AF615884

The University of Wisconsin

PUBLICATIONS IN MEDIEVAL SCIENCE

PUBLICATIONS IN MEDIEVAL SCIENCE

1

The Medieval Science of Weights (Scientia de ponderibus): Treatises Ascribed to Euclid, Archimedes, Thabit ibn Qurra, Jordanus de Nemore, and Blasius of Parma.
Edited by Ernest A. Moody and Marshall Clagett.

2

Thomas of Bradwardine His Tractatus de proportionibus: *Its Significance for the Development of Mathematical Physics.*
Edited and translated by H. Lamar Crosby, Jr.

3

William Heytesbury: Medieval Logic and the Rise of Mathematical Physics.
By Curtis Wilson.

4

The Science of Mechanics in the Middle Ages.
By Marshall Clagett.

5

Galileo Galilei: On Motion *and* On Mechanics.
Translated by I. E. Drabkin and Stillman Drake.

6

Archimedes in the Middle Ages, Volume I: The Arabo-Latin Tradition.
By Marshall Clagett.

Archimedes in the Middle Ages

VOLUME I

ARCHIMEDES

in the Middle Ages

VOLUME I

THE ARABO-LATIN TRADITION

MARSHALL CLAGETT

The University of Wisconsin Press

MADISON, 1964

Published by The University of Wisconsin Press
430 Sterling Court, Madison 6, Wisconsin

Printed in The Netherlands by N.V. Drukkerij G. J. Thieme, Nijmegen

Library of Congress Catalog Number 62-7218

To Sue

Preface

All the preceding volumes of this series have been concerned with mechanics in the Middle Ages or with its modifications in early modern times. In this volume (and the one to succeed it) an effort is made to present certain mathematical material available to medieval scholars, namely, medieval Archimedean texts.

Volume One is devoted to texts which are in the Arabo-Latin tradition and texts which are closely allied to that tradition, while Volume Two will consist almost exclusively of William of Moerbeke's translation of the corpus of Archimedes' works, made from the Greek in 1269, and of later texts that made use of that translation. No major text has been omitted in Volume One so far as I can ascertain by many years of study of the manuscripts in European libraries. The texts presented here either are previously unpublished ones which I have discovered or are new versions of earlier efforts made by me or by others but based on a wider study and collation of the manuscripts. They have all been translated into English for the first time.

It will be evident to the reader on examining these texts that medieval mathematicians with their rather elementary knowledge of geometry based largely on Euclid's *Elements* felt obliged to paraphrase and elaborate the terse and more mature Archimedean mathematics. It will be further evident from a comparison of Volume One with Volume Two that far less of Archimedes' corpus became available by means of the Arabo-Latin tradition than through the Moerbeke translations, but that on the whole more use was made of the more elementary and incomplete Arabo-Latin works than of the Moerbeke translations. The detailed comparison of the two traditions cannot be accurately made, however, until the presentation of both sets of texts is complete.

Concerning the texts published here, one final warning is pertinent. This volume was in preparation and publication over the course of a number of years. Consequently, new manuscripts and versions continued to turn up right through the last stages of publication. I have tried to include in their appropriate places references to such new information as became available. However, some parts of the text had gone through final stages of proof when new information appeared. In such cases I was unable to add pertinent remarks in every appropriate place. For example, when I say on page 30 that Gerard of Cremona's translation was the source of some fourteen versions of Propositions I, II, and III of the *De mensura circuli*, that reference is to the fourteen versions published in Chapters III and V and thus does not include recently discovered versions briefly noted in other parts of the volume (e.g., see page 38; page 80 *n*1; and page 96 *n*6). These additional versions will be discussed and published in the second volume.

Once more I must acknowledge the financial help of many organizations over the past years and in particular that of the following organizations: the National Science Foundation for research grants in 1955–56 and 1962–63 and for a grant-in-aid of publication, the Institute for Advanced Study at Princeton for a grant and membership in 1958–59 and for membership in 1963, the Research Committee of the University of Wisconsin for research grants in 1958 and 1959–60, and the Trustees of the William F. Vilas Estate for its support of me as Vilas Research Professor in the History of Science since 1962.

I must also acknowledge the help of a number of scholars—past and present—whose studies have been important to me. I have occasionally criticized the textual efforts of M. Curtze in this volume but such criticism should not obscure the debt that I and all students of medieval mathematics owe to his untiring efforts to discover and publish texts. Furthermore, the works of J. L. Heiberg, T. L. Heath, and, more recently, E. J. Dijksterhuis have been most useful to me in understanding the background of Greek mathematics.

Particular thanks are due my friends Professor Carl Boyer of Brooklyn College and Professor John Murdoch of Harvard University for reading and criticizing the typescript in a helpful manner, and Dr. Shlomo Pines of the Hebrew University of Jerusalem for critically checking my translations of the Arabic variant readings in Chapters Two and Four. I am also appreciative of the readiness with which my friends Richard Hunt of the Bodleian Library and Marie-Thérèse d'Alverny, formerly of the Bibliothèque Nationale, have checked manuscripts in their respective collections for me; needless to say, if I had not been offered the hospitality of the principal manuscript libraries of Europe, my work could not have been completed.

A final word of thanks must be said concerning the assistance of my secretary, Mrs. Loretta Freiling, who not only typed this volume (and parts of it more than once) but also made the preliminary copies of its many diagrams.

M.C.

Institute for Research in the Humanities
University of Wisconsin
September 1963

Contents

Preface ix
Note on Textual Procedures xv
General List of Manuscripts and Their Sigla xix
Illustrations of Manuscripts, *following page* 144
Chapter One: The Impact of Archimedes on Medieval Science . . 1
Chapter Two: Translations of the *De mensura circuli* from the Arabic 15
1. The Translation Perhaps by Plato of Tivoli 16
2. The Translation of Gerard of Cremona 30

Chapter Three: Emended Versions of the *De mensura circuli* . . . 59
1. The Cambridge Version 63
2. The Naples Version 80
3. The Florence Versions 91
4. The Version of Gordanus 142
5. The Corpus Christi Version 165
6. The Munich Version 193

Chapter Four: The *Verba filiorum* of the Banū Mūsā 223
Chapter Five: Further Versions of the *De mensura circuli* 368
1. The Pseudo-Bradwardine Version *(Versio Vaticana)* 370
2. The Abbreviated Version of the Pseudo-Bradwardine Text . . 388
3. Albert of Saxony's *Question on the Quadrature of the Circle* . . . 398

Chapter Six: Archimedes' *De sphaera et cylindro* in the Latin West before 1269 433
1. A Latin Fragment of the *De sphaera et cylindro* 433

2. The *Liber de curvis superficiebus* of Johannes de Tinemue . . . 439
3. Two Propositions Interposed in the Text of the *Liber de curvis superficiebus* 520
4. Three Propositions Added to the *Liber de curvis superficiebus* . . 530
5. An Anonymous Comment on Proposition Seven of the *Liber de curvis superficiebus* 547

Chapter Seven: The Arabo-Latin Tradition of Archimedes in Retrospect 558

Appendix I: Some Non-Archimedean Treatments of Quadrature . 567
1. The Theorum of Jordanus 567
2. Two Anonymous Quadrature Proofs of the Thirteenth Century . 576
3. The *Quadratura circuli* Attributed to Campanus 581

Appendix II: The *Quadratura circuli per lunulas* 610

Appendix III: Some Medieval Latin Citations of Archimedes . . 627

Appendix IV: A Medieval Treatment of Hero's Theorem on the Area of a Triangle 635

Appendix V: A Version of Philo's Solution of the Problem of Two Mean Proportionals 658

Appendix VI: Jordanus and Campanus on the Trisection of an Angle 666
1. Jordanus' Solution 666
2. The Solution Attributed to Campanus 678

Bibliography 684

A Selective Index of Latin Geometrical Terms 691

An Index of the Citations of Euclid's *Elements* 709

An Index of Latin Manuscripts Cited 711

General Index 713

Note on Textual Procedures

My evident intention in this work is to include all of the known Latin Archimedean texts stemming from the medieval period. In general, my procedure has been to examine and collate all of the known manuscripts of each text; but usually I have decided which are the best manuscripts and accordingly based the text on those manuscripts. Hence, often when it is clear that a late manuscript was copied from an earlier and better one used by me in the establishment of the text, I have omitted the variant readings of the later manuscript. I have done this so that the significant variant readings of earlier and better manuscripts will not be obscured by the wholesale inclusion of corrupt and senseless readings.

My main objective throughout has been to present texts that are easily readable as well as ones that faithfully represent the original works. For this reason I have been free with punctuation and capitalization. Thus I have added commas, periods, and other marks of punctuation when the meaning seemed to demand them. In the case of capitalization, I have consistently capitalized the letters used to mark geometrical quantities even though small letters are generally used in the manuscripts. I have also capitalized the enunciations of propositions in order that I might represent the common medieval practice of using a larger hand for the enunciations. Furthermore, I have always capitalized the first letter of each word beginning the sentence, although the medieval practice is irregular—capital letters sometimes being used and sometimes not.

In general, I have left the medieval orthography of the manuscripts as it is. For example, the classical *ae* is always written as *e* in the manuscripts, and I have followed the medieval practice. Particular and peculiar spellings have been noted in the Introductions preceding the texts. I have also followed the medieval authors in their careless procedure of reversing letters which designate geometrical magnitudes. That is, an author will often use AB and BA to designate the same line (even in the same sentence) without having any geometrical objective in so reversing the letters.

The diagrams are taken from the manuscripts. However, on some occasions I have added a reconstructed diagram where the figure in the manuscript is obscure or incorrect.

I have tried to make the English translations literal but easily understandable. Occasionally I have been arbitrary. Thus *curva superficies*, which is literally rendered by "curved surface," I have often translated as "lateral surface" or "lateral area" when the latter translations seemed appropriate. Similarly, I have translated *hypothenusa* when applied to cones as "slant height" rather than as "hypotenuse." Sometimes the various forms of *resecare* have been rendered by a form of the word "exhaust" in the proofs using the so-called "exhaustion method." I am, of course, perfectly aware that in this technique no actual exhaustion is intended but only that by taking away successively more and more of a magnitude there will remain ultimately a quantity smaller than any assigned quantity.

In the translations I have often used the fractional form to represent ratios, although this is in some respects anachronistic. I do not mean to imply by this practice that the authors in question were employing a modern system of real numbers; I use it only for the sake of convenience—whatever concept of ratio, magnitude, and number the particular author might have had. Incidentally, when rendering the rhetorical Latin of a geometrical argument by modern symbols, I have often omitted translating the word *linea* so as to keep the modern expression compact and uncluttered. I have done this only when it is perfectly clear that the quantity specified is indeed a line. Thus, *quadratum linee AB est equale quadrato linee BC et quadrato linee AC* would be translated as $AB^2 = BC^2 + AC^2$. Furthermore, the equal sign is used to render *est equalis*, *erit equalis*, *equatur*, *valet*, and so on, when the argument is translated by the use of modern symbols.

Three forms of bracketing have been employed. The first is parentheses, ( ), which I have used in the text or in the variant readings to enclose an editorial comment or sign and/or an alternate reading. In addition, parentheses have occasionally been used in the translation to enclose factors to be multiplied together, for example, $(A + B) \cdot (B + D)$. The second form is square brackets, [ ], which have been employed in the text and in the variant readings to enclose additions to the text made either by me as editor or by some later scribe for the purpose of clarification—but additions which were probably not present in the original text. The third form is angle brackets, ⟨ ⟩; these have been used to enclose additions that I have made to the text because I was reasonably certain they were present in the original text though omitted in the extant manuscripts. In the translations, the signs $>$ and $<$ have been used to represent "greater than" and "less than." At first glance they might be confused with angle brackets, but the context will always show the reader their particular use. Incidentally, in the translations I have often used a raised period, ·, to indicate multiplication, for example, $AB \cdot BC$, since the authors sometimes used two or more letters to represent a single geometric magnitude and sometimes used only one letter. The use of the dot eliminates the ambiguity inherent in this vacillating practice.

The system used to indicate variant readings is a common one. Preferred readings are stated first and separated from variant readings by a colon (for example, ipse: ille *B*—"ipse" being the readings of all manuscripts but *B*, which has "ille"). The following abbreviations have been used in the variant readings:

add. = addidit
cf. = confer
cod. = codex
corr. = correxi, correxit
del. = delevi, delevit
ed. = edidit, editio, editor
hab. = habet
inser. = inseruit
iter. = iteravit
lac. = lacuna
lib. = liber
m. rec. = manus recentior
MS = codex manuscriptus
mg. = in margine
om. = omisit, omiserunt, omisi
supra scr. = supra scripsit
*tr.** = transposuit, transtulit

* This abbreviation is used to indicate two different practices: (1) the mutual transposition of two words (e.g., "recta linea *tr. H*" means that in *H* the phrase appears as "linea recta"), and (2) the transfer of a single word or phrase (in this case it always appears with *ante* or *post*).

A General List of Manuscripts and Their Sigla

The manuscripts listed here are those actually employed in the establishment of the texts presented in this volume. In general, I have used successive letters as sigla in order to group manuscripts of the same library together, but this has not always been possible. In any event, the list is short enough so that all of the manuscripts from a single library can be easily located. Special orthographic and other characteristics peculiar to the sections of the various codexes used in this volume are included in the introductions and in the sigla lists preceding each of the texts. However, I have given here references to more general discussions of the codexes found in catalogues and elsewhere. I have confined my discussion of codex dates to the sections used in this volume. Specimens of the most important of the manuscripts are included in the plates.

1. *A** = Naples, Biblioteca Nazionale, VIII. C.22

Date: 13c.
Sections used: 57r–60r, *De curvis superficiebus* (Chapter Six, Section 2); 65v–66v, *De mensura circuli*—Naples Version (Chapter Three, Section 2).
Description: Not described in Cataldus Iannelius' catalogue.

2. *Aa** = Vatican Library, Vat. lat. 3102

Date: 14c.
Section used: 111v–112v, the Vatican Version of the Pseudo-Bradwardine tract on quadrature (Chapter Five, Section 1).

* A specimen of this MS is included in the plates, following page 144.

3. *B** = Oxford, Bodleian Library, Auct. F.5.28

Date: middle of 13c (for the sections used here).

Sections used: 101v–102v, *De mensura circuli*—Gerard translation (Chapter Two, Section 2);
111r–116r, *De curvis superficiebus* (Chapter Six, Sections 2);
116r, *Quadratura circuli per lunulas* (Appendix II).

Description: F. Madan and H. H. E. Craster, *Summary Catalogue of Western Manuscripts in the Bodleian Library*, vol. *2* (Oxford, 1922), pp. 706–707.

4. *Bc** = Oxford, Corpus Christi College 234

Date: 15c.

Section used: 170r–172v, *De mensura circuli*—Corpus Christi Version (Chapter Three, Section 5).

Description: H. O. Coxe, *Catalogus codicum MMS qui in collegiis aulisque Oxoniensibus hodie adservantur*, vol. *2* (Oxford, 1852), Corpus Christi Section, p. 97.

5. *Bd** = Oxford, Bodleian Library, Digby 147

Date: 14c.

Section used: 89r–91v, *Quadratura circuli*, attributed to Campanus (Appendix I, Section 3).

Description: G. D. Macray, *Catalogi codium manuscriptorium Bibliothecae Bodleianae. Pars nona, Codices a...Kenelm Digby...donatos, complectens* (Oxford, 1883), cc. 144–46; cf. L. Thorndike, *History of Magic and Experimental Science*, vol. *3* (New York, 1934), p. 143*n*; vol. *2* (New York, 1923), p. 500*n*.

6. *C** = Oxford, Bodleian Library, Digby 174

Date: 13c. Macray's catalogue seems to imply that the sections used are from the twelfth century. However, this is unlikely since the same thing is implied for the item preceding these. But this item is the *Elementa de ponderibus*, which does not appear to be of the twelfth century.

Sections used: 133v–134v, *De mensura circuli*—Gerard translation (Chapter Two, Section 2, Sigla);
136v–137r, other quadrature propositions (Appendix I, Section 2);
174v–178r, *De curvis superficiebus* (Chapter Six, Section 2).

Description: G. D. Macray, *Pars nona...complectens*, cc. 184–185.

7. *Ca** = Oxford, Corpus Christi College 251

Date: 13c.

Sections used: 83v–84r, *Quadratura circuli* [*per lunulas*] (Appendix II); 84v, proposition on quadrature from Jordanus' *De triangulis* (Appendix I, Section 1).

Description: H. O. Coxe, *Catalogus codicum...adservantur*, vol. *2*, p. 104.

8. *D** = Florence, Biblioteca Nazionale, Conv. Soppr. J.V.30

Date: 14c.

Sections used: 1r–4v, *De curvis superficiebus* (Chapter Six, Sections 2, 3); 9v–12v, *De mensura circuli*—Florence Versions (Chapter Three, Section 3).

Description: A. A. Björnbo, "Die mathematischen S. Marcohandschriften in Florenz," *Bibliotheca mathematica*, 3. Folge, vol. *4* (1903), pp. 241–45.

9. D_1 = Paris, Bibliothèque Nationale, Fonds latin 11247

Date: late 15 or 16c. Copied from *D*.

Sections used: 2r–25v, *De curvis superficiebus* (Chapter Six, Sections 2, 3); 51r–66r, *De mensura circuli*—Florence Versions (Chapter Three, Section 3).

Description: *Bibliothèque de l'École de Chartes*, vol. *24* (1863), p. 225.

10. *E* = London, British Museum, Harleian 625

Date: 14c(?).

Sections used: 123r–130r, *De triangulis* of Jordanus (Appendix I, Section 1, and Appendixes V and VI); 137r–139v, *De curvis superficiebus* (Chapter Six, Sections 2, 5).

Description: *A Catalogue of the Harleian Manuscripts*, vol. *1* (London, 1808), p. 391. This catalogue does not date the codex and it misnames the author of the *De curvis superficiebus:* "Liber Mahumedis...." For an estimate of the date, see L. Thorndike in *Isis*, vol. *50* (1959), p. 37*n31*.

11. *Ea** = Florence, Biblioteca Nazionale, Conv. Soppr. J.IX.26

Date: late 15c (about 1500).

Section used: 49v–50v, an abbreviated version of the Pseudo-Bradwardine tract on quadrature (Chapter Five, Section 2).

Description: A. A. Björnbo, "Die mathematischen S. Marcohandschriften," *Bibliotheca mathematica*, 3. Folge, vol. *12* (1911–12), pp. 97–99. But Björnbo fails to note this work on 49v–50v.

12. *Eb* = Erfurt, Stadtbibliothek, Amplon. F.178

Date: middle of 14c.

Section used: 138r–139v, *Quadratura circuli*, attributed to Campanus (Appendix I, Section 3).

Description: W. Schum, *Beschreibendes Verzeichnis der Amplonianischen Handschriften-Sammlung zu Erfurt* (Berlin, 1887), pp. 236–38.

13. *Ec** = Erfurt, Stadtbibliothek, Amplon. Q.361

Date: first half of 14c.

Section used: 79v–80r, *Quadratura circuli*, attributed to Campanus (Appendix I, Section 3).

Description: W. Schum, *Beschreibendes Verzeichnis...Erfurt*, pp. 601–606.

14. *Er* = Erfurt, Stadtbibliothek, Amplon. Q.385

Date: late 14c.

Section used: 51r–53r, *Quadratura circuli*, attributed to Campanus (Appendix I, Section 3).

Description: W. Schum, *Beschreibendes Verzeichnis...Erfurt*, pp. 641–44.

15. *F** = Vienna, Nationalbibliothek, cod. 5303

Date: 15–16c.

Sections used: 11r–18v, *De curvis superficiebus* (Chapter Six, Section 2);
19r–21v, *De mensura circuli*—Gerard translation (Chapter Two, Section 2, Sigla);
19r, 21v, other quadrature proofs (Appendix I, Section 2).

Description: *Tabulae codicum manuscriptorum...in Bibliotheca Palatina Vindobonensi asservatorum*, vol. *4* (Vienna, 1870), p. 93.

16. *Fa* = Vienna, Nationalbibliothek, cod. 5257

Date: 1390.

Sections used: 64v–67r, *Questio de quadratura circuli* of Albert of Saxony (Chapter Five, Section 3);
67r–69r; *De mensura circuli*—Gordanus Version (Chapter Three, Section 4).

Description: *Tabulae codium...asservatorum*, vol. *4*, p. 77.

17. *G* = Cambridge University Library, Mm.III.11 (= 2327)

Date: 15c.

Section used: 196r–198v, *De curvis superficiebus* (Chapter Six, Section 2).

Description: *A Catalogue of the Manuscripts Preserved in the Library of the University of Cambridge*, vol. *4* (Cambridge, 1861), p. 181.

18. *Ga* = Cambridge University Library, Ee.III.61 (= 1017)

Date: 15c.

Section used: 176v–177v, *Quadratura circuli*, usually attributed to Campanus but here attributed to Franco of Liège (Appendix I, Section 3).

Description: *A Catalogue of the Manuscripts...Cambridge*, vol. *2* (Cambridge, 1857), pp. 114–20.

19. *H** = Basel, Öffentliche Bibliothek der Universität, F.II.33

Date: middle of 14c.

Sections used: 116v–122r, *Verba filiorum* of the Banū Mūsā (Chapter Four); 151r–153r, *De curvis superficiebus* (Chapter Six, Section 2).

Description: A. A. Björnbo and S. Vogl, *Alkindi, Tideus und Pseudo-Euklid, Abhandlungen zur Geschichte der mathematischen Wissenschaften, 26.* Heft (1912), pp. 124–29, 171–72; cf. A. Lejeune, *L'Optique de Claude Ptolémée* (Louvain, 1956), p. 39.

20. *Ha* = Oxford, Bodleian Library, Digby 190

Date: 13c (for section used).

Section used: 87v, *Quadratura circuli per lunulas* (Appendix II).

Description: G. D. Macray, *Pars none...complectens*, c. 203.

21. *I** = Dresden, Sächs. Landesbibliothek, Db. 86

Date: early 14c.

Sections used: 50r–61v, *De triangulis* of Jordanus (Appendix I, Section 1, and Appendixes V and VI);
175v–176v, 178r, *De mensura circuli*—Gerard translation (Chapter Two, Section 2);
178r–v, a medieval version of Hero's theorem for the area of a triangle (Appendix IV);
188r–194v, *De curvis superficiebus* (Chapter Six, Section 2).

Description: M. Curtze, "Über eine Handschrift der Köningl. öffentl. Bibliothek zu Dresden," *Zeitschrift für Mathematik und Physik*, vol. *28* (1883), Hist.-lit. Abtheilung, pp. 1–13; cf. A. A. Björnbo and S. Vogl, *Alkindi,...Abhandlungen...Wissenschaften, 26.* Heft (1912), p. 130.

22. *J** = Berlin, Deutsche Staatsbibliothek (MS now at Marburg, Westdeutsche Bibliothek), Q.150

Date: 13c.

Sections used: 89r–v, *De mensura circuli*—Gerard translation (Chapter Two, Section 2);
90r–94v, *De curvis superficiebus* (Chapter Six, Section 2);
94v, *Quadratura circuli per lunulas* (Appendix II).

Description: This manuscript is identical with Libri manuscript 665, described in the *Catalogue of the Sale of Libri Manuscripts* (London, 1859), pp. 145–48, although there it is misdated as twelfth century.

23. *K** = Paris, Bibliothèque Nationale, Fonds latin 11246

Date: 13c (so dated by Curtze and Libri in the works noted below and by Mlle. Marie-Thérèse d'Alverny in a letter to me, but ascribed by Delisle and Bubnov to the fifteenth century).

Section used: 37v–39r, *In quadratum circuli*—translation of the *De mensura circuli* Perhaps by Plato of Tivoli (Chapter Two, Section 1).

Description: M. Curtze, *Der "Liber Embadorum" des Savasorda... Abhandlungen zur Geschichte der mathematischen Wissenschaften, 12*, Heft (1902), pp. 3–4; G. Libri, *Historie des sciences mathematiques en Italie*, vol. *2*, pp. 480–86; L. Delisle, in *Bibliothèque de l'École de Chartes*, vol. *24* (1863), p. 225; N. Bubnov, *Opera Gerberti* (Berlin, 1898), pp. 302–35.

24. *Ka* = Oxford, Bodleian Library, Dibgy 153

Date: 14c.

Section used: 184r, *Quadratura circuli per lunulas* (Appendix II).

Description: C. D. Macray, *Pars non... complectens*, cc. 152–54; L. Thorndike, *History of Magic and Experimental Science*, vol. *2*, p. 726*n*.

25. *L** = Oxford, Bodleian Library, Arch. Seld. B.13

Date: 15c (for section used; much of the rest of the manuscript is 13c). Concerning the date, R. Hunt, Keeper of the Western Manuscripts, writes me: "At the end of the Archimedes there is a long erased inscription of 7 lines going right across the page.... Only the date at the very end is clear: anno domini Mccccxlii die... novembris." Hunt also notes that on fol. 3r the name of Magister Nicholaus and the date 28 Aug. 1469 (or 1489) appear.

Section used: 2r–v, *De mensura circuli*—Gerard translation (Chapter Two, Section 2).

Description: F. Madan and H. H. E. Craster, *Summary Catalogue... Bodeleian Library*, vol. *1*, p. 618.

26. *M** = Florence, Biblioteca Nazionale, Conv. Soppr. J.V.18

Date: 14c.

Sections used: 33r, *Quadratura circuli per lunulas* (Appendix II);
92r–v, *De mensura circuli*—Cambridge Version (Chapter Three, Section 1);
92v–96v, *De curvis superficiebus*—paraphrase (Chapter Six, Sections 2, 4).

Description: A. A. Björnbo, in *Bibliotheca mathematica*, 3. Folge, vol. *12* (1911–12), pp. 218–22 (but Björnbo omits mention of tracts on 92r–96v).

27. *Ma** = Paris, Bibliothèque Mazarine, 3637 (1256)

Date: late 14c. The source of manuscript *R*.

Section used: 1r–13v, *Verba filiorum* of the Banū Mūsā (Chapter Four, Sigla).

Description: A. Molinier, *Catalogue des manuscrits de la Bibliothèque Mazarine*, vol. *3* (Paris, 1890), pp. 149–50.

28. *Mi* = Milan, Bibliotheca Ambrosiana, H. 144 Inf.

Date: 15c. Amelli dates part of the codex as 1431.

Section used: 145r–146r, *Quadratura circuli*, attributed to Campanus (Appendix I, Section 3).

Description: A. M. Amelli, "Indice dei codici manoscritti della Biblioteca Ambrosiana," *Rivista delle biblioteche e degli archivi*, vol. *21* (1910), p. 144.

29. *N** = Cambridge, Gonville and Caius College 504/271

Date: 13c.

Section used: 108v–109v, *De mensura circuli*—Cambridge Version (Chapter Three, Section 1).

Description: M. R. James, *A Descriptive Catalogue of the Manuscripts in the Library of Gonville and Caius College*, vol. *2* (Cambridge, 1908), p. 574.

30. *O* = Paris, Bibliothèque Nationale, Fonds latin 7224

Date: 16c. Copied from *K*.

Section used: 63r–65r, *In quadratum circuli*—translation of the *De mensura circuli* perhaps by Plato of Tivoli (Chapter Two, Section 1).

Description: M. Curtze, in *Der "Liber Embadorum"... Abhandlungen... Wissenschaften*, *12*. Heft (1902), p. 4.

31. *Oa** = Paris, Bibliothèque Nationale, Fonds latin 7434

Date: 14c.

Section used: 84v–87v, Propositions IV.12–IV.28 of the *De triangulis* of Jordanus (Appendix I, Section 1, and Appendixes V and VI).

32. *P** = Paris, Bibliothèque Nationale, Fonds latin 9335

Date: 14c.

Sections used: 28v–29v, *De mensura circuli*—Gerard translation (Chapter Two, Section 2);
55v–63r, *Verba filiorum* of the Banū Mūsā (Chapter Four).

Description: A. A. Björnbo and S. Vogl, in *Alkindi,... Abhandlungen... Wissenschaften*, *14*. Heft (1902), pp. 137–38, and *26*. Heft (1912), pp. 138, 171; P. Tannery, in *Bibliotheca mathematica*, 3. Folge, vol. *2* (1901), pp. 46–47, and A. A. Björnbo, *ibid.*, 3. Folge, vol. *3* (1902), pp. 63–75; L. Delisle, in *Bibliothèque de l'École de Chartes*, vol. *23* (1862), pp. 304–305.

33. *Q** = Paris, Bibliothèque Nationale, Fonds latin 7378A

Date: 14c.

Sections used: 18r–v, *Quadratura circuli*, attributed to Campanus (Appendix I, Section 3);
19r–v, *De mensura circuli*—Gerard translation (Chapter Two, Section 2);
40r–v, a medieval version of Hero's theorem for the area of a triangle (Appendix IV).

Description: L. Thorndike, *History of Magic and Experimental Science*, vol. *3*, p. 304.

34. *R* = Paris, Bibliothèque Nationale, Fonds latin 7225A

Date: early 16c. Mlle. Marie-Thérèse d'Alverny writes "First half of the 16c." This manuscript was copied from *Ma*.

Section used: 2r–31r, *Verba filiorum* of the Banū Mūsā (Chapter Four).

35. *S** = Oxford, Bodleian Library, Digby 168

Date: 14c (for sections used; some other parts of the manuscript are 13c).

Sections used: 122r (old page 121r), a fragment of the *On the Sphere and the*

Cylinder of Archimedes (Chapter Six, Section 1); also the so-called Leonardo Proof of the two mean proportionals problem (Appendix V);

124r–v (old page 123r–v), a fragment from the *Verba filiorum* of the Banū Mūsā (Macray catalogue puts this on 123r–v; see Chapter Four).

Description: G. D. Macray, *Pars nona... complectens*, cc. 172–177; L. Thorndike, *History of Magic and Experimental Science*, vol. *3*, p. 261.

36. *Sl* = London, British Museum, Sloane 285

Date: 14c.

Section used: 80r–92v, *De triangulis* of Jordanus (Appendix I, Section 1, and Appendixes V and VI).

Description: E. J. L. Scott, *Index to the Sloane Manuscripts in the British Museum* (London, 1904), p. 390.

37. *T* = Thorn (Torún), Gymnasialbibliothek, R 4° 2

Date: 14c.

Section used: pp. 73–79, *Verba filiorum* of the Banū Mūsā (Chapter Four).

Description: M. Curtze, "Über die Handschrift R.4°.2, Problematum Euclidis explicatio der Königl. Gymnasialbibliothek zu Thorn," *Zeitschrift für Mathematik und Physik*, vol. *13*, Suppl. (1868), pp. 45–104.

38. *U* = London, British Museum, Addit. 17368

Date: late 13 or early 14c (the catalogue does not mention the item in question but dates the Euclid *Optica et catoptrica* preceding it as "written at the end of the XIIIth cent.").

Section used: 69v, *De mensura circuli*—Gerard translation (Chapter Two, Section 2).

Description: *A Catalogue of Additions to the Manuscripts in the British Museum in the Years MDCCCXLVIII–MDCCCLIII* (London, 1868), p. 10.

39. *V** London, British Museum, Royal 12E.25

Date: ca. 1300.

Sections used: 150v, *Quadratura circuli per lunulas* (Appendix II);

150v–151v, *De mensura circuli*—Cambridge Version (Chapter Three, Section 1).

Description: *Catalogue of Western Manuscripts in the Old Royal and King's Collections*, vol. *2* (London, 1921), pp. 59–61.

40. *W* = Dublin, Trinity College D.2.9

Date: 1565, written by Franceso Barozzi.

Section used: 54r–55r, *De circuli dimensione*—translation of the *De mensura circuli* perhaps by Plato of Tivoli (Chapter Two, Section 1).

Description: T. K. Abbot, *Catalogue of the Manuscripts in the Library of Trinity College, Dublin* (Dublin, 1900), p. 60 (cod. no. 390).

41. *X** = Vatican Library, Pal. lat. 1389

Date: 15c.

Section used: 108r–111v, *De mensura circuli*—Gordanus version (Chapter Three, Section 4).

42. *Xa* = Vatican Library, Reg. suev. 1261

Date: 14c (ca. 1350–1375).

Section used: 57v–58r, a medieval version of Hero's theorem for the area of a triangle (Appendix IV).

Description: A. A. Björnbo and S. Vogl, in *Alkiudi,... Abhandlungen... Wissenschaften, 14.* Heft (1902), pp. 146–50.

43. *Y** = Munich, Bayrische Staatsbibliothek, cod. 56

Date: 15c (1434–1436).

Section used: 182r–186v, *De mensura circuli*—Munich Version (Chapter Three, Section 6).

Description: *Catalogus codium latinorum Bibliothecae Regiae Monacensis*, 2d ed., vol. *1*, pars 1 (Munich, 1892), p. 12.

44. *Ya** = Munich, Bayrische Staatsbibliothek, cod. 234

Date: 15c.

Section used: 105v–108v, a medieval version of Hero's theorem for the area of a triangle (see Appendix IV).

Description: *Catalogus codium... Monacensis*, 2d. ed., vol. *1* (Munich, 1892), p. 58.

45. *Z** = Bern, Bürgerbibliothek, A.50

Date: 15c.

Sections used: 168r–169r, *Quadratura circuli per lunulas*, Version II (Appendix II);

169r–172r, *Questio de quadratura circuli* of Albert of Saxony (Chapter Five, Section 3).

Description: H. Suter, "Der Tractatus 'De quadratura circuli' des Albertus de Saxonia," *Zeitschrift für Mathematik und Physik*, vol. *29* (1884), Hist.-lit. Abtheilung, pp. 83–85.

46. *Zm* = Madrid, Biblioteca Nacional 10010

Date: 14c.

Sections used: 77v–83r, *Verba filiorum* of the Banū Mūsā (Chapter IV, Sigla);
83v–84v, the "Leonardo" version of a proposition on finding two mean proportionals (Appendix V);
84r, a fragment of Archimedes' *De sphaera et cylindro* (Chapter VI, Section 1).

Description: J. M. Millas Vallicrosa, *Las traducciones orientales en los manuscritos de la Biblioteca Catedral de Toledo* (Madrid, 1942), pp. 210–11.

47. *Zo* = Vatican Library, Ottob. lat. 1870

Date: 15c.

Section used: 151v–157v, *De mensura circuli*—Florence versions (Chapter Three, Section 3).

Archimedes in the Middle Ages

VOLUME I

Chapter one

The Impact of Archimedes on Medieval Science*

The importance of the role played by Archimedes in the history of science can scarcely be exaggerated. He was emulated and admired in his own day and at successive periods in later times. His name appears on the pages of the works of the great figures who fashioned the beginnings of modern mechanics. For example, Galileo owed a not inconsiderable debt to Archimedes—both direct and indirect. Galileo mentions Archimedes by actual count over one hundred times[1] and in almost Homeric hyperbole, using such expressions as *suprahumanus Archimedes*, *inimitablilis Archimedes*, *divinissimus Archimedes*, and so on. Archimedes' significance for these founders of early modern science lay in the use of mathematics in the treatment of physical problems[2] as well as in the originality and fertility of his mathematical techniques. But all of this is well known.

* This chapter is based on a paper given at Oxford in May 1956. It was printed in substantially the same form as given here in *Isis*, vol. *50* (1959), pp. 419–29.

1 See the extensive list of citations of Archimedes' name by Galileo in the index volume of the national edition of his works, *Opere di Galileo Galilei*, Ediz. Naz., vol. *20* (Florence, 1909), pp. 69–70. For example, see vol. *1*, p. 300: "His responderem, me sub suprahumani Archimedis (quem nunquam absque admiratione nomino) alis memet protegere."

2 One place where Galileo was particularly influenced by Archimedes was in the formation of his concept of a rectilinear inertial motion to be used in the analysis of certain physical problems. I have briefly noted the relevant passages in Galileo's works for the enunciation of his inertial doctrine in chapter XI of my *Science of Mechanics in the Middle Ages* (Madison, 1959), pp. 668–71, n. 132. Fundamentally, Galileo recognized three kinds of movement: natural, violent, and circular inertial motion continually at right angles to radii extending to the center of the earth. In physical problems, however, where the trajectory of horizontal motion is insignificant in comparison to the radius of the earth, Galileo assumed that the horizontal trajectory over which the in-

Perhaps less well known is the role played by Archimedes in the Middle Ages. The problem of the medieval Archimedes is a classic one in the transmission of Greek learning. In some ways it has an advantage for study over other similar problems of transmission, since one is not overwhelmed by the abundance of materials as in the case of Euclid, and even more so in the case of Aristotle. That is to say, one has hopes of both defining the limits of the investigation and presenting a substantially complete solution of the problem. The basic problem of the impact of Archimedes on medieval science can be subdivided into three questions:

1. How much of the corpus of Archimedes' writings in their actual form—or some closely modified state—came directly into Latin?

2. How much of Archimedes' methods and results was made available by translations from the Arabic or Greek of works which were influenced (but not so directly) by Archimedes?

3. What use was made by medieval authors of the works and ideas available in Latin translation?

I hope in this work to present in detailed fashion the answers to these questions by giving the medieval Latin versions of the works of Archimedes and of the works influenced by Archimedean techniques. In this introductory chapter, I shall merely give in short compass the highlights of the answers as I conceive them.

Before answering the questions, two brief observations ought to be made. The first concerns the temporal limits of the problem—the twelfth to fifteenth century. The latter terminus is arbitrary; the former is set by the state of the materials, for it must be realized that there is virtually no Archimedes before the twelfth century in the Latin tongue. (This has to be qualified by reference to some half a dozen brief indications of Archimedean ideas in the early medieval period. They mostly concern the density problem and I shall omit them here, since I have treated them elsewhere in some detail.[3])

ertial motion takes place is a straight line rather than a circle. In his *Discorsi e dimostrazioni matematiche intorno a due nuove scienze* he notes that this assumption is similar to that made by Archimedes when he postulated that weights on a balance arm act at right angles to the arm (*Le Opere...*, vol. *8*, pp. 274-75).

[3] I have discussed the various passages concerning hydrostatics available to early medieval Schoolmen in my *Science of Mechanics in the Middle Ages*, chapter II. I do not know to which Archimedean work Cassiodorus refers when in Theodoric's name he praises Boethius for having restored in Latin the mechanician Archimedes to the Sicilians (Cassiodorus, *Liber variorum*, Bk. I, Ep. 45, "Mechanicum etiam Archimedem Latialem Siculis reddisti"). Certainly no Archimedean work has

The second preliminary observation concerns itself with the state of the Greek and Arabic texts of Archimedes available at the time of the translations. The Greek text extant in the Byzantine period consisted of at least three manuscripts, whose fates we can trace.[4] Manuscript A, the principal manuscript, contained all of the works now known except the *On Floating Bodies*, the *On the Method*, the fragmentary *Stomachion*, and the *Bovine Problem*. This was one of the two manuscripts available to Moerbeke. It is the source of all Renaissance copies of Archimedes—but it has been lost. Manuscript 𝔅 included the mechanical works—*On the Equilibrium of Planes*, *On Floating Bodies*—and the *On Spiral Lines* and the *Quadrature of the Parabola*. It too was available to Moerbeke. But it drops out of history after a reference to it in the early fourteenth century. Finally, we can mention manuscript C, not available to the Latin West in the Middle Ages, or in modern times until its identification by Heiberg in 1906 in Constantinople. It contains fragments of *On the Sphere and the Cylinder*, *On Spiral Lines*, *Measurement of the Circle*, *On the Equilibrium of Planes*, *Stomachion*, most of the Greek text of *On Floating Bodies*, and the brilliant work *On the Method* not present in the other manuscripts.

The apparent paucity of Greek manuscripts during the Byzantine period is probably the cause of the relative scarcity of Archimedean texts available to the Arabs. It now seems quite unlikely that they had access to any manuscript as complete as Greek manuscript A—the basis of most of our modern texts of Archimedes. The Arabic Archimedes contains the following works:[5] the *On the Sphere and the Cylinder* and at least a portion of Eutocius' commentary on it; (2) the *Measurement of the Circle*—with perhaps Eutocius' commentary; (3) a fragment of the *On Floating Bodies* (consisting of the enunciations without proofs of seven of the nine propositions of the first book and the first proposition of the second book—apparently the Arabs were not interested in the abstruse problems of the stability of

been found in any early medieval Latin codex.

[4] The standard account of the Greek manuscripts of Archimedes is still that of J. L. Heiberg, *Archimedis opera omina*, vol. *3* (Leipzig, 1915), Prolegomena. Cf. J. L. Heiberg, "Le Rôle d'Archimède dans le développement des sciences exactes," *Scientia*, vol. *20* (1916), pp. 81–89; and E. J. Dijksterhuis, *Archimedes* (Copenhagen, 1956), chapter II. The manuscript designations A, 𝔅, and C are those employed by Heiberg.

[5] For the works of Archimedes translated by the Arabs, see H. Suter, "Die Mathematiker und Astronomen der Araber," *Abhandlungen zur Geschichte der mathematischen Wissenschaften*, 10. Heft (1900), pp. 36–37, 52; E. Wiedemann, "Beiträge zur Geschichte der Naturwissenschaften III," *Sitzungsberichte d. Phys.-med. Soz. in Erlangen*, vol. *37* (1905), pp. 234, 247–50, 257.

parabaloidal segments found in Book II); (4) some indirect material from the *On the Equilibrium of Planes* found in other mechanical works translated into Arabic (such as Hero's *Mechanics*; the so-called Euclid *On the Balance*; *Liber karastonis*, etc.); (5) works perhaps based on originals of Archimedes now lost[6]—such as the *Lemmata* (*Liber assumptorum*), a *Book of Triangles*, *On the Seven-part Division of a Circle*, *On Touching Circles*, *On Parallel Lines*, *On Data*, *On Properties of a Right-angled Triangle*, and *On a Water-clock*. Distinguished by their absence are the *On Spiral Lines*, the *Quadrature of the Parabola*, the *Conoids and Spheroids*, the *Sandreckoner*, and the *On the Method*. One final observation must be made concerning the Arabic Archimedes. Even though a number of the best treatises are missing, the Arabs nevertheless mastered the techniques that mark Archimedes' work, and in such a way as to show that they had made them their own—in fact they did so much more readily on the whole than did the Latin mathematicians in the thirteenth and fourteenth centuries.

Now for our three questions. The first concerned the translation of the actual texts (or texts very close to actual ones) of Archimedes. Here we first should observe the importance of the date 1269, when the Flemish Dominican William of Moerbeke rendered virtually the whole corpus of Archimedes into Latin from the Greek text. Before that time the texts of Archimedes known to the Latin Schoolmen came from the Arabic and were limited in number, although not in circulation. The first work of Archimedes translated into Latin was the *Measurement of the Circle*. It was translated from the Arabic twice in the twelfth century (see Chapter Two). The first translation, which I have argued (but not with any finality) was done by Plato of Tivoli, was most inferior; just three manuscripts are known, only one of which is medieval. Apparently not long after this first translation the great translator Gerard of Cremona again used the Arabic text and rendered the *Measurement of the Circle* into Latin. This time the translation was quite accurate, and so before 1187 a faithful version of this short but important treatise became available. We are fortunate that this version was included in MS Bibliothèque Nationale, Fonds latin 9335, a handsome codex of Gerard translations. (Incidentally, this manuscript is one of the best examples of intelligent copying of scientific works. It has marginal variant readings which cite alternate traditions. The drawings are carefully made. Still more important, the transcription of numbers —even of six places—is almost perfect.)

[6] This list of works is largely drawn from al-Qiftī's account of Archimedes, as noted in Wiedemann, *op. cit.*, p. 247 (see note 5 above). Also see page 38 below.

Gerard's translation of the *Measurement of the Circle* seems to have some importance for the textual study of the original Greek text. The extant Greek version is clearly far removed from the text as composed by Archimedes. While the Gerard translation is not much closer, it does include a corollary (giving the area of a sector of a circle in terms of the length of the arc of the sector and its radius) that seems to go back to the original Archimedean text—although it is missing in the extant Greek text. Hero, who can be dated much earlier than our extant text, specifically states that Archimedes proved this corollary in his *Measurement of the Circle*.[7]

Of the Gerard version there are at least twelve manuscripts, and no doubt others which I have overlooked (see Chapter Two). Furthermore, it was the point of departure for some seven different paraphrases or reworkings of Proposition I, and influenced at least three others. Proposition I, you will recall, relates the area of the circle to a right triangle whose two sides including the right angle are equal respectively to the circumference and to the radius. These reworkings centered on elaborating by the use of Euclid's *Elements* the geometric steps only implied in the text of Archimedes. For example, by reference to the tenth book of the *Elements* (Proposition I) they demonstrate in a manner similar to that found in Proposition XII.2 of the *Elements* that as we continually double the number of sides of a regular polygon inscribed in a circle, more than half of the unexhausted area of the circle is thereby exhausted at each step—or, to put it in modern terms, they prove that the area of the inscribed polygon converges as close as we like toward that of the circle as we continually double the number of sides (see Chapter Three, page 605, footnote 1). These various reworkings of the Gerard version of Proposition I ought to be classified as a part of our answer to the third principal question on the use of Archimedes. Obviously, at least the first proposition of Archimedes' *Measurement of the Circle* was standard mathematical fare for the geometers of the thirteenth, fourteenth, and fifteenth centuries since it appears in one form or another in so many of the mathematical codices that date from these centuries. One ought to remark that on occasion the reworkings of the Gerard version appear in more than one manuscript, as the paraphrases themselves became standard for a given school or area.

Proposition II of the Gerard translation was also paraphrased along with some of these reworkings of Proposition I. Proposition III, involving

[7] Hero of Alexandria, *Metrica*, ed. H. Schöne (Leipzig, 1903), Bk. I, Chap. 37, p. 86.

the calculation of π, seems to have frightened the scribes, no doubt because of the extensive arithmetical calculations given there. At any rate, I know of only six independent versions of Proposition III other than the discussion by the Banū Mūsā (see Chapter Four) and the initial translation by Gerard. The first of these, by Johannes de Muris, appears to be based on the Moerbeke translation (see Volume Two). The second is in a manuscript that is so chaotic as to the numbers used that it would have been unintelligible (see Chapter Three, Section 3, footnote 6). However, the third of these versions—given in a Florence manuscript (Bibl. Naz., Con. Soppr. J.V. 30, ff. 9v–12v)—is most interesting, for it attempts to work out the calculations beyond the bare figures given in the Archimedean text; it does this somewhat in the manner of Eutocius in his commentary to the *Measurement of the Circle* (see Chapter Three, Section 3). The fourth is a brief fragment (see the Commentary to Chapter Three, Section 3, Proposition III, lines 147–67). For the other two, see Chapter Three, Section 3, footnote 6.

The second work of the pre-1269 period which has a more or less direct relationship with a text of Archimedes is the so-called *Liber de curvis superficiebus Archimenidis* (see Chapter Six, Section 2). This work is a paraphrase and reworking of some of the propositions of Book I of the *On the Sphere and the Cylinder*; it was done by a mysterious figure known, in Latin, as Johannes de Tinemue. It probably was composed in late antique or early Byzantine times, as I argue in Chapter Six below. This tract has the same objective as that of the *On the Sphere and the Cylinder*, namely, the finding of the surface area and volume of a sphere. It has a number of propositions closely akin to the propositions in the *On the Sphere and the Cylinder*, with demonstrations that are completely Archimedean in character—i.e., they employ the exhaustion procedure. These propositions—like Proposition I of the *Measurement of the Circle*—were to serve as models to medieval mathematicians when they attempted to apply this exhaustion procedure to entirely different propositions. The *Liber de curvis superficiebus* was a work that was quite popular, being cited almost as often as the *De mensura circuli*. And hardly an important geometer of the high and late Middle Ages failed to cite this treatise. Thus, in the fourteenth century it was used by such men as Bradwardine, Albert of Saxony, Johannes de Muris, etc. In my text of this treatise I have cited some twelve manuscripts of it, and, in some of these, additional and original medieval propositions with Archimedean type proofs were added —further evidence of the use of Archimedes by medieval authors.

In addition to the *De mensura circuli* and the *Liber de curvis superficiebus* there also circulated before 1269 a short fragment of the *On the Sphere and the Cylinder* (see Chapter Six, Section 1). This fragment was probably translated by Gerard of Cremona; it consists of some propositions which are cited in the introductions to the first and second books of the *On the Sphere and the Cylinder*—but no proofs were added. I have as yet found no citation of this fragment by later authors.

One further work translated from the Arabic by Gerard is quite close to the spirit of Archimedes' works. This is the so-called *Verba filiorum* or *Liber de geometria* of the eminent Arabic mathematicians, the Banū Mūsā, i.e., the sons of Mūsā ibn Shākir (see Chapter Four). This tract contains the first and third propositions of the *Measurement of the Circle* and several propositions deriving from the *On the Sphere and the Cylinder*—all with proofs in the Archimedean manner. It likewise contained solutions of two of the most fruitful problems of Greek geometry, the finding of two continually proportional means between two given quantities, and the trisection of any angle by a mechanical method. It further has the first known demonstration in Latin of Hero's formula for the area of a triangle in terms of its sides. (Incidentally, this formula is attributed to Archimedes by an Arab mathematician.[8])

When one appreciates the maturity of the geometry involved in the *Verba filiorum*, he must realize the great gulf that separates the elementary geometry of the early Middle Ages and the geometry of the period following the translations. The *Verba filiorum* was used by the famous mathematician Leonardo of Pisa, by Roger Bacon, and by others, as I shall show in my edition of this work. Furthermore, the so-called Pseudo-Bradwardine Version of the *De mensura circuli* (see Chapter Five, Section 1), as well as Albert of Saxony's treatment of quadrature (Chapter Five, Section 3), perhaps made use of it.

In the foregoing remarks I have touched upon the highlights of direct translation before 1269. In that year, as I said earlier, Moerbeke made his translation from the Greek (see also Volume Two). It included all of the texts known today with the exception of the *Sandreckoner*, the *On the Method*, the fragmentary *Stomachion*, and the brief *Bovine Problem*. It also included two of the three known commentaries of Eutocius, with only the commentary on the *Measurement of the Circle* being missing. The complete

[8] Al-Bīrūnī in a treatise *On Finding Sines in Circles* labels this theorem as one by Archimedes. See H. Suter, "Das Buch der Auffindung der Sehnen im Kreise...," *Bibliotheca Mathematica*, 3. Folge, vol. *11* (1910–11), p. 39.

Moerbeke translation exists in only one codex, a Vatican manuscript, Ottob. lat. 1850. Although the name of the translator is nowhere mentioned in this codex, consideration of historical circumstances and translating techniques convinces us that it was done by Moerbeke, as I shall show in Volume Two.

While a number of mistakes in translation occur, for the most part it is a careful and understandable rendering, though quite literal. The places that are particularly poor are usually those where the Greek text was itself at fault. It would appear to me that Moerbeke went to quite some trouble to master the mathematics involved. This must have been particularly difficult since there was so little Latin medieval precedence for most of the mathematics, say, of the *Conoids and Spheroids*, or the *On Spiral Lines*, or the *Quadrature of the Parabola*. His achievement is even more remarkable when we think of the variety of subjects involved in the various works of Aristotle and others that he translated. And Roger Bacon's comment that "William the Fleming knew nothing worth while in the sciences or in languages"[9] seems grossly unfair. I shall speak shortly of the use made of this translation in the answer to the third question.

The answer to the second question, concerning works translated that showed less direct influence, I shall make brief. These works were particularly important in the fields of hydrostatics and statics. In hydrostatics a pseudo-Archimedean *On Floating Bodies*, reflecting some of the propositions of the orginal treatise, was composed in the thirteenth century and was probably based on an Arabic translation. It circulated under the title of *De ponderibus Archimenidis* or *De insidentibus in humidum*.[10] We find in this treatise the first expression in Latin of the celebrated Principle of Archimedes.[11] It contains the first use of the expression "specifically heavier"—although, of course, it was not the first to use the concept.[12]

In the field of statics several short works were translated; particularly

[9] R. Bacon, *Compendium studii philisophiae*, in *Opera quaedam hactenus inedita*, ed. J. S. Brewer (London, 1859), p. 472.

[10] I have discussed this tract and its antecedents in my *Science of Mechanics in the Middle Ages*, chapter II. The text was published in E. A. Moody and M. Clagett, *Medieval Science of Weights* (Madison, 1952), pp. 35–53.

[11] Moody and Clagett, *Medieval Science of Weights*, p. 42, "Omnis corporis pondus in aere quam in aqua maius est per pondus aque sibi equalis in magnitudine."

[12] *Ibid.*, p. 42, "Duorum equalium corporum, altero gravius esse specie cuius pondus maiori calculorum numero adequatur."

worth noting were an anonymous *De canonio* from Greek and a *Liber karastonis* of Thābit ibn Qurra, from Arabic, by Gerard of Cremona.[13] These are both concerned with the problem of the steelyard or Roman balance (i.e., the balance of unequal arm lengths). These works show the influence of Archimedes in three respects. (1) They have adopted the Archimedean type of geometrical demonstration of statical theorems and the geometrical form implicit in weightless beams, weights that are really only geometrical magnitudes, etc. (2) They give reference in geometrical form to the law of the lever—and at least the reference in the *De canonio* is connected directly with Archimedes.[14] (3) They indirectly reflect the centers-of-gravity doctrine so important to Archimedes, in that in both treatises we find the practice employed of substituting for a segment of a material beam a weight equal in weight to this segment but hung at the middle point of a weightless segment used to replace the material segment.[15] These treatises, so Archimedean in form (even if combined with Aristotelian ideas), were certainly to prove vastly stimulating to the development of the statics in the thirteenth century associated with the name of Jordanus. And if one grants the importance of the Jordanus treatises, as I believe one must certainly do, then once again we have evidence of the importance of Archimedes in the growth of science, in an area where his importance is not widely realized.

In suggesting the answers to the first and second questions posed at the beginning of this chapter, I have partly answered my third question as to the use made by medieval mathematicians of these translations. An example of the original use of the *Measurement of the Circle* and the *Curved Surfaces* is found in a little-known treatise of the thirteenth-century geometer Gerard of Brussels, entitled *Liber de motu*.[16] It is devoted to kinematic problems arising from the rotation of geometrical figures. It has as its objective reducing the varying curvilinear velocities of such rotating figures to

[13] In *ibid.*, pp. 57–117, the texts and English translations of these works are given. A further discussion is found in chapter II of my *Science of Mechanics in the Middle Ages*.

[14] The author of the *De canonio* assumes that the law of the lever has been demonstrated "ab Euclide et Archimede et aliis." See Moody and Clagett, *Medieval Science of Weights*, p. 66. An earlier Latin association of Archimedes with the balance is found in Gerard of Cremona's translation of the *Epistola Ameti filii Iosephi de proportione et proportionalitate*. For the text of the reference by Ametus filius Iosephi to Archimedes, see page 629.

[15] Moody and Clagett, *Medieval Science of Weights*, pp. 64–67, 102–109.

[16] I have published the text of this work in my study, "The *Liber de motu* of Gerard of Brussels and the Origins of Kinematics in the West," *Osiris*, vol. *12* (1956), pp. 73–175.

uniform movements of translation. Not only does Gerard of Brussels cite the results of the two Archimedean treatises,[17] but what is even more important he borrows the basic kind of indirect proof found in these treatises, using it for the proof of wholly different theorems. Similarly, the mathematician Leonardo of Pisa makes original usage of the Archimedean type of proof, which he gained in part from reading the *Verba filiorum* of the Banū Mūsā (see Chapter Four).

Less fruitful was the use of the hydrostatic ideas of the pseudo-Archimedean treatise *De insidentibus in humidum*.[18] The vague, nonmathematical, nonexperimental character of the Aristotelian treatment of the density problem dominated the scholastic discussion of this problem in the fourteenth century. However, certain Schoolmen like Albert of Saxony and Blasius of Parma were not uninfluenced by the pseudo-Archimedean treatise. But at the crucial point when one would expect them to use the Principle of Archimedes, so succinctly expressed in the *De insidentibus*, to determine the specific gravities of bodies heavier than the fluid in which they are immersed, they fall back on the primitive method of comparing specific gravities by comparing (but without measuring) the speeds of descent of the bodies in the liquid. However, a somewhat more faithful student of the pseudo-Archimedean hydrostatics was Johannes de Muris at Paris in the 1340's. He repeats much of the earlier treatise in his *Quadripartitum numerorum*.[19]

I have spoken earlier of the flowering of statics in the thirteenth century under Jordanus, and I have indicated this was in part indirectly influenced by the Archimedean statical tradition which joined with the more philosophical Aristotelian tradition to produce brilliant proofs of the law of the lever for straight and bent levers, the equilibrium of weights on oppositely inclined planes, etc. This combined Aristotelian-Archimedean statics pro-

[17] *Ibid.*, p. 113, line 29, and p. 130, line 125 (where the *De quadratura circuli* is cited), and p. 132, line 7 (where the *De curvis superficiebus* is called *De piramidibus*; cf. p. 121, variant readings for lines 8–10). If one studies the analysis of contents given in the third part of the paper on Gerard, he will see the affinities in procedure between Gerard and Archimedes. One remarkable technique, somewhat incorrectly used by Gerard, involves the comparison of the movements of two rotating figures by comparing the motions of two corresponding line elements of the figures. This has some similarity with the balancing of corresponding line or surface elements of two figures by Archimedes in his *On the Method*. Needless to say, the *On the Method* was not available to Gerard.

[18] I have discussed the influence of the *De insidentibus in humidum* in chapter II of my *Science of Mechanics in the Middle Ages*.

[19] The main part of the hydrostatic section of the *Quadripartitum numerorum* has been given as "Document 2.3" of the *Science of Mechanics in the Middle Ages*.

duced in the Middle Ages was to circulate in the sixteenth century along with the more purely Archimedean statics of the *Equilibrium of Planes* and was to influence some of the early modern mechanicians in their approach to statics.

We have left one final problem on the use of Archimedes in the Middle Ages. To what extent did the new translation of the whole Archimedean corpus by Moerbeke influence the course of medieval mathematics? Heiberg felt that the Moerbeke translation exerted no influence at all in the Middle Ages. Put so baldly this is certainly not correct, as I shall show in Volume Two. The number of extant manuscripts of the Moerbeke translations at first glance seems to bear out Heiberg.[20] In addition to the single holograph manuscript of the whole corpus at the Vatican (Ottob. lat. 1850) there are only two other manuscripts with parts of the Moerbeke corpus. One is a Vatican manuscript of the early fourteenth century (Reg. suev. 1253, 14r–33r), which contains only the *On Spiral Lines*. The other is a badly done manuscript of the early fifteenth century (Italian in origin but now at Madrid, BN 9119) which contains the *Measurement of the Circle*, the *Equilibrium of Planes*, the *Quadrature of the Parabola*, the *On Floating Bodies*, and Eutocius' commentary on the *Equilibrium of Planes*.

However, these three manuscripts do not tell the whole story. For in 1340 there was produced at Paris (or at least circulated at Paris) a hybrid treatise entitled *De quadratura circuli*, which is compounded of a number of propositions of the *On Spiral Lines* with the first proposition of the *Measurement of the Circle* (for the text of this treatise, see Volume Two). This showed some real ingenuity on the part of the compositor. For he realized that the real difficulty in the first proposition of the *Measurement of the Circle* is not in the exhaustion proof showing that the circle is equal to the right triangle whose sides including the right angle are equal to the circumference and the radius of the circle. The proof is sound enough. But the true difficulty is one that precedes the proof; it is whether we can construct such a triangle, i.e., find straight lines equal to the curved lines. The first eighteen propositions of the *On Spiral Lines* lead to the construction of such a triangle. So our compositor borrowed most of these propositions with their proofs from the *On Spiral Lines* in the Moerbeke translation and, having constructed such a triangle, then proceeded to

[20] I have discussed the various manuscripts in "The Use of the Moerbeke Translations of Archimedes in the Works of Johannes de Muris," *Isis*, vol. 43 (1952), pp. 236–42, and more thoroughly in Volume Two of this work.

give the first proposition of the *Measurement of the Circle* to find its area. The celebrated Nicole Oresme knew this hybrid treatise and perhaps more of the original Vatican autograph of Moerbeke.[21] For it is at least possible that Oresme was the continuator who finished off Johannes de Muris' *Ars mensurandi* after the latter's death. In the continuation of this treatise —whether by Oresme or some other fourteenth-century figure is for our present purposes immaterial—the continuator included many propositions from the *On the Sphere and the Cylinder* and the *Conoids and Spheroids*, both in the Moerbeke translation. Further, there are a number of manuscripts including this continuation, which indicates that these Archimedean propositions were circulating.

So our conclusion—for fourteenth-century Paris at least—must be that the following works in the Moerbeke translation were known: *Measurement of the Circle*, *On Spiral Lines*, *On the Sphere and the Cylinder*, and *Conoids and Spheroids*. These were coupled with all of the earlier works translated from the Arabic which can be shown to have been used not only at Paris but in other schools of the fourteenth century.

As we approach the Renaissance and the end of our trail of medieval Archimedean texts and ideas, we ought to note the considerable influence exercised by the Moerbeke translation in the Renaissance. Heiberg is rightly convinced that the fifteenth-century translation of the works of Archimedes done in *ca.* 1450 by Jacobus Cremonensis was made with an eye on the Moerbeke translation.[22] (Incidentally, none of the hitherto known manuscripts[23] of this translation actually affirms Jacobus' authorship, but one recently noted copy, Paris, BN Nouv. acquis. lat. 1538, 1r–168v, ascribes it to Francesco Cereo de Borgo S. Sepulchri—"Interprete Franci‹s›co Cereo de Burgo Sancti Sepulchri."[24] I shall discuss the autorship of this version in Volume II). The Moerbeke translation was

[21] M. Clagett, "Three Notes," *Isis*, vol. *48* (1957), pp. 182–83.

[22] J. L. Heiberg, "Neue Studien zu Archimedes," *Zeitschrift für Mathematik und Physik*, vol. *34* (1890), Suppl. p. 93.

[23] A manuscript copy of this translation corrected by Regiomontanus was brought to Nuremberg and was published by Venatorius (Basel, 1544) with the first Greek edition; it is, I take it, still at Nuremberg, Stadtbibliothek(?), with a number Cent. V. 15. Other manuscript copies are at Venice (Bibl. Naz. San Marco f. a. lat. 327), Paris (BN lat. 7220, 7221), Florence (Bibl. Riccard. 106), and the Vatican (Urb. lat. 261).

[24] But at the end of Urb. lat 261 we read, "finis librorum Archimedis, quos transcribi iussit dominus Franciscus Burgensis." For the attribution to Jacobus, see J. L. Heiberg, ed., *Archimedis Opera omnia*, vol. *3* pp. LXIX–LXX, LXXIV–LXXV.

also known and cited by Leonardo da Vinci.[25] Gaurico and Tartaglia pilfered it in the sixteenth century,[26] and actually the first translations of Archimedes to appear in print were the medieval translations of Moerbeke published at Venice by Gaurico in 1503 (the *Measurement of the Circle* and the *Quadrature of the Parabola*), and by Tartaglia in 1543 when he published directly from Gaurico's edition the same treatises and in addition, from the Madrid manuscript, the Moerbeke translation of the *Equilibirium of Planes* and Book I of the *On Floating Bodies*. Both books of Moerbeke's translation of the *On Floating Bodies* were published twice in 1565, the first time by Curtius Troianus in Venice and the second time by Federigo Commandino in Bologna. Since the Greek text in MS 𝔅 of the *On Floating Bodies* used by Moerbeke was by this time lost, Moerbeke's translation remained the only version of this important work until the discovery of manuscript C in the twentieth century.

Of the sixteenth-century Latin versions of the works of Archimedes, Commandino's translations published at Venice in 1558 are among the most important.[27] They perhaps demonstrate some acquaintance with the Moerbeke translation. A second translation (or, rather, a paraphrase translation of Archimedes' works made by Francesco Maurolico over the period 1534–1549) was published in an edition of 1570 which was almost completely destroyed and was then reprinted in 1685.[28] That Maurolico knew of both the Gerard of Cremona translation of the *Measurement of the Circle* as well as that of Moerbeke can be shown. At least one further translation of Archimedes' works was made in the sixteenth century, that of Antonius de Albertis (see Volume Two), who studied with Marsiglio Ficino's successor at the Florentine Academy, Francesco da Diacetto (1466–

[25] W. Schmidt, "Zur Textgeschichte der 'Ochumena' des Archimedes," *Bibliotheca Mathematica*, 3. Folge, vol. *3* (1902), pp. 176–79.

[26] Clagett, "The Use of the Moerbeke Translations..." p. 237.

[27] On the whole, Commandino's translation is most original and mathematically perceptive. His *Archimedis opera nonnulla in latinum conversa et commentariis illustra* (Venice, 1558) contains translations of the *Measurement of the Circle*, *On Spiral Lines*, *Quadrature of the Parabola*, *Conoids and Spheroids*, and the *Sandreckoner*. His correction (without a Greek text) of the Moerbeke translation of the *On Floating Bodies* (Bologna, 1565) was brilliant, containing as it did considerable reconstruction of the text already much corrupted by the time of the writing of MS 𝔅 which Moerbeke used.

[28] *Admirandi Archimedis Syracusani monumenta omnia mathematica, quae extant... ex traditione doctissimi viri D. Francisci Maurolici* (Palermo, 1685). The original dates of the various works are given in this edition. Since MS 𝔅 was already lost, no franslation of the *On Floating Bodies* was made by Maurolico.

1522).[29] This translation is found in a Viennese codex;[30] it appears unrelated to any known translation.

As we glance back over this brief outline of the role of Archimedes in the Middle Ages, I think we must conclude that he played a modest but nevertheless important part in that thin tradition of Greek mathematics and physics that trickled down through the Middle Ages into modern times.

[29] A. della Torre, *Storia dell'Accademia Platonica di Firenze* (Florence, 1902), pp. 832–33. Cf. O. Kristeller, *Renaissance Thought and Letters* (Rome, 1956), pp. 322n–23n.

[30] Vienna, Nationalbibliothek, cod. 10701. We read, f. 1r, "Archimedis de quadratura paraboles liber e graeco tralatus Antonio de Albertis interprete." On the bottom of f. 1r we find the scribe's name, "Mich. Creten (?)," with a date which I read as "1562" and a place "Bonon" (Bologna?). The table of contents gives these works: *Quadrature of the Parabola*, *On the Sphere and the Cylinder*, *On Spiral Lines*, *Conoids and Spheroids*, *Measurement of the Circle*, *On the Equilibrium of Planes*, and *Sandreckoner*. But in the present manuscript, there is nothing after the middle of Proposition 22 of the *Conoids and Spheroids* (p. 356, line 10, of Heiberg's edition). Cf. Heiberg, *Arch. Op. omnia*, p. LXXXIV, where in addition to this translation of Antonius, Heiberg also notes still another translation: "cod. Riccard. 107: Archimedis opera latine per Iosephum Averianum (autogr.), cum Eutocio. in prima pag. legitur: 'haec scribebam succisivis horis primo anno quo Pisas me contuli ad perdiscendam iurisprudentiam.'" I have not seen this last codex, but I shall discuss it in Volume Two.

Chapter two

Translations of the *De mensura circuli* from the Arabic

As I indicated briefly in the first chapter, direct contact of the Latin Schoolmen with the works of Archimedes began with the translation from the Arabic of the short tract *Measurement of the Circle*. Early Latin mathematicians were familiar with the quadrature problem in only the vaguest way, it having been briefly discussed in Boethius' *Commentary on the Categories of Aristotle*.[1] There were several non-Archimedean discussions of quadrature before the twelfth century and a few even after the appearance of the work of Archimedes.[2] But with the translation of the

[1] Boethius, *In Categorias Aristotelis*, Bk. II (ed. in Migne, *Patrologia latina*, vol. *64*, pp. 230–31). See below, Commentary to lines 2–3 of Section 3 of Appendix I.

[2] Among the eleventh-century considerations of quadrature was Franco of Liège's *De quadratura circuli*, edited by Dr. Winterberg, *Abhandlungen zur Geschichte der Mathematik*, 4. Heft (Leipzig, 1882), pp. 137–90. Franco accepts as a fact that the surface area of a circle is exactly equal to the square of the diameter multiplied by 11/14, a formula transmitted by the Roman *agrimensores*. Franco takes a circle with a diameter of 14 feet and first constructs a rectangle 14 by 11. He then seeks in vain to construct a square equivalent to that rectangle, since he does not know how to find a line which is a geometrical mean between two given lines. For Franco's treatise and other efforts of this sort, see P. Tannery, *Mémoires scientifiques*, vol. *5* (Paris, Toulouse, 1922), pp. 86–90, 233, 257–62, 294. In the thirteenth century Jordanus, the eminent mathematician, treated the quadrature question in a non-Archimedean manner in his *De triangulis* (see Appendix I, Section 1). Later in the century the Euclid commentator Campanus of Novara perhaps was the author who misinterpreted the problem (see Appendix I, Section 3). Compare with these treatments that incorrectly attributed to Johannes de Chinemue (in Appendix I, Section 2). Also see Ramón Lull's *De quadratura et triangulatura circuli* (ed. by J. E. Hofmann, "Cusanus-Studien VII. Die Quellen der Cusanischen Mathematik

De mensura circuli in the twelfth century, Archimedes' solution (although no doubt not in its pristine form[3]) became generally available and began to dominate discussions of the question of quadrature, as Chapter Three amply demonstrates.

1. The Translation Perhaps by Plato of Tivoli

Of the various versions of the *De mensura circuli* that circulated in the twelfth and thirteenth centuries only two appear to be direct translations from the Arabic. The first translation, i.e., the one I judge to be the older, is contained in a thirteenth-century manuscript of the Bibliothèque Nationale at Paris (MS *K*). This copy was twice copied in the sixteenth century (MS *O* and MS *W*). In view of the fact that these last two manuscripts were in all probability rendered from the thirteenth-century manuscript and that that manuscript is very clear and easy to read, only the medieval copy is essential to the establishment of the text. The text here published is a slight revision of the text which I published in 1952.[4]

That the translation was done from the Arabic cannot be doubted. For it opens with the traditional praise to God or *bismillah*: "In nomine domini misericordis et miseratoris," which translates the Arabic phrase, "Bismillāhi al-rahmanī al-rahīm." Furthermore, it will be noticed that an Arabic form of the name of Archimedes is used: "Incipit liber Ersemidis in quadratum circuli." The form *Ersemides* is similar to the forms *Arsamithes* and *Arsamides* used by Gerard of Cremona in translating from the Arabic.[5]

I: Ramon Lulls Kreisquadratur," *Sitzungsberichte der Heidelberger Akademie der Wissenschaften. Philosophisch-historische Klasse*, Jahrgang 1941–42. 4. Abhandlung, 1–37), and also Lull's *Nova geometria* as studied by J. M. Millás Vallicrosa, *Estudios sobre historia de la ciencia española* (Barcelona, 1949), pp. 359–68. A different but quite interesting discussion of quadrature (also translated from the Greek) was the discussion from Simplicius' *Commentary on the Physics of Aristotle*, translated and paraphrased by Robert Grosseteste under the title *Quadratura circuli* or *Quadratura circuli per lunulas* (see Appendix II).

3 A. F. von Pauly and G. Wissowa, *Real-encyclopädie der classischen Altertumswissenschaft*, Supplementband III (Stuttgart, 1918), cols. 145–46.

4 "Archimedes in the Middle Ages: The *De mensura circuli*," *Osiris*, vol. *10* (1952), pp. 599–605.

5 *Arsamithes* is the form used by Gerard in his translation of the *De mensura circuli* (see the text published in Section 2 of this chapter) as well as in his translation of Anaritius' commentary on the *Elements* of Euclid; see Euclid, *Opera omnia*, Supplementband (Leipzig, 1899), p. 6, line 2; p. 24, line 30; p. 28, line 18; p. 162, line 23.

It is possible that this translation was done by Plato of Tivoli. It follows Plato's translation of the *Liber embadorum* of Abraham bar Ḥiyya (Savasorda) from the Hebrew in BN lat. 11246 and seems to use the same kind of terminology as the latter work.[6] Plato's translations can be dated between 1134 and 1145 and were made for the most part at Barcelona.[7] The translation, or at least the medieval copy we have, is quite corrupt. As the reader glances through the text (and pays particular attention to the variant readings), he will notice a number of errors in rendering numbers. These errors might have arisen from the ineptness of the scribe rather than from the translator. But even so, when one compares this version with the Arabic text of the *De mensura circuli*, which to be sure exists only in the later version of al-Ṭūsī,[8] he will find a certain awkwardness of translation even in the parts that do not involve numbers. Furthermore, this Latin version, as we have it, is defective. Only the first half of the third proposition involving the calculation of π is included.

My text is based almost entirely on manuscript *K*, and diverges from that copy only where an obvious scribal error has been committed. In those cases where I have altered the reading of the manuscript I have, of

But in his translation of the *Epistola Ameti filii Iosephi de proportione et proportionalitate* he uses *Arsamides* (MS BN lat. 9335, 66r). In his translation of the *Verba filiorum* of the Banū Mūsā he uses *Archemides* (*ibid.*, 58r), as well as *Archimenides*. H. Suter lists as possibilities: Arsamides, Ersemides, Arschimenides, Arsimenides, Archimenides, and Arsamites. In the thirteenth and fourteenth centuries *Archimenides* was by far the most popular (see the various versions of Chapter Three).

[6] For example, both use *embadum* for "area," and they use many words and phrases that are common and do so in the same situations: the manifold use of the verb *abscido* and its derivatives, the common use of the word *colligitur*, the rendering of polygon by *multiangula figura*, and many others. See the text of the *Liber embadorum* published by M. Curtze, *Urkunden zur Geschichte der Mathematik im Mittelalter und der Renaissance* in *Abhandlungen zur Geschichte der mathematischen Wissenschaften*, 12. Heft (Leipzig, 1902), pp. 1–183.

[7] Charles H. Haskins, *Studies in the History of Mediaeval Science*, 2 ed. (Cambridge, Mass., 1927), p. 11. Cf. B. Boncompagni, "Delle versioni fatte da Platone Tiburtino ...," *Atti dell' Accademia Pontificia dei Nuovi Lincei*, vol. 4 (1851), pp. 248–86, and J. M. Millás Vallicrosa, *Nuevos estudios sobre historia de la ciencia española* (Barcelona, 1960), pp. 109–110.

[8] I have examined a number of copies of the Arabic text and they all seem to be Thābit ibn Qurra's translation as edited by the thirteenth-century mathematician Nasīr al-Dīn al-Ṭūsī; for example, BN Fonds arabe 2467, 139v–42v, Bodleian Arabic MSS, Marsh 709, 35v–38r, Arch. Seld. A45, 83r–85r, and John Rylands Library, Arabic MS 350, 120v–25r, 341r–47r. Others that I have spotted in the catalogues of Arabic manuscripts also appear to be of the al-Ṭūsī edition. For the Arabic text of al-Ṭūsī, see the special variant readings from the Arabic added to the Gerard of Cremona text in Section 2 of this chapter.

course, given the original reading in the variant readings. On occasion I have used the corrections of *W* (executed by Francesco Barozzi in 1565) to correct *K*, but I have not given as variants the many modern readings of *W*. To do so would be of no assistance in establishing the original text, since *W* is copied from *K* (or possibly from *O*, which was itself copied from *K*).

It ought to be noticed that both *W* and *O* have the same gross numerical errors as *K*, including even the error of line 105 where the scribe of *K* apparently changed the letters *CL*, which designated a geometric quantity, to the numerals 150. The repetition of the errors of *K* and particularly of the last mentioned error establishes without doubt the close affinity of the sixteenth century copies with *K*. While I have generally not included the variants from *W*, I can here note briefly some of the modernizations of *W*. Thus, the Arabic *bismillah*, mentioned above, has been omitted. The title has been altered to read: "Incipit liber Ersemidis De dimensionibus. De circuli dimensione." Obviously the circulation of the Greek text has had its effect on Barozzi. Other characteristic alterations made by Barozzi in the text of *K* are *segmenta* for *portiones* or *abscisiones*, *area* for *embadum*, *ratio* for *proportio*, and so on. Barozzi has also added some marginal comments. The propositions, unnumbered in *K* and *O*, are numbered in *W*.

The Note on Textural Procedure should be consulted in regard to general practice in presenting these texts. In addition to this general note, it should be observed that the scribe of *K* used the Indo-Arabic numerals throughout. This practice was probably the exception rather than the rule in these medieval geometric manuscripts, at least before the middle of the fourteenth century. And it is evident that our scribe of the Paris manuscript was copying from a manuscript that had used Roman numerals which he converted to Arabic numerals, for (as I have noted) in line 105 he made the error of converting the *CL* that designated a line to the Arabic numerals 150 (see the variant reading for this line).

The marginal folio numbers are from MS *K*.

Sigla of Manuscripts

K = Paris, BN lat. 11246, 37v–39r, 13c.
O = Paris, BN lat. 7224, 63r–65r, 16c.
W = Dublin, Trinity College D.2.9, 54r–55r, 1565.

On the Square of the Circle

In Quadratum Circuli

37v / In nomine domini misericordis et miseratoris,
incipit liber Ersemidis in quadratum circuli.

⟨I.⟩ OMNIS CIRCULUS TRIANGULO ORTOGONIO, CUIUS UNUM EX DUOBUS LATERIBUS RECTO ADIACENTIBUS 5 ANGULO DIMIDIO DIAMETRI EIUSDEM CIRCULI, ALTERUM VERO LATUS LINEE CIRCUMFERENTI EQUATUR, EXISTIT EQUALIS.

Sit ergo circulus *ABCD* [Fig. 1] triangulo *E* in supradictis omnibus equus. Dicemus igitur ipsius embadum embado illius equum existere. 10 Quod si ei equalis non fuerit, erit itaque circuli embadum eo maius vel minus. Facta autem primus positione quod sit maius, intra circulum quadratus *AC* describatur. Eo igitur descripto ex circulo *ABCD* plus dimidio intra quadratum continebitur. De hinc arcum *AB*, et ceteros arcus, in duo equa supra punctum *R*, et super alia puncta, abscidamus. 15 Post hoc alias duas lineas *AR*, *RB* producamus. A reliquis igitur sectionibus circuli plus earumdem medietate separabitur. Cumque hoc

14 alia puncta *correxi ex* alicuius punctos *in KO et ex* aliqua alia puncta *in W*

On the Square of the Circle

In the name of the most merciful God, the Book of Archimedes on the Square of the Circle begins.

I. EVERY CIRCLE IS EQUAL TO A RIGHT TRIANGLE, ONE OF WHOSE TWO SIDES ADJACENT TO THE RIGHT ANGLE IS EQUAL TO THE RADIUS OF THE SAME CIRCLE, WHILE THE OTHER SIDE EQUALS THE CIRCUMFERENCE.

Hence let circle *ABCD* [see Fig. 1] be equal to △ *E* in all the abovementioned respects. Therefore, we shall say that the area of the one is equal to the area of the other. For if the area of the circle is not equal to

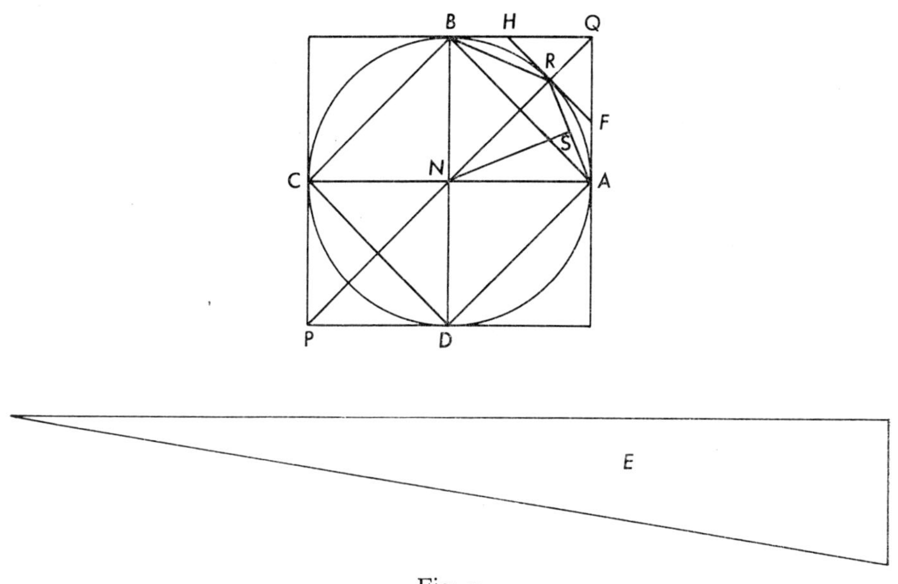

Fig. 1

it, then it will be either greater than or less than it. Having assumed first that it is greater, let the square *AC* be inscribed in the circle. Hence within the square there will be contained more than half of circle *ABCD* in which it is inscribed. Then we bisect arc *AB* at point *R*, and the other arcs at other points. After this, let us draw two other lines, *AR* and *RB*. Hence more than half of the [area of the] remaining segments of the circle will

frequenter fecerimus abscisiones minores eo in quo predictus circulus prefatum triangulum excedit in constante remanebunt. Rectilinea itaque figura multiangula ab ipso circulo circumdata triangulo maior existit. Positoque circuli centro puncto *N* ca/thetum *NS* protrahemus. Linea igitur *NS* minor est quovis illorum laterum trianguli que recto adiacent angulo; et multiangule figure circumductio reliquo latere minor existit. Quodque ex multiplicatione alterius lateris trianguli recto angulo adiacentis in alterum, scilicet, latus recto adiacens angulo, quod est duplum totius aree trianguli colligitur. Maius est eo quod ex multiplicatione *NS* in circumdatione multiangule figure coadunatur quod est duplum aree multiangule figure. Dimidium itaque dimidio maius existit. Maior est ergo triangulus figura multiangula. Superius autem eum minorem esse monstravimus. Igitur hoc est impossibile.

38r 21 25

Item si predictus circulus minor triangulo *E* et circa ipsum quadratus *PQ* describatur, abscisum est itaque ex quadrato *PQ* plus sui ipsius dimidio, quod est circuli quantitas. Hiis itaque peractis arcum *BA* in duo equa supra punctum *R*, ceteros etiam arcus super alia puncta, in duo divideres equalia. A portionum punctis lineas ipsum circulum contingentes abstrahemus. Linea igitur *FH* in duo equa supra punctum *R* absciditur. Super lineam itaque *FH* linea *NR* perpendiculariter erigitur. Idem et de ceteris lineis ostendetur. Verum quia *QF* et *QH* sunt maiores *HF*, erit dimidium maius dimidio. Linea igitur *QH* maior est linea *HR*, que est equalis linee *HB*. Triangulus itaque *RQH* maior est dimidio trianguli *QRB*. Multo igitur maior dimidio figure *RQB* lineis *BQ*, *QR*, et arcu *BR* contente. Similiter et triangulus *QRF* maior est *RFA*. Totum igitur *QHF* dimidio figure *AQB* maius existit. Eodemque modo alii trianguli maiores dimidio aliarum abscitionum fore probantur. Cumque hoc frequenter factum fuerit, abscisiones semper extra circulum remanebunt que in unum collecte minus eo in quo triangulus *E* circulum *ABCD* superat apparebunt. Facta namque positione quod portio *RFA* alieque portiones supersint, figuram rectilineam circa ipsum circulum descriptam triangulo *E* minorem fore pronunciabimus; quod esse non potest, ipso enim maior existit. Circulus itaque triangulo *E* minor esse non potest. Supra vero probatum est ipsum ipso maiorem nullatenus esse. Necesse est igitur

30 35 40 45 50

19 multiangula *W* multiangulo *KO*
20 cathetum *corr. ex* chatetum *in KO*
23 trianguli *W* trangulo *KO*
24 recto *corr. ex* recti *in KO*
33 ceteros *W* curtos *KO*
40 maior est: maiorem (?) *O*
41 lineis: sub lineis *O*
42 AQB: a quibus *O*

be extracted. And when we shall have done this a number of times, there will unalterably remain segments [whose total area is] less than the amount by which the aforesaid circle exceeds the previously mentioned triangle. Consequently the polygon circumscribed by that circle is larger than the triangle. With the center of the circle placed at point *N*, let us draw the perpendicular *NS*. Therefore, line *NS* is less than one of those sides of the triangle adjacent to the right angle, and the perimeter of the polygon is less than the remaining side. The product of [1] one of the sides of the triangle adjacent to the right angle and [2] the other side adjacent to the right angle, which product is twice the area of the whole triangle, is obtained. It is greater than the product of *NS* and the perimeter of the polygon, which product is double the area of the polygon. And so its half is greater than the half of the latter product. Therefore, the triangle is greater than the polygon. But we have demonstrated above that it is less. Therefore, this is impossible.

Also, if the aforesaid circle is less than △ *E*, and around it square *PQ* is circumscribed, then the area of the circle has cut out more than half of square *PQ*. With these things done, you should bisect arc *BA* at point *R*, and the other arcs at other points. We shall draw lines tangent to the circle at the points [making equal divisions] of the arcs. Thus line *FH* is bisected at point *R*. And so line *NR* is erected perpendicularly on line *FH*. The same thing will be shown regarding the other lines. It is true that, since *QF* and *QH* are together greater than *HF*, the half [of *QF* and *QH* together] is greater than the half [of *HF*]. Therefore, line *QH* is greater than line *HR*, which is itself equal to line *HB*. And so △ *RQH* is greater than half of △ *QRB*. Hence it is much greater than half of figure *RQB* contained by lines *BQ*, *QR*, and arc *BR*. Similarly △ *QRF* is greater than *RFA*. Therefore the whole of *QHF* is greater than half of the figure *AQB*. In the same way the other triangles are proved to be greater than half of the other parts cut off. When this is done repeatedly, there will always result outside of the circle [some totality of] cut off portions which joined together will be less than the amount by which △ *E* exceeds circle *ABCD*. For it having been posited that portion *RAF* together with the other portions constitute the excess [beyond the circle], we shall assert that the rectilinear figure circumscribed about the circle would be less than △ *E*, which cannot be so, for it is greater than it. And so the circle cannot be less than △ *E*; and above it was proved that it was in no wise greater than it. It is necessary, therefore, that they are mutually equal. Similarly it will be proved that the area of △ *E* is equal

eos sibimet invicem equos existere. Similiter probabitur embadum
38v trianguli *E* ei quod procedit ex mul/tiplicatione sui perpendicularis
in dimidium sue basis equum existere; eiusdem quoque perpendicu-
55 laris equus est dimidio diametri circuli *ABCD*; ipsiusque basis linee
circumferenti eiusdem circuli existit equalis. Illud igitur quod ex multiplicatione dimidii diametri in dimidium circumferentie *ABCD* provenit embado trianguli est equale. Et hoc volumus.

⟨II.⟩ PROPORTIO AREE OMNIS CIRCULI AD QUADRA-
60 TUM SUI DIAMETRI EST UT PROPORTIO 11 AD 14.

Sit itaque linea *AB* circuli diametrus supra quem quadratum *CG*
constituamus [Fig. 2]. Sitque linea *DC* linee *DE* dimidium. Sit etiam
EF septima pars linee *DC*. Quia igitur proportio trianguli *ACE* ad
⟨tri⟩angulum *ACD* est ut proportio 21 ad 7 et proportio *ACD* ad
65 *AEF* est ut proportio 7 ad 1, erit proportio trianguli *ACF* ad
⟨tri⟩angulum *ACD* sicut proportio 22 ad 7. Proportio autem quadrati
CG est quadrupla triangulo *ACD* et triangulus *ACF* circulo *AB*
equus existit, eo quod perpendicularis *AC* linee que ab huius circuli
centro ad circumferentiam protrahitur est equalis et basis *CF* equa
70 est linee circumferenti. Linea enim circumferens maior est triplo
diametri septimaque superaddita fere. Probatum est igitur ex hoc
quod dictum est, quod circuli *AB* proportio ad quadratum *CG* est
ut proportio 11 ad 14, et id est quod volumus, ut in subscripta figura
monstratur.

75 ⟨III.⟩ OMNIS LINEA CIRCUMFERENS SUPPERADDIT VEL
MAIOR EST TRIPLO DIAMETRI MINUS SEPTIMA PLUSQUE
DECEM PARTIBUS EX 71 DIAMETRI PARTIBUS.

Sit itaque linea *AC* diametrus circuli cuius centrum est punctus *E*
et linea *DF* eundem circulum contingat [Fig. 3]. Angulus quoque
80 *FEC* duas unius anguli recti partes contineat. Proportio igitur *EF*
ad *FC* est ut proportio 306 ad 153, et proportio *EC* ad *FC* maior
est proportione 265 ad 153. Hiis ita constitutis angulum *FEC* in duo
equa cum linea *GE* dividemus. Igitur ⟨pro⟩portio duarum linearum
FE, *EC* simul collectarum ad *FC* est ut ⟨pro⟩portio *EC* ad *CG*. Erit

57 diametri *corr. ex.* diametrum *in KO*/ *post* ABCD *add. KO* quod *sed delevi*
61 diametrus *corr. ex* diametrum *in KO*
64, 66 ⟨tri⟩angulum *corr. W ex* angulum *in KO*
68 ab *corr. W ex* ad *in KO*
78 diametrus *corr. ex* diametrum *in KO*
83, 84, 85 ⟨pro⟩portio *corr. O ex* portio *in K*

to the product of its altitude and one half of its base, its altitude being equal to the radius of circle *ABCD* and its base being equal to the circumference of the same circle. Therefore, the product of the radius and one half of the circumference of circle *ABCD* is equal to the area of the triangle. And this we wish.

II. THE RATIO OF THE AREA OF ANY CIRCLE TO THE SQUARE OF ITS DIAMETER IS AS THE RATIO OF 11 TO 14.

And so let line *AB* be the diameter of the circle on which we construct square *CG* [see Fig. 2]. And let line *DC* be half of line *DE*. Further let *EF* be a seventh part of line *DC*. Because, therefore, the ratio of △ *ACE* to △ *ACD* is as the ratio of 21 to 7 and the ratio of *ACD* to *AEF* is as the ratio of 7 to 1, the ratio of △ *ACF* to △ *ACD* will be as the ratio of 22 to 7. Moreover, the ratio of square *CG* to △ *ACD* is as 4 to 1, and △ *ACF* is equal to circle *AB* because the altitude *AC* is equal to the line drawn from the center of the circle to the circumference and the base *CF* is equal to the circumference. For the circumference is more than three times the diameter by an amount that is almost one seventh [of the diameter]. Therefore, from what has been said, it has been proved that the ratio of circle *AB* to square *CG* is as the ratio of 11 to 14. And this is what we wish, as is shown in the figure [Fig. 2] drawn below.

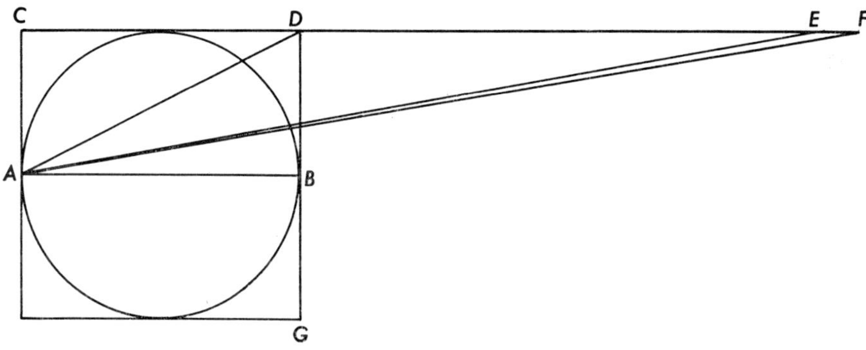

Fig. 2

III. EVERY CIRCUMFERENCE EXCEEDS OR IS GREATER THAN THREE TIMES THE DIAMETER BY AN AMOUNT LESS THAN ONE SEVENTH AND GREATER THAN TEN SEVENTY-FIRST PARTS OF THE DIAMETER.

And so let line *AC* be the diameter of the circle whose center is point *E*, and let line *DF* be tangent to the same circle [see Fig. 3]. Also let ∠ *FEC* contain two [of the six?] parts of one right angle. Therefore, *EF*/*FC* =

85 itaque ⟨pro⟩portio *CE* ad *CG* maior proportione 571 ad 153. Quare proportio *EG* in fortitudine ad *CG* in fortitudine est ut proportio ⟨34⟩9 milium quadringentorum ⟨et 50⟩, i.e., ad ⟨vigenti⟩ tria milia quadringenta novem. Proportio vero quam eadem *EG* ad eandem *CG* habet in longitudine maior est proportione 591 ⟨ et unius octave⟩
90 ad 153. Simili quoque modo angulo *CEG* in duo equa cum linea *EH* diviso, probabitur quod proportio *EC* ad *CH* maior est proportione 1162 et unius octave ad 153. Proportio igitur *HE* ad *CH* est maior proportione 1172 et unius octave ad 153. Item si angulum *HEC* in duo equa cum linea *EK* diviserimus, erit proportio *EC* ad *CK* maior
95 proportione 2334 et unius quarte ad 153. Proportio igitur *EK* ad *CK* maior est proportione 2339 uniusque quarte ad 153. Similiter diviso angulo *KEC* in duo equalia cum linea *LE*, erit proportio *EC* ad *CL* in longitudine maior proportione 4673 et semissis ad 153. Et quoniam
39r angulus *FEC* duas unius anguli recti partes continet, erit / angulus
100 *LEC* una pars ex 48 partibus unius anguli recti. Supra punctum etiam *E* si fecerimus angulum *CEM* angulo *LEC* equalem, angulus *LEM* unam partem ex 24 partibus unius anguli recti continebit. Linea igitur *LM* recta est latus multiangule figure circa ipsum angulum designate que 96 equos angulos amplectitur. Et quia probatum est quod pro-
105 portio *EC* ad *CL* sit maior proportione 4673 et semissis ad 153 et duplum linee *EC* est linea *AC*, duplum quoque *CL* est linea *LM*, erit proportio *AC* ad circumductionem figure multiangule maior proportione 4673 et semissis ad 14688, quod est plus triplo 667 et semissis. Figura⟨m⟩ igitur multiangula⟨m⟩ circa ipsum circulum descriptam
110 maiorem triplo diametri in quantitate unius septime partis totius diametri fore necesse est.

[*Ends without second half of proof.*]

87 ⟨34⟩ *om KWO*/⟨et 50⟩ *om KWO*/⟨vigenti⟩ *om KWO*
89 591 *corr. ex* 571 *in KWO*/⟨et... octave⟩ *om. KWO*
91 CH *corr. ex* GH *in KWO*
92 1162 *corr. ex* 1062 *in KWO*
93 octave *corr. ex* quarte *in KWO*
96 Similiter *corr. W ex* simile *in KO*
97 CL *corr. ex* DK *in KWO*
98 4673 *corr. ex* 4693 *in KWO*
100 una pars *corr. ex* unius partibus *in KWO*
102 24 *corr. mg. W ex* 23 *in KWO*
105 CL *corr. ex* 150 *in KWO*
105, 108 4673 *corr. ex* 4693 *in KWO*
106 CL *corr. ex* GL *in KWO*
108 667 : 567 *O*
109 ⟨m⟩[1, 2] *om. KWO*

306/153, and $EC/FC > 265/153$. With these things established, we shall bisect $\angle FEC$ with line GE. Therefore, $(FE + EC)/FC = EC/EG$. And so $CE/CG > 571/153$. Therefore, $EG^2/CG^2 = 349450/23409$, while $\frac{EG}{CG} > \frac{591\frac{1}{8}}{153}$. Again in the same way, with $\angle CEG$ bisected by line EH, it will be proved that $\frac{EC}{CH} > \frac{1162\frac{1}{8}}{153}$. Hence $\frac{HE}{CH} > \frac{1172\frac{1}{8}}{153}$. In like manner, if we bisect $\angle HEC$ with line EK, then $\frac{EC}{CK} > \frac{2334\frac{1}{4}}{153}$. Hence $\frac{EK}{CK} > \frac{2339\frac{1}{4}}{153}$. With $\angle KEC$ bisected in the same way by line LE, $\frac{EC}{CL} > \frac{4673\frac{1}{2}}{153}$.

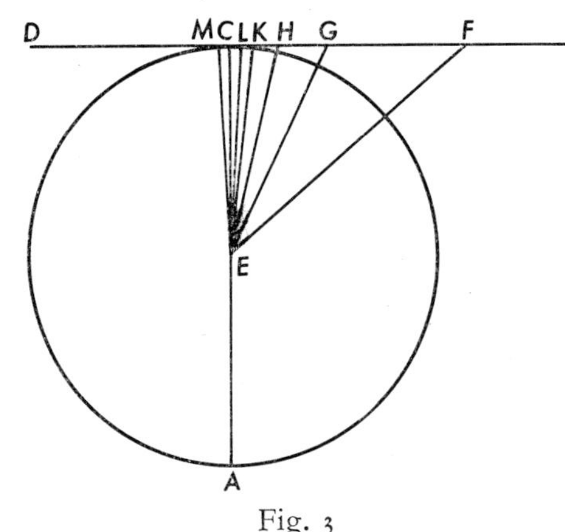

Fig. 3

Since $\angle FEC$ contains two [of the six?] parts of one right angle, hence $\angle LEC$ will be 1/48 of a right angle. If on point E we make $\angle CEM$ equal to $\angle LEC$, then $\angle LEM$ will contain 1/24 of a right angle. Therefore, straight line LM is the side of a polygon drawn about that angle (i.e., circle) which includes 96 equal angles. And since it was proved that $\frac{EC}{CL} > \frac{4673\frac{1}{2}}{153}$ and line $AC = 2\,EC$ and $LM = 2\,CL$, the ratio of AC to the perimeter of the polygon is greater than the ratio of $4673\frac{1}{2}$ to 14688, which ratio is greater than 3/1 by $667\frac{1}{2}$. Therefore, it is necessary that the [perimeter of the] polygon described about that circle would be more than triple the diameter by an amount of [less than] one seventh part of the whole diameter.

COMMENTARY

As I pointed out in the introduction to Section 1, the translation was in all probability defective, although to be sure the translator might have had a poor Arabic text to translate from. While the text on the whole is like the translation of Gerard of Cremona given in Section 2 of this chapter, it diverges in enough places for us to be sure that the two translators did not use precisely the same Arabic text. Particularly important is the fact that this text lacks the corrollaries to Proposition I that are given in the Gerard translation. This text is also defective in the second half of the third proposition. The mistakes in the rendering of numbers in the third proposition (indicated in the variant readings) are perhaps mistakes of the scribe rather than of the translator, since at least some figures in the long numbers of four or five places are given correctly.

12–13 "Eo...continebitur." The comment to the effect that the square exhausts more than half of the circle is missing from the Greek text, which throughout this work I employ in the edition of J. L. Heiberg, Archimedes, *Opera omnia cum commentariis Eutocii*, vol. *1* (Leipzig, 1910), pp. 232–43. See Chapter Three, Section 2, footnote 1, for comments on the "exhaustion" procedure.

22–28 "Quodque.... multiangula." The Greek text lacks the obvious multiplication of the two sides of the triangle which is given in this text. The main conclusion is presented here in reverse form from that of the Greek text, which states that "the polygon is less than triangle *E*," while here it is stated that "the triangle is greater than the polygon."

31–32 "abscisum... quantitas." The comment to the effect that the circle exhausts more than half of the circumscribed figure is missing in the Greek text.

34–51 "A.... esse." The elaborate proof here presented of the statement that more than half of the difference between the polygon and the circle is exhausted at each step of doubling the number of the sides of the circumscribed polygon is only briefly intimated in the Greek text. See Chapter Three, page 60, footnote 1.

52–58 "Similiter....equale." The further elaboration contained in these lines is completely absent from the Greek text. It should be noticed that in the passage from Greek to Arabic to Latin there has been considerable altering of the letters used on the figures of this proposition.

59–74 The second proposition as given in this translation is close to the reading of the Greek text. It is almost universally agreed that this whole proposition was probably not in the pristine version of this treatise or, if it was in the original version, that it came after the present third proposition, for the approximation of π calculated in the third proposition is used in the second proposition. It is of interest to note that when al-Ṭūsī edited Thābit ibn Qurra's translation he reversed the order of the second and third propositions.

75–111 The first half of the third proposition as given in this translation follows the Greek text closely, except that—as I have said—the numbers are incorrectly given in a number of places. Notice that like an emendation to the Moerbeke translation (see Volume Two of this work) but unlike the main Greek manuscripts we read here in lines 81–82 that EC/FC is *greater* than 265/153 (and not *equal to* that ratio, as in the Greek text). The reading of this translation is obviously the correct one mathematically. Compare the *Verba filiorum*, Proposition VI, lines 45–46.

103 "circa ipsum angulum" appears to be an error of translation (or copying) and it should read "circa ipsum circulum."

104 "angulos" is given both here and in the translation of Gerard of Cremona. It is missing in the Greek manuscripts but in its stead Heiberg adds *pleura* on Wurm's authority.

109–111 "Figura ⟨m⟩...est." Like the Greek text, this translation says that the "polygon is more than triple the diameter..." when it should read that the "perimeter of the polygon is more than triple the diameter...." In line 110, instead of "in quantitatie unius septime...," it should read "in quantitate minore septima una...." Furthermore, this translator not only omits the whole second half of the proof, as I have already said, but he omits the conclusion of the first half to the effect that not only is the perimeter less than $3\frac{1}{7}$ the diameter, but so too is the circumference of the circle since it is less than the perimeter. (Cf. the similar omission in Proposition VI of the *Verba filiorum* of the Banū Mūsā, in Chapter IV below.)

2. The Translation of the *De mensura circuli* of Archimedes by Gerard of Cremona

In contrast to the translation of Archimedes' *Measurement of the Circle* which has been discussed in the first part of this chapter, the translation done by the celebrated translator Gerard of Cremona was quite popular. As I pointed out in the first chapter, it is extant in at least twelve manuscripts in its original form (for a list of these manuscripts, see the sigla of manuscripts preceding the text). Furthermore, it appears to have been the major source of seven different versions of Proposition I (as well as a supplementary source for three others), at least three of Proposition II, and one of Proposition III. The texts and discussion of these emended versions will be given in Chapters Three and Five below.

Granting for the present the popularity of this translation, we must first examine the evidence that links it with Gerard of Cremona; for it happens that in no manuscript is the translation directly attributed to him. But the translation is included in a manuscript of Paris, BN lat. 9335, which is quite evidently a codex devoted to the translations of Gerard of Cremona, as Wüstenfeld argued more than eighty years ago.[1] Furthermore, it reveals the same kind of terminology and translating techniques as do other works translated by Gerard of Cremona.[2] And it seems reasonable to conclude that the reference in Gerard's *Vita*[3] to a translation listed as

[1] F. Wüstenfeld, "Die Übersetzungen arabischer Werke in das Lateinische," *Abhandlungen der K. Gesellschaft der Wissenschaften der Göttingen*, vol. *22* (1877), p. 59.

[2] The repeated fondness for this form of the Q.E.D., "Et illud est quod voluimus declarare" (see the text of the *Verba filiorum* in Chapter Four, passim), the phrase "hoc vero contrarium est et impossibile," the general use of the word "declarare" for "prove," the persistent use of "narrare" where "dicere" might be expected, the written forms of the numbers —these are but a few of the common usages found in this and other works translated by Gerard of Cremona. The passages in the *Verba filiorum* that render the third proposition of the *De mensura* are closely similar to those of the *De mensura circuli*. For Gerard's translating techniques, see A.A. Björnbo, "Gerhard von Cremonas Übersetzung von Alkhwarizmis Algebra und von Euklids Elementen," *Bibliotheca mathematica*, 3. Folge, vol. *6* (1905), pp. 239–48, and A. A. Björnbo and S. Vogl, "Alkindi, Tideus, und Pseudo-Euklid," *Abhandlungen zur Geschichte der mathematischen Wissenschaften*, *26*. Heft (1912), p. 149.

[3] Wüstenfeld, *op. cit.*, in note 1, p. 59. Cf. B. Boncompagni, "Della vita e delle opere di Gherardo cremonese etc.," *Atti dell'Accademia Pontificia de' Nuovi Lincei*, vol. *4* (1851), pp. 387–493 (see particularly p. 389 where Archimedes' treatise is given as the seventh item in the list of Gerard's translations).

Archimenidis tractatus I is a reference to the *De mensura circuli*, since Gerard translated no other work of Archimedes, except perhaps certain isolated propositions from the introductions of Books I and II of the *Sphere and the Cylinder* (see Chapter Six, Section 1).

Assuming the translation to be by Gerard of Cremona, we have, of course, no idea as to the exact date of translation. But we do know that Gerard's activity at Toledo dominated the third quarter of the twelfth century and continued until his death in 1187.[4]

There are several points concerning this translation that should be emphasized. First, and above all, is its accuracy regarding numbers, particularly in the first tradition of the text, which tradition forms the point of departure for the text I have given here. While most of the manuscripts have an occasional error in the writing of the numbers, manuscripts *P* and *L* have no errors in the main body of the translation, although in common with all the manuscripts they omit one ratio that is in the Greek text (see the Commentary for line 90). As indicated in the sigla of manuscripts preceding the text, manuscripts *P* and *L* belong to the first tradition of the text, a tradition that comprises only these two extant manuscripts but which is closer to the Arabic and Greek texts than the second, more popular tradition.[5] The second tradition is, however, very good and diverges only slightly from the first. My text almost always follows the first tradition and so makes important corrections of the published text of J. L. Heiberg which was based on a single manuscript of the second tradition.[6] The present text is itself a corrected version of the text I published in 1952 and is based on a wider study of the extant manuscripts.[7]

In both copies of the first tradition of this translation it bears the title *Liber Arsamithis de mensura circuli*. In the second tradition we find some variation in the title given. In MS *B* two statements of the title are found. In the hand of the principal scribe of the manuscript we simply read *De quadratura circuli tractatus*. A later cursive hand has added *Tractatus Archimenidis de quadratura circuli*. In MS *I* the title appears as *Liber Archimenidis de comparatione figurarum circularum ad rectilineas*. In MS *J* we read *Incipit de quadratura circuli*. In *C* a late sixteenth-century hand has scrawled

[4] C. H. Haskins, *Studies in Medieval Science*, 2 ed., Cambridge, Mass. (1927), pp. 14–15. Cf. Boncompagni, *op. cit.*, in note 3 above.

[5] See the Commentary following this section.

[6] J. L. Heiberg, "Beiträge zur Geschichte der Mathematik im Mittelalter," *Zeitschrift für Mathematik und Physik*, vol. *35* (1890), Hist.-lit. Abtheilung, pp. 41–48.

[7] M. Clagett, "Archimedes in the Middle Ages: The *De mensura circuli*," *Osiris*, vol. *10* (1952), pp. 599–607.

Archimedis tetragonismus. MS *V* has *Liber Archimedis de mensura circuli*. Notice also the variations in title given in the emended versions published in Chapter Three, which of course depended ultimately on Gerard's translation. The most popular title was simply *Liber de quadratura circuli* as taken from the second tradition of Gerard's translation. Incidentally, in the popular *De isoperimetris*, translated from the Greek in the thirteenth century, Archimedes' work was entitled *In mensuratione circuli*.[8]

In a very general way we can say that Gerard's translation is based on a text that is not too different from the extant Greek text. However, it is clearly a more elaborate and less summary version than that of the Greek text. This is particularly evident in the first proposition where it is made clear, for example, that as we double the number of the sides of the inscribed regular polygon we exhaust at each doubling more than half of the area contained between the perimeter of the polygon and the circumference of the circle. This is left unsaid in the Greek text. Furthermore, in including the corollaries to the first proposition, and particularly Corollary II on the sector of a circle, Gerard's translation—or rather the Arabic text from which it was made—seems to indicate access to a Greek tradition somewhat more complete than the Byzantine tradition on which the extant Greek text is based. I have already noted in the first chapter that the Gerard translation gives in Corollary II a statement attributed to Archimedes by Hero but absent from the extant Greek text.[9] There is also some elaboration of the Greek text in the third proposition, as the Commentary to Section 2 reveals.

As we study Gerard's translation, our conclusion must be that its readers (in either of the two traditions) had before them a very clear rendering of the treatise. Even so, as the emended versions given in the next chapter show, the medieval geometers attempted to spell out in greater detail the arguments used by Archimedes. Not only did those who prepared these emended versions use and speculate on the text of Archimedes, but other authors made at least passing reference to it. Among the earliest citations were those of the little known mathematician Gerard of Brussels, who in his *Liber de motu* based his equation of curvilinear and rectilinear velocities on the assumption that a rectangle equal to a circle can be found, and it is to the first proposition of the *De quadratura circuli* (i.e., the *De mensura circuli*) that Gerard of Brussels turns for his authority.[10] The *De curvis*

[8] See Appendix III, paragraph 7.

[9] See above, Chapter One, note 7; compare Hero's wording with the Latin text of Gerard and with the Arabic text (Bodl. MS Arab. Arch. Seld. A.45, 83v).

[10] See Chapter One, note 17.

superficiebus, which we shall edit in Chapter Six, also depended in a very fundamental way on the *De mensura circuli*; but since it now appears that the work was itself a translation, it is hardly possible that the author got his knowledge of the *De mensura circuli* from the translation of Gerard of Cremona. Still, its wide circulation undoubtedly helped in stimulating interest in the problem of quadrature. The same thing can be said regarding the conclusions concerning quadrature to be found in the *Practica geometrie*, written in 1220 by the well-known mathematician Leonardo Fibonacci Pisano. These conclusions perhaps depend rather on the treatment of quadrature found in the *Verba filiorum* of the Banū Mūsā (see Chapter Four) than on Gerard's translation of the *Measurement of the Circle*.

It is of interest to note that apparently neither of the other two famous mathematicians of the thirteenth century—Jordanus de Nemore and Campanus de Novara—used the Archimedean treatment of the quadrature question (see Appendix I). But the celebrated mathematician and theologian of Oxford, Thomas Bradwardine, in the first part of the fourteenth century took up quadrature in his short *Geometria* (or *Geometria speculativa*). He did not prove Proposition I of the *De mensura circuli*, but he was acquainted with the Archimedean treatise, presumably in the Gerard version or in one of the emended versions which were circulating. His discussion is of some interest to us and so I quote it extensively:[11]

[11] For the text of this passage I have used Vat. lat. 3102, 103v–104r (*V*) and Florence, Bibl. Naz. Conv. Soppr. J.IV.29, 91v–93r (*F*). Occasional reference in the variant readings is also made to the edition of Paris, 1495, 14v–15r (*Ed*):

"Post predictam consequens est tangere de quadraturis. Est autum aliquam figuram quadrare aream quadratam equalem invenire. Causa autem in quadraturis est quia figura quadrata certioris est mensure quam quecunque alia figura. Cum enim habes quod superficies data sit duorum pedum quadratorum vel quattuor vel secundum alium numerum, iam certificatus es de mensura eius certificatione ultima.... Ponam ergo conclusiones paucas de quadraturis, et incipiam a superficiebus similioribus quadratis et reducam considerationem usque ad circulos. Et sit prima conclusio de figura altera parte longiore que est quadrato similius talis: (1) Figura altera parte longior per medie rei inventionem et eius ductum in seipsum in quadratum reducitur.... (2) Area trianguli equilateri vel ysozelis equalis

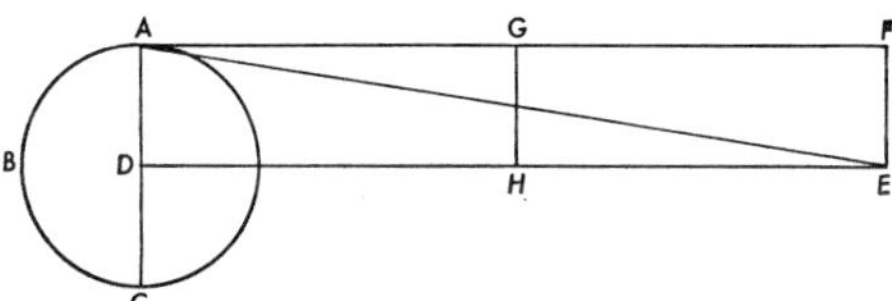

est tetragono contento sub duabus lineis, quarum una est medietas basis, altera vero dividens basim angulumque basi oppositum totumque triangulum per medium.... (3) Area trianguli omnium laterum inequalium equalis est medio tetragoni contenti sub duabus lineis, quarum una est latus maximum eiusdem trianguli, altera vero a maximo angulo eius super

After the aformentioned, it is necessary to touch upon quadrature. To square some figure is to find an equal squared area. The cause in quadrature is this, that a square figure is of a more certain measure than any other figure. For when you find that a given surface is of two square feet—or of four or some other

maximum latus veniens perpendiculariter in se ductis.... (4) Omne poli-
35 gonium per resolutionem factam in triangulos et per quadraturas ipsorum triangulorum et demum per perscriptiones gnomonicas in formam quadrati reduci possibile est.... (5) Area
40 omnis circuli equalis est tetragono sub medietate circumferentie et medietate diametri contento. Supponam unam propositionem Archimenidis de mensura circuli, et erit mihi petitio-
45 nem quia eam demonstrare maiorem requiret quam sit istud totum, et est ista: Omnis circulus triangulo orthogonio equalis est, ciuus unum duorum laterum rectum angulum continen-
50 tium semidyametro circuli equatur, alterum vero ipsorum linee continenti circulum. Est autem proportio linee continentis circulum ad dyametrum tripla sesquiseptima, ita quod circum-
55 ferentia continet dyametrum et ultra hoc septimam eius partem, quod habetur in predicto libello. Verbi gratia, in circulo *ABC* sit *AC* dyameter et eius semidyameter sit *AD* et a puncto
60 *D* ducatur orthoganliter *DE* usque ad equalitatem cum circumferentia circuli, et ducatur *AE* linea, perficiens triangulum *ADE*. Est ergo tunc intentio Archimenidis, quod triangulus
65 *AED* est equalis circulo et hoc demonstrat certissime. Ex quo patet intentum, ducam enim *AF* equedistantem *DE*, et ducatur *EF* tetragonum perficiens. Habeo ita parallelogram-
70 mum *AFDE* divisum in duos triangulos per lineam dyagonalem *AE*. Sed illi trianguli sunt equales per penultimam capituli de triangulis, et circulus est uni eorum equalis, ut dicit
75 propositio Archimenidis. Ergo circulus est equalis medietati illius tetragoni. Dividam tetragonum hunc in duos tetragonos equales per lineam *GH*, et erit circulus alterutri eorum
80 equalis. Sed quilibet eorum continetur sub medietate circumferentie et medietate dyametri. Ergo circulus erit equalis tetragono sub semicircumferentia et semidyametro contento. Qua-
85 dretur ergo tetragonus per primam huius capituli, et sic circulus quadratus.

Aristoteles secundo priorum de deductione sumit tale argumentum, quod circulus potest quadrari quia
90 omne quod est equale figure rectilinee quadrari potest. Omnis circulus est equalis alicui rectilinee figure; ergo omnis circulus quadrari potest. Et maior docetur in quattuor primis de-
95 monstrationibus huius presentis capituli. Minor habetur in fine huius capituli per sententiam Archimenidis. Et sic videtur hoc totum capitulum tendere ad hanc conclusionem, quod cir-
100 culus potest quadrari.

1 consequenter *V* 2 quadraturis *F* quadratura *V* 3–4 quadrati *F* 4 autem *om. V* 5 quadraturis *F* quadratis *V* 6 certioris *F* certior *V* mensure *F* mensura *V* 7 enim habes tr. *V* 8 data *F* diametri *V* 11 es *V* est *F* eius certificatione *F* et eius certificatio *V* 13 quadratura *V* 14 similioribus quadratis *F* similibus quaratorum *V* 15 reducam *F* deducam *V* 16 sit *om. V* 17 figura *F* quadratura *V* longiore *corr. ex* longiorum *V* longioris *F* est *F* est in *V* 18 talis *F* hec *V* *et post* hec *add. V* Sequitur conclusio Figura *om.F* 20 ductam in seipsam *V* 22 ysozelis *F* ysochelis *V* 24 una *om. F* basis *om. F* 26 totumque *F* et totum *V* 27 *post* medium *add. V* in se ductum 29 contento *V* 31 eiusdem *om. V* 32 vero *F* vero linea *V* angulo eius *tr. V* 33 latus *F* eiusdem latus *V* 34 ductis *F* ductum *V*

number—you are assured of its measure with ultimate certitude. Hence I pose a few conclusions regarding quadrature, and I begin with surfaces which are more like squares and lead up to the consideration of circles. The first conclusion concerns the quadrature of a rectangle, which is like a square: (1) A rectangle is reduced to a square by finding the mean proportional line [between the two sides] and [using that line as] the quantity to multiply into itself [or to square].... (2) The area of an equilateral or isosceles triangle is equal to the rectangle contained by two lines, one of which is half of the base of the triangle and the other is the line bisecting the base, the angle opposite the base, and the whole triangle.... (3) The area of a scalene triangle is equal to half of the rectangle contained in the multiplication of two lines, one of which is the longest side of the triangle while the other is a line drawn from the largest angle perpendicularly to the longest side.... (4) Every polygon can be reduced to the form of a square by its resolution into triangles, by the quadrature of those triangles, and then by using gnomonic rules.... (5) The area of every circle is equal to a rectangle contained by half of the circumference and half of the diameter. I posit a proposition from Archimedes' *On the Measurement of the Circle*, and it will be a postulate for me, since to demonstrate it would require more than this whole [chapter]. The supposition is this: Every circle is equal to a right triangle, one of whose two sides containing the right angle is equal to the radius of the circle and the other to the circumference. Furthermore, the ratio of the circumference to the diameter is three and one seventh, so that the circumference contains the diameter three times and in addition a seventh part of it, as it is had in the aforesaid little book [of Archimedes].... Aristotle in the second book of the *Prior Analytics* takes up an argument to the effect that a circle can be squared because every thing which is equal to a rectilinear figure can be squared. Every circle is equal to some rectilinear figure; therefore every circle can be squared. The major is taught in the first four demonstrations of this chapter, while the minor is had by means of the opinion of Archimedes. And so it is seen that the whole chapter tends to this conclusion, that the circle can be squared.

35 per *om. V* 37 triangulorum *V* quadratorum *F* 41 sub *V* ex *F* 44 mensura *FEd* quadratura *V* 45 quia *F* quoniam *V* 46 quam *F* tractatum quam *V* istud totum *F* totum istud opus *V* istud capitulum *Ed* 46–47 est ista *F* sit suppositio ista *V* 48 equalis est *tr. V* 50 semidyametro *F* medietate dyametri *V* equatur *F* est equale *V* 51 alterum vero *F* et alterum *V* ipsorum *om. F* 52–57 Est.... libello *om. F* 56–57 quod habetur *FV* ut habetur ab eodem Archimenide *Ed* 58–59 et eius *V* cuius *F* 61 cum circumferentia *V* circumferentie *F* 62 linea *om. F* 63 ADE *om. F* Est ergo *F* erit igitur *V* 66 Ex quo patet *V* Tunc ex illo sequitur *F* 67 equedistantem *F* equidistanter *V* 68 ducatur *om. F* tetragonum *V* tetragonismum *F* 69 Et habeo *F* ita *om. F* 73 capituli *om. V* et *V* sed *F* 74 eorum *V* illorum *F* 74–75 ut dicit propositio *V* per propositionem *F* 76 medietate *V* 77 tetragonum hunc *V* ergo hunc tetragonum *F* 82–84 erit.... contento *V* est tertragonizatus *F* 84–86 Quadretur.... sic *V* Si ergo quadretur ille tetragonus erit *F* 87-100 Aristoteles.... quadrari *om. F*

I shall show in Chapter Five that the reworking of Proposition I of the *De mensura circuli* which is found attached to a Vatican manuscript of Bradwardine's *Geometria* is probably not by Bradwardine, since so far as I know it appears in no other copy of the *Geometria*. The little tract on quadrature that does appear with the printed editions of the *Geometria* is a tract of the thirteenth century usually ascribed to Campanus de Novara (see Appendix I); it is not Archimedean in character.

Other Schoolmen of the fourteenth century took up the conclusions of the *De mensura circuli*. These were Schoolmen who simply cited the tract without reworking it in the manner of Albert of Saxony and those unknown authors of the various versions given in Chapter Three. Among those who mention the *De mensura circuli* we can cite Bernardus Pistoiensis, who copied Bradwardine's conclusions verbatim,[12] as well as the anonymous author of a *Practica geometrie* that circulated in the fourteenth century.[13]

A final word remains to be said about the text of Gerard's translation. In following the more accurate copy of Tradition I, namely *P*, I have made one change. In manuscript *P* the numbers are written out in rhetorical fashion, while in manuscript *L*, also of Tradition I, we find that the numbers are usually in the Indo-Arabic numerals, and in the manuscripts of Tradition II there is a mixture of Roman numerals and rhetorical expressions. For ease of reading I have changed all numbers (except fractions) to Indo-Arabic numerals. Fractions I have left in their rhetorical expressions. Actually, the system I have adopted is used in manuscript *L* throughout, and in the margin at least once by our most careful scribe, the scribe who wrote *P*. To be sure, *L* often modifies this system by using common abreviations for the fractions, such as 4^a for quarta, 7^a for septima, etc. In my opinion, however, Gerard originally used simply the

[12] *Geometrie liber*, British Museum Addit. 17420, 22r: "Area omnis circuli est equalis tetragono ex medietate circumferentie et medietate dyametri conscripto. Supponam unam propositionem Archindi (!) de mensura circuli, et erit mea petitio.... Et est ista: omnis circulus triangulo orthogonio est equalis, cuius unum duorum laterum rectum angulum continentium est semidiameter circuli, alterum vero linee continentis circulum."

[13] *Practica geometrie*, Vat. Reg. suev. 1261, 294v: "Probatio huius propositionis habetur in Archimede qui docet invenire orthogonium triangulum equalem circulo...." Cf. the questions given in the fourteenth-century MS BN lat. 7377B, 35r: "Nunc vero sequitur de mensuratione circuli cuius exemplum ubi sit circulus aliquis cuius diameter sit 10. Cum autem quantitatem ipsius voluero invenire, multiplicabo diametrum in se; de quadrato cuius minuam tres 14as que equatur septime et medietati septime et residuum equabitur circulo, quia proportio aree omnis circuli ad quadratum diametri eius est sicut proportio 11 ad 14 ex Archimenidi."

rhetorical expressions and the reader should keep this in mind. I have carefully compared and collated all of the extant manuscripts, but for the text given here I give only the variant readings of the first two manuscripts of each tradition, MSS *L* and *P* for Tradition I and *B* and *I* for Tradition II. For these four manuscripts are clearly the best manuscripts, although MS *J* in Tradition II is very good also. Citing the careless errors of the other manuscripts would not contribute to our efforts toward the establishment of a sound text. The drawings in the various copies are generally quite good. I have followed the drawings in *P*, my only change being to correct the proportions of the lines representing the numbers given in the text. The marginal folio numbers are from MS *P*.

Sigla of Manuscripts

Tradition I

P = Paris, BN lat. 9335, 28v–29v, 14c.
L = Oxford, Bodl. Arch. Seld. B. 13, 2r–v, 15c (for this folio).

Tradition II

B = Oxford, Bodl. Auct. F. 5.28, 101v–102v, 13c.
I = Dresden, Sächs. Landesbibliothek, Db. 86, 175v–176v, 178r, early 14c. (I have also used the edition of this done by J. L. Heiberg in the *Zeitschrift für Mathematik und Physik*, Bd. 35 [1890], Hist.-lit. Abtheilung, pp. 41–48.)
J = Berlin, Deutsche Staatsbibliothek (*MS* now at Marburg, Westdeutsche Bibliothek), Q. 150, 89r–v, 13c.
C = Oxford, Bodl. Digby 174, 133v–134v, 13c.
A = Naples, Bibl. Naz. VIII. C. 22, 65v–66v, 13c. (Proposition II and the enunciation of Proposition III—both on folio 66v—are in the form of the original translation. For Proposition I, see Chapter Three, Section 2.)
Q = Paris, BN lat. 7378A, 19r–v, 14c.
U = London, British Museum Addit. 17368, 69v, late 13c or early 14c.
F = Vienna, Nationalbibliothek cod. 5303, 19r–21v, 15–16c.
[As this volume went to press my attention was called by Mr. John Murdoch to another manuscript of the Gerard translation in Paris, BN lat. 16649, 1r–4r. This piece is dated 1519 by the scribe (fol. 4r). The scribe had access to both traditions of the text, for he sometimes follows

the one and sometimes the other. The manuscript is of no significance for the establishment of the text. Incidentally, the scribe labels the second half of Proposition III as a new Proposition IV. This reminds us somewhat of another manuscript (Vat. lat. 4275, 81v–83v) which contains, in addition to a paraphrase of the Naples Version of Proposition I, paraphrases of Gerard's translation of Propositions II and III, and where the scribe inserted a new enunciation between the two halves of Proposition III (see Chapter Three, Section 2, footnote 1). In the very last stages of page proof still another manuscript has come to my attention: Glasgow University Library MS BE 8-y.18 of the late fourteenth or early fifteenth century. It includes a collection of tracts on the quadrature of the circle (ff. 202r–214r). Among them is a copy of Gerard's translation of the *De mensura circuli* in the second tradition (ff. 202r–203v). I shall discuss the remaining items in the first appendix to Volume Two. However, it can be remarked here that this manuscript includes a further paraphrase or version of all three propositions of the *De mensura circuli* in the manner of the versions given in Chapters Three and Five below. It also includes copies or versions of the quadrature items I have presented below in Appendixes I and II.]

Arabic Text of al-Ṭūsī

Ar = Nasīr al-Dīn al-Ṭūsī, *Majmūᶜ al-Rasāʾil*, vol. 2 (Hyderabad, 1940), as an addition to the *Sphere and Cylinder*, pp. 127–33. [In additon to the publication of these two genuine works of Archimedes, there has appeared at Hyderabad in 1948 a small volume entitled *Rasāʾil Ibn Qurra* containing two of the probably spurious works of Archimedes: *Book of the Elements of Geometry* (no doubt identical with the *Book of Triangles*, see page 4 above) and the tract *On Touching Circles*. These works are apparently translations from the Greek by Thābit ibn Qurra.]

On the Measurement of the Circle

De Mensura Circuli

28v c. 1 / Liber Arsamithis de mensura circuli.

I. OMNIS CIRCULUS TRIANGULO ORTHOGONIO EST EQUALIS, CUIUS UNUM DUORUM LATERUM RECTUM CONTINENTIUM ANGULUM MEDIETATI DIAMETRI CIRCULI EQUATUR ET ALTERUM IPSORUM/LINEE CIRCU (c. 2) 6 LUM CONTINENTI.

Sit itaque circulus *ABGD* [Fig. 4] triangulo *E* equalis, secundum quod ante narravimus in propositione. Dico ergo quod eius mensura ipsius mensure equatur. Quod si non ita fuerit, tunc circulus aut maior 10 aut minor eo erit. Sit itaque primo maior. Faciam autem in circulo quadratum *ABGD*. Iam ergo separatum est ex circulo *ABGD* plus medietate ipsius et est quadratus *ABGD*. Secabo autem arcum *AB* in duo media supra punctm *F* et arcus ei similes similiter, et copulabo

1 For variants of title see introduction to text.
2 triangulo orthogonio *tr. BI*
8 ergo *PL* itaque *BI*/mensura: vel area *mg. P*
9 mensure: vel aree *mg. P*
9–10 aut.... minor: aut minor aut maior *L*
10 eo *om. P*
11 ABGD[1]: ADGB *PL*
11–12 Iam.... ABGD *PL om. BI* (*sed cf. Arab. tex.*)
12 ABGD *corr. ex* ADGB *in PL*
13 supra: super *BI*/similes: vel relatos *mg. P* similes et relatos et *L*

---

1 *post* circuli *add. Ar.* وهى ثلاثة اشكال
(And there are three figures) ((figures = propositions))
4–5 circuli: تلك الدائرة
(*of this circle*) ((Cf. "eiusdem circuli" in line 5 of the Plato translation))
6 *post* continenti *add. Ar.* والحاصل انها تساوى سطح نصف قطرها فى الخط المساوى لنصف محيطها
(And the product is equal to the area arising from one half its diameter and the line equal to half its circumference.)
7 triangulo E equalis: المثلث المذكور مثلث ه
(the mentioned triangle is triangle E) *((*rendering ه by *E))*
8–9 Dico...equatur *om. Ar.*
10 Faciam: نرسم *(We describe)*
11 ABGD: ا ب ج *(ABG)*
12 et...ABGD *om. Ar.*
13 arcus...similiter: وهكذا القسى الاربع
(and in this way the four arcs)
13–14 copulabo...similes: نصل الاوتار
(we join its chords)

On the Measurement of the Circle

The Book of Archimedes on the Measure of the Circle.

I. EVERY CIRCLE IS EQUAL TO A RIGHT TRIANGLE, ONE OF WHOSE TWO SIDES CONTAINING THE RIGHT ANGLE IS EQUAL TO THE RADIUS OF THE CIRCLE AND THE OTHER OF THEM TO THE CIRCUMFERENCE OF THE CIRCLE.

And so let circle *ABGD* be equal to △ *E*, according as we have just stated in the proposition [see Fig. 4]. I say, therefore, that the measure of the one equals the measure of the other. For if this is not so, then the

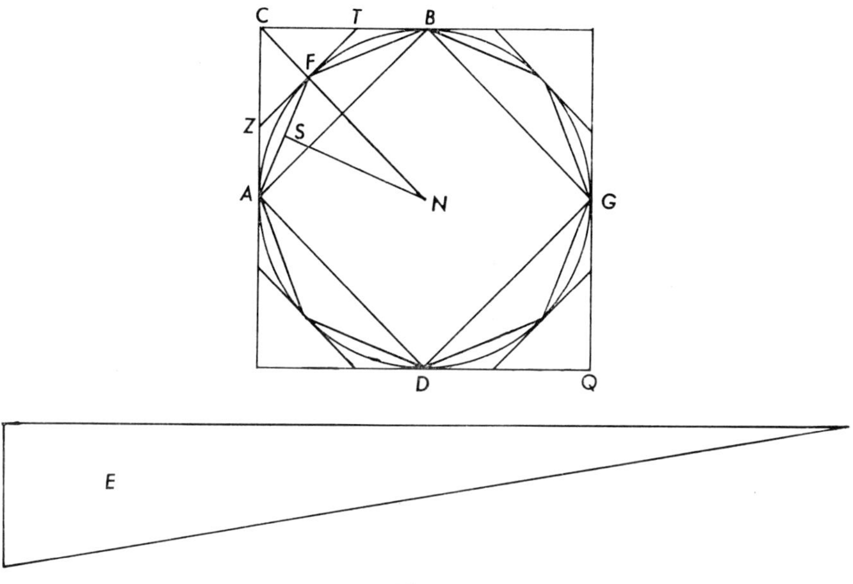

Fig. 4

circle will either be greater or less than it. And so in the first place let it be greater. I shall inscribe square *ABGD* in the circle and so square *ABGD* has thereby exhausted more than half of circle *ABCD*. Further, I shall bisect arc *AB* at point *F*, and in the same way the arcs similar to it, and I shall connect *AF* and *FB* and the lines similar to it. Thus by *AFB*

AF et *FB* et ei similes. Iam ergo separatum est etiam ex residuis
15 portionibus circuli *ABGD* plus medietate ipsarum et est *AFB* et sibi similes. Cum ergo fecerimus ita secundum illud quod sequitur, remanebunt portiones que erunt minores quantitate eius quod circulus addit super triangulum *E*. Et figura tunc rectilinea poligonia, quam continet circulus, erit maior triangulo. Sit itaque figura illa *AFB* et
20 eius similes. Ponam autem centrum circuli *N* et producam perpendicularem *NS*. Linea igitur *NS* est minor uno duorum laterum trianguli continentium rectum angulum. Et linea circumdans poligonium est minor reliquo latere ipsorum, quoniam ipsa etiam est minor circumferentia circuli. Quod autem fit ex multiplicatione unius duorum
25 laterum trianguli continentium rectum angulum in alterum, et est duplum aree trianguli, est plus aggregato ex *NS* in lineam circumdantem poligonium, et est duplum poligonii. Cum igitur illud ita sit, tunc triangulus est maior poligonio. Sed iam fuerat minor, et hoc quidem est contrarium et impossibile.

30 Sit etiam circulus minor triangulo *E*, si fuerit illud possibile. Describam autem supra ipsum quadratum continentem ipsum. Sitque quadratus *QC*. Et iam quidem separatum est ex quadrato *QC* plus medietate eius, et est circulus. Dividam autem arcum *BA* in duo media et arcus sibi similes in duo media. Linee ergo que transeunt per
9r c. 1 puncta sectionum contingunt circulum. / Tunc linea *ZT* iam divisa
36 est in duo media supra *F*. Et linea *CF* est perpendicularis super *ZT*,

14 ei: eis *I*
18 E *PL om. BI* (*cf. Arab. tex.*)/Et: et etiam *BI*/polligonia *L hic et ubique*
23 etiam *om. L*
26 aree *om. BI* / *post* NS *add. BI* linea
27 igitur: ergo *L* / *ante* sit *del. P* erit
28 fuerat: fuit *L*
31 ipsum: circumlum ipsum *BI*
34 sibi similes *tr. L*
36 super: supra *BI*

14–16 Iam...similes: فتفصل المثلثات الحادثه اعظم من نصف القطع لما مربيانه
(*And so the new triangles separate out* more than half of the portion* [*which is in excess*] *according as we showed it.*) ((*Changing فتفصل to فنفصل))
18 E *om. Ar*
19–20 Sit...similes *om. Ar.*
21 *ante* uno *add. Ar.* ن ص (*NṢ*) ((equal to the radius and thus distinct from ن س))
22 continentium rectum angulum: ه
23 reliquo...minor *om. Ar.*
24 *post* circuli *add. Ar.*
المتساوى للضلع الاخر من مثلث — ه —
(*equal to the other side of triangle E*)

and those [other triangles] similar to it there has been exhausted more than half of the remaining segments of circle *ABGD*. When we shall have continued this procedure successively, there will remain [at some time] segments which [in sum] will be less than the quantity by which the circle exceeds △ *E*. And then the polygon, which the circle contains, will be greater than the triangle. And so let it be the figure [including] *AFB* and those [triangles] similar to it [that is the desired end of the successive doubling of sides]. Further I shall posit *N* as the center of the circle and draw the perpendicular *NS*. Therefore line *NS* is less than one of the two sides of the triangle containing the right angle. And the perimeter of the polygon is less than the other of these sides, since it is also less than the circumference of the circle. Further, the product of [1] one of the two sides of the triangle containing the right angle and [2] the other is double the area of the triangle and is greater than the product of *NS* and the perimeter of the polygon, which latter product is double the [area of the] polygon. Since this is so, therefore the triangle is greater than the polygon. But it had been [shown to be] less, and indeed this is a contradiction and is impossible.

Again, let the circle be less than △ *E*, if that is possible. Further, I shall describe a square about it (the circle), and let the square be *QC*. Now indeed the circle has exhausted more than half of the square *QC*. And I shall bisect arc *BA* and the other arcs similar to it. Hence the lines which intersect the points bisecting the arcs are tangent to the circle. So line *ZT* has accordingly been bisected at point *F* and line *CF* is perpendicular to *ZT*, and the same thing holds for the other lines similar to it. Since $(ZC + CT) > ZT$ and $1/2\ (ZC + CT) > 1/2\ ZT$, so line $CT >$

24–29 Quod....impossibile: – فسطح – ن س
فى محيط الشكل اعنى ضعف مقدار الشكل اصغر
من ضعف المثلث فالشكل اصغر من المثلث وكان
اعظم منه هذا خلف

(And so the product of NS and the perimeter of the figure—i.e., double the quantity of the figure—is less than double the triangle, and so the figure is less than the triangle though it was greater than it; this is contradictory.)

30 E...possibile *om. Ar.*

32 QC: ع ق

33 *post* BA *add. Ar.* على ف *(at F)*

34 et...media *om. Ar.*

34–37 Linee....similes:
ونخرج – ز ف ط –
مماسا للدائرة على – ف – ويكون نصف قطر
– ن ف – عمودا عليه وهكذا نعمل فى سائر القسى

(And we draw ZFT tangent to the circle at F and the radius NF will be perpendicular to it, and we proceed in the same way in the remaining arcs.)

et similiter linee ei similes. Et quoniam *ZC* et *CT* sunt maius *ZT* et est earum medietas maior medietate ipsius, tunc linea *CT* est maior *TF*, que est equalis *TB*. Ergo triangulus *FCT* est maior medietate
40 trianguli *FCB*. Et multo plus illo erit maior medietate figure *FCB*, que continetur duabus lineis *FC*, *CB*, et arcu *BF*. Et similiter erit triangulus *CFZ* maior medietate figure *CFA*. Ergo totus *TCZ* est maior medietate figure *AFBC*, que continetur duabus lineis *AC*, *CB* et arcu *AFB*. Et similiter sunt trianguli sibi similes plus medietate
45 portionum aliarum sibi similium. Cum ergo fecerimus illud in eo quod sequitur, remanebunt portiones supra circulum, que cum aggregabuntur erunt minus augmento trianguli *E* super circulum *ABGD*. Remaneat ergo portio *FZA* et portiones sibi similes. Figura igitur tunc rectilinea que circulum continet erit minor triangulo *E*. Sed
50 hoc quidem est impossibile, quoniam fuit maior. Et illud ideo, quoniam *NF* equatur catheto trianguli et linea continens poligonium est maior reliquo latere trianguli quod continet rectum angulum, eo quod sit maior linea circumdante circulum. Illud ergo quod fit ex multiplicatione *FN* in lineam continentem figuram poligoniam est maius eo
55 quod fit ex multiplicatione unius duorum laterum trianguli continentium rectum angulum in alterum. Non est igitur circulus minor triangulo *E*. Et iam quidem ostensum fuit in hiis que premissa sunt, quod ipse non est maior eo. Circulus igitur *ABGD* est equalis triangulo *E*.

37 ZC: linee ZC *BI*
40 trianguli *PL* figure *BI*
44 sibi similes *tr. I*
46 super *I*
47 E: C(?) *L* / supra *BI*
48 remanet *BI* / igitur: ergo *L*
51 catecho *L*
55 multiplicationem *I*
58 igitur: ergo *L*

37–64 Et quoniam....E:
ولأن – ق ب – ق ا –
متساويان وكذلك – ط ب – ط ف – ز ف – ز ا –
الاربعة متساوية يكون – ط ق – ق ز – متساويين
وممامعًا اطول من – ط ز – فق ط – اطول من
– ب ط – فمثلث – ق ف ط – اظم من مثلث
– ط ف ب – الذى هو اعظم قطعة – ط ف ى ب –
الخرجة من الدائرة وكذلك فى البواقى فالمثلثات
الاربعة التى على زوايا المربع تفصل من باق
المربع بعد نقصان الدائرة اعظم من النصف وينصف
القسى هكذا مرة بعد الخرى وتخرج الخطوط المماسة
للدائرة الى ان تبقى قطع خارجة من الدائرة مجموعها
اصغر من زيادة مثلث – ه – على الدائرة ويكون
الشكل الكثير الاضلاح الذى على الدائرة اصغر من
مثلث – ه – ولكن سطح – ن ف – نصف القطر فى
محيط الشكل الذى على الدائرة اعنى ضعف مقدار
الشكل اعظم من ضعف المثلث لكون محيط الشكل
اعظم من محيط الدائرة فالشكل اعظم من المثلث

TF; and *TF* is equal to *TB*. Hence △ *FCT* > 1/2 △ *FCB*. And a fortiori it will be greater than 1/2 the figure *FCB* contained by the two lines *FC* and *CB* together with the arc *BF*. In the same way, △ *CFZ* > 1/2 the figure *CFA*. Therefore, the whole *TCZ* > 1/2 the figure *AFCB* contained by the two lines *AC* and *CB* together with the arc *AFB*. In the same way, the triangles similar to it (*TCZ*) are greater than one half of the figures similar to *AFB*. When we shall have continued this procedure successively, there will [at some time] remain portions beyond the circle, which when totaled will be less than the excess of △ *E* beyond circle *ABGD*. Let there remain [as such portions] portion *FZA* and the portions similar to it. Therefore, the polygon which contains the circle will be less than △ *E*. But indeed this is impossible, since it was greater. And this follows since *NF* equals the altitude of the triangle and the perimeter of the polygon is greater than the other side of the triangle which contains the right angle due to the fact that it (the perimeter) is greater than the circumference. Therefore, the product of *FN* and the perimeter of the polygon is greater than the product of [1] one of the two sides of the triangle containing the right angle and [2] the other. Hence the circle is not less than △ *E*. And it was already shown in the first part that the circle was not greater than the triangle. Therefore, circle *ABGD* is equal to △ *E*.

وكان اصغر ميه هـذا خلف فاذا الدائـرة مساويـة
بمثلث — ه — فسطح نصف القطر فى نصف المحيط
مساو لسطح الدائرة وذلك ما اردناه

(And because QB, QA are equal, and similarly the four TB, TF, ZF, and ZA are equal, TQ and QZ will be equal, and both of them together are greater than TZ. And so QT is greater than BT. And so triangle QFT is greater than triangle TFB, which is itself greater than the segment TFYB outside of the circle; and similarly for the rest. And so the four triangles which are at the angles of the square separate out more than half of that which remains of the square after the circle has been subtracted from it. And the arcs are successively bisected and lines are drawn tangent to the circle until a portion remains outside of the circle which all together is smaller than the excess of triangle E over the circle. And so the polygonal figure which is around the circle is smaller than triangle E. But the product of radius NF and the perimeter of the figure which is about the circle—i.e., double the quantity of the figure—is greater than double the triangle. For the perimeter of the figure is greater than the circumference of the circle. And so the figure is greater than the triangle, though it was less than it; this is contradictory. And hence the circle is equal to triangle E. And so the product of the radius and half the circumference is equal to the area of the circle, and this is what we proposed.)

[Corollarium I:] Et etiam quia area trianguli *E* est equalis ei quod
60 fit ex multiplicatione perpendicularis sue in medietatem basis ipsius, et eius perpendicularis est equalis medietati diametri *ABGD*, et basis eius equalis circumferentie circuli *ABGD*, tunc quod fit ex multiplicatione eius in medietatem sectionis circumferentie est area figure accepta equalis aree trianguli *E*.

65 [Corollarium II:] Et propter hoc erit multiplicatio medietatis diametri in medietatem portionis circumferentie area figure que continetur ab illa portione et duabus lineis egredientibus a duabus extremitatibus portionis ad centrum. Et illud est cuius voluimus declarationem.

II. PROPORTIO AREE OMNIS CIRCULI AD QUADRATUM
70 DIAMETRI IPSIUS EST SICUT PROPORTIO 11 AD 14.

Exempli causa, sit linea *AB* [Fig. 5] diametrus circuli, et super ipsam quidem faciam quadratum *HG*. Sitque *DG* medietas *DE*, et sit linea *EZ* septima *GD*. Et quia proportio trianguli *AGE* ad triangulum *AGD* est sicut proportio 3 ad 1 et proportio trianguli *AGD*
75 ad triangulum *AEZ* est sicut proportio 7 ad 1, tunc propter illud
c. 2 / fit proportio trianguli *AGZ* ad triangulum *AGD* sicut proportio 22 ad 7. Quadratus vero *GH* est quadruplus trianguli *ADG*, et triangulus *AGZ* est equalis circulo *AB*, quoniam perpendicularis *AG* est equalis linee que egreditur a centro circuli ad lineam ipsum circum-
80 dantem, et basis *GZ* est equalis circumferentie circuli, quoniam est plus triplo diametri ipsius et septima diametri fere. Iam igitur verificatum est quod diximus, quod proportio circuli *AB* ad quadratum *GH* est sicut proportio 11 ad 14. Et illud est quod voluimus declarare.

68 illud est *tr. BI* / volimus (!) *L*
73 GD: DG *L* / quia: vel sic, et quia GD est septupla ad EZ et DE dupla DG, erit ergo GE continens EZ vicesies et semel. Ergo proportio GZ totalis ad EZ est sicut proportio XXII ad I. Ergo proportio trianguli AGZ ad triangulum AEZ est sicut XXII ad I *mg. BI*
74–75 et.... 1 *om. L*
76 proportio[1] *tr. BI post.* AGZ
77 quadratum *BI* / quadruplum *BI*
79 a: e *BI*
80–81 est plus *tr. BI*
81 igitur: ergo *L*

---

69 *ante* proportio *add. Ar.*
اذا كان محيط الدائرة
ثلاثه امثال القطر وسبعة وهى نسبة تقريبة اصطلح
عليه المساحون

(And since the circumference of the circle is 3 1/7 times the diameter and this is an approximate ratio to which the measurements conform)

[Corollary I:] And also, because the area of △ *E* is equal to the product of its altitude and half of its base, and its altitude is equal to the radius of *ABGD* while its base is equal to the circumference of circle *ABGD*, then the product of the altitude and the semicircumference is the area of a [rectangular] figure accepted as equal to the area of △ *E*.

[Corollary II:] And accordingly the product of the radius and half of an arc of the circumference is equal to the area of a figure (i.e., a sector) which is contained by that arc and the two radii drawn from the extremities of the arc. And this is what we wished to prove.

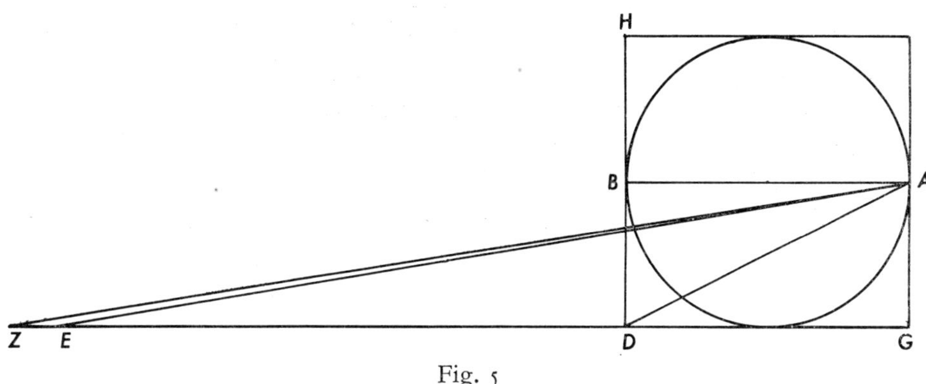

Fig. 5

II. THE RATIO OF THE AREA OF ANY CIRCLE TO THE SQUARE OF ITS DIAMETER IS AS THE RATIO OF 11 TO 14.

For example, let line *AB* [see Fig. 5] be the diameter of a circle, and on it I shall produce square *HG*. Let *DG* be 1/2 *DE* and *EZ* be 1/7 *GD*. Now because △ *AGE*/△ *AGD* = 3/1, and △ *AGD*/△*AEZ* = 7/1, then accordingly △ *AGZ*/△ *AGD* = 22/7. But square *GH* is quadruple △ *ADG*, and △ *AGZ* is equal to circle *AB*, since altitude *AG* is equal to the radius and the base *GZ* is equal to the circumference of the circle—*GZ* being almost $3\frac{1}{7}$ times the diameter. Hence our assertion has now been verified, namely, that circle *AB*/ square *GH* = 11/14. And that is what we wished to show.

74 3 ad 1: احد وعشرين الى سبعه
(*21 to 7*)
78 perpendicularis *om. Ar.*
80 basis *om. Ar.*
80–83 circumferentie.... 14:
بالتقريب المحيط
فنسبه مربـع القطـر الى سطح الدائـرة نسبـة ثمانية وعشرين الى اثنيى وعشرين بـل نسبـة اربعة عشر الى احد عشر
(*approximately to the circumference, and so the ratio of the square of the diameter to the area of the circle is as the ratio of 28 to 22. But this is the ratio of 14 to 11.*)
87 AG^2 *om. Ar.*

III. OMNIS LINEA CONTINENS CIRCULUM ADDIT SUPER
85 TRIPLUM DIAMETRI IPISIUS MINUS SEPTIMA ET PLUS 10
PARTIBUS 71 PARTIUM DIAMETRI.

Exempli causa, sit linea *AG* diametrus circuli *AG* [Fig. 6]. Sitque eius centrum *E*, et linea *DZ* sit contingens circulum, et sit angulus *ZEG* tertia anguli recti. Ergo proportio *EZ* ad *ZG* est sicut proportio
90 306 ad 153. Dividam autem angulum *ZEG* in duo media linea *HE*. Ergo proportio *ZE* ad *EG* est sicut proportio *ZH* ad *GH*. Ergo proportio *ZE* et *EG* coniunctarum ad *ZG* est sicut proportio *EG* ad *GH*. Fit ergo proportio *EG* ad *GH* maior proportione 571 ad 153. Ergo proportio *EH* in potentia ad *HG* in potentia est plus propor-
95 tione 349450 ad 23409. Ergo proportio eius ad ipsam in longitudine est maior proportione 591 et octave ad 153. Angulum quoque *HEG* dividam in duo media linea *ET*. Ergo secundum similitudinem eius quod diximus, declaratur quod proportio *EG* ad *GT* est maior pro-
29v c. 1 portione 1162 et oc/tave ad 153. Ergo proportio *TE* ad *TG* est maior
100 proportione 1172 et octave ad 153. Angulum quoque *TEG* in duo

84 *mg. BIC* Sudor Archimenidis
85 triplum diametri: diametrum *L*
89 *mg. PL* vult ut cum ZE posita fuerit trecenta et sex sit ZG eius medietas semper, quoniam est medietas lateris exagoni cadentis in circulo cuius latus ZE est medietas diametri, propterea quod angulus ZEG est tertia recti.
90 *mg. BI* quia linea EZ est dupla ad lineam GZ ex 4 et 32 et 6 primi Euclidis protracta GD ad equalitatem DZ./dua *P*
91 *supra* ad EG est *add. B* ex III sexti
94 *mg. P* Propterea quod facta est proportio EG ad GH maior proportione quingentorum et septuaginta unius ad centum et quinquaginta tria, tunc cum tu posueris ZG centum et quinquaginta tria, erit EG plus quingentis et septuaginta uno. Et fiet propter illud HE in potentia trecenta milia et quadraginta novem milia et quadringinta et quinquaginta. Et in longitunine plus quingentis et nonaginta uno et octava unius. / *mg. BI* ex dulk (i.e., I. 47) et VIII. quinti
95 *mg. P* quoniam si tu permittas ea, deindeponis proportionem unius antecedentium ad unum consequentium sicut omnium antecedentium ad omnia consequentia.
96 maior *corr. P ex* sicut / et angulum *I*

---

89 *post* recti *add. Ar.* اعنى نصف زاوية مـن زوايا المثلث المتساوى الاضلاع
(i.e., half of one of the angles of an equilateral triangle)
90 *ante* 306 *add. Ar.*
الاثنين الى الواحد ولتكن كنسبة
(2 to 1 and let it be as the ratio)
90 *post* 153 *add. Ar.* واذا الفنيا مربع العدد الذى بـازاء – ز ج – مـن مربـع العدد الذى بـازاء – ه ز – واخدنـا جـذر الباق كان – ه ج – بذلك المقدار اكثر من (٢٧٠) بكسر ما
(And when the square of the number representing ZG has been subtracted from

III. EVERY CIRCUMFERENCE OF A CIRCLE EXCEEDS THREE TIMES ITS DIAMETER BY AN AMOUNT LESS THAN ONE SEVENTH AND MORE THAN 10 PARTS OF 71 PARTS OF THE DIAMETER.

For example, let line AG be the diameter of the circle AG [see Fig. 6]. Let its center be E and let line DZ be tangent to the circle. Let $\angle ZEG$ be 1/3 of a right angle. Therefore, $EZ/ZG = 306/153$. Now I bisect $\angle ZEG$ by line HE. Therefore, $ZE/EG = ZH/GH$. Therefore, $(ZE + EG)/ZG = EG/GH$. Hence $EG/GH > 571/153$. Therefore, $EH^2/HG^2 > 349450/23409$. Therefore, $\frac{EH}{HG} > \frac{591\frac{1}{8}}{153}$. I also bisect $\angle HEG$ by line ET. Therefore, in the same way that we said before it is shown that $\frac{EG}{GT} > \frac{1162\frac{1}{8}}{153}$. Hence $\frac{TE}{TG} > \frac{1172\frac{1}{8}}{153}$. I further bisect $\angle TEG$ by line EK. Therefore, $\frac{EG}{GK} > \frac{2334\frac{1}{4}}{153}$. I also bisect $\angle KEG$ by line LE. Therefore, $\frac{EG}{GL} > \frac{4673\frac{1}{2}}{153}$. And because $\angle ZEG$ was 1/3

the square of the number representing ZE and we have taken the square root of the remainder, EG was a quantity greater than 265 by some fraction.) ((Cf. lines 81–82 of the Plato translation.))

90 *post* ZEG *add. Ar.* على – ح – *(at H)*

91 *post* GH *add. Ar* واذا ركبنا و ابدلنا
(And when we have substituted [numbers] and conjoined [terms])

93 Fit...153: فاذا جمعنا العددين اللذين بازاء – ز ه – ه ج – كان اكثر من (٥٧١) فنجعله بازاء – ه ج – ويصير الذى بازاء – ج ح – بهذا المقدار (١٥٣)
(And when we have joined the two numbers representing ZE and EG, the result was greater than 571; and so we put it in place of EG, and that which is in place of GH will be the quantity 153.)

94–96 Ergo...153: واذا جمعنا مربعهلا واحذنا جذرهما كان – ه ح – بهذا المقدار اكثر من (٥٩١) وثمن
(And when we added the square of the two and took the square root of this addition, the result EH was a quantity greater than 591 1/8.)

96 *post* HEG *add. Ar.* على – ط – *(at T)*

98–100 declaratur...153: نسبة – ح ه – ه ج – الى – ح ج – كنسبة – ه ج – الى – ج ط – واذا جمعنا عددى – ح ه – ه ج – وجعلناهما بازاء – ه ج – كان – ه ج – اكثر من (١١٧٢) وثمن و – ط ج – بذلك المقدار (١٥٣) ويكون بمثل مامر – ه ط – بذلك المقدار اكثر من (١١٧٢) وثمن
(the ratio of $\frac{HE + EG}{HG} = \frac{EG}{GT}$, and when we have added the two quantities HE and EG and have put them in the place of EG, EG was greater than $1162\frac{1}{8}$ where TG is that quantity 153. And in the same way as before ET will be a quantity greater than $1172\frac{1}{8}$.)

100 *post* TEG *add. Ar.* على – ك – *(at K)*

media dividam linea *EK*. Proportio igitur *EG* ad *GK* est maior proportione 2334 et quarte ad 153. Et angulum etiam *KEG* dividam in duo media linea *LE*. Proportio igitur *EG* ad *GL* in longitudine est maior proportione 4673 et medietatis ad 153. Et quia angulus *ZEG*
105 fuit tertia anguli recti, oportet ut sit angulus *LEG* quadragesima octava pars anguli recti. Faciam autem supra punctum *E* angulum equalem angulo *LEG*, sitque angulus *GEM*. Angulus igitur *LEM* est vicesima quarta pars recti anguli. Linea ergo recta *LM* est latus figure poligonie continentis circulum habentis 96 angulos equales.
110 Et quoniam iam declaravimus quod proportio *EG* ad *GL* est maior proportione 4673 et medietatis ad 153, et duplum *EG* est linea *AG*, et duplum *GL* est linea *LM*, sequitur ut sit proportio *AG* ad lineam circumdantem figuram poligoniam 96 angulorum maior proportione 4673 et medietatis ad 14688. Et illud quod est plus triplo 667 et me-
115 dietatis, cuius proportio ad 4673 et medietatem est minor septima. Oportet ergo ut sit figura poligonia continens circulum plus triplo diametri ipsius per id quod est minus septima diametri et plus diminutione linee continentis circulum a triplo diametri eius et septima.

Et sit circulus cuius diametrus sit *AG* [Fig. 7]. Describam autem
120 in ipso latus exagoni, quod sit *GB*. Angulus igitur *GAB* est tertia recti. Ergo proportio *AB* ad *BG* est minor proportione 1351 ad 780, propterea quod proportio *AG* ad *GB* est sicut proportio 1560 ad
c. 2 780, quoniam *AG* est dupla *GB*. Dividam autem an/gulum *GAB*

101 dividam *tr. BI ante* in / igitur: ergo *L*
103 igitur: ergo *BI*
105 quadragessima *P*
108 vicessima *P* / recti anguli *tr. BL*
109 *post* circulum *add. BI* et
112 LM: LRA(?) *I* / sequitur: vel oportet *mg. P*
114 *post* triplo *add. BI* eius secundum quantitatem (*sed om. Ar.*)
117 per: vel secundum *mg.P*
119 Et: at *I* / sit² *om. BI*
120 igitur: vero *L*

101–102 Proportio...153: وتكون نسبة – ط ه – ه ج – الى – ط ج – كنسبة – ه ج – الى – خط – ج ك – فتصير هذه النوبة بازاء – ه ج – اكثر من (٢٣٣٤) وربع وثمن وبازاء – ج ك – (١٥٣) ويكون – ه ك – بهذا المقدار اكثر من (٢٣٣٩) وربع وثمن

(*And* $\frac{TE + EG}{TG} = \frac{EG}{GK}$. *And that which is put in place of EG is greater than* $2334 + \frac{1}{4} + \frac{1}{8}$* *and that in place of GK, 153. And EK will be quantity greater than* $2339 + \frac{1}{4} + \frac{1}{8}$*.*)
((*note the erroneous additions of $\frac{1}{8}$)).

102 *post* KEG *add. Ar.* – ل – على *(at L)*

of a right angle, it is necessary that $\angle$ *LEG* be 1/48 of a right angle. Now I shall construct $\angle$ *GEM* on point *E* and let it be equal to $\angle$ *LEG*. Therefore, $\angle$ *LEM* is 1/24 of a right angle. Therefore, straight line *LM* is a side of the polygon circumscribing the circle and having 96 equal angles. And since we have already shown that $\frac{EG}{GL} > \frac{4673\frac{1}{2}}{153}$ with $AG = 2\ EG$ and $LM = 2\ GL$, it follows that the ratio of *AG* to the perimeter of the polygon having 96 angles is greater than the ratio of $4673\frac{1}{2}$ to 14688. And the ratio of 14688 to $4673\frac{1}{2}$ is greater than 3/1 by $667\frac{1}{2}$, and $\frac{667\frac{1}{2}}{4673\frac{1}{2}} < \frac{1}{7}$. It is therefore necessary that the [perimeter of the]polygon circumscribing the circle is greater than triple its

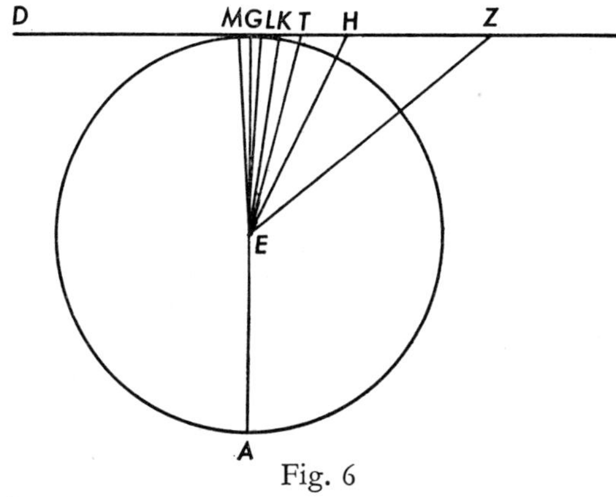

Fig. 6

diameter by an amount which is less than 1/7 the diameter, and [consequently] the circumference of the circle is less than $3\frac{1}{7}$ the diameter by an even greater amount.

Now let there be a circle whose diameter is *AG* [see Fig. 7]. I inscribe in it the side of a hexagon, which side is *GB*. Hence $\angle$ *GAB* is 1/3 a right angle. Therefore, $AB/BG < 1351/780$ because, with $AG = 2\ GB$, then $AG/GB = 1560/780$. Further, I bisect $\angle$ *GAB* by line *AH*. And

106 *post* E *add. Ar.* من خط – ج ه –
(by line GE)

109 angulos: الاضلاع و الزوايا
(sides and angles)

110–114 Et....quod: فاذا ضربنا العدد الذى بازاء – ل م – فى ستة و تسعين بلغ ضعف هذا العدد (١٤٤٨٨) ويكون القطر بذلك المقدار ضعف (٤٦٧٣) ونصف فالذى بازاء محيط الشكل
(And when we multiply the number representing LM by 96, the result is 2 × 14488 ((! 14688?)) and the diameter is 2 × $4673\frac{1}{2}$, and so that which represents the primeter of the figure)

114 *post* triplo *add. Ar.* الذى بازاء القطر
(which represents the diameter)

115 4673 et medietatem: عدد القطر
(quantity of the diameter)

120 in...exagoni *om. Ar.*

in duo media linea *AH*. Et quia angulus *BAH* est equalis angulo
125 *HAG*, angulo *AHG* communi, erunt anguli trianguli *AHG* equales angulis *ABZ*. Ergo proportio *AH* ad *HG* est sicut proportio *AB* ad *BZ*, et sicut proportio *AG* ad *GZ* et sicut proportio *GA*, *AB* coniunctarum ad *BG*. Et ex eo declaratur quod proportio *AH* ad *HG* est minor proportione 2911 ad 780, et quod proportio *AG* ad *HG* est
130 minor proportione 3013 et medietatis et quarte ad 780. Dividam autem angulum *GAH* in duo media linea *AT*. Declarabitur ergo ex eo quod premisimus, quod proportio *AT* ad *TG* est minor proportione 5924 et medietatis et quarte ad 780. Et illud est sicut proportio 1823 ad 240, quoniam proportio cuiusque duorum numerorum primorum
135 ad suum relativum duorum numerorum postremorum est sicut proportio 3 et quarte ad 1. Fit ergo proportio *AG* ad *GT* minor proportione 1838 et novem undecimarum partium unius ad 240. Et etiam dividam angulum *TAG* in duo media linea *AK*. Ergo proportio *AK* ad *KG* est minor proportione 3661 et novem undecimarum unius ad
140 240 et illud est sicut proportio 1007 ad 66, quoniam proportio cuiusque duorum numerorum primorum ad suum relativum duorum numerorum postremorum est sicut proportio 40 ad 11. Ergo proportio *AG* | [30r c. 1]
ad *KG* est minor proportione 1009 et sexte ad 66. Angulum quoque *KAG* dividam in duo media linea *AL*. Ergo proportio *AL* ad *LG*
145 est minor proportione 2016 et sexte ad 66. Ergo proportio *AG* ad *GL*

125 *de* angulo AHG communi *scr. P mg.* Et quia est equalis angulo ABG, quoniam sunt duo recti. / AHG[1]: HAG *BI*
126 angulus *P*
127 proportio[2] *om. I*
128 coniunctarum: vel simul *mg. P* / *mg. BI* quoniam proportio AB ad BZ est tanquam AG ad GZ per 3 sexti et permutatim / *mg. P* ex tertio 6
129 HG: GH *B*
130 3013: vel dicit 3073 *mg. L* / autem: ergo *PL*
132 premissimus *P*
133 -24: XIIII *B*
135 posteriorum *BI*
136 Fit: sit *L*
139 -661: in alio, sexcentorum et septingentorum *mg. P*
143 minor proportione: sicut proportio *BI* / sexte: 6[a] *L*

---

124 *post* AH *add. Ar.* ونصل – ج ح –
(and we join GH)
124–126 Et...ABZ: ولان فى مثلثات – اح ج – ج ح ز – اب ز – زوايا – ح اج – ح ج ز – ب از – متساوية وزوايا – ح – ب – قائمة تكون المثلثات متشابهة
(And because in triangles AHG, GHZ, ABZ, the angles HAG, HGZ, BAZ are equal and the angles at H and B are right angles, the triangles are similar.)
128–30 AH....780: وعددا – اج – اب – جميعا اقل من (٢٩١١) وعدد – ج ب – (٧٨٠)

because ∠ BAH is equal to ∠ HAG, and ∠ AHG is common (i.e., being a right angle is equal to ∠ ABG), the angles of △ AHG are equal to the angles of ABZ. Therefore, $AH/HG = AB/BZ = AG/GZ = (GA + AB)/BG$. From this it is shown that $AH/HG < 2911/780$ and that $AG/HG < (3013 + \frac{1}{2} + \frac{1}{4})/780$. Now I bisect ∠ GAH by line AT. Hence it will be shown from what we put forth before that $\frac{AT}{TG} < \frac{5924 + \frac{1}{2} + \frac{1}{4}}{780}$, that is, $AT/TG < 1823/240$, since [in the case of geometric proportion] the ratio of each of the two antecedent

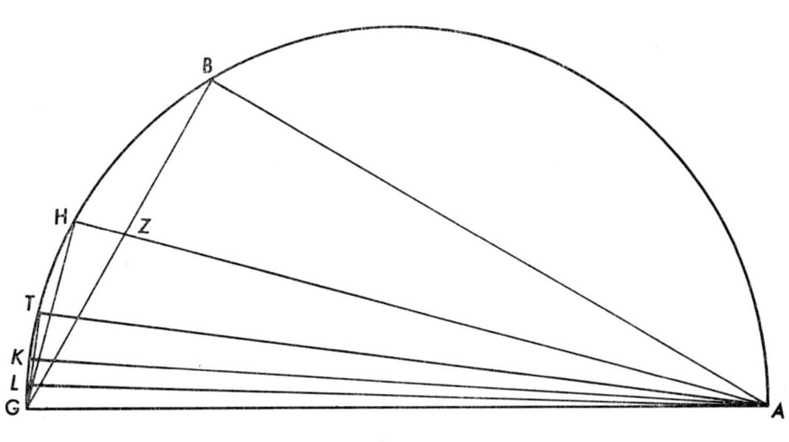

Fig. 7

numbers to its respective consequent of the two consequent numbers is as $3\frac{1}{4}$ to 1 (i.e., 13/4). Therefore, $\frac{AG}{GT} < \frac{1838\frac{9}{11}}{240}$. I also bisect ∠ TAG by line AK. Therefore, $\frac{AK}{KG} < \frac{3661\frac{9}{11}}{240}$, that is, $AK/KG < 1007/66$, since the ratio of each of the two antecedent numbers to its respective consequent of the two consequent numbers is as 40 to 11. Therefore, $\frac{AG}{KG} < \frac{1009\frac{1}{6}}{66}$. I further bisect ∠ KAG by line AL. Hence,

فاذا جعلناهما بازاء – ا ح – ح ج – كان – ا ج – بذلك المقدار اقل من (٣٠١٣) ونصف وربع
(And $AG + AB < 2911$ and $GB = 780$; and since we have put these two numbers in place of AH and HG, hence $AG < 3013 + \frac{1}{2} + \frac{1}{4}$.)

131 *post* AT *add.* *AR* ونصل – ط ج –
(And we join TG)

133 et[1]...quarte *om.* *Ar.*

142–43 Ergo...66 *om.* *Ar.*

est minor proportione 2017 et quarte ad 66. Cum ergo composuerimus, fiet proportio linee continentis figuram poligoniam ad diametrum maior proportione 6336 ad 2017 et quartam. Sed proportio 6336 est plus triplo 2017 et quarte secundum plus 10 partibus 71 partium unius.
150 Ergo linea continens figuram poligoniam habentem 96 angulos, quam circulus continet, addit super triplum diametri eius plus 10 partibus 71 partium. Fit ergo linea continens circulum plus triplo diametri sui secundum id quod est plus 10 partibus 71 partium et fit augmentum eius super hanc quantitatem plus augmento laterum figure poligonie.
155 Linea ergo continens circulum addit super triplum diametri eius minus septima ipsius et plus 10 partibus 71. Et illud est quod declarare voluimus.

146–47 composuerimus: convertimus *BI* in alio, convertimus *mg. P*
147 *post* poligoniam *add. BI* cuius unumquodque laterum est latus exagoni (*not in Greek text*)/ad diametrum *om. L*
148 *post* quartam *add. P et mg. L* Glossa. (*om. L*) Quod sic probatur. Accipe id in quo 6336 sunt plus triplo 2017 (2013 *P*) et quarta una que sunt 284 (2848 *P*) et quarta una, et multiplica ea in 71 et proveniet 20181 et tres quarte, que (et *L*) divide per 2017 et unam quartam, et provenient 10 et remanebunt (remanent *L*) in communi 9 et una quarta, et remanentia in communi (*mg. P textu L* scilicet, 284 que adduntur [addunt *L*] super triplum) si proportionaverimus ad 2017 et quartam erunt minus 1/7. Ergo provenient 10 et minus 1/7. Et provenit plus 10 partibus 7 unius (*i.e.*, 71) partium, nam 10 sunt 10 partes 7 unius (*i.e.*, 71) partium. / quartam: 4ª *L*
149 secundum: vel per id quod est *mg. P*
152 *ante* Fit *add. I* et fit augmentum eius et *supra eum scr.* vacat / Fit ergo *tr. L* / sui *om. BI*
157 *post* voluimus *add. L* Explicit / *Post tertiam propositionem add. BCJ quartam propositionem*: Omnis trianguli in semicirculo cadentis unius duorum laterum in alterum multiplicatio est equalis multiplicationi diametri in perpendicularem que cadit (*om. J*) super basim trianguli. *Et add. BJ*: Ratio patiens est (*om. J*) [*supra B* per VII. 1 et] per primam partem 15ᵉ (*or* 19ᵉ, 16ᵉ *J*) sexti (*om. J*) euclidis. Sicut enim diametrus ad alterum laterum ita reliquum ad perpendicularem.

$\frac{AL}{LG} < \frac{2016\frac{1}{6}}{66}$. Hence $\frac{AG}{GL} < \frac{2017\frac{1}{4}}{66}$. Hence by composition the ratio of the perimeter of the polygon to the diameter will become greater than 6336 to $2017\frac{1}{4}$. But 6336 is greater than triple $2017\frac{1}{4}$ by an amount that is more than 10/71 of 1. Therefore, the perimeter of the polygon having 96 angles, which the circle contains, exceeds triple its diameter by an amount that is more than 10/71 [of the diameter]. Therefore, the circumference of the circle is more than triple its diameter by an amount that is more than 10/71 [of the diameter], and the circumference's excess [over the diameter] is greater than that of the perimeter of the polygon.

Therefore, the circumference of the circle exceeds triple its diameter by an amount that is less than one seventh of it and more than 10/71. And this is what we wished to show.

146–48 Cum....6336[2]: واذا ضربنا ستة وستين في ستة وتسعين صار جميع اضلاع الشكل ذى الستة و التسعين ضلع الذى على الدائرة (٦٣٣٦) وهو

(And since we have multiplied 66 by 96, the sum of the sides of the figure having 96 sides, and which is in the circle, is 6336, and this)

150 96 angulos: المساوى الاضلاع و الزوايا

(of equal sides and angles)

152–54 Fit...poligonie: و محيط الدائرة اعظم منه

(And the perimeter of the circle is greater than it.)

Note: al-Ṭūsī, following upon his version of Thābit's translation, adds other proofs not present in any of the Latin translations, and so I shall omit them.

COMMENTARY

2–6 "Omnis . . . continenti." The wording of Gerard's translation follows what Heiberg in his edition of the Greek text calls the "more correct" reading as given by Eutocius.

2 "triangulo orthogonio" of Tradition I reflects the more literal translation of the Greek text than does "orthogonio triangulo" in Tradition II.

9–10 "Quod . . . erit" clearly supplies what is an unfortunae lacuna in the Greek text.

11–12 "Iam . . . *ABGD*" is absent from the Greek text. It constitutes the first step in the argument that will ultimately show that more than half of the space between the perimeter of the polygon and the circle is exhausted with each doubling of the number of sides of the polygon.

12–19 "Secabo. . . . triangulo" constitutes an elaboration of lines 8–11 (p. 232) of the Greek text. The elaboration may well represent something closer to what Archimedes originally had than does the extant Greek text. Granted that Archimedes often does not elaborate his proofs in detail, yet a number of the additions to the text of the sort found here seem to be necessary. Heiberg's comment at this point of the Greek text is of interest: "omnino in toto hoc opusculo genus dicendi et exponendi brevitate tam neglegenti laborat, ut manum excerptoris potius quam Archimedis agnoscas."

21–22 "uno duorum laterum trianguli continentium rectum angulum" is preferable to the abbreviated reading in the Greek text.

24–29 "Quod. . . . impossibile" is also an improved elaboration of the Greek text (lines 4–5, p. 234).

30–48 "Describam. . . . similes" is a very considerable elaboration of the Greek text (lines 5–13, p. 234), and certainly something like this elaboration may have been in the original text.

48–58 "Figura. . . . *E*" details the argument only suggested in lines 13–17, p. 234, of the Greek text.

57–58 "Et iam . . . . *E*" is a much more formal and symmetrical presentation of the conclusion than is evident in the Greek text.

59–68 Both corollaries are missing in the Greek text. Corollary I is quite obscurely presented. The intent seems to be to hold that with the main proposition assumed—namely, that the circle is equal to the given right triangle—the circle is also equal to a rectangle contained by lines equal respectively to the semicircumference and the radius of the circle. Corollary II, as we have already seen on page 32 of this section, was attributed to Archimedes by Hero.

73–74 "proportio . . . 1" is given as "21 to 7" in the Greek text.

80 "est equalis circumferentie circuli" is missing in the Greek text but certainly ought to be there so long as the phrase "perpendicularis . . . circumdantem" of lines 78–80 is considered a part of the text.

88 The designation of *DZ* for the tangent is preferable to *GLZ* in the Greek text, since *L* is not introduced until later.

89–118 These lines are virtually identical with the Greek text.

90 Gerard's text leaves out the statement found in the Greek text to the effect that $EG/GZ = 265/153$. This seems to have been an oversight, for at least a modified form of the ratio was in the Plato translation, namely, $EC/FC > 265/153$. See the Commentary to Section 1, lines 75-111.

109 "angulos" See the Commentary to Section 1, line 104.

119–20 "Describam . . . *GB*" is not in the Greek text.

123 "quoniam . . . *GB*" is not in the Greek text.

125 "angulo *AHG* communi" appears to be an error, or at least to be incomplete. It should read "et communis est angulus *AHG* rectus." (Note that Tradition II wrongly has *HAG*.) The more correct reading is suggested by the marginal comment of the first tradition: "Et quia est equalis angulo *ABG*, quoniam sunt duo recti." The statement in the Greek text concerning the equality of the third angles, i.e., of angle *AGH* to angle *HZG*, is missing in Gerard's text.

126–27 Instead of "*AB* ad *BZ*" the Greek text has "*GH* ad *HZ*."

138–40 "Ergo . . . 240" is missing in the Greek text, but is present in Eutocius's commentary, edition of J. L. Heiberg, Archimedes, *Opera omnia*, vol *3* (Leipzig, 1915), p. 254.

143 "minor proportione" in Tradition I is preferable to "sicut proportio" in Tradition II, as being not only correct but also in the Greek text.

147 I have followed the Greek text in preferring the simple reading "poligoniam" rather than adopting the added phrase of Tradition II (see the variant readings).

148 The "glossa" after "quartam" in Tradition I (see the variant readings)

might be somewhat confusing to the modern reader and so I have restated it in fractional form:

$$\frac{6336}{2017\frac{1}{4}} = 3 + \frac{284\frac{1}{4}}{2017\frac{1}{4}} \qquad (1)$$

and

$$\frac{284\frac{1}{4} \times \frac{71}{71}}{2017\frac{1}{4}} = \frac{\frac{20181\frac{3}{4}}{71}}{2017\frac{1}{4}} = \frac{10\frac{9\frac{1}{4}}{2017\frac{1}{4}}}{71}. \qquad (2)$$

The marginal phrase "scilicet . . . triplum" has evidently been put in the wrong place. It should follow "multiplica ea."

Chapter three

Emended Versions of the *De mensura circuli*

I have already mentioned in the preceding chapters the fact that a number of emended versions of the *De mensura circuli* circulated in the thirteenth, fourteenth, and fifteenth centuries. The principal question which arises on examining these various reworkings of the *De mensura circuli* is this: Are they further translations from the Arabic—or even from the Greek? Or are they original versions done by medieval geometers by applying Euclid to the text of the Gerard translation and/or to the discussion of the quadrature question by the Banū Mūsā in their *Verba filiorum*? (See Chapter Four.) I think that there can be very little doubt that the various versions are Latin in origin. Since I shall argue the case for each of the versions in the various sections of this chapter, I shall not repeat these arguments at this point. But we can suggest some of the criteria used to show their Latin origin: the use of the wording of the enunciations and proofs found in the Gerard translation, the evidence of citations from one of the Adelard translations of the *Elements* of Euclid, the citation of Latin works such as the commentaries on the *Elements* of Campanus and the *Commentarius in somnium Scipionis* of Macrobius, and so on.

The various emended versions here discussed can in general be grouped together in three main classes. The first class contains those elaborations which are quite close to the original translation of Gerard of Cremona but which show some geometric proliferation by an appeal to Euclid. In this group are the Cambridge, Naples, and Florence versions. The second class depends on the first and is only distinguished from it by the excessive elaboration present in the treatises, an elaboration in part similar to the organization found in scholastic tracts of the fourteenth century.

In this class we can place the Gordanus, Corpus Christi, and Munich versions. The third class includes treatises that depend on the *De mensura circuli* in one or another of its elaborations but at the same time diverge from its proof in a fundamental way and perhaps depend also on the treatment of quadrature found in the *Verba filiorum* of the Banū Mūsā. For this reason I have placed the three works of this class—the Pseudo-Bradwardine Version, the *Versio abbreviata*, and the *Questio de quadratura circuli* of Albert of Saxony—in Chapter Five following the presentation in Chapter Four of the *Verba filiorum*.

Before discussing each version of the first two groups in detail, I should point out that certain of the versions have some interesting features in common. The major point of methodological similarity found in most of the versions is the specific application of Proposition X.1 of the *Elements* of Euclid, which latter proposition holds that "With two unequal magnitudes proposed, if more than half is subtracted from the greater, and then from the remainder again more than half is subtracted, and this subtraction takes place continuously, there will finally remain a magnitude less than the lesser of the proposed magnitudes."[1] Archimedes' aim in the

[1] In the translation of the *Elements*, which elsewhere I have called Adelard II and which was by far the most popular of the translations of the *Elements*, Proposition X.1 runs (MS Brit. Museum Add. 34018, 38v): "Si duabus quantitatibus inequalibus positis maius dimidio ⟨a⟩ maiori detrahatur itemque de reliquo maius dimidio dematur deinscepsque eodem modo, necesse est ut tandem minore positarum minor quantitas relinquatur." It ought to be remarked in the first place that the use of the term "exhaustion" for this procedure of Archimedes as elaborated by the various versions on the basis of Proposition X.1 is something of a misnomer, although it is now almost universally accepted. For this procedure, unlike the pre-Archimedean procedure associated with the name of Antiphon, does not intend a complete exhaustion (and in fact it was invented to obviate the illogic felt to exist in complete exhaustion); rather, it intends a successive extraction until a quantity less than an assigned quantity remains. Professor John Murdoch writes me in connection with the extraction procedure used by the various paraphrasers of the Archimedean proof as follows: "Proposition XII.2 of the *Elements*, which of course also uses Proposition X.1, has a note attached to it in the Gerard of Cremona translation: 'Et huic questioni est propositio quod quando cadit in circulo quadratum equilaterum, tunc ipsum est plus medietate eius; et si cadit in eo octogonium, tunc sectiones (sectores triangulorum *in Vat. lat 7299 and Vat. Ross. 579*) qui accidunt in eo sunt plus medietate eius. Et similiter si cadat in eo habens sedecim latera, et usque ad infinitum.'" This comment is in the text in BN lat. 7216, 91r, in the margin of Vat. lat. 7299, 114r, and Vat. Ross. 579, 107v, and is missing in Vat. Reg. suev. 1268. Incidentally, one should point out that all of the Arabic versions of the *Elements* as well as most of the Latin versions (except for that of Campanus and the translation from the Greek in BN lat. 7373) lack the corollary to Proposition X.1, which corollary achieves the same objective as X.1 by

first proposition of the *De mensura circuli* (an aim sketched very briefly in the extant Greek text) was to show that if the circle is said to exceed the right triangle, there will result some inscribed polygon which is at the same time greater than and less than the given triangle, an obvious contradiction. A similar contradiction concerning some circumscribed polygon ensues if we assume the circle to be less than the given triangle. In elaborating this proof not only do most of the versions specifically cite Proposition X.1 of Euclid, as I have said, but they also cite many other propositions of Euclid, as my detailed commentaries on these various versions will show. (Incidently, although the form and substance of the argument is similar to Proposition XII.2 of Euclid's *Elements*, this proposition is generally not cited.) Furthermore, in the proof of Proposition I they tend to assign a literal value (such as D in the Cambridge Version) to any given magnitude by which the circle is said to exceed the triangle in the first part of the proof and to be exceeded by the triangle in the second part of the proof. It will be obvious to anyone examining either the Greek text or the two translations from the Arabic given in Chapter Two above that such literal specification was not given in these earlier versions. After the assignment of such a quantity, the emended versions then go on to prove by the use of Euclid's *Elements* that if we keep doubling the number of sides of the inscribed regular polygon, having started with an inscribed square, we are each time extracting more than half of the space remaining between the perimeter of the polygon and the circumference of the circle. Hence, if we continue this process indefinitely, we shall by Proposition X.1 of the *Elements* finally arrive at some remaining quantity that is less than the literally designated quantity by which the circle exceeds the triangle. For the sake of brevity it is said that we have found that remaining quantity when we have inscribed an octagon in the circle. At this point the logical inference is drawn that the regular polygon is greater than the triangle. But, the emended versions go on to say, it can be shown that the polygon is actually less than the triangle, for the polygon is inscribed in a circle and

taking half of each remainder. Most of the authors, starting with Archimedes' proof where the "more-than-half" procedure is implied and with Euclid XII.2 where the same procedure is evident, settled for it, even to the point of commencing with a square that was "more than half" of the circle. Mr. Murdoch remarks that the "Arabs were particularly troubled with the 'more-than-half' requirement of exhaustion proofs, a disturbance occasioned, or at least abetted, by their lack of the corollary to Proposition X.1." Cf. similar statements in his paper, "The Medieval Language of Proportions," in A. Crombie, ed., *Scientific Change* (New York, 1963), pp. 244-45.

hence the two quantities which measure double its area—namely, the perimeter of the polygon and the perpendicular drawn from the center of the circle to one of the sides of the polygon—are respectively less than the two quantities which measure double the area of the triangle—that is, the circumference and the radius of the circle. Hence the assumption from which the contrary inference was drawn must be false, and so the circle cannot be said to be greater than the triangle. The argument refuting the assumption that the circle is less than the given triangle is also spelled out in some detail in these emended versions. It is of further interest that many of the emended versions make specific reference to the logical structure of the argument, as, for example, pointing out just where the logical fallacy lies or indicating the form that the argument will take.

A further common point of many of the elaborated versions is that after they have shown that the designated right triangle is equal to the circle, they then indicate that the triangle must be converted to a square by first converting it to a rectangle and then finding the mean proportional line between the altitude and base of the rectangle (cf. also the treatment in Bradwardine's *Geometry*; see Chapter Two, Section 2, note 11, above).

A minor but interesting terminological point ties together five of the seven emended versions of this chapter. It centers on the misuse of the term *lunula* for a segment of a circle (or even on occasion for a figure contained by two straight lines and an arc). The correct usage of this term (for an area contained between two intersecting arcs) was, however, evident in the translation and paraphrase of Simplicius' treatment of quadratures by means of lunes (see Appendix II). But sometime in the course of the thirteenth century—perhaps from Averroes' *Commentary on the Physics* (see page 569n, below)—the term was taken up and misapplied to segments of circles. In addition to appearing in this way in the Cambridge and Naples versions, it also appears incorrectly in the two versions of Florence and in the Corpus Christi Version. As a matter of fact, in the Corpus Christi Version we see it beginning to be replaced by the word commonly used for small segments of circles, that is, *portiuncula*, which appears also in the Gordanus Version.

While I have emphasized some of the common features of these various emended versions, the individual differences are manifold and it will be the purpose of this chapter not only to give texts and translations of these various versions but also to emphasize their peculiarities.

1. The Cambridge Version

Certainly one of the oldest of the emended versions is the one which I have called the Cambridge Version from its principal manuscript (manuscript *N*; see sigla). The reader will notice immediately that the author of this version had decided that certain postulates were necessary before embarking on the proofs of the two propositions which constitute this version. The first postulate assumes that there can be found a straight line equal to a curve, and vice versa. As Eutocius noted in his commentary to the *Measurement of the Circle* (ed. of J. L. Heiberg. 1915, p. 230), Archimedes did not directly point this out. It is of course crucial to the whole treatment of the quadrature problem and is at least tacitly assumed in any version of Archimedes' treatise. Our author could scarely have framed this postulate as the result of Eutocius' statement, which almost certainly was not available to him. He might, however, have drawn it from the similar statement found in the first proposition of the *De curvis superficiebus* of Johannes de Tinemue (see Chapter Six, Proposition I, lines 15–17). It will be noticed that the Corpus Christi Version of the *De mensura circuli* given in Section 5 of this chapter makes this postulate its second assumption (line 8). See also the Gordanus tract, page 143, note 4.

The second postulate of the Cambridge Version assumes that an arc is greater than its chord. Such an assumption was embraced by the more general postulate given by Archimedes at the head of his *On the Sphere and the Cylinder*:[2] "Of lines which have the same extremities the straight line is the least." The Cambridge postulate was also repeated in the Corpus Christi Version (lines 5–7) and in the text of the version of Gordanus (line 109) and of Munich (lines 128–30).

The last of the three postulates of the Cambridge Version assumes that the perimeter of an including figure is greater than the perimeter of the included figure. While not making precise what constitutes "including" and "included," this assumption obviously has the same purpose as the conclusion drawn by Archimedes from his postulates in the *On the Sphere*

[2] Archimedes, *Opera omnia*, vol. *1* (Leipzig, 1910), p. 8. Interestingly enough this postulate not only was known to the Latin geometers in the Moerbeke translation of 1269 (see Volume Two of this work), but was also known earlier in Gerard of Cremona's translation of the commentary of Anaritius on the *Elements* of Euclid, where it is given in two forms (see Appendix III, Passages 1–2). Cf. H. G. Zeuthen, "Ueber einige archimedische postulate," *Archiv für die Geschichte der Naturwissenschaften und der Technik*, vol. *1* (1909), pp. 320–27.

and the Cylinder:[3] "If a polygon is inscribed in a circle, it is evident that the perimeter of the inscribed polygon is less than the circumference of the circle," together with the first proposition of that work,[4] "If a polygon is circumscribed about a circle, the perimeter of the circumscribed polygon is greater than the perimeter of the circle." As a matter of fact, these two conclusions of Archimedes form the first two propositions of the Corpus Christi Version (see that text), lines 35–49, and are similar to a supposition made by Albert of Saxony in his *Questio de quadratura* (see Chapter Five below, Section 3, lines 198–200).

I have already indicated that the Cambridge Version was one of the many versions to make specific citations to Euclid's *Elements*. I shall give the details of these citations below in the Commentary, but I may say here that the *Elements* is cited nine times in the proof of the first proposition and three times in the proof of the second proposition. The Cambridge Version was also one of those versions which made X.1 of the *Elements* the heart of the proof of the first proposition. The phrasing in lines 19–21 where X.1 is cited obviously reflects one of the translations of the *Elements* of Adelard of Bath (see note 1 above) and confirms our conclusion that we have here an original Latin version rather than a new translation. Incidentally, other points confirming the Latin origin of this version are (1) the almost exact verbal correspondence between the enunciations of Propositions I and II in this version and in the translation of the *De mensura circuli* by Gerard of Cremona, and (2) the citation of the Latin author Macrobius in the second proposition (see lines 87–88).

In detailing the proof of the first proposition, the author of the Cambridge Version assigns the letter *D* to the quantity by which the circle exceeds the triangle in the first part of the argument. This perhaps has some significance for the problem of dating the Cambridge Version. It would seem that our principal manuscript of this version, manuscript *N*, is of the thirteenth century, although it might be a bit later. The thirteenth-century manuscript, Bodleian, Auct. F.5.28, which we used in the establishment of the text of the Gerard of Cremona translation, includes a drawing for the first proposition that has quantity *D* specified, even though the text makes no mention of *D*. This seems to show that the scribe had at hand not only a copy of the text of the Gerard of Cremona translation but also a manuscript that included the Cambridge Version. If my reasoning is correct, this would substantiate the thirteenth-century origin of the Cambridge Version.

[3] Archimedes, *op. cit.*, p. 10. [4] *Ibid.*

An interesting feature of the Cambridge Version is the use throughout of single letters to designate various geometric magnitudes. Although this introduces some ambiguity, the argument becomes transparently easy to follow once these letters have been properly assigned and understood. It is also of interest that the Cambridge Version goes one step beyond the Gerard translation by squaring the triangle found to be equal to the circle. This was to be characteristic of the emended versions of the *De mensura circuli*.

The reader should be reminded that the author of the Cambridge Version apparently reworked only the first two propositions of the *De mensura circuli*. At least there is no extant manuscript of this version that includes Proposition III. Of the three manuscripts of this version, manuscript *N*, as I have said, is the oldest and the best. The scribe of *N* had, however, at least two blind spots so far as orthography was concerned. During the copying of the first proposition, he originally often used the form *linule* rather than the correct form *lunule* before finally using the correct form. (I have already commented on the fact that the author of the Cambridge Version was one of those authors who misused the term *lunule* for segments of circles.) The other spelling error that is found several times in the first part of Proposition I in manuscript *N* is *ex ducta* for *ex ductu*. Once again, the scribe later employed the correct form.

After *N* the next best manuscript is *V*. It appears to have been copied from *N* sometime close to 1300. It makes the same errors as *N* but leaves out many crucial phrases, particularly in Proposition II, where both scribes have some trouble with the numbers. Both *N* and *V* mix Indo-Arabic numerals and Roman numerals with rhetorical expressions for numbers.

Manuscript *M* is a hopelessly confused copy and I have but rarely reported its variant readings. Its confusion is illustrated by the continual replacement of the words *octogonium* and *poligonium* by the word *ortogonium*. It leaves out crucial phrases and quantities. Furthermore, it includes only the first proposition and, as a matter of fact, leaves off the conclusion of that proposition. One interesting fact about *M* is that its scribe makes Proposition I precede the *De curvis superficiebus*. I have already commented on the possible dependence of the author of the Cambridge Version on the *De curvis superficiebus*. It is not improbable that the original author of the Cambridge Version composed his version of the *De mensura circuli* with an eye on the *De curvis superficiebus* and that he wrote his version in a codex that included the latter work.

The figures for Proposition I have been taken from *N*. All of the figures are missing in *V*, while such figures as are found in *M* are quite incorrect. The figure for Proposition II has been reconstructed from the text, it being absent from both *N* and *V*, the two manuscripts which include Proposition II. The marginal folio numbers are from MS *N*.

Sigla of Manuscripts

N = Cambridge, Gon. & Caius College 504/271, 108v–109v, 13c.
V = London, British Museum Royal 12E.25, 150v–151v, ca. 1300.
M = Florence, Bibl. Naz. Conv. Soppr. J.V.18, 92r–v, 14c.

On the Quadrature of the Circle

De Quadratura Circuli

108v
c. 1

/ Liber Archimenidis de quadratura circuli

[Petitiones]

[1.] Cuilibet recte linee aliquam curvam esse equalem et cuilibet curve aliquam rectam.

5 [2.] Item cordam quamlibet esse minorem arcu.

[3.] Item ambitum cuiuslibet figure includentis esse maiorem ambitu figure incluse.

[Propositiones]

I. OMNIS CIRCULUS TRIGONO ORTOGONIO EST EQUALIS, CUIUS UNUM DUORUM LATERUM RECTUM ANGULUM CONTINENTIUM MEDIETATI DIAMETRI EQUATUR ET ALTERUM IPSORUM LINEE CONTINENTI CIRCULUM.

Sit datus circulus *O*, linea recta equalis circumferentie *A*, semidiameter *B* et applicetur illi ad rectum angulum et subtendatur basis [Fig. 8]. Et sit ille triangulus *C*. Dico ergo quod *C* triangulus est equalis *O* circulo. Si dicatur quod circulus sit maior, ergo triangulus cum aliquo excessu est equalis *O* circulo. Sit ille excessus *D*, et non refert cuius forme sit. Inde sic, *O* est equalis *C* et *D*. Ergo est maior *D*. Ergo si ab *O* subtrahatur maius medietate et iterum a residuo maius medietate, iuxta primam X Euclidis tandem occurret minus minore predictarum, scilicet *D*. Item inscribatur quadratum circulo *O* et fiat diameter que sit diameter et circuli et quadrati, et ad quantitatem diametri eius intelligatur fieri quadratum quod erit minus circulo per tertiam petitionem. Quadratus ergo inscriptus *O* est medietas

1 Liber...circuli *om.* *VM*
5 quamlibet *om.* *V*
6 ambitu *om.* *N*
14 ad *om.* *V*
15 ille *N* iste *V* / C triangulus *tr.* *V* / est *N* sit *V*
17 excessu *VM* excessio *N*
17–19 et...D *NM om.* *V*
20 X Euclidis *N* ex eius *V* / occurret *NM* occuret *V*
20–21 minus...D: quantitas minor D et hoc per primam decimi *M*
21 predictorum *V* / scilicet *N* C *V*
22 que...quadrati *N* et circuli quadrati que sit diameter *V*
23 dyametri *V* / minus *V* maius *N*

On the Quadrature of the Circle

The Book of Archimedes on the Quadrature of the Circle

[Postulates]

[1.] There is some curved line equal to any straight line and some straight line equal to any curved line.

[2.] Any chord is less than [its] arc.

[3.] The perimeter of any including figure is greater than the perimeter of the included figure.

[Propositions]

I. EVERY CIRCLE IS EQUAL TO A RIGHT TRIANGLE, ONE OF WHOSE TWO SIDES CONTAINING THE RIGHT ANGLE IS EQUAL TO THE RADIUS OF THE CIRCLE AND THE OTHER OF THEM TO THE CIRCUMFERENCE OF THE CIRCLE.

Let O be the given circle, A a straight line equal to the circumference, B one equal to the radius, and let A be applied to B at a right angle; let the base be drawn [see Fig. 8]. Let C be that triangle. I assert therefore that $\triangle\, C$ is equal to circle O. If it is said that the circle is greater, then the triangle plus some excess is equal to circle O. Let that excess be D—and its shape is of no matter. Thence, $O = (C + D)$. Therefore, $O > D$. Hence if a quantity greater than one half [of O] is subtracted from O, and again from the remainder more than half is subtracted [and this is done continuously], then according to X. 1 of [the *Elements* of] Euclid, there will finally result a quantity less than the lesser of the aforesaid quantities, that is, less than D. [In explanation of this,] also let a square be inscribed in circle O and a diameter is produced which is the diameter of both the circle and the square. And by reference to the quantity of its diameter, let it be understood that a square is produced which will be less than the circle, by the third postulate. Therefore, the square inscribed in O is one

25 exterioris per penultimam primi. Ergo est maior medietate circuli. Ergo si ab illo circulo subtrahatur illud quadratum, subtrahetur maius medietate. Item inscribe circulo octogonium equilaterum. Probo quod c. 2 quilibet triangulus qui est ex latere quadrati / et duobus lateribus octogonii est maius medietate lunule cui inscribatur, quod in uno 30 appareat. Ducatur itaque a puncto contactus duorum laterum linea equidistans basi illius trianguli et fiat parallelogramum. Ille triangulus et illud parallelogramum sunt inter duas lineas equidistantes super eandem basim. Ergo triangulus est medietas illius per XL primi. Ergo ille triangulus est maior medietate lunule cum parallelogramum 35 sit maius lunula. Subtracta igitur sic semper maiori parte tandem relinquetur superficies ex parvis lunulis minor *D*. Inde sic, illud poligonium et ille relicte lunule sunt equales *D* et *C*, que lunule sunt minores *D*. Ergo poligonium est maius triangulo *C*. Sed probo quod

25 medietate *bis V*
26 subtrahetur *NM* subtractum *V*
28 ex *N* in (?) *V*
29 octogonium *V* / linule *N* / uno *N* unum *V*
30 itaque *N* ergo *V*
31 equedistans *V* / parallelogramum *hic et ubique corr. ex* parallellogramum *in MSS*
32 equedistantes *V*
34 lunule *corr. ex* linule *N* linee *V* simile *M*
35 lunula *N* linula *V*
36 relinquentur *V* / D *N* sit (?) D *V*
36–37 Inde...equales *om. V*
37 et *om. V*
38 est *N* existens (?) *V* / *post* maius *add. N* medietate

half of an exterior [circumscribed square], by the penultimate [proposition] of [Book] I [of the *Elements*]. Therefore it (the inscribed square) is greater than half of the circle. Hence if the square is subtracted from the circle, more than half is subtracted. Then inscribe in the circle a regular octagon. I prove that any triangle which is composed of a side of the square and two sides of the octagon is greater than half of the lune* (*actually* segment) in which it is inscribed, which we let be evident by a single example.

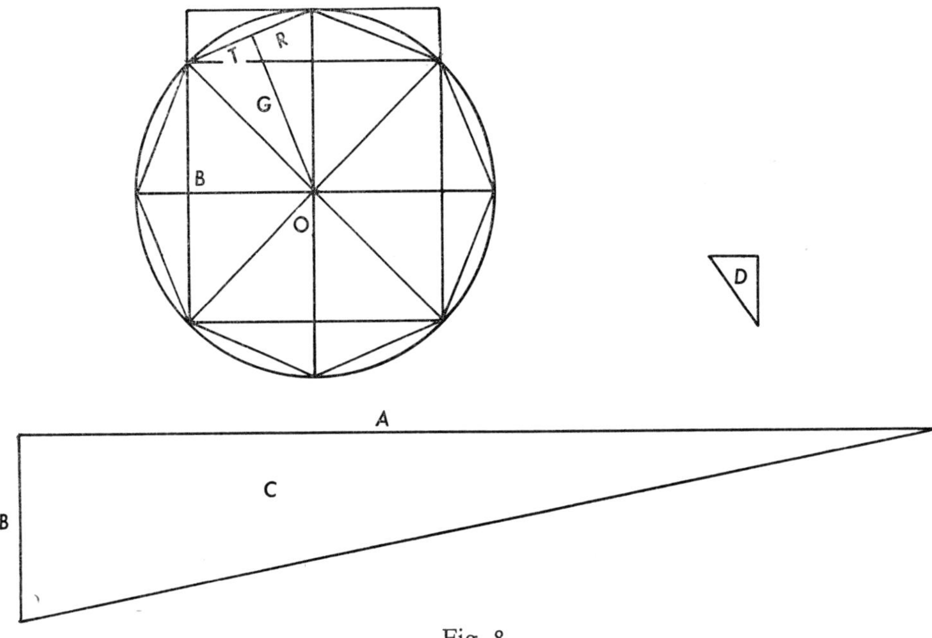

Fig. 8

And so let there be drawn from the point of contact of the two sides [of the octagon] a line parallel [and equal] to the base of the triangle, and let a parallelogram be constructed [by joining the ends of the lines]. Hence the triangle and the parallelogram are on the same base and between two parallel lines. Therefore, the triangle is one half of the parallelogram, by I.40 (I.41, *Greek text*) [of the *Elements*]. Hence the triangle is more than one half of the segment, since the parallelogram is greater than the segment. Therefore, with the greater part always subtracted in this manner, finally there will remain a surface consisting of small segments which is less than D. Proceed as follows: The polygon and those segments which remain are equal to $(D + C)$, but the segments are less than D. Hence the polygon is greater than $\triangle C$. But I prove that it is less. For a perpen-

* For use of the word *lunula* for *segment*, see the Introduction to this chapter, p. 62.

est minus. Ducatur ergo catetus a centro *O* ad latus poligonii dividens
40 illud per duo equa et faciet angulos rectos per tertiam tertii. Et ducantur
a terminis illius lateris poligonii ad centrum due linee. Inde sic, illud
quod fit ex ductu *G* in *T* est duplum ad illum parvum triangulum et
illud quod fit ex ductu *G* in *R* est duplum ad reliquum partialem. Ergo
quod fit ex ductu *G* in *TR* est duplum ad totum triangulum per XL
45 primi. Et ita illud quod fit ex ductu *G* in totalem ambitum poligonii
est duplum ad poligonium. Sed illud quod fit ex ductu *B* in *A* est
duplum ad *C* trigonum, et *G* est minor quam *B* quia est minor semidiametro, et ambitus poligonii est minor ⟨*A*⟩ quia est minor circulo per
tertiam petitionem. Ergo duplum trianguli est maius duplo poligonii.
50 Ergo triangulus est maior poligonio per XV quinti. Relinquitur ergo
quod *O* non est maior.

Si *O* est minor *C* triangulo, ergo *O* et aliqua superficies sunt equales
109r c. 1 *C*. Sit illud *D*. Quadretur ergo diameter circa circulum *O*. / Inde sic
[Fig. 9], ambitus huius quadrati est maior *A*. Ergo maius est illud
55 quod fit ex ductu *B* in ambitum quadrati quam quod fit ex ductu *B*
eius in *A*. Vocentur ergo superficies que sunt excessus quadrati ad
O, *F*. Inde ergo *O* et *D* sunt equales *C*, et *O* et *F* sunt maiores *C*,
et *O* equale sibi ipsi. Ergo *F* est maius *D*. Ergo si ab *F* detrahatur
maius medietate et iterum a residuo maius medietate, tandem relin-
60 quetur quantitas minor; quod fiet hoc modo. Sit *NC* latus octogonii
equilateri et ducantur ad centrum *O* linee a terminis *NC*. Deinde ab
angulo quadrati ducatur per punctum contactus ad centrum *O* linea
X. Deinde ab eodem puncto contactus ad punctum contactus lateris

39 ergo *om.* *V*
40 equa *NM* equalia *V* / faciet *N* facies *V*
41 due linee *om.* *V* / Inde *N* Deinde *V*
42 ducta *N hic et alibi* / in *N* et *V* / illum *corr.* *N ex* illud / parvum *om.* *V*
43 ducta *N* / reliquum *om.* *V*
48 ambitu *N*
51 non est *bis* *V*
53 illud *om.* *V* / O *om.* *V*
55 quam *N* quamvis *V*
56 vocantur (?) *V* / ergo *om.* *V*
58 F[1] *om.* *V*
59 et...medietate *om.* *V*
60 NC *corr.* *ex* N *in NM et ex* enim *in V*
61 equilateris *V* / ducantur *tr.* *V post* O / NC *corr.* *ex* N *in N et ex* eius *in V* / Deinde *N* ducta *V*
63 ad punctum *N* a puncto *V*

dicular (G) is drawn from the center O to a side of the polygon, and it bisects the side and forms right angles, by III. 3 [of the *Elements*]. Then from the termini of that side of the polygon are drawn two lines to the center. Proceed then as follows: The product of G and T is double the small triangle and the product of G and R is double the remaining partial [triangle]. Therefore, the product of G and $(T + R)$ is double the whole triangle, by I.40 (I.41) [of the *Elements*]. And so the product of G and the whole perimeter of the polygon is double the polygon. But the product of B and A is double $\triangle$ C, and G is less than B because it is less than the radius, while the perimeter of the polygon is less than A because it is less than the [circumference of the] circle by the third postulate. Therefore, double the triangle is greater than double the polygon and so the triangle is greater than the polygon, by V.15 [of the *Elements*]. It results, therefore, that O is not greater [than $\triangle$ C].

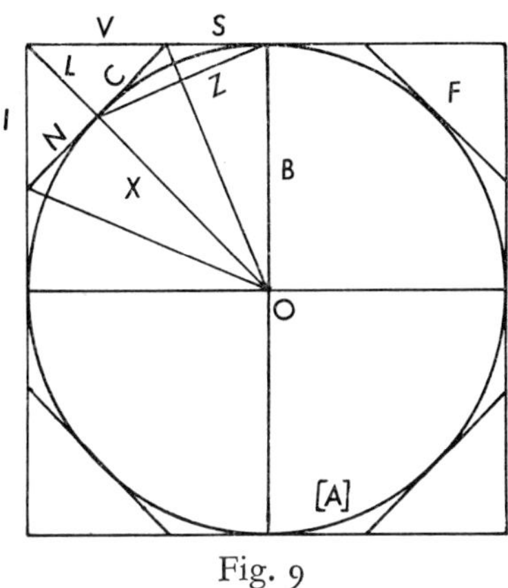

Fig. 9

If O is less than $\triangle$ C, hence O plus some surface is equal to C. Let that be D. Let the diameter be squared about circle O [see Fig. 9]. Proceed then as follows: The perimeter of the square is greater than A. Therefore, the product of B and the perimeter of the square is greater than the product of B and A. Therefore, let the surfaces which constitute the excess of the square over O be called F. Thus $(O + D) = C$, and $(O + F) > C$, and O is equal to itself. Therefore, $F > D$. Therefore, if from F is subtracted more than its half, and again more than half is subtracted from the remainder, [and this is done continuously,] finally there will remain a lesser quantity. This is done in this way: Let $(N + C)$ be the side of a regular octagon and let lines be drawn from the ends of $(N + C)$ to the center O. Then from the angle of the square let line X be drawn through the

quadrati ducatur *Z*; suntque latera duorum triangulorum *I* et *N* et *C*
65 et *V* et *L*. Inde sic, *V* respicit ad angulum rectum propter lineam contingentem. Ergo ⟨quadratum *V*⟩ valet quadrata *L*, *C*. Sed *C* est equale *S*. Ergo *V* est maius *S*. Sed *C* est equale *S* per angulos sub basi et angulos supra basim. Inde sic, triangulus cuius *V* est basis et ille cuius ⟨*S*⟩ est basis sunt eiusdem altitudinis. Ergo que est proportio
70 *V* ad *S* eadem est trianguli ad triangulum secundum primam sexti. Sed *V* est maior *S*. Ergo triangulus est maior triangulo. Ergo multo magis est aliqua ex circulo lunula que est pars eius; a simili de reliquo triangulorum. Ergo ab illa totali superficie que constat ex quatuor figuris subtracta eius maior est pars medietate; relictum ergo sic minus
75 est *D*, tandem enim necesse est pervenire ad minus. Inde sic, *O* et *D* sunt equalia *C*, et *Q* minus *D*. Sit *Q* illud ultimum relictum. Ergo *O* et *Q* sunt minus *C*. Sed *O* et *Q* sunt poligonium. Ergo poligonium illud est minus *C*. Sed contra, ambitus poligonii maior est ⟨*A*⟩. Ergo
c. 2 maius est illud quod fit ex ductu *B* in ambitum poligonii quam illud /
80 quod fit ex ductu *B* in *A*. Ergo maius est duplum poligonii quam duplum *C*. Ergo poligonium est maius *C*. Relinquitur ergo quod *C* sit equale *O*. Sed dato trigono equum quadratum invenire per ultimam secundi; ergo dato circulo quadratum equale; et hoc erat propositum.

II. PROPORTIO AREE OMNIS CIRCULI AD QUADRATUM
85 DIAMETRI EST SICUT PROPORTIO UNDECIM AD QUATUORDECIM.

Et probatur iuxta illud Macrobii [quod] diametros triplicata cum adiectione septime partis est equalis circumferentie circuli. Sit circulus *O*, cuius diameter triplicata cum adiectione septime partis [Fig. 10].
90 Sit *ABD* ita quod septima pars sit *D*. Semidiameter illius, scilicet *C*, applicetur *A* ad rectum angulum. Per proximam ergo trigonus

66 Ergo *N* Deinde ergo *V* / C[1] *N* et C *V*
67 V *N* N *V* / C *V* V *N*
68–69 et...basis *om.* *V*
70 eadem *VM* ad eandem *N* / primam sexti *M* primam *NV*
72 magis *VM* maius *N* / pars *om.* *V*
72–73 reliquo triangulorum *N* triangulo reliquo *V*
73 Ergo *N* constet ergo *V*
74 eius maior est *N* est maior *V*
75 est[1] *om.* *V* / enim *om.* *V*
76 equalia C *N* equales *V* / Sit Q *N* sitque *V*
77 minus C *N* inequalia *V*
77–78 Sed...contra *om.* *V*
79–80 ambitum...B *om.* *V*
80 duplum poligonii *N* poligonium *V* / quam duplum *N* quadruplum *V*
81 est *om.* *V*
82 O *N* D *V*
83 equale *N* equum describere *V* / et hoc *N* quod *V*
84 II: 2[a] *mg.* *N* *om.* *V*
85 dyametri *N*
87 Et *N* Hec *V*
87, 89 triplicatus *V* (see Commentary)
88–89 equalis...partis *om.* *V*

point of tangency to the center O. Then from that same point of tangency [of side $(N + C)$ and circle O] line Z is drawn to the point of tangency of the side of the square. And the sides of the two triangles are I and N, C and V, and L. Proceed then as follows: V is opposite the right angle formed as the result of line [C] being tangent. Therefore $V^2 = C^2 + L^2$. But $C = S$. Therefore, $V > S$. But C is equal to S because of the equality of the angles under the base and the equality of the angles above the base. Thence the triangle of which V is the base and that of which S is the base are of the same altitude. Hence the ratio of triangle to triangle is as the ratio of V to S, according to VI.1 [of the *Elements*]. But $V > S$; therefore the triangle [with base V] is greater than the triangle [with base S]. Therefore, it is much greater than some segment of the circle which is its part. The same thing is true for the remaining triangles. Hence from that total surface consisting of the four figures has been subtracted more than half of it. The [final] remainder, therefore, [when this practice is continued,] is thus less than D, for ultimately it is necessary to arrive at a lesser quantity. Proceed thus: $(O + D) = C$ and $Q < D$. Let Q be the ultimate remainder. Therefore, $(O + Q) < C$. But $(O + Q)$ equals the polygon. Hence the polygon is less than C. But the contrary [can be shown]: The perimeter of the polygon is greater than A. Therefore, the product of B and the perimeter of the polygon is greater than the product of B and A. Therefore, double the polygon is greater than double C. Therefore, the polygon is greater than C. It remains, therefore, that C equals O. But a square can be found which is equal to the given triangle by the last proposition of [Book] II [of the *Elements*]. Therefore, we have found a square which is equal to the given circle, and this was proposed.

II. THE RATIO OF THE AREA OF ANY CIRCLE TO THE SQUARE OF ITS DIAMETER IS AS THE RATIO OF 11 TO 14.

And this is proved, assuming the conclusion of Macrobius to the effect that triple the diameter with the additon of a seventh part is equal to the circumference of the circle. Let O be the circle whose diameter is tripled with the addition of a seventh part [see Fig. 10]. Let $(A + B + D)$ be such that D is $1/7$ [of A, and $B = 2A$]. Let the radius C be applied to A at a right angle. By the first proposition the triangle made up from $(A +$

ABD, *C* est equalis *O*. Inde sic, que est proportio *A* ad *D* ea est septinarii ad unitatem. Ergo que est proportio *AB* ad *D* ea est 21 ad unitatem. Ergo que est proportio *ABD* ad *D* ea est 22 ad unitatem.
95 Sed que est basis ad basim ea est trianguli ad triangulum per primam sexti. Ergo que est proportio *ABD* trigoni ad *D* trigonum ea est 22 ad unitatem. Ergo que est circuli ad *D* ea est 22 ad unitatem. Item *A* est diameter; *C* est semidiameter. Ergo illud quod fit ex ductu *C* in *A* est subduplum ad quadratum *A*. Sed triangulus *A* est subduplus ad
100 illud quod fit ex ductu C in *A*. Ergo trigonus est subquadruplus ad quadratum *A*. Sed que est proportio *A* trigoni ad trigonum ⟨*D*⟩ est 7 ad unitatem. Ergo que est quadrati ad trigonum *D* ea est 28 ad unitatem. Disponantur ergo numeri et quadratum et circulus et trigonus *D*. Sic ergo que est proportio 28 ad 1 ea est quadrati *A* ad *D*.
105 Sed que est 22 ad unitatem ea est circuli *O* ad *D* trigonum. Ergo que
29v c. 1 est 28 ad 22 ea est quadrati *A* ad circulum *O* per 11 / quinti. Ergo que est quadrati ad circulum ea est 14 ad 11 per tertiam quinti. Ergo permutatim que est circuli ad quadratum ea est 11 ad 14.

92 ABD, C *corr. ex* ABCD *in NV*
93 ea *N* eadem *V* / 21 *N* tertiarii *V*
94 22 *om. V*
95 basem *V*
97 22 *om. V*
98 illud *corr. ex* cd *in N; om. V*
100 quod *V* et (?) *N* / ex ductu *om. V*
101 A trigoni *N* 11 tertragoni *V*
102 7 *corr. ex* 11 *in NV* / ea *om. N* / 28 ad unitatem *N* a D ad C *V*
103 Disponatur *N* / et[1] *N* ad *V*
104 28 *N* 8 *V* / 1 *N* D.1 *V*
105 22 *om.* V / circuli O *om. V et lacunam habet*
106 28 ad 22 *om V* / est[2] *om. V* / *A om. V*
107 14 *N* X[or] (?) *V* / tertiam *N* primam *V*

$B + D$) and C is equal to O. Then proceed as follows: $A/D = 7/1$. Hence $(A + B)/D = 21/1$. Therefore, $(A + B + D)/D = 22/1$. But the triangle is to the triangle as the base is to the base, by VI.1 [of the *Elements*]. Hence $\triangle\ ABD/\triangle\ D = 22/1$. Hence circle $O/\triangle\ D = 22/1$. Now A is the diameter and C is the radius; therefore, the product of

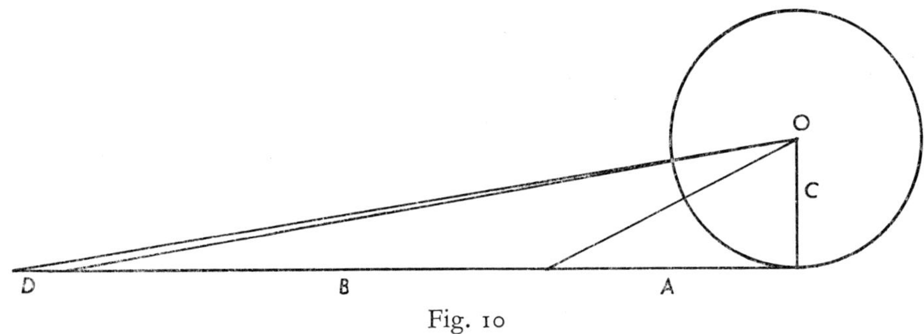

Fig. 10

C and A is one half the square of A. But $\triangle\ A$ is one half of the product of C and A. Therefore, triangle [A] is one fourth the square of A. But $\triangle\ A/\triangle\ D = 7/1$. Therefore, square of $A/\triangle\ D = 28/1$. Hence let the square, the circle, $\triangle\ D$, and their numbers be disposed [in proportion]. Thus $28/1 =$ square of $A/\triangle\ D$, and $22/1 =$ circle $O/\triangle\ D$. Therefore, square A/circle $O = 28/22 = 14/11$, by V.11 and V.3 [of the *Elements*]. Hence, permutatively, the circle is to the square of the diameter as is 11 to 14.

COMMENTARY

9 "trigono." "triangulo" is used in the Gerard translation. However, these words are completely interchangeable among medieval geometers.

10–11 "angulum continentium." These words are transposed in the original Gerard translation.

12 "continenti circulum." These words are also transposed in the Gerard translation.

19–21 "subtrahatur...predictarum." This phrasing should be compared to that given on page 60, note 1, for the Adelard II translation of X.1 of the *Elements*. As a matter of fact, the *predictarum* of manuscript *N* is probably feminine because of the *positarum* of the Adelard tradition. It is for this reason that I have retained it instead of the apparently better reading of *predictorum* in *V*.

25 "penultimam primi." This is a reference to the Pythgorean theorem (I.46 in the Adelard II translation; I.47 in the Greek text). It follows immediately from the Pythagorean theorem that the exterior square which is constructed on the diagonal of the interior square is twice the area of the interior square. The theorem as given in the Adelard II translation of the *Elements* reads (MS Brit. Museum Add. 34018, 7r): "In omni triangulo rectangulo quadratum quod a latere recto angulo opposito in se ipsum ducto describitur equum est duobus quadratis que ex duobus reliquis equis lateribus conscribuntur."

33 "XL primi" should read "XLI primi," for in both the Adelard II translation and the Greek text the forty-first proposition is the relevant proposition. In the Adelard translation it runs (*ms. cit.*, 6v): "Si paralellogramum (!) triangulusque in eadem basi atque in eisdem alternis lineis fuerint constituta, paralellogramum triangulo duplum esse conveniet."

40 "tertiam tertii." In the Adelard II translation we read (*ms. cit.*, 11r): "Si lineam intra circulum preter centrum collocatam alia a centro ducta per equalia secet ortogonaliter super eandem insistere, et si in eam ortogonaliter steterit, [eam] per equalia dividere necesse est."

44–45 "XL primi" again should probably be "XLI primi," as in line 33.

48–49 "per tertiam petitionem." The third postulate is the only one that is specifically cited in the course of the proofs.

50 "XV quinti." In the Adelard translation this proposition runs (*ms. cit.*, 20r): "Si fuerint aliquibus quantitatibus eque multiplicationes* assignate, erit ipsarum multiplicium atque submultiplicium proportio una."

58–60 "detrahatur...minor." Compare with the wording of X.1 of the *Elements* as given on page 60, note 1, above.

70 "primam sexti." In the Adelard II translation we read (*ms. cit.*, 21v): "Si duarum rectilinearum superficierum equidistantium laterum sive triangulorum fuerit altitudo una, tanta erit alterutra earum ad alteram quanta sua basis ad basim alterius."

82–83 "ultimam secundi." In Adelard II (*ms. cit.*, 10r): "Dato triangulo equum quadratum describere."

87–88 "Et...circuli." The reference is to Macrobius, *Commentarius in somnium Scipionis*, Book I, Chapter XX, section 16 (ed. of F. Eyssenhardt, Leipzig, 1893, p. 567; cf. ed. of J. Willis [Leipzig, 1963], p. 81): "Item omnis diametros cuiuscumque orbis triplicata cum adiectione septimae partis suae mensuram facit circuli, quo orbis includitur." Notice in lines 87, 89, I have accepted the reading *triplicata* from *N* because it jibes with the quotation from Macrobius. Three forms of the word "diameter" are used in medieval manuscripts: *diametros*, *diametrus*, *diameter* (in addition to the same three forms with *dya*). When the Greek form *diametros* is used, it is taken as feminine. The other two are often considered as masculine, but other times are considered to be feminine.

95–96 "primam sexti." See the Commentary, line 70 above

106 "11 quinti." In the Adelard II translation (*ms. cit.*, 19v): "Si fuerint quantitatum proportiones alicui uni equales, ipsas quoque proportiones sibi invicem esse equales [necesse est]."

107 "tertiam quinti." In Adelard II (*ms. cit.*, 18v): "Si fuerint primum secundi et tertium quarti eque multiplicia, ad primum vero et tertium multiplicationes* sumantur equales, erit multiplex primi ad secundum atque multiplex tertii ad quartum eque multiplicia."

*Or multiplices; see Companus edition of Ratdolt (Venice, 1482).

2. The Naples Version

One other reworking of the *De mensura circuli* can with some surety be placed in the thirteenth century. This is the version which I have called the Naples Version from its principal manuscript (see MS *A*, in Siglum below).[1] This so-called Naples Version consists of the first two propositions of the *De mensura circuli* and the enunciation of the third. However, it ought to be remarked that only Proposition I has been reworked. Proposition II and the enunciation of Proposition III are simply taken directly from Gerard of Cremona's translation. I shall accordingly only give the text of Proposition I in this section.

It will be noticed from my text of Proposition I below that the wording of the enunciation does not so closely parallel the Gerard of Cremona translation as does that of the Cambridge Version. But in view of its association with Propositions II and III taken verbatim from the Gerard translation, I feel some confidence in asserting that the ultimate source of the Naples Version was in fact the Gerard of Cremona translation. Further, the similarity of the title in the Naples Version (*De circuli quadratura*) to the title as given in Tradition II of Gerard's *De mensura circuli* suggests that the Naples Version had its origin in that tradition.

Of considerable interest is the fact that the author of the Naples Version immediately announces that the proof is to be *per impossibile*, that is, by reduction to absurdity. This announcement is reminiscent of the similar announcements made by Gerard of Brussels in his very original *Liber de motu*.[2] In this connection it is of further interest that the *Liber de motu* immediately precedes the *De circuli quadratura* in the Naples manuscript (60v–65v). These two facts raise the possibility that it was Gerard of Brussels who produced this Naples Version of the *De mensura*

[1] As this volume was going to press, my attention was called to a paraphrase of the Naples Version to be found in Vat. lat. 4275, 81v–83v, 15c. In addition to the paraphrase of the Naples Version of Proposition I of the *De mensura circuli*, the Vatican manuscript includes also paraphrases of Propositions II and III of the *De mensura circuli* in the Gerard translation; and incidentally, between the first and second half of Proposition III the scribe inserts the enunciation of a fourth proposition that occasionally circulated with the *De mensura circuli* (for this enunciation, see the variant reading after line 157 of Gerard of Cremona's translation in Chapter Two, Section 2). I shall edit this version and include it as an appendix to Volume Two.

[2] M. Clagett, "The *Liber de motu* of Gerard of Brussels and the Origins of Kinematics in the West," *Osiris*, vol. 12 (1956), p. 120, lines 283–84; p. 130, line 140; p. 142, lines 10–11.

circuli and that the scribe of *A* copied both of these works from a Gerard of Brussels codex. If this is so, then the Naples Version must, like the *Liber de motu*, date from the first half of the thirteenth century.[3]

The Naples Version has a number of points of similarity with the Cambridge Version. It too bases the proof of Proposition I on specific reference to X.1 of the *Elements*. It also cites numerous other propositions from the *Elements* (actually, ten such propositions). Many of the propositions are those cited in the Cambridge Version. The Naples Version also specifically mentions the fact that one can continually produce smaller and smaller segments to find some remainder less than a specified quantity *Y* by which the circle is said to exceed the given triangle. It adds, however, a phrase only implied in the Cambridge Version to the effect that "for the sake of brevity" it will be assumed that the eight segments between the perimeter of an inscribed regular octagon and the circumference of the circle are less than *Y*. Furthermore, the Naples Version also misuses the term *lunula*. For the author of this version the term means not only a segment of a circle but in fact any mixed figure contained by an arc and one or more straight lines. Thus, in the second part of the proof he applies the term to the triangular figures resulting from the circumscription of regular polygons about the circle and contained by arcs of the circle and successive pairs of the sides of the polygons. This last usage of the term is distinguished by the author from its use for segments of a circle by the addition of the term "exterior"—*lunula exterior*. In my translation I have rendered *lunula* in the first half of the proof by "segment" and in the second half by "arcal figure."

In my text of the Naples Version I have found it necessary in a few instances to correct the reading of *A*. The original readings are given in the variant readings. One peculiarity of orthography needs to be mentioned. The form *hee* is several times used (see lines 26, 28, and 77) where *he* (=*hae*), or perhaps on occasion *ee* (=*eae*), is demanded. Incidentally *he* is one time used in a similar phrase (see line 25). The use of *hee* for *he* or *ee* is not uncommon in Latin manuscripts.[4] The marginal folio numbers are, of course, from *A*.

Siglum of Manuscript

A = Naples, Bibl. Naz. VIII.C.22, 65v–66v, 13c.

[3] *Ibid.*, pp. 104–107.

[4] See H. Ziegel, *De is et hic pronominibus quatenus confusa sint apud antiquos* (Marburg, 1897), pp. 45–47.

De Circuli Quadratura

/ Archimenides de circuli quadratura

I. OMNIS CIRCULUS EST EQUALIS TRIGONO ⟨ORTHOGONIO⟩, CUIUS UNUM LATERUM RECTUM ANGULUM CONTINENTIUM PERIFERIE EST EQUALE, RELIQUUM 5 SEMIDIAMETRO CIRCULI EQUALE.

Hoc probatur per impossibile. Si enim trigonus non sit equalis 66r circulo, erit vel maior vel minor. Sit primo maior, et sit / equalis *Z* trigono et *Y* quantitati proposite [Fig. 11]. Ergo est maior *Z*. Sed quoniam propositis duabus quantitatibus inequalibus a maiore est 10 abscindere maius medietate et iterum maius medietate donec supersit minus reliqua per primam decimi, igitur a circulo subtrahatur maius medietate hoc modo: Inscribatur circulo quadratus *ABCD*, qui pro-

1 Archimenides...quadratura *mg. A*
3 laterum *corr. ex* latus
8 Z *corr. ex* Y

On the Quadrature of the Circle

On the Quadrature of the Circle of Archimedes

I. EVERY CIRCLE IS EQUAL TO A RIGHT TRIANGLE, ONE OF WHOSE SIDES CONTAINING THE RIGHT ANGLE IS EQUAL TO THE CIRCUMFERENCE, WHILE THE OTHER SIDE IS EQUAL TO THE RADIUS OF THE CIRCLE.

This is proved by absurdity. For if the triangle is not equal to the circle, it will be either greater or less. In the first place let it be greater, and let it be equal to △ *Z* plus a proposed quantity *Y* [see Fig. 11]. Therefore, it

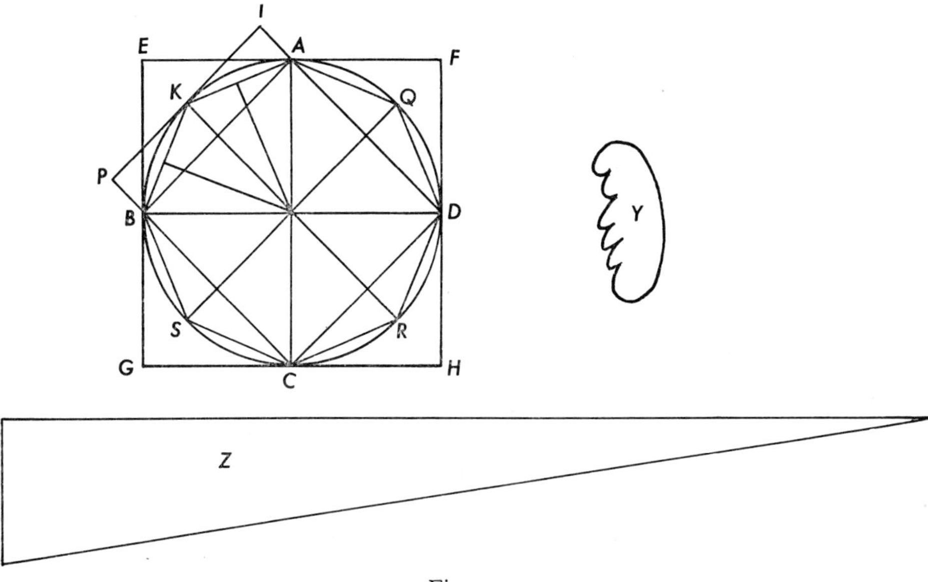

Fig. 11

is greater than *Z*. But because, with two unequal quantities proposed, it is [possible] to cut more than half from the greater and again to cut more than half [from the remainder, and to repeat this process continually] until the lesser [of the original unequal quantities] exceeds the remainder [of the successive cuttings], by X.1 [of the *Elements*], therefore, let more than half be subtracted from the circle in this way: Let square *ABCD* be inscribed in the circle. The square is proved to be more than half of the

batur maior medietate circuli, quia *ABD* triangulus medietas est quadrati *ABCD*. Sed *EFBD* parallelogramum et *ABD* triangulus
15 sunt inter duas lineas equidistantes super bases equales. Ergo parallelogramum est duplex ad triangulum. Quare *ABD* est equalis *AEB* et *AFD*. Triangulus ergo maior *AKB* et *AQD* lunulis. Eadem ratione *BCD* triangulus maior est *BSC* et *DRC* lunulis. Quare *ABCD* quadratus maior est quatuor lunulis residuis; quare maior medietate
20 circuli. Sed si item ille 4 lunule sunt maiores *Y*, secetur *AB* arcus per equalia in *K*. Et fiat *AKB* tringulus, qui et *AIBP* sunt inter lineas equidistantes. Quare ut prius *AKB* triangulus maior est *AK* et *KB* lunulis; similiter de *AQD* triangulo et de ceteris triangulis ad lunulas sic sumptas. Et ita abscidisti maius medietate residui. Rursum, si
25 etiam he 8 lunule non sint minores *Y*, fac ut prius donec occurrat quantitas minor *Y*. Sed sit, causa compendii, quod hee 8 lunule sint minores *Y*. Age igitur, circulus est equalis *Z*, *Y*; et una pars circuli que est hee 8 lunule est minor *Y*. Ergo reliqua, que est octogonum *AKBSCRDQ*, maior est *Z* triangulo; ergo duplum huius maius
30 duplo illius. Quod probatur esse falsum: Dividantur enim latera octogoni per equalia, et ad media puncta ducantur linee a centro perpendiculares, ex 3ª tertii, que probabuntur esse equales ex 8ª et 3ª primi. Ductis etiam lineis ad angulos a centro, fiunt igitur ibi partiales trianguli sedecim, quorum quilibet subduplus est ad id quod fit ex ductu
35 unius linearum ad medium a centro ductarum in medietatem basis octogoni; quod est videre ex 4ª primi si compleatur figura. Erunt ergo duo triangulorum subdupli ad id quod fit ex ductu eiusdem linee in unam basim octogoni. Cumque coacervaveris, erunt omnes illi trianguli, pariter accepti, subdupli ad id quod fit ex ductu unius pre-
40 dictarum linearum in omnia latera octogoni. Sed illi trianguli equales octogono. Quare id quod fit ex ductu linee predicte in omnia latera octogoni duplum est octogono. Sed quelibet linearum dividentium latera minor est semidiametro, et latera octogoni, pariter accepta, minora periferia, quia quelibet corda minor suo arcu. Ergo duplum octo-
45 goni, quod fit ex ductu linee minoris semidiametro in latera octogoni, que minora sunt periferia, minus est eo quod fit ex ductu semidiametri

20 Sed *corr. ex* Quod (?)
35 a *corr. ex* ad / in medietatem *corr. ex* ad medium
39 subdupli *corr. ex* subduplum
42 dividentium *corr. ex* dividiantium

circle, because △ *ABD* is half of the square *ABCD*. But parallelogram *EFBD* and △ *ABD* are [constructed] between two parallel lines and on equal bases. Therefore, the parallelogram is double the triangle. Therefore, $ABD = AEB + AFD$. Hence the triangle [*ABD*] is greater than segments *AKB* and *AQD*. By the same reasoning △ *BCD* is greater than segments *BSC* and *DRC*. Consequently square *ABCD* is greater than the four segments left over; therefore it is greater than half of the circle. But if these four segments together are greater than *Y*, let arc *AB* be bisected in *K*. And let △ *AKB* be constructed, which triangle and [parallelogram] *AIBP* are [constructed] between parallel lines. Therefore, as before, △ *AKB* is greater than segments *AK* and *KB*. The same thing holds for △ *AQD* and for the other triangles related to segments in the same way [as before]. You have thus exhausted more than half of the remainder. Again if these eight segments are also not less than *Y*, proceed as before until a quantity less than *Y* results. But, for the sake of brevity, let these eight segments be less than *Y*. Proceed therefore: The circle is equal to $(Z + Y)$, and one part of the circle—namely, the eight segments—is less than *Y*. Therefore, the rest of it—the octagon *AKBSCRDQ*—is greater than triangle *Z*; therefore, double the octagon is greater than double the triangle. But this is proved to be false: For let the sides of the octagon be bisected and to the middle points let perpendiculars be drawn from the center—by III.3 [of the *Elements*]. These lines are proved to be equal, by I.3 and I.8 [of the *Elements*]. With lines also drawn from the center to the angles, sixteen partial triangles are produced. Any one of these [triangles] is one half of the product of [1] one of the lines drawn from the center to the middle [of the sides of the octagon] and [2] one half of a side of the octagon. This is seen [to follow] from I.4 [of the *Elements*], if the figure is completed. Hence two of the triangles will equal one half of the product of this same line and a side of the octagon. And when you have accumulated them, all of these triangles taken together are equal to one half of the product of one of the aforesaid lines and all of the sides of the octagon. But these triangles equal the octagon. Therefore, the product of the aforesaid line and the perimeter of the polygon is double the octagon. But any of these lines bisecting the sides is less than the radius, and the sides of the octagon taken together are less than the circumference, because any chord is less than its arc. Therefore, double the octagon—which results from the multiplication of a line less than the radius by the sides of the octagon, which are less than the circumference—is less than the product of the radius and the circumference, which product

in periferiam, quod est duplum *Z* trianguli. Et sic constat per impossibile, circulum non esse maiorem *Z* triangulo.

Sit circulus minor *Z*, ita quod *Z* sit equalis circulo et *Y* quantitati.
50 Circulo igitur circumscribatur quadratus *MNE'G* [Fig. 12], qui quadratus maior est *Z* triangulo, quoniam duplum eius maius est duplo trianguli. Duplum enim quadrati est quod fit ex ductu semidiametri in omnia sua latera, que maiora sunt periferia. Duplum vero trianguli ex ductu periferie in semidiametrum. Cum igitur circulus et *Y* sint
55 equales *Z*, et *MNE'G* maior *Z*, erit quadratus maius circulo et *Y*. Sed circulus est pars quadrati. Ergo residuum pars quadrati quod est 4 lunule. Dematur ergo ab illo residuo maius medietate et iterum maius medietate donec supersit minus *Y* hoc modo: Ducatur a centro linea ad *N* angulum, et ubi ipsa secabitur periferiam duc contingentem
60 ad circulum, que rectos angulos faciet cum ducta a centro ad *N*, ex 17ª tertii. Subtende etiam duas cordas medietatibus arcus quem secuisti. Age igitur, *C* angulus rectus est; ergo *E* totalis rectus, per 13 primi. Similiter *D* totalis rectus. Sed etiam *E*, *D* partiales equales ex 5ª primi. Ergo residui partiales qui interiacent *E*, *C* et *D*, *A* sunt equales. Quare
65 per 6 primi respiciunt equalia latera *AB* et *BC*. Sed *BC* minor est *BN*, quoniam *BN* rectum angulum respicit in triangulo *BNC*. Ergo *AB*

is double △ Z. And so it is clear by absurdity that the circle is not greater than △ Z.

Let the circle be less than Z, so that Z is equal to the circle plus quantity Y. Hence let the square $MNE'G$ be circumscribed about the circle [see Fig. 12]. This square is greater than △ Z, since its double is greater than double the triangle. For the double of the square arises from the product of the radius and all of its sides, which [sides together] are greater than the circumference, while double the triangle arises from the product of the circumference and the radius. Since, therefore, the circle and Y [together] equal Z, and $MNE'G$ is greater than Z, the square is greater than the

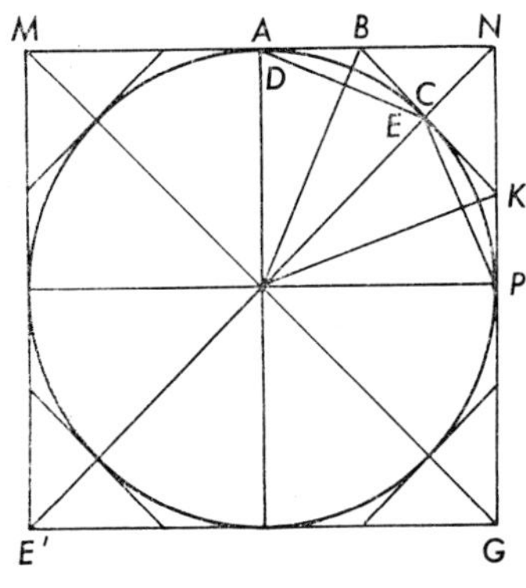

Fig. 12
Note: I have added the prime sign to E'.

circle and Y. But the circle is part of the square. Therefore, the remainder is that part of the square constituted out of the four [curvi-rectilinear] arcal figures. Hence let there be subtracted from that remainder more than [its] half and again more than half [successively] until the [final] remainder is less than Y, and do it in this way: Let a line be drawn from the center to ∠ N. Where that line intersects the circumference, draw a tangent to the circle which makes right angles with the line drawn from the center to N, by III.17 (III.18, *Greek text*) [of the *Elements*]. Also draw chords to the two halves of the arc which you have bisected. Proceed then as follows: ∠ C is a right angle. Hence the whole angle at E is a right angle, by I.13 [of the *Elements*]. Similarly, the whole angle at D is a right angle. But the partial angles at E and D are equal, by I.5 [of the *Elements*]. Therefore, the partial remainders, ($C — E$) and ($A — D$), are equal. Hence by I.6 [of the *Elements*] these angles face equal sides, AB and BC. But $BC < BN$, since BN faces the right angle in △ BNC. Hence $AB < BN$. But

minor est quam *BN*. Sed que est proportio *AB* ad *BN* ea est *ABC* trianguli ad *BNC* triangulum, ex prima sexti, quoniam eorum est altitudo una in *C* puncto. Ergo *ABC* triangulus minor est *BNC*
70 triangulo. Multo magis *AC* lunula ⟨exterior⟩ minor est eodem *BNC*. Et eadem ratione *CKP* triangulus minor est *CKN* triangulo. Multo magis *PC* lunula exterior minor est *KCN*. Ergo *AC* et *CP* exteriores lunule minores sunt *BNK* triangulo, et sic abscidisti a lunula totali maius medietate, quod est *BNK* triangulus. Et similiter fac de aliis, et
75 hoc modo de minoribus, donec proveniat minus *Y*. Sed, causa compendii, sint hee lunule 8 minores *Y*. Ergo iste circulus et hee lunule sunt minus *Z* triangulo. Sed iste circulus et hee 8 lunule sunt equales hoc octogono. Ergo hoc octogonum minus est triangulo. Ergo duplum octogoni minus duplo trianguli, quod sic improbatur: Du-
80 plum trianguli est id quod fit ex ductu semidiametri in periferiam; duplum vero octogoni ex ductu semidiametri in latera octogoni, que sunt maiora periferia. Ergo duplum octogoni maius est duplo trianguli. Sed dictum erat quod minus. Quod autem duplum octogoni sit quod fit ex ductu semidiametri in latera octogoni proba ut superius. Cum
85 igitur circulus propositus triangulo proposito nec sit maior nec minor, erit eidem equalis, quod propositum est.

73 minores *corr. ex* minus   77 equales *corr. ex* maiores

$AB/BN = \triangle ABC/\triangle BCN$, by VI.1 since they (the triangles) have the same altitude in point C. Hence $\triangle ABC < \triangle BNC$. [Thus] the exterior arcal figure [contained by arc] AC [and lines AB, BC] is even smaller than that same [$\triangle$] BNC. By the same reasoning $\triangle CKP < \triangle CKN$. [Thus] the exterior arcal figure [contained by arc] PC [and lines CK, KP] is even smaller than KCN. Therefore, the exterior arcal figures on [arcs] AC and CP are [together] less than $\triangle BNK$, and so with $\triangle BNK$ you have exhausted more than half of the total arcal figure [$ACPN$]. The same thing holds for the other [corner arcal figures and triangles]. And in this way [one can proceed] with smaller [circumscribed polygons] until there results a quantity less than Y. But, for the sake of brevity, let these eight arcal figures be less than Y. Therefore, this circle and these eight arcal figures are less than $\triangle Z$. But this circle and these eight arcal figures are equal to the octagon. Therefore, the octagon is less than the triangle. Therefore, double the octagon is less than double the triangle, which is refuted in this way: The double of the triangle arises from the product of the radius and the circumference, while the double of the octagon arises from the product of the radius and the perimeter of the octagon. But the perimeter of the octagon is greater than the circumference. Therefore, double the octagon is greater than double the triangle. However, it was said that it was less. Moreover, that the double of the octagon arises from the product of the radius and the perimeter, prove as above.

Since, therefore, the proposed circle is neither greater than nor less than the proposed triangle, it will be equal to it; which was proposed.

COMMENTARY

2 "trigono ⟨orthogonio⟩." The necessity of adding *orthogonio* is apparent. Notice that here *trigono* is used rather than the *triangulo* of the Gerard of Cremona translation.

3 "unum laterum," instead of the "unum duorum laterum" of the Gerard translation.

3–4 "angulum continentium." Like the Cambridge Version, the Naples Version transposes the words of the Gerard translation.

4–5 "perifierie...equale." Here the order of the quantities to which the two sides of the triangle are equal is reversed from the Gerard translation. Notice the use of the word *periferia* for circumference. This, of course, is a transliteration of the Greek word (see Eutocius, *Commentarius in dimensionem circuli*, ed. of J. L. Heiberg in Archimedes, *Opera omnia*, vol. 3, Leipzig, 1915, p. 230, line 10). Notice also that *semidiametro* replaces the *medietati diametri* of the Gerard translation.

8–11 "Sed...decimi." Compare this wording with that of X.1 of the Adelard II translation as given on page 60, note 1, of this chapter.

32 "3ª tertii." See the Commentary, line 40, of the Cambridge Version.

32 "8ª et 3ª primi." See the Adelard II translation (*ms. cit.*, 2v): "Omnium duorum triangulorum quorum duo latera unius duobus lateribus alterius fuerint equalia basique unius basi alterius equalis, duos angulos equis lateribus contentos equales esse necesse est." And for I.3 see the *ms. cit.*, 1v–2r: "Propositis duabus lineis inequalibus de longiore earum equalem breviori abscindere." There is perhaps a mistake in these numbers, for these two propositions do not seem to be the most appropriate to cite, although one could say that I.8 leads ultimately to the conclusion that the triangles on the various sides are congruent and thus their altitudes are equal.

36 "4ª primi." See the Adelard II translation (*ms. cit.*, 2r): "Omnium duorum triangulorum quorum duo latera unius duobus lateribus alterius fuerint equalia duoque anguli illis equis lateribus contenti equales, erit alter alteri equalis lateraque illorum sese respicientia equalia; reliqui vero anguli unius reliquis angulis alterius equales."

44 "quia...arcu." This is the second postulate of the Cambridge Version. It is of interest that here it is explicitly quoted, thus illustrating the tendency of these versions to mention every step and authority.

47–48 "per impossibile." This constitutes a formal reiteration that this first half of the proof has been demonstrated by the method of reduction to absurdity.

61 "17ª tertii." This is III.17 of the Adelard II translation but is III.18 of the Greek text. In the former it runs (*ms. cit.*, 13r): "Si circulum recta linea contingat ⟨et⟩ a contactu ad centrum linea recta ducatur necesse eam super lineam contingentem perpendicularem esse."

62 "13 primi." See the Adelard II translation (*ms. cit.*, 3r): "Omnis linee recte super rectam lineam stantis duo utrobique anguli aut recti sunt aut duobus rectis equales."

63 "5ª primi." See the Adelard II translation (*ms. cit.*, 2r): "Omnis trianguli

duum equalium laterum angulos qui supra basim sunt equales esse necesse est. Quod (Et?) si duo equa latera eius directe protrahantur, fient quoque sub basi duo anguli sibi invicem equales."

65 "6 primi." See the Adelard II translation (*ms. cit.*, 2v–3r): "Si duo anguli alicuius trianguli fuerint equales, duo quoque latera eius angulos illos respicientia erunt equalia."

68 "prima sexti." See the Commentary, line 70, of the Cambridge Version.

70 "⟨exterior⟩." I have added this word because of the usage in line 72.

84 "proba ut superius." Consult lines 30–40.

3. The Florence Versions

Not unlike the versions of Cambridge and Naples are the versions of the *De mensura circuli* found in the Florence manuscript, Biblioteca Nazionale, Con. Soppr. J.V.30, 9v–12v (MS *D*). In speaking of the versions of Florence, I mean to suggest that more than one author was involved in the preparation of the four items found in this manuscript. In the order that they are given in the manuscript these items are: (1) folios 9v–11v, Proposition III in an extended commentary; (2) folio 12r, a reworking of Proposition I (which I have called F.IA); (3) folios 12r–v, Proposition II; and (4) folio 12v, a different reworking of Proposition I (here designated as F.IB). It is quite evident that at least two authors were involved in the preparation of these various items, for it hardly seems likely that the same man composed F.IA and F.IB, which—while similar in over-all structure—are completely different in language and detail. Accepting two authors, we could then say that the scribe of the Florence manuscript found in one codex a version of the *De mensura circuli* that contained all three propositions and in another codex a different version of Proposition I. As possible evidence that the first three items do belong to a single author we can note the explicit statement in Proposition II which cites both Proposition I and Proposition II (i.e., the "prior proposition" and the "following proposition"; see Proposition II below). Furthermore, the scribe distinguishes F.IB from the other propositions by writing in the margin next to its enunciation the word "interposed (*interposita*)."[1] Of course, if all three

[1] On the other hand, one might suggest that F.IB rather than F.IA belongs to the same author as that of the commentary of Proposition III; for, when the author of

items are from a single version of the *De mensura circuli*, one wonders why the scribe placed Proposition III before the other two propositions. We can reject the explanation that the leaves of codex *D* have been improperly bound, for F.IB and the beginning of a treatise *De isoperimetris* follow Proposition II on folio 12v. One could say, however, that the leaves of the codex from which the scribe of *D* made his copy had been incorrectly bound. Another possible explanation is that a third author was involved: that is, that after the scribe of *D* came across this version of Proposition III and copied it, he found another codex with F.IA and Proposition II in it and still a third codex including F.IB. We do not have, however, any earlier manuscript including any of these items, although all four items are found in the same order in two later manuscripts, *Zo* and D_1 (see Sigla). *Zo* was copied from *D*, as is evident from the fact that where *Zo* diverges from *D* it is almost always where *D* is difficult to read and in almost every case one can see how the poor writing of *D* is the precise source of the error of *Zo*.[2] Furthermore, *Zo* follows almost all of the erroneous readings of *D*, even where *D* is clearly using the wrong word,

F.IB has completed the first half of the proof and he wishes to return to the second alternative stated at the beginning of the proposition, he uses the word *redeamus*. Similarly, in several different places the author of Proposition III, after completing incidental proofs, tells us that we must turn back to the main line of the proof; in such places he too employs *redeamus*.

[2] For example, in line 18 of Proposition F.IA, the scribe of *D* has added over the letter "L" a bar. This provoked the scribe of *Zo* to write *in*, mistaking, as he did, the "L" for an "I". Also *D* in line 106 of Proposition III makes a 8^e that looks something like a g^e, and indeed *Zo* carefully makes it g^e. Again, in line 262 of the same proposition *D* writes AK, which is correct, but the tail of the K is somewhat obscured, making it look like AB, and *Zo* writes AB. These examples could be multiplied many times. In several cases, *Zo* omits lines or parts of lines which are in *D* and are necessary for the argument. For example, lines 246–47 in Proposition III ("Ergo...57600) are omitted by *Zo*. Incidentally, *Zo* copies from *D* the long marginal note to Proposition III, including the same signs to mark the rewritten passage (see variant reading to lines 112–67). The scribe of *Zo* is obviously not a trained mathematician. He is much confused by the term *lunule*, which sometimes he writes as *linule* and sometimes as *limule*. Incidentally, *Zo* uses the form *octogonius* in F.IB where *D* has *octogonus*. A further argument that seems to support the conclusion that *Zo* copied from *D* is that *Zo* has included a marginal note found in *D* which does not bear on the Archimedes tract but rather is an addition to a tract *De proportionibus* which precedes the *De quadratura circuli* in *D* and ends on the same page (9v). Apparently the scribe of *Zo* found this marginal note at the bottom of this page in *D* (9v) and thus thought it belonged to the *De quadratura* and so copied it in the margin of his manuscript at the bottom of the page (151v) (where, incidentally, only the *De quadratura* appears without the *De proportionibus*).

the wrong case, or the wrong gender.[3] Thus, although *Zo* is beautifully written and I have collated it completely with *D*, I have generally not included its variant readings since they do not serve to establish the text. Nor do I give the variant readings of D_1. It is a particularly careless copy of *D*, or perhaps of *Zo*, and dates from the sixteenth century.

The dependence of at least three of the four Florence items on Gerard of Cremona's translation of the *De mensura circuli* is evident; item number 4 (F.IB) apparently diverges further from the original text than the other items. At the outset it should be noticed that the title found in *D* clearly comes from the second tradition of the Gerard of Cremona translation; it runs *Liber de quadratura circuli Archimenidis*. Furthermore, in F.IA and in Proposition II and III we find precisely the same wording for the enunciations as in the Gerard of Cremona translation. And the wording of the proof in Proposition II and parts of Proposition III almost exactly follows that of the Gerard translation (although to be sure in Proposition III the pristine statements of the Gerard translation are accompanied by a large amount of commentary). In F.IB the paraphrasing is complete and nothing remains of Gerard's language, but I still would suppose that the ultimate inspiration was Gerard's translation.[4] The principal change in form of

[3] Thus *Zo* repeats from *D* all the wrong numbers in Proposition III (see the variants for lines 34, 35, 255, 271, 273, 284) In Proposition F.IA, the scribe of *Zo*, like that of *D*, writes *factum* for *falsum* in line 31 and *proportionem* for *propositionem* in lines 81, 85. Similarly, *Zo* takes from *D* the form *dimium* instead of the correct form *dimidium* in line 73 of Proposition F.IA, as well as the unmathematical and ungrammatical sentence *octogonus sit maiores triangulo* instead of something like the phrase I restored, namely, *octo sunt minores QTSO* (see F.IB, lines 21–22). In the same proposition both scribes write *concurrunt* for *continent* (see line 49). See also the variant of Proposition F.IA, line 32. For examples of wrong case endings appearing in both manuscripts, see the variants for Proposition F.IA, lines 4, 34; Proposition III, lines 20, 168, etc. For a wrong gender, see the variant of Proposition III, line 196. There are several other examples, which the reader can note by studying the many corrections that I have suggested in the texts. In virtually every case of correction, the incorrect reading not only appears in *D* but in *Zo*. The only clear-cut case where *Zo* corrects *D* occurs in line 57 of F.IB where *Zo* corrects the confused reading of *propor̄tum* in *D* to *propositum*. This correction appears to be an obvious one. The point that is surprising is that *Zo* did not correct other mistakes of *D* that seem to be equally obvious. In conclusion, we should note that although the scribe *Zo* generally expands the abbreviations of *D*, in some cases where he is perhaps unsure of the proper words he follows *D* slavishly. Thus, for example, when *D* has *d mr in ao* for *ductu MR in ambitum ocotgoni* (F.IB, line 53), *Zo* repeats the abbreviated phrase.

[4] If the reasoning in note 1 is significant and the same person composed both F.IB and Proposition III, then we can be sure that the author of F.IB was acquainted not only with some emended version of

F.IB from F.IA is that, like the Cambridge Version, F.IB makes the final objective of the first proposition not just the equation of a right triangle and a circle but rather the finding of a square equal to the circle. This is done, of course, by first showing that the right triangle is equal to the circle and then converting the triangle into a square by using the last proposition of the second book of the *Elements*. This is similar to the procedure suggested by Thomas Bradwardine in his *Geometria* (see Chapter Two, Section 2, note 11) and followed in the Gordanus, the Corpus Christi, and the Munich version—all given below in this chapter—and in the Albert of Saxony Version, given in Chapter Five.

Both F.IA and F.IB exhibit the same tendencies toward elaboration and specification of the geometric steps found in the Cambridge and Naples versions. However, F.IA cites in the text the *Elements* of Euclid only twice, while F.IB cites it eleven times[5]. Both versions recognize that Euclid's Proposition X.I is the heart of the proof of Proposition I. Both specify the quantity by which the circle is said to exceed, or to be exceeded by, the right triangle (Z and X [actually 8] in F.IA; $QTSO$ in F.IB). Each spells out the procedure by which the circle is to be exhausted until a quantity less than the specified quantity remains. Both apply the word *lunula* to segments of a circle, although neither applies it in the second half of the proof to those mixed figures included by arcs of the circle and the sides of the circumscribed polygon. Both versions specifically state that for the sake of brevity we can use the inscribed or circumscribed octagon as the polygon which will allow us to find the desired remainder less than the specified quantity (that is, F.IA uses the expression *causa brevitatis*, while F.IB expresses the same idea by saying *non est procedere in infinitum*). Actually, the expression of F.IB seems to be highlighting more explicitly that one does not actually proceed to infinity in this extraction or "exhaustion" process (no true exhaustion being intended

the Gerard translation but with the Gerard translation itself, since he quotes verbatim from that translation in Proposition III.

[5] Apparently the scribe of *D* had taken it upon himself to supply citations to the *Elements* where he felt that the text was deficient in such citations. For in the margin of F.IA (see Commentary, lines 7–37, 38–69, 71) he notes six citations for the first half of the proof and sixteen for the second half of the proof (two of which are not to Euclid); for the corollary, he has one citation. For Proposition II (see Commentary, lines 74–85) he gives three citations, two of which are to the *Elements*. For Proposition III, twenty-one marginal citations are given for the first half of the proposition (see Commentary, lines 4–167) and nineteen citations for the second half (see Commentary, lines 168–352). He gives no marginal citations for F.IB, since the text itself was well supplied with such citations.

by the procedure) but only that one proceeds until a quantity less than an assigned quantity is found. Hence one might as well use an octagon rather than some figure with so many sides and lines that the proof will be unduly long and the figure hard to see. While both F.IA and F.IB recognized that the proof is by reduction to absurdity, of particular interest is the explicit statement in F.IA at the end of the first half of the proof (lines 36–37) of the formal conclusion: "This cannot stand, nor can the premise from which it followed, namely, that the circle is greater than the triangle." Incidentally, the authors used different forms of the nominative case of octagon: *octogonium* in F.IA and *octogonus* in F.IB. A third form, *octogonum*, was employed in the Naples Version.

As I have suggested, Proposition II in this Florence codex follows almost verbatim the Gerard of Cremona translation. Its most interesting additions are the references to the first and third propositions, references that are absent from the Gerard translation. In relating this proof to Proposition III the author tells us that because "the excess of line *ZG* [equaling $3\frac{1}{7}$ the diameter] beyond the circumference is almost imperceptible, we have said that it is equal to the circumference, which will be evident from the following proposition."

Proposition III is easily the most interesting of the items contained in the Florence manuscript. For, while starting from the Gerard of Cremona translation, it includes in addition detailed efforts to explain or confirm the figures in the text. It does this somewhat in the manner of the Greek commentary of Eutocius. But I would suppose that the Latin commentator who composed this version did not have at hand the commentary of Eutocius; for not only is there no evidence of a Latin translation of this commentary of Eutocius, but Eutocius' method of calculation is basically different from that of the Latin commentator. The Latin commentator tends to convert all of the numbers containing fractions to their improper fractional form before proceeding to the arithmetical operations of multiplication, division, or taking the square root. Eutocius on the other hand operates with the mixed forms of integers and fractions. Unfortunately, in neither case are we told exactly how to find the square roots of numbers which are not perfects squares. In fact, the principal purpose of both expositions is merely to show that the roots given by Archimedes are close roots. The same difference of procedure that distinguishes the Latin commentary on Proposition III and the commentary by Eutocius also distinguishes the Latin commentary from the account of Proposition III found in the *Verba filiorum* of the Banū Mūsā translated by Gerard of

Cremona. Our author did not follow the *Verba filiorum* with any exactitude if he knew of that work.

The major point of interest in the Florence Version of Proposition III centers on the efforts made to check the accuracy of the figures of the original text and to fill in the bare skeleton of figures given in the original text. In this connection it is of interest to note that the principal scribe of manuscript *D*, who probably dates from about the beginning of the fifteenth century, was himself a competent calculator, for in copying Proposition III he realized that the original commentator had made some serious calculating errors (see lines 112–167 with the variant readings). He accordingly gives the corrected reading in the margin, introducing it by the following statement: "It seems to me that here the commentator has erred in multiplying, for it is based on a false root. Accordingly I have placed a [corrected] reading [in the margin] where the sign # is given" (see variant reading for lines 112–67). Following the reading he adds: "Note that I have interposed this reading, because the comment seemed to me to be based on a false root. However, I have not dared to correct [the text]; and I have placed a sign # at the beginning and the sign ◁| at the end" (see the variant reading for lines 166–67). The reader will notice by consulting the variant readings that I have followed the corrected reading of the scribe in my text, leaving the errors of the original version in the variant readings, except that I have retained the conventional spelling of *dyametri* followed in the main text rather than the marginal reading of *diametri*.

In the manner of the compositor of F.IA, the commentator of Proposition III is sparing in his citation in the text of the *Elements*. In fact, only once does he allude to the *Elements* and then under the title of *Liber geometrie*. He or the scribe of *D* has, however, generously compensated for this lack of citations in the text by adding citations in the margins (see Commentary, lines 4–167, 168–352).

Before leaving the question of the treatment of Proposition III in the Florence manuscript it should be remarked that there exists at the end of the codex (57v) a fragment which includes a small part of the final calculation of the first half of the proposition. This fragment is quite distinct from the main text of Proposition III and no doubt is from some different version of Proposition III (see Commentary, lines 147–67).[6]

[6] A corrupt fragment of a still different version of Proposition III that ought to be compared with our Florence version is found in Bruges, Stadsbibliotheek 530, 6v–7r, 14c. This fragment follows directly after Proposition IV.11 of Jordanus' *De*

It is impossible to know with any exactitude the dates of the four items in *D*. It is apparent that manuscript *D* stems from about the turn of the fifteenth century. But the scribe has clearly copied this material; that is to say, he was not the original author of any of this material, except perhaps the marginal additions. Thus all we can say is that it is probably fourteenth century in origin. As I have pointed out, F.IB is further removed from the original Gerard translation than F.IA and so might have been composed at a later date. Neither F.IA nor F.IB shows evidence of the scholastic form apparent in the other versions of the *De mensura circuli* that are given in the following sections in this chapter. In short, I would say that the form and content of F.IA and F.IB point to a period no later than the early fourteenth century. The techniques of calculation with fractions in Proposition III, which on first glance seem promising as an aid to dating, do not in fact give us any sure evidence for dating that item, since they are the methods of algorism which were commonly employed in both the thirteenth and fourteenth centuries.

In connection with my texts of the Florence items, certain points are to be noted. Throughout the copy in *D*, Indo-Arabic numerals are in

triangulis (which in this manuscript breaks off at this point). It consists in a paraphrase of only the first half of Gerard's translation. Because of its corrupt nature and incompleteness, I shall not give the full text of it, hoping that in the future a better copy will turn up. But we can give the *incipit* and *explicit*: "Circonferentia addit super triplum dyametri minus septima et plus decem partibus septuaginta unius parcium dyametri. Sit *E* circulus et centrum, *AG* dyameter, *DZ* contingens, *ZEG* angulus tercia recti divisus in duo equalia *EB* (!), angulus quoque *HEG* divisus in duo equalia, et *TEG* angulus divisus in equalia *EK*, *KEG* quoque divisus in equalia. Elige. Ratio. Ex tercia quinti Euclidis, que est proporcio *EZ* ad *EG* que *ZH* ad *DH*.... (7r)...Sed 14688 superant triplum numerus (*!*) 4673 et medietatis secundum quantitatem 667 et medietatem quod est minus 7ª numeri 4673 et medietatis; continet circonferentia triplum dyametri et minus septima." Although taken basically from Gerard's translation, there is a similarity of procedure with the treatment of the Archimedean proposition found in the *Verba filiorum* of the Banū Mūsā (see Chapter Four, Proposition VI). The similarity is shown by the fact that in the Bruges fragment, as in the *Verba filiorum*, the successive bisections of the angles used are done before any calculation is started, while in the Gerard translation each bisection follows the completion of the calculation arising with the whole angle. Incidentally, may I underscore the fact that in this fragment the numbers are often omitted or given incorrectly. There is considerable citation of the *Elements*, and the Pythagorean theorem is several times cited as "dulk," standing for *dulkarnan* or *dulcarnon* (which appears to come from the Arabic *dhu'l-qarnain*, meaning "possessor of two horns," a reference to the two squares sticking up as horns; see the Euclid manuscript, Bodl. Auct. F.3.13, 3v; cf. T. L. Heath, *Euclid, The Elements*, vol. 1 [Annapolis, 1947], p. 418). After this volume went to Press, two other versions of Proposition III were discovered in Vat. lat. 4275 and Glasgow Library BE 8-y. 18. They will be discussed and compared with the Florence Version in Volume II.

overwhelming preponderance. I have accordingly used those numerals in my texts, and indeed have altered to such numerals the rare instances of the use of Roman numerals. Like so many Greek, Arabic, and Latin treatises, no effort was made in these Florence items to use the same order for the letters in a given designation. That is, the scribe sometimes used *AB* and sometimes *BA* for the same quantity, and did so even in the same line of the text. In this matter I have followed the manuscript. As was usual in late medieval manuscripts, *D* uses *c*'s and *t*'s interchangeably before *i* and an additional vowel. Thus we very often find *proporcio*. I have, however, been consistent in my text and have used *proportio* throughout, whereas in later texts based on one manuscript (like the Corpus Christi and Munich versions), where an author consistently used the *cio*, *cia*, or *cie* forms, I have used those forms in the texts.

It will be observed that F.IA in the second half of the proof rather oddly employs the numerals 8, 3, 4, and 9 to mark the points of geometrical magnitudes or a magnitude itself. This seems quite peculiar since the rest of the points and magnitudes are marked by letters. I have accordingly altered these numerals, using instead the letters *X*, *E*, *I*, and *H*, respectively. The places where these changes have been made are indicated in the variant readings.

Finally, we should observe that I have changed the order of the Florence items in my texts below, giving them in the order of F.IA, II, F.IB, and III. The marginal folio numbers are from MS *D*.

Sigla of Manuscripts

D = Florence, Biblioteca Nazionale, Con. Soppr. J.V.30, 9v–12v, 14c.
D_1 = Paris, BN lat. 11247, 51r–66r, 16c.
Zo = Vatican, Ottob. lat. 1870, 151v–157v, 15c.

On the Quadrature of the Circle

De Quadratura Circuli

Liber de quadratura circuli Archimenidis

12r ⟨I.⟩ / (F.IA) OMNIS CIRCULUS ORTHOGONIO TRIANGULO EST EQUALIS, CUIUS UNUM DUORUM LATERUM CONTINENTIUM ANGULUM RECTUM MEDIETATI DYAMETRI 5 CIRCULI EQUATUR ET ALTERUM IPSORUM LINEE CIRCULUM CONTINENTI.

Sit itaque circumferentia circuli *ABGD* equalis *PM* lateri trianguli orthogonii *PMT* et semidyameter *FH* equalis lateri *PT* [Fig. 13]. Dico ergo circulum *ABGD* esse equalem triangulo *PMT*. Quod si 10 ita non fuerit, aut maior est circulus aut minor triangulo. Si maior, sit excessus quantitas *Z*. Et circulo inscribatur quadratum *ABDG*, quod constat maius esse medietate circuli, cum superficies equidistantium laterum constituta super dyametrum *AG* ad altitudinem *HB* sit maior semicirculo *ABG* et dupla ad triangulum *ABG*. Et sit medietas 15 quadrati maior medietate circuli. Quare linee quatuor cum quadrato circulum constituentes sunt minores medietate circuli. Secentur iterum quatuor arcus *AB*, *BG*, *GD*, *DA* in suas medietates per puncta *F*, *N*, *R*, *L*, protractis *AF*, *BF*, *BN*, *NG*, *GR*, *RD*, *DL*, *LA* includentibus octogonium inscriptum circulo. Et sic a circulo quatuor supra- 20 dictis lunulis auferatur maius medietate illarum pariter acceptarum per quatuor triangulos *AFB*, *BNG*, *GRD*, *DLA*, supradicta ratione si attendantur parallelograma constitua supra bases *AB*, *BG*, *GD*, *DA* ad altitudinem perpendicularis ab uno angulorum descendentis ad alkaydem. Eodem modo, si octo lunularum octo arcus dividantur in 25 suas medietates, ut predicto modo poligonium fiat ab octo lunulis per octo triangulos maius medietate earum, scilicet pariter acceptarum. Et in tali continua detractione necesse est qualibet quantitate proposita relinqui minus per primam decimi Euclidis. Causa ergo brevitatis ille octo lunule constituantur iam minores, pariter accepte, *Z* quantitate,

1 Liber...Archimenidis *in angulis superioribus D*
4 medietati *corr. ex* medietate
8 PT *supra scr. D*
11 *post* sit *del. D* circulus
16 minores *corr. ex* maiores (?)
26 scilicet...acceptarum *corr. ex* spal'r

On the Quadrature of the Circle

The Book of Archimedes on the Quadrature of the Circle

I. (F. IA) EVERY CIRCLE IS EQUAL TO A RIGHT TRIANGLE, ONE OF WHOSE TWO SIDES CONTAINING THE RIGHT ANGLE IS EQUAL TO THE RADIUS OF THE CIRCLE AND THE OTHER OF THEM TO THE CIRCUMFERENCE OF THE CIRCLE.

And so let the circumference of circle *ABGD* be equal to side *PM* of right △ *PMT* and radius *FH* be equal to side *PT* [see Fig. 13]. I assert, therefore, that circle *ABGD* is equal to △ *PMT*. Because if it is not so, the circle is either greater than or less than the triangle. If it is greater, let the excess be quantity *Z*. And let square *ABDG* be inscribed in the circle. Evidently the square is greater than half the circle, since the rectangle constructed on diameter *AG* with altitude *HB* is greater than the semicircle *ABG* and is twice △ *ABG*. And half the square is greater than half the circle. Hence the four lines that make up the circle and the [lines of the] square [enclose a space] less than half the circle. Again let the four arcs *AB*, *BG*, *GD*, and *DA* be bisected through points *F*, *N*, *R*, and *L*, with lines *AF*, *BF*, *BN*, *NG*, *GR*, *RD*, *DL*, and *LA* drawn to include an octagon inscribed in the circle. And thus let more than half of the above-mentioned segments of the circle—taken together—be exhausted by the four triangles *AFB*, *BNG*, *GRD*, and *DLA*. That this is so is shown by the same reasoning as above when we consider the rectangles constructed upon bases *AB*, *BG*, *GD*, and *DA*, each of which has an altitude equal to a perpendicular drawn from one of the angles to its base. In the same way, if the eight arcs of the eight segments are bisected, a [regular] polygon is formed within the eight segments by eight triangles fashioned as before [and these triangles constitute] more than half of the segments taken together. And by such continual exhaustion, it is necessary by X.1 [of the *Elements*] of Euclid that there will [ultimately] remain a quantity less than any proposed quantity. Therefore, for the sake of brevity let these eight segments taken together be already constituted as less than the quantity *Z*, and so the residual

30 et ita poligonium residuum circuli erit maius triangulo *PTM*, quod convincitur esse falsum. Ducta perpendiculari *HS* a centro circuli, que, cum sit minor semidyametro *FH*, que respicit maiorem angulum in triangulo *SFH*, ducta in lineam ambientem octogonium minorem circumferentia ambiente eum producit minus duplo trianguli et sic 35 octogonium minus triangulo *PTM*, quod tamen prius dicebatur esse maius. Hoc autem stare non potest; quare nec illud ex quo hoc sequebatur, scilicet quod circulus sit maior triangulo.

Si minor fuerit circulus *ABGD* triangulo *PMT*, excessus trianguli supra circulum *X* [Fig. 14]. Et inscribatur circulus quadrato *CKQP'*, 40 et sic separatum est a quadrato maius medietate sua per circulum. Protrahatur ab *N* ad *C* linea *CN*, qua constat et arcum *AB* et cordam eius separare in duo equalia. Cum anguli *CAB* et *CBA* sint equales propter semidyametros equales *AN* et *BN* et angulos rectos *CAN* et *CBN*, inde colligitur omnes duas partes linearum contingentium 45 existentes inter puncta contactuum earum et circuli et punctum suum commune esse equales. Protrahatur item contingens a puncto communi linee *CN* et arcus *AB* ex utraque parte donec reperiat puncta *Z*, *T*. Et simili modo protrahantur *OEY*, *MIV*, *SHR*. Sic ergo *FT* est equalis *TB*. Et subtiliter intuenti constabit illas octo lineas constituere

31 falsum *corr. ex* factum / perpendiculari *Zo*
32 maiorem *corr. ex* minorem
34 circumferentia ambiente *corr. ex* circumferentiam ambientem
39, 63 X *corr. ex* 8
41 qua *corr. ex* quam (?)
48, 56–57, 58–59, 63 -E- *corr. ex* 3 / -I- *corr. ex* 4 / -H- *corr. ex* 9

polygon of the circle will be greater than △ *PTM*; which is demonstrated to be false. Since the perpendicular *HS* drawn from the center of the circle is less than the radius *FH* which is opposite the larger angle in △ *SFH*, the product of [1] *HS* and [2] the perimeter of the octagon —which perimeter is less than the circumference surrounding it— results in a quantity less than double the triangle. And so the octagon is less than △ *PTM*, although earlier it was said to be greater. This cannot stand, nor can the premise from which it followed, namely, that the circle is greater than the triangle.

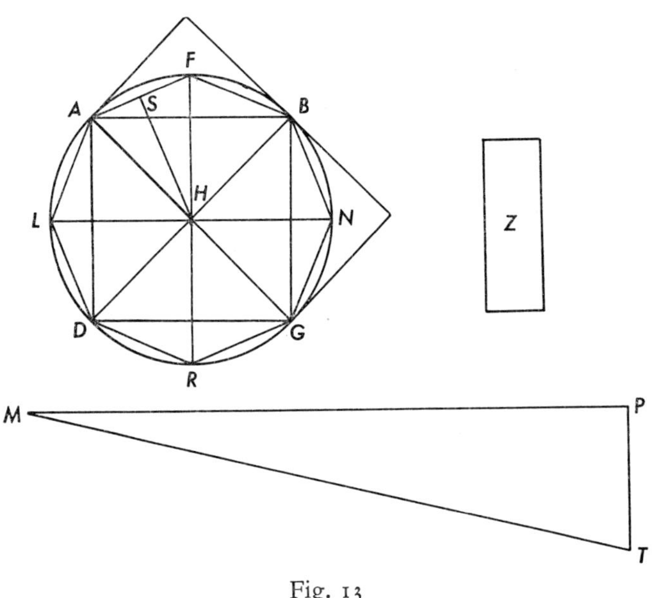

Fig. 13

If circle *ABGD* is less than △ *PTM* [Fig. 13], let the excess of the triangle over the circle be *X* [see Fig. 14]. And let the circle be inscribed in square *CKQP'*. Thus more than its half has been exhausted from the square by the circle. Let line *CN* be drawn from *N* to *C*. It is evident that by this line the arc *AB* and its chord have been bisected. Since angles *CAB* and *CBA* are equal because of the equal radii *AN* and *BN* and the equal right angles *CAN* and *CBN*, so it is inferred that the two parts of the tangent lines which lie between the points of tangency of these lines with the circle and their common intersection are equal. Also let a tangent be drawn from the intersection of line *CN* and arc *AB* in each direction until it reaches points *Z* and *T*. In the same way, let *OEY*, *MIV*, and *SHR* be drawn. Therefore, *FT* is equal to *TB*. And it will be evident to

50 octogonium ⟨quo⟩ circumscribitur circulus et quamlibet illarum dividi in suas partes medietates ad medias lineas. Et cum *CT* sit maior *FT* quia respicit maiorem angulum, est et maior *TB*. Quare per primam sexti Euclidis triangulus *FBT* est minor triangulo *CFT*; quare pars eius extra circulum *FTB* multo magis minor eodem triangulo *CFT*.
55 Et similiter *AZF* minor triangulo *CZF* et ita *BFT* et *FZA* sunt minores triangulo *ZCT*, et alie consimiles partes *BOE*, *EYG*; *GMI*, *IVD*; *DSH*, *HRA*, sunt minores triangulis sibi suprapositis, scilicet *OKY*, *MQV*, *SP'R*. Eodem modo a portionibus *FTB*, *BOE*, *EYG*, *GMI*, *IVD*, *DSH*, *HRA*, *AZF* supra circulum existentibus per con-
60 tingentes potest subtrahi maius medietate earum. Et in tali continua detractione qualibet quantitate proposita occurret minus. Causa brevitatis ergo portiones octo iam relicte supra circulum constituantur minores *X* excessu. Constat ergo poligonium *FTBOEYGMIVDSHRAZ* omne minus triangulo *PMT*. Sed linea ambiens poligonium maior
65 circumferentia, que est equalis *PM* lateri trianguli, et *BN* semidyameter equalis *PT*. Maius est ergo duplum poligonii quod fit ex ductu semidyametri in lineam ambientem poligonium duplo trianguli *PTM*. Quare et poligonium maius triangulo. Sed prius dicebatur esse minus. Constat ergo circulum, cum nec sit maior nec minor, esse
70 equalem triangulo. Et hoc est quod declarare volumus.

him who considers it subtly that these eight lines constitute an octagon by which the circle is circumscribed and that any of them is bisected by the middle lines. And since $CT > FT$ because it is opposite the larger angle, $CT > TB$. Hence by VI.1 [of the *Elements*] of Euclid $\triangle FBT < \triangle CFT$. Accordingly, the part *FTB* outside the circle is less than $\triangle CFT$ by an even greater amount. Similarly $AZF < \triangle CZF$. And so [the arcal figures] *BFT* and *FZA* [together] are less than $\triangle ZCT$, and the other similar parts *BOE* and *EYG*, *GMI* and *IVD*, *DSH* and *HRA* are less than the triangles situated above them, namely, [triangles] *OKY*, *MQV*, and *SP′R*. In the same way, from the figures *FTB*, *BOE*, *EYG*, *GMI*, *IVD*, *DSH*, *HRA*, and *AZF* existing beyond the circle, there can be

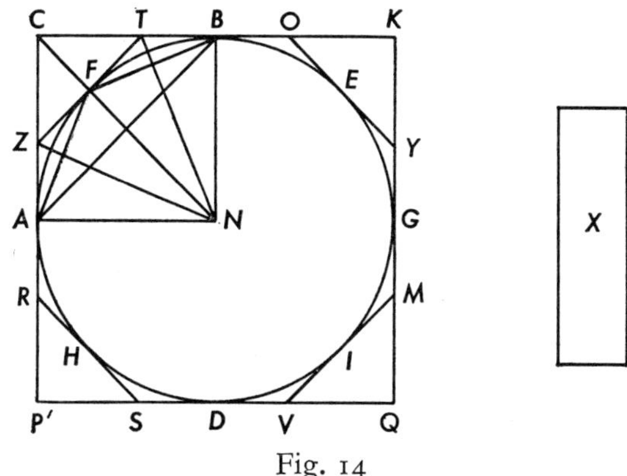

Fig. 14

Note: I have added the prime sign to *P′* here and in the text to distinguish it from *P* in Figure 13.

extracted by means of the tangents an amount greater than their half. And by such continual extraction there [ultimately] results [a quantity] less than the proposed quantity. Hence for the sake of brevity let those eight figures already remaining beyond the circle constitute [a quantity] less than the excess *X*. It is evident, therefore, that the whole polygon *FTBOEYGMIVDSHRAZ* is less than $\triangle PMT$ [Fig. 13]. But the perimeter of the polygon is greater than the circumference of the circle, which latter is equal to side *PM* of the triangle, and radius *BN* is equal to [side] *PT* [Figs. 13 and 14]. Therefore, double the polygon, which arises from the product of the radius and the perimeter of the polygon, is greater than double $\triangle PTM$. Therefore, the polygon is also greater than the triangle. But it was said before to be less. Hence it is evident that the circle, since it is neither more nor less, is equal to the triangle. And this is what we wish to show.

[Corollarium:] Colligitur etiam ex dictis omnem aream circuli esse equalem quadrangulo quod fit ex multiplicatione medietatis dyametri in dimidium circumferentie.

⟨II.⟩ PROPORTIO AREE OMNIS CIRCULI AD QUADRA-
75 TUM DYAMETRI IPSIUS EST SICUT PROPORTIO 11 AD 14.

Exempli causa, sit linea *AB* dyameter, cuius quadratum sit *GDHC*
12v [Fig. 15]. Sit item *GD* medietas *ED* et *ZE* septima / *GD*. Et quia triangulus *ADG* ad *AZE* que 7 ad 1, tota proportio *AZG* ad *ADG* sicut proportio 22 ad 7. Cum ergo quadratum sit quadruplum ad
80 triangulum *ADG*, et triangulus *AZG* equalis aree circuli per priorem propositionem, erit proportio quadrati ad circulum sicut proportio 28 ad 22; quare sicut medietas ad medietatem, scilicet 14 ad 11. Attende tamen quod propter excessum *ZG* linee ad circumferentiam fere imperceptibilem diximus esse eam equalem circumferentie, quod ma-
85 nifestum erit per sequentem propositionem.

⟨I.⟩ (F.IB) CIRCULUM QUADRARE EO QUOD OMNIS TRIANGULUS ORTHOGONIUS, CUIUS UNUM LATUS EQUATUR CIRCUMFERENTIE, RELIQUUM LATUS SEMIDYAMETRO, EQUALIS EST IPSI CIRCULO.

5 Sit enim *AB* equalis *CD* semidyametro et *AF* circumferentie [Fig. 16]. Dico quod *ABF* triangulus est equalis circulo. Detur enim quod sit minor circulo. Quantitas ergo que sit *QTSO* que cum triangulo sit equalis circulo. Describatur ergo *NRMK* quadratum in circulo, per sextam quarti. Sed probatur in prima decimi quod possumus
10 omnem quantitatem maiorem alia resecare relinquendo maius medietate quousque sit minor minore. Relinquitur ergo per positionem

72 *post* equalem *del. D* triangulo
73 dimidium *corr. ex* dimium
81, 85 propositionem *corr. ex* proporcionem
1 *mg. habet D* interposita

[Corollary:] It is inferred from [these] statements that the area of any circle is equal to a rectangle arising from the product of the radius and the semicircumference.

II. THE RATIO OF THE AREA OF ANY CIRCLE TO THE SQUARE OF ITS DIAMETER IS AS THE RATIO OF 11 TO 14.

For example, let line *AB* be the diameter whose square is *GDHC* [see Fig. 15]. Also let *GD* be one half of *ED* and *EZ* one seventh of *GD*. Now because △ *ADG*/△ *AZE* = 7/1 [therefore] *AZG*/*ADG* = 22/7. Therefore, since the square is quadruple △ *ADG*, and since △ *AZG* equals the area of the circle by the prior proposition, the ratio of the square to the circle will be as 28 to 22. Hence their halves are as 14 to 11. Observe, however, that because the excess of line *ZG* beyond the circum-

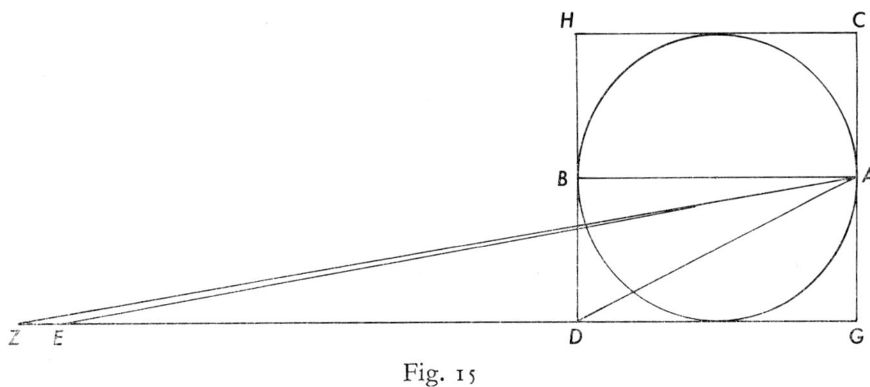

Fig. 15

ference is almost imperceptible, we have said that it is equal to the circumference; which will be evident by the following proposition (that is, Proposition III).

I. (F.IB) TO SQUARE A CIRCLE, FROM THE FACT THAT EVERY RIGHT TRIANGLE, ONE SIDE OF WHICH IS EQUAL TO THE CIRCUMFERENCE [AND] THE REMAINING SIDE TO THE RADIUS, IS EQUAL TO THIS CIRCLE.

For let *AB* be equal to the radius *CD* and *AF* to the circumference [see Fig. 16]. I say that △ *ABF* is equal to the circle. For let it be given that it is less than the circle. Hence let *QTSO* be the quantity which together with the triangle is equal to the circle. Therefore, let square *NRMK* be inscribed in the circle, by IV.6 [of the *Elements*]. But it is proved in X.1 [of the *Elements*] that we can [continually] cut away, by remaindering, more than half of the greater of two quantities until it is less than the lesser [of those quantities]. Hence by [this] proposition let the circle be [so]

circulus donec sit minor. Sed *NRMK* quadratum est maius reliquis quatuor lunulis circuli, quoniam utraque eius medietas est maior duabus lunulis circa ipsam. Et probetur hoc per 41^am^ primi, quoniam
15 parallelogramum duplum est ad triangulum. Sic ergo relinquitur maius medietate. Aut ergo ⟨residuum⟩ minus est quadrato [*QTSO*] aut non. Si non, dividatur in duo equalia quodlibet latus quadrati [*NRMK*], et a quolibet puncto sectionis ducatur linea recta directe ad circumferentiam. Et fiat octogonus sicquod quatuor anguli a supradictis qua
20 tuor lunulis secentur, et sic lunule que iam facte sunt minores triangulis relictis, ut probetur per 41^am^ primi. Detur ergo hee lunule quod octo sunt minores *QTSO*, quoniam sic non est procedere in infinitum. Ergo ⟨octogonus⟩ est maior *ABF*. Sed contra: Illud quod fit ex ductu *CE* in *ND* cordam duplum est ad *NDC* triangulum. Et sic
25 illud quod fit ⟨ex ductu⟩ *EC* in totalem ambitum octogoni duplum est ad octogonum per secundam secundi. Sed *CD* linea est maior *EC* et circumferentia est maior ambitu octogoni. Ergo quod fit ex ductu *CD* in circumferentiam est maius duplo octogoni. Sed illud quod fit ex ductu *CD* in circumferentiam duplum est ad *ABF* per 41^am^ primi.
30 Ergo *ABF* est maior octogono, et dictum est quod erat minor.

21–22 octo...QTSO *corr. ex* octogonus sit maiores triangulo
23 *ante* est *del. D* lunule sunt minores
25 illud *corr. ex* id

remaindered until it is less. But square *NRMK* is greater than the remaining four segments of the circle, since each half of it is greater than the two segments which are [circumscribed] about it. And let this be proved by I.41 [of the *Elements*,] since a parallelogram is double a triangle [with the same altitude and base]. Hence more than half is exhausted as follows: Either the remainder is less than square [*QTSO*] or not. If not, let each side of square [*NRMK*] be bisected, and from each point of [such] bisection let a straight line be drawn to the circumference. And let an octagon be

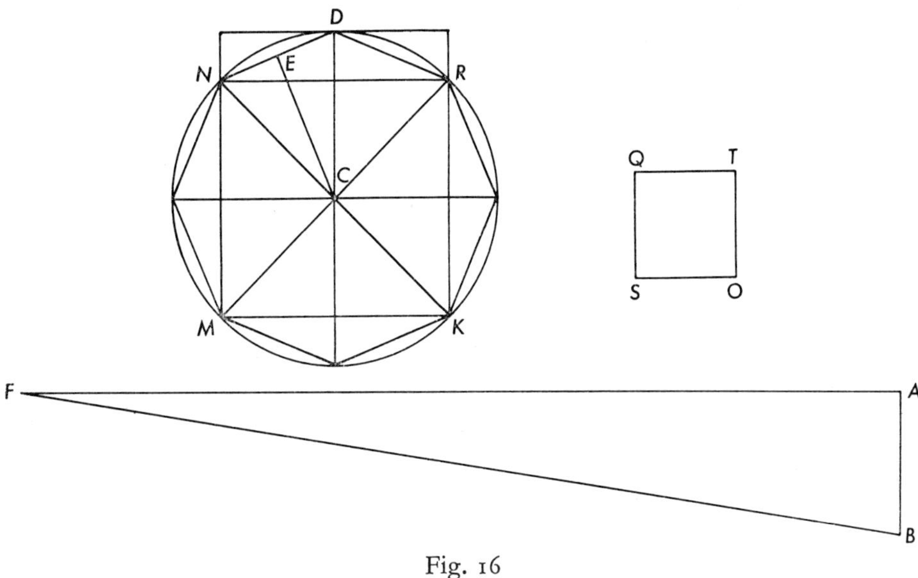

Fig. 16

constructed such that four angles are cut out of the aforesaid four segments. Thus the [eight] segments which have been so constructed are less than the triangles remaining [from the aforesaid segments], as one can prove by I.41 [of the *Elements*]. Therefore, let it be given that the eight segments are less than *QTSO*, since one does not have to proceed in this way to infinity. Therefore, the octagon is greater than [$\triangle$] *ABF*. But the opposite can be shown: For $CE \cdot ND = 2 \triangle NDC$. Thus the product of *EC* and the perimeter of the octagon is double the octagon, by II.2 [of the *Elements*]. But line $CD > EC$ and the circumference is greater than the perimeter of the octagon. Therefore, the product of *CD* and the circumference is greater than double the octagon. But the product of *CD* and the circumference is double [$\triangle$] *ABF*, by I.41 [of the *Elements*]. Therefore, *ABF* is greater than the octagon, and it has been said that it was less.

Redeamus, et dicatur quod *ABF* sit maior circulo. Sumatur ergo quantitas que sit *QTSO*, et illa cum circulo sit equalis trigono. Et describatur quadratum circa circulum per septimam quarti. Et probetur quod *DCEN* quadratum [Fig. 17] sit maius *ABF* triangulo,
35 quoniam, ut patet ex predictis, illud quod fit ex ⟨ductu⟩ *MR* in ambitum quadrati duplum est ad quadratum et est maius duplo trianguli, quod fit ex ductu *MR* in circumferentiam. Ergo quadratum est maius *QTSO* et circulo. Sed circulus est equalis sibi. Ergo partes quadrati residue a circulo sunt maiores *QTSO*. Resecentur ergo per primam
40 decimi donec sint minores *QTSO*. Ducatur ergo contingens a quolibet puncto ubi duo diametri *DE*, *CN* secant circulum ita quod quelibet contingens ducatur ad duo quadrati sibi proxima latera, et sic erit octogonus circa circulum. Sed probatur quod sic resecatum est plus medietate, quod enim dicetur de uno intelligatur de omnibus. Quanta
45 est *DK* ad *KR* tanta proportio est *DBK* ad *KRB* per primam sexti. Sed *DK* est maius *KR*, quoniam *KR* est equalis *KB* per quartum librum; sunt enim contingentes et concurrunt. Sed *KB* est minus *DK* per penultimam primi. Ergo *DKB* est maius *KBR*. Ergo multo magis est maius eo quod *BK* et *KR* et *BR* pars circumferentie continent. Si
50 ergo nondum dicatur quod partes octogoni extrinsece a circulo sunt minores *QTSO*, iterum resecentur secundum ante. Sed dentur hee alie minores propter evidentiam, et sic octogonus est minor. Sed contra:

41 DE, CN *corr. ex* DC, EN
49 continent *corr. ex* concurrunt
51 ante *corr. ex* artem

Let us go back and say that ABF is greater than the circle. Let the quantity $QTSO$ be assumed, which quantity together with the circle is equal to the triangle. And let a square be described about the circle, by IV.7 [of the *Elements*]. And let it be proved that square $DCEN$ [see Fig. 17] is greater than $\triangle$ ABF [Fig. 16], since, as is obvious from what we have said, the product of MR and the perimeter of the square is double the square and is more than double the triangle, which arises from the product of MR and the circumference. Therefore, the square is greater than the sum of $QTSO$ [Fig. 16] and the circle. But the circle is equal to itself. Therefore, the parts of the square remaining beyond the circle are greater than $QTSO$. Let the cutting away continue by X.1 [of the *Elements*]

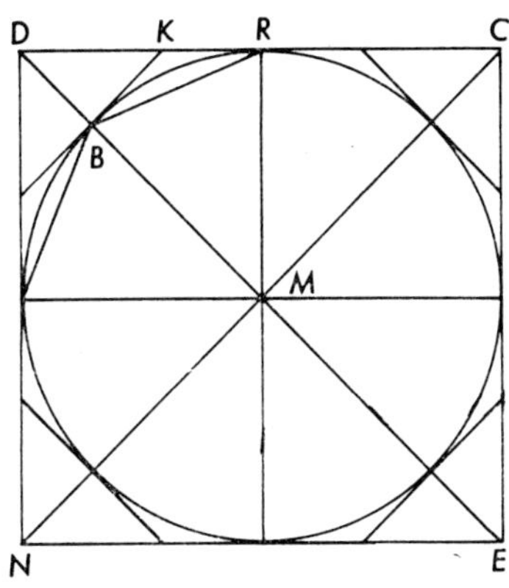

Fig. 17

until the [residual] parts are less than $QTSO$. So let a tangent be drawn from each point where the two diagonals DE and CN intersect the circle, so that each tangent is drawn to two adjacent sides of the square; and so there will be an octagon about the circle. But it is proved that there has been thus exhausted more than half, for what will be said concerning one is understood for all. [Now] $DK/KR = DKB/KRB$, by VI.1 [of the *Elements*]. But $DK > KR$, since $KR = KB$, by the fourth book [of the *Elements*], for KR and KB are tangent and meet. But $KB < DK$, by the penultimate of [Book] I [of the *Elements*]. Therefore $DKB > KBR$. Therefore, a fortiori DKB is greater than the figure included by BK, KR, and arc BR. Hence if it is said that the extrinsic parts of the octagon beyond the circle are not yet less than $QTSO$ [Fig. 16], let them again be exhausted as before. But let it be that these other parts are less—for the sake of clarity—and so the octagon is less [than the triangle]. But the opposite

Illud quod fit ex ductu *MR* in ambitum octogoni duplum est ad octogonum et est maius duplo trianguli, i.e., eo quod fit ex ductu *MR* 55 in circumferentiam. Ergo octogonus est maior triangulo. Fac ergo quadratum equale triangulo per ultimam secundi, et sic concludes propositum. Explicit.

9v ⟨III.⟩ / OMNIS LINEA CONTINENS CIRCULUM ADDIT SUPER TRIPLUM DYAMETRI IPSIUS MINUS SEPTIMA ET PLUS 10 PARTIBUS 71 UNIUS PARTIUM DYAMETRI.

Exempli causa, sit linea *AG* ⟨dyameter circuli *AG*⟩ [Fig. 18]. Sitque 5 eius centrum *E*, et linea *DZ* sit contingens circulum, et sit angulus *ZEG* tertia anguli recti. Ergo *ZG* est subdupla *ZE*, quia circumscripti circuli *ZGE* triangulo *ZE* est dyameter, cum *ZGE* angulus sit rectus. Quare que est proportio *GEZ* anguli ad *GZE* angulum eadem est arcus corde *ZG* ad arcum corde *GE* et sic arcus corde *ZG* subduplus 10 est arcus corde *GE*. Quare est tertia pars semicircumferentie, ergo sexta pars totalis circumferentie circumscripte triangulo *ZGE*. Sic ergo corda *ZG*, cum sit latus exagoni inscripti illi circulo, est medietas *ZE*. Ergo proportio *EZ* ad *ZG* sicut proportio 306 ad 153. Dividam autem angulum *ZEG* in duo media *EH*. Ergo proportio *ZE* ad *EG* sicut 15 *ZH* ad *GH* per tertiam propositionem sexti libri geometrie. Quare permutatim que est proportio *ZE* ad *ZH* eadem est *GE* ad *HG*. Ergo que est proportio *ZE* ad *HZ* eadem est *ZE* et *EG* coniunctarum ad *ZHG*. Quare que est *GE* ad *GH* eadem est proportio *ZE* et *EG* coniunctarum ad *ZG*. Ergo cum proportio *ZE* et *EG* coniunctarum 20 ad *GZ* sit maior proportione 571 ad 153, erit et maior proportio *GE* ad *HG* proportione 571 ad 153. Quod maior sit *EZ* et *EG* coniunctarum etc., sic declaratur: Sumatur quadratum numeri 306, quod est 93636, facile enim invenire poteris si numerum 306 per se ipsum

53 ambitum octogoni *corr. ex* ao
54 ductu *corr. ex* d in d
57 propositum *Zo* propotum *D*
1 *mg. manu recentiori* De quadratura circuli; D_1 *habet* Liber de quadratura circuli; *et habet Zo* Incipit liber Archimenidis de Quadratura circuli
4 ⟨dyameter circuli AG⟩ *supplevi, cf. lin. 168; sed in D_1 et mg. D* continens circulum
6 *de* tertia anguli recti *scr. mg. D mg. Zo* Inscribe circulo duodecagonum et trahendo lineas a centro ad quoslibet angulos quodlibet latus eius subtendetur angulo in centro qui erit tertia pars anguli recti, erunt enim in centro 12 anguli qui omnes simul per correlarium 15 primi valent 4 rectos. Ergo quilibet valet tertiam partem recti.
11 sexta *corr. ex* vi
20 proportione *corr. ex* proportio

can be shown: The product of MR and the perimeter of the octagon is double the octagon and is more than double the triangle, for the latter [i.e., double the triangle] arises from the product of MR and the circumference. Therefore, the octagon is greater than the triangle. Therefore, make a square equal to the triangle by the last [proposition of Book] II [of the *Elements*], and you will thus conclude that which was proposed. The end.

III. EVERY CIRCUMFERENCE OF A CIRCLE EXCEEDS THREE TIMES ITS DIAMETER BY AN AMOUNT LESS THAN ONE SEVENTH AND MORE THAN 10 PARTS OF 71 PARTS OF THE DIAMETER.

For example, let line AG be the diameter of circle AG [see Fig. 18]. Let its center be E and let DZ be tangent to the circle. Let ZEG be 1/3 of a right angle. Therefore, $ZG = 1/2\ ZE$, because ZE is the diameter of a circle circumscribed about $\triangle\ ZGE$ since ZGE is a right angle. Hence,

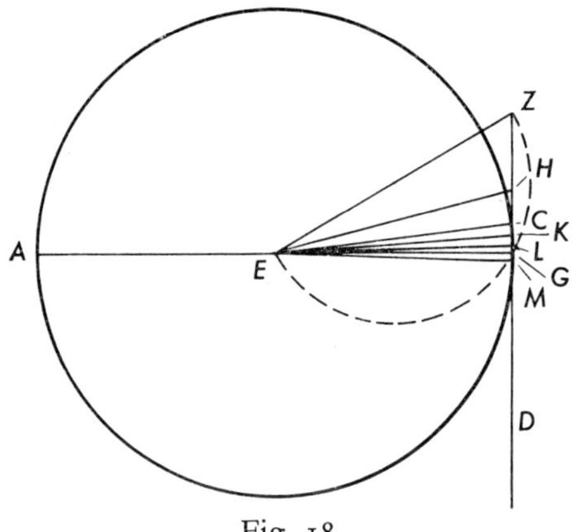

Fig. 18
Note: I have added the dotted semicircle.

$\angle\ GEZ/\angle\ GZE$ = arc ZG/arc GE. Thus arc $ZG = 1/2$ arc GE. Therefore, arc $ZG = 1/3$ semicircumference and arc $ZG = 1/6$ circumference circumscribed about $\triangle\ ZGE$. Hence chord ZG, since it is the side of a hexagon inscribed in that circle, is $1/2\ ZE$. Therefore, $EZ/ZG = 306/153$.

Now I bisect $\angle\ ZEG$ by [line] HE. Therefore, $ZE/EG = ZH/GH$, by VI.3 of the *Book of Geometry* (i.e., the *Elements*). Therefore, permutatively $ZE/ZH = GE/HG$. Therefore, $ZE/HZ = (ZE + EG)/(ZH + HG)$. Hence, $GE/GH = (ZE + EG)/ZG$. Hence, since $(ZE + EG)/GZ > 571/153$, also $GE/HG > 571/153$. That $(EZ + EG)/GZ > 571/153$ is proved as follows: Let 306^2 be taken; it is 93,636, for you will be able

multiplicaveris. Inveniatur iterum quadratum numeri 153, quod est
25 23409. Subtrahe ergo quadratum 23409 a quadrato 93636 et remanebit numerus 70227, qui est equalis quadrato *GE* linee, cum 93636, quod est quadratum *ZE* linee que est quasi 306, sit equale 23409 quadrato *ZG* linee que est quasi 153 et quadrato linee *EG*. Item numero 70227, scilicet quadrato *EG* linee, subtrahe radicem que, nisi superesset 2,
30 esset iste numerus 265. Sume ergo radicem proximi minoris quadrati, cum ipse careat vera radice, que est hec 265. Addatur itaque hec radix 306 et proveniet ille numerus 571, quem numerum constat se habere in minori proportione ad 153 proportione *EZ*, *EG* coniunctarum ad 26 ⟨5⟩, in quanto 265 radix proximi minoris esset minor vera radice
35 numeri 70227 qui est equalis quadrato *EG*, si veram haberet. Constat ergo quod maior est proportio *GE* ad *GH* proportione 571 ad 153. Ergo maior est proportio quadrati *GE* ad quadratum *HG* linee proportione quadrati 571, quod est 326041, ad quadratum 153, quod est 23409. Ergo coniunctim maior est proportio quadrati *EG* et quadrati
40 *HG* ad quadratum *HG* proportione quadrati 571 et quadrati numeri 153 coniunctorum, scilicet que coniuncta sunt 349450, ad quadratum 153. Et quadratum *EH*, quia respicit rectum angulum, valet quadratum *GE* et quadratum *HG* coniuncta. Ergo proportio quadrati *HE* ad quadratum *HG* linee est maior proportione numeri 349450 ad 23409.
45 Ergo maior est proportio *HE* ad *HG*, cum *HE* et *HG* sunt vere radices quadrati *HE* et quadrati *HG*, proportione vere radicis numeri 349450, si haberet, ad 153. Cum ergo numerus 591 cum octava unius minor sit vera radice eius, quod postea probabitur, multo magis maior est proportio *HE* ad *HG* proportione 591 cum octava unius ad 153.

50 Probatur autem hoc modo: Duc 591 in octavas, i.e., multiplica eos
10r per 8 et excrescunt / minuta eiusdem denominationis, i.e., octave, equivalentia integris 591, que sunt 4728. Addita ergo octava proveniunt hec minuta 4729, que equivalent integris 591 cum octava unius. Multiplica ergo hec minuta per se et provenient minuta que sunt
55 sexagesime quarte, scilicet hec 22363441. Has iterum sexagesimas quartas divide per suam denominationem, scilicet 64, et provenient hec integra 349428, que cum istis minutis 49 remanentibus post divisionem equivalent omnibus istis minutis pariter acceptis. At illorum

34 26 ⟨5⟩ *corr. ex* 26
35 70227 *corr. ex* 23409
37 HG *corr. D ex* GHG
40 ad quadratum HG *mg. D*
42 Et quadratum EH *corr. ex* ad quadratum ATCH
48 sit *supra scr. D*
49 ad[2] *corr. ex* a

to find it easily if you multiply 306 by itself. Let 153^2 be found; it is 23,409. Hence subtract 23,409 from 93,636, and 70,227 will remain. Now $70{,}227 = GE^2$, since 93,636 (equal to ZE^2 when $ZE = 306$) is equal to 23,409 (equal to ZG^2 when $ZG = 153$) plus EG^2. If out of the number 70,227 (equal to GE^2) you extract the root, it would be 265, except that it (70,227) would exceed [265^2] by 2. So take the root of the next smaller square [below 70,227], i.e., 265, since 70,227 lacks a true [rational] root. And so this root [265] is added to 306 and 571 results. Thus, $571/153 < (EZ + EG)/265$ by the amount that 265, the root of the next smaller square, would be less than the true root—if it had a true root—of the number 70,227, which is equal to EG^2. It is evident, therefore, that $GE/GH > 571/153$. Hence $(GE)^2/(HG)^2 > 571^2/153^2$, i.e., $(GE)^2/(HG)^2 > 326{,}041/23{,}409$. Hence by composition, $(EG^2 + HG^2)/(HG)^2 > 349{,}450/23{,}409$. And—because EH is opposite a right angle—$EH^2 = GE^2 + HG^2$. Therefore, $(HE)^2/(HG)^2 > 349{,}450/23{,}409$. Therefore, since HE and HG are the true roots of HE^2 and HG^2, $HE/HG > \sqrt{349{,}450}/153$. Since $591\frac{1}{8}$ is less than the true root of 349,450, which will be proved afterwards, a fortiori $\dfrac{HE}{HG} > \dfrac{591\frac{1}{8}}{153}$.

It is proved moreover in this way: Turn 591 into eighths, i.e., multiply it by 8/8, and there will result 4,728/8, which equals 591 in integers. Hence with 1/8 added, we get 4,729/8, equivalent to $591\frac{1}{8}$. Multiply 4,729/8 by itself and 22,363,441/64 results. Divide this through by 64 and there results in integers $349{,}428 + \frac{49}{64}$, which is equivalent to all these sixty-fourths taken together. And so the root of both the expression in

minutorum radix est 4729, quare et istorum integrorum 349428 cum
60 istis sexagesimis quartis 49 remanentibus. Sic constat hec integra
349428 cum istis minutis 49 esse minora 349450 fere in quantum 50
excedunt 28. "Fere" dico propter minuta 49. Constat ergo quod vera
radix 349450, si haberet, maior esset hiis octavis 4729, que sunt radix
integrorum cum minutis 49 et equivalentia hiis integris 591 cum octava
65 unius. Hoc declarato ad principale redeamus propositum.

Maior est proportio *HE* ad *GH* proportione 591 cum octava unius
ad 153, et, ut dictum est, maior est *GE* linee ad *HG* lineam proportione 571 ad 153. Ergo maior est proportio *HE* et *GE* coniunctarum
ad *GH* proportione numeri 591 cum octava unius et numeri 571
70 coniunctorum, qui coniuncti producunt 1162 cum octava unius, ad
153.

Angulum autem *HEG* dividam in duo equalia linea *CE*. Et que
erit proportio *HE* et *GE* coniunctarum ad *GH* eadem erit *GE* ad *GC*.
Quare maior est proportio *GE* ad *GC* proportione 1162 et octave ad
75 153. Ergo maior est proportio *GE* linee quadrati ad quadratum *GC*
linee proportione quadrati numeri 1162 et octave, quod est 86434209—
quia ducantur 1162 in octavas, i.e., multiplicentur per 8 et excrescunt
hec minuta 9296 equiparentia illi integro; apponatur iterum octava
et erunt 9297 que valent 1162 et octavam; et iterum multiplicentur hec
80 minuta 9297 et provenient iste sexagesime quarte 86434209, et sic
quadratum 1162 et octave integrorum—ad quadratum 153, quod
quadratum valet istas sexagesimas quartas 1498176, que fiunt ex multiplicatione ⟨in se⟩ istarum octavarum 1224, que equivalent isti numero
153. Ergo coniunctim maior est proportio quadrati *EG* et quadrati
85 *GC* coniunctorum ad quadratum *GC* proportione quadrati numeri
1162 et octave et quadrati numeri 153 coniunctorum, que coniuncta
sunt 87932385, ad quadratum numeri 153, quod est, ut sunt iste
sexagesime quarte, 1498176. Ergo maior est proportio quadrati *EC*
ad quadratum *GC* proportione numeri 87932385 ad numerum 1498176.
90 Ergo maior est proportio linee *EC* ad lineam *GC*, cum sint vere
radices quadrati *EC* et quadrati *GC*, proportione vere radicis 87932385,
si haberet, ad 153. Constat ergo quod maior est proportio linee *EC*
ad lineam *GC* proportione numeri 1172 et octave unius, cum sit
minor vera radice 87932385, ad 153. Quod 1172 cum octava sit minor
95 vera radice illius, constat. Ducatur enim iste 1172 in octavas et hee

76 quadrati *mg. D*
84 *ante* 153 *del. D* Item ergo
95 *ante* in *del. D* et

fractions and of those integers, i.e., $349{,}428 + \frac{49}{64}$, is 4,729/8. Thus it is evident that $349{,}428 + \frac{49}{64}$ is less than 349,450 almost by the amount that 50 exceeds 28. I say "almost" because of the $\frac{49}{64}$. It is clear, therefore, that the true root of 349,450, if it were had, would be greater than 4,729/8, the root of the integers plus the $\frac{49}{64}$, and thus would be greater than the equivalent in integers, namely, $591\frac{1}{8}$. With this shown, we return to the main proposal.

[Now] $\frac{HE}{GH} > \frac{591\frac{1}{8}}{153}$ and, as was said, $GE/HG > 571/153$. Therefore, by composition, $\frac{HE + GE}{GH} > \frac{591\frac{1}{8} + 571}{153}$, i.e., $\frac{HE + GE}{GH} > \frac{1{,}162\frac{1}{8}}{153}$.

Moreover, I bisect $\angle HEG$ by line CE. And $(HE + GE)/GH = GE/GC$. Therefore, $\frac{GE}{GC} > \frac{1{,}162\frac{1}{8}}{153}$. Therefore, $\frac{GE^2}{GC^2} > \frac{(1{,}162\frac{1}{8})^2}{(153)^2}$, or $\frac{GE^2}{GC^2} > \frac{86{,}434{,}209/64}{1{,}498{,}176/64}$. [The numerator arises] by turning 1,162 into eighths, i.e., by multiplying by 8/8 and thus getting 9,296/8, equivalent to the integers. Then 1/8 is added and we get 9,297/8, equivalent to $1{,}162\frac{1}{8}$. Then multiply 9,297/8 by itself and we get 86,434,209/64. So much for the square of $1{,}162\frac{1}{8}$. [Now the denominator arises from the square of 153] which is equivalent to 1,498,176/64 and results from multiplying 1,224/8 (equal to 153) by itself. Therefore by composition, $\frac{EG^2 + GC^2}{GC^2} > \frac{(1{,}162\frac{1}{8})^2 + 153^2}{153^2}$, or $\frac{EG^2 + GC^2}{GC^2} > \frac{87{,}932{,}385/64}{1{,}498{,}176/64}$, Therefore, $\frac{EC^2}{GC^2} > \frac{87{,}932{,}385/64}{1{,}498{,}176/64}$. Therefore, since line EC and line GC are the true roots of EC^2 and GC^2, $\frac{EC}{GC} > \frac{\sqrt{87{,}932{,}385/64}}{153}$. It is evident, therefore, that $\frac{EC}{GC} > \frac{1{,}172\frac{1}{8}}{153}$, since $1{,}172\frac{1}{8}$ is less than the true root of 87,932,385/64. That $1{,}172\frac{1}{8}$ is less than its true root is evident. For turn 1,172 into eighths and there results 9,376/8. An eighth is added, making

erunt octave 9376. Apponatur iterum octava et erunt 9377 que in se multiplicate producunt istas sexagesimas quartas 87928129. Cum ergo istorum numerorum radix sit 1172 cum octava et ista minuta excedunt scilicet 87932385 in quantum 32385 excedunt 28129, minor est 1172
100 cum octava vera radice 87932385. Redeamus ergo ad principale propositum.

Maior est proportio *EC* ad *GC* proportione 1172 et octave ad 153, et, ut dictum est, maior est proportio *GE* ad *GC* proportione 1162 et octave ad 153. Ergo ut prius maior est proportio *EC* et *EG* coniunc-
105 tarum ad *GC* proportione numeri 1172 et octave et numeri 1162 et octave coniunctorum, qui coniuncti procreant 2334 et quartam unius ad 153.

Dividam autem angulum *CEG* in duo equalia linea *EK*. Ergo secundum tenorem dictorum maior est proportio *EG* ad *GK* proportione
110 2334 et quarte unius ad 153. Ergo maior est proportio quadrati linee *EG* ad quadratum linee *GK* proportione quadrati numeri 2334 et quarte unius ad quadratum numeri 153, que duo quadrata sunt in sextisdecimis 87179569, 374544, quippe ducantur hec integra 2334 in quartas, i.e., multiplicentur per 4 et provenient hee quarte 9336.
115 Apponatur quarta unius et provenient hec minuta 9337, equivalentia illis integris 2334 cum quarta, que etiam multiplicata per se producunt predicta minuta 87179569. Quadratum vero 153 est 23409; multiplicetur per 16 et provenient hee sextedecime 374544. Ergo coniunctim maior est proportio *EK* linee quadrati ad quadratum *GK* linee porpor-
120 tione illorum duorum quadratorum, que coniuncta faciunt in sextisdecimis 87554113, ad quadratum numeri 153. Ergo maior est proportio *EK* in longitudine ad *GK*, cum sint vere radices, proportione

105–106 et[2]...octave *mg. D*

112–67 in....theorematis *mg. D mg. Zo*; *in altera margine de lectura textus add. D et mg. Zo* Hic videtur mihi quod commentator erraverit in multiplicando, fundat enim se in falsa radice, ut mihi videtur, propter quod apposui lecturam ubi est tale signum #. (*cf. lineam 167*) *In lineis 112–67 lecturam marginis do, quia est melior, et hic varias lectiones ex textu addo*

112–13 in sextisdecimis *om. Dt* (= *textus D*)

113 87179569, 374544: 87198244 et 1498176 *Dt*

114 4: quartas 4 *Dt* / proveniunt *Dt* / 9336: 9337 *Dt*

115 9337: 9338 *Dt* / equivalentia: equiparentia *Dt*

116 etiam: etiam minuta *Dt*

117 87179569: scilicet 87198244 *Dt*

117–18 Quadratum...374544: quare 1498176 sit quadratum 153 ut prius dictum est *Dt*

118 sextedecime *corr. ex* sextisdecime

120 duorum quadratorum: quadratorum duorum coniunctorum *Dt* / faciunt in sextisdecimis: sunt *Dt*

121 87554113: 87563888 *Dt* / 153 *Dt* 151 *mg D*

9,377/8. Multiply 9,377/8 by itself and the result is 87,928,129/64. Since the root of this number is $1{,}172\frac{1}{8}$ and 87,932,385/64 exceeds 87,928,129/64 by the amount that 32,385/64 exceeds 28,129/64, thus $1{,}172\frac{1}{8}$ is less than the true root of 87,932,385/64. Hence let us return to the main proposal. $\frac{EC}{GC} > \frac{1{,}172\frac{1}{8}}{153}$ and, as was said, $\frac{GE}{GC} > \frac{1{,}162\frac{1}{8}}{153}$. Therefore, as before, $\frac{EC + EG}{GC} > \frac{1{,}172\frac{1}{8} + 1{,}162\frac{1}{8}}{153}$, or $\frac{EC + EG}{GC} > \frac{2{,}334\frac{1}{4}}{153}$.

Moreover, I bisect $\angle$ *CEG* by line *EK*. Therefore, according to the [prior] line of reasoning, $\frac{EG}{GK} > \frac{2{,}334\frac{1}{4}}{153}$. Therefore, $\frac{EG^2}{GK^2} > \frac{(2{,}334\frac{1}{4})^2}{153^2}$, or $\frac{EG^2}{GK^2} > \frac{87{,}179{,}569/16}{374{,}544/16}$. For we turn 2,334 into fourths, i.e., multiply it by 4/4, and there arises 9,336/4. One fourth is added and the result is 9,337/4, equivalent in integers to $2{,}334\frac{1}{4}$. Then we multiply 9,337/4 into itself and there results 87,179,569/16. Now the square of 153 is 23,409, which when multiplied by 16/16 produces 374,544/16. Therefore, by composition, $\frac{EK^2}{GK^2} > \frac{87{,}554{,}113/16}{374{,}544/16}$. Therefore, since *EK* and *GK* are true roots [of EK^2 and GK^2], $\frac{EK}{GK} > \frac{2{,}339\frac{1}{4}}{153}$, since $2{,}339\frac{1}{4}$ is less than

2339 numeri cum quarta unius ad 153, cum 2339 cum quarta sit minor vera radice istarum sedecimarum 87554113, quia ductis hiis integris 125 2339 in quartas hee proveniunt quarte 9356. Apposita quarta sunt 9357. Que quarte per se multiplicate procreant has sextasdecimas 87553449, que sunt minora istis sextisdecimis 87554113. Sic ergo 10v maior est / proportio *EK* linee ad *GK* lineam proportione numeri 2339 cum quarta unius ad* 153; et, ut dictum est, maior est proportio *GE* 130 linee ad *GK* lineam proportione numeri 2334 cum quarta unius ad 153. Quare maior est proportio *EK* et *EG* simul ad *GK* proportione duorum numerorum 2339 cum quarta et 2334 cum quarta simul, qui coniuncti faciunt 4673 et medietatem unius, ad 153.

Angulum autem *GEK* dividam per equalia linea *EL*, eritque maior 135 proportio *EG* ad *GL* proportione numeri 4673 cum medietate unius ad 153. Et quia angulus *ZEG* fuit tertia anguli recti, oportet ut sit angulus *LEG* quadragesima octava pars anguli recti. Faciam autem supra punctum *E* angulum equalem angulo *LEG*. Sit autem angulus *GEM*. Angulus igitur *LEM* est 24[a] pars anguli recti. Ergo linea recta 140 *LM* est latus poligonie figure continentis circulum et habentis 96 angulos equales in centro circuli quem continet, quoniam in centro circuli sunt 4 anguli recti, qui, si dividatur quilibet per 24, erunt partiales anguli 96 et totidem latera poligonii respicientia illos angulos. Et quoniam iam declaravimus quod proportio *EG* ad *GL* maior est 145 proportione 4673 et medietatis unius ad 153 et duplum *EG* sit *AG* et duplum *GL* sit *ML*, maior est proportio *AG* ad *ML* proportione 4673 et medietatis unius ad 153. At eadem est proportio *ML* ad tota-

* Marginal version continues on folio 10v with this word.

---

123 2339 numeri: numeri 2348 *Dt* / cum quarta: et quarte *Dt* / 2339: 2348 *Dt*
124 istarum...87554113: 87963888 *Dt*
125 2339: 2346 *Dt* / 9356: 9392 *Dt corr. mg. D ex* 9336
125–26 sunt 9357: erunt 939 *Dt*; *corr. mg. D ex* sunt 9337
126–27 has...87553449: 8822849 *Dt*
127 que sunt *om. Dt* / istis...87554113: 87563888 *Dt*
128 numeri 2339: 2348 *Dt*
129 cum quarta unius: et quarte *Dt*
129–30 GE linee *tr. Dt*
130–35 numeri....numeri *om. Dt*
135 4673...medietate: 4682 et medietatis *Dt*
139 ergo *tr. Dt post* recta
141 quem *corr. ex* quam
141–42 in[1]...sunt: qui cum sint illi *Dt*
142 qui *om. Dt* / dividatur *corr. ex* dividantur / quilibet: quilibet illorum *Dt* / per 24: in 24 partes *Dt*
143 *post* anguli *del. Dt* 26 / respicientia *om. Dt*
145, 147 4673: 4682 *Dt*
145 sit *om. Dt*
147 proportio *om. Dt*

the true root of 87,554,113/16. For if we turn 2,339 into fourths, the result is 9,356/4, and, with a fourth added, 9,357/4. When 9,357/4 is multiplied into itself there results 87,553,449/16, which is less than 87,554,113/16. So, therefore, $\frac{EK}{GK} > \frac{2{,}339\frac{1}{4}}{153}$ and, as was said, $\frac{GE}{GK} > \frac{2{,}334\frac{1}{4}}{153}$. Hence $\frac{EK+EG}{GK} > \frac{2{,}339\frac{1}{4} + 2{,}334\frac{1}{4}}{153}$, or $\frac{EK+EG}{GK} > \frac{4{,}673\frac{1}{2}}{153}$.

Now I bisect $\angle\ GEK$ by line EL, and $\frac{EG}{GL} > \frac{4{,}673\frac{1}{2}}{153}$. And because $\angle\ ZEG$ was 1/3 of a right angle, it is necessary that $\angle\ LEG$ is 1/48 of a right angle. Now I construct at point E $\angle\ GEM$ equal to $\angle\ LEG$. Hence $\angle\ LEM$ is 1/24 of a right angle. Therefore, straight line LM is a side of a [regular] polygon containing the circle and having 96 equal angles in the center of the circle, since in the center of the circle there are four right angles, which, with each divided into 24 parts, will include 96 angles, and there are just as many sides of the polygon opposite those angles. And since we have already shown that $\frac{EG}{GL} > \frac{4{,}673\frac{1}{2}}{153}$, while $AG = 2\,EG$ and $ML = 2\,GL$, then $\frac{AG}{ML} > \frac{4{,}673\frac{1}{2}}{153}$. But $\frac{ML}{\text{perimeter of polygon}} = \frac{153}{14{,}688}$, because if you multiply 153 by 96 there results 14,688, in which 153 is contained as many times as ML is in the perimeter of the polygon. So

lem ambitum poligonii que est 153 ad 14688, quia 153 multiplica per 96 et proveniet iste numerus 14688, in quo totiens est 153 quotiens
150 *ML* in ambitu poligonii. Sic ergo habemus quod maior est proportio *AG* dyametri ad ambitum poligonii proportione numeri 4673 et medietatis ad 14688. Ergo econverso minor est proportio ambitus poligonii ad dyametrum *AG* proportione istius numeri 14688 ad 4673 et medietatem unius. At numerus 14688 non continet triplum et septi-
155 mam numeri 4673 et medietatis unius. Ergo multo fortius ambitus poligonii non est triplum et septima dyametri. Ad illud autem probandum: Dupla numerum 4673 et medietatem unius et habebis has dimidias 9347. Dupla iterum numerum 14688 et habebis has dimidias 29376. Subtrahe ergo numerum 9347 ter a numero 29376 et remane-
160 bunt 1335, qui est minor septima parte numeri 9347, quia in dividendo numerum 9347 per 7 redit in numero quotiens 1335 et remanent due dimidie indivise. Constat igitur quod numerus 14688 maior est triplo minus septima numeri 4673 et medietatis unius. Ergo multo fortius ambitus poligonii maior est triplo dyametri minus septima dyametri.
165 Ergo multo magis circumferentia, cum sit minor ambitu poligonii, est maior triplo dyametri minus septima. Sic ergo patet prima pars nostri theorematis.

Sit autem *AG* dyameter circuli *AG* et describatur in illo circulo latus exagoni *BG*, faciens angulum cum *AG* dyametro [Fig. 19].
170 Subtendatur basis angulo *AGB*, scilicet *AB*. Erit ergo *GAB* angulus tertia recti, cum respiciat latus exagoni, et proportio *AG* dyametri

148 hambitum *Dt* / est *om. Dt*
149 proveniet: procedit *Dt*
150 hambitu *Dt*
151 hambitum *Dt* / numeri 4673: 4682 *Dt*
153 istius numeri *om. Dt*
153–54 4673...medietatem: 4682 cum medietate *Dt* / At numerus: Sed *Dt*
54–63 At....fortius: Sed 14688 est maior tripla 4682 cum medietate unius septima eius, ad quod probandum multiplica 14688 per 2, duc medietates 4682 et procedit iste numerus 9364. Apponatur iterum medietas et erit ille numerus 9365. Divide ergo 29376 per istum 9365 et invenies quod divisor cum tribus vicibus in diviso et quod post divisionem remanebit ille numerus 1241, quem constat esse minorem septima 9365. Divide enim 9365 per numerum remanentem, scilicit 1241, et videbis quod septies continebit eum et insuper numerum qui relinquitur post divisionem, scilicet 9398. Constat igitur quod maior est triplo eius minus septima, quia multo magis *Dt Zot* (= *textus Zo*)
164 est *om. Dt*
165 Ergo et *Dt*
166 est maior *tr. Dt*
166–67 Sic...theorematis *om. Dt Zot* / *post lecturam in margine add. D et mg. Zo* Nota quod istam lecturam interposui, quia commentum videbatur mihi fundare se super falsam radicem. Non tamen fui ausus corrigere, et posui in principio tale signum # et in fine tale ◁|.
168 dyameter *corr. ex* dyametrum

we have it, therefore, that $\frac{\text{diameter } AG}{\text{perimeter of polygon}} > \frac{4{,}673\frac{1}{2}}{14{,}688}$. Conversely, $\frac{\text{perimeter of polygon}}{\text{diameter } AG} < \frac{14{,}688}{4{,}673\frac{1}{2}}$. But 14,688 does not contain 4,673½ three and one seventh times. Hence a fortiori the perimeter of the polygon is not 3 1/7 times the diameter. For the proof of this, turn 4,673½ into halves and you will have 9,347/2. Turn 14,688 into halves and you

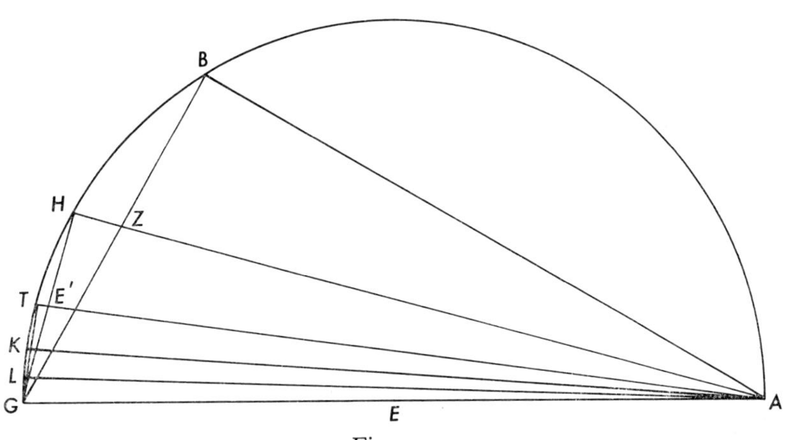

Fig. 19
Note: I have added the prime sign to E'.

will have 29,376/2. Hence subtract 9,347/2 three times from 29,376/2 and there will remain 1,335/2, which number is less than 1/7 of 9,347/2. For if you divide 9,347/2 by 7, the quotient will be 1,335/2 plus an undivided remainder of 2 halves. It is evident, therefore, that 14,688 is more than three times 4,673½ by an amount less than 1/7 of 4,673½. Therefore, a fortiori the perimeter of the polygon is more than three times the diameter by an amount less than 1/7 of the diameter. And still further, the circumference, since it is less than the perimeter of the polygon, is more than three times the diameter by an amount less than 1/7 of the diameter. Thus the first part of our theorem is evident.

Now let AG be the diameter of circle AG and let BG be the side of a hexagon described in that circle and making an angle with diameter AG [see Fig. 19]. Let side AB subtend $\angle AGB$. Therefore, $\angle GAB$ will be one third of a right angle, since it is opposite the side of a hexagon. And the ratio of diameter AG to its half, namely, GB, is as 1,560 to its half,

ad suum subduplum, scilicet *GB*, que 1560 ad 780, suum subduplum. Constat ergo si diviseris 1560 per duo relinqui numerum 780. Quare est medietas eius. Ergo minor est proportio *AB* ad *GB* proportione 1351
175 ad 780. Hec conclusio sic constat:

Sume quadratum 1560, quod est 2433600; sume iterum quadratum numeri 780, quod est 608400. Subtrahe ergo 608400 a quadrato 243360 et remanebit numerus equalis quadrato *AB*, scilicet 1825200. Cum 2433600 sit quadratum *AG* linee respicientis rectum angulum,
180 quia radix ipsius, scilicet 1560, et 608400 sit quadratum *BG* linee, est quasi *AB* linea 1351. Item quere radicem huius numeri, scilicet 1825200; si diligenter inspicis, videbis, quod totus numerus superior consumptus erit antequam prima figura numeri suppositi per se ipsam multiplicetur, quia vel erit supra primam scilicet 1 residuum, vel et
185 ille erit numerus suppositus 2700. Quare non deest aliquid superiori numero ad habendam veram radicem nisi 1. Dupla ergo 1351 et postea divide istum numerum 2702 per duo et habebis radicem quadrati maioris, scilicet 1351. Cum ergo 1825200 valeat quadratum *AB* linee et numerus 1351 sit maior vera radice eius, minor est proportio *AB*,
190 cum sit vera radix quadrati *AB*, ad *GB* proportione 1351 ad 780 numerum suum quasi lineam *BG*.

Dividam iterum angulum *GAB* in duo media linea *AH*. Ergo minor est proportio *AB* ad *BZ* proportione 2911 ad 780; [quia, quoniam] quecunque sit proportio *AG* ad *GB* eadem sit 1560 ad 780 et minor
195 *AB* ad *GB* proportione 1351 ad 780, maior est proportio numeri 1560 et numeri 1351 coniunctorum, qui coniuncti sunt 2911, ad 780 proportione *AG* et *AB* coniunctarum ad *BG*, et que est illarum coniunctarum ad *BG* eadem est *AB* ad *BZ*, ut sepe probatum est; maior ergo proportio 2911 ad 780 proportione *AB* ad *BZ*. Subtende ergo
200 basim angulo *GAH*, scilicet *HG*, et erunt trianguli isti similes *ABZ*, *AHG*, quia *ABZ* angulus et *AHG* angulus sint recti, quia uterque supra circumferentiam, et angulus *ZAB* equalis angulo *GAH* ex ypothesi; ergo tertius tertio, scilicet *AZB* equalis *AGH*. Constat ergo quod latera continentia angulos equales sunt proportionalia.

179 *ante* Cum *delevi* linee
180–81 608400...1351 *corr. ex* est quasi AB linea 1351, et 608400 sit quadratum BG linee
181 1351 *mg. D*
184 vel[1]: $\frac{l}{u}$ (?) *D* nihil *Zo* / vel[2]: $\frac{i}{u}$ (ubi?) *Zo*
185 2700 *corr. ex* 2701
186 veram *corr. ex* vera / 1351 *mg. D* et *in textu habet* 1; *cf. Zo* 1351.1
192 minor *corr. ex* maior
196 qui coniuncti *corr. ex* que coniuncta
201 et AHG *corr. D ex* Ḥ et A^H^G (*et habet Zo* H et AHG)

780. For if you divide 1,560 by 2 you will evidently get 780; wherefore, it is half of it. Hence $AB/GB < 1{,}351/780$. This conclusion is evident as follows:

Take the square of 1,560, which is 2,433,600. Further, take the square of 780, which is 608,400. Then subtract 608,400 from 2,433,600 and a number equal to AB^2 will remain, namely, 1,825,200. With AG opposite a right angle, if it is equal to 1,560 and its square is 2,433,600 and 608,400 equals BG^2, AB is almost 1,351. For seek the root of this number, i.e., 1,825,200. *If you diligently consider it, you will see that the whole superior number (1,825,200) will be consumed before the first figure of the supposed number (1,351) is multiplied by itself, because [if the supposed number is multiplied by itself the square of] it (1,825,201) exceeds the first number (1,825,200) by 1, while if the supposed number [were 1,350, its square (1,822,500) would be exceeded by 1,825,200 by] 2,700. Therefore, the superior number lacks only 1 of having a true root. Therefore, double 1,351 and afterwards divide this number 2702 by 2, and you will have the root of the [next] larger square, namely, the root 1,351.* Therefore, 1,825,200 equals AB^2,and 1,351 is greater than its true root. Since AB is the true root of AB^2 and GB is as 780, hence $AB/GB < 1{,}351/780$.

Again I bisect $\angle GAB$ by line AH. Therefore, $AB/BZ < 2{,}911/780$. For $AG/GB = 1{,}560/780$; $AB/GB < 1{,}351/780$; $(1{,}560 + 1{,}351)/780 > (AG + AB)/BG$, or $2{,}911/780 > (AG + AB)/BG$; $(AG + AB)/BG = AB/BZ$, as has often been proved; and therefore, $2{,}911/780 > AB/BZ$. Therefore, subtend side HG to $\angle GAH$. Then these triangles ABZ and AHG will be similar. For angles ABZ and AHG are right angles since each is inscribed in a semicircle, and $\angle ZAB$ is equal to $\angle GAH$ by hypothesis; therefore, the third $\angle AZB$ is equal to the third $\angle AGH$. It is evident, therefore, that the sides containing equal angles are pro-

... This whole passage appears to be corrupt; see the Commentary, lines 181–88.

205 Quare que est proportio *AB* linee ad *BZ* eadem est *AH* ad *HG*. Ergo maior est proportio 2911 ad 780 proportione *AH* linee ad *GH* lineam. Ergo maior est proportio quadrati 2911, quod est 8473921, ad quadratum numeri 780, quod est 608400, proportione quadrati *AH* ad quadratum *HG*. Ergo maior est proportio illorum numerorum con-
210 iunctorum, qui coniuncti sunt 9082321, ad 608400 proportione quadratorum linearum coniunctarum ad quadratum *HG* linee; quare maior proportione quadrati *AG* linee ad quadratum *HG*. Ergo maior est proportio vere radicis 9082321, si haberet, ad 780 proportione *AG* linee ad *HG*. Ergo cum 3013 cum medietate et cum quarta sit maior
11r vera radice / eius, maior est proportio 3013 et medietatis et quarte
216 ⟨ad⟩ 780 proportione *AG* ad *HG*.

Quod maior sit vera radice eius sic habetur: Duc 3013 in quartas, i.e., per quatuor multiplica, et proveniunt hec minuta 12052, que quarte sunt. Iterum hiis quartis adiunge medietatem unius et quartam,
220 que sunt tres quarte, et habebis hec minuta 12055. Multiplica iterum hec minuta ⟨per se⟩ et habebis has sextasdecimas 145323025, quas iterum si dividas per suam denominationem, scilicet 16, habebis hec integra 9082689; nec relinquitur post divisionem de minutis his que una sola sexagesima, scilicet 1. Constat ergo, cum 9082689 sit maior
225 9082321, cum 3013 cum medietate et cum quarta sit radix huius numeri 9082689 cum una sextadecima, quod maior sit vera radice numeri huius 9082321. Insistamus igitur principali proposito.

Maior est proportio 3013 et medietatis et quarte ad 780 proportione linee *AG* ad *HG*. Ergo maior est proportio 5924 et medietatis et
230 quarte, cum sit coniunctus ex 2911 et 3013 et medietate et quarta, ad 780 proportione *AG* et *AH* coniunctarum ad *HG*.

Dividam ergo angulum *GAH* in duo media linea *AT*, et subtendam basim *GT*. Ergo maior est 5924 et medietatis et quarte ad 780 proportione *AH* linee ad *E'H* lineam. Ergo *AHE'* et *AGT* triangulo-
235 rum simultudine minor est proportio *AT* ad *GT* proportione 5924 et medietatis et quarte ad 780. Causa brevitatis sumantur numeri minores se habentes in eadem proportione, scilicet 1823, 240. Quod se habeant in eadem proportione sic probabis: Duc maximum in minimum et unum reliquorum in reliquum. Et si idem numerus pro-
240 venerit, hinc inde erunt proportionales. Cum ergo ita verum sit, minor est proportio *AT* ad *TG* proportione 1823 ad 240. Ergo minor est proportio quadrati *AT* linee ad quadratum *TG* linee proportione

225 huius numeri *corr. ex* huiusmodi   241 240 *mg. D*

portional. Hence $AB/BZ = AH/HG$. Therefore, $2{,}911/780 > AH/GH$. Therefore, $2{,}911^2/780^2 > AH^2/HG^2$, or $8{,}473{,}921/608{,}400 > AH^2/HG^2$. Therefore, by composition, $9{,}082{,}321/608{,}400 > (AH^2 + HG^2)/HG^2$.

Therefore, $9{,}082{,}321/608{,}400 > AG^2/HG^2$. Therefore, $\frac{\sqrt{9{,}082{,}321}}{780} > \frac{AG}{HG}$. Hence, since $3{,}013 + \frac{1}{2} + \frac{1}{4}$ is greater than the true root of $9{,}082{,}321$, $\frac{3{,}013 + \frac{1}{2} + \frac{1}{4}}{780} > \frac{AG}{HG}$.

That $3{,}013 + \frac{1}{2} + \frac{1}{4}$ is greater than that true root is proved as follows. Turn 3,013 into fourths, i.e., multiply it by 4/4, and there results 12,052/4. Add 1/2 and 1/4, i.e., 3/4, to these fourths, making altogether 12,055/4. Multiply 12,055/4 by itself and you will have 145,323,025/16. Reduced to integers by dividing through by the denominator 16, you will have, after division, in integers $9{,}082{,}689 + \frac{1}{16}$. It is clear, therefore, that, since 9,082,689 is greater than 9,082,321, then $3{,}013 + \frac{1}{2} + \frac{1}{4}$, the root of $9{,}082{,}689\frac{1}{16}$, is greater than the true root of 9,082,321. Let us press on to the principal proposal.

Now $\frac{3{,}013 + \frac{1}{2} + \frac{1}{4}}{780} > \frac{AG}{HG}$. Therefore, by composition, $\frac{5{,}924 + \frac{1}{2} + \frac{1}{4}}{780} > \frac{AG + AH}{HG}$, since $5{,}924 + \frac{1}{2} + \frac{1}{4} = 2{,}911 + 3{,}013 + \frac{1}{2} + \frac{1}{4}$.

Hence I bisect $\angle\, GAH$ by line AT and I draw side GT. Therefore, $\frac{5{,}924 + \frac{1}{2} + \frac{1}{4}}{780} > \frac{AH}{E'H}$. Therefore, because of the similarity of triangles AHE' and AGT, $\frac{AT}{GT} < \frac{5{,}924 + \frac{1}{2} + \frac{1}{4}}{780}$. For the sake of brevity, smaller numbers in the same ratio are assumed, namely, the numbers 1,823 and 240. That these numbers are related in the same ratio you will prove as follows: Cross multiply the four numbers and, if the products are the same, the numbers will be proportional. Hence, since such is true, $AT/TG < 1{,}823/240$. Hence, $AT^2/TG^2 < 1{,}823^2/240^2$, or $AT^2/TG^2 <$

quadrati numeri 1823, quod est 3323329, ad quadratum 240, quod est 57600. Ergo minor est proportio quadrati *AG* ad quadratum *GT* linee
245 proportione numeri coniuncti ex 1823, 240 quadratis, qui numerus est 3380929 ad 57600. Ergo minor est *AG* ad *GT* proportione vere radicis 3380929, si haberet, ad 240, cum sit vera radix 57600. At numerus 1838 cum novem undecimis unius est maior vera radice eius, si haberet, quod sic constat:

250 Duc 1838 in undecimas, que undecime cum novem undecimis erunt hee 20227 equivalentes 1838 cum novem undecimis. Has iterum multiplica per se et habebis has centesimas vicesimas primas 409131529. Has iterum divide per suam denominationem et habebis hec integra 3381252, et post divisionem remanebunt hec minuta 37. Cum ergo
255 constat quod iste numerus 3381252 sit maior 3380929, constat etiam quod maior est 1838 cum novem undecimis vera radice 3380929. Sic ergo habemus quod minor est proportio *AG* ad *GT* proportione 1838 cum novem undecimis ad 240. Ergo minor est ⟨proportio⟩ *AG* et *AT* coniunctarum ad *GT* proportione 3661 cum novem undecimis,
260 quia coniunctus est ex 1838 cum novem undecimis et 1823, ad 240.

Dividam ergo angulum *GAT* in duo media linea *AK*. Ergo ut prius minor est proportio *AK* ad *GK* proportione 3661 cum novem undecimis ad 240, et hoc est sicut 1007 ad 66. Ad hoc probandum debes ducere 3661 in undecimas, hoc est multiplicare per 11, et postea minu-
265 tis modo provenientibus adicere novem undecimas, et erunt hee 40280; multiplicantur gratia consorcii 66 per 11 et habebis 726, iterum 1007 et provenient minuta 11077; 240 multiplica per 11 et erunt hec minuta 2640. Cum ergo ex multiplicatione maximi in minimum, scilicet 40280, 726, provenit iste numerus 29243280, et ex multiplicatione
270 11077 in 2640 proveniat idem, erunt predicti numeri proportionales. Minor ergo est proportio *AK* ad *KG* proportione 1007 ad 66. Ergo minor est proportio quadrati *AK* ad quadratum *KG* proportione quadrati 1007, quod est 1014049 ad quadratum 66, quod est 4356. Ergo minor est proportio *AG* quadrati ad quadratum *GK* proportione
275 illorum duorum numerorum coniunctorum, qui coniuncti sunt 1018405 ad 4356. Ergo cum 1009 cum sexta sit vera radix eius vel maior, minor est proportio *AG* ad *KG* proportione 1009 cum sexta ad 66.

246–47 Ergo...57600 *om. Zo*
251 1838 *mg. D*
255 3381252 *corr. D ex* 3380929
257 minor *mg. D*
260 et *mg. D*
266 multiplicantur *corr. ex* multiplicaverunt (?)
271, 273 1077 *corr. ex* 11007

3,323,329/57,600, and so [by composition], $AG^2/TG^2 < 3{,}380{,}929/57{,}600$. Therefore, since 240 equals the true root of 57,600, $\frac{AG}{GT} < \frac{\sqrt{3{,}380{,}929}}{240}$.

But $1{,}838\frac{9}{11}$ is greater than the true root of 3,380,929, which is evident as follows:

Turn 1,838 into elevenths, which, with 9/11 added, will equal 20,227/11, equivalent to $1{,}838\frac{9}{11}$. Again, multiply 20,227/11 by itself and you will have 409,131,529/121. Divide by the denominator and you will have in integers $3{,}381{,}252 + \frac{37}{121}$. Since, therefore, this number 3,381,252 is greater than 3,380,929, it is evident also that $1{,}838\frac{9}{11}$ is greater than the true root of 3,380,929. And so we have it, therefore, that $\frac{AG}{GT} < \frac{1{,}838\frac{9}{11}}{240}$. Therefore $\frac{AG + AT}{GT} < \frac{3{,}661\frac{9}{11}}{240}$, since $3{,}661\frac{9}{11} = 1{,}838\frac{9}{11} + 1{,}823$.

Hence I bisect $\angle$ *GAT* by line *AK*. Therefore, as before, $\frac{AK}{GK} < \frac{3{,}661\frac{9}{11}}{240}$, or $AK/KG < 1{,}007/66$. To prove the identity of $\frac{3{,}661\frac{9}{11}}{240}$ and 1,007/66, turn 3,661 into elevenths, i.e., multiply by 11/11, and, adding 9/11 to the result, there will be 40,280/11. For the sake of obtaining common fractions, multiply 66 by 11/11 and you will have 726/11. Do the same to 1,007 and there will be 11,077/11; and multiply 240 by 11 and you will have 2,640/11. Therefore, since $40{,}280 \times 76 = 29{,}243{,}280 = 11{,}077 \times 2{,}640$, the aforesaid numbers will be proportional. Therefore, $AK/KG < 1{,}007/66$. Hence $AK^2/KG^2 < 1{,}007^2/66^2$, or $AK^2/KG^2 < 1{,}014{,}049/4{,}356$. Therefore, $AG^2/KG^2 < 1{,}018{,}405/4{,}356$. Therefore, since $1{,}009\frac{1}{6}$ is the true root or [actually] is more than the true root of 1,018,405, $\frac{AG}{KG} < \frac{1{,}009\frac{1}{6}}{66}$.

Quod sit maior vera radice sic probatur: Duc 1009 in sextas et hiis que proveniunt appone sextam et erunt hee 6055. Has in se 280 multiplica et proveniunt hec minuta 36663025. Divide iterum hec minuta per suam denominationem et hec habebis integra 1018417 cum 10v hiis minutis remanentibus 13. / Cum istorum integrorum cum istis minutis vera radix sit 1009 cum sexta, et illa integra sunt maiora istis 1018405, maior est vera radice 1018405. Redeamus ergo ad pro- 285 positum.

Minor est proportio *AG* ad *GK* proportione 1009 et sexte ad 66, et, ut dictum est, minor est proportio *AK* ad *KG* 1007 ad 66. Ergo minor est proportio *AG* et *AK* coniunctarum ad *KG* proportione 2016 cum sexta, cum sit coniunctus ex illis duobus numeris, 290 ad 66.

Dividam ergo *GAK* angulum in duo media linea *AL*. Ergo si priorum memineris, minor est proportio *AL* ad *LG* proportione 2016 cum sexta ad 66. Ergo minor est proportio quadrati *AL* linee ad quadratum *LG* linee proportione quadrati 2016 cum sexta ad quadra- 295 tum numeri 66. Ergo minor ⟨est⟩ proportio quadrati *AG* linee ad quadratum *GL* proportione quadratorum illorum numerorum coniunctorum ad quadratum numeri 66. Ergo minor est proportio *AG* ad *LG* proportione vere radicis numeri coniuncti ad 66. Ergo cum 2017 cum quarta sit maior vera radice numeri coniuncti, minor est 300 proportio *AG* ad *GL* proportione 2017 cum quarta ad 66.

Ad probandum quod 2017 cum quarta sit maior vera radice numeri produc in sextas 2016 et sextis illis adiunge unam sextam et erunt hee sexte 12097 equiparentes integris cum sexta. Postea multiplica per se et habebis tricesimas sextas has 146337409. Item quadratum 66, quod 305 est 4356, duc in tricesimas sextas, i.e., multiplica per 36, et proveniunt hec minuta 156816. Adiunge iterum hec minuta prioribus, scilicet 146337409, et provenient hec minuta 146494225, que sunt tricesime sexte, et valent numerum coniunctum ex quadrato numeri 2016 cum sexta et ex quadrato numeri huius 66. Divide postea ultima minuta 310 per suam denominationem et habebis hec integra 4069284, una sola tricesima sexta scilicet remanente post divisionem; que integra cum illo remanente valent eadem minuta, scilicet 146494225. Cum hoc feceris, duc 2017 in quartas et quartis illis postea adiunge quartam et hee erunt quarte 8069. Has statim per se multiplica et provenient ex

284 1018405 *corr. D ex* 1018417

295 *post* linee *add. mg. D* et quadrati GL simul

That $1,009\frac{1}{6}$ is greater than the true root of 1,018,405 is proved as follows: Turn 1,009 into sixths, and, adding 1/6, there will result 6,055/6. Multiply 6,055/6 by itself and the result is 36,663,025/36. Complete the division by the denominator and you will have in integers $1,018,417\frac{13}{36}$. Since the true root of $1,018,417\frac{13}{36}$ is $1,009\frac{1}{6}$ and 1,018,417 is greater than 1,018,405, so $1,009\frac{1}{6}$ is greater than the true root of 1,018,405. Let us return to that which was proposed.

Now, $\frac{AG}{GK} < \frac{1,009\frac{1}{6}}{66}$ and, as was said, $AK/KG < 1,007/66$. Therefore, $\frac{AG + AK}{KG} < \frac{2,016\frac{1}{6}}{66}$, since $2,016\frac{1}{6} = 1,009\frac{1}{6} + 1,007$.

Hence I bisect $\angle\ GAK$ by line AL. Hence, if you keep in mind the prior procedures, $\frac{AL}{LG} < \frac{2,016\frac{1}{6}}{66}$. Therefore $\frac{AL^2}{LG^2} < \frac{(2,016\frac{1}{6})^2}{66^2}$. Therefore, $\frac{AG^2}{GL^2} < \frac{(2,016\frac{1}{6})^2 + 66^2}{66^2}$. Hence $\frac{AG}{LG} < \frac{\sqrt{(2,016\frac{1}{6})^2 + 66^2}}{66}$.
Hence, since $2,017\frac{1}{4}$ is greater than the true root of $[(2,016\frac{1}{6})^2 + 66^2]$, $\frac{AG}{GL} < \frac{2,017\frac{1}{4}}{66}$.

To prove that $2,017\frac{1}{4}$ is greater than the true root [as noted above], turn 2,016 into sixths, and, with the addition of 1/6, there will result 12,097/6, equivalent to $2,016\frac{1}{6}$. Afterwards multiply 12,097/6 by itself and you will have 146,337,409/36. Also turn 66^2, i.e., 4,356, into thirty-sixths, i.e., multiply by 36/36, and 156,816/36 results. Add this latter fraction to the former, evidently, 146,337,409/36, and the result is 146,494,225/36. This is equivalent to the sum: $(2,016\frac{1}{6})^2 + 66^2$. Complete the division of the last fraction by dividing through by the denominator and you will have in integers $4,069,284 + \frac{1}{36}$, equivalent to 146,494,225/36. When this is completed, turn 2,017 into fourths, and with the addition of 1/4 you will have 8,069/4. Immediately multiply 8,069/4

315 illa multiplicatione hee sextedecime 65108761, que iterum divide per suam denominationem. Hec procreabunt integra 4069297; residua erunt hec minuta 9 post divisionem. Constat ergo, cum hec integra sint maiora hiis 4069284, quod 2017 cum quarta maior vera radice numeri coniuncti ex quadratis duorum numerorum predictorum. Igi-
320 tur redeamus ad propositum principale.

Minor est proportio *AG* ad *GL* proportione 2017 cum quarta ad 66. Ergo maior est proportio *GL* ad *AG* proportione 66 ad 2017 cum quarta. At *GL* est latus poligonii circumscripti circulo et nonaginta sex angulorum, quia, cum angulus *GAL* sit 48[a] pars anguli recti et
325 angulus *GEL* sit duplus ad ipsum, quia cum sit extrinsecus est equalis istis duobus equalibus *EAL*, *ALE*, est ergo *GEL* 24[a] pars anguli recti. At constat, [quod] si diviseris quemlibet quatuor istorum angulorum rectorum in 24 partes, habebis nonaginta sex angulos et totidem latera poligonii respicientia illos angulos. Postea multiplica 66 per 96
330 et procedit iste numerus 6336, in quo totiens est 66 quotiens *GL* in ambitu poligonii. Cum ergo maior sit proportio *GL* ad *AG* proportione 66 ad 2017 cum quarta et eadem sit *GL* ad ambitum poligonii et 66 ad 6336, ergo maior est proportio ambitus poligonii ad *AG* dyametrum proportione 6336 ad 2017 cum quarta. At 6336 continet
335 ter illum et plus 10 partibus 71 partium 2017 cum quarta.

Ad hoc probandum duc 6336 in quartas, que sunt hee 25344. Postea duc 2017 in quartas et illis appone quartam, que sunt hee 8069. Divide 25344 per 8069 et videbis quod ter continet illum et remanebunt hee quarte post divisionem 1137. Postea duc 8069 in septuagesimas primas,
340 i.e., multiplica per 71, et erunt hee 572899. Constat ergo numerus 8069 est septuagesima prima pars huius numeri 572899, quia multiplicatur per 71 ⟨et⟩ procreat illum. Multiplica iterum 8069 per 10 et habebis numerum qui est 10 septuagesime prime huius 572899, qui est 80690. Divide postea 80690 per 71 et reduces in has quartas 1136,
345 remanentibus hiis minutis 34 post divisionem. Cum ergo hee quarte 1137 sint plures hiis 1136 cum hiis minutis remanentibus, scilicet 34, in quantum deest hiis minutis ad constituendam quartam, constat ergo 6336 continet ter 2017 cum quarta et plus 10, 71 partibus partium eius. Quare etiam multo magis linea ambiens poligonium addit super
350 triplum dyametri plus 10, 71 partibus partium dyametri. Ergo et multo magis linea continens circulum est maior triplo dyametri excessu qui

324 quia *corr. ex* qui / 48[a] *corr. ex* 48
326 24[a] *corr. ex* 24
338 quod *corr. ex* quia
349 Quare *corr. ex* quia

by itself, and the result of this multiplication is 65,108,761/16. You again divide through by the denominator and there results in integers $4{,}069{,}297 + \frac{9}{16}$. It is evident, hence, that since 4,069,297 is greater than 4,069,284, so $2017\frac{1}{4}$ is greater than the true root of $[(2{,}016\frac{1}{6})^2 + 66^2]$.

Hence let us return to the main proposal. Now, $\frac{AG}{GL} < \frac{2{,}017\frac{1}{4}}{66}$. Hence $\frac{GL}{AG} > \frac{66}{2{,}017\frac{1}{4}}$. But GL is the side of a [regular] polygon circumscribed by the circle and having 96 angles. For since GAL is 1/48 of a right angle and $\angle\ GEL = 2\ GAL$ (because GEL being extrinsic is equal to the two equal angles EAL and ALE), therefore GEL is 1/24 of a right angle. But it is evident that if you divide each of the four right angles [at the center] into 24 parts, you will have 96 angles and as many sides of the polygon opposite those angles. Then $66 \times 96 = 6{,}336$; 66 is contained in 6,336 as many times as GL is in the perimeter of the polygon. Since, therefore, $\frac{GL}{AG} > \frac{66}{2{,}017\frac{1}{4}}$ and $\frac{GL}{\text{perimeter of polygon}} = \frac{66}{6{,}336}$, then $\frac{\text{perimeter of polygon}}{AG \text{ diameter}} > \frac{6{,}336}{2{,}017\frac{1}{4}}$. But 6,336 contains $2{,}017\frac{1}{4}$ three times and in addition more than 10/17 of $2{,}017\frac{1}{4}$.

To prove this, turn 6,336 into fourths and there results 25,344/4. Then turn 2,017 into fourths, and with the addition of 1/4 you will have 8,069/4. Divide 25,344 by 8,069 and you will see that the former contains the latter three times plus 1,137/8,069. Afterwards turn 8,069 into seventy-first parts, i.e., multiply it by 71/71 and there results 572,899/71. It is clear, therefore, that 8,069 is 1/71 of 572,899, for the latter is produced by multiplying 8,069 by 71. Multiply 8,069 by 10 and the result is 80,690, which is 10/71 of 572,899. Then divide 80,690 by 71 and it becomes 1,136 + 34/71. Since 1,137 is greater than $1{,}136\frac{34}{71}$ by the amount that the fraction 34/71 is less than 1, it is evident that 6,336 contains $2{,}017\frac{1}{4}$ three times and in addition more than 10/71 of it. Therefore, a fortiori the perimeter of the polygon exceeds triple the diameter by more than 10/71 of the diameter. Therefore, even more does the circumference exceed triple the diameter by an excess which is greater than 10/71 of the

est maior 10, 71 partibus partium dyametri. Constat ergo quod linea continens circulum addit super triplum dyametri minus septima et plus 10, 71 partibus partium dyametri. Et hoc est quod voluimus
355 probare.

diameter. It is evident, therefore, that the circumference exceeds triple the diameter by an amount less than 1/7 and more than 10/71 of the diameter. And this is what we wished to prove.

COMMENTARY

F.IA

2 "orthogonio triangulo." This is the order of words found in Tradition II of the Gerard of Cremona translation.

7–37 In connection with the first half of the proof the scribe of *D* lists in the margin (12r) [cf. *Zo*, 156r] the following citations of Euclid's *Elements* in the Adelard II–Campanus Version:

2 (!1?) —
- per 6.4ⁱ.e (per sextam quarti Euclidis, i.e. IV.6)
- per 41.pⁱ.e (I.41)
- per 1.10ⁱ.e (X.1)
- per 18. pⁱ.e (I.18)
- per 41. pⁱ.e (I.41)
- per 1.secundi.e (II.1)

15–16 "Quare...circuli." I give my reconstruction of the meaning of this sentence in the translation. I am not at all happy with the reconstruction, since it involved substituting *minores* in line 16 for what may be *maiores* in the manuscript. I am, however, at a loss to explain its meaning in any other way.

24 "alkaydem." This is a transliteration of the Arabic word *al-qā'idat*, meaning "base." We find it used extensively in the so-called Adelard I translation of the *Elements* of Euclid; see M. Clagett, "The Medieval Latin Translations from the Arabic of the *Elements* of Euclid," *Isis*, vol. *44* (1953), p. 20.

28 "primam decimi Euclidis." See page 60, note 1, above.

31–34 "Ducta...trianguli." The Latin of this passage is very awkward, but the meaning as reflected in my translation is clear.

38–69 In connection with the second half of the proof the scribe of *D* adds in the margin (12r) as a unit, without specifying to which passages they refer, the following citations to the *Elements* of Euclid in the Adelard II–Campanus version [cf. *Zo*, 156v]:

- per 7.4ⁱ.e (IV.7)
- per 17.3ⁱ.e (III.17 = Gr. III.18)
- per 5.pⁱ.e et dif. Cⁱ (I.5 et diffinitio circuli)
- per communem scientiam
- per 6.pⁱ.e (I.6) [Should this be VI.1?]
- per 8.pⁱ.e (I.8)
- per 25.3ⁱ.e (III.25 = Gr. III.26)
- per 4.pⁱ.e (I.4)

2. —
- per 26.pⁱ.e (I.26)
- per cor. 15.3ⁱ.e (corollarium III.15 = Gr. porism III.6)
- per 18.pⁱ.e (I.18)
- per communem scientiam
- per 1. sexti.e (VI.1)
- per 1.decimi.e (X.1)
- per 41.pⁱ.e (I.41)
- per 1. secundi.e (II.1)

This list of citations does not seem to fit the steps taken in lines 38–69 exactly. This makes it difficult to know to what axioms the two references *per communem scientiam* correspond. But I think it plausible that both refer to the first of the additional axioms added in the Adelard II Version (and as a matter of fact in Adelard I as well), which was taken up by Campanus and in the latter's version reads: "Si duae quantitates aequales ad quamlibet tertiam eiusdem generis comparentur, simul erunt ambae illa tertia, aut aeque maiores, aut aeque minores, aut simul aequales" (ed. of Basel, 1546, p. 3). That this axiom is the one cited seems to be clearer in the first instance. There it perhaps refers to the step of lines 51–52: with $FT = TB$, if $CT > FT$, then $CT > TB$. Incidentally, I have also given in parentheses the citations to the Greek text of the *Elements* where it differs.

51 "ad medias lineas." This expression is unclear. It is either redundant with "in suas partes medietates" or it means that any of the sides of the octagon is bisected by the radius intersecting it.

71 The scribe of *D* has added one marginal citation for the corollary: "corollarium: per primam sexti Euclidis" (cf. *Zo*, 157r).

Proposition II

74–77 "Proportio... GD^2." Note the similarity to Gerard, lines 69–73.

74–85 For Proposition II, the scribe of *D* (12r, top right) adds the following marginal citations (cf. *Zo*, 157r).

per 1.sexti.e (VI.1)
per premissam
per 41.p^{i}.e (I.41)

The citation *per premissam* probably refers to the assumption that *ZE* is one seventh of *GD* (line 77).

77 Notice that the commentator has omitted from the Gerard translation the statement that triangle *AGE* is to triangle *AGD* as 3 is to 1.

80–81 "per priorem propositionem." Our commentator substitutes this for the longer clause of the Gerard translation: "quoniam...circuli" (lines 78–80).

82–85 "Attende...propositionem" is a comment reflecting the phrase "quoniam est plus triplo diametri ipsius et septima diametri fere" in the Gerard translation (lines 80–81).

F.IB

1–4 "Circulum...circulo." In the introduction to the texts of the Florence versions I have commented on the fact that F.IB abandons the wording of the Gerard translation in favor of a paraphrase. Note that the author of F.IB fails to say that the sides of the triangle which are equal to the circumference and radius are those sides including the right angle.

9 "sextam quarti." In the Adelard II translation (MS British Museum Add. 34018, 16r) this proposition runs: "Intra datum circulum quadratum describere." "prima decimi." See page 60, note 1, above.

10 "relinquendo." This term with this author became the technical term for the process of taking a series of divisions in which in each step more than half of the remainder is cut away. To maintain the spirit of the Latin I have translated it by "remaindering."

14, 21, 29 "41^{am} primi." See the Commentary to the Cambridge Version, line 33.

26 "secundam secundi." See the Adelard II translation of the *Elements*, which reads (*ms. cit.*, 8r): "Si fuerit linea in partes divisa, illud quod fit ex ductu totius linee in se ipsam equum erit hiis que fiunt ex ductu eiusdem in omnes partes."

33 "septimam quarti." See Adelard II translation of the *Elements*, which reads (*ms. cit.*, 16r): "Circa propositum circulum quadratum designare."

39–40 "primam decimi." See page 60, note 1, of this chapter.

45 "primam sexti." See the Commentary to the Cambridge Version, line 70.

46–47 "quartum librum." This is probably a wrong reference. That two tangents drawn to a circle from a common point are equal is proved in the course of the penultimate proposition of Book III. In the Campanus version (III.35, Venice, 1482) we read: "Nota etiam quod si a quolibet puncto extra circulum signato due linee contingentes ad circulum ipsum ducantur, ipse erunt adinvicem equales."

48 "penultimam primi." See the Commentary to the Cambridge Version, line 25.

56 "ultimam secundi." See the Commentary to the Cambridge Version, line 82.

Proposition III

4–167 The scribe of *D* gives the following marginal citations (10r) for the first half of the proposition (cf. *Zo*, 152v); the citations, it is clear, are from the Adelard II–Campanus version of the *Elements* of Euclid:

Prima pars —
- per illud quod probatum est in 16 primi euclidis (I.16)
- per correlarium 15.p^{i}.e (Cor. I.15)
- per 17.3^{i}.e (III.17 = Gr. III.18)
- per 5. quarti.e (IV.5)
- per 32. primi.e (I.32)
- per ultimam sexti.e (VI.33)
- per conversam 30.3^{i}.e (III.30 = Gr. III.31)
- per diffinitionem diametri
- per correlarium 15 quarti (Cor. IV.15)
- per 9.p^{i}.e (I.9) — quotiens oportet
- per 3.6^{i}.e (VI.3) — quotiens oportet
- per 16.5^{i}.e (V.16) — quotiens oportet
- per 13.5^{i}.e (V.13 = Gr. V.12) — quotiens oportet
- per penultimam p^{i}.e (I.46 = Gr. I.47) — quotiens oportet
- per primam partem 8.5^{i}.e (V.8) — quotiens oportet
- per interpositam — quotiens oportet
- per 18.5^{i}.e — quotiens oportet
- per 3 interpositam 5^{i}.e (Campan. comm. V. def. 3) — quotiens oportet
- per 15. quinti e (V.15)
- per 1. interpositam 5^{i}.e (Campan. comm. V. def. 1)
- per conversam diffinitionis minoris inproportionalitatis

The principal difficulty in this list lies with the proper identification of the three items which have *interpositam* in them. I have taken them to be references to the comments of Campanus to the definitions of Book V, but I am not sure I see the pertinence of those comments to the specific steps in the Archimedean proof, and so perhaps some other explanation should be sought. Incidentally, in connection with the practice of citing the "converse" of definitions, Mr. Murdoch writes by letter: "The annotator reflects the growing concern (from the thirteenth century) with the logical structure of Euclidian definitions. They take, as did some Arabs before them, Euclid to have set down 'if...then' assertions in definitions, etc., rather than 'if and only if' assertions (as he certainly intended). Hence the necessity of filling in the imagined logical gaps by reference to converses of definitions and first principles." The *quotiens oportet* simply means that these propositions are to be used as many times as is necessary in the various divisions of the angle.

4 "⟨dyameter circuli AG⟩." I have added this phrase not only because at this point the commentator is quoting the Gerard of Cremona translation where the phrase is found, but also because of its presence in line 168. I have accordingly rejected the marginal addition of "continens circulum."

6–12 "Ergo....*ZE*." This is an elaboration of the marginal comment found in manuscripts *PL* of the Gerard translation (see variant readings of Gerard translation, line 89). It is quite different from the comment of Eutocius (ed. of J. L. Heiberg, *Archimedis opera omnia*, vol. *3* (Leipzig, 1915), from p. 232, line 20, to p. 234, line 6).

15 "tertium...geometrie." In the Adelard II translation the appropriate first part of VI.3 runs (*ms. cit.*, 22r): "Si ab aliquo angulorum trianguli linea adducta [ad basim] angulum illum per equalia dividat, in (*! omit*?) duas partes ipsius basis reliquis eiusdem trianguli lateribus proportionales esse...necesse comprobatur." This is the only specific citation to a theorem of the *Elements* in Proposition III and similarly Eutocius only cites this one proposition (*ed. cit.*, 234, lines 18–20).

21–36 "Quod....153." Compare this commentary with that of Eutocius (*ed. cit.*, from p. 234, line 17, to p. 236, line 7).

31 "careat vera radice." Throughout the commentary this author speaks of the "true root" (*vera radix*) of numbers that are lacking exact roots. He often uses the expression "si veram haberet" (e.g., line 35) in connection with such roots. The Greek commentary of Eutocius in

these similar places says that the squares of these approximate roots are less than the "exact" squares (Eutocius, *ed. cit.*, p. 234, lines 14–15; p. 236, lines 15–16; p. 238, lines 16–17).

37–65 "Ergo....propositum." Cf. Eutocius, *ed. cit.*, p. 236, lines 8–19. The technique of reducing mixed numbers to fractions of the same denominator before beginning operations which is used by the Latin commentator is, as I have said, different from the procedure of Eutocius in which the mixed numbers themselves are manipulated with.

51 "denominationis." This word was used in the algorist tradition to distinguish fractions by their numerical denominators, as in "thirds," "fifths," etc. See Clagett, *Giovanni Marliani and Late Medieval Physics* (New York, 1941), p. 155. I point this out only because there is a connected but more subtle use of the term in medieval tracts on proportion where the term expresses the essential aspect of a ratio even where the ratio is between incommensurable quantities and the ratio is said to be "denominated" only "mediately" by numbers rather than "immediately." See E. Grant, "Nicole Oresme and his *De proportionibus proportionum*." *Isis*, vol. *51* (1960), p. 301.

66–75 "Maior....153." Cf. Eutocius, *ed. cit.*, p. 236, line 20, to p. 238, line 9.

75–101 "Ergo....propositum." Cf. Eutocius, *ed. cit.*, p. 238, lines 10–18.

102–110 "Maior...153." Cf. Eutocius, *ed. cit.*, p. 240, lines 1–11.

110–29 "Ergo....153." Cf. Eutocius, *ed. cit.*, p. 240, lines 12–18.

129–36 "et...153." Cf. Eutocius, *ed. cit.*, p. 242, lines 1–14.

136–67 "Et....theorematis." Cf. Eutocius, *ed. cit.*, from p. 242, line 15, to p. 244, line 29.

147–67 "At....theorematis." A fragment at the end of codex *D* (57v) includes part of a different version of Proposition III: "Et illud, scilicet 14688, est plus triplo 4673 et medietatis in 667 et medietate et istius proportio ad 4673 et medietatem est minor septima. Oportet ergo ut sit linea figure poligonie continentis circulum minor triplo dyametri et eius septima, et hoc ideo quia dyameter *AG* est maior 4673 et medietate. Cum ergo linea figure poligonie non contineat 4673 et medietatem ⟨ter⟩ et insuper eius septimam, multo minus continebit dyametrum ter et insuper eius septimam. Sed linea figure poligonie est maior circumferentia circuli. Ergo nec circumferentia circuli dyametrum ter et eius septimam. Et linea figure poligonie est plus diminutione linee continentis circulum a triplo dyametri eius et septima, i.e., cum linea continens circulum diminuat a triplo dyametri eius et septima; linea figure poligonie minus diminuit quia ipsa est maior circumferentia circuli. Hoc est necessarium in Archimenide."

168–352 The scribe of *D* adds in the margin (10v) for the second part of the proof the following citations from Euclid (cf. *Zo*, 153r):

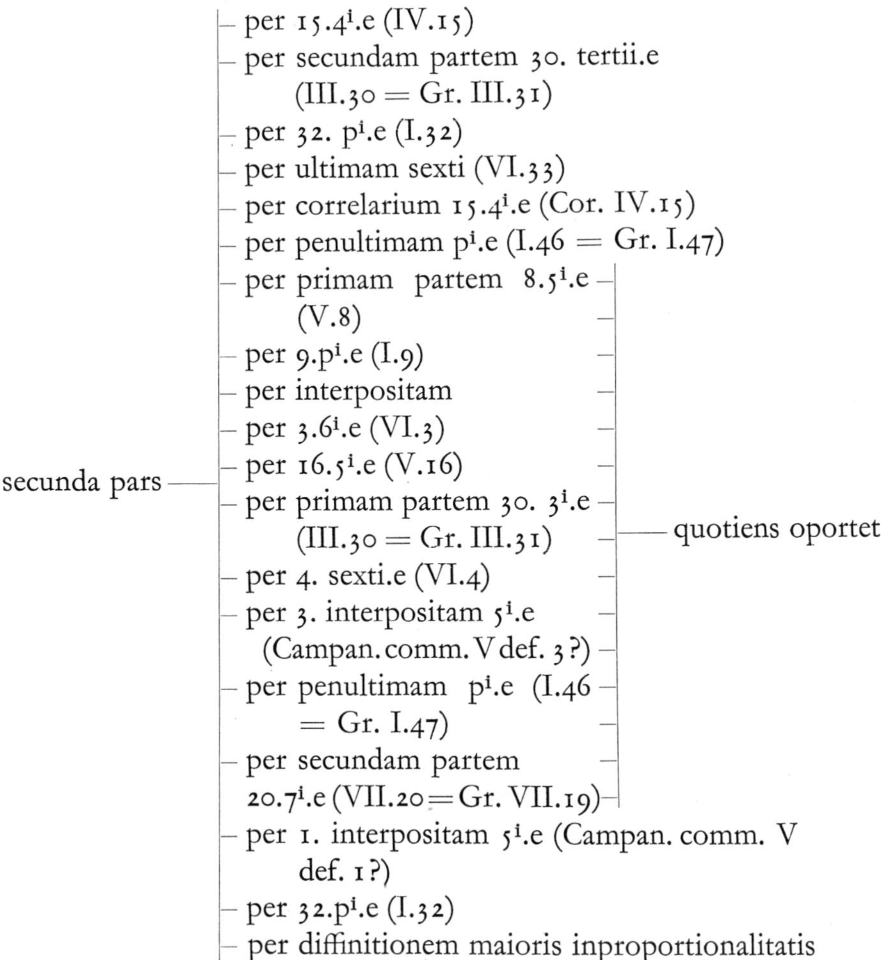

See the Commentary, lines 4–167 above.

168–91 "Sit.... *BG*." Cf. Eutocius, *ed. cit.*, from p. 244, line 30, to p. 246, line 19.

181–88 "Item.... 1351." There appears to be something wrong with the text here. As my translation indicates, I believe that the commentator is merely trying to show that the next largest perfect square beyond 1825200, namely, 1825201, yields a much better approximate root (1351) than does the next smaller square. But the Latin text beginning with *quia* in line 184 and ending with "2700" in line 185 by no means says this clearly. Furthermore, I am not confident of my reading of *vel* as it appears twice in line 184. Also puzzling is the sentence "Dupla... 1351" (lines 186–188). The "1351" in line 186 has been added from the margin, but as the sentence stands it makes little sense. I attempted to connect it with the known methods of extracting square roots in the Middle Ages but without success.

192–227 "Dividam....proposito." Cf. Eutocius, *ed. cit.*, from p. 248, line 1, to p. 250, line 4.

228–58 "Maior....240." Cf. Eutocius, *ed. cit.*, from p. 250, line 5, to p. 252, line 10.

258–84 "Ergo....propositum." Cf. Eutocius, *ed. cit.*, p. 254, lines 1–18.

286–320 "Minor....principale." Cf. Eutocius, *ed. cit.*, p. 256, lines 1–14.

321–52 "Minor....dyametri." Cf. Eutocius, *ed. cit.*, p. 258, lines 1–14.

4. The Version of Gordanus

At some unspecified time during or not long after the late thirteenth century still another effort was made to expand Archimedes' proof of Proposition I of the *De mensura circuli*. This appeared in an omnibus astronomical and geometrical work entitled, in a fifteenth-century manuscript of the Vatican (Pal. lat. 1389, MS *X*), *Compilacio quorundam canonum in practicis astronomie et geometrie*. The author, whose name is merely given as Gordanus—and this in a hand later than that in which the manuscript is written—is unknown to me. Needless to say, I had considered the possibility of Gordanus being the celebrated mathematician Jordanus de Nemore, but the calendaric tables that precede the work and yet appear to be part of the work seem to preclude this identification. The first table (9v), which is for finding the Arabic year when the Christian year is given, goes from 1232 to 1440; a second table (10r) giving the number of days in the Christian years and a third table (11r) converting Arabic to Christian years both go from 1260 to 1468. Furthermore, on folio 12r there is another conversion table which runs from 1272 to 1292. At the bottom of this page we read: "potest tabula extendi ad placitum," which I take to mean that the table is good for the next few years after the time of its composition but that it can be easily expanded if one wishes.

Another note on this same folio tells us "annus Arabum 689 incipit feria sexta." The present tense *incipit* may have some significance for the time of composition of the table. According to the table, the Islamic year 689 overlaps the Christian years 1290–1291.[1] Following these tables the *Compilacio* proper begins and is in the same hand. Of course, the *Compilacio* could be earlier than the preceding tables, for it could have been copied from an earlier manuscript and added to the tables, or it could have been very much later.[2] Fortunately we can place a later limit of 1390 on the tract, for the part on quadrature was copied intact as a part of a geometrical treatise appearing in a Viennese codex almost certainly written and composed by the Franciscan Wigandus Durnheimer (MS *Fa* in the Sigla below). This geometry is the first work in the Durnheimer codex, which also includes an arithmetic and an optics. Two items of the codex bear the name of Wigandus and the date 1390; and the very last item was completed at Paris in that year.[3]

The section of the *Compilacio* that concerns us here occurs within the eighth part of the work devoted to "In mensura figurarum." It is Chapter 23 and bears the title: "Circulum demonstrative quadrare." The introduction to the chapter is of some interest. It tells us that the purpose of the chapter is "to treat of and explain further the proof of Archimedes." This is in accordance with an earlier promise in Chapter 18 ("Circuli aream concludere").[4]

[1] On folio 10r at the top appears the date 1324 twice, but I am not sure in what connection this date is given. Interestingly, following the preface to the *Compilacio*, which begins on folio 13r, there is a table *ad latitudinem 48 graduum*, which is about the latitude of Munich or Vienna.

[2] At the close of the eighth part of the *Compilacio* (117r), which contains the chapter of interest to us, appears in what I judge to be the same bold hand as the rest of the part: "Et hec de mensura figurarum sufficiant 1461." I assume that this is the date of copying. On folio 157r there is scribbled in another hand the date 1482.

[3] There is little doubt that the geometrical treatise that occupies folios 1r–89v of Vienna, Nat.-bibl. 5257, was composed (and probably written) by Wigandus Durnheimer. In style, form, and terminology it is completely like the remaining parts of the codex, which bear the name of Wigandus Durnheimer. A second treatise on arithmetical subjects has this colophon (f. 118v): "Perfectus est iste libellus et compilatus a fratre wigando durnheimer ordinis minorum anno domini 1390, die 22 mai, indictione 12ª, luna 11ª. Deo gratias. Explicit." The last treatise ends as follows: "Perfectus est iste liber a fratre wigando durnheimer ordinis minorum Parisius anno domini 1390...."

[4] On folio 107r we read: "Probatio huius talis est: concesso quod linea curva sit equalis recte fiat circulus super centrum *C*, protrahatur semidyameter *CG*, item ex puncto *F* per 16 tertii Euclidis ducatur linea ad circulum contingens ipsum in puncto. Igitur hec linea sit equalis circumferentie. Deinde ducta *EF* claudatur triangulus ortogonius, quod secundum

In fulfilling his purpose, Gordanus gives a proof that is even more elaborate than those of the emended versions we have discussed so far. As in the case of the preceding versions, the author leans heavily on Euclid's *Elements*, citing it some thirty times, although many of the citations are to the same propositions. Thus every step is proved by a proper citation to the *Elements*. A case in point occurs when the author proves in a detailed and obvious manner that the four triangles formed by the corner angles of the circumscribed square and the appropriate sides of the circumscribed octagon are more than half of the space between the circumference of the circle and the sides of the square (see lines 156–77). The author's attempt to reduce a geometrical argument to numbers (lines 68–72) is another example of his tendency to elaborate the obvious.

There are a good many points of similarity between the Version of Gordanus and the other versions already discussed. It is for this reason that we suspect that Gordanus, if he did not consult directly the translation of Gerard of Cremona, had as his model one or more of the various emended versions of the Gerard translation. Like these other versions, Gordanus specifically makes X.1 of the *Elements* the central point of his proof. He too makes a literal specification of the quantity by which the circle is said to exceed or to be less than the triangle. He assumes this quantity to be triangular in shape, as is the case of the drawing accompanying the Corpus Christi Version, treated below. The reader will recall that in the Cambridge Version it was stated that the form of this excess quantity is of no concern (see lines 17–18). In the Naples Version the form is not specified, although it is represented in the drawing as a small curled figure (see Fig. 11). Similarly, in Florence Version F.IA its shape is unspecified in the text but is drawn as a rectangle (see Fig. 13), while in F.IB it is mentioned as a square (line 16). As in the case of all the emended versions, Gordanus stops his exhaustion process with the inscription and circumscription of regular octagons. He states that he is doing this "for the sake of brevity, lest the oppression (*pressura*) of many lines impede the demonstration" (see lines 64–66). Like F.IA of Florence, our author explicitly reminds us of the logical structure of the argument by reduction to absurdity, as, for example, when he says toward the end of the first half of the proof

Archimenidem equalis est circulo; sed huius trianguli area ex precedentibus est nota, ergo area circuli est nota. Quod autem iste triangulus sit equalis isti circulo inferius demonstrative probatur ubi circuli quadratura docetur. Est autem propositio quam Archimenides probat: talis omnis trigonus ortogonius contentus semidyametro circuli et linea recta equali periferie illius circuli, que due linee faciunt illum rectum angulum illius trianguli ortogonii equalis est circulo...."

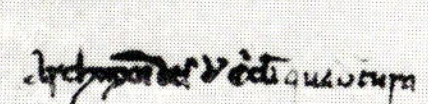

MS *A* = Naples, Biblioteca Nazionale, VIII. C.22, 65v

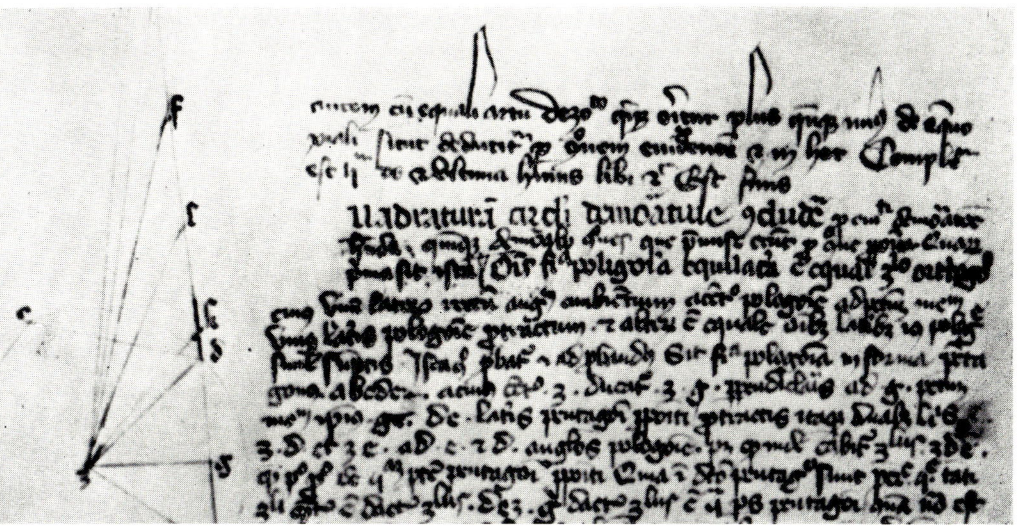

MS *Aa* = Vatican Library, Vat. lat. 3102, 111v

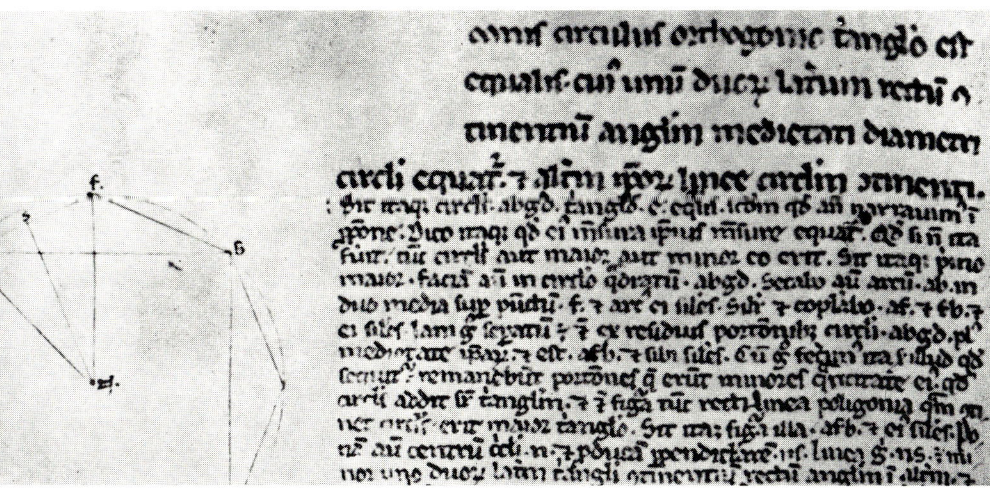

MS *B* = Oxford, Bodleian Library, Auct. F.5. 28, 101v

MS *Bc* = Oxford, Corpus Christi College 234, 170r

MS *Bd* = Oxford, Bodleian Library, Digby 147, 89r

MS *C* = Oxford, Bodleian Library, Digby 174, 137r

MS *Ca* = Oxford, Corpus Christi College 251, 83v

MS *D* = Florence, Biblioteca Nazionale, Conv. Soppr. J.V.30, 10r

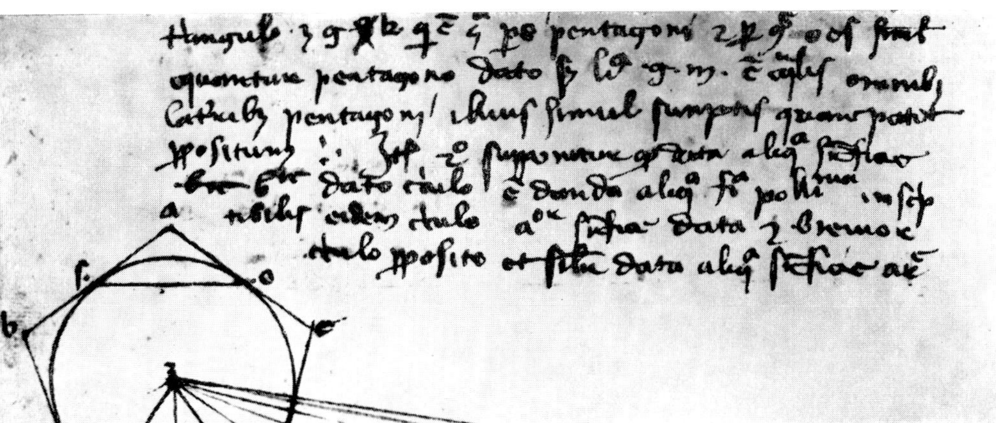

MS *Ea* = Florence, Biblioteca Nazionale, Conv. Soppr. J.IX.26, 49v

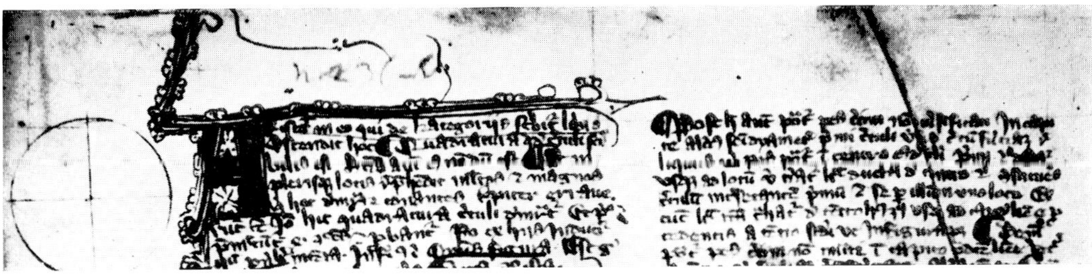

MS *Ec* = Erfurt, Stadtbibliothek, Amplon. Q.361, 79v

MS *F* = Vienna, Nationalbibliothek, cod. 5303, 19r

MS *H* = Basel, Öffentliche Bibliothek der Universität, F.II.33, 116v

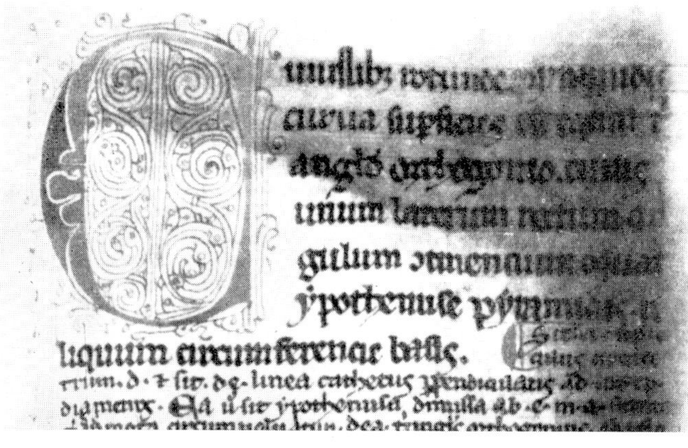

MS *I* = Dresden, Sächs. Landesbibliothek, Db.86, 188r

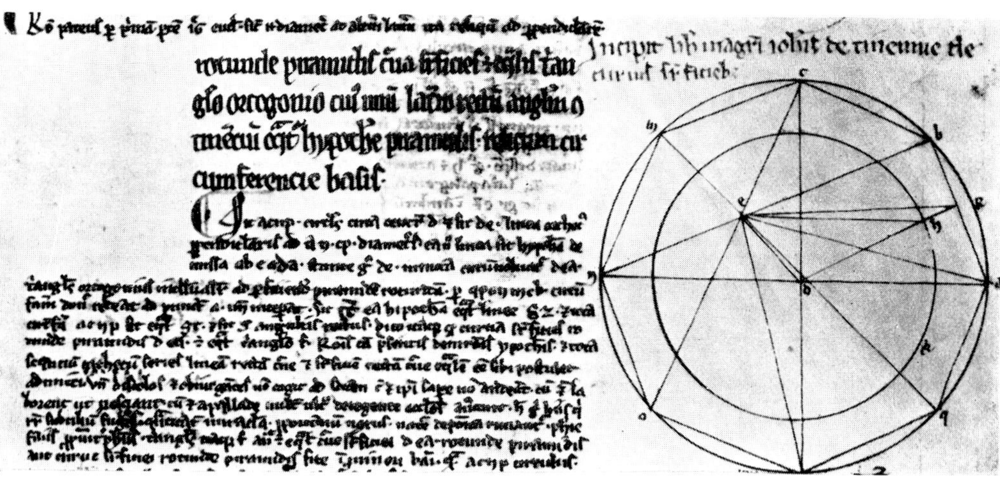

MS *J* = Berlin, Deutsche Staatsbibl. (now at Marburg, Westd. Bibliothek), Q.150, 90r

MS *K* = Paris, Bibliothèque Nationale, Fonds lat. 11246, 37v

MS *L* = Oxford, Bodleian Library, Arch. Seld. B.13, 2r

MS *M* = Florence, Biblioteca Nazionale, Conv. Soppr. J.V.18, 95v

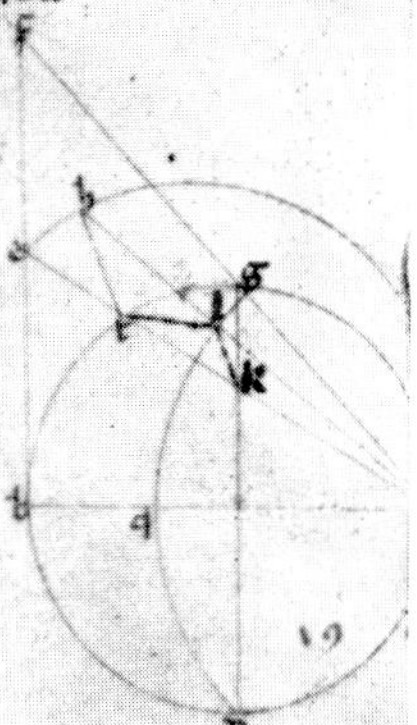

MS *Ma* = Paris, Bibliothèque Mazarine, 3637 (1256), 11v

MS *N* = Cambridge, Gonville and Caius College 504/271, 108v

MS *Oa* = Paris, Bibliothèque Nationale, Fonds lat. 7434, 85r

MS *P* = Paris, Bibliothèque Nationale, Fonds lat. 9335, 55v

MS *Q* = Paris, Bibliothèque Nationale, Fonds lat. 7378A, 18r

MS *S* = Oxford, Bodleian Library, Digby 168, 121r

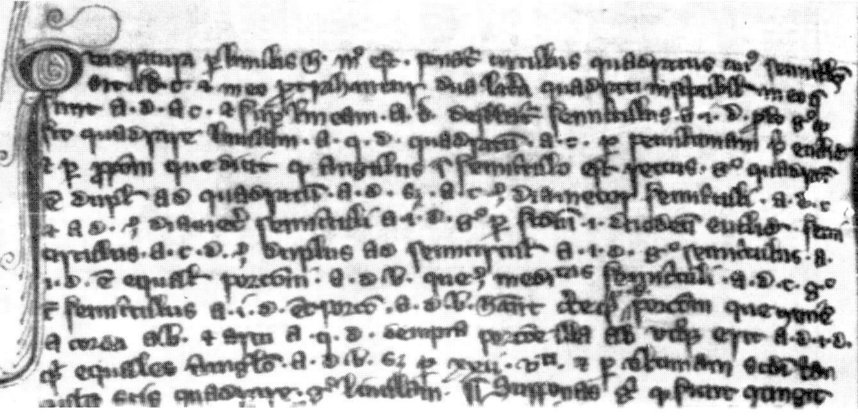

MS *V* = London, British Museum, Royal 12E. 25, 150v

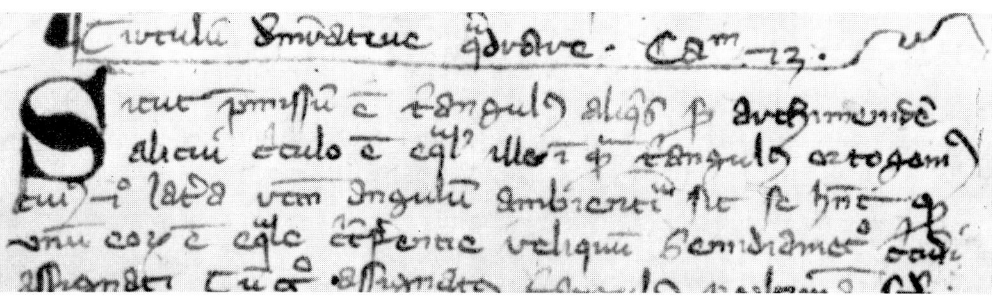

MS *X* = Vatican Library, Pal. lat. 1389, 108r

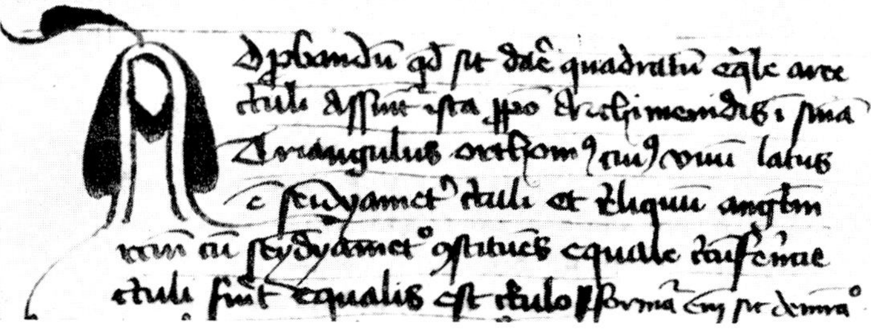

MS *Y* = Munich, Bayrische Staatsbibliothek, cod. 56, 182r

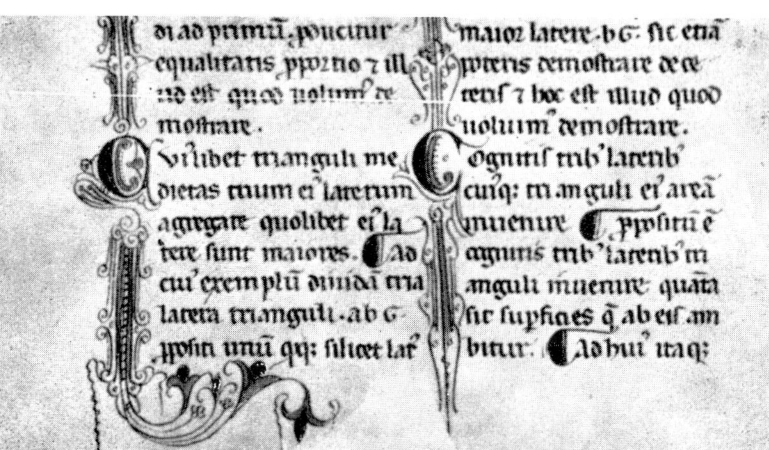

MS *Ya* = Munich, Bayrische Staatsbibliothek, cod. 234, 105v

MS *Z* = Bern, Bürgerbibliothek, A.50, 169r

(lines 103–105): "This moreover is false and impossible. Therefore also [false and impossible is] that from which it follows, namely, that the octagon is greater than the product of the radius and half the circumference." Immediately after this statement he adds another explicit comment on his method of reasoning (lines 106–110): "By the same line of reasoning it would follow that, if *OT* were multiplied by the eight sides of the octagon, the product would be greater than the product of the radius and the whole circumference, which constitutes the same contradiction. For it is implied that any chord is greater than its arc, which again is impossible." Incidentally, at the end of this statement, when he remarks on the impossibility of the chord being greater than its arc, he puts in negative form the postulate given in the Cambridge and Corpus Christi versions. A further point of similarity with the Corpus Christi Version is the author's use of the word *portiuncula*. This word replaces the more common *lunula* both for the small segments formed by the circumference of the circle and the perimeter of the inscribed octagon and for the mixed triangular figures formed by the circumference and the perimeter of the circumscribed octagon. Actually, the author of the Corpus Christi Version uses *portiuncula* only for the segments of the circle and not for the mixed triangular figures. The Version of Gordanus also resembles a number of the other versions by noting that the right triangle found to be equal to the circle must be converted to a square by using Proposition II.14 of the *Elements* of Euclid.

One interesting feature can be remarked concerning the author's use of letters to designate geometrical magnitudes. If the author is talking about two quantities that have a common point, such as the two sides *HA* and *AB* of the octagon, he writes in the Greek manner *HAB*; similarly for triangles *HOA* and *OAB*, which have a common side *AO*, he writes *HOAB*. In the case of each of the many instances of such usage I have altered the manuscript readings to the modern form, although I have, of course, given the original readings in the variant readings. This practice of syncopation has not been followed in any other version. Incidentally, the letters used by Gordanus are quite different from those found in the other versions.

It should be clear from my discussion that Gordanus concerns himself with Proposition I only. He does assume a value of π equivalent to $3\frac{1}{7}$ but without discussing Proposition III.[5]

[5] The value of $3\frac{1}{7}$ for π is implied in Chapters 20 and 24, as follows: (107v) "Circuli maiuraturam (?) cognoscere. Ca^m 20. Dyameter ducatur in se; productum multiplicetur per 22. Et summa excrescens dividatur per 7. Quantitas

Little need be said about my text of the Gordanus Version. It depends almost exclusively on MS *X*, although some variant readings have been added from Durnheimer's copy (MS *Fa*). Actually, Durnheimer has added a number of citations to other chapters of his own treatise. These I have not attempted to include in the variant readings. Nor have I noted the numerous word transpositions of *Fa*, the occasional orthographic variations, the frequent substitution of *isto* for *illo*, *igitur* for *ergo*, and so on. *Fa* reads the letter designating the excess of the triangle over the circle (and vice versa) as *n*, while *X* has what I read to be *H*, but which perhaps could be a capital *N*. I have retained the very common medieval spelling *paralellogramum*, since in the cases where it is written out in both manuscripts this form is followed. Most of the time the scribe of *X* has written *semidyameter* but occasionally has *semidiameter*. I have retained both spellings, according to the use of the scribe. The drawings are clear and accurate in both *X* and *Fa*. The marginal folio numbers are from *X*.

Sigla of Manuscripts

X = Vatican, Pal. lat. 1389, 108r–111v, 15c.
Fa = Vienna, Nat.-bibl. cod. 5257, 67r–69r, 1390.

exiens docebit circuli maiuraturam que proprie est quadrupla ad aream ipsius. Vel aliter ducatur dyameter in circumferentiam et habebitur propositum. Hoc probatur ratione numeri, ut si dyameter est 7, maiuratura erit 154....(111v) Quadrati circulationem assignare. Ca^m^ 24^m^. Latus quadrati ducatur in se; productum multiplicetur per 14. Summa excrescens dividatur per XI; denominationis radix erit dyameter circuli equalis quadrato proposito. Istud probatur per illud theoreuma quo superius area circuli concluditur."

The Version of the *Measurement of the Circle* by Gordanus

Versio Gordani *De mensura circuli*

108r / Circulum demonstrative quadrare. Capitulum 23.

Sicut premissum est triangulus aliquis secundum Archimenidem alicui circulo est equalis; ille inquam triangulus ortogonius cuius duo latera rectum angulum ambientia sic se habent quod unum eorum est
5 equale circumferentie, reliquum semidiametro circuli assignati. Cum ergo assignatus triangulus per ultimam secundi Euclidis quadretur, sequitur necessario quod circulus illi equalis quadretur. Itaque probationem ipsius Archimenidis pertractare et magis declarare volentes ostendimus dictum triangulum memorato circulo nec maiorem esse
10 nec minorem et ita propositum reputabimus nos habere.

Sit triangulus iste *KPO* [Fig. 20]; circulus vero sit super *O* centrum descriptus. Itaque si dixeris circulum esse triangulo maiorem, sit

1–10 Circulum....habere *om. Fa* 4 ambientia. *corr. ex* ambientium

The Version of the *Measurement of the Circle* by Gordanus

Chapter 23. To Square a Circle Demonstratively

It was premised [in Chapter 18] that according to Archimedes some triangle is equal to some circle. This triangle, I say, is a right triangle whose two sides including the right angle are so related that one of them is equal to the circumference and the other to the radius of the given circle. Since a given triangle is squared by the last [proposition] of [Book] II

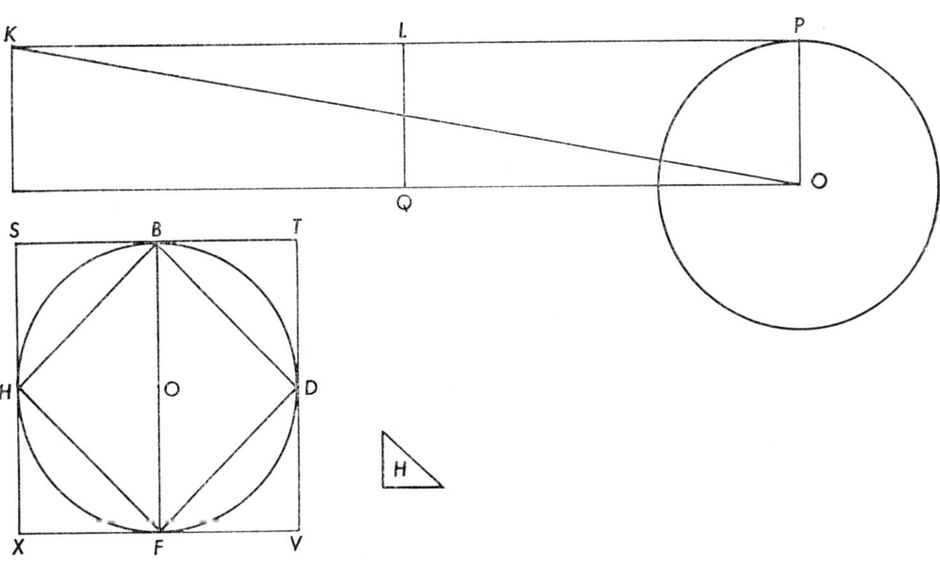

Fig. 20

[of the *Elements*] of Euclid, it follows necessarily that the circle equal to that triangle is squared. And so wishing to treat and explain further the proof of Archimedes, we demonstrate that the said triangle is neither more nor less than the mentioned circle, and thus we believe that we have that which has been proposed.

Let the triangle be *KPO* [see Fig. 20], while the circle is described about *O* as a center. And so if you say that the circle is greater than the triangle,

excessus triangulus *H*. Sequitur ergo quod dictus *KPO* triangulus et triangulus *H* simul sumpti circulo proposito adequantur. Tunc ex
15 10 Euclidis per primam eius propositionem detrahatur a circulo maius medietate et iterum a residuo plus sua medietate et sic semper donec ex ipso circulo remaneat minus quantitate *H*. Et primum detrahatur *HBDF* quadratum eidem circulo per 6tam 4ti Euclidis inscriptum. Quod autem hoc quadratum sit maius medietate circuli probatur sic:
108v Per 7am 4ti Euclidis *STVX* / quadratum circumscribitur circulo et
21 est maius ipso. Sed quadratum quod est inscriptum est medietas illius quod est extra scriptum, quod probabo. Ergo *HBDF* quadratum, cum sit medietas maioris, erit utique maius medietate minoris, scilicet circuli, et hoc est propositum. Probo quod quadratum inscriptum est
25 medietas extra scripti; ex penultima primi Euclidis apparet quod quadratum linee *BF* valet duo quadrata, scilicet *BD* et *DF*. Sed illa duo quadrata sunt equalia, cum habeant radices equales, ut patet. Ergo quadratum *BF* est duplum ad unum illorum, verbi gratia, ad quadratum *BD*, quod hic est inscriptum. Ergo quadratum *BF* du-
30 plum est ad quadratum *HBDF* hic inscriptum. Sed linea *SX* est equalis ipsi *BF* ratione equidistantie per 34tam primi Euclidis. Ergo quadratum ipsius *SX*, quod extra designatur, est duplum ad quadratum *HBDF*. Igitur istud est medietas illius. Erit ergo, ut prius dictum est, maius medietate circuli, et hoc est propositum.

35 Isto itaque quadrato subtracto, scilicet 4or residue portiones circuli fuerint maiores quantitate *H*. Detrahatur adhuc ab eis maius medietate sua hoc modo: Per 29 tertii Euclidis *HAB* arcus in duo equalia dividatur in puncto *A* [Fig. 21], et ex illo due recte ducantur ad terminos *HB* lateris quadrati eritque triangulus *HAB*. Eodem modo fiat in
40 aliis tribus portionibus et erunt 4or trianguli ex 4or lateribus quadrati et 8 lateribus octogoni constituti. Ductis itaque a singulis octogoni angulis ad singulos oppositos 4or dyametris, que sint *AE*, *BF*, *CG*, *DH*, constituentur 8 trianguli totalem octogonum componentes, qui 8 trianguli per descriptionem circuli, 28 tertii, 8 et 4tam primi Euclidis
45 equales esse probantur. Itaque 4or trianguli iam dicti, scilicet qui su-
109r per latera quadrati constituuntur a 4or portionibus circuli supra / dictis, detrahantur et remanebunt 8 portiuncule in circulum. Sunt autem isti

13 H: n *Fa hic et ubique*
18 per...Euclidis *om. Fa*
20 Per...Euclidis *om. Fa*
36 fuerint *bis Fa*
37 Per...Euclidis *om. Fa*
42 dyametris *Fa* dyametros *X*
43–44 totalem...trianguli *om. Fa*
44–45 28...probantur: equales *Fa*
46 constituunt *Fa*

let the excess be △ *H*. It follows, therefore, that the said △ *KPO* and △ *H* taken together are equal to the proposed circle. Then by Proposition X.1 [of the *Elements*] of Euclid more than half is extracted from the circle, and from the remainder again more than its half is extracted, and so on continually until there remains of this circle a quantity less than quantity *H*. First let the square *HBDF*, inscribed in the same circle by IV.6 [of the *Elements*] of Euclid, be extracted. That this square is greater than half of the circle is proved as follows:

By IV.7 [of the *Elements*] of Euclid, square *STVX* is circumscribed about the circle and [thus] is greater than it. But the inscribed square is one half the circumscribed square, which I shall prove. Therefore, square *HBDF*, since it is one half the larger quantity, will certainly be greater than one half of the lesser quantity, i.e., the circle, and this is what was proposed. I prove that the inscribed square is one half the circumscribed square [thus]: From the penultimate [proposition] of [Book] I [of the *Elements*] of Euclid it is apparent that $BF^2 = BD^2 + DF^2$. But $BD^2 = DF^2$, since $BD = DF$, as is evident. Therefore, BF^2 is double each of them, e.g., it is double BD^2, which is here the inscribed square. Therefore BF^2 is double the inscribed square *HBDF*. But line $SX = BF$ because of parallelism, by I.34 [of the *Elements*] of Euclid. Therefore, the circumscribed square, SX^2, equals twice the inscribed square, *HBDF*. Therefore, the latter is one half of the former. It will, therefore, as was said before, be greater than one half of the circle, and this is what was proposed.

And so with the square subtracted the four residual segments of the circle are greater than the quantity *H*. At this point let more than one half be extracted from these segments in this way: By III.29 (III.30, *Greek text*) [of the *Elements*] of Euclid, arc *HAB* is bisected in point *A* [see Fig. 21]. Let two straight lines be drawn from that point to the termini of side *HB* of the square and there results △ *HAB*. We proceed in the same way in the other segments and there will result four triangles constituted out of the four sides of the square and the eight sides of the octagon. Draw from the individual angles of the octagon to the angles opposite them the four diameters *AE*, *BF*, *CG*, and *DH*, and there are [thus] formed eight triangles comprising the whole octagon. These eight triangles are proved to be equal by the definition of the circle and by III.28, I.8, I.4 [of the *Elements*] of Euclid. And so let the aforesaid four triangles, i.e., those which are constructed on the four sides of the square in the four above-mentioned segments of the circle, be extracted and there will remain

4^{or} trianguli plus medietate illarum 4^{or} portionum, quod probatur sic: Sint equidistantes *FD*, *LM* inter *FL*, *DM*; per 8^{vam} primi Euclidis
50 et per 4^{tam} eiusdem patet quod *FEZ*, *EZD* trianguli partiales inter se sunt equales, et *EZD* angulus est rectus. Tunc arguo quod [illud quod] fit ex ductu *EZ* in *ZD* per 41 primi Euclidis duplum est ad triangulum *EZD*, et hoc est paralellogramum *ZM*. Est ergo paralellogramum *ZM* equale toti triangulo *FED*. Consequenter probatur
55 quod fit ex ductu *EZ*, sive *DM* que ei est equalis ratione equidistantie, in totam *FD* erit duplum ad totalem triangulum *FED* per 41 primi Euclidis. Est ergo *FED* triangulus medietas paralellogrami *LD*. Ex quo patet quod maior est medietate *FED* portionis, cum sit medietas maioris quantitatis, ut ostensum est. Similiter intellige de aliis tribus
60 triangulis a reliquis tribus portionibus detrahendis.

Detractis itaque premissis 4 triangulis, si adhuc dixerit quas residuas 8 portiunculas esse maiores quantitate *H*, fiat per modum similem detractio ulterior donec de necessitate occurrat et relinquatur quantitas minor quam *H*. Sed ad presens gratia compendii, ne pressura multarum
65 linearum demonstrationem impediat, concedantur ille 8 portiuncule esse minores quantitate *H*. Sequitur ergo quod iste octogonus est maior triangulo *KPO*, quoniam ex ypothesi idem triangulus cum quantitate *H* acceptus circulo equipollet. Hec autem illatio patet magis in numeris, ut si totus circulus sit 6 ita quod octogonus sit 5, 8 vero
70 portiuncule sint unum, unde et *KPO* triangulus cum *H* valebit 6 ratione ypothesis. Sit ergo triangulus 4^{or}, *H* vero duo, et ita patet

50 FEZ, EZD *corr. ex* FEZD
51 quod: sic quod *Fa*
54 probatur: autem *Fa*
62 portiunculas: portiones *Fa*
64–65 ne...impediat *om. Fa*
69 quod: ut *Fa*

eight small segments of the circle. Moreover these four triangles are greater than half of the four [larger] segments, which is proved thus: Between *FL* and *DM* let there be parallel lines *FD* and *LM*. By I.8 and I.4 [of the *Elements*] of Euclid it is evident that partial triangles *FEZ* and *EZD* are equal to each other and that ∠ *EZD* is a right angle. Then I argue by I.41 [of the *Elements*] of Euclid that the product of *EZ* and *ZD* is double △ *EZD* and constitutes rectangle *ZM*. Therefore, the rectangle *ZM* equals the whole △ *FED*. Consequently, the product of [1] *EZ*—or *DM* equal to it by reason of parallelism—and [2] the whole of *FD* will equal double the whole △ *FED* by I.41 [of the *Elements*] of Euclid. Therefore, △ *FED* is one half of rectangle *LD*. From which

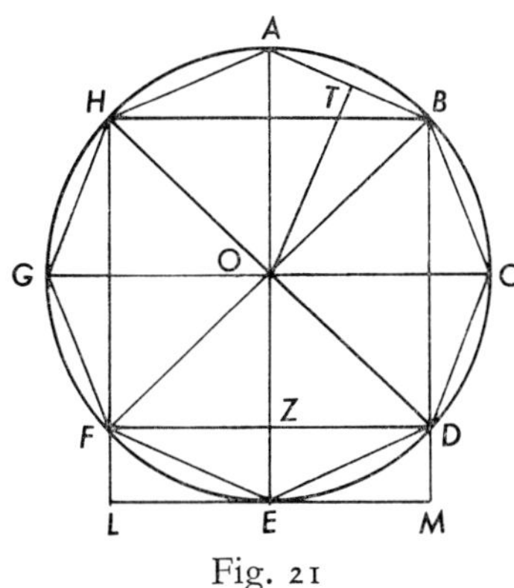

Fig. 21

it is evident that △ *FED* is greater than one half segment *FED*, since △ *FED* is one half of a greater quantity, as was demonstrated. Understand the same thing for the other three triangles to be subtracted from the three remaining segments.

With the four triangles subtracted, if one still says that the eight residual small segments are greater than quantity *H*, further subtraction is performed by a similar method until by necessity there results and remains a quantity less than *H*. But at present for the sake of brevity, lest the oppression of many lines impede the demonstration, these eight small segments are conceded to be less than quantity *H*. It follows, therefore, that this octagon is greater than △ *KPO*, since by hypothesis the sum of △ *KPO* and *H* equals the circle. This reasoning is more evident in numbers: If the total circle is 6 so that the octagon is 5, while the eight small segments are 1, then by reason of the hypothesis △ *KPO* plus *H* will equal 6. Therefore, let the triangle be 4, while *H* is 2, and thus is evident the above

illatio iam facta. Sed *KPO* triangulus [Fig. 20] est equalis ei quod fit ex ductu semidiametri, sive *PO* quod idem est, in medietatem circumferentie, sive in *PL* quod idem valet, quoniam *KP*, que equalis
75 est toti circumferentie, in puncto *L* per medium dividitur. Ergo *PQ* paralellogramum dicto triangulo est equale, quod probatur sic:

Ducatur *PO* in *PK* et proveniet paralellogramum *KO*, quod per 41 primi Euclidis est duplum ad triangulum *KPO*. Ergo eius medietas, que est paralellogramum *PQ*, erit eidem triangulo equalis. Sequitur
109v ergo quod predictus octogonus est maior illo / quod fit ex ductu
81 semidyametri in medietatem circumferentie. Sed quod hoc falsum sit probo. *AB* latus octogoni per 10^{am} primi Euclidis dividitur in puncto *T* et *OT* ducitur [Fig. 21]. Itaque per circuli descriptionem et per 8 et 4^{tam} primi Euclidis patet quod *AOB* triangulus in duos partiales
85 et equales dividitur et quod *OTB* angulus sit rectus. Inde sit quod fit ex ductu *OT* in *TB* per 41 primi Euclidis est duplum ad triangulum *OTB*. Ergo est equale totali triangulo *AOB*. Sequitur ergo quod si eadem *OT* in totalem *AB*, que est latus octogoni, ducatur, cum ipsa *AB* sit dupla ad *TB*, productum erit duplum ad totalem *AOB* trian-
90 gulum. Igitur ductus *OT* in *AB* duos triangulos valebit, que sunt *HOA*, *OAB* quoniam ipsi inter se sunt equales, ut patuit supra. Similiter si eadem *OT* ducatur in duo latera octogoni, que sunt *HA*, *AB*, productum valebit 4^{or} triangulos que sunt medietas octogoni. Et ita per consequens, si dicta *OT* ducatur in 4^{or} latera octogoni, que sunt *GH*,
95 *HA*, *AB*, *BC*, productum valebit totalem octogonum. Si ergo octogonus maior est illo quod fit ex ductu semidyametri in medietatem circumferentie, ut supra conclusum est, sequitur quod etiam istud quod fit ex ductu *OT* in 4^{or} latera octogoni erit maius illo eodem, scilicet, quod provenit ex ductu semidyametri in medietatem circum-
100 ferentie. Sed *OT* est minor semidyametro et 4^{or} latera octogoni sunt
110r minora medietate circumferentie, ut patet. Ergo / quod fit ex ductu minoris in minus maius est illo quod fit ex ductu maioris in maius. Hoc autem est falsum et impossibile. Ergo et illud ex quo sequitur, scilicet, quod octogonus sit maior illo quod fit ex ductu semidyametri
105 in medietatem circumferentie.

Per eundem rationis decursum sequeretur quod si *OT* duceretur

82 per...Euclidis *om. Fa*
83 ducatur *Fa*
90–91 HOA, OAB *corr. ex* HOAB
92 HA, AB *corr. ex* HAB
94–95 GH, HA, AB, BC *corr. ex* GHABC
101 *ante* ut *add. Fa* quia corda est minor arcu
106 discursum *Fa*

produced reasoning. But △ *KPO* [see Fig. 20] is equal to the product of [1] the radius—or of *PO* since it is the same—and [2] one half of the circumference—or *PL* since it is the same—; for *KP*, which equals the whole circumference, is bisected at point *L*. Therefore, rectangle *PQ* is equal to the said triangle, which is proved thus:

Let *PO* be multiplied into *PK* and rectangle *KO* will result. By I.41 [of the *Elements*] of Euclid, rectangle *KO* is double △ *KPO*. Therefore, the half of rectangle *KO*, which is rectangle *PQ*, will be equal to △ *KPO*. It follows, therefore, that the aforesaid octagon is greater than the product of the radius and half the circumference. But I shall prove that this is false. *AB*, a side of the octagon, is bisected in point *T* [see Fig. 21], by I.10 [of the *Elements*] of Euclid. Line *OT* is drawn. And thus by the definition of a circle and by I.8, and I.4, [of the *Elements*] of Euclid it is evident that △ *AOB* is divided into two partial and equal triangles and that ∠ *OTB* is a right angle. Hence, by I.41 [of the *Elements*] of Euclid, the product of *OT* and *TB* is double △ *OTB*. Therefore, it is equal to the whole △ *AOB*. It follows, therefore, that if the same *OT* is multiplied by the whole of *AB*, which is the side of the octagon, the product will be double the whole △ *AOB*, since $AB = 2\,TB$. Therefore, the product of *OT* and *AB* will equal the two triangles *HOA* and *OAB*, since $HOA = OAB$, as was evident above. Similarly if the same *OT* is multiplied by the two sides of the octagon *HA* and *AB*, the product will equal the four triangles which are half the octagon. And so consequently, if the said *OT* is multiplied by the four sides of the octagon *GH*, *HA*, *AB*, and *BC*, the product will equal the whole octagon. If, therefore, the octagon is greater than the product of the radius and half the circumference, as was concluded above, it follows also that the product of *OT* and the four sides of the octagon will be greater than the product of the radius and half the circumference. But *OT* is less than the radius, and the four sides of the octagon are less than half the circumference, as is evident. Therefore, the product of a lesser quantity and a lesser quantity is greater than the product of a greater quantity and a greater quantity. This moreover is false and impossible. Therefore also [false and impossible is] that from which it follows, namely, that the octagon is greater than the product of the radius and half the circumference.

By the same line of reasoning it would follow that, if *OT* were multiplied by the eight sides of the octagon, the product would be greater than the

in 8 latera octogoni productum esset maius illo quod fit ex ductu semidyametri in totam circumferentiam, quod est idem inconveniens; implicatur enim quod quelibet corda sit maior suo arcu, quod iterum
110 est impossibile. Non est igitur dictus octogonus maior illo quod fit ex ductu semidyametri in medietatem circumferentie. Unde cum illud productum sit equale *KPO* triangulo, ut supra probatum est, sequitur quod non est octogonus maior illo triangulo, quod tamen oportet necessario, ut superius patuit, si propositus circulus esset maior eodem
115 triangulo, ex quo residue 8 portiones sunt minores quantitate *H*. Relinquitur itaque de necessitate quod circulus non sit maior triangulo.

Consequenter ostendendum est quod non sit minor. Si detur quod sit minor, igitur sit *H* quantitas qua triangulus *KPO* excedit circulum memoratum. Secundum hoc ergo ipse circulus cum *H* acceptus ade-
120 quabitur triangulo. Tunc per 7^{mam} 4^{ti} Euclidis designetur circa circulum *HBDF* quadratum [Fig. 22]. Quo facto autem ille 4^{or} portiones que sub angulis quadrati et partibus circumferentie continentur simul sumpte sunt equales quantitati *H*, aut maiores aut minores; hoc est dicere, ipsum totale quadratum esse equale triangulo *KPO*,
125 aut maius aut minus.

Si dicatur minus, ducetur ad inconveniens hoc, scilicet, quod illud quod fit ex ductu semidyametri in 4^{or} latera quadrati circulo circumscripti erit minus illo quod fit ex ductu eiusdem semidyametri in circumferentiam, quod est impossibile. Decursus rationis erit ducendo
130 semidyametrum *OZ* in latera quadrati omnino sicut supra processum est ducendo *OT* lineam in latera octogoni ibi inscripti.

120 Per...Euclidis *om. Fa*/signetur *Fa* 131 ibi *om. Fa*
129 discursus *Fa*

product of the radius and the whole circumference, which constitutes the same contradiction. For it is implied that any chord is greater than its arc, which again is impossible. Therefore, the said octagon is not greater than the product of the radius and half the circumference. Whence, since that product is equal to △ *KPO*, as was proved above, it follows that the octagon is not greater than the triangle, which however it must necessarily be by the evident reasoning above if the proposed circle would be greater than the triangle, because of the fact that the residual eight segments are less than quantity *H*. And so it necessarily remains that the circle is not greater than the triangle.

Consequently it must be shown that it is not less. If it is given that it

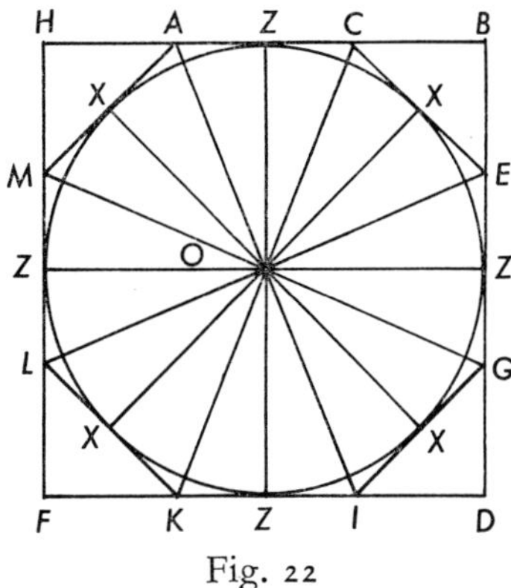

Fig. 22

is less, then let *H* be the quantity by which △ *KPO* [Fig. 20] exceeds the mentioned circle. Hence accordingly the circle together with *H* will equal the triangle. Then, by IV.7 [of the *Elements*] of Euclid, let square *HBDF* be described about the circle [see Fig. 22]. Then the sum of the four [mixed] figures which are contained by the angles of the square and the parts of the circumference equals the quantity *H*, or is greater than or less than it. That is to say, the total square is equal to △ *KPO*, or is greater than or less than [it].

If it is said to be less [than it], this contradiction is inferred: the product of the radius and the four sides of the square circumscribed about the circle will be less than the product of the radius and the circumference, which is impossible. The line of reasoning will be by multiplying the radius *OZ* by the sides of the square in completely the same way as was done above when *OT* [Fig. 21] was multiplied by the sides of the octagon, which was there an inscribed octagon.

Si vero dixerit idem quadratum esse equale aut maius *KPO* triangulo, et hoc est dicere, illas 4^or portiones simul sumptas esse equales aut maiores quantitate *H*, fiat detractio per primam 10^i Euclidis hoc
135 modo: Detrahatur primum a totali quadrato circulus eidem inscriptus, qui utique est maior medietate illius quadrati, quod patet cum sit maior quadrato sibi intelligibiliter inscripto, quod quidem quadratum
110v inscriptum est medietas / quadrati circumscripti, ut probatum est prius Iterum a residuo totalis quadrati detrahatur maius sua medietate hoc
140 modo:

In circumferentia circuli signentur per *X* et *Z* 8 puncta angularia octogoni intelligibiliter inscripti. Et ab ipsis punctis per 17 tertii Euclidis 8 linee contingentes in utramque partem trahantur donec concurrant et constituetur octogonus circumscriptus; ad cuius angu-
145 los singulos ab *O* centro linee recte ducantur, et erunt 8 trianguli totalem octogonum componentes, quem octogonum probabis esse equiangulum et equilaterum et per consequens omnes illos triangulos 8 super latera octogoni constitutos equales probabis. Inquam per corollarium ultime tertii et per 17 et 26 eiusdem et per 4 vel 8 et 26
150 et 5^tam primi Euclidis et per communem conceptionem—quorum dimidia sunt equalia ipsa equa esse—argumentum formabis sicut in probatione duodecime quarti Euclidis formatur. Hoc facto detracti sunt 4^or trianguli qui sunt *MHA*, *CBE*, *GDI*, *KFL*. Quod ergo isti trianguli simul sumpti sint plus medietate predictarum 4^or portionum
155 probatur sic:

Sit *PH* linea medietas lateris quadrati circumscripti [Fig. 23], *HQ* medietas alterius, et sit *PXQ* quarta pars circumferentie, *MA* latus octogoni circumscripti, quod in puncto *X* per equalia dividitur. *HX* ducitur; et *XS* equidistans ipsi *PH* protrahitur. Similiter *XR* ducitur
160 in equidistantia ad *HQ*. Item due linee *PX*, *XQ* ducuntur. Adhuc super *A* centrum pede circini posito et occupato *Q* semicirculus super lineam *HQ* describitur. Similiter super *M* centrum posito circino et occupato *P* super lineam *HP* alius semicirculus designatur. Tunc argue sic: Isti duo trianguli *HXA*, *XAQ* sunt unius altitudinis, quia
111r inter duas lineas equidistantes consistunt. Ergo per primam / 6^i Eucli-

132 dixeris *Fa*
142–43 per...Euclidis *om. Fa*
147 illos: illorum *Fa*
148 constitutas *Fa*
149–50 collarim...et^1: 4 primi euclidis et primam capituli de triangulis item per secundam eiusdem capituli et per *Fa*
151–52 argumentum...formatur *om. Fa*
153 ergo: autem *Fa*
160 PX, XQ, *corr. ex* PXQ
162 describatur *Fa*
164 HXA, XAQ *corr. ex* HXAQ

If one says that the same square is equal to, or is greater than, △ *KPO* [Fig. 20], that is to say, that the four figures taken together are equal to or greater than quantity *H*, a subtraction is performed, by X.1 [of the *Elements*] of Euclid, in this way: First let the circle inscribed in it be subtracted from the whole square. The circle is certainly greater than half of the square. This is evident, since the circle is greater than a square inscribed in it in the understood manner. For indeed the inscribed square is half the circumscribed square, as was proved earlier. From the remaining part of the total [circumscribed] square subtract once more a quantity greater than its half in this way:

On the circumference of the circle let there be designated by *X* and *Z* the eight angular points of an octagon inscribed in the understood way. Then by III.17 [of the *Elements*] of Euclid from these points eight tangents are drawn in both directions until they meet and a circumscribed octagon will be formed. To each of its angles is drawn a straight line from the center *O*. There will result eight triangles which comprise the octagon. You will prove that this octagon is equiangular and equilateral and hence you will prove that all eight of these triangles formed on the sides of the octagon are equal. You will fashion an argument, I say, in the manner of IV.12 [of the *Elements*] of Euclid by means of the corollary to the last [proposition] of the third book,* III.17, III.26, I.4 (or I.8), I.26, and I.5—[all from the *Elements*] of Euclid, and by means of this axiom: those things whose halves are equal are themselves equal. With this done, the four triangles *MHA*, *CBE*, *GDI*, and *KFL* are subtracted. That therefore these triangles taken together are more than half of the aforesaid four figures is proved as follows:

Let *PH* be half a side of the circumscribed square [see Fig. 23], *HQ* half the other [adjacent] side; and let *PXQ* be a quadrant of the circumference, *MA* a side of the circumscribed octagon which is bisected in point *X*. *HX* is drawn; and *XS* is drawn parallel to *PH*. Similarly *XR* is drawn parallel to *HQ*. Also draw the two lines *PX* and *XQ*. Further, with one foot of the compass placed on *A* as a center and [the other] situated at *Q*, a semicircle is described on line *HQ*. Similarly with [one foot of] the compass placed on *M* as a center and [the other] situated at *P*, another semicircle is described on line *HP*. Then argue as follows: These two triangles *HXA* and *XAQ* are of the same altitude because they are contained between two parallel lines. Therefore, by VI.1 [of the *Elements*]

* See Commentary, lines 149–50.

166 dis que proportio basis ad basim eadem est trianguli ad triangulum. Sed *HA* basis maior est *AQ* basi, quia totum maius est parte, et per circuli descriptionem; ergo *HXA* triangulus maior est *XAQ* triangulo. Eodem modo ex parte altera argumentum facies in *HXM*, *XMP*
170 triangulos. Patet ergo quod totalis *MHA* triangulus maior est *PMX*, *XAQ* triangulis, qui partem circumferentie continent. Multo ergo fortius maior est idem *MHA* triangulus quam ille due portiuncule que a dictis duobus triangulis et extra circumferentiam continentur. Et sic patet quod detractum est plus medietate illius portionis que sub *MHA*
175 angulo quadrati et *PXQ* quarta parte circumferentie continetur si totalis *MHA* triangulus detrahatur. Eodem modo intellige in reliquis tribus portionibus in precedenti figura distinctis.

Ille itaque 8 portiuncule residue que sub angulis octogoni circumscripti et partibus circumferentie continentur, si negentur esse minores
180 quantitate *H*, fiat ab eis detractio predicto modo, et sic semper donec de necessitate occurrat quantitas minor quam *H*. Sed gratia brevitatis concedantur esse minores quam *H*, ipse, dico, simul sumpte. Est ergo secundum hoc totalis ille octogonus minor *KPO* triangulo assignato. Sed cum ille triangulus sit equalis illi quod fit ex ductu *OZ*
185 semidyametri in medietatem circumferentie, sequitur quod idem octogonus est minor illo quod fit ex ductu semidyametri in medietatem
111v circumferentie. Hoc autem, quia falsum est, refellitur / ducendo semidyametrum in omnia latera octogoni circumscripti, omnino sicut supra dictum est ducendo *OT* lineam in latera octogoni ibi inscripti.

167 est[2]: sua *Fa*
169 HXM, XMP *corr. ex* HXMP
170 triangulos: triangulis *Fa*
170–71 PMX, XAQ *corr. ex* PMXAQ
187 refellitur *Fa* refellatur (?) *X*
189 dictum: ostensum *Fa*

of Euclid, triangle/triangle = base/base. But base *AH* is larger than base *AQ* because the whole is greater than the part and by the definition of a circle. Therefore, △ *HXA* is greater than △ *XAQ*. You will fashion the same argument on the other side in connection with triangles *HXM* and *XMP*. It is evident, therefore, that the whole △ *MHA* is greater than triangles *PMX* and *XAQ* which contain part of the circumference. Therefore, even more is the same △ *MHA* greater than the two small figures contained by the two mentioned triangles and outside of the circumference. And thus it is evident that more than half of the figure contained by ∠ *MHA* [i.e., by lines *PH* and *QH*] and by [arc] *PXQ*, a quarter of the

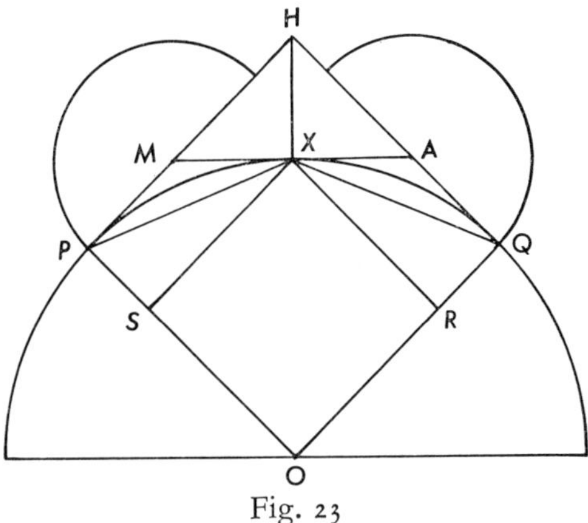

Fig. 23

circumference, has been subtracted when the whole △ *MHA* is subtracted. Understand the same procedure in regard to the three remaining [triangular] portions evident in the preceding figure [Fig. 22].

If it is denied that the residual eight small figures contained by the angles of the octagon and the parts of the circumference are less than quantity *H*, let subtraction from them be done in the aforesaid way, and continually so, until there occurs of neccesity a quantity less than *H*. But, for the sake of brevity, let these [eight figures], taken together, I say, be conceded to be less than *H*. Accordingly the whole octagon is less than the designated △ *KPO*. But since that triangle is equal to the product of radius *OZ* and half the circumference, it follows that the octagon is less than the product of the radius and half the circumference. But since it is false, this is refuted by multiplying the radius by all of the sides of the circumscribed octagon in completely the same way as it was said above when line *OT* was multiplied by the sides of the octagon, which was there an inscribed octagon [Fig. 21].

190 Et sequitur inconveniens hoc, scilicet, quod illud quod fit ex ductu semidyametri in omnia latera octogoni circumscripti minus erit illo quod fit ex ductu semidyametri in circumferentiam et ita illud quod fit ex ductu alicuius in maius minus erit illo quod fit ex ductu eiusdem in minus, quod est impossibile. Non est ergo ille octogonus minor 195 *KPO* triangulo aut illo quod fit ex ductu semidyametri in medietatem circumferentie, quod tamen oporteret si circulus propositus minor esset illo triangulo, ut supra patuit, ex quo residue portiones sunt minores *H*.

Relinquitur ergo necessario ex omnibus predictis quod circulus 200 propositus assignato triangulo est equalis, cum probatum sit quod nec maior est nec minor. Tunc, arguo, triangulus quadratur, ergo et circulus ei equalis.

194 in *om. Fa*
196 oportet *Fa*
197 *ante* ut *del. X* quod
201–202 Tunc...equalis *om. Fa*

And this contradiction follows: the product of the radius and the perimeter of the circumscribed octagon will be less than the product of the radius and the circumference, and thus the product of some quantity and the greater [of two other quantities] will be less than the product of that same quantity and the lesser [of the two other quantities], which is impossible. Therefore, the octagon is not less than △ *KPO* or less than the product of the radius and half the circumference. But such would, however, be necessary if the proposed circle were less than the triangle, as was evident above, because the residual figures are less than *H*.

From all that has been said above, it necessarily results, therefore, that the proposed circle is equal to the designated triangle, since it has been proved that it is neither greater nor less [than it]. Then, I argue, the triangle is squared and, therefore, [so] is the circle which is equal to it.

COMMENTARY

5–7 "Cum...quadretur." As in the case of Bradwardine's *Geometria* and the Cambridge and F.IB Versions, Gordanus' text takes the added step of squaring the right triangle shown to be equal to the circle.

6 "ultimam secundi." See the Commentary to the Cambridge Version, lines 82–83.

14–15 "ex 10...propositionem." See page 60, note 1, above, for the Adelard II text of X.1 of the *Elements*.

18 "6^{tam} 4^{ti}." See the Commentary to the Florence Versions, F.IB, line 9.

20 "7^{tam} 4^{ti}." See the Commentary to F.IB, line 33

25 "penultima primi." See the Commentary to the Cambridge Version, line 25.

31 "34^{tam} primi." In the Adelard II translation of the *Elements*, I.34 runs (MS Birt. Mus. Add. 34018, 5v): "Omnis superficies equidistantibus contenta lateribus lineas atque angulos ex adverso collocatos habet equales, dimaetro dividente eam per medium."

37 "29 tertii." III.29 (= III.30 of the Greek text) in the Adelard II translation is as follows (*ms. cit.*, 14r): "Datum arcum per equalia dividere." After *dividere* the MS has *res postulat*.

44 "28 tertii 8 et 4^{tam} primi." III.28 (= III.29 of the Greek text) in the Adelard text is as follows (*ms. cit.*, 14r): "Circulorum equalium equos

arcus, equas cordas habere necesse est." For I.8 and I.4 see the commentary to the Naples Version, lines 32 and 36.

47 "portiuncule." Gordanus uses this term instead of the term "lunule" throughout.

49–50 "8vam...eiusdem." For I.8 and I.4, see the Commentary to the Naples Version, lines 32 and 36.

52, 56, 77–78 "41 primi." See the Commentary to the Cambridge Version, line 33.

68–69 "Hec...numeris." The author's confidence in the greater clarity of the argument as expressed in numbers is similar to a statement made by Johannes de Muris in his *Quadripartitum numerorum* while discussing a hydrostatic proposition: "Posses declarare per litteras alphabeti; sed mihi sunt numeri clariores." (See Clagett, *Science of Mechanics in the Middle Ages*, p. 129.)

82 "10am primi." I.10 in the Adelard II translation runs as follows (*ms. cit.*, 3r): "Proposita linea recta eam per equalia dividere."

83–84 "8...primi." See the Commentary to the Naples Version, lines 32 and 36.

86 "41 primi." See the Commentary to the Cambridge Version, line 33.

120 "7mam 4ti." See the Commentary to F.IB, line 33.

134 "primam 10i." See page 60, note 1, above.

142 "17 tertii." See the Commentary to the Naples Version, line 61

149–50 "corollarium...primi." There is no corollary to the last proposition of the Book III of the *Elements*. Perhaps Gordanus meant to cite the corollary to III.15 (= III.16 in the Greek text) which reads in the Adelard II translation (*ms. cit.*, 31): "Unde etiam manifestum est omnem lineam rectam a termino diametri cuiuslibet circuli ortogonaliter ductam circulum contingere." For III.17 (= III.18 in the Greek text), see the Commentary to the Naples Version, line 61. III.26 (= III.27 in the Greek text) in the Adelard II translation is as follows (*ms. cit.*, 14r): "Si in circulis equis equi sumantur arcus, in illos cadentes angulos sive super centra eorum sive super circumferentias constituantur equos esse." For I.4 and I.8, see the commentary to the Naples Version, lines 32 and 36. I.26 reads (*ms. cit.*, 4v): "Omnium duorum trigonorum quorum duo anguli unius duobus angulis alterius, uterque se respicienti, equales fuerint, latusque* unius lateri alterius equale fueritque latus illud inter duos angulos equales aut uni eorum oppo-

* The *que* is written above and after *fuerint* in this manuscript, but it should follow *latus*, as it does in other manuscripts.

situm, erunt quoque duo unius reliqua latera duobus reliquis alterius trigoni lateribus, unumquodque se respicienti, equalia, angulusque reliquus unius angulo reliquo alterius equalis." For I.5 see the Commentary to the Naples Version, line 63.

152 "duodecimi quarti." IV.12 in the Adelard II translation reads as follows (*ms. cit.*, 16v): "Circa propositum circulum pentagonum equilaterum atque equiangulum designare."

165 "primam 6ti." See the Commentary to the Cambridge Version, line 70.

5. The Corpus Christi Version

In a fifteenth-century manuscript of Oxford (Corpus Christi College 234, 170r–172v) appears still another version of the *De mensura circuli*. In structure and form this version is further removed from the original translations than the other versions we have so far discussed. For in this version some of the fundamental propositions needed for the proof of the main quadrature theorem are singled out for prior enunciation and proof. This constitutes still another step in the increasing elaboration of the pristine version of the *De mensura circuli*.

It is difficult to date the composition of the Corpus Christi Version. It is free from the scholastic form that is apparent in the other versions of the middle and late fourteenth century. For this reason it could be argued that it precedes those versions and dates perhaps from the late thirteenth or early fourteenth century. But such reasoning is hazardous, and without further evidence we must set aside the question of a precise dating for the *terminus ante quem* any earlier than the fifteenth century. On the other hand, some knowledge, direct or indirect, of the *On the Sphere and the Cylinder* of Archimedes on the part of the author of this version is probable, as I shall show. But the only known available source of the *On the Sphere and the Cylinder* in Latin before the middle of the fifteenth century was the translation by Moerbeke in 1269. This date then should be the *terminus post quem* of the version's composition.

Regardless of its date, the Corpus Christi Version can be solidly connected with Gerard of Cremona's translation of the *De mensura circuli*. Gerard's wording of the enunciations of Propositions I and II is used by the Corpus Christi author in his Propositions IV and V. As a matter of fact, our author, in taking over Gerard's enunciation of Proposition I, follows the second tradition of Gerard's translation since he gives the order of words followed in the second tradition (*orthogonio triangulo*) rather than that of the first tradition (*triangulo orthogonio*).

As I have said, the author of the Corpus Christi Version presents some postulates and propositions before considering the first two propositions of the *De mensura circuli.* As in the case of the Cambridge Version, the author first gives us three postulates (*petita*) which, although postulated, he says are "in reality principles (*principia*) and are known per se." The first postulate (lines 5–7) is that "an arc is greater than its chord." I have indicated in my previous discussions that the Cambridge Version (line 5) and the Gordanus Version (line 109) made explicit reference to a similar postulate. As I have already noted in my discussion of the Cambridge postulate above, this postulate is included in the more general first assumption of Archimedes' *On the Sphere and the Cylinder*: "Of lines which have the same extremities the straight line is the least." The Corpus Christi author even goes so far as to illustrate the postulate (see Figure 24a).

The second postulate (line 8) to the effect that a curved line can be equal to a straight line was perhaps suggested by the first postulate in the Cambridge Version (see that text, lines 3–4), although the wording of the Cambridge postulate is somewhat different and is, as a matter of fact, to be preferred. In the Corpus Christi Version we have two cases given for confirmation of this postulate. The first one—strictly physical—is that one can take a hair or silk thread and have it either extended in a straight line or bent around to form a circumference; and so, the author asserts, "who will doubt—unless he is hare-brained—that the hair or thread is the same whether it is bent circumference-wise or extended in a straight line and is just as long the one time as the other." The second illustration of this postulate concerns a circle which, standing perpendicular to a plane, is rolled along the plane until the point originally in contact with the plane is again in contact with the plane. In such an operation, the author says, "it can be scarcely doubted by anyone with a mind in his head that the circumference will have described a straight line equal to the circumference." This is reminiscent of the procedure used by Hero in his *Mechanics* when he discusses the paradox known as "the wheel of Aristotle."[1] I am not suggesting, however, that the medieval author knew Hero's work since there is no evidence that it was known in the Latin West. Much more

[1] Hero, *Mechanics*, ed. of L. Nix (Leipzig, 1900), book I, chap. 7. English translation in my *Science of Mechanics in the Middle Ages* (Madison, 1959), pp. 40, 48.

likely is the possibility that, if the author was influenced by the *rota Aristotelis*, he learned about it from Part I, Chapter 2, of Algazel's *Metaphysics* (ed. of J. T. Muckle, Toronto, 1933, p. 13).

The third Corpus Christi postulate is of considerable interest since, as I shall demonstrate, it shows some ultimate influence of the *On the Sphere and the Cylinder* of Archimedes. It states (lines 20–22): "Any curved line sharing two termini of a circumferential arc and including it in the direction of the convexity of the arc is greater than the arc." This should be compared with the second assumption of Archimedes' *On the Sphere and the Cylinder:* "Of other lines in a plane and having the same extremities, [any two] such are unequal whenever both are concave in the same direction and one of them is either wholly included between the other and the straight line which has the same extremities with it, or is partly included by, and is partly common with, the other; and that [line] which is included is the lesser [of the two]."[2] The general similarity of the two statements is evident. Furthermore, the Corpus Christi author goes on to indicate that "curved lines" include bent lines composed of straight lines and thus is similar to Archimedes' καμπύλαι γραμμαί, rendered, I might add, by Moerbeke in his translation of the *On the Sphere and Cylinder* as *curve linee*.[3] Notice also that Archimedes uses the expression "concave in the same direction," while our medieval author speaks of "the direction of the convexity of the arc." And so, from the over-all similarity of the two statements, one can conclude that the general tenor (if not the precise statement) of the Archimedean assumption had become available to our author.

It will be noticed by the reader that the third postulate of the Corpus Christi Version is quite different from the third postulate of the Cambridge Version which held that the perimeter of any "including" figure is greater than the perimeter of the "included" figure. (Compare also Albert of Saxony's *Questio*, Chapter Five, Section 3, lines 198–200.) But when the first two propositions of the Corpus Christi Version are examined, it will be evident that the Corpus Christi author has reached a similar conclusion. For he tells us in Proposition I that "the perimeter of any polygon inscribed in a circle... is less than the circumference" and in Proposition II that "the perimeter of a polygon circumscribed about a circle is greater than the circumference." Again, Archimedes' *On the Sphere and the Cylinder* appears to be the source of these propositions as they are both found in

[2] T. L. Heath, *The Works of Archimedes* (Cambridge, 1897), p. 4.

[3] *Ibid.*, p. 2. For Moerbeke's translation, see MS Vat. Ottob. lat. 1850, 23v, col. 1.

that work (See Section I, notes 3 and 4). One might argue, although not correctly in my opinion, that these first two propositions were suggested by the wording of the second conclusion of the Pseudo-Bradwardine Version (see Chapter Five, Section 1), which reads: "The perimeter of any polygon circumscribed about a circle constitutes a line longer than the circumference of the aforesaid circle [the circumference] having been transformed into a straight line (*rectificata*)." It is obvious that this is much farther removed from the statement of the Corpus Christi Version than is the conclusion of Archimedes, which is almost identical with the statement in the Corpus Christi Version. Furthermore, the Pseudo-Bradwardine Version has no complementary statement regarding the relationship of the perimeter of an inscribed polygon with the circumference of the circle, as do both Archimedes' *On the Sphere and the Cylinder* and the Corpus Christi Version.

Before passing on to the actual propositions drawn from the *De mensura circuli*, the Corpus Christi author singles out the proposition to the effect that the product of the perimeter of an inscribed or circumscribed regular polygon and the line drawn from the center to the middle point of one of the sides of the polygon is double the area of the polygon. The author actually breaks this proposition up into two parts, one (Proposition IIIA) for the inscribed polygon and the other (Proposition IIIB) for the circumscribed polygon. Since this theorem represents an important step in the proof of Proposition I of the *De mensura circuli*, it is not surprising that many of the authors of the emended versions thought it necessary to include the proposition with some proof. The only distinctive feature of the Corpus Christi treatment as compared with the other versions we have already treated is that the proposition is proved before the main proposition is commenced. Such a proposition, although in slightly different form, is also a preliminary proposition for Albert of Saxony (see his text below, lines 141–47), the *Verba filiorum* (see that text, Proposition I, lines 1–4), and the so-called Pseudo-Bradwardine Version (*Versio Vaticana*, lines 4–10); and *Versio Abbreviata* (lines 1–7).

As in the cases of almost all of the emended versions of the *De mensura circuli*, the author of the Corpus Christi Version quotes Euclid's *Elements* extensively—in fact, thirty-one times. As in the Florence Versions, the *Elements* is cited under the title *Liber geometrie*. Because of its importance Proposition X.1 is not only cited but quoted in full. In the course of asserting that if the right triangle is not equal to the circle it is greater or less than it, the geometrical form of the magnitude representing the excess is not specified. However, it is drawn as a small triangle (see *G* in Fig. 27) in the first part of Proposition IV and as a small square (see *O* in Fig. 28) in the second part of that proposition. In detailing the proof of Proposition IV, the Corpus Christi author is, like all the other authors, satisfied with the inscription and circumscription of octagons. For the designation of the small segments formed by the perimeter of the inscribed octagon and the circumference of the circle he uses both the common term *lunule* and the term used by Gordanus, *porciuncule*. However, unlike Gordanus he does not use *porciuncule* for the mixed triangular figures formed by the perimeter of the circumscribed polygon and the circumference. Still one further point of similarity of this version with the others is the conclusion that once the circle has been shown to be equal to the right triangle, that triangle must be converted to a square (see lines 224–42). This conversion is made by first finding a rectangle equal to the triangle and then finding the mean proportional line between the base and altitude of the rectangle. The only interesting point of distinction here is that the porism to Proposition VI.8 of the *Elements* is used to find the mean proportional rather than the commonly employed Proposition II.14. Finally, we should note that the author of the Corpus Christi Version is also concerned with the logical form of the proof. He uses the logical device of *falsigraphus* (lines 217–19, and the variant reading, line 218), as do so many of the medieval logical and mathematical discussions.[4] *Falsigraphus* is for him the opponent who asserts a position which will be shown to lead to a contradiction. *Falsigraphus* is, of course, the medieval form of Aristotle's ψευδογράφος, whose role in geometrical arguments is described in the *De sophisticis elenchis* (171ᵇ 34–172ᵃ 7; see the Commentary to the version of Albert of Saxony, Chapter Five, Section 3, lines 3–4).

My text of the Corpus Christi Version follows the same procedures used in the previous texts The figures are taken from the manuscript without significant change. The marginal folio numbers are of course from MS *Bc*.

Siglum of Manuscript

Bc = Oxford, Corpus Christi College 234, 170r–72v, 15c.

[4] Notice, for example, the use of *falsigraphus* in the text of the *De curvis superficiebus* in Chapter Six, Section 2, below, Proposition I, lines 33, 72, 76, 84, 94; Proposition III, line 13; Proposition VI, line 39; *et al.* See also the introduction to what I have called the Adelard III translation or version of Euclid ("King Alfred and the Elements of Euclid," *Isis*, vol. *45* [1954], p. 274): "Instantie dissolutio est cum falsigraphus insistit non sic vel aliter accidere quam geometer affirmat."

Versio Corpus Christi Collegii De Mensura Circuli

170r / Tria petimus que ad demonstracionem quadrature circuli et ad demonstracionem curvarum superficierum et spericarum et ad astronomiam et ad perfeccionem doctrine sunt necessaria. Licet autem petamus tria, sciendum tamen est quod revera sunt principia et per se nota.

5 [1.] Trium petitorum primum est, quod arcus sit maior corda. Hoc petimus, licet per se notum sit et cuilibet sane mentis indubitabile. Verbi gratia, arcus *ABC* est maior corda *AC* [Fig. 24*a*].

[2.] Secundum petitorum est, quod linea curva sit equalis recte. Hoc petimus, licet sit principium et per se notum et cuilibet sani 10 capitis cognitum; si enim capillus vel filum sericum in plana superficie circumferencialiter circumflectatur, deinde idem in directum in eodem plano extendatur, quis nisi cerebrosus dubitet eundem esse capillum et idem filum sive circumflectatur sive in directum extendatur et tantum esse quantum prius. Rursus, si circulus super planam superficiem 15 circumrotetur circumferencia planam superficiem tangente et in directum circumvoluta super planam superficiem ab uno sui puncto donec perveniat ad idem punctum circumferencie quibus nisi mente capitis dubitabile est quin circumferencia descripserit lineam rectam equalem circumferencie.

11 idem *supra scr. Bc*

The Corpus Christi College Version of the *Measurement of the Circle*

We postulate three things which are necessary for the demonstration of the quadrature of the circle and for the demonstration [of the quadrature] of curved and spherical surfaces, as well as for astronomy and for the perfection of learning. Although we postulate the three things, still it should be realized that they are in reality principles and are known per se.

[1.] The first of the three postulates is that an arc is greater than [its] chord. We postulate this, although it is known per se and is not doubted by any one of sound mind. For example, arc *ABC* is greater than chord *AC* [see Fig. 24*a*].

[2.] The second of the postulates is that a curved line is equal to a straight line. We postulate this although it is a principle known per se and recog-

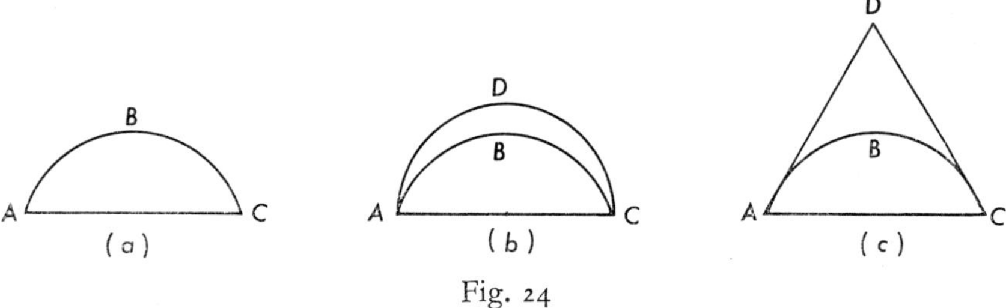

Fig. 24

nized by anybody with a sound head. For if a hair or silk thread is bent around circumference-wise in a plane suface and then afterwards is extended in a straight line in the same plane, who will doubt—unless he is hare-brained—that the hair or thread is the same whether it is bent circumference-wise or extended in a straight line and is just as long the one time as the other. Again, if a circle is rolled on a plane with the circumference tangent to the plane surface and if it is rolled in a straight line from one point of the circumference until it arrives at the same point of the circumference, it can scarcely be doubted by any one with a mind in his head that the circumference will have described a straight line equal to the circumference.

20 [3.] Tercium petitorum tale est: Quelibet linea curva duobus terminis arcus circumferencialis conterminata ex parte convexitatis arcus arcum ambiens maior est illo arcu. Revera licet hoc petamus, tamen principium est per se notum, nec alicui rationali creature discrecionem habenti debet trahi in dubium. Verbi gratia: Esto *ABC* arcus circum-
25 ferencialis [Fig. 24*b*]. Deinde protrahatur linea curva a puncto *A* ad punctum *C* ex parte convexitatis arcus transiens per punctum *D*; cui ergo vel quibus dubitabile est quin linea curva *ADC* sit maior arcu *ABC*?

Eodem modo si a duobus terminis arcus circumferencialis *ABC*
30 [Fig. 24*c*] protrahantur due linee recte ex parte convexitatis arcus concurrentes in puncto uno, scilicet *D*, quis posset dubitare lineam curvam *ADC* ex duabus rectis lineis compositam maiorem esse arcu *ABC*.

Constent igitur tria petita principia.

35 [I.] EX PRIMO PRINCIPIO MANIFESTUM EST QUOD CUIUSLIBET POLIGONII CIRCULO INSCRIPTI LINEA CURVA
170v AMBITUS POLIGONII, I.E., LINEA CURVA AMBIENS / POLIGONIUM, MINOR EST CIRCUMFERENCIA, CUM OMNES PARCIALES ARCUS SINT MAIORES SUIS CORDIS.

40 Verbi gratia: Describatur circulus *ABCDEF*, cui inscribatur exagonus *ABCDEF* [Fig. 25]. Manifestum est quod sex parciales arcus, scilicet arcus *AB* et arcus *BC* et arcus *CD* et arcus *DE* et arcus *EF* et arcus *FA*, sunt maiores sex cordis. Sic necessarium est quod omnis

[3] The third of the postulates is as follows: Any curved line sharing the two termini of a circumferential arc and including it in the direction of the convexity of the arc is greater than the arc. Although we are actually postulating this, still it is a principle known per se, nor ought it be doubted by any rational creature having discretion. For example, let *ABC* be the circumferential arc [see Fig. 24*b*]. Then let a curved line be drawn from point *A* to point *C* through point *D* and in the direction of the convexity of the arc. Who then will doubt that the curve *ADC* is greater than the arc *ABC*?

Similarly, if from the two termini of the circumferential arc *ABC* two straight lines are drawn in the direction of the convexity of the arc until

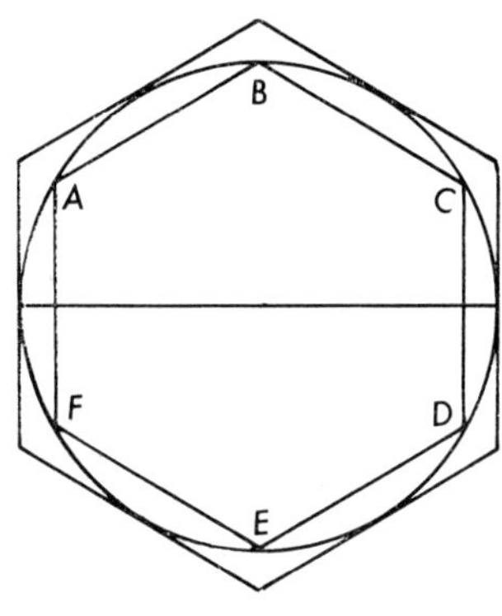

Fig. 25

they meet in one point, namely, *D*, who could doubt that the bent line *ADC* composed of the two straight lines is greater than arc *ABC* [see Fig. 24*c*]?

Let, therefore, the three postulated principles be evident.

[1] FROM THE FIRST PRINCIPLE IT IS MANIFEST THAT THE BENT LINE* [OR] PERIMETER OF ANY POLYGON INSCRIBED IN A CIRCLE, I.E., THE BENT LINE* ENCOMPASSING THE POLYGON, IS LESS THAN THE CIRCUMFERENCE, SINCE ALL THE PARTIAL ARCS ARE GREATER THAN THEIR CHORDS.

For example, let circle *ABCDEF* be described and let hexagon *ABCDEF* be inscribed in it [see Fig. 25]. It is manifest that the six partial arcs *AB*, *BC*, *CD*, *DE*, *EF*, and *FA* are greater than the six chords. Thus it is

* See Commentary, line 20.

circumferencia maior sit quolibet ambitu poligonii circulo inscripti.

45 [II.] EX TERCIA PETICIONE MANIFESTUM EST QUOD OMNIS AMBITUS POLIGONII CIRCULO CIRCUMSCRIPTI MAIOR EST CIRCUMFERENCIA, CUM PARCIALES LINEE RECTE EXTERIORES MAIORES SINT ARCUBUS CIRCUMFERENCIE.

50 Ad evidenciam itaque subsequencium quedam sunt demonstranda que ad sequencium demonstracionem sunt necessaria.

[IIIA.] DICIMUS ERGO QUOD SI ALIQUOD POLIGONIUM CIRCULO INSCRIBATUR ET A CENTRO ILLIUS CIRCULI DUCATUR PERPENDICULARIS AD UNUM LATERUM PO-

55 LIGONII SIVE AD MEDIUM PUNCTUM EIUSDEM LATERIS, QUOD IDEM EST EX TERCIA TERCII, QUOD ID QUOD FIT EX DUCTU ILLIUS PERPENDICULARIS IN TOTALEM LINEAM CONTINENTEM SIVE AMBIENTEM POLIGONIUM EST DUPLUM AD POLIGONIUM.

60 [IIIB.] PRORSUS EODEM MODO ⟨SI⟩ ALICUI CIRCULO CIRCUMSCRIBATUR POLIGONIUM ET A CENTRO ILLIUS CIRCULI AD MEDIUM PUNCTUM UNIUS LATERUM ILLIUS POLIGONII DUCATUR LINEA, DICIMUS QUOD ID QUOD FIT EX DUCTU ILLIUS LINEE IN TOTALEM AMBIENTEM

65 POLIGONIUM EST DUPLUM AD IPSUM POLIGONIUM.

Hoc autem facillime demonstrabitur per primam secundi et per 41 primi. Verbi gratia: Inscribatur circulo poligonium octogonium quod vocetur *ABCDEFGH*; centrum autem circuli vocetur *K* [Fig. 26]. Deinde a centro *K* ducatur linea ad medium punctum lateris poligonii

70 *AB*, que linea vocetur *KM*. Igitur linea *KM* erit linea perpendicularis ad latus *AB* ex 3 tercii. Quo constante a centro *K* ducantur due linee ad puncta *A* et *B* Ex hiis sic: quod fit ex ductu *KM* linee in *MB* lineam est duplum ad triangulum *KMB* ex 41 primi. Prorsus eodem modo quod fit ex ductu *KM* linee in *AM* lineam est duplum ad trian-

75 gulum *KMA* ex eadem primi. Ergo ex prima secundi quod fit ex ductu *KM* linee in duas lineas, scilicet *AM* lineam et *MB* lineam, est duplum ad duos triangulos, scilicet triangulum *KMA* et triangulum *KMB*; ergo ad triangulum *KAB*. Prorsus eodem modo prosequaris de aliis lateribus poligonii et de similibus triangulis super ipsa latera consti-

46 circum- *supra scr. Bc*

necessary that every circumference be greater than the perimeter of any polygon inscribed in the circle.

[II.] FROM THE THIRD POSTULATE IT IS MANIFEST THAT EVERY PERIMETER OF A POLYGON CIRCUMSCRIBED ABOUT A CIRCLE IS GREATER THAN THE CIRCUMFERENCE, SINCE THE PARTIAL STRAIGHT LINES LYING OUTSIDE ARE GREATER THAN THE ARCS OF THE CIRCUMFERENCE.

And so for the clarification of what follows immediately certain things must be demonstrated—things which are necessary for the demonstration of what follows [later].

[IIIA.] WE SAY, THEREFORE, THAT IF SOME [REGULAR] POLYGON IS INSCRIBED IN A CIRCLE AND IF A PERPENDICULAR IS DRAWN FROM THE CENTER OF THAT CIRCLE TO ONE OF THE SIDES OF THE POLYGON—OR THAT WHICH IS THE SAME THING, IF, BY III.3 [OF THE *ELEMENTS* OF EUCLID], A LINE IS DRAWN TO THE MIDDLE POINT OF THE SAME SIDE—THEN THE PRODUCT OF THAT PERPENDICULAR AND THE WHOLE LINE CONTAINING OR ENCOMPASSING THE POLYGON IS DOUBLE THE POLYGON.

[IIIB.] FURTHER, IN THE SAME WAY, IF ANY [REGULAR] POLYGON IS CIRCUMSCRIBED ABOUT A CIRCLE FROM WHOSE CENTER A LINE IS DRAWN TO THE MIDDLE POINT OF ONE OF ITS SIDES, WE SAY THAT THE PRODUCT OF THAT LINE AND THE WHOLE PERIMETER OF THE POLYGON IS DOUBLE THE POLYGON.

This will be very easily demonstrated by II.1 and I.41 [of the *Elements*]. For example, let a octagonal polygon designated as *ABCDEFGH* be inscribed in a circle and let the center of the circle be called *K* [see Fig. 26]. Then from center *K* let a line be drawn to the middle point of side *AB* of the polygon. This line we let be called *KM*. Therefore, line *KM* will be perpendicular to side *AB*, by III.3 [of the *Elements*]. With this evident, let two lines be drawn from center *K* to points *A* and *B*. Accordingly, the product of line *KM* and line *MB* is double △ *KMB*, by I.41 [of the *Elements*]. Further, in the same say, the product of line *KM* and line *AM* is double △ *KMA*, from the same [proposition] of [Book] I. Hence, from II.1 [of the *Elements*], the product of line *KM* and the [sum of the] two lines *AM* and *MB* is double the [sum of the] two triangles *KMA* and *KMB*; and, therefore, it is double △ *KAB*. Further, you will proceed in the same way in regard to the other sides of the polygon and to the

80 tutis. Ergo quod fit ex ductu *KM* linee in totalem lineam ambientem poligonium est duplum ad ipsum poligonium ex prima secundi.

Prorsus eodem modo demonstrabis et eisdem theoreumatibus quod id quod fit ex ductu linee exeuntis a centro *K* ad medium unius lateris poligonii exterioris circulo circumscripti in totalem lineam ambientem 85 exterius poligonium circulo circumscriptum est duplum ad ipsum poligonium, que demonstrare proposuimus.

Accedamus itaque ad demonstracionem quadrature circuli idem ad demonstracionem inveniendum quadratum equale circulo proposito. Dicit ergo Archimenides:

90 [IV.] OMNIS CIRCULUS ORTOGONIO TRIANGULO EST EQUALIS, CUIUS UNUM DUORUM LATERUM ANGULUM RECTUM CONTINENCIUM MEDIETATI DIAMETRI CIRCULI EQUATUR ET ALTERUM IPSORUM LINEE CIRCULUM CONTINENTI:

95 Describatur circulus *ABCD*, cuius diameter vocetur *AC*, centrum 171r vero vocetur *E* [Fig. 27]. Deinde des/cribatur triangulus ortogonius, cuius cathetus sit equalis semidiametro *EA*, basis vero sit equalis circumferencie dati circuli, que vocetur *EF*, et continent iste due linee angulum rectum. Ypothenusa vero trianguli vocetur *AF*. Ha-100 bemus ergo triangulum ortogonium et habemus circulum propositum. Est ergo propositum demonstrare quod ille triangulus est equalis circulo proposito; quod sic demonstramus:

Circulus propositus aut est maior triangulo aut minor aut equalis.

90 *supra* ortogonio *scr. Bc* i. [e.] habenti angulum rectum / triangulo *supra scr. Bc*
92–93 *supra* circuli *scr. Bc* est equale
93–94 *supra* alterum...continenti *scr. Bc* laterum angulum rectum continentium est equale circumferencie

similar triangles formed on those sides. Therefore, the product of line *KM* and the total perimeter of the polygon is double the polygon from II.1 [of the *Elements*].

Furthermore, you will demonstrate in the same way and by the same theorems that the product of [1] the line going out from the center *K* to the middle of one side of the exterior polygon circumscribed about the circle and [2] the total perimeter of the polygon circumscribed about the circle is double that polygon. This is what we proposed to demonstrate.

And so let us approach the demonstration of the quadrature of the circle, which is the same as demonstrating that there is to be found a square equal to the proposed circle.

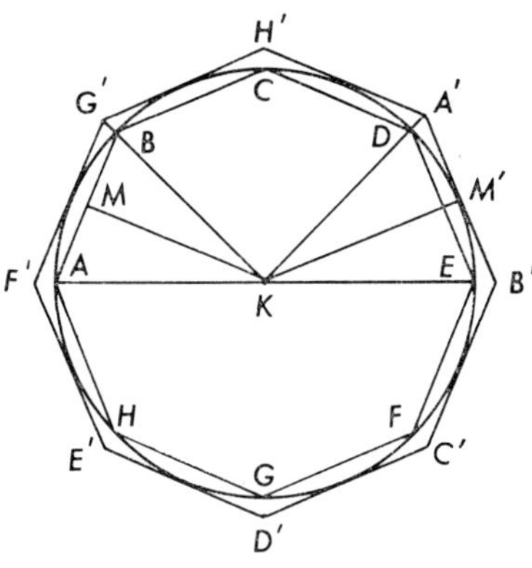

Fig. 26
Note: I have added the primes to the outer ring of letters.

Therefore, Archimedes says:

[IV.] EVERY CIRCLE IS EQUAL TO A RIGHT TRIANGLE ONE OF WHOSE TWO SIDES CONTAINING THE RIGHT ANGLE IS EQUAL TO THE RADIUS OF THE CIRCLE AND THE OTHER TO THE CIRCUMFERENCE OF THE CIRCLE.

Let circle *ABCD* be described; let its diameter be called *AC* and its center *E* [See Fig. 27]. Then let a right triangle be drawn whose altitude is equal to radius *EA* and whose base, designated by *EF*, is equal to the circumference of the given circle. These two lines contain the right angle, while the hypotenuse of the triangle let us call *AF*. We have, therefore, both the right triangle and the proposed circle. It is proposed, therefore, to demonstrate that that triangle is equal to the proposed circle. This we demonstrate as follows:

The proposed circle is greater than the triangle, less than it, or equal

Esto primo quod sit maior. Inde sic: circulus datus est maior triangulo
105 *AEF*. Ergo aliquanto est maior eo. Hec est enim communis animi
concepcio: quod si aliquod totum est maius alio, quod eo est maius
aliquanto excessu. Esto itaque *G* excessus quo circulus est maior
triangulo. Sic ergo datus circulus est equalis duobus, scilicet triangulo
et *G*; ergo est maior *G*. Age itaque per primum theoreuma decimi libri
110 geometrie, quod tale est: Si duabus quantitatibus inequalibus positis
maius dimidio [a] maiori detrahatur, itemque de reliquo maius dimidio
dematur donec minus eo relinquatur, deincepsque eo modo, necesse est
ut tandem minore positarum minor quantitas relinquatur. Dematur itaque a dato circulo maius medietate et iterum a residuo maius medietate
115 eius et iterum a residuo maius medietate eius et tandem occurret
minus *G*. Quod sic fiat: Dato circulo inscribatur quadratum *ABCD*.
Cum ergo illud quadratum sit medietas quadrati diametri, cum quadratum diametri sit duplum ad quadratum coste ex penultima primi,
et quadratum diametri sit maius circulo dato tanquam totum maius
120 sua parte, subtrahatur predictum quadratum a circulo dato et sic subtrahetur maius medietate circuli et remanebunt 4^{or} circuli porciones

115 et...eius *mg. Bc*

to it. In the first place let it be greater. From this, then, the given circle is greater than △ *AEF*. Therefore, it is greater than it by some amount. For this is a common axiom, that if some whole is greater than another, it is greater by some excess. And so let *G* be the excess by which the circle is greater than the triangle. So, therefore, the given circle is equal to two quantities, namely, the triangle and *G*. Hence it is greater than *G*. And so proceed by X.1 of the *Book of Geometry* [i.e., *Elements* of Euclid] which is as follows: "With two unequal quantities posited, if more than half is subtracted from the greater and again from the remainder more

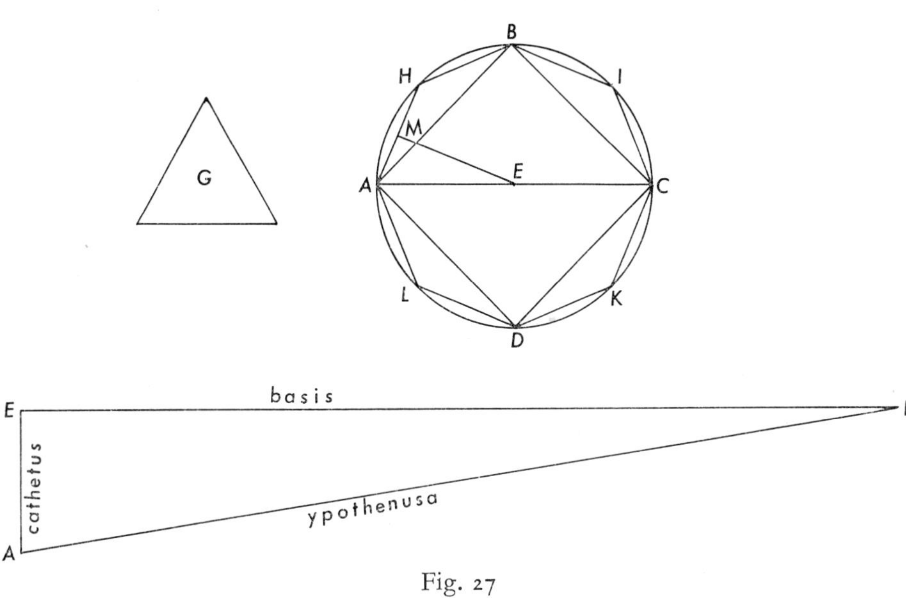

Fig. 27

than half is taken until something less than it remains, and this is done continuously, it is necessary that finally a quantity less than the lesser of the [two] posited quantities remains." And so let more than half be taken from the given circle and again more than half from its remainder and again more than half from its remainder, and finally there will occur a quantity less than *G*. This is done in the following way:

Let square *ABCD* be inscribed in the given circle. Since that square is one half of the square of the diameter—the square of the diameter being double the square of a side by the penultimate [proposition] of [Book] I [of the *Elements*]—and since the square of the diameter is greater than the given circle just as a whole is greater than its part, let the aforesaid square be subtracted from the given circle and thus more than half of the circle is

contente 4^{or} arcubus et 4^{or} cordis. Si ergo ille quatuor circuli porciones non sunt minores *G*, fiat iterum subtraccio hoc modo:

Quatuor arcus predictarum porcionum circuli dividantur per equa- 125 lia ex 29 tercii libri geometrie in punctis *H* et *I* et *K* et *L*. Deinde a quatuor punctis *A* et *B* et *C* et *D* protrahantur 8 corde ad quatuor puncta, hoc est *H* et *I* et *K* et *L*. Habebimus itaque quatuor triangulos, quorum primus est *AHB*, secundus est *BIC*, tercius est *CKD*, quartus est *DLA*. Isti 4^{or} trianguli sunt maiores medietate 4^{or} predictarum 130 porcionum circuli, quod manifestissime patebit ex 41 primi libri geometrie si triangulorum parallelograma describantur, illa enim parallelograma erunt tota ad predictas porciones circuli. Subtrahantur ergo isti quatuor trianguli a predictis 4 circuli porcionibus et remanebunt octo lunule quasi octo circuli porciuncule. Si ergo ille 8 lunule non 135 sunt minores *G*, fiat consimilis detraccio a predictis octo lunulis et a residuis et tandem relinquentur porciuncule circuli minores *G* secundum primum theoreuma decimi libri geometrie. Esto tamen causa brevitatis et facillioris intelligencie quod octo prefate lunule sint minores *G*. Ergo residuum de circulo, scilicet poligonium octogonium, 140 est maius prefato triangulo *AEF*; cuius contrarium demonstramus sic:

Protrahatur linea a centro *E* ad medium punctum unius lateris 171v octogonii, quod latus vocatur *AH*, que linea vocetur *EM*. / Est itaque *EM* linea perpendicularis ad lineam *AH* ex tercia tercii libri geometrie. Ex hiis sic: linea *EM* est minor linea *EA*. Eodem modo 145 totalis linea ambiens octogonium est minor linea *EF*, cum sit minor circumferencia ex prima peticione. Ergo quod fit ex ductu linee *EM* in totalem lineam ambientem octogonium est minus eo qoud fit ex ductu *EA* linee in *EF* lineam. Sed quod fit ex ductu *EM* linee in totalem lineam ambientem poligonium est duplum poligonii, ut prius demon- 150 stratum est. Et similiter quod fit ex ductu *EA* linee in *EF* lineam est duplum trianguli ex 41 primi libri geometrie. Ergo duplum poligonii est minus duplo trianguli. Ergo poligonium est minus triangulo. Concessum est autem quod sit maius. Relinquitur itaque quod circulus datus non est maior triangulo prefato, cum ex eo sequatur impossibile 155 per se.

Esto igitur quod circulus datus est minor prefato triangulo. Procede

142 *post* EM *del. Bc* in totalem

subtracted and there will remain four segments of the circle—segments contained by four arcs and four chords. If, therefore, these four segments are not less than G, let another subtraction be performed in this way:

Let the four arcs of the aforesaid segments of the circle be bisected, by III.29 of the *Book of Geometry* [of Euclid], in points H, I, K, and L. Then from the four points A, B, C, and D let eight chords be drawn to the four points H, I, K, and L. And so we shall have four triangles, of which the first is AHB, the second BIC, the third CKD, and the fourth DLA. These four triangles are more than half of the aforesaid four segments of the circle. This will be very manifestly evident from I.41 of the *Book of Geometry* [of Euclid] if parallelograms are described on the triangles; for these parallelograms will be [double the triangles and] all together [more than] the aforesaid segments of the circle. Therefore, let these four triangles be subtracted from the aforesaid 4 segments of the circle and there will remain 8 "lunules" in the form of 8 small segments of the circle. Then if these 8 "lunules" are not less than G, let a similar subtraction from the aforesaid 8 "lunules" be performed and [also] from their remainders, and finally there will remain small segments of the circle which are less than G, by X.1 of the *Book of Geometry* [of Euclid]. However, for the sake of brevity and easier understanding let the eight designated "lunules" be less than G. Therefore, the rest of the circle, namely, the octagonal polygon, is greater than the designated $\triangle$ AEF. [But] we demonstrate the contrary of this as follows:

Let a line be drawn from the center E to the middle point of one side of the octagon, which side is called AH and which line we call EM. And so EM is a line perpendicular to line AH, by III.3 of the *Book of Geometry* [of Euclid]. Then accordingly line EM is less than line EA. Similarly, the whole perimeter of the octagon is less than line EF since it is less than the circumference by the first postulate. Therefore, the product of line EM and the whole perimeter of the octagon is less than the product of line EA and line EF. But the product of line EM and the whole perimeter of the polygon is double the polygon, as was demonstrated before. And similarly the product of line EA and line EF is double the triangle, by I.41 of the *Book of Geometry* [of Euclid]. Therefore, double the polygon is less than double the triangle. Therefore, the polygon is less than the triangle. But it was conceded that it is greater. And so it remains that the given circle is not greater than the designated triangle, since from it follows something that is per se impossible.

Hence let it be that the given circle is less than the designated triangle.

ergo sic: Circulus datus est minor triangulo. Ergo aliquanto est minor triangulo et esto *O* excessus quo triangulus est maior circulo [Fig. 28]. Unde triangulus est equalis dato circulo et *O*. Circumscribatur itaque
160 quadratum diametri dato circulo, quod vocetur *HIKL*, et necessarium est quod tale quadratum sit maius triangulo; quod sic demonstramus:

Totalis linea ambiens quadratum diametri est maior circumferencia ex tercia peticione. Ergo quod fit ex ductu *EA* linee in totalem lineam ambientem quadratum diametri est maius quam id quod fit ex ductu
165 *EA* linee in *EF* lineam, cum *EF* linea sit equalis circumferencie ex ypothesi. Sed quod fit ex ductu *EA* linee in totalem lineam ambientem quadratum diametri est duplum quadrati diametri ex 41 primi libri geometrie. Similiter quod fit ex ductu *EA* linee in *EF* lineam est duplum trianguli. Ex eadem ergo duplum quadrati diametri est maius
170 duplo trianguli. Ergo quadratum diametri est maius triangulo. Sed triangulus est equalis circulo et *O*. Ergo quadratum diametri est maius circulo et *O*. Ergo dempto circulo residuum de quadrato diametri est maius *O*. Sed residuum de quadrato diametri dempto circulo sunt 4 trianguli, quorum quilibet habet duo latera, scilicet duas lineas rec-
175 tas, et basim arcum ex parte trianguli convexum. Cum igitur isti 4 trianguli sint maiores *O*, per primam decimi libri geometrie subtrahatur ab eis maius medietate et de residuo maius medietate et tali con-

174 scilicet *supra scr.* *Bc*

Therefore, proceed as follows: The given circle is less than the triangle. Therefore, it is less than the triangle by some amount, and let *O* be the excess by which the triangle is greater than the circle [see Fig. 28]. Whence the triangle is equal to the given circle and *O*. And so let the square of the diameter, square *HIKL*, be circumscribed about the given circle. It is necessary that such a square be greater than the triangle. This we demonstrate as follows:

The whole perimeter of the square of the diameter is greater than the circumference, by the third postulate. Hence the product of line *EA* and the whole perimeter of the square of the diameter is greater than the product of line *EA* and line *EF* [Fig. 27], since line *EF* is equal to the circumference by hypothesis. But the product of line *EA* and the whole

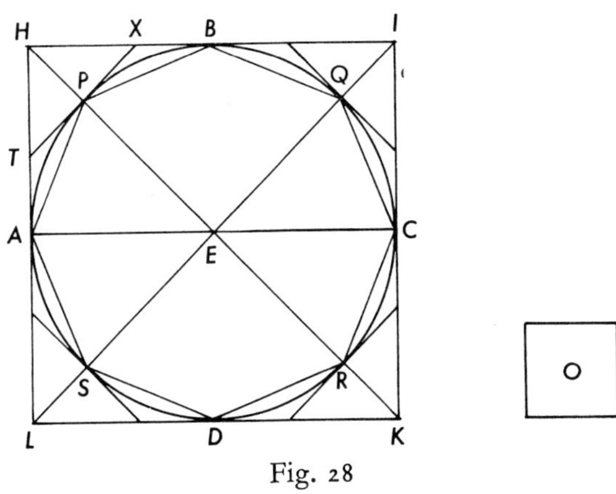

Fig. 28

perimeter of the square of the diameter is double the square of the diameter, by I.41 of the *Book of Geometry* [of Euclid]. Similarly the product of line *EA* and line *EF* is double the triangle. In the same way, therefore, double the square of the diameter is greater than double the triangle. Therefore, the square of the diameter is greater than the triangle. But the triangle is equal to the circle and *O*. Therefore, the square of the diameter is greater than the circle and *O*. Therefore, with the circle subtracted, the rest of the square of the diameter is greater than *O*. But the rest of the square of the diameter after the circle has been subtracted consists of the 4 triangles, each of which has straight lines as its two sides and as its base an arc convex in the direction of the triangle. Therefore, since these 4 triangles are greater than *O*, by means of X.1 of the *Book of Geometry* [of Euclid] let more than half be subtracted from them and [also] more than half from their remainder, and with continued subtraction of this sort finally there

tinua detraccione facta tandem relinquetur minus *O*, scilicet minimi trianguli consimiles predictis. Fiat itaque detraccio maioris medietate 180 hoc modo:

Signentur media puncta 4 arcuum circuli, scilicet arcus *AB* et arcus *BC* et arcus *CD* et arcus *DA*, per puncta *P*, *Q*, *R*, *S*. Deinde ab istis 4 punctis protrahantur ex utraque parte 4 linee contingentes ad circulum, et ex utraque parte protrahantur donec concurrant cum lateri- 185 bus quadrati diametri. Et gratia maioris evidencie linea protracta a puncto *P* concurrens cum latere quadrati diametri *HL* vocetur *PT*. Reliqua vero linea protracta a puncto *P* concurrens cum latere quadrati 172r *HI* vocetur *PX*. / Quibus peractis protrahatur linea a puncto *A* ad punctum *P* et iterum a puncto *P* ad punctum *B* et deinde a puncto *E* 190 per punctum *P* ad punctum *H*. Ex hiis sic: trianguli *TXH* duo latera sunt maiora tercio ex 19 primi libri geometrie, et sic latus *TH* et latus *HX* sunt maiora linea *TX*. Cum ergo linea *TH* et linea *HX* sunt equales ex 6 primi libri geometrie, cum etiam linea *TP* et linea *PX* sunt equales ex 4 primi libri geometrie, manifestum est quod linea 195 *TH* est maior linea *TP*. Sed linea *TP* est equalis linee *TA* ex 6 primi. Ergo linea *TH* est maior linea *TA*. Ergo ex prima sexti libri geometrie triangulus *THP* est maior triangulo *ATP*. Eodem modo demonstrabis quod triangulus *XHP* est maior triangulo *XBP*. Ergo totalis triangulus *THX* est maior duobus parvis triangulis, scilicet triangulo *ATP* 200 et triangulo *XBP*. Constat itaque quod triangulus *THX* est maius medietate illius trianguli cuius duo latera sunt *AH* et *HB* et basis est arcus *AB*. Consimilis subtraccio fiat in consimilibus tribus triangulis. Habemus ergo 4 triangulos rectilineos ortogonios habentes 4 angulos quadrati diametri, qui quatuor sunt maius dimidio prefatorum 205 quatuor triangulorum quorum bases sunt arcus convexi.

Facta itaque consimili detraccione maioris medietate a residuis tandem relinquentur parvi trianguli quorum bases erunt arcus, qui omnes parvi trianguli erunt minus excessu *O* per primam decimi, qui parvi trianguli cum circulo constituunt poligonium ⟨circum⟩scriptum circulo. 210 Ergo illud poligonium est minus circulo et *O*. Ergo est minus triangulo. Quod autem illud poligonium sit maius triangulo patet ex premissis sic:

Totalis linea ambiens poligonium circumscriptum circulo est maior circumferencia ex tercia peticione huius. Ergo quod fit ex ductu *EA*

192 ergo *supra scr. Bc*

will remain a quantity less than O consisting of very small triangles similar to the aforementioned ones. And so let the subtraction of more than half be performed in this way:

Let points P, Q, R, and S be designated as midpoints of 4 circular arcs AB, BC, CD, and DA. Then from these 4 points let there be drawn in each direction 4 lines tangent to the circle, and let these be protracted in each direction until they meet the sides of the square of the diameter. For greater clarity let line PT be drawn from point P and meeting side HL of the square of the diameter. The remaining line drawn from P and meeting side HI of the square we let be called PX. With these things done, let a line be drawn from point A to point P, another from point P to point B, and then one more from point E through point P to point H. Then accordingly two sides of $\triangle$ TXH are greater than the third side, from I. 19 of the *Book of Geometry* [of Euclid], and so $(TH + HX) > TX$. Hence, since line TH and line HX are equal from I.6 of the *Book of Geometry* [of Euclid] and since line TP and line PX are equal from I.4 of the *Book of Geometry* [of Euclid], it is manifest that line TH is greater than line TP. But line TP is equal to line TA from I.6 [of the *Elements*]. Therefore, line TH is greater than line TA. Therefore, from VI.1 of the *Book of Geometry* [of Euclid] $\triangle$ THP is greater than $\triangle$ ATP. In the same way you will demonstrate that $\triangle$ XHP is greater than $\triangle$ XBP. Therefore, the whole $\triangle$ THX is greater than the two small triangles ATP and XBP. And so it is clear that $\triangle$ THX is greater than half of the triangle whose two sides are AH and HB and whose base is arc AB. Let a similar subtraction be performed in the three similar triangles. We have, therefore, 4 right triangles having the 4 angles of the square of the diameter. These 4 triangles are greater than half of the 4 designated triangles whose bases are the convex arcs.

With a similar subtraction made of more than half from the remainders, there will finally remain small triangles whose bases are arcs and which all together are less than the excess O, by X.1 [of the *Elements*]. These small triangles together with the circle constitute a [regular] polygon circumscribed about the circle. Therefore, that polygon is less than the circle and O [together]. Therefore, it is less than the triangle. But that the polygon is greater than the triangle is obvious from the premises as follows:

The whole perimeter of the polygon circumscribed about the circle is greater than the circumference, by the third postulate of this [work]. Therefore, the product of line EA and the whole perimeter is greater than

215 linee in lineam totalem est maius eo quod fit ex ductu *EA* linee in *EF* lineam. Ergo duplum poligonii est maius duplo trianguli. Ergo illud poligonium est maius triangulo. Concessum est autem quod minus, que duo sunt incompossibilia que secuntur ex eo quod dicit falsigraphus quod circulus est minor triangulo. Ergo circulus non est minor
220 triangulo. Concessum est autem quod minor sit. Igitur cum circulus sit maior vel minor vel equalis triangulo, sed non est maior, ut demonstratum est, nec est minor, ut demonstratum est, ergo propositus circulus est equalis illi triangulo, quod demonstrare debuimus.

Hiis demonstratis ducatur *EA* linea in medietatem *EF* linee, i.e.,
225 ducatur semidiameter circuli dati in semicircumferenciam eiusdem et habebis quadrangulum rectangulum equale circulo sive triangulo ex 41 primi libri geometrie. Si enim quod fit ex ductu *EA* linee in *EF* lineam est duplum ad triangulum, quod fit ex ductu *EA* linee in medietatem *EF* linee est equale triangulo, et sic habemus quadrangulum
230 equale circulo. Vocetur itaque medietas *EF* linee *EZ* [Fig. 29]. Deinde iste due linee *EA* et *EZ* sibi directe applicentur in plana superficie, ut ex illis duabus fiat una linea recta, que est *ZA*, quam lineam, scilicet *ZA*, dividas per equalia ex 10 primi libri geometrie in puncto *Y*. Post hoc fiat *Y* centrum et ad quantitatem diametri *ZA* describatur
235 semicircumferencia. Quibus peractis a puncto *E* extrahatur perpendicularis usque ad semicircumferenciam, que vocetur *EM*. Manifestum

218 *mg Bc* Hic falsigraphia est ex eo quod circulus triangulo est improportionalis; ideo nec maior neque minor etiam nec equalis

the product of line EA and line EF [Fig. 27]. Therefore, double the polygon is greater than double the triangle. Therefore, that polygon is greater than the triangle. However it was conceded that it was less. [Hence] these are two contradictory statements following from the fact that the pseudographer* says that the circle is less than the triangle. Therefore, the circle is not less than the triangle even though it was conceded that it is less. Therefore, since the circle must be greater than, less than, or equal to the triangle, but it is not greater, as has been demonstrated, nor is it less, as has been demonstrated, therefore the proposed circle is equal to the triangle, which is what we had to demonstrate.

With these things demonstrated, line EA is multiplied by 1/2 line EF, i.e., the radius of the given circle is multiplied by its semicircumference,

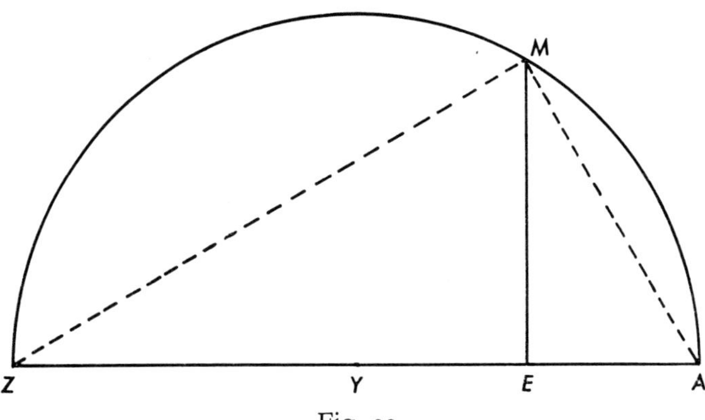

Fig. 29
Note: I have added the dotted lines.

and the result is a rectangle equal to the circle or triangle, by I.41 of the *Book of Geometry* [of Euclid]. For if the product of line EA and line EF is double the triangle, then the product of EA and 1/2 EF is equal to the triangle, and thus we have a rectangle equal to the circle. And so let 1/2 EF be called line EZ [see Fig. 29]. Then these two lines EA and EZ are directly applied to one another in a plane surface, so that from the two of them a single straight line ZA is formed. ZA is bisected at point Y, by I.10 of the *Book of Geometry* [of Euclid]. After this let Y be made the center and let a semicircumference be described with a diameter ZA. When these steps have been completed, let a perpendicular EM be drawn from point E up to the semicircumference. And so it is manifest from the corollary to VI.8 of the *Book of Geometry* [of Euclid] that line EM is a

* See Commentary, lines 218–19.

est itaque ex corollario octavi theoreumatis sexti libri geometrie quod linea *EM* est medio loco proporcionalis inter lineam *EA* et lineam *EZ*. 172v Ergo ex 13 sexti quadratum linee *EM* est equale rectangulo / quod 240 fit ex ductu *EA* linee in *EZ* lineam, et sic quadratum *EM* linee est equale circulo proposito, et sic perfecte invenimus quadratum equale circulo proposito.

[V.] PROPORCIO AREE OMNIS CIRCULI AD QUADRATUM DIAMETRI IPSIUS EST SICUT PROPORCIO UNDECIM AD 245 QUATUORDECIM.

Describatur circulus *ABCD*, cuius diameter sit *AC* [Fig. 30]. Deinde circumscribatur circulo quadratum diametri, quod vocetur *EFGH*. Dico ergo quod que est proporcio plane superficiei circuli ad planam superficiem quadrati diametri eadem est undecim ad quatuordecim; 250 quod sic demonstramus: Protrahatur linea *FG* ex parte *G* in continuum et directum, linea dupla ad lineam *FG*, que vocetur *GK*. Deinde ulterius protrahatur linea ad equalitatem septime partis diametri, que vocetur *KM*. Quibus peractis a puncto *A* protrahantur tres linee, prima ad punctum *G*, secunda ad punctum *K*, tercia ad punctum *M*. 255 Ex hiis sic: linea *FG* est septupla ad lineam *KM* ex ypothesi. Sed linea *GK* est dupla ad lineam *FG* ex ypothesi. Ergo linea *GK* est quatuordecim vicibus maior linea *KM*. Ergo linea *FK* est 21 vicibus maior linea *KM*. Ergo linea *FM* est 22 vicibus maior linea *KM*. Procede ergo sic:

260 Triangulus *AFM* est equalis circulo proposito ex proxima. Est enim linea *FM* tanquam circumferencia circuli propositi. Sed ex prima sexti que est proporcio linee *FM* ad lineam *KM* eadem est trianguli *AFM* ad triangulum *AKM*. Ergo triangulus *AFM* est maior triangulo *AKM* 22 vicibus. Ergo circulus propositus est maior triangulo 265 *AKM* 22 vicibus. Quo constante procedamus iterum sic:

Que est proporcio linee *FG* ad lineam *KM* eadem est trianguli *AFG* ad triangulum *AKM*. Sed linea *FG* est septuplo maior linea *KM* ex

255–56 Sed...ypothesi *mg. Bc* 258 Ergo...KM *mg. Bc*

mean proportional between line EA and EZ. Therefore, from VI.13 (VI.14, *Greek text*) [of the *Elements*], the square of line EM is equal to the rectangle arising from the multiplication of line EA by line EZ, and thus the square of line EM is equal to the proposed circle. And thus we have completely found a square equal to the proposed circle.

[V.] THE RATIO OF THE AREA OF ANY CIRCLE TO THE SQUARE OF ITS DIAMETER IS AS THE RATIO OF 11 TO 14.

Let circle $ABCD$ be described and let its diameter be AC [see Fig. 30]. Then let a square of the diameter be circumscribed about the circle. Let the square be called $EFGH$. I say, therefore, that the ratio of the plane

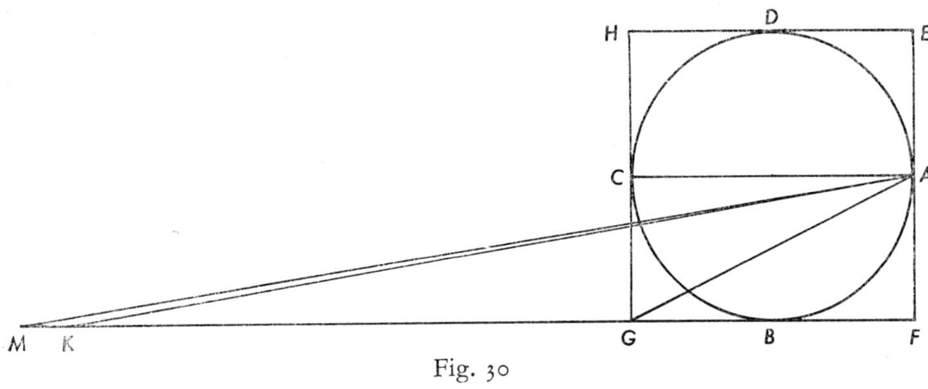

Fig. 30

surface of the circle to the plane surface of the square of the diameter is as 11 to 14. This we demonstrate as follows:

Let line FG be continuously and directly protracted in the direction of G by line GK equal to 2 FG. Then the line is protracted further by a line KM equal to 1/7 of the diameter. With these things done, let three lines be drawn from point A, the first to point G, the second to point K, and the third to point M. From these things we go on to the following: Line $FG = 7$ line KM, by hypothesis. But line $GK = 2$ line FG, by hypothesis. Therefore, line $GK = 14$ line KM. Therefore, line $FK = 21$ line KM. Therefore, line $FM = 22$ line KM. Therefore, proceed as follows:

Triangle AFM is equal to the proposed circle, from the preceding proposition; for FM is as the circumference of the proposed circle. But from VI.1 [of the *Elements*], line FM/line KM = $\triangle AFM / \triangle AKM$. Therefore, $\triangle AFM = 22 \triangle AKM$. Therefore, the proposed circle is 22 times greater than $\triangle AKM$.

With this evident, let us proceed further as follows: Line FG/lineKM =

ypothesi. Ergo triangulus *AFG* est septuplo maior triangulo *AKM*. Sed quadratum diametri est quadruplum ad triangulum *AFG*, cum
270 triangulus *AFG* sit quarta pars ipsius quadrati diametri. Ergo quadratum diametri est 28 vicibus maius triangulo *AKM*. Collige ergo sic: Que est proporcio trianguli *AFM* ad triangulum *AKM* eadem est 22 ad unum. Sed que est proporcio trianguli *AKM* ad quadratum diametri eadem est unius ad 28. Ergo a principio que est proporcio trianguli
275 *AFM* ad quadratum diametri eadem est 22 ad 28 ex 20 quinti libri geometrie. Sed, ut demonstratum est, triangulus *AFM* est equalis circulo. Ergo ex 11[a] quinti libri geometrie que est proporcio circuli ad quadratum diametri eadem est 22 ad 28. Ergo ex 15 quinti libri geometrie que est proporcio circuli ad quadratum diametri eadem
280 est undecim ad quatuordecim, cum multiplicium et submultiplicium sit eadem proporcio; quod demonstrare debuimus.

$\triangle$ *AFG*/$\triangle$*AKM*. But line *FG* = 7 line *KM* by hypothesis. Therefore, $\triangle$ *AFG* = 7 $\triangle$ *AKM*. But the square of the diameter is quadruple $\triangle$ *AFG*, since $\triangle$ *AFG* = 1/4 the square of the diameter. Therefore, the square of the diameter = 28 $\triangle$ *AKM*. Hence, reason as follows: $\triangle$ *AFM*/$\triangle$*AKM* = 22/1. But $\triangle$ *AKM*/sq of diam = 1/28. Therefore, by principle, $\triangle$ *AFM*/sq of diam = 22/28, and by V.20 of the *Book of Geometry* [of Euclid]. But, as has been demonstrated, $\triangle$ *AFM* is equal to the circle. Therefore, from V.11 of the *Book of Geometry* [of Euclid], the circle/sq of diam = 22/28. Therefore, by V.15 of the *Book of Geometry* [of Euclid], the circle/sq of diam = 11/14, since the ratio of multiples and submultiples is the same. This is what we had to demonstrate.

COMMENTARY

1–2 "demonstracionem curvarum superficierem." This may be an indirect reference to *De curvis superficiebus Archimenidis*, particularly since that work emphasized the importance of at least one of these postulates for its theorems (see Chapter Six, Section 2, the text of the *De curvis*, Proposition I, lines 15–17).

20 "linea curva." As I remarked in the introduction to the text of the Corpus Christi Version, this expression includes bent lines composed of straight lines (see particularly lines 31–32, 36, 37, where *linea curva* is employed for such bent lines).

56, 71 "tercia tercii." See the Commentary to the Cambridge Version, line 40.

66–67 "per primam...primi." For II.1 of the Adelard II translation of the *Elements*, see MS Brit. Mus. Add. 34018, 7v–8r: "Si fuerint due linee quarum una in quotlibet partes dividatur, illud quod ex ductu unius earum in alteram fiet equum erit hiis que ex ductu linee indivise in unamquamque partem linee particulatim divise rectangula producentur." For I.41 see the Commentary to the Cambridge Version, line 33.

73 "41 primi." See the Commentary to the Cambridge Version, line 33.

75 "eadem primi." *Ibidem*.

75, 81 "primum secundi." See above in this Commentary, lines 66–67.

109–113 "prima...relinquatur." Compare this to the reading of X.1 on page 60, note 1, above. The latter omits "donec minus eo relinquatur," found here in line 112.

118 "penultima primi." See the Commentary to the Cambridge Version, line 25.

125 "29 tercii." See the Commentary to the Gordanus Version, line 37.

130 "41 primi." See the Commentary to the Cambridge Version, line 33.

137 "primum...decimi." See above, lines 109–113.

143 "tercia tercii." See above, line 71.

149–50 "ut...est." This was demonstrated as proposition IIIA.

151 "41 primi." See the Commentary to the Cambridge Version, line 33.

153–55 "Relinquatur...se." This is another case of an author taking the trouble to point out the logical structure of the proof.

162–63 "Totalis...peticione." It is quite true that this follows from the third postulate, but it follows more immediately from the author's first proposition.

167 "41 primi." See the Commentary to the Cambridge Version, line 33.

176 "primam decimi." See above, lines 109–113.

191 "19 primi." For the Adelard translation of I.19, see the *ms. cit.*, 3v: "Omnis trianguli maior angulus longiori lateri oppositus est." In the Campanus text (Venice, 1482), I.19 reads: "Omnis trianguli maiori angulo longius latus oppositum est." This is I.18 in the Adelard text.

193, 195 "6 primi." See the Commentary to the Naples Version, line 65.

194 "4 primi." See the Commentary to the Naples Version, line 36.

196 "prima sexti." See the Commentary to the Cambridge Version, line 70.

208 "primam decimi." See above, lines 109–113.

218–19 "dicit falsigraphus." I have mentioned the use of *falsigraphus* in my introduction to the text. But I did not comment on the marginal note (see variant reading, line 218), which is of some interest. It reads in translation: "Here there is a false position from this, that the circle is unrelatable by ratio to the triangle and hence it is neither more, nor less, nor even equal." This "false position" is reflective of the Aristotelian and pre-Archimedean belief that a curved line is not really relatable to a straight line, since they are of different species of figure (see the *Physics* of Aristotle, 248a–249a).

227 "41 primi." See the Commentary to the Cambridge Version, line 33.

233 "10 primi." See the Commentary to the Gordanus Version, line 82.

237 "corollario...sexti." For the porism to VI.8 in the Adelard II translation, see the *ms. cit.*, 22v–23r: "Unde etiam manifestum est, quod

in omni triangulo rectangulo, si ab eius angulo recto ad basim perpendicularis ducatur, ipsa perpendicularis inter duas sectiones basis proportionalis erit."

239 "13 sexti." *Ibid.*, 23v: "Si due superficies equidistantium laterum quarum unius cuius angulus uni angulo alterius equalis equales fuerint, latera duos [equos] angulos continentia mutekefia esse; si vero latera duos earum equos angulos continentia mutekefia fuerint, duas superficies equales esse necesse est." This is VI.14 in the Greek text. It is the second half of the proposition which is used here, and in modern parlance we would render it: "If the sides about equal angles of two equiangular parallelograms are reciprocally proportional, the parallelograms are equal."

261–62 "prima sexti." See above, line 196.

275 "20 quinti." See the *ms. cit.*, 20v: "Si fuerint quotlibet quantitates alieque secundum earum numerum atque queque due priorum secundum proportionem duarum postremarum, necesse est in proportionalitate quidem equalitatis, ut si fuerit priorum prima ultima maior et posteriorum primam ultima esse maiorem, quod si minor et minorem, si vero equalis et equalem." This takes the wording from V.20 and V.22 in the Greek text.

277 "11 quinta." *Ibid.*, 19v: "Si fuerint quantitatum proportiones alicui uni equales, ipsas quoque proportiones sibi invicem esse equales [necesse est]."

278 "15 quinti." *Ibid.*, 20r: "Si fuerint aliquibus quantitatibus eque multiplicationes assignate, erit ipsarum multiplicium atque submultiplicium proportio una."

6. The Munich Version

In the beginning of this chapter I suggested that certain of the emended versions of the *De mensura circuli* seemed to show a self-conscious attention to the logical structure of the proof of the initial proposition of the *De mensura circuli.* This is particularly true of the version I have called the Munich Version from its unique manuscript. In fact, the author of the Munich Version is so self-conscious of the structure of his argument that he exhibits the form often found in logical tracts of the fourteenth century where the *consequentie* are laid out and the major, minor, and conclusion are specified. It is for this reason that I assign this version of the *De mensura circuli* to what might be called a scholastic *genre* of these treatises. And because of this logical structure, I am inclined to date it in the fourteenth century in spite of the fact that the one manuscript is of the fifteenth century. That this manuscript is a copy rather than a holograph is evident from the state of confusion of the figures added to the text and the appearance of copying errors, such as the silly substitution of *ypostasis* where *ypothesis* was clearly in the original. The only point that we can be sure of in connection with its dating is that it cites and thus postdates Campanus' version of the *Elements* which was composed some time toward the end of the third quarter of the thirteenth century.

The Munich Version consists of a quite extensive elaboration of Proposition I of the *De mensura circuli*. It bears the title associated with the second tradition of Gerard of Cremona's translation, namely, *De quadratura circuli*. However, when the author specifically cites the *De mensura circuli*, his citation is a paraphrase rather than an exact rendering of Gerard of Cremona's translation. His paraphrase is similar to the wording found in Version F.IB in that it says that the right triangle is equal to the circle rather than that the circle is equal to the right triangle, as it is given in the translation and in most of the versions.

The second paragraph of the Munich Version is a scholastic laying out of the form of the proof, breaking it down into two propositions which are to be proved: [1] the specified right triangle is equal to the given circle, and [2] a square can be found which is equal to the right triangle of Proposition. [1] The conclusion from these two propositions and the first axiom of the *Elements* is that the square and the circle are equal. The third paragraph in turn lays out the form of the argument needed to prove the first part of the first proposition specified in the preceding paragraph. The steps of the proof are indicated as follows: (a) If the triangle is less than the circle, then the circle, on the authority of Aristotle's *Physics*, can be divided into a "quantity which is exceeded" and a "quantity by which it (the circle) exceeds." The excess of the circle over the triangle is designated as P and is added to the triangle. (b) With such a quantity P assumed, it would then be possible to inscribe a regular polygon within the circle which would be greater than the triangle. But this is impossible. Therefore, (a) is impossible. Having shown the form of the argument, the author then expounds (a) and (b) and shows their necessary connection. Incidentally, like the authors of so many of these emended versions, he cites the

Elements of Euclid heavily; and, as I have noted, he also cites Campanus. The second half of the proof is treated in a similar manner. Here we find, in addition to citations to the *Elements*, references to Aristotle's *De caelo* (which, in the medieval manner, he misnames *De caelo et mundo*) and to Ptolemy's *Almagest*. Upon completing the proof of Proposition I, he takes up Proposition II. Its proof is briefly indicated by attention to Proposition II.14 of Euclid's *Elements*. Finally the treatise ends by an extract from Campanus to show that two tangents from the same point outside of a circle are equal, a proposition that entered into the earlier proof of Proposition [1].

There are not many features distinctive to this version of the *De mensura circuli*. While, as my commentary indicates, a few new propositions of Euclid are cited, the over-all form of the proof is much like the other elaborations and depends on X.1 of the *Elements*. The author was also familiar with XII.2 of the *Elements*, which makes use of X.1 in the same manner as the various versions of Proposition I of the *De mensura circuli*. In fact, it is quite evident that the procedure of exhausting the circle by inscribing polygons, which is only indicated in Archimedes' text, was taken in its detailed description by the various authors of the emended versions from XII.2, as I have already indicated. But the other authors did not specifically cite this proposition; and even the author of the Munich Version only cites it in an incidental fashion. Like all the other versions, the Munich Version specifies the quantity by which the circle is said to exceed the triangle or the triangle the circle. It is quantity *P* and is represented by a triangular addition to the right triangle *A*. The author of the Munich Version has freed himself from the use of *lunula* to represent a segment of a circle or the mixed triangles composed of straight lines and an arc. Nor does the author list postulates at the beginning of the treatise as do the Cambridge and Corpus Christi versions, although on numerous occasions he calls attention to basic axioms which he labels as *communis sciencia*, *communis animi concepcio*, or simply *communis concepcio*. These are either axioms drawn directly from Euclid or inferred from Euclidian axioms (or, on occasion, from Aristotle). He does specify (line 130) that the chord is the least line between two points—an Archimedean axiom appearing in somewhat different form among the postulates of the Cambridge and Corpus Christi versions.

My text is a new transcription of manuscript *Y*. Curtze's text of this work has also been consulted, although it is defective in some respects. As a matter of fact, its chief defect is that Curtze made numerous changes in the text and drawings without specifying that these changes had been made and without indicating what the original readings were. I have followed the medieval spelling for the most part. For example, I have left the *ci* form before a vowel instead of converting it to *ti*. I have also retained the spelling *paralellogrammum*, which is the form almost exclusively used by the author. The author vacillates between *polligonium* and *poligonium*, as well as between *semidyameter* and *semydyameter*, but I have retained the spelling *semidyameter* and *poligonium* throughout. The illustrations have been reconstructed slightly. I have restored the orientation given the drawings in the manuscript, which orientation Curtze had altered. But I have kept the relation of circumscribed square to inscribed square of Fig. 31 suggested by Curtze and demanded by the text, for the text tells us that line *BG* and line *LM* are parallel lines. Yet in two drawings (one on folio 182r and the other on folio 183r) *BG* is so drawn that if extended it would intersect *LM*. On Fig. 31, I have kept point *V* as suggested by at least one drawing (183v) and by the text; thus I have rejected Curtze's change to *S* as unnecessary. However, I have accepted on Fig. 32 Curtze's emendation of *V* to *R*, since without this emendation *V* would be repeated twice on the same figure (and indeed on the same line). But I have seen no reason to follow Curtze in reversing the positions of *R* and *Q*. This is another incidence of Curtze's easy tampering with the text when such tampering is not necessary. I have followed Curtze in placing *S* in Fig. 32, for the copyist clearly mixed up *F* and *S*, and surely the original author meant to have *S* be in the position I have now assigned it, as the text clearly reveals. I have not followed the scribe in repeating the same drawings from folio to folio. Fig. 33 is taken from the Campanus text of Euclid, for it was quite erroneously given by the scribe.

Siglum of Manuscript

Y = Munich, Bay. Staatsbibl. cod. 56, 182r-186v, 15c (1434–36). (Cf. the text of Curtze, *Bibliotheca Mathematica*, 3. Folge, vol. 2 [1901], pp. 48–56.)

The Munich Version of the Quadrature of the Circle

Versio Monacensis De Quadratura Circuli

182r / De quadratura circuli

Ad probandum quod sit dare quadratum equale aree circuli assumitur ista proposicio Archimenidis in summa: "Triangulus orthogonius, cuius unum latus est semidyameter circuli et reliquum angu-
5 lum rectum cum semidyametro constituens equale circumferencie circuli fuerit, equalis est circulo."

Formatur enim sic demonstracio: Si triangulus *A* ex circumferencia et semidyametro dati circuli rectum angulum ambientibus et tertio latere opposito recto angulo constitutus [est], dico hunc triangulum
10 esse equalem circulo dato. Sed huic triangulo *A* dabo quadratum equale per ultimam 2ⁱ libri Euclidis, et tunc erunt circuli dati et huius quadrati superficies equales uni et eidem triangulo *A*. Ergo erunt inter se equales per communem animi concepcionem et cetera. Sunt igitur due proposiciones accepte ad probandum, quarum prima est, quod
15 triangulus constitutus ex circumferencia, scilicet dati circuli, et eius [semi]dyametro est equalis circulo dato; et hic triangulus vocetur *A*. Secunda proposicio: quod triangulo *A* erit dare quadratum equale. Et tunc sequitur conclusio, scilicet quod circulus datus et quadratum illud inter se adequantur. Prima sic probatur:

20 Describatur circulus circa centrum *E* [Fig. 31] circulus *BCDGH* secundum quantitatem semidyametri *EB*. Tunc sic triangulus *A* est equalis circulo dato, aut maior aut minor. Si primum, habetur propositum. Si autem idem triangulus sit minor circulo dato, tunc possibile est quantitatem circuli dividi in quantitatem que excellitur et in
25 eam qua excellit, que est communis concepcio quam ponit Aristoteles 4° physicorum capitulo de vacuo. Et equale huius in quo excellit circulus triangulum addatur triangulo, et hoc equale excessui triangulo additum vocetur *P*. Tunc ergo triangulus *A* cum illa quantitate

3–4 orthogonius *corr. ex* othonius
17 triangulo *corr. ex* angulo
20 BCDGH *corr. ex* ABCDGH
21 EB *corr. ex* EBC (?)
24 que *corr. ex* qua

The Munich Version of the Quadrature of the Circle

On the Quadrature of the Circle

For proving that there can be given a square equal to the area of a circle this proposition of Archimedes in its completeness is assumed as follows: "A right triangle, one side of which is the radius of a circle and the other [side] of which makes a right angle with the radius and is equal to the circumference of the circle, is equal to the circle."

The demonstration is fashioned as follows: If △ *A* has been constructed from the circumference and radius of the given circle as the sides including the right angle and with a third side opposite the right angle, I say that this triangle is equal to the given circle. But I shall give a square equal to this △ *A* by the last [proposition] of [Book] II [of the *Elements*] of Euclid, and then the sufaces of the given circle and this square will be equal to one and the same △ *A*; therefore, they will be equal to each other by axiom [1 of Book I of the *Elements*] etc. Hence there are two propositions accepted for proof. The first of them is that the triangle constructed from the circumference and radius of the given circle is equal to the given circle; and let this triangle be called *A*. The second proposition is that a square equal to △ *A* can be given. And then follows the conclusion, namely, that the given circle and the square are equal to each other. The first proposition is proved as follows:

Let circle *BCDGH* be described about center *E* with radius *EB* [see Fig. 31]. So then △ *A* is equal to the given circle, or is greater or less [than it]. If the first, the proposition is had. But if the same triangle is less than the given circle, then it is possible for the quantity of the circle to be divided into a "quantity which is exceeded" and a "quantity by which it (the circle) exceeds." This is an axiom Aristotle posited in [Book] IV of the *Physics* in the chapter "On a Vacuum." And the amount by which the circle exceeds the triangle is added to the triangle. And let this amount, equal to the excess which is added to the triangle, be called *P*. Therefore, △ *A* together with quantity *P* will be equal to the given circle. But with

P erit equalis circulo dato. Sed hoc posito quod circulus sit maior
30 triangulo *A* in illa quantitate *P* sequitur quod esset poligonium infra
eundem circulum inscribi maius triangulo *A*, quod est impossibile.
Ergo est impossibile quod circulus datus sit maior triangulo *A* in
aliqua quantitate, que dicatur *P*, quantacunque sit illa, et cetera.

182v / Nunc ergo ostendendum est quod sequitur hoc impossibile, et
35 postea quod hoc quod sequitur sit impossibile, ut dicitur. Quod autem
sequatur poligonium posse inscribi circulo maius triangulo *A*, sic
demonstratur:

A maiori quantitate possum demere maius suo dimidio, et iterum
de residuo possum demere maius suo dimidio, et hoc semper donec
40 relinquatur minor quantitas qualibet proposita. Et hoc posse fieri
demonstrat prima decimi Euclidis. Dematur ergo de dato circulo isto
modo donec relinquatur minor quantitas de circulo quantitate *P*.
Dematur autem hoc modo: Inscribatur quadratum infra datum circulum, secundum quod docet 6ª quarti Euclidis. Inscribi autem quid
45 sit, docet prima diffinicio posita in principio 4ⁱ Euclidis. Sit ergo dato
circulo quadratum inscriptum *BDGH*, et sic quadratum est maius

30 *post* P *del.* *Y* erit equalis / poligonium *corr. ex* paralellogrammum

it posited that the circle is greater than △ *A* by the quantity *P*, it follows that there would be a polygon inscribed in the same circle which would be greater than △ *A*, which is impossible. Therefore, it is impossible that the given circle is greater than △ *A* by some quantity *P* regardless of that quantity's magnitude, etc.

Therefore, it must now be shown that the [first] impossibility follows and then afterwards that the [second] impossibility dependent on it follows,

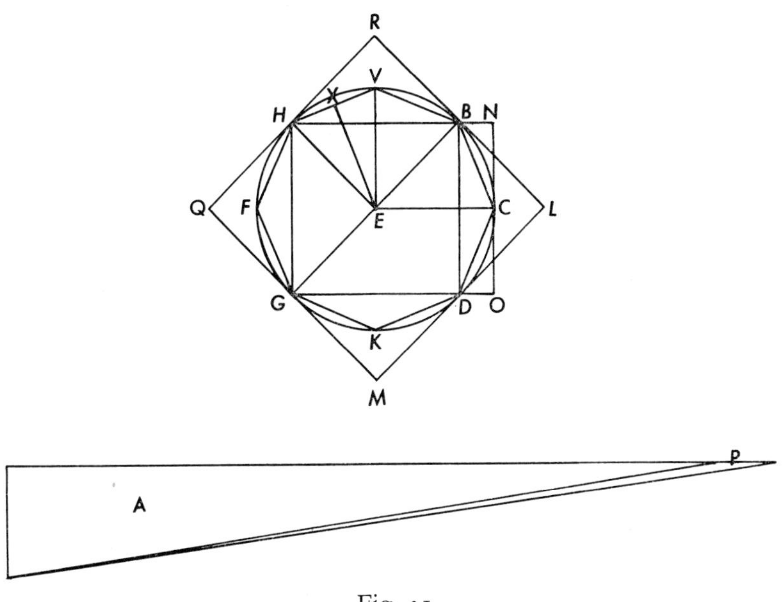

Fig. 31

just as it is stated [above]. But that it follows that a polygon greater than △ *A* can be inscribed in the circle is demonstrated thus:

From a greater quantity I can take away more than its half, and again from the remainder I can take away more than its half, and I can continue to do this until there remains a quantity less than any proposed quantity; and [Proposition] X.1 [of the *Elements*] of Euclid demonstrates that this can be done. Therefore, let us take away quantities from the given circle in this manner until there remains of the circle a quantity less than quantity *P*. Let the subtraction be done in this way. Let a square be inscribed in the given circle as IV.6 of [the *Elements* of] Euclid teaches; moreover, the first definition posited in the beginning of [Book] IV of [the *Elements* of] Euclid teaches us what it is "to be inscribed." Therefore, let square *BDGH* be inscribed in the given circle; and so the square is

dimidio circuli dati, quod sic patet: Quia quadratum *BDGH* est
medietas quadrati *LMQR* circumscripti ipsi dato circulo per penultimam primi Euclidis, quia latera *BD* et *DG* trianguli *BDG* sunt
50 equalia per 6^{am} primi; sed quadratum linee *BG* est equale duobus
quadratis duarum linearum *BD* et *DG* per illam penultimam primi;
ergo idem quadratum linee *BG* est duplum ad alterum illorum duorum quadratorum, quia illa duo sunt equalia. Sed quadratum inscriptum circulo est super lineam *BD*, quia cetera sunt isti equalia quia
55 omnia sunt invicem equalia per diffinicionem quadrati et inscripcionis
et per 6^{am} primi et per primam partem 30^{e} Euclidis 3^{ii}; quadratum
autem circumscriptum eidem circulo est super lineam *BG* quia linea
BG est equalis alteri *LM* quadrati circumscripti quia sunt inter se
equedistantes. Patet igitur quod quadratum inscriptum est dimidium
60 quadrati circumscripti eidem circulo. Hoc eciam docet Campanus in
demonstracione 2^{e} 12^{i}. Sed quadratum circumscriptum est maius circulo dato cui circumscribitur per communem concepcionem que dicit
quod omne totum est maius sua parte. Est ergo quadratum inscriptum
maioris quantitatis medietate quam sit circuli; est ergo idem quadratum
65 inscriptum maius medietate eiusdem circuli dati. Demptum est igitur
secundum quod prima 10^{i} iniungebat.

Porciones ergo relicte simul sumpte minus sunt medietate circuli.
183r Demam iterum de ultimis / porcionibus plus medietate, quod faciam
isto modo: Ab angulis porcionum ducam lineas ad punctum in medio
70 arcus signatum. Primum ad *C* in medio arcus *BCD* signatum ducam
lineas *BC* et *DC*. Similiter faciam in aliis porcionibus omnibus. Et
demam triangulum *BCD* ab ipsa porcione *BCD*. Tunc scio me demp-
sisse a porcione eadem plus medietate porcionis, quia idem triangulus
est medietas maioris quantitatis quam sit porcio. Ergo est maior me-
75 dietate porcionis. Hoc autem quod sit maioris quantitatis medietas
quam sit porcio patebit, quia si super lineam *BD* constituas paralellogrammum cuius altitudinem attingat punctus *C* trianguli *BCD*,
hoc eundem paralellogrammum constat esse duplum ad triangulum
BCD. Sit autem paralellogrammum hoc *BDON*. Cum enim triangu-
80 lus *BCD* et paralellogrammum *BDON* sint in equedistantibus lineis
et super eandem basim *BD*, erit paralellogrammum duplum ad triangulum per 41^{am} primi. Similiter demam de aliis porcionibus similes

47 circuli *corr. ex* circulo
51 BD *corr. ex* $\frac{\mathrm{B}}{\mathrm{KD}}$
60 docet *corr. ex* dicit (?) (*cf. lin. 45, 196*)
64 circuli *corr. ex* circulus (?)
78 eundem *corr. ex* enim (?)
80 *ante* BCD *del.* *Y* BD

greater than half of the given circle, which is evident as follows: Square *BDGH* is one half of square *LMQR*, which is circumscribed about the given circle, by the penultimate [proposition] of [Book] I of [the *Elements* of] Euclid, since sides *BD* and *DG* of △ *BDG* are equal by I.6 [of the *Elements*]; but $BG^2 = BD^2 + DG^2$, by that penultimate [proposition] of [Book] I; therefore, that same square of line *BG* is double each of the two squares since these two squares are equal. But the square inscribed in the circle is on line *BD*, since the other [sides] are equal to it—all [the sides] being mutually equal by the definition of a square and that of inscription and by I.6 and the first part of III.30 of [the *Elements* of] Euclid. Moreover, the square circumscribed about the same circle is on *BG*, because line *BG* is equal to the other side *LM* of the circumscribed square, *BG* and *LM* being parallel lines. It is evident, therefore, that the inscribed square is half of the square circumscribed about the same circle. Campanus also teaches this in the demonstration of XII.2 [of the *Elements*]. But the circumscribed square is greater than the given circle about which it is circumscribed, by the axiom which says that the whole is greater than its part. Therefore, the inscribed square is of greater quantity than half of the circle; therefore, the same inscribed square is greater than half of the same given circle. Hence the subtraction has been performed as X.1 [of the *Elements*] enjoined.

Therefore, the segments left over [in the circle], when taken together, are less than half of the circle. I shall again take away from these last segments more than half, which I shall do in this way: From the angles of the segments I shall draw lines to a point marked in the middle of the arc. First to point *C*, marked in the middel of arc *BCD*, I shall draw lines *BC* and *DC*. I shall do the same thing in all the other segments. And I shall take △ *BCD* away from that segment *BCD*. Then I know that I have subtracted from the same segment more than half of the segment, for the same triangle is half of a quantity which is greater than the segment; therefore, it is more than half of the segment. That it is a quantity more than half of the segment will be evident. For, if you construct a [rectangular] parallelogram on line *BD* with an altitude which point *C* of △ *BCD* terminates, this parallelogram is clearly double △ *BCD*. Now let this parallelogram be *BDON*. Since △ *BCD* and parallelogram *BDON* are between parallel lines and are on the same base *BD*, the parallelogram will be double the triangle, by I.41 [of the *Elements*]. I shall take away in

triangulos. Quo facto, plus dempsero medietate porcionum. Et si illud quod relinquitur de eisdem porcionibus, demptis triangulis pre-
85 dictis, [non] fuerit minus quantitate *P*, demam simili modo triangulos a porcionibus residuis; quod similiter faciam ut feci predictis, scilicet ducendo lineas ab angulis porcionum [ad punctum] in medio arcus eiusdem porcionis signatum et cetera. Quo peracto, quod relinquitur, si sit [minus] quantitate *P*, habeo quod quero. Quod si non est minus,
90 demam semper modo simili sicut feci. Qua operacione demendi completa, dum quod relinquitur erit minus quantitate *P* tunc poligonium factum intra circulum continet quantitatem tocius circuli dati excepto residuo, quod est minus quantitate *P*. Cum ergo circulus datus continet quantitatem trianguli *A* et quantitatem *P*, erit poligonium in-
95 scriptum eidem circulo maius triangulo *A* in tanto quanto ipsius poligonii residuum in circulo est minus quantitate *P*. Conclusum est igitur quod fuit premissum, scilicet quod si triangulus *A* esset minor circulo dato in quantitate aliqua que dicta est *P*, erit dare poligonium intra circulum eundum inscriptum, quod erit maius triangulo *A*, quod
100 erat probandum.

Sequitur ergo demonstrare hoc esse impossibile, scilicet quod intra datum circulum est dare poligonium maius triangulo *A*. Demonstratur autem sic, sumpta ista communi sciencia, quod [quando] quecunque dupla sunt maiora ad aliud ipsorum, media sunt maiora ad me-
183v dium alterius. Si duplum trianguli *A* est maius duplo / poligonii infra
106 circulum inscripti, tunc ipsum medium maioris, scilicet triangulus *A*, est maius dimidio minoris, scilicet poligonio predicto. Quod autem duplum trianguli *A* sit maius duplo poligonii dicti sic patet. Quia [ex] lineis maioribus utriusque angulum rectum in paralellogrammo
110 ambientibus maius fit paralellogrammum per diffinicionem primam positam in principio secundi Euclidis. Sed linee due trianguli *A*, quarum altera equalis est circumferencie dati circuli ex ypothesi et altera semidyametro, ex quibus paralellogrammum duplum [fit] triangulo, sunt maiores lineis ex quibus fit duplum poligonii predicti. Ergo et
115 duplum trianguli *A* est maius duplo poligonii predicti. Quod autem ex illis duabus lineis trianguli rectum angulum constituentibus in triangulo *A* fiat paralellogrammum, patere potest per preallegatam diffinicionem paralellogrammi positam in principio secundi Euclidis. Quod autem idem paralellogrammum sit duplum ad triangulum *A*

92, 101 intra: inter (?) *Y*   112 ypothesi *corr. ex* hypostasi

the same fashion the similar triangles from the other segments. With this done, I shall have taken away more than half of the segments. And if that which remains from the same segments after the aforesaid triangles have been subtracted is not less than quantity P, I shall in the same way subtract triangles from the residual segments. This I shall do in the same way as I did before, i.e., by drawing lines from the angles of the segments [to a point] marked in the middle of the arc of the same segment, etc. With this done, that which remains, if it is [less] than quantity P, is what I seek. But if it is not less, I shall continue to do more subtracting in the same way as I have done. With the operation of subtraction completed, so long as that which remains is less than quantity P, then the polygon inscribed in the circle contains the quantity of the whole given circle excepting for a remainder which is less than quantity P. Therefore, since the given circle contains the quantity of $\triangle A$ and quantity P, the polygon inscribed in the same circle will be more than $\triangle A$ by the amount that the excess of the circle over the polygon is less than quantity P. Therefore, we have concluded what was premised, namely, that if $\triangle A$ were less than the given circle by some quantity designated as P, there can be given a polygon inscribed in the same circle, which will be greater than $\triangle A$, which was to be proved.

Therefore, we must next demonstrate that this is impossible, namely, that inscribed within the same given circle there can be given a polygon greater than $\triangle A$. This is demonstrated as follows, with the axiom assumed, that when doubles are greater than some quantity their halves are greater than the half of the quantity. If double $\triangle A$ is greater than double the polygon inscribed in the circle, then the half of the larger quantity, i.e., $\triangle A$, is greater than half of the lesser quantity, i.e., the aforesaid polygon. That double $\triangle A$ is greater than double the said polygon is evident as follows: For a larger parallelogram results when it has longer sides about the right angle, by the first definition posited in the beginning of [Book] II of [the *Elements* of] Euclid. But the two sides of $\triangle A$, which by hypothesis are equal to the circumference and the radius of the given circle and from which is produced a [rectangular] parallelogram double the triangle, are greater than the lines from which is produced [a rectangle] double the aforesaid polygon. Therefore, double $\triangle A$ is greater than double the aforesaid polygon. That moreover from these two sides containing the right angle in $\triangle A$ a [rectangular] parallelogram is produced can be evident by the previously mentioned definition of a [rectangular] parallelogram posited in the beginning of the second [Book] of Euclid. And, moreover,

120 patet per 41am primi. Quod autem ille linee trianguli *A* sunt maiores lineis que faciunt duplum poligonii predicti, sic constat. Sit enim poligonium octogonium *BCDKGFHV*, cuius centrum sit *E*, et ducatur linea orthogonaliter a centro *E* ad unum suorum laterum, quod gracia exempli sit *HV*, et hec linea egrediatur a puncto *E* in lineam 125 *HV*. Tunc dico quod ex ductu linee *EX* in omnes [lineas] laterales poligonii constituitur duplum poligonii huius. Sed omnes linee laterales non adequantur circumferencie circuli, que est alterum latus trianguli *A*. Quod autem non adequatur patet, quia quelibet corda et arcus ducuntur ab eodem puncto ad eundum, recta autem linea 130 que est corda brevior que esse potest inter illa duo puncta. Ergo omnes corde que sunt latera poligonii inscripti circulo sunt minores ipsa circumferencia. Similiter linea *EX* brevior est semidyametro, quia non procedit ad circumferenciam. Patet ergo quod utraque est brevior in poligonio sua correlativa in triangulo *A*. Quod autem duplum 135 poligonii proveniat ex ductu linee *EX* in omnes lineas laterales poligonii, sic demonstratur. Ex ductu enim *HX* in *XE* fit paralellogrammum per primam diffinicionem positam in principio 2i Euclidis. Sed illud paralellogrammum est duplum ad triangulum *EHX* per 41am primi. Eodem modo provenit duplum trianguli *EVX* ex 184r ductu *VX* in *EX*; ergo ex ductu *EX* / in totam *HV* provenit du- 141 plum utriusque trianguli *EHX* et *EVX* per primam diffinicionem 2i Euclidis. Similiter ex ductu linee equalis *EX* in similia latera poligonii provenit duplum tocius poligonii in eosdem triangulos resoluti. Patet ergo quod querebatur, scilicet quod ex ductu *EX* in omnia 145 latera poligonii provenit duplum poligonii. Sed hoc duplum, ut patuit, est minus duplo trianguli *A*. Ergo et poligonium inscriptum in circulo dato esse maius triangulo *A* est impossibile. Hoc tamen impossibile concludebatur ex ypothesi illa que dicebat triangulum cuius duo latera sunt circumferencie et semidyametro circuli dati 150 equalia esse minorem circulo dato. Non est ergo possibile quod triangulus ille sit minor circulo dato et cetera.

Restat nunc ostendere quod non sit maior idem triangulus *A*, cuius

135 EX *corr. ex* CX
140 EX2 *corr. ex* CX / HV *corr. ex* $\frac{V}{HB}$ 148 ypothesi *corr. ex* ypostasi

that the same parallelogram is double △ *A* is evident by I.41 [of the *Elements*]. Further, that the sides of △ *A* are greater than the lines producing [the parallelogram] double the aforesaid polygon is clear as follows: For let there be an octagonal polygon *BCDKGFHV* [see Fig. 31], whose center is *E*, and let a line be drawn perpendicularly from center *E* to one of its sides, e.g., *HV*, and [so] let this line go out from point *E* to line *HV*. Then I say that the product of *EX* and all the sides of the polygon is double the polygon. But the sum of all the sides is not equal to the circumference of the circle, which circumference is equal to one side of △ *A*. That it is not equal is evident, for any chord and its arc are drawn between the same points and the straight line which is the chord is the least line which can be drawn between these two points. Therefore, all the chords which are the sides of the polygon inscribed in the circle are [together] less than the circumference. Similarly line *EX* is less than the radius because *EX* does not go up to the circumference. It is evident, therefore, that each [of the two lines] in the polygon is less than its corresponding line in △ *A*. That moreover [a rectangular parallelogram] double the polygon arises from the product of *EX* and the sum of the sides of the polygon is demonstrated as follows: A [rectangular] parallelogram arises from the product of *HX* and *XE* by the first definition posited in [Book] II of [the *Elements* of] Euclid. But that parallelogram is double △ *EHX* by I.41 [of the *Elements*]. In the same way [a rectangular parallelogram] double △ *EVX* arises from the product of *VX* and *EX*. Hence from the product of *EX* and the whole *HV* arises [a rectangular parallelogram] double both triangles *EHX* and *EVX*, by the first definition of [Book] II of [the *Elements* of] Euclid. Similarly, from the product of a line equal to *EX* and the similar [other] sides of the polygon arises [a parallelogram] double the whole polygon resolved into those same triangles. Therefore, what was sought is now evident, namely, that from the product of *EX* and the sum of all of the sides of the polygon arises [a parallelogram] double the polygon. But, as was obvious, this double is less than double △ *A*. Therefore, it is impossible that the polygon inscribed in the given circle is greater than △ *A*. However, just this impossibility was inferred from the hypothesis which said that the triangle whose two sides are equal to the circumference and the radius of the circle is less than the given circle. Therefore, it is not possible that the triangle is less than the given circle, etc.

It remains now to show that the same △ *A*, whose two sides including

duo latera angulum rectum ambientia sunt circumferencie et semidyametro circuli dati equalia.

155 Circumscribatur idem circulus dato quadrato *BCDE* [Fig. 32]. Describatur in hoc quadratum poligonium 8 angulorum *FGHKLMNO* extra circulum datum. Dico hoc quadratum circumscriptum esse maius triangulo *A* quia est maius poligonio intra se inscripto, quia est totum respectu ipsius tanquam partis. Contentum enim poligo-160 nium a quadrato pars est quadrati cuiusdam. Sed idem poligonium maius est triangulo *A*, quia duplum poligonii maius est duplo trianguli *A* quia duplum poligonii constituitur ex maioribus lineis quam duplum trianguli *A*. Ita ex istis arguitur. Duplum trianguli *A* per 41^{am} primi constituitur ex duobus lateribus, quorum aliud per ypothe-165 sim circumferencie circuli dati, aliud vero eiusdem dati circuli semidyametro est equale. Duplum vero poligonii constituitur per eandem 41^{am} primi et per primam 2^{i} Euclidis ex semidyametro eiusdem circuli ducto in omnia latera poligonii. Omnia autem latera poligonii sunt quantitates maiores quam circumferencia, cui est equale aliud trianguli 170 latus. Ergo quamvis semidyameter utrobique sit idem vel eius equale in ductu facto in triangulo *A* et in poligonio, tamen reliquum in quod ducitur est maius in poligonio quam in triangulo *A*. Quod autem maioris sint quantitatis omnia latera poligonii quam circumferencia, 184v cui est circumscriptum poligonium, / liquet, quod maius continet 175 spacium poligonium quam circulus idem, quia circulus continetur a poligonio. Si ergo continet maius spacium ipsum poligonium circulo

156 FGHKLMNO *corr. ex* FGHLMNO
160 pars *corr. ex* par / quadrati *corr. ex* quadrato
164–65 ypothesim *corr. ex* ypostasim
167 semidyametro *corr. ex* dyametro

the right angle are equal to the circumference and radius, is not greater [than the circle].

Let the same circle be inscribed in a given square *BCDE* [see Fig. 32]. Let there be described in this square—but outside the given circle—a [regular] polygon of eight angles, *FGHKLMNO*. I say that this circumscribed square is greater than △ *A* because it is greater than the polygon inscribed in it, for it is as a whole is related to its part since the polygon contained by the square is a part of the same square. But the same polygon is greater than △ *A* because double the polygon is greater than double △ *A* since double the polygon is produced from greater lines than double △ *A*. The argument is as follows: By I.41 [of the *Elements*], double △ *A*

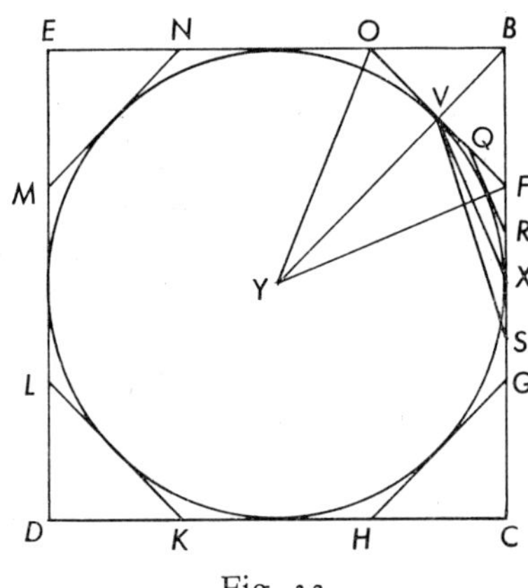

Fig. 32
Note: For triangle *A*, see Fig. 31.

is produced from the two sides, which by hypothesis are equal to the circumference and the radius of the given circle. But by the same I.41 and II.1 [of the *Elements*] double the polygon arises from the product of the radius of the same circle and the sum of all of the sides of the polygon. Moreover, all of the sides of the polygon are [together] greater than the circumference, to which one side of the triangle is equal. Therefore, although in each case the radius is the same or its equal in the product from which arises the triangle and that from which arises the polygon, still the other term by which it is multiplied is greater in the case of the polygon than in that of △ *A*. That, moreover, all of the sides of the polygon [together] are of greater quantity than the circumference about which the polygon has been circumscribed is clear, because the polygon contains more space than does the circle since the circle is contained by the polygon. If, therefore, the polygon contains more space than the given circle, there

dato, ergo et in lineis ambientibus est maior quantitas ponenda, quia circularis linea equalis plus continet de spacio quam recta, ut dicit Aristoteles primo celi et mundi, et Ptolemeus primo libro almagesti
180 dicit capitulo 3° sic: "Et est circulus maior figuris superficialibus et est spera maior figuris corporeis." Conclusum est ergo quod duplum poligonii maius est duplo trianguli *A*. Ergo poligonium est maius triangulo *A* per communem scienciam: quorum dupla sunt maiora, ipsa sunt maiora. Multo forcius ergo quadratum circumscriptum cir-
185 culo, quod continet idem poligonium octogonium, est maius triangulo *A* per communem scienciam: quod est maius maiore est maius minore.

Hoc habito circumscribatur idem poligonium octogonium ipsi circulo dato per hunc modum: A puncto *F* signato in latere *BC* qua-
190 drati circumscripti ducatur linea contingens in punctum *O* lateris quadrati *BE*, fiuntque partes *BF* et *BO* equales, que resecate sunt de predictis lateribus. Simili modo ducantur linee contingentes circulum intra cetera latera, ut a puncto *G* lateris *BC* ducatur linea *GH* ad punctum *H* lateris *CD* et a puncto *K* lateris *CD* ad punctum *L* lateris
195 *DE* ducatur linea *KL* et cetera, eritque poligonium 8 angulorum *FGHKLMNO*. Quid autem sit linea contingens circulum docet diffinicio secunda in principio 3ⁱ Euclidis posita.

Hiis ergo sic dispositis demonstratur quod triangulus *A* datus non sit maior circulo dato, cuius trianguli duo latera talia sunt quod aliud
200 est equale circumferencie et aliud semidyametro circuli dati. Fiatque talis consequencia: si triangulus *A* est maior circulo dato, sequitur quod erit dare poligonium circumscriptum circulo dato, quod quidem poligonium erit minus triangulo *A*. Sed hoc est impossibile; ergo et illud ex cuius posicione sequitur hec conclusio. Non ergo est possibile
205 quod triangulus sit maior circulo dato.

Primo ostendatur consequencia secunda; quod quidem sit impossibile sic demonstratur. Resumatur secundum per ypothesim quod triangulus *A* est maior circulo dato. Sit hoc ergo in aliqua quantitate que vocetur *P*. Nunc enim est dividi quod excellens et cetera, ut
210 supra. Hic autem excessus *P* minor est excessu in quo quadratum
185r circumscriptum circulo / excedit circulum, quia et triangulus *A* est minor quadrato circumscripto, ut demonstratum fuit. Possis autem

177 ponenda: pon^me (?) *Y*
184 circum- / *corr. ex* in-
189 signato *corr. ex* signo
193 intra: int᷑ *Y* / GH *corr. ex* BH
201 circulo *corr. ex* semicirculo
207 ypothesim *corr. ex* ypostasim

is to be posited a greater quantity within the enclosing lines [i.e., the perimeter of the polygon]. For an equivalent circular line contains more space than a straight line, as Aristotle says in [Book] I of the *Heavens and World*; and Ptolemy says in [Book] I, Chapter 3, of the *Almagest*: "The circle is the greatest of surface figures and the sphere the greatest of corporeal figures." It has been concluded, therefore, that double the polygon is greater than double △ *A*. Therefore, the polygon is greater than △ *A* by the axiom: those whose doubles are greater are themselves greater. Hence a fortiori the square circumscribed about the circle—i.e., the square which contains the same octagonal polygon—is greater than △ *A* by the axiom: That which is greater than the greater [of two quantities] is greater than the lesser [of those two quantities].

With this accepted, let the octagonal polygon be circumscribed about the given circle by this method: From point *F* marked in side *BC* of the circumscribed square a tangent [to the circle] is drawn to point *O* in side *BE* of the square, such that parts *BF* and *BO*, which have been cut out from the aforesaid sides, are made equal. In the same way, tangents to the circle are drawn between the other sides, so that, from point *G* in side *BC*, line *GH* is drawn to point *H* in line *CD* and, from point *K* in side *CD*, line *KL* is drawn to point *L* in side *DE*, and so on; and there will [thus] result a polygon of 8 angles, *FGHKLMNO*. Moreover, the second definition posited in the beginning of [Book] III of [the *Elements* of] Euclid teaches [us] what a tangent to a circle is.

Hence with these things so disposed, it is demonstrated that the given △ *A* is not greater than the given circle—the triangle having two sides such that one is equal to the circumference and the other is equal to the radius of the given circle. And let the following inference be drawn: If △ *A* is greater than the given circle, it follows that there can be given a polygon circumscribed about the circle which indeed will be less than △ *A*. But this is impossible, and therefore the position from which this conclusion follows [is also impossible]. Therefore, it is not possible that the triangle is greater than the given circle.

In the first place let the second inference be shown. That indeed it is impossible is demonstrated as follows: Let it again be assumed by hypothesis that △ *A* is greater than the given circle by some quantity which we call *P*, for now there can be a division into "that which [it] exceeds" and "that by which it exceeds," as above. Moreover, this excess *P* is less than the excess by which the square circumscribed about the circle exceeds the circle, because △ *A* is less than the circumscribed square, as was shown. But you can

demere de excessu qua excedit quadratum circulum, quia hoc docet prima 10^{i}, donec relinquat minus ipso *P*, quod faciam isto modo:
215 Ducam a centro ad angulum *B* quadrati circumscripti [lineam] que secat lineam *FO* contingentem in puncto *V*. Tunc dico quod latus *BF* trianguli *BVF*, quod est oppositum angulo recto, maius est latere *FV* eiusdem per 14^{am} primi. Sed latus *FV* est equale linee *FX* per 16^{am} proposicionem tercii. Est autem *X* punctus in quo latus *BC*
220 quadrati *BCDE* contingit circulum datum. Sed cum angulus *BVF* trianguli *BVF* sit rectus quia 4 anguli qui sunt ad *V* punctum contingencie sunt recti per 17^{am} 3^{ii}, ergo *BF* erit ipso *FV* maior per 18^{am} primi. Igitur *BF* est maior *FX*. Ducatur igitur *VX*. Dico quod triangulus *FBV*, qui est super *FB* basim maiorem, maior est triangulo
225 *FVX*, qui habet *FX* basim minorem inter latera quadrati inscripti equedistancia per conversam 38^{e} primi. Si enim inter equedistantes trianguli equalium basium fuerint, sunt equales; ergo qui sunt equales trianguli inter equedistantes habent equales bases. Igitur qui non habent equales bases non sunt equales. Si enim ille qui habet minorem
230 basim sit equalis habenti maiorem, ut quod triangulus *FXV* qui habet minorem basim, ut probatum est, sit equalis triangulo *BFV*, tunc producatur *FX* linea usque ad *S* punctum, ubi sit equalis *BF*, *SF*, et ducatur linea *SV*; tunc dico quod, cum triangulus *BFV* habeat basim *BF* [et] triangulus *FVS* habeat *FS* basim equalem basi *BF* et in
235 eadem linea, igitur triangulus *FSV* est equalis triangulo *BFV* per 38^{am} primi. Sunt enim inter equedistantes per 41^{am} primi. Igitur si eidem *BFV* est equalis, ergo inter se sunt equales illi duo, scilicet *FXV* et *FSV*, scilicet pars et totum, quod est impossibile. Verum est igitur quod triangulus *BFV* est maior triangulo *FXV*. Ergo idem
240 triangulus *BFV* multo a forciori est maior parte trianguli *FXV*, que est porcio contenta ex arcu *XV* et inter duas lineas *FV* et *FX*. Dempto ergo triangulo *BFV* a porcione, que erat inter lineas *BV*
185v et *BX* et extra arcum circuli dati inter / illas contentum, demptum erit maius medietate eiusdem porcionis. Similiter si dempsero trian-
245 gulum *BVO* de alia simili porcione, demptum erit maius dimidio eiusdem porcionis. Similiter operabor circa alias porciones inter reliquos quadrati angulos contentas. Quo peracto, si quod relinquitur inter latera poligonii circumscripti et arcus circuli lateribus subiectos

214 relinquat *corr. ex* reliquam
218 FV[2] *corr. ex* $^{VR}_{FR}$
232 S *corr. ex* B / SF *corr. ex* SX
238 FXV *corr. ex* SXV

make subtractions from the excess by which the square exceeds the circle—since X.1 [of the *Elements*] teaches it—until there remains a quantity less than *P*. This I shall do in the following way: I shall draw from the center to ∠ *B* of the circumscribed square [a line] which cuts the tangent *FO* in point *V*. Then I say that side *BF* of △ *BVF*, which side is opposite the right angle, is greater than side *FV* of the same [triangle] by I.14 (I.18?) [of the *Elements*]. But side *FV* is equal to line *FX* by Proposition III.16* [of the *Elements*]. Further, *X* is the point at which side *BC* of square *BCDE* is tangent to the given circle. But since ∠ *BVF* of △ *BVF* is a right angle because the four angles at *V*, the point of tangency, are right angles by III.17 [of the *Elements*], therefore *BF* will be greater than *FV* by I.18. Hence *BF* is greater than *FX*. Hence let *VX* be drawn. I say that △ *FBV*, which is on the greater base *FB*, is greater than △ *FVX*, on the smaller base *FX*, [for these triangles are constructed] between the parallel sides of the inscribed [and circumscribed] square, by the converse of I.38 [of the *Elements*]. For if triangles of equal bases are constructed between parallel lines, they are equal; therefore, those which are equal triangles and are constructed between parallel lines have equal bases. Therefore, those which do not have equal bases are not equal. For if that which has a lesser base is equal to that which has a larger base, so that △ *FXV*, which was proved to have a lesser base, is equal to △ *BFV*, then let line *FX* be continued to point *S* where *BF* = *SF* and let line *SV* be drawn. Then I say that, since △ *BFV* has base *BF* and △ *FVS* has base *FS* equal to base *BF* and in the same line, therefore △ *FSV* is equal to △ *BFV* by I.38 [of the *Elements*], for they are between parallel lines by I.41 [of the *Elements*]. Therefore, if each is equal to the same *BFV*, these two triangles—*FXV* and *FSV*—are equal, i.e., the part and the whole, which is impossible. It is true, therefore, that △ *BFV* is greater than △ *FXV*. Therefore, the same △ *BFV* is even greater than the part of △ *FXV* consisting of the segment contained by arc *XV* and the two lines *FV* and *FX*. Therefore, with △ *BFV* taken away from the figure contained between lines *BV* and *BX* and outside of the arc of the given circle which is limited by these lines, there will have been subtracted more than half of that figure. Similarly, if I take away △ *BVO* from the other similar figure, more than half will have been subtracted from that same figure. I shall operate in the same way in regard to the other figures contained in the remaining angles of the square. With this done, if what remains between the sides of the circumscribed polygon and the

* See Commentary, lines 308–310.

hiis demptis minus est quantitate *P*, tunc habeo quod quero. Si non
250 sit minus, adhuc demam similiter sicut feci. De lateribus enim poligonii circa angulum *F*, que latera sunt *FO* et *FG*, sumam equalia circa *F* et ducam lineam *RQ* ita quod *R* sit in latere *FG* et *Q* sit in latere *FO* sic quod linea *RQ* contingens sit circulum. Tunc similiter probo ut prius triangulum *FRQ* esse maiorem dimidio porcionis prius relicte.
255 Sicque probo in ceteris porcionibus. Et hos triangulos cum dempsero, subtractum erit maius dimidio porcionum relictarum. Sicque faciam donec excessus qui est inter latera poligonii circumscripti et circumferenciam circuli, quo poligonium ultimo formatum excedit circulum datum, minor sit quantitate *P*. Tunc igitur cum circulus cum quantitate
260 *P* equaretur triangulo *A*, sequitur quod poligonium in quo est circulus datus cum minore quantitate quam sit *P* sit minus ipso triangulo *A*. Conclusum est igitur quod est dare poligonium extra circulum datum, quod est minus ipso triangulo *A*, quod dixi esse impossibile. Et per hoc posicio erit impossibilis.

265 Nunc superest demonstrare quod hoc consequens sit impossibile, quod scilicet possit circumscribi circulo dato poligonium minus triangulo. Ducantur linee a centro circuli dati ad singulos angulos poligonii. Ponatur enim gracia exempli quod poligonium istud sit ipsum octogonium *FGHKLMNO* [Fig. 32]. Ducantur insuper linee a centro
270 ad singula puncta contactus laterum poligonii, de quibus primo pro exemplo ea que egreditur a centro *Y* ad *V* punctum contactus lateris *FO* facit angulos rectos ad eandem lineam *FO* per 17^{am} 3^{ii} Euclidis. Similiter et alie ducte ad alia puncta contactuum faciunt angulos rectos. Hec autem ducam in punctum *V* si ducatur in latus *FV* faciet per
275 41^{am} primi duplum trianguli *FYV*. Ponatur enim in centro *Y*. Similiter si ducatur eadem linea *YV* in latus *VO* faciet duplum trianguli
186r *OYV* per eandem / 41^{am} primi. Est enim paralellogrammum rectangulum hoc quod fit ex *YV* in *VO* per diffinicionem primam positam in principio secundi Euclidis. Similiter si eadem *YV* vel sibi equalis
280 ducatur in quamlibet ceterarum, faciet duplum trianguli cuiuslibet, in cuius latus ducetur. Ergo si ducatur *YV* in omnia latera poligonii, faciet duplum omnium sive duplum tocius poligonii per primam secundi. Tunc sic duplum poligonii circumscripti circulo dato maius est duplo

252 RQ *corr. ex* VQ / R *corr. ex* V
253 RQ *corr. ex* VQ
254 FRQ *corr. ex* FVQ
269 FGHKLMNO *corr. ex* FGHKMNO
271 ea *corr. ex* eam
273 alie *corr. ex* alia
276 duplum *corr. Y ex* triangulum
277 paralellogrammum *corr. ex* poligonium

arcs subjoined to these sides after the subtraction has been made is less than quantity *P*, then I have what I seek. If it is not less, I make a further subtraction in the same way as I have done [above]. For from sides *FO* and *FG* about ∠ *F*, I take equal lines from *F*, and I draw line *RQ* so that *R* is in line *FG* and *Q* is in line *FO* and such that line *RQ* is tangent to the circle. Then in the same way as before I prove that △ *FRQ* is greater than half of the figure left over before. This I also prove in regard to the other figures. And when I take away these triangles, more than half of these residual figures will have been subtracted. I shall continue to proceed in this way until the excess which is between the sides of a circumscribed polygon and the circumference of the circle and by which the polygon finally constructed exceeds the given circle is less than quantity *P*. Hence, since the circle with *P* would be equal to △ *A*, it follows that [the sum of] the polygon in which the given circle lies and a quantity less than *P* is less than this △ *A*. It has been concluded, therefore, that there can be given a polygon outside of the given circle which is less than △ *A*. This I have said to be impossible. Accordingly the position [holding the triangle to be greater than the circle] will be impossible.

Now it remains to demonstrate that this consequent is impossible, namely, that a polygon less than the triangle can be circumscribed about the given circle. Let lines be drawn from the center of the given circle to each of the angles of the polygon—it is posited as an example that this polygon is the octagon *FGHKLMNO* [see Fig. 32]. In addition, let lines be drawn from the center to each of the points of tangency of the sides of the polygon. Of these lines, our first example is the line which extends from the center *Y* to *V* the point of contact with side *FO*. This line makes right angles with that same line *FO*, by III.17 of [the *Elements* of] Euclid. And in the same way other lines drawn to the other points [of contact would] form right angles. If the line drawn to *V* is multiplied by side *FV*, the product will be double △ *FYV*, by I.41 [of the *Elements*]; for let the line start from center *Y*. Similarly, if the same line *YV* is multiplied by side *VO*, the product will be double △ *OYV*, by the same I.41. For it is a rectangular parallelogram which arises from the product of *YV* and *VO*, by the first definition posited in the beginning of [Book] II of [the *Elements* of] Euclid. Similarly, if the same *YV*, or a line equal to it, is multiplied by any one of the sides, the product will be double the triangle whose side was used in the multiplication. Hence if *YV* is multiplied by all of the sides of the polygon, the product will be double all [the triangles], or double the whole polygon, by II.1 [of the *Elements*]. Thus, double the

trianguli *A*, ergo poligonium est maius triangulo *A* per communem
285 scienciam: quorum dupla sunt maiora ipsa sunt maiora. Quod autem
duplum poligonii sit maius duplo trianguli *A* patet, quia, ut dictum
est supra et probatum in ista demonstracione, omnia latera poligonii
circumscripti sunt maioris quantitatis ipsa circumferencia, cum poligonium contineat maius spacium circulo. Sed maius latus trianguli *A*
290 est equale circumferencie circuli dati, et minus latus eiusdem trianguli
est equale linee *YV*, quia est semidyameter. Cum ergo aliud latus in
poligonio, scilicet quantitas omnium laterum eius, sit maius altero
latere trianguli *A*, reliquis existentibus equalibus quia sunt equalia
semidyametro relique duo, fiet maius paralellogrammum ex ductu
295 semidyametri in latera poligonii quam ex ductu eiusdem semidyametri,
qui est latus in triangulo *A*, in aliud latus trianguli eiusdem. Universaliter ergo verum est omne poligonium circumscriptum dato circulo
esse maius triangulo, cuius aliud latus est equale circumferencie et
aliud latus semidyametro, que due latera ambiunt rectum angulum.
300 Sequebatur autem quod esset minus hoc poligonium predicto triangulo ex ypothesi que ponebat triangulum predictum esse maiorem
circulo dato; ergo ypothesis illa est falsa. Non est ergo maior triangulus talis circulo dato, nec minor, ut superius est probatum; ergo
est equalis et cetera.

305 Invenitur autem quadratum quod huic triangulo sit equale per
doctrinam 14[e] proposicionis 2[i], que est ultima eius 2[i] Euclidis; ad
quod eciam faciunt 42 aut 44 primi et 5[ta] secundi et cetera.

186v / Proposicio autem [ex] 36 vel 16 tercii est: "Puncto extra circulum
signato si ab eo ducantur due linee circulum contingentes, ipse sunt
310 sibi invicem equales." Hanc enim sic demonstrat Campanus super
penultimam tercii Euclidis. Sit punctus *A* extra circulum *BCD* [Fig. 33],
cuius centrum est *E*, et ab ipso ducantur due linee *AB* et *AD* contingentes circulum in punctis *B*, *D*. Dicam ipsas esse equales. Producam enim lineas *EB* et *ED* et *EA*, eruntque per 17[am] 3[ii] anguli *B*
315 et *D* sibi invicem equales. Ducatur insuper linea *BD*, eritque angulus
EBD per 5[am] primi equalis angulo *EDB*, quia latera *EB* et *ED* sunt
equalia; ita et duo residui anguli *ABD* et *ADB* sunt equales per
communem scienciam: si ab equalibus equalia demantur et cetera.

301 ypothesi *corr. ex* ypostasi
302 ypothesis *corr. ex* ypostasis
314 EB et ED et EA *corr. ex* CB et CD et CA

polygon circumscribed about the given circle is more than double △ *A* [see Fig. 31]; therefore, the polygon is greater than △ *A* by the axiom: those whose doubles are greater are themselves greater. That moreover double the polygon is greater than double △ *A* is obvious, because, as was said and proved above in this demonstration, all the sides of the circumscribed polygon [together] are of greater quantity than the circumference, since the polygon contains more space than the circle. But the larger side of △ *A* is equal to the circumference of the given circle and the smaller side of the same triangle is equal to line *YV* because it is the radius. Since, therefore, one multiplier in the polygon, i.e., the sum of all of its sides, is greater than one side of △ *A*, while the remaining multiplicands [respectively of the polygon and △ *A*] are equal since they both equal the radius, a greater [rectangular] parallelogram arises from the product of the radius and the sides of the polygon than from the product of the same radius—as a side of △ *A*—and the other side of the same triangle. Therefore, it is universally true that every polygon about the given circle is greater than the triangle whose sides are equal respectively to the circumference and the radius—the sides being those which include the right angle. It followed, however, that this polygon would be less than the aforesaid triangle from the hypothesis which posited the aforesaid triangle to be greater than the given circle. Therefore, that hypothesis is false. Hence such a triangle is not greater than the given circle, nor less, as was proved above; therefore, it is equal, and so on.

Now a square which is equal to this triangle is found by the teaching of Proposition II.14 [of the *Elements*], which is the last proposition of [Book] II of Euclid. Use also for this I.42, or I.44, and II.5 [of the *Elements*], and so on.

There is a further proposition [following from] III.36 or III.16: "If from a designated point outside of a circle two lines are drawn tangent to the circle, these lines will be equal." For Campanus demonstrates this [in the commentary] on the penultimate [proposition of Book] III of [the *Elements* of] Euclid. Let point *A* be outside of circle *BCD*, whose center is *E* [see Fig. 33]. And from *A* let lines *AB* and *AD* be drawn tangent to the circle at points *B* and *D*. I say that these lines are equal. For I draw lines *EB*, *ED*, and *EA*; and ∠ *B* will be equal to ∠ *D*, by III.17 [of the *Elements*]. In addition, line *BD* is drawn; and ∠ *EBD* = ∠ *EDB*, by I.5 [of the *Elements*]; for sides *EB* and *ED* are equal. And thus the two residual angles *ABD* and *ADB* are equal by the axiom: if equals are

Igitur duo latera *AD* et *AB* sunt equalia per 6am primi, quod volebam. Patet igitur, cum in nostro paralellogrammo* a puncto *F* ducantur linee *FV* et *FX* [Fig. 32] circulum contingentes, quod ipse sunt equales. Sed cum angulus *BVF* trianguli *BVF* sit rectus quia 4 anguli qui sunt ad *V* punctum contingencie sunt recti per 17am 3ii, igitur *FV* erit minor ipso [*BF*].

320

* See Commentary, line 320.

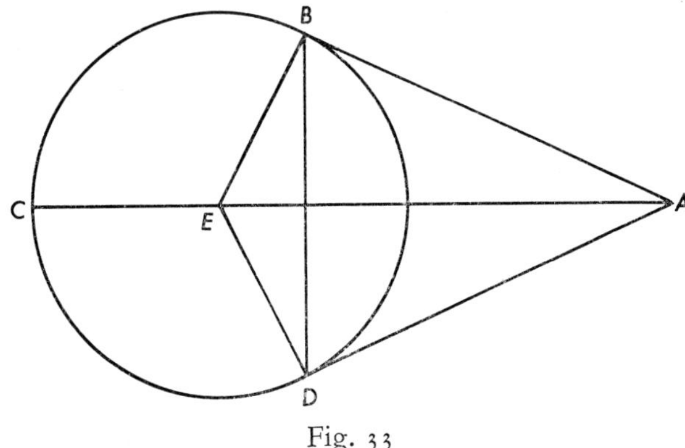

Fig. 33

subtracted from equals, and so on. Therefore, the two sides AD and AB are equal, by I.6 [of the *Elements*], which is what I wished.

It is obvious, therefore, that, since in our parallelogram (? proof ?) [see Fig. 32] lines FV and FX are drawn from point F as tangents to the circle, they are equal. But since $\angle BVF$ of $\triangle BVF$ is a right angle —all four angles at V being right by III.17 [of the *Elements*]—therefore FV will be less than it [i.e., BF].

COMMENTARY

3 "summa." This reading does not completely satisfy me. Curtze's text has *Mensura*, which makes sense, but it is clear that this is not justified by the manuscript. Another possible reading is *sententia*. See the specimen of MS *Y* in the plates.

13 "communem animi concepcionem." This is the first axiom of Euclid's *Elements*. For various medieval Latin versions of this axiom, see M. Clagett, "Medieval Latin Translations of Euclid's *Elements* from the Arabic," *Isis*, vol. 44 (1953), pp. 31, 39. It is of interest to note that the Munich author also uses the shortened form "communis concepcio" and the expression "communis sciencia" for axiom.

25–26 "communis . . . vacuo." Aristotle's *Physics*, Bk. IV, chap. 8, 215b; medieval translation (of Moerbeke?) printed with the exposition of Aquinas (Rome, 1954, p. 255): "necesse est enim dividi excedens (*or* excellens) in excedentiam (*or* excellentiam) et in id quod exceditur (*or* excellitur)." (The variant readings in parentheses come from the copy of this translation included in Walter Burley, *Super Aristotelis libros de physica auscultatione...commentaria*, Venice, 1589, 464.) Cf. Thomas Bradwardine, *De continuo*, Supposit. 1, MS Thorn, R 4° 2, p. 156: "Omne maius posse dividi in equale et in differentiam qua excedit...prima supponitur 4to Physicorum, capitulo de vacuo, differentiam et excessum."

41 "prima decimi." See page 60, note 1, above.

44 "sexta quarti." See the Commentary to Section 3 above, Version F.IB, line 9.

45 "prima...quarti." Defenition 1, Book IV, runs in the Adelard II Version (BM Add.34018, 15v) as follows: "Figura intra figuram dicitur [in]scribi quando ea que inscribitur eius in qua [in]scribitur latera unoquoque suo angulo ab interiore sui parte contingit."

48–49, 51 "penultimam primi." See the Commentary to the Cambridge Version, line 25.

50, 56 "6am primi." See the Commentary to the Naples Version, line 65.

56 "primam...3ii." For the first part of III.30 [III.31 in the Greek text], see the *ms. cit.*, 14r: "Si rectilineus angulus in semicirculo supra arcum constituat rectus est."

60–61 "Campanus...12i." See the E. Ratdolt edition (Venice, 1482), Prop. XII.2, "...quadratum enim quod est duplum ad ipsum [inscriptum] est circulum circumscribens, ut patet ex penultima primi et 7 quarti."

62–63 "communem...parte." See Adelard II (*ms. cit.*, 1r) where this axiom is quoted in exactly these words.

66 "prima 10i." See page 60, note 1, above.

82 "41am primi." See the Commentary to the Cambridge Version, line 33.

103–105 "communi...alterius." This is not included among the axioms—genuine or false—that accompany the Adelard II version of the *Elements*. One could say that it is embraced by Definition 5, Book V.

110–11, 118, 137, 141 "difficionem...secundi." For the Adelard II version of Def. 1, Book II, see the *ms. cit.*, 7v: "Omne parallelogramum rectangulum sub duabus lineis rectum angulum continentibus dicitur contineri."

120, 139, 164, 167 "41am primi." See the Commentary to the Cambridge Version, line 33.

128–30 "quelibet...puncta." Cf. note 2, Section 1, of this chapter for the similar statement of Archimedes.

167 "primam secundi." See the Commentary to the Corpus Christi Version, lines 66–67.

179 "Aristoteles...mundi." I cannot find any such statement in the *De caelo*, but the author might have interpreted the argument of Bk. II, Chap. 4 (287ª), along these lines.

179–81 "Ptolemeus...corporeis." Cf. the edition of the *Almagestum* (Venice 1515), Bk. I, Chap. 3, 3r: "Et quia figure plurium laterum que sunt in circulis equalibus plures habentes angulos sunt eis maiores, est circulus maior figuris superficialibus et est sphera maior figuris corporeis." Cf. *Tractatus de ysoperimetris* (MS Bodl. F.5.28, 106v): "...circulus omnium ysoperimetrarum figurarum maximus est.... Omnium ysoperimetrorum solidorum maximum est spera."

183–84 "communem...maiora." See the Commentary, lines 103–105, above.

186–87 "communem...minore." This is not a formally stated axiom in the *Elements*. However, it clearly follows from the last axiom of Book I of the medieval Euclid, namely, that a whole is greater than its part.

197 "diffinicio...3ⁱ." See the Campanus text of Ratdolt, Bk. III, Def. 2: "Circulum linea contingere dicitur, que cum circulum tangat in utramque partem eiecta, circulum non secat."

214 "prima decimi." See the Commentary, line 41, above.

218 "14ᵃᵐ primi." This is in error; it should, I believe, be I.18 (*ms. cit.*, 3v): "Omnis trianguli longius latus maiori angulo oppositum est."

219 "16ᵃᵐ...tercii." III.16 [III.17 in Greek text] in Adelard II (*ms. cit.*, 13r) runs: "A puncto dato ad datum circulum lineam contingentem ducere."

222 "17ᵃᵐ tercii." See the Commentary to the Naples Version, line 61.

222–23 "18ᵃᵐ primi." See the Commentary, line 218, above.

226 "conversam 38ᵉ primi." I.38 in the Adelard II version (*ms. cit.*, 6r) runs: "Si trianguli super bases equales atque inter lineas equidistantes ceciderint, equales eos esse necesse est."

236 "38ᵃᵐ primi." See the Commentary, line 226, above.

236, 275, 277 "41ᵃᵐ primi." See the Commentary to the Cambridge Version, line 33.

272 "17ᵃᵐ 3ⁱⁱ." See the Commentary to the Naples Version, line 61.

278–79 "diffinicionem...secundi." See the Commentary, lines 110–11, above.

284–85 "communem...maiora." See the Commentary, lines 103–105, above.

306 "14... 2^{i}." See the Commentary to the Cambridge Version, line 82–83.

307 "42...secundi." I.42 in the Adelard II version (*ms. cit.*, 6v) runs: "Equidistantium laterum superficiem designare cuius angulus sit angulo assignato equalis, ipsa vero superficies triangulo assignato equalis." I.44 (*ibid.*, 7r): "Proposita linea recta super eam superficiem equidistantium laterum designare cuius angulus angulo dato equalis, ipsa vero superficies triangulo assignato equalis." II.5 (*ibid.*, 8v): "Si recta linea per duo equalia duoque inequalia secetur, quod sub inequalibus sectionibus rectangulum continetur cum eo quadrato quod ab ea describitur que inter utrasque est sectiones, equum est ei quadrato quod a dimidio totius linee describitur."

308–310 "36...equales." For III.16, see the Commentary, line 219, above. Actually Campanus states and proves this extra theorem in the manner given here in his comment on III.35 [III.36 in the Greek text]. I am puzzled by the reference to III.36 (unless it is an error for III.35), for it hardly seems likely that the author was referring to the Greek text since all of his references are to the Adelard II version and the commentary of Campanus based on that version.

311 "5^{am} primi." See the Commentary to the Naples Version, line 63.

319 "6^{am} primi." See the Commentary to the Naples Version, line 65.

320 "paralellogrammo." Another possible reading is *paralogismo*. Neither word makes very good sense. Of course, the whole Figure 32 is inclosed by a square and, in view of the looseness of the use of the word *paralellogrammum* in medieval texts, the reference might be to a square. The manuscript reading does look more like *paralogismo* but the ordinary meaning of that term is "logical fallacy." Perhaps it might be used loosely with the meaning of "proof," as I have suggested in my translation.

323 "17^{am} 3^{ii}." See the Commentary to the Naples Version, line 61.

Chapter four

The *Verba filorum* of the Banū Mūsā

1. Content and Authors

We have seen that Gerard of Cremona was responsible for the most accurate and popular translation of the *De mensura circuli* which circulated in the Latin West. He was also responsible for the first knowledge of some of the conclusions of Archimedes' *On the Sphere and the Cylinder*. This came through his translation of a short but important treatise entitled in Latin *Verba filiorum Moysi filii Sekir, i.e. Maumeti, Hameti, Hasen.*[1] This Arabic treatise on the measurement of the areas and volumes of certain plane and solid curved figures was composed by the celebrated Arabic mathematicians of the ninth century, the Banū Mūsā ibn Shākir, three brothers whose contribution to Islamic mathematics was of great significance.

The importance of this work for Western geometry is twofold: it gave the formulas for the area of the circle, the area and volumes of a sphere, and so on (as did certain practical manuals available to the medieval Schoolmen); and it also presented demonstrations of an Archimedean character of these formulas. In fact, we can single out the following particular contributions of this treatise: (1) A proof of Proposition I of the *De mensura circuli* somewhat different from the proof of Archimedes but still fundamentally based on the so-called exhaustion principle. (2) A determination of the value of π drawn from Proposition III of Archimedes' *De mensura*

[1] Listed in the medieval list of Gerard's translations as *Liber trium fratrum* (see F. Wüstenfeld, "Die Übersetzungen arabischer Werke in das Lateinische," *Abhandlungen der K. Gesellschaft der Wissenschaften zu Göttingen*, vol. 22 (1877), p. 59.

circuli but with further calculations similar to those found in Eutocius' commentary on the *De mensura*. (3) Hero's theorem for the area of a triangle in terms of the sides (that is, $A^2 = s(s-a)(s-b)(s-c)$, where A is the area, s the semiperimeter, and a, b, and c are the sides) with the first demonstration of that proposition appearing in Latin. (4) Theorems for the surface area and volume of a cone, again with geometrical demonstrations. (5) Theorems for the area and volume of a sphere with demonstrations of an Archimedean character. (6) A use of a form of the formula for the area of a circle equivalent to $A = \pi r^2$ in addition to the more common Archimedean form $A = 1/2\, cr$. Instead of the modern symbol π the authors use the expression "the quantity which when multiplied by the diameter produces the circumference." (7) The first introduction into the West of the problem of finding two mean proportionals between two given quantities. In this treatise we find two solutions presented: (a) the solution attributed by the Banū Mūsā to Menelaus and by Eutocius to Archytas, and (b) the solution presented by the Banū Mūsā as their own but attributed by Eutocius to Plato. (8) The first solution in Latin of the famous Greek problem of trisecting an angle, a solution to some extent reminiscent of the one found in the so-called *Lemmata* (or *Liber assumptorum*) attributed to Archimedes. (9) A method of approximating cube roots to any desired limit.

As I have said in the first chapter, the *Verba filiorum* presents an impressively rich fare when compared with the geometric diet of the centuries immediately preceding it, that is, with the geometry produced by a Gerbert or one familiar to Schoolmen like Hugh of St. Victor. I think it just to say that this tract played an important role in the gradual spread of the knowledge of Archimedean geometry. In the first place, the famous mathematician Leonardo Fibonacci of Pisa in his revised *Practica geometrie* (1220) borrows heavily (and often in verbatim fashion)[2] from the *Verba filiorum* in his propositions relating to the area of a circle, Hero's formula for the area of a triangle, the area and volume of a cone, the area and volume of a sphere, and the finding of two continually proportional means between two given quantities. And, of course, Leonardo's work was itself quite well known down to the time of the Renaissance.

Leonardo's contemporary, the equally important Jordanus de Nemore, in his *De triangulis* takes from the *Verba filiorum* one solution of the problem of finding two means (see Appendix V) and the solution of the problem of trisecting an angle (see Appendix VI). It was possibly from

[2] B. Boncompagni, ed., *Scritti di Leonardo Pisano* (*Practica geometrie*), vol. 2 (Rome, 1862), pp. 40–42, 87–91, 153–58, 178–87.

Jordanus' treatment of the trisection problem that the solution was further altered and joined to Campanus of Novara's version of the *Elements* (see Appendix VI).

The *Verba filiorum* was also known in the thirteenth century to Roger Bacon, who gives it in a list of geometrical works under the title *Liber trium fratrum*.[3] Somewhat later, in the fourteenth century, the author of some geometric questions which are in a manuscript now at Paris was clearly influenced by the treatise of the Banū Mūsā and quotes it in connection with the area and volume of a sphere.[4] Furthermore, at about the same time Thomas Bradwardine (or, more probably, one of his contemporaries) was perhaps aware of Proposition IV from the *Verba filiorum* when constructing his demonstration of the general proposition relative to the area of a circle in terms of the radius and circumference of the circle. (That demonstration is published in Chapter Five under the name of Pseudo-Bradwardine.) Furthermore, from the late fourteenth or early fifteenth century stems an anonymous treatise *De inquisicione capacitatis figurarum* which cites the *Verba* as *Geometria trium fratrum*.[5]

Before passing on to the text and its construction, I should like to say something about the activity of the Banū Mūsā, who played a central role as patrons and students of Greek science in Baghdad in the ninth century.

[3] Roger Bacon, *Communia mathematica*, ed. of R. Steele (Oxford, 1940), p. 44, lines 20–25: "Et hee practice [partes geometrie] omnes traduntur in libris propriis secundum numerum earum, et notum est illis qui eas sciunt, quorum nomina sunt imposita secundum proprietatem earum, ut *liber de Ysoperimetris*, et *de Replentibus Locuum*, et *de Curvis Superficiebus* Archimenidis, et *liber Trium Fratrum*...."

[4] MS Paris, BN lat. 7377B, 37v: "In mensuratione autem spere primo per quantitatem diametri ipsius invenis quantitatem circuli maioris eius, quam postea multiplicas in 4 et provenit superficies totius spere, quod probatur in libro trium fratrum. Deinde superficiem spere multiplicas in sextam diametri et provenit soliditas totius spere, quia in libro trium fratrum probatur quod multiplicatio medietatis diametri spere in tertiam superficiei ipsius equatur soliditati totius spere." The *Liber trium fratrum* is also cited (folio 33r) for the solution of the area of a triangle in terms of its sides.

[5] *De inquisicione capacitatis figurarum*, ed. of M. Curtze in *Abhandlungen zur Geschichte der Mathematik*, *8*. Heft (1898), pp. 37–38. The following propositions of the Banū Mūsā treatise are used and cited by the unknown author: Proposition IV and its corollary, Proposition VI (cited as Proposition 7 by the author), Proposition XIV (cited as Proposition 15*?* [or 17?] by the author; Curtze reads it as 15 and then gives proposition 15 in a note, which proposition refers to the volume of the sphere, but it is clear from the context that Proposition XIV is the required proposition). The fact that Proposition VI is cited as 7 suggests that the author was using a copy of the *Verba filiorum* in the tradition of *MaR* where VI is given as 7 (see Sigla). On the other hand, XIV is given by *MaR* as 17, not as 15 which this work seems to have (cf. another manuscript of the *De inquisicione* in Vienna, Nat.-bibl. 5277, 102r).

They were sons of Musā ibn Shākir, who is variously described as a reformed bandit and an astronomer or astrologer of the Caliph al-Maᶜmūn.[6] He is said to have given his sons to the astronomer Yaḥya ibn Abī Manṣūr for instruction in mathematics. At any rate, they joined the circle of mathematicians that grew up in Baghdad at the time of al-Ma'mūn (fl. 813–33) and his successors. The names of the three brothers were Abū Jaᶜfar Muḥammad, Abū 'l-Kāsim Aḥmad, and al-Ḥasan. They devoted their energies and resources to the acquisition of Greek scientific manuscripts and the propagation of their contents either by translation or by independent works.[7] They sent agents or went themselves on trips into the Byzantine provinces to search for and purchase manuscripts. It was probably on one such trip that Muḥammad (or, as some say, Aḥmad) met the mathematician and translator Thābit ibn Qurra of Ḥarran and persuaded him to come to Baghdad and join their circle. As a matter of fact, there is a distinct possibility that Thābit ibn Qurra had something to do with the preparation of the work which we are here editing, since Thābit is credited with the writing of a tract having almost the same title: *On the Measure of Plane Figures and Other Surfaces and Corporeal Figures*.[8]

While devoting their talents to science and learning, the Banū Mūsā were also involved in the political quarrels and court fighting of the day. Their hostility to the famous philosopher al-Kindī is said to have stemmed from the fact that it was he rather than they who was picked to educate the son of the Caliph al-Muᶜtaṣim, and they later actively opposed the accession of that son to the caliphate.

There is little information which makes individuals of the three brothers. Some would say that Muḥammad, the eldest of the three, was the most important of the brothers.[9] He was learned in the works of Euclid, Ptolemy, and other mathematicians and astronomers. He is said by the *Fihrist* (edition of G. Flügel, p. 271) to have died in the year 259 a.H. (873 A.D.). Aḥmad is described as having been particularly interested in mechanics[10] and is named by the *Fihrist* (*loc. cit.*) as the author of a treatise on mechanics that is elsewhere attributed to all of the brothers. Al-Ḥasan's

[6] H. Suter, "Die Mathematiker and Astronomer der Araber und Ihrer Werke," *Abhandlungen zur Geschichte der Mathematik*, *10*. Heft (1900), p. 20. See also *Encyclopedia of Islam*, 1st ed., "Banu Musa" (J. Ruska). Cf. *Abhandlungen...*, *7*. Heft (1892), p. 24.

[7] M. Meyerhof, "New Light on Hunain ibn Ishaq and his Period," *Isis*, vol. *8* (1926). pp. 714–15.

[8] A. G. Kapp, "Arabische Übersetzer und Kommentatoren Euklids," *Isis*, vol. *23* (1935), p. 61.

[9] M. Casiri, *Bibliotheca Arabico-Hispana Escurialensis*, vol. *1* (Madrid, 1790), p. 418.

[10] *Ibid.*, pp. 418–19.

special interest was geometry.[11] Among the many works attributed to the brothers, we can single out, in addition to the treatise on mechanics already mentioned, a treatise on the *Ellipse* by al-Ḥasan; a new edition of or commentary on Apollonius' *Conics*, a *Book on the Movement of the First Sphere* by Muḥammad; a book by Aḥmad demonstrating that there is no movement of the ninth sphere beyond the sphere of the fixed stars, and of course the work which we are here editing and which is attributed to all three brothers. In the *Fihrist* its title is given as *The Book of the Measurement of the Sphere, the Trisection of an Angle, and the Finding of Two* Quantities Between Two Quantities Such That All of the Quantities are Continually Proportional*—which sounds more like a partial enumeration of the contents than a title. But in the extant manuscripts of the thirteenth-century revision by al-Ṭūsī the work is entitled *The Book of the Knowledge of the Measurement of Plane and Spherical Figures by the Sons of Moses: Muḥammad, and al-Ḥasan, and Aḥmad* (see Arabic variant readings below). This seems more probable as the original title. The word *Verba* used by Gerard of Cremona in his shortened title is no doubt a rendering of كتاب since, of course, *Verba* can be used in the sense of "a discourse."

2. The Text of the *Verba filiorum*

I know of eight manuscripts of Gerard of Cremona's translation of the *Verba filiorum* of the Banū Mūsā (see Sigla below). Of these eight, five are complete copies of the text: *P* (early fourteenth century), *Zm* (fourteenth century), *H* (fourteenth century), *Ma* (late fourteenth or early fifteenth century), and *R* (late fifteenth century?). Manuscripts *S* and *T*, both of the fourteenth century, are much truncated copies and are of little help for textual construction. For the fragmentary eighth copy, see page 236.

Manuscript *P* is the best manuscript,[1] although *Zm* is very good indeed.

[11] *Ibid.*

* Corrected from "one quantity" in the *Fihrist*.

[1] The reader will remember that the best tradition of Gerard's translation of the *De mensura circuli* is also contained in *P* (see Chapter Two, Section 2). Its excellence for other texts of Gerard's translations has been noted by P. Tannery, A. A. Björnbo, and H. Suter. Tannery, "Sur le 'liber augmenti et diminutionis' etc.," *Bibliotheca Mathematica*, 3. Folge. vol. *2* (1901), p. 45, says: "le Lat. 9335 [= *P*] semble bien voisin de l'an 1300. Ce manuscrit, in-fol., qui a appartenu à Ismaël Boulliau, mais où je n'ai trouvé aucun indice sur les

It is practically identical with *P* and fails to give good readings in only a very few places (e.g., see the variant reading for Proposition XVII, line 48). I do not think that *Zm* was copied directly from *P*, since *P* has in three places brought into the text readings which he labels as alternate readings, at the same time suppressing the original textual readings, while *Zm* leaves the alternate readings in the margin and gives the original readings not present in *P*. In view of the excellence of *P* and *Zm* I have almost always preferred their readings to those of the other manuscripts. However, when the traditions of *Ma* and *H* agree with one another against *P* and *Zm* (or particularly when *Ma* and *H* both agree with *Zm* against *P* alone), I have on accasion followed *Ma* and *H*. I have done this because of the possibility that *Ma* (of which *R* is merely a faithful copy) was copied not from *P*, *Zm*, or *H* but from some other copy or even from the original version.[2] In the cases where *H* varies from a reading found in the other four complete manuscripts, I have of course rejected *H*'s reading. Similarly, I have also rejected the manifold cases where the tradition of *MaR* differs from *PZm* and *H* together, although in two cases where *MaR* agrees with *Zm* against *H* and *P*, I have followed *ZmMaR*.[3] I have reported the variant readings of *P*, *Zm*, and *H* throughout the whole text, but I have given all of the variant readings of *MaR* only though Proposition

possesseurs antérieurs, est remarquablement soigné, écrit avec très peu d'abréviations, et bien lisible." Björnbo, "Über zwei mathematische Handschriften aus dem vierzehnten Jahrhundert," *Ibid.*, *3* (1902), 63 *et seq.*, compares *P* with our manuscript *H* (Basel F.II.33) as follows: "Beide sind vielfach verwendet worden, obwohl sie keineswegs gleichwertig sind; denn der Baseler Codex (B) [= *H*] ist nur mit Vorsicht zu benutzen, während der Pariser Codex (P) in jeder Beziehung zuverlässig ist." He then goes on to compare several works common to the two manuscripts: Theodosius, *De habitationibus*, Alkindi, *De aspectibus*, Pseudo-Euclid, *De speculis*, and the *Verba filiorum* of the Banū Mūsā. For the first work he finds that *P* can be the base of the text, while *H* is worthless. And in the two optical works *H* can only be used with the greatest of caution. As for the last work, after showing a few improved readings of *P*, he laments that Curtze depended on *H* rather than on *P* for his edition of the *Verba filiorum*. And finally Björnbo notes the excellence of *P* in comparison with other fourteenth-century manuscripts for a text of Menelaus' *Spherics*. Suter, "Über die Geometrie der Söhne des Mûsâ ben Schâkir," *Ibid.*, vol. *3* (1902), p. 259, echoes Björnbo's conclusions. Incidentally, a student of mine, Sister St. Martin Van Ryzin, discovered a further manuscript of Gerard of Cremona's translation of Euclid's *Elements*, Vat. Ross. 579, which I am convinced is in the same Italian hand as *P* and is executed with the same care. It follows the spelling characteristics of *P*.

[2] It is obvious from my variant readings of the text that *R* was copied from *Ma*, for many peculiar variants of *Ma* are duplicated in *R*.

[3] Proposition VI, line 43, where *Zm MaR* has the geometrically correct reading of "BZ", and Proposition VII, line 60, where *ZmMaR*'s reading of *remanet* seems preferable.

III. After Proposition III, I have given *Ma*'s (and occasionally *R*'s) readings only where *PZm* and *H* differ so that we might decide whether *PZm* or *H* is correct.[4] To have reported throughout the whole text all of the errors of *MaR*, such as wrong case endings, transpositions, nonagreements of subject and predicate—trivial errors that run into the hundreds—would certainly not have served our purpose of reconstructing the text as Gerard of Cremona composed it. Inclusion of these inconsequential errors would have merely obscured the significant variant readings. While *S* and *T* are abbreviated versions, they do occasionally give the pristine text. And since they give relatively little of the text, I have been able to include all of their readings without unduely extending the variant readings. On the whole, *S* and *T* appear somewhat closer to *H* than to *PZm*.

By using manuscripts *P* and *Zm* as the principal sources I am thus able to present a text that is very much sounder than the text which Curtze edited on the basis of manuscript *H*. The reader will notice from an examination of the variant readings how often *H* erred (cf. the remarks of note 1 on the relative value of *P* and *H*, and *Zm* is of course virtually identical with *P*). Unfortunately, Curtze often misread and altered *H* even where *H* had a correct reading.[5] The result is that Curtze's text is quite corrupt and thoroughly unreliable. It must also be remarked that Curtze

[4] Thus, in the cases where I am using the four manuscripts *P*, *Zm*, *H*, and *Ma* and I employ the form "equidistantes: extremitates *H*," this indicates that I have checked *Zm* and *Ma* and they agree with *P* against *H*.

[5] M. Curtze, "Verba Filiorum Moysi, Filii Sekir, id est Maumeti, Hameti et Hasen. Der Liber trium fratrum de Geometria, nach der Lesart des Codex Basileensis F.II.33 mit Einleitung und Commentar." *Nova Acta der Ksl. Leop.-Carol. Deutschen Akademie der Naturförscher*, vol. 49, (Halle, 1885), pp. 109–67. While I am appreciative of the important pioneer work of Curtze, I must say that in the case of this text Curtze commits almost every possible kind of error in setting up a text: using the wrong manuscript as a base, altering readings without proper indication that he is doing so, making incorrect alterations when the reading of the manuscript is correct, misreading the manuscript, misinterpreting the nature of the marginal alternate readings, faulty punctuation, and so on. I see no point in repeating all of these faults in detail since they run to many pages, but I can give a few examples: Proemium, line 28, Curtze misreads *intelligere* as *intrare*; Proemium, line 31, Curtze accepts the absurd reading of *infinita* in *H* for the correct *in figura* of *P* although he purportedly used *P* for collation in the Proemium; similarly, he accepts in Proemium, line 66, the nonsensical *superficialis* of *H* instead of the correct *super alias* of *P*; Proemium, line 72, Curtze badly corrects *mensuratur* in *H* to *mensuracio* (while *P* has the correct reading of *mensuratis*); Proemium, line 75, Curtze misreads *magis* as *ingenium*; Proemium, line 87, Curtze misreads *sine* as *super*; Proemium, line 87, Curtze misreads *rem* as *cum*; Proposition II line 6, Curtze changes "T" to "E" (and elsewhere in the proposition) without noting the change; Proposition II, lines 11–12, "TH...BG," Curtze erroneously changes to *EK in medieta-*

was quite wrong in seeing *H* as a more complete copy than *P*.[6] Everything that is in *H* is also in *P*. Curtze was thrown off the track by *H*'s practice of including the alternate readings, no doubt given by Gerard of Cremona as marginal notes (see below), in the body of the text. *H*'s scribe often put them in the text in the wrong places. This is particularly true of Proposition VII. Examination of the variant readings will show how confused the scribe of *H* became when he attempted to add the marginal readings to the body of the text.[7] A similar confusion is evident in the inclusion of the marginal proof of $A = \pi r^2$ to the text in Proposition XIII.[8]

tem linee KG; Proposition II, line 12, Curtze changes "TBG, BHG" (in both *H* and *P*) to read "EHG, BHE"; Proposition VI, line 111, "AT ad TQ," here Curtze wrongly corrects the incorrect reading of *H* (see variants) to read in each case "AB ad BG", which makes no sense; Proposition VII, *H* once has "M" for "N" (actually "U" in *P*) and so Curtze replaces "N" throughout with "M"; Proposition VII, line 98, Curtze corrects *mensurationi* into *mensurationis* without indicating it; Proposition VIII, line 9, he adds *recta* without brackets even though it is missing in *H*; Proposition IX, line 54, Curtze changes correct reading of "ML" to "OL" without indicating the change; Proposition XI, line 94, he misreads *si* as *sive*; Proposition XII, line 25, Curtze misreads "ZL" as "ZB"; Proposition XII, line 45, he omits "*super... media*" without indication; Proposition XII, line 47, he misreads "BL" as "HL" (and thus seriously changes the geometry); Proposition XII, line 50, his misreading of "EDB" as "EDH" similarly makes hash of the geometry; Proposition XIII, Curtze changes all the letters of the subproof of lines 97–118 without indicating the changes; Proposition XVII, here the readings of *H* are quite faulty and Curtze goes quite astray in trying to correct them. Although the above list of errors is but a small number of the total, it does represent the various kinds of errors committed by Curtze.

[6] Curtze, *op. cit.* (see note 5 above).

[7] The marginal note of *P* given as a variant for line 83 of Proposition VII appears to be an alternate account in specific form of the argument of lines 78–106. But it was taken (along with the note given in the variant to line 90) by the scribe of *H* (and thus by Curtze) as the continuation of the text after "trianguli" in line 83 and thus as the conclusion of the first proof of Proposition VII. Then lines 83–106, which appear in the text of *P* as the preferred conclusion of the first proof, are transferred by *H* to form a conclusion of the second proof where a lacuna had been left by *P*. It is my opinion that this transference is not proper and that Gerard's text also had a lacuna due to the defectiveness of the Arabic texts he had (see the Commentary below, lines 83, 90). The marginal note of line 83 I believe either to be an explanation of lines 78–106 added by Gerard or some other Latin author, or an alternate form in another Arabic exemplar used by Gerard (see the further remarks in the Commentary). The culminating effect of comparing the readings of *P* with the other manuscripts, and particularly with *H*, is one inspiring great confidence in *P* and open scepticism as regards *H* and the other manuscripts. In short, *P* appears to have followed the original text of Gerard very closely.

[8] In the case of his marginal notes, *P* usually puts a sign on the first word of the statement or phrase in the text to which the marginal note refers. In the case of the marginal note to Proposition XIII, *H* apparently took such a mark or sign in the

The relation of manuscript *S* to *T* is not completely clear. Both start from the same place in the middle of the Proemium. But in the remainder of the Proemium, *T* abbreviates more radically than does *S*. However, in the proofs of propositions, *T*—so far as it goes down through part of Proposition VI—is fairly close to the text while *S*, where it gives proofs, gives only brief paraphrases. Furthermore, *S*, in spite of the fact that it leaves out some of the propositions entirely (see Sigla below), does not break off at Proposition VI but goes through Proposition XIX, which it reports in the briefest fashion. It is thus clear that *S* did not copy from *T* nor *T* from *S*, since each includes textual matter not in the other. Evidently the scribes of *S* and *T* had access to some copy, perhaps like *H*, but which unlike *H* had the beginning and the end of the treatise cut off (but the scribe of *S* must also have known a version with the beginning of the treatise; see Commentary, Proemium, lines 4–34). Incidentally, at first glance it might seem, because the Arabic text of al-Ṭūsī also begins in the middle of the proemium in almost the same place as *S* and *T*, that these two manuscripts are copies of a translation of the al-Ṭūsī text rather than of the Gerard of Cremona translation of the original Banū Mūsā text. But this is not so. Where *S* and *T* give the text, they obviously agree with the Gerard of Cremona translation instead of the Arabic text of al-Ṭūsī; and the agreement is a verbatim agreement that can only be explained by their having as their ultimate source the Gerard of Cremona translation.

An interesting feature of the text as given by *P* and *Zm* is the presence

copy he was using as an invitation to insert the whole marginal note after the word which has the marginal sign above it. This resulted in inserting the proof of the formula before the statement of the formula! *H* further compounded the confusion by following the proof of $A = \pi r^2$ with lines 91–96 in capital letters as if to indicate that these lines constituted the next (i.e., fourteenth) proposition. But of course what follows these lines is in fact really a continuation of the proof of the main thirteenth proposition, that is, it is a proof of the second half of the proposition to the effect that: Area of circumscribed body $> 2\pi r^2$. Thus it makes absolutely no sense to do as *H* does and include the conclusion of the first half of the proposition (i.e., Area of included body $< 2\pi r^2$) in large letters as if it is the enunciation of the second part of the proposition. It should be noticed that Curtze follows *H* in this hopeless confusion. Actually, I believe it can be doubted that the marginal proof in Proposition XIII was included as an original part of Gerard's text. Its terminology differs from that of the main text as a ready comparison of the note (now included in brackets as lines 97–118) with the rest of the text. Compare, for example, Gerard's commonly used phrase "illud quod agregatur" with the phrase of the note "eius quod provenit" used in the note in the same context; and compare Gerard's "linea circumdans" used over and over again with the word "circumferentia" employed in the same place in the note.

of a number of marginal notes. Curtze apparently misunderstood the nature of the notes, taking them to be a part of the Latin text. This resulted from his use of *H* where the notes are put into the body of the text. The fact is that the notes (with very few exceptions) go back to Gerard's original translation and for the most part appear to refer to the Arabic text. The marginal notes are of three kinds.

1. The first kind appear to be references by Gerard to alternate ways of translating the original Arabic term. They are often introduced by *vel*. For example, *planiciem* is given as a marginal alternate for *extensionem* in line 54 of the Proemium. Both of these terms can be used to translate the term انبساط as found in the extant Arabic text of al-Ṭūsī. On the other hand, there is little possibility of the two Latin words being confused by scribes. Similarly, in line 62 of the Proemium we have the word *comparatur* in the text and *mensuratur* in the margin. Both are acceptable ways to translate يقاس in the Arabic text.

2. A second type of marginal note occurs when Gerard wished to indicate that there were alternate readings in the Arabic text. Apparently Gerard had at least two copies of the text (or one copy with alternate readings indicated). This second type of note is introduced by the phrase *in alio*. An example is *scientia* in line 60 of the Proemium which has the marginal reference *in alio*, *operatio*. In the extant Arabic text we have العمل, which can be rendered by the alternate reading, *operatio*. On the other hand, it is apparent that Gerard had another reading which he preferred, namely, العلم, which is translatable by *scientia*. Clearly, in the two different Arabic readings there had been an interchange of *lam* and *mim*, a type of interchange that is a most frequent occurrence in Arabic manuscripts. It is quite inconceivable that *scientia* could be misread in Latin for *operatio*, while the suggested shift in Arabic orthography is a commonplace. This is the kind of evidence that has convinced me that many of the marginal notes do indeed go back to Gerard and refer to the Arabic and thus do not stem from later Latin scribes.

3. A third type of note has neither *vel* nor *in alio* before it. The shorter examples of this type consist of words or phrases that *P* omitted in the copying process. These additions clearly belong to the text (e.g., "centrum sit," Proposition VI, line 26). Somewhat more difficult to classify are the long notes added to Proposition VII (see variant and Commentary for line 83), Proposition XI (see text, lines 52–71), and Proposition XIII (see text, lines 97–118). The first of these notes is either an alternate passage from another Arabic exemplar available to Gerard, or it is an addition to

the text omitted by Gerard in his first translating effort or added by him or someone else to make the text more specific. The second, as I explain in the text, I believe to be an addition by Gerard. The third of the long notes also appears to me to have been not a part of the original translation of Gerard but rather a later explanatory addition (see note 8). The short note given in the variant to lines 14–15 of the Proemium does not have a *vel* or an *in alio* before it. It may be a part of the text originally omitted by Gerard and then added by him later, or it may be an explanation of the text given by Gerard.

As in the case of the *De mensura circuli* given above in Chapter Two, I have here included a quite complete set of variant readings found in the Arabic text of the *Verba filiorum* as edited by al-Ṭūsī. It will be evident to the reader that al-Ṭūsī has been quite free in his treatment of the original text of the Banū Mūsā as represented by the Gerard translation. Al-Ṭūsī's objective was to give a mathematically sound text rather than to preserve the original readings. He shortens the majority of the propositions by omitting the definition or specification (*diorismos*) of the proposition in terms of the specific, concrete figures, while, as is evident from the Gerard of Cremona text, which of course precedes the thirteenth-century edition of al-Ṭūsī, the Banū Mūsā had originally followed the formal construction of propositions in the Greek manner. It will also be noticed that al-Ṭūsī omitted much of the introduction and that he radically abbreviated the last part of the treatise. Suter has stated that al-Ṭūsī's re-edition is a sounder text than the version of the text of the Banū Mūsā used by Gerard.[9] But this judgement—based as it was on Curtze's defective text of Gerard's translation of the original text of the Banū Mūsā as one term of his comparison—is certainly not true.

Notice that Gerard's translation is generally quite accurate, if almost painfully literal. One of the peculiarities of his translation, however, is the free use of *quadratum* to translate words that sometimes mean "square," and sometimes "area" in the sense of "product" or "multiplication," and sometimes "quadrilateral." This will be particularly evident if the reader compares the readings of the Arabic text with Gerard's wording of the seventh proposition, and particularly lines 65–66, where the first two different usages appear in the same sentence. Cf. also Proposition XII, line 38, where it means "quadrilateral." Note also the loose use of the word *circulus* to mean either the whole circle or its circumference. This reflects a

[9] Suter, *op. cit.*, p. 260 (see note 1, division 2, above).

similar ambiguity in the Arabic. I have retained the ambiguity in the English translation.

The drawings are, for the most part, based on those given in *P*. But comparison has been made with the drawings in the other manuscripts and the printed Arabic text. It will be noticed that in a number of cases I not only have given the drawing as it existed in the medieval versions of the text but also have added a modern reconstruction where I thought this might be helpful to the reader.

I have followed *P* in orthography even to the point of preserving odd spellings like *agregatur*, *sexagessima*, *millia*, etc., and variable spellings like both *columna* and *columpna*, *diameter* and *diametrus*, and I have also tended to follow it in capitalization (except that as usual I have capitalized letters refering to geometric magnitudes although *P* has not done so). I have felt free to punctuate as I thought the meaning demanded. Orthographic peculiarities and other particular characteristics of each manuscript are noted below in the list of manuscripts given with the Sigla.

In the variant readings for this text the only special point to note is that when the long variant passages from *S* and *T* are quoted, I have punctuated them to make sense, although my general policy on short variants is to omit punctuation. I have also placed a comma after "in alio" when *P* and *Zm* give an alternate reading in the margin, although no such mark is actually used in the MMS. Often *Zm* deletes words and phrases that have been put out of order or repeated twice. I have not indicated these deletions. The marginal folio numbers refer to manuscript *P*.

Sigla and Characteristics of Manuscripts

P = Paris, BN lat. 9335, 55v–63r, 14c. For the reliability of this manuscript, see note 1 of division 2 of this chapter. *P*'s orthographic characteristics are revealed in the text, but we can note a few of them here: *agregatur* for *aggregatur*, both *columna* and *columpna* are used, both *diameter* and *diametrus*, *millia*, *quattuor*.

Zm = Madrid, Bibl. Nac. 10010, 77v–83r, 14c. A good copy, very close to *P*. The orthography generally is the same as *P*, although the scribe vacillates between *milia* and *millia*, *columna* and *columpna* (although generally preferring the latter). The form *sexagesima* appears instead of *P*'s *sexagessima* (and similarly for the other numbers). In the calculations of Proposition VI, we sometimes find the numbers expressed rhetorically, sometimes with Arabic numerals, and often with a mixture of both rhetoric and Arabic numerals. This manuscript includes all of the marginal notes found in *P* except the long note included in the text of Proposition XIII, lines 97–118. On the other hand, it has two other long notes on Proposition XIII not found in *P* (see the variant readings for line 74 and lines 133–49). These two notes are added to the bottom of folio 84r, two pages from the end of the *Verba filiorum* tract. Still another note (included in the text of Proposition XI, lines 52–71) had been added by the scribe of *Zm* to the end of the *Epistola de proportione et proportionalitate* of Ametus filius Iosephi (folio 77r). Incidentally, this same note was added to the Ametus text in a Florence manuscript, Biblioteca Medicea-Laurenziana, San Marco 184, 112r–v.

H = Basel, Univ. Bibl. F.II.33, 116v–122r, 14c. See note 1 of division 2 of this chapter. Some orthographic characteristics: both *dyameter* and *dyametrus*, *equedistans* for *equidistans* in *P*, *existimo* for *estimo* in *P*, the use of -N for -U in marking geometric quantities (note -U is used in both *P* and the Arabic text), the use of *ci* for *ti* before vowels, *milia* instead of *P*'s *millia*, *quatuor* for *P*'s *quattuor*, *sexagesima* for *P*'s *sexagessima* (and similarly for other numbers). Hindu-Arabic numerals are used through line 109 in Proposition VI. After that point, numbers are written out as *P* does throughout. Note from the variants in Proposition VI the erroneous method of writing numerals often used in *H* (e.g., in Proposition VI, lines 53–54, 326000 is written "300 et 26000"). *H* adds marginal notes to text, sometimes in the wrong place (see notes 7, 8, of division 2 of this chapter). *H* was the source of Curtze's unsound text of the *Verba filiorum* (see note 5 of division 2 of this chapter).

Ma = Paris, Bibl. Mazarine 3637 (1256), 1r–13v, late 14c. *Ma* is the source of *R*. Orthographic characteristics: *alliquis*, *set* for *sed*, *ci* instead of *ti* before vowels (e.g., *cognicionem*), *narare*, *demitat*, occasionally *seste* for *sexte*, *disscretionis*, *embade* for *embadus*. For the most part the marginal notes of *P* are inserted in the text in *Ma*, sometimes enclosed by a box (e.g., see variant reading for VIII, line 24). The propositions are numbered with Arabic numerals in *Ma* but there is considerable confusion: before "Iam" in line 35 of Proposition IV, *Ma* inserts "5," and thus V becomes 6 and VI becomes 7 in *Ma*; before "Incipiamus" in line 24 of Proposition VI, *Ma* puts "8." Number "9" is placed before "proportione" (for "proportionem") in line 178 of Proposition VI rather than before "Volo" in Proposition VII, line 1. And so Propositions VIII, IX, X, and XI are labeled as 10, 11, 12, and 13 in *Ma*. Before "Iam" in Proposition XI, line 78, *Ma* puts "14." Thus Proposition XII becomes 15 in *Ma*, and so on for the succeeding propositions until at the end Proposition XIX (unnumbered in *P* and the other manuscripts, except *R*) becomes 22. As I have noted in the introduction, this manuscript and its copy *R* are replete with wrong case endings, transpositions, nonagreements of subject and predicate, and other trivial errors. I have given all of its readings only through Proposition III (see division 2 of Introduction to this chapter).

R = Paris, BN lat. 7225A, 2r–31r, early 16c. Copied almost exactly from *Ma*. Orthographic characteristics: *inpossibile*, *aggregatur* for *P*'s *agregatur*, *quatuor* for *quattuor* in *P*, occasionally *milibus* for *P*'s *millibus*, and in addition most of the peculiarities of spelling found in *Ma*. Also like *Ma*, *R* on occasion puts the notes that are included in the text in boxes. It follows *Ma* exactly in its curious errors in the numbering of the propositions.

T = Thorn (Torún), Gymnasialbibliothek R 4° 2, pp. 73–79, 14c. An abbreviated version, beginning in Proemium, line 35, and ending in Proposition VI, line 45. Some orthographic characteristics: *dyameter* for *diameter* in *P*, *punctus* for *punctum* in *P*, *sicud* for *sicut*, occasionally *capud* for *caput*, *ortogonaliter* for *orthogonaliter*.

S = Oxford, Bodl. Digby 168, 124r–v (123r–v, old pag.), 14c. Another abbreviated copy. Begins in Proemium, line 35, and ends in Proposition XIX, line 35. When the enunciations are given, they are given quite correctly, but the proofs when given are quite truncated or paraphrased (e.g., Propositions I, III, IV, VI, XIX). In a number of propositions *S* has only the enunciation or only the enunciation and a sentence or so (e.g., II, V, VII–XI, XIV–XV). *S* omits entirely XII–XIII, XVI–XVIII. Some orthographic characteristics: *dyameter* for *diameter* in *P*, *aggregatur* for *agregatur* in *P*, *ci* for *ti* before vowels (e.g., *operacio*, *fraccionibus*, etc.).

[As this volume neared publication, an eighth manuscript, Vienna Nat.-bibl., cod. 5277, 297v–98v, 16 c, was discovered. It includes only the Proemium and the text through line 23 of Proposition III. It is a careless copy and of no use for the establishment of the text.]

Arabic Text of al-Ṭūsī

Ar = Nasīr al-Dīn al-Ṭūsī, *Majmū*[c] *al-Rasā'il*, vol. 2 (Hyderabad, 1940); the second text of volume 2 is the tract of the Banū Mūsā, each tract being numbered separately. I compared this text with a manuscript, Paris, BN arabe 2467, 58v–68r. For the variants of this text, see the Arabic variants. Consult Suter, *op. cit.* in note 1 of division 2 of Introduction to this chapter, for a discussion of this text. He gives a German translation of some parts of the Arabic text.

55r
c. 2

Verba Filiorum Moysi Filii Sekir, I. [E.] Maumeti, Hameti, Hasen

[Proemium]

Propterea quia vidimus quod conveniens est necessitas scientie men-
5 sure figurarum superficialium et magnitudinis corporum, et vidimus quod de rebus sunt, quarum scientia necessaria est in hac specie scientie, cuius cognitionem non apprehendit aliquis usque ad hoc nostrum tempus, secundum quod apparet nobis; et de eis sunt de quibus consecuti sumus quod quidam de antiquis qui preterierunt
10 consecuti sunt cognitionem eius, verumtamen scientia illius non pervenit ad nos, neque aliquis de illis, quibus atestati sumus, comprehendit; et de eis sunt, quarum scientiam aliqui precedentium sapientum comprehenderunt, et eam in suis scripserunt libris, verumtamen earum scientia pervenit ad proprietatem eorum, qui sunt nostri temporis sine
15 communitate—tunc propter illud visum est nobis ut componamus librum in quo ostendamus illud, cuius scientia necessaria est de eo,

1 *mg. H. manu rec.* Liber trium fratrum/ Moysi *PH Zm om. S* Moysy *MaR*/ Sekir *PS* Sakir *MaR* sekyr *T* sehir *H*
1–4 Verba... vidimus: Ex libro intitulatur verba filiorum sekir, i., mauemeti, hasen, qui sic incipit "Propterea quia vidimus et cetera" *S*
2 Hameti: Ameti *MaR om. S*/Hasen: et hasen *H* Asatht. Rubrica *MaR*
4–35 Propterea....illud[1] *om. TS* (*sed cf. titulum S supra*)
6 quorum *MaR*
7 scientie *om. MaR*
9 quidam *P Zm* quidem *H* quia *MaR*
10 consecuti *Zm HMaR* consequuti *P*/ cognitione *MaR*/- tantum *MaR*
11 sumus *om. MaR*
12 sunt *Zm HMaR* est *P*/ quorum *MaR*/ aliqui: antiqui *H*
13 comprehendent *H*/eam: etiam *MaR*
14 eorum: eius *H*
14–15 *de* sine communitate *scr. P mg et Zm supra* (*et add. MaR ante* sine): i.e., ad quosdam proprie et non ad omnes sapientes
16 ostendemus *MaR*

1–2 Verba...Hasen: كتاب معرفة مساحة الاشكال البسيطة و الكرية لبنى موسى محمد والحسن واحمه — ثمانية عشر شكلا (*The Book of the Knowledge of the Measurement of Plane and Spherical Figures by the Sons of Moses: Muḥammad, and al-Ḥasan, and Aḥmad. Eighteen Propositions*)
4–34 Propterea....corporis. Et *om. Ar.*

The Discourse of the Sons of Mūsā Ibn Shākir: Muḥammad, Aḥmad, and Ḥasan

[Preface]

Because we have seen (1) that there is fitting need for the knowledge of the measure of surface figures and of the volume of bodies, and we have seen (2) that there are some things, a knowledge of which is necessary for this field of learning but which—as it appears to us—no one up to our time understands, and [(3) that] there are some things we have pursued because certain of the ancients who lived in the past had sought understanding of them and yet knowledge has not come down to us, nor does any one of those we have examined understand, and [(4) that] there are some things which some of the early savants understood and wrote about in their books but knowledge of which, although coming down to us, is not common in our time—for all these reasons it has seemed to us that we ought to compose a book in which we demonstrate the necessary part of this knowledge that has become evident to us.

quod nobis manifestum est de hac scientia. Et si viderimus aliquid eorum, que posuerunt antiqui et quorum scientia publicata est in hominibus nostri temporis, quo indigeamus ad testificandum super
20 aliquod eorum que ponemus in libro nostro, dicemus illud rememorando tantum; et non erit nobis necessarium narrare illud in libro nostro, cum sit eius scientia publicata; propterea quia querimus abbreviationem. Et si viderimus aliquid eorum que posuerunt antiqui de illis, quorum rememoratio non est famosa, et non est exquisita
25 eius scientia, cuius narratione indigeamus in hoc nostro libro, ponemus illud in eo, et proportionabimus illud eius auctori. Et declarabitur ex eo quod narrabimus de compositione huius nostri libri quod oportet ei qui vult legere et intelligere ipsum, ut sit bene instructus in libris geometrie publicatis in hominibus nostri temporis.

30 Proprietas communis omnis superficiei est quod est habens longitudinem et latitudinem tantum. Sed proprietas in figura corporea est quod est habens longitudinem et latitudinem et altitudinem. Et longitudo et latitudo et altitudo sunt quantitates que terminant magnitudinem omnis corporis. Et longitudo est prima quantitatum que
35 terminant illud. Et est illud quod extenditur secundum rectitudinem

17–18 viderimus...que: videmus aliquem illorum qui (que *Ma*) *MaR*
18 est *om MaR*
19 nostris (!) *Ma*
20 que ponemus: quod ponimus *MaR*
22 cum: tamen *R* (?) *Ma* / sit *om. MaR* / publica *Zm*
24 quarum *H* / est[1] *Ma om. R* / exquesita *H*
26 declarabitur: declinabitur *H*
27 nostri libri *tr. H*
28 ei: illum *MaR* / et intelligere *om. MaR*
29 temporis: operis *H*
30 omni *MaR*
31 in figura: infinita *H*
32 et[1] *om. H* / altitudinem et latitudinem *MaR*
35 Et: Longitudo *S* (*cf. T*) / extenditur: *MaR*
35–95 Et....volumus: Longitudo est illud quod extenditur secundum rectitudinem in duas partes simul tantum. Sed cum hec (*corr. ex* huius) extenditur latitudinaliter scilicet preter (*corr. ex* partem) in partem suam [et] suam rectitudinem dicitur latitudo ista extensio. Unde patet error dicentis latitudinem esse lineam continentem superficiem in parte alia a longitudine sua. Altitudo [est] extensio superficiei in partem aliam a longitudine et latitudine sua, scilicet in altum. Unde altitudo non est linearum. Scientia amplitudinis et magnitudinis in mensurando est per unum superficiale quo ad superficierum mensurationem, cuius longitudo est una et latitudo est una, cuius anguli sunt recti, et per unum corporale quo ad magnitudines quod est corpus cuius longitudo est una, latitudo una, et cuius altitudo est una, et elevatio superficierun eius quarundam super alias est super angulos rectos. Cuius ratio quia oportet quod quantitas qua mensurentur super-

35 illud[1]: الاشكال (*figures*)

And if we consider some of those things which the ancients posed and the knowledge of which has become public among men of our time but which we need for the proof of something we pose in our book, we shall merely call it to mind and it will not be necessary for us in our book to describe it [in detail], since knowledge of it is common; for this reason we seek only a brief statement. On the other hand, if we consider something which the ancients posed and which is not well remembered nor excellently known but the explanation of which we need in our book, then we shall put it in our book, relating it to its author. It will be evident from what we shall recount concerning the composition of our book that one who wishes to read and understand it must be well instructed in the books of geometry in common usage among men of our time.

The common property of every surface is the possession of length and breadth alone, while the property of a corporeal figure is the possession of length, breadth, and height. Length, breadth, and height are quantities which delimit the magnitude of every body. Length is the first of the quantities which delimit the body and it is that which is extended in a straight line in both directions simultaneously. For nothing except length alone arises from it. When length is extended latitudinally, that is, in

55v c. 1 in duas / partes simul. Nam non fit ex eo nisi longitudo tantum. Et cum extenditur longitudo latitudinaliter, scilicet preter in partem suam et suam rectitudinem, tunc illa extensio est latitudo. Et tunc provenit superficies. Et latitudo quidem non est sicut estimant plures hominum, 40 scilicet quod est linea que continet superficiem in parte alia a longitudine sua. Et si esset illud sicut dicunt, non esset superficies habens longitudinem et latitudinem tantum, et esset tunc latitudo longitudo etiam. Et illud est quoniam latitudo in eorum estimationibus est linea et linea est longitudo. Et Euclides quidem iam sapienter dixit illud 45 ubi dixit quod linea est longitudo tantum, et superficies est habens longitudinem et latitudinem. Altitudo vero est extensio superficiei in partem que est preter longitudinem et latitudinem, scilicet extensio eius in altum. Et illi quidem qui estimaverunt quod latitudo est linea estimaverunt iterum quod altitudo est linea. Et declaratio erroris eorum 50 in illo est equalis.

Iam ergo ostensum est quid sit longitudo et quid latitudo et quid altitudo. Et declaratur cum hoc quod iste tres quantitates, scilicet longitudo, latitudo et altitudo, determinant magnitudinem omnis corporis et extensionem omnis superficiei. Et declaratur iterum quod 55 non est aliquid corporum indigens quantitate alia quarta qua eius

ficies et corpora sit talis ut cum duplantur continuentur ad invicem taliter ne dimittat in vacuitatibus aliquod de superficie et corpore mensuratis super quod non veniat. Et est necessarium cum hoc ut sit illud super quod venit mensuratio de superficie aut corpore facile dum non prohibetur eius mensuratio. Et illud non est repertum nisi in quadrato et in tali figura quia quando duplatur alteratur eius quantitas sed remanet eius quadratura et iterum continuatio unius cum altero quando duplatur est continuatio non dimittens in vacuitatibus huius quod mensuratur super (*corr. ex* sicud) quod non veniat, et illud velocius sit in corporibus et superficiebus per quadratum orthogonium quia ipsum est maius aliis. Igitur et cetera. *T*

36 fit: sit *MaR*

38 est: dicitur *S* (*cf. T*) / pervenit *HMaR*

39 quidem: que *MaR* / extimant *MaRZm*

40 scilicet *om. MaRS* / a: in *H*

41 illud *om. H* / dicunt: illi dicunt *H*

42 latitudo *om. MaR*

43 Et...estimationibus: quia latitudo ut dicunt *S* / extimationibus *MaRZm*

44 Et[2]...illud: sicut dixit euclides *MaR* / quidem *om. S* / iam: tam (?) H / sapienter: sapiens *H*

46 Altitudo vero: et eadem ratione altitudo non est linea sed *S* / vera *MaR*

47 parte *MaR* / longitudine et latitudine *MaR*

48 quidem *om. MaR*

48–61 Et....nisi: illa mensuratio corporum et superficierum est *S*

48, 49 extimaverunt *MaRZm*

49 erroris eorum *tr. MaR* / eorum *om. H*

52 declaratur cum: declaratum est *H*

53 *post* longitudo *add. PZm* et; *sed om. HMaR*

54 *de* extensionem *scr. P mg. et Zm supra* vel planiciem / extensionem : extensionem vel pleniorem *R* vel pleniorem et extensionem *Ma* / declaratum *H*

55 *ante* quantitate *add. H* tam (?)

other than its own direction and the direction of it as a straight line, then that extension is breadth, and then a surface is formed. Breadth is not, as many people believe, the line which contains the surface in a direction other than its length. If it were that, as they say, a surface would not be that which "has length and breadth only," but [in fact] breadth would then be length as well. This is so because breadth in their judgement is a line and a line is length. Euclid has already wisely stated this when he has said that a line is only length and a surface is that which has length and breadth. Height, in truth, is the extension of a surface in a direction which is neither that of its length nor that of its breadth; evidently, it is extension upward. Those who thought that breadth is a line also believed that height is a line. The revelation of their error in doing so is the same [as before].

Therefore, it has now been shown what length, breadth, and height are. It is declared in addition that these three quantities, i.e., length, breadth, and height, delimit the magnitude of every body and the extension of every surface. It is further declared that there is no body requiring another [or] fourth quantity to delimit its magnitude. Hence after we have

37 longitudo: السطح (*surface*)

37–38 preter....rectitudinem: في غير جهة الطول (*in other than the direction of its length*)

38–39 Et[1]...superficies *om. Ar.* ((except as suggested in variant for line 37))

46 *post* latitudinem *add. Ar.* فقط (*only*) / superficiei *om. Ar.*

47–48 scilicet...altum *om. Ar.*

51–52 Iam...quod *om. Ar.*

52–53 scilicet...altitudo *om. Ar.*

54 extensionem: انبساط (*extension*) ((See the introduction for a discussion of this.))

54–60 Et....corporum *om. Ar.*

magnitudo terminetur. Postquam ergo ostensum est illud quod narravimus, tunc oportet ut incipiamus ostendere illud cuius volumus narrationem in hoc nostro libro. Et quoniam nos nolumus significare per illud nisi super scientiam amplitudinis superficierum et magnitu-
60 dinis corporum et scientia in mensuratione quantitatis illius non comparatur nisi per unum superficiale et per unum corporale et unum superficiale per quod comparatur superficies est superficies cuius longitudo est una et cuius latitudo est una et cuius anguli sunt recti, et unum corporale quo comparatur corpus est corpus cuius longitudo
65 est una et cuius latitudo est una et cuius altitudo est una et elevatio quarundam superficierum eius super alias est super rectos angulos, tunc propter illud oportet ut narremus causam quare posite sunt iste due quantitates quibus comparentur amplitudo superficierum et magnitudo corporum. Causa vero in hoc est, quoniam oportet ut quantitas
70 qua mensurantur superficies et corpora sit talis ut cum duplantur continuentur ad invicem taliter ne dimittat in vacuitatibus aliquid de superficie et corpore mensuratis super quod non veniat. Et est necessarium cum hoc ut sit illud super quod venit mensuratio de superficie aut corpore facile dum non prohibetur eius mensuratio. Et neque est
75 aliquid magis ultimum in facilitate discretionis illius quam ut sit iudicium unius quo comparatur superficies aut corpus in singularitate

57 ut: quod *MaR*
57–58 incipiamus...narrationem: ostendamus illud de quo volumus narare (narrare *R*) *MaR*
57, 58 voluimus *H*
60–62 et... superficies[1]: et per unum corporale et unum superficiale et alia (in alio *Ma*) operatio et scientia in mensuratione quantitatis illius non comparatur nisi per unum superficiales (!) per quod mensuratur comparatur *R*
60 *de* scientia *scr. P Zm mg.* in alio, operatio
61 per[2] *om. S* / et unum: unum autem *S*
62 *de* comparatur *scr. P mg. et Zm supra* vel mensuratur / comparatur: mensurantur et comparantur *S*
63 *post* una[2] *add. H* et cuius altitudo est una / anguli: angeli *R*
65 cuius[1] *om. MaR*
66 quorundam *MaR* / eius *om. S* / super alias: super alia *MaR* superficialis *H*
67–69 tunc...corporum *om. S*
67 narrem *H*
69 quoniam oportet *om. H*
70 qua: que *R* / *post* corpora *add. S* ad invicem / cum duplantur: conduplantur *MaR*
71 continetur *H* / taliter *om. H* / in: de *S* / *de* vacuitatibus *scr. PS mg. et Zm supra* vel (*om. S*) in toto illo / vacuitatibus: vel in toto illo vacuitatibus *MaR* / aliqui *R* alliqui *Ma*
72 mensuratur *H*
73 illud: talis (?) *H*
73 superficiei *R*
74 *de* dum... mensuratio *scr. P mg. et add. H MaRS in textu* (*ante* facile): in (*om. S.Zm?*, vel *H*) alio (*om. S Zm?*) taliter (*om. S*) quod non veniat super ipsum mensuratio eius (*om. Zm*) (*et add. S in text.*: secundam aliam literam)
75 aliquid: aliqui *Ma R* sicut *H*

established what we have described, then it is necessary for us to begin to establish that whose exposition we desire in our book. Now since (1) we do not wish to signify by "that" anything except a knowledge of the amplitude of surfaces and the volume of bodies, and [since] (2) a knowledge of the mensuration of such quantities is compared only by means of a unit surface and by means of a unit body—and a unit surface for the comparison of surfaces is a surface whose length is one, whose breadth is one, and whose angles are right angles, while a unit body for the comparison of bodies is a body whose length is one, whose breadth is one, whose height is one, and wherein the elevation of the certain surfaces upon each other is at right angles—for these reasons, then, it is necessary for us to discuss the cause as to why these two [unit] quantities have been posited for the comparison of the amplitude of surfaces and the volume of bodies. The reason is because it is necessary that, when the quantity by which the surface and body are measured is continuously repeated and one unit placed beside another, it be of such a nature that there must not be in the suface and body uncovered spaces over which the measure does not come. It is necessary in addition that it be easy [to distinguish] that part of the surface or body being measured, so that its measurement is not hindered. And nothing is better for easily distinguishing this than a unit measure for the comparison of a surface or a body which maintains the same character whether used singularly or in duplication, so that by a single

60 scientia: العمل (*procedure*) ((Cf. with the introduction, where it is noted that Gerard gives *operatio* as an alternate reading; presumably the first tradition Gerard used had العلم, which would indeed be rendered by *scientia*.))

60–61 comparatur: يتبين بالقياس (*is investigated by comparison*).

67–69 tunc....est *om. Ar.*

71 in vacuitatibus: فى خلله (*in its gaps*) ((This is the first tradition followed by Gerard of Cremona and not that of his marginal notation; see the Latin variants. Presumably the alternate tradition read فى كلّه—*in its whole*.))

71–72 de superficie et corpore *om. Ar.*

73–74 sit...mensuratio: تحتاج مع ذلك الى ان يكون تميز ما اتى عليه للتقدير مما لم يأت عليه سهلا

(*And it is necessary in addition that the distinction of that which its measure covers from that which it does not cover is easy*) ((cf. the alternate reading))

c. 2 sua in sua duplatione iudicium unum, ut sit labor in discernendo / illud super quod cadit mensuratio ab eo quod non mensuratur unus. Et hoc quidem non est repertum in aliqua figurarum nisi in quadrato. Et
80 illud est quoniam quando duplatur alteratur eius quantitas sed remanet eius quadratura. Et iterum continuatio unius cum altero quando duplantur est continuatio non dimittens in vacuitatibus suis aliquid de eo quod mensuratur per ipsum super quod non veniat.

Iam ergo manifestum est propter quam causam usi sunt uno quadra-
85 to de superficiali et corporali loco quantitatis qua comparetur omnis quantitas superficierum et corporum. Causa autem in utendo orthogonio sine aliis non est nisi quoniam mensurans rem oportet ut sit quantitas qua mensuratur veniens super eam et continens eam velociter. Et non est aliqua figurarum quadratarum velocius continens illud
90 quod cum ea mensuratur quam orthogonia, quoniam est maior earum.

Iam ergo manifestum est propter quam causam ponitur quadratum orthogonium ex superficiebus et corporibus esse quantitas qua comparantur superficies et corpora. Et ita verificatur sermo in eo cuius narrationem voluimus in hoc nostro libro. Incipiamus ergo nunc
95 narrare illud quod volumus.

[I.] OMNIS FIGURE LATERATE CONTINENTIS CIRCULUM MULTIPLICATIO MEDIETATIS DIAMETRI CIRCULI IN MEDIETATEM OMNIUM LATERUM FIGURE CONTINENTIS CIRCULUM EST EMBADUM FIGURE LATERATE.

5 Verbi gratia, sit circulus *ZDH* contentus a figura *ABG* [Fig. 34].

77 sua[1] *om. R* / duplicatione *S*
79–80 Et illud est *om. S*
80 duplicatur *S* / *post* duplatur *add. P* ne *sed om. Zm HMaRST* / alterantur *MaR*/ quantitates *MaR*
81 quadraturam *MaR* / unius: eius *R*
81–82 duplatur *H* duplentur *Zm*
83 veniat: omnia *R* venia *Ma*
84–89 Iam.... Et: utuntur autem orthogonio quia *S*
84 quam causam: quartam *MaR*
85 qua: que *MaR* / comparentur *H*
87 quoniam mensurans *tr. H* / mensuras *MaR*
88 qua: que *MaR* / venies *MaR*
89 Et *om. H* / continet *MaR*
90 mensuraretur *MaR* / quam: que *R* / hortogonia *R*
91–95 Iam.... volumus *om. S*
91 propter: ob (?) *H* / quam: qua *MaR* / ponitur *ZmHMaR* potest ? *P*
92 hortogonio *R* / qua: que *R*
94 nostro *om. H* / nunc: nam *H*
95 narrare *PZmR* manifestum *H* narare *Ma*
1 omnes *H*
2 multiplicatio *ZmHMaRTS mg. P*
2–3 medietatem: medietate *MaR hic et ubique* ((I have not noted this elsewhere))
5–21 Verbi.... voluimus: Probatur resolvendo figuram continentem circulum in triangulos, ut si fuerit triangulus in 3

effort one can distinguish the part covered by the measure from the part not covered. And this is not found in any figure but the square. This is because when it is duplicated, although its quantity is altered, its squareness remains. And furthermore the continuous juncture of one unit with another in the duplication process is such as not to leave any vacant spaces uncovered by the measure.

It has thus become clear why the unit square has been employed as the unit quantity for the comparison of surfaces and bodies. The reason for using a right-angular unit rather than one with other angles is only because, in measuring something, it is necessary that the quantity used for measuring cover and encompass it quickly. And no other figure of a square-like nature [with equal sides] encompasses that which it measures more quickly than one with right angles [i.e., a square or a cube], since it is the largest of such figures [with equal sides].

Hence it has now become evident why, of surfaces and bodies, the right-angled square-like quantity [i.e., a square or a cube] is posited as the quantity by which surfaces and bodies are compared. And so the terminology for that which we wish to recount in our book is verified. Hence let us now begin our desired narrative.

[I.] IN THE CASE OF EVERY [REGULAR] POLYGON CONTAINING A CIRCLE THE MULTIPLICATION OF THE RADIUS OF THE CIRCLE BY HALF THE PERIMETER OF THE POLYGON CONTAINING THE CIRCLE IS THE AREA OF THE POLYGON.

For example, let circle ZDH be contained by figure ABG [see Fig. 34].

81 eius quadratura: تربيعه (*its squareness*)
81–95 Et....volumus: واظم الاشكال اطربعة احاطه هو القائم الزوايا فهذا هو العلة فى جعل ذلك معيارادون غيره
(*And the greatest of figures with a quadrangular perimeter is that which is right angular and this is the reason for using this measure rather than some other one.*) ((Cf. particularly lines 89–90: Et....earum.))
3–4 continentis circulum *om. Ar.*
4 figure laterate *om. Ar.*

Et centrum circuli sit punctum *E*. Et sit linea *EH* medietas diametri eius. Dico ergo quod multiplicatio linee *EH* in medietatem omnium laterum figure *ABG* est embadum superficiei *ABG*. Cuius hec est demonstratio.

10 Protraham duas lineas *BED*, *GEZ*. Ergo linea *EH* est erecta super lineam *BG*. Ergo est perpendicularis trianguli *BEG*. Ergo multiplicatio *EH* in medietatem linee *BG* est embadum trianguli *BEG*. Et per huiusmodi proprietatem sciemus quod multiplicatio medietatis diametri circuli *ZDH* in medietatem linee *AB* aut in medietatem 15 linee *AG* est embadum duorum triangulorum *GEA*, *AEB*. Iam ergo ostensum est quod multiplicatio medietatis diametri circuli *ZDH* in medietatem linearum *BA*, *AG*, *GB* agregatarum est embadum figure *ABG*. Et iam scitur ex eo quod narravimus quod omnis corporis continentis speram multiplicatio medietatis diametri spere 20 in tertiam embadi superficiei corporis continentis speram est embadum magnitudinis corporis. Et illud est quod declarare voluimus.

triangulos et ducendo (*bis*) a centro circuli ad loca contactus circuli cum figura illa lineas quarum quelibet est semidyameter circuli et cadit perpendiculariter super latus figure scilicet trianguli et utendo hac propositione: embadus cuiuslibet trianguli consurgit ex ductu perpendicularis in medietatem basis super quam cadit, multociens, scilicet secundum numerum triangulorum in quos resolvitur illa figura. *S*
5 Verbi gratia *om. MaRT* / a *om. T* / figure *T*
6 punctus *T*
6 EH: HE *H*
7 eius: ipsius *T*
7, 10 ergo: igitur *T*
8 figure: eius figure *H*
8–9 hec...demonstratio: demonstratio est *T*
11 BG *om. T* / Ergo...BEG *om. MaR*
13 huiusmodi proprietatem: hunc modum *T* / *de* proprietatem *scr. P mg. et Zm supra* vel modum / multiplicati *MaR*
14 aut: igitur *T*
15 embade *MaR* / GEA : GET *R*
15–16 Iam...quod: Unde *T*
17 agragatarum *R*
18 Et: etiam *MaR* / iam...eo: sequitur *T*
18–19 omnis...continentis *corr. ex* omne corpus continens *in MSS*
19 speram *P* speram ideo *H* speram quod *MaRTZm*
20 embade *R*
20–21 embadum magnitudinis: embadum magnitudinis *Zm*
21 est *om. H* / declarare voluimus *PZmR* declarandum erat *H* demonstrare voluimus *T*

7–11 Dico....Ergo[1]: ونصل – ١٥ – ٥ ب – ٥ ج – فظاهر ان – ٥ ح –
(*And we join EA, EB, EG. And so it is apparent that EH*) ((I have followed Gerard in rendering – ٥ – by "E" since "H" is employed for ح.))

12–18 Et....ABG: وكذلك الحكم فى مثلثيى – ا ٥ ب – ا ٥ ج – فاذا نصف قطر الدائرة فى نصف جميع الاضلاع هو مساحة مثلث – ا ب ج
(*And in the same way [we have] knowledge concerning the two triangles AEB, AEG. And so [the multiplication] of half the*

And let the center of the circle be point E. And let line EH be its radius. I say, therefore, that the multiplication of line EH by half the perimeter of polygon ABG is the area of surface ABG.

Demonstration: I shall draw two lines BED, GEZ. Hence line EH has been erected on line BG. Therefore, it is the altitude of $\triangle BEG$. Hence, $EH \cdot 1/2\ BG = \triangle BEG$. In the same way we know that $EH \cdot 1/2\ AB = \triangle AEB$ or that $EH \cdot 1/2\ AG = \triangle GEA$. Hence, it has now been shown that $EH \cdot 1/2\ (BA + AG + GB) = \triangle ABG$.

[Corollary:] And it is now known from what we have narrated that in the case of every [regular] body containing a sphere, the multiplication of the radius of the sphere by one third the surface area of the body containing the sphere is the volume of the body. And this is what we wished to prove.

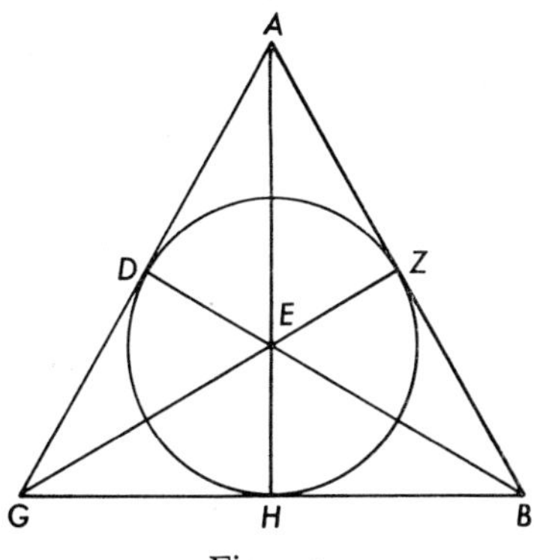

Fig. 34

diameter of the circle by half of the sum of the sides is the area of triangle ABG.)

19 medietatis *om. Ar.* ((in printed edition, but included in Paris MS of Arabic text))

21 *post* corporis *add. Ar.* وهو اعظم من تكسير الكرة (*And it is greater than the measure* ((i.e., volume)) *of the sphere.*) ((Note the meaning "volume" which seems to be demanded by the sense, although this is certainly not a common meaning of تكسير. But see R. Dozy, *Supplement aux dictionnaires arabes*, vol. 2, p. 465.)) / Et...voluimus *om. Ar.* There is a final sentence in Arabic which was probably added by al-Ṭūsī:

اقول هذا انما يتبين بتوهم قسمة الجسم بمخرو طات رؤسها مركز اكرة وقواعدها قواعد الجسم ويكون نصف قطر الكرة اعمدة على قواعدها فتكون مساحته مساحة تلك المخروطات

(*I say that this can only be demonstrated by supposing the division of the body into pyramids whose vertices are at the center of the sphere and whose bases are the bases* ((i.e., faces)) *of the body and the radius of the sphere will be the altitudes on these bases. And so its volume is the* [*combined*] *volume of these pyramids.*)

[II.] MEDIETATIS DIAMETRI CIRCULI MULTIPLICATIO IN MEDIETATEM OMNIUM LATERUM OMNIS FIGURE IN CIRCULO CONTENTE EST MINOR EMBADO SUPERFICIEI CIRCULI.

5 Verbi gratia, sit figura *ABG* contenta a circulo *ABG*. [Fig. 35] 56r c. 1 / Et sit centrum circuli punctum *T*. Dico ergo quod multiplicatio medietatis diametri circuli *ABG* in medietatem laterum figure *ABG* agregatorum est minor embado circuli *ABG*. Cuius hec est demonstratio.

Protraham duas lineas *TB*, *TG*. Et protraham lineam *TK* perpen- 10 dicularem super lineam *BG*, et producam eam usque ad *H*. Et protraham duas lineas *BH*, *HG*. Ergo multiplicatio linee *TH* in medietatem linee *BG* est embadum duorum triangulorum *TBG*, *BHG*. Et per modum similem huic scitur quod multiplicatio medietatis diametri circuli *ABG* in medietatem laterum *AG*, *BG*, *AB* est minor embado 15 circuli *ABG*. Iam ergo declaratum est quod multiplicatio medietatis diametri circuli *ABG* ⟨in medietatem omnium laterum figure⟩ est minor embado circuli *ABG*. Iam ergo ostensum est quod multiplicatio medietatis diametri circuli in medietatem omnium laterum figure est minor embado circuli. Et scitur ex eo quod narravimus, quod multi-

3 embado: ab amd' *MaR*
4 circuli *om. MaR*
5–19 Verba....circuli: Hanc probat per precedentem *S*
5 Verbi gratia *om. T et tr. R post* sit *et super scr. Ma* / *ante* ABG[1] *del. P* A / a circulo: ad circulum *MaR* circulo *T*
6 ergo *om. T*
7 medietatem: medietate *MaR hic et ubique*
8 aggregatarum *R* / hec est *om. T tr. Ma*
9 T-: E- (?) *T hic et ubique in hac propositione* / TK : EAH *T*
12 BG...Et *om. T* / TBG: RBG *R*
13 scitur: scilicet *T*
15–17 Iam...ergo: Unde *T*
16 ⟨in...figure⟩ *supplevi*
17 minor *om. MaR*
18 medietatis *mg. P*
19 scitur ex *PR* est ex *H* scilicet *T*

6 T: – ه – (*E*) ((i.e., ordinarily Gerard represents ه by "E," as he did in the first proposition.))
6–8 "Dico....demonstratio *om. Ar.*
9 TB, BG: ه ج – ه ب – (*EB*, *EG*) / TK: – ه د – (*ED*) ((Note: د of Arabic edition is clearly ك in Paris MS, which thus is like Gerard's copy of it.))
10 H: – ز – (*Z*) ((ز is ر in accompanying figure in edition and also in text of Paris MS.))
11 BH, HG: – ز ج – ب ز – (*BZ*, *GZ*) / TH: – ه ز – (*EZ*)

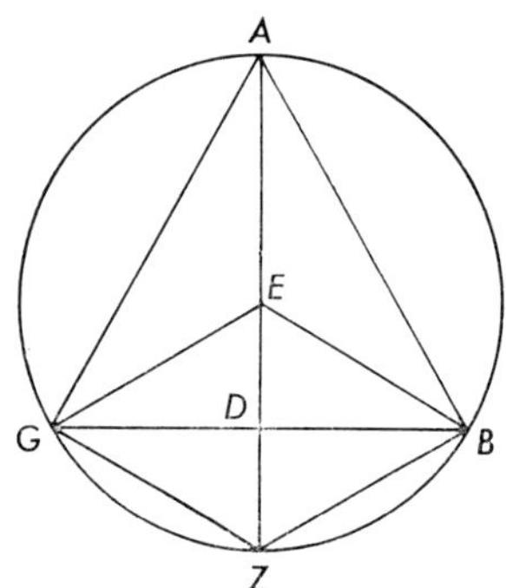

Note: The letters have been transliterated from the Arabic

[II.] THE MULTIPLICATION OF THE RADIUS OF A CIRCLE BY HALF THE PERIMETER OF EVERY [REGULAR] POLYGON CONTAINED IN A CIRCLE IS LESS THAN THE AREA OF THE CIRCLE.

For example, let polygon *ABG* be contained by circle *ABG* [see Fig. 35]. And let the center of the circle be point *T*. I say, therefore, that the multiplication of the radius of circle *ABG* by half the perimeter of polygon *ABG* is less than the area of circle *ABG*.

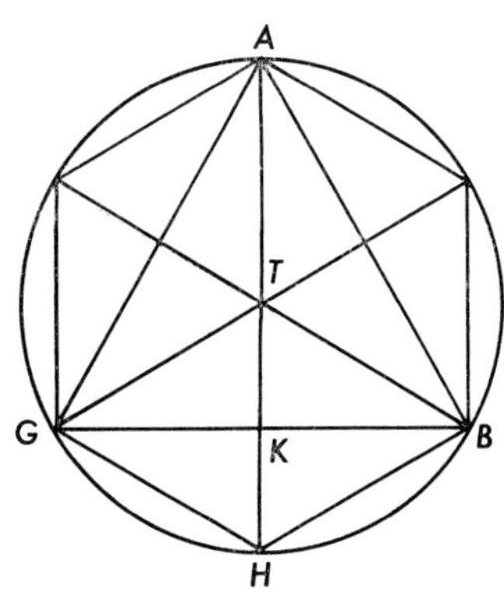

Fig. 35

Demonstration: I shall draw two lines, *TB*, *TG*, and I shall draw line *TK* perpendicular to line *BG*, producing it [*TK*] to *H*. And I shall draw the two lines *BH*, *HG*. Hence, $TH \cdot 1/2\ BG = (\triangle\ TBG + \triangle\ BHG)$. By a method similar to this it is known that $TH \cdot 1/2\ (AG + BG + AB) <$ circle *ABG*. Hence it has now been demonstrated that the multiplication of the radius of circle *ABG* by half the perimeter of the polygon is less than the area of circle *ABG*. Hence it has now been demonstrated that the multiplication of the radius of a circle by half the perimeter of an inscribed polygon is less than the area of the circle.

[Corollary:] And it is known from what we have recounted that the

12 TBG, BHG: — ه ب ج — ز ب ج
(*EBG*, *ZBG*)
12–19 Et....circuli: وهو اقل من مساحة قطلع
— ه ب زج — واعظم من مساحة مثلث — ه ب ج
— و بمثله تبين فى باقى الشكل و تبين ان مساحة
الدائرة اعظم كثىرا من مساحة مثلث — ا ب ج —

(And it is less than the area of the sector EBZG and greater than the area of the triangle EBG. And in the same way you demonstrate [it] for the remaining figures and you demonstrate that the area of the circle is much greater than the area of triangle ABG.)

20 plicatio medietatis diametri spere in tertiam embadi superficiei omnis figure corporee contente a spera est minor embado magnitudinis spere. Et illud est quod declarare voluimus.

[III.] SI FUERIT OMNIS LINEA TERMINATA ET CIRCULUS, TUNC SI FUERIT LINEA TERMINATA BREVIOR LINEA CONTINENTE CIRCULUM, TUNC POSSIBILE EST UT FIAT IN CIRCULO FIGURA LATERATA ET ANGULATA QUAM 5 CONTINEAT CIRCULUS, ET SINT LATERA EIUS CONIUNCTA LONGIUS LINEA TERMINATA. ET SI FUERIT LINEA TERMINATA LONGIOR LINEA CONTINENTE CIRCULUM, TUNC POSSIBILE EST UT FIAT SUPER CIRCULUM FIGURA LATERATA ET ANGULATA CONTINENS EUM ET ERUNT 10 LATERA EIUS AGREGATA BREVIUS LINEA TERMINATA.

Verbi gratia, sit linea terminata, linea *HU*, et circulus, circulus *ABG* [Fig. 36]. Et ponam lineam *HU* in primis breviorem linea *ABG*, que est circumferentia circuli. Dico ergo quod possibile est ut faciamus in circulo *ABG* figuram lateratam quam circulus contineat, et sint latera 15 eius agregata longius linea *HU*, quod sic probatur.

Linea *HU* est minor linea *ABG*. Sit itaque linea *DZE* continens circulum *DZE* equalis linee *HU* recte. Et faciam in circulo *ABG*

21 continente *Ma* / a *om. MaR* / embado magnitudinie *Zm*
22 Et...voluimus *om. S* / illud...declarare: demonstrare *T* / declarare *P MaR* demonstrare *H Zm*
1 *de* omnis *scr. P mg.* vel quelibet / *post* omnis *add. MaR Zm* vel quelibet
3 continentes *MaR* / ut: quod *T*
4, 9 et angulata *om. T*
5 circulos *H* / sunt *T* / latera eius: lateratis *MaR*
6–7 Et...terminata *om. MaR*
9 eum: eam *H*
11–36 Verbi....figure: Probatio prime partis: Sit HU linea terminata et data primo minor BAG circulo. Sit igitur equalis DEZ circulo minori. Describam igitur intra A circulum figuram lateratam tangendo per angulos suos A sed nullo modo E. Hec igitur figura laterata est maior E circulo, ergo est maior suo equali, scilicet HU, per positionem. Igitur et cetera [*add. mg.*: ut patet ex predictis]. Probatio secunde partis: Si enim sit maius A circulo, sit equalis F circulo maiori A. Describam igitur extra A figuram lateratam non tangendo F, que erit minor HU quia includitur, et erit minus suo equali, scilicet F. *S*.
11 Verbi gratia *om. TMaR* / HU *Zm PMaRSAr* HN *HT hic et ubique* / circulus circulus *Zm PMaR* circulus *HT*
12 Et...breviorem: Sit linea HN primo brevior *T*
13 ergo *om. MaR*
14 ABG *om. H* / quam...contineat: continentam circulo *R* / sit *R*
15–16 quod....minor: et est maior *R*
15 probatur: proponatur *Ma*
16 minor: maior *MaR* / DEZ *H*
17 ZDE *MaRH* / faciamus *MaR* / in circulo *om. MaR*

multiplication of the radius of a sphere by one third the surface area of every [regular] body contained by the sphere is less than the volume of the sphere. And this is what we wished to prove.

[III.] GIVEN SOME DEFINITE LINE AND A CIRCLE, IF THE DEFINITE LINE IS SHORTER THAN THE CIRCUMFERENCE OF THE CIRCLE, THEN IT IS POSSIBLE TO CONSTRUCT IN THE CIRCLE A [REGULAR] POLYGON WHICH THE CIRCLE CIRCUMSCRIBES AND WHOSE PERIMETER IS LONGER THAN THE DEFINITE LINE. AND IF THE DEFINITE LINE IS LONGER THAN THE CIRCUMFERENCE OF THE CIRCLE, THEN IT IS POSSIBLE TO CONSTRUCT ON THE CIRCLE A [REGULAR] POLYGON WHICH CIRCUMSCRIBES THE CIRCLE AND WHOSE PERIMETER IS SHORTER THAN THE DEFINITE LINE.

For example, let there be a definite line, line *HU*, and a circle, circle *ABG* [see Fig. 36]. First I shall posit line *HU* as shorter than line *ABG*, the circumference of the circle. I say, therefore, that we can construct in *ABG* a [regular] polygon which the circle circumscribes and whose perimeter is longer than *HU*.

Proof: Line *HU* is less than line *ABG*. And so let line *DZE*, the circumference of circle *DZE*, be equal to straight line *HU*. I shall construct*

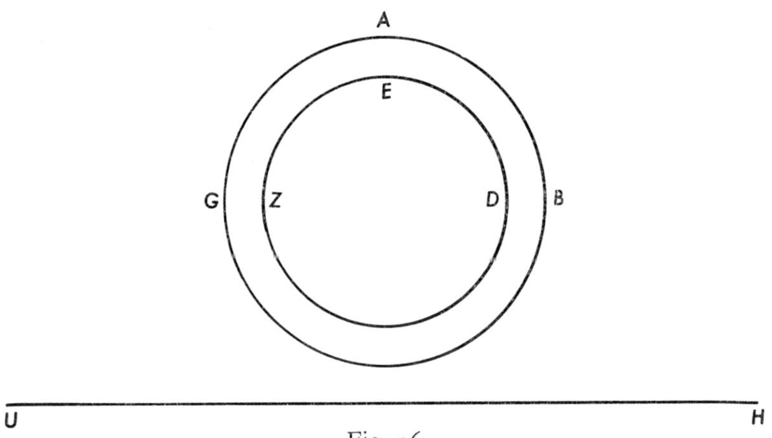

Fig. 36

* The polygon is not actually added to the figure.

1 omnis *om. Ar.*
4 Figura laterata et angula: شكل مضلع (*the sided figure*) ((i.e., regular polygon))
13–16 Dico....ABG *om. Ar.*

figuram lateratam et angulosam non contingentem circulum *DZE*. Tunc latera figure facte agregata sunt longius linea *EDZ*. Sed linea
20 *EDZ* est equalis linee *HU*. Iam ergo ostensum est quod possibile est ut faciamus in circulo *ABG* figuram lateratam et angulosam et latera eius agregata sint longius linea *HU*. Et illud est quod declarare voluimus.

Deinde ponemus lineam *HU* longiorem linea continente circulum
c. 2 *DZE*. Di/co ergo quod possibile est ut faciamus super circulum *EDZ*
26 figuram continentem ipsum, et sint latera eius agregata brevius linea *HU*. Cuius hec est demonstratio.

Linea *HU* est longior linea *EDZ*. Sit ergo linea *ABG* continens circulum *ABG* equalis linee *HU*. Fiat ergo in circulo *ABG* figura
30 laterata non contingens circulum *EDZ*. Dico ergo quod latera figure facte agregata sunt brevius linea *ABG*. Et cum fit super circulum *EDZ* figura laterata et non tangit circulum *ABG* similis figure facte in circulo *ABG*, erunt latera figure contingentis circulum *EDZ* agregata brevius plurimum linea *ABG*. Et linea *ABG* est equalis
35 linee *HU*. Ergo latera figure facte sunt brevius linea *HU*. Et illud est quod declarare voluimus. Et hec est forma figure.

18 figuram: 2am 12i (?) *add. P mg* / et *om. MaR* / contingente *MaR*
19 latam *MaR*
20 Iam ergo: igitur *T* / quod *om. T*
20–21 est ut faciamus: facere (?) *T*
21 ABG *om. T* / et angulosam *om. T* / et[1] *om. MaR* / et[2]: cuius *T*
22 eius *om. T* / est...declarare: demonstrare *T* / declarare: demonstrare *H*
24 ponemus lineam: sit linea *T* ponamus linea *R* ponemus linea *Ma* / longior *T* / linea: lineam *MaR* / *ante* linea *add. T* quam / continentem *MaR*
25 DEZ *H*
26 continentis *T* / et...agregata: cuius latera aggregata sunt *T*
27 hec *om. T* / est demonstratio *tr. T*
28 Linea...longior: longior est linea HN *H*
30 Dico ergo *om. R* / ergo *om. Ma* / latere *MaR*
30–31 figure facte *tr. T*
31 agregata *tr. T post* brevius / brevis *R* / *ante* linea *add. T* quam / fit: sit *H*
31–32 Et...similis: similiter *MaR*
34 agregata *om. T* / Et linea ABG *om. H*
35–36 Et...est[1] *om. T*
36 declarate *PMaR Zm* demonstrare *HT* / Et...figure *PZm om. HTMaR*

19–23 Sed....voluimus: اعنى من خطّ – ح و – (*i.e.*, [*longer*] *than line HU*)
25–28 Dico....EDZ *om. Ar.*
34–35 Et....HU[2]: اعنى من خطّ – ح و – (*i.e.*, [*shorter*] *than line HU*)
36 Et...figure *om. Ar.* / *post* voluimus *add. Ar.* ((no doubt by al-Ṭūsī))

اقول هذا مبنى على وجود دائرة يساوى محيطها اى خطّ محدود يفرض و هذا مما لم يتبين فى موضع
(*I say this is based on the existence of a circle whose circumference is equal to a postulated fixed line, but this is never demonstrated in any place.*)

in circle *ABG* a [regular] polygon not touching circle *DZE*. Then the perimeter of the polygon is longer than line *EDZ*. But line *EDZ* is equal to line *HU*. Hence it has now been demonstrated that we can construct in circle *ABG* a [regular] polygon whose perimeter is longer than line *HU*. And this is what we wished to show.

Then we posit line *HU* to be longer than the circumference of circle *DZE*. I say, therefore, that we can construct on circle *EDZ* a polygon circumscribing the circle whose perimeter is shorter than line *HU*.

Demonstration: Line *HU* is longer than line *EDZ*. Hence let line *ABG*, the circumference of circle *ABG*, be equal to line *HU*. Therefore, let us construct in circle *ABG* a [regular] polygon not touching circle *EDZ*. I say, therefore, that the perimeter of the figure so constructed is shorter than line *ABG*. And since there is formed on circle *EDZ* a polygon which does not touch circle *ABG* and which is similar to the figure constructed in circle *ABG*, the perimeter of the polygon touching circle *EDZ* will be very much shorter than line *ABG*, and line *ABG* is equal to line *HU*. Therefore, the perimeter of the figure constructed is shorter than line *HU*. And this is what we wished to show. And this is the form of the figure [Fig. 36].

[IV.] MEDIETATIS DIAMETRI OMNIS CIRCULI MULTIPLICATIO IN MEDIETATEM LINEE CONTINENTIS IPSUM EST EMBADUM SUPERFICIEI EIUS.

Verbi gratia, sit circulus *ABG*, cuius centrum sit punctum *E*
5 [Fig. 37]. Et sit medietas diametri eius linea *EG*. Dico ergo quod multiplicatio linee *EG* in medietatem linee *ABG* est embadum circuli *ABG*, quod sic demonstratur.

Si non fuerit ita, tunc erit multiplicatio linee *EG* in quantitatem breviorem aut longiorem medietate linee *ABG* ipsum embadum cir-

2 in *om. H*
3 eius: ipsius *T*
4–40 Verbi....arcus[1]: Probatio per impossibile, quia GE semidyametri circuli A multiplicatio nec in maiorem nec minorem medietate linee continentis circulum A confacit embadum. Sit enim primo quod multiplicatio GE in H facit embadum et sit enim minor mediatate linee [continentis] A. Igitur duplum H est minor ABG [i.e., circumferentia circuli A]. Igitur per primam partem precedentis fiat figura laterata contenta ab A maior dupla H. Igitur per secundam multiplicatio GE in medietatem omnium laterum huius figure est minor embado superficiei circuli A. Sed H est minor aggregato es hiis medietatibus, ut sequitur ex hypothesi (?) et prima parte precedentis. Igitur multiplicatio GE in H est multo minor embado A, sed positum erat equalis A. Hoc est contrarium. Sequitur, si dicatur quod H, ex eius multiplicatione per GE provenit embadum A, sit maior linea continente medietatem linee A, quod per secundam partem precedentis fiat figura laterata continens A, cuius latera aggregata sint brevius duplo linee H, quia duplum linee H est maior toto A. Sed medium est maius medio. Ergo medietates omnium laterum eius erunt brevius H. Sed per primam huius multiplicatio GE in medietates omnium laterum huius figure aggregatas facit embadum eius. Igitur embadum est maius embado A. Igitur multiplicatio GE in H est multo maior embado superficiei circuli A. Iam vero fuit ei equalis, quod est contrarium, quia quod est maius maiore est maius minore. Corellarium: et ex hoc sequitur quod cum sumitur ex circulo A arcus aliquis, *S*

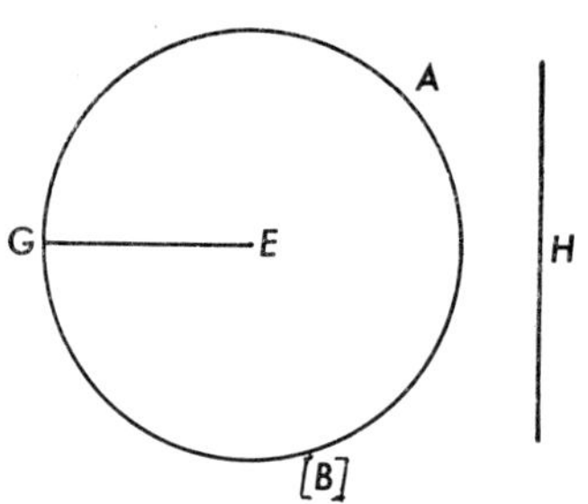

4 Verbi gratia *om. TMa* / cuius *om. TR* / punctum *om. T*
5 Et sit *om. T* / ergo *om. T*
7 quod...demonstratur: cuius demonstratio est *T*
8 ita: sic *T* / tunc *PMaTZm* unde *H* / linee *PRTZm* e linee *H*
9, 11, 22, 26 medietati (?) *H*
9 medietate *om. T* / linea *T*

5–8 Dico....ita: – فان لم يكن سطح – ه ج – فى نصف محيط – ا ب ج – مساويا لمساحة الدائرة

(*And so if the multiplication of EG by one half the circumference AB is not equal to the area of the circle*)

[IV.] THE MULTIPLICATION OF THE RADIUS OF ANY CIRCLE BY HALF ITS CIRCUMFERENCE IS THE AREA OF ITS SURFACE.

For example, let there be a circle ABG, whose center is point E [see Fig. 37]. And let its radius be line EG. I say, therefore, that

$$EG \cdot 1/2 \text{ circum } ABG = \text{area of circle } ABG.$$

Proof: If it is not so, then the multiplication of EG by a quantity either shorter or longer than 1/2 line ABG will be the area of circle ABG.

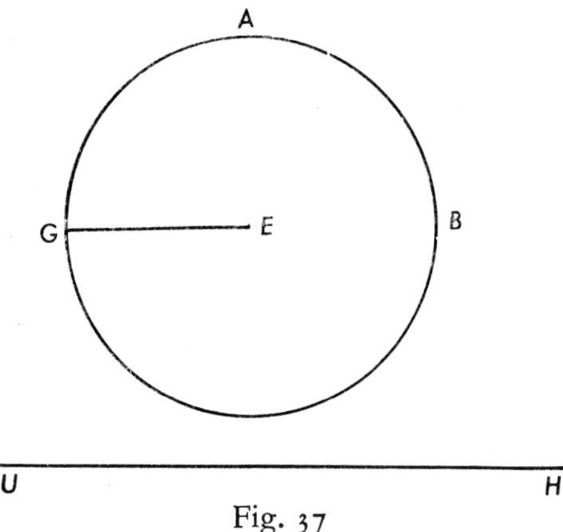

Fig. 37

10 culi *ABG*. Si possibile fuerit, sit itaque multiplicatio in primis in minorem quantitatem medietate linee *ABG*. Ponam autem quantitatem illam lineam *HU*. Ergo multiplicatio linee *EG* in lineam *HU* est embadum circuli *ABG*. Et duplum eius est brevius linea *ABG* continente. Ergo possibile est ut faciamus in circulo *ABG* figuram late-
15 ratam cuius latera agregata sint longius duplo linee *HU*. Cum ergo fiet hec figura in circulo *ABG*, erit medietas agregationis laterum eius longius linea *HU*. Et erit multiplicatio medietatis diametri circuli *ABG* in medietatem agregationis laterum figure facte in eo minor embado circuli *ABG*. Ergo multiplicatio linee *HU* in *EG* est multo minor
20 embado circuli *ABG*. Et iam fuit ei equalis. Hoc vero contrarium est et impossibile.

Et sit multiplicatio linee *EG* in quantitatem longiorem medietate linee *ABG* ipsum embadum circuli *ABG*, si fuerit possibile. Et ponam quantitatem illam lineam *HU* iterum. Ergo multiplicatio linee *EG* in
25 lineam *HU* est embadum circuli *ABG*. Sed linea *HU* est longior medietate linee *ABG*. Et duplum eius est longius linea *ABG*. Ergo possibile est ut fiat super circulum *ABG* figura laterata cuius latera
56v c.1 agregata sint brevius duplo linee *HU*. Et cum fiet hec figura / super circulum *ABG* erit medietas omnium laterum eius brevior linea *HU*.
30 Sed multiplicatio medietatis diametri circuli *ABG* in medietatem agregationis laterum figure facte super ipsum est embadum figure continentis circulum *ABG*. Et embadum eius est maius embado circuli *ABG*. Ergo multiplicatio medietatis diametri circuli *ABG* in lineam *HU* est multo maior embado superficiei circuli *ABG*. Iam

10 ABG *om. T* / *ante* sit *add. Zm, Pmg.* in alio
10–12 Si...HU[1]: Sit ergo multiplicatio in breviorem sicud lineam HN que est brevior medietate linee ABG contingens (*!*) circulum *T*
10–11 sit...ABG *ZmHMa, mg. P*
11 *post* ABG add. *HZm, mg. P* que sit linea HU et est (est ei et est *H*) quantitas brevior medietate linee ABG continentis circulum (*istud totum om. Ma*)
12 lineam[2] *om. T*
13 *post* brevius *add. T* eius
13–14 continente *om. T*
15–16 cum...erit: et huius figure *T*
16 hec *PR Zm om. H* / erit: est duplum *Zm et Zm supra scr.* in alio, medietas
19 Ergo: sed *H*
20 embado circuli *tr. H* / Hoc...est: quod est contrarium *T*
22 Et sit *PR Zm* si vero *T* et si *H* / quantitatem *om. T*
23 ipsum *om. T*
23–24 si...iterum: Sit igitur huius linea longior HN *T*
27 fiat *tr. T post* ABG
29 circulum *om. H*
30 in *om. H*
31 figure facte *tr. T* / super ipsum *om. T*
32, 33, 34 ABG *om. T*
32 *post* embado *del. Zm* superficiei

If it is possible, first let the multiplication be by a quantity less than 1/2 line *ABG*. Moreover, I shall posit that quantity to be line *HU*. Therefore, $EG \cdot HU =$ circle *ABG*. And $2\ HU <$ circum *ABG*. Therefore, we can construct* in circle *ABG* a [regular] polygon whose perimeter is longer than 2 *HU* [by Proposition III]. When, therefore, this polygon is constructed in circle *ABG*, 1/2 its perimeter will be longer than *HU*. And the multiplication of the radius of circle *ABG* by 1/2 the perimeter of the polygon constructed in it will be less than the area of circle *ABG*. Therefore, $(HU \cdot EG)$ is much less than the area of circle *ABG*. But it was already [posited as being] equal to it. This indeed is a contradiction and is impossible.

And let the multiplication of line *EG* by a quantity longer than 1/2 line *ABG* be the area of circle *ABG*, if it is possible. And I shall again posit that this quantity is line *HU*. Therefore, $EG \cdot HU =$ circle *ABG*. But $HU > 1/2$ line *ABG*. Therefore, $2\ HU >$ line *ABG*. Therefore, it is possible to construct upon circle *ABG* a [regular] polygon whose perimeter is less than double line *HU*. And when this polygon is formed on circle *ABG*, 1/2 its perimeter will be less than line *HU*. But the multiplication of the radius of circle *ABG* by 1/2 the perimeter of the polygon constructed on it equals the area of the polygon containing circle *ABG*, and its area is greater than the area of circle *ABG*. Therefore, the multiplication of the radius of circle *ABG* by line *HU* is much greater than the surface area of circle *ABG*. But it was already [posited as being]

* As in Fig. 36, this polygon is not actually constructed.

---

10 Si...fuerit *om. Ar.* | *post* multiplicatio *add. Ar.* – ه ج – (*EG*)
15–16 Cum...ABG *om. Ar.*
17 circuli ABG: – ه ج – (*EG*)
19, 20 ABG *om. Ar.*
21 et impossibile *om. Ar.*
26 linea ABG: محيط الدائرة (*the circumference of the circle*)
28–29 cum...ABG *om Ar.*
30 circuli ABG: – ه ج – (*EG*)
31–32 facte...eius *om. Ar.*
33 ABG[1] *om. Ar.* | medietatis...ABG: – ه ج – (*EG*)

35 vero fuit ei equalis. Et hoc quidem est contrarium et impossibile. Iam ergo declaratum est quod multiplicatio medietatis diametri omnis circuli in medietatem linee continentis ipsum est embadum superficiei circuli. Et illud est quod demonstrare voluimus.

Et iam scitur ex eo quod narravimus quod cum sumitur ex circulo 40 *ABG* arcus, quicunque arcus sit, et protrahuntur ex duabus extremitatibus eius due linee ad centrum circuli, est embadum huius trianguli, quem continet iste arcus et due linee que protracte sunt ab extremitatibus eius ad centrum, illud quod fit ex multiplicatione medietatis diametri circuli *ABG* in medietatem arcus assumpti ex eo. Et illud est 45 quod voluimus declarare. Et hec est forma figure.

[V.] PROPORTIO DIAMETRI OMNIS CIRCULI AD LINEAM CONTINENTEM IPSUM EST UNA.

Verbi gratia, sint duo circuli diversi, qui sint duo circuli *ABG*, *DEZ*. Et sint diametri eorum, scilicet diameter circuli *ABG* linea 5 *BG* et diameter circuli *DEZ* linea *EZ* [Fig. 38]. Dico ergo quod proportio diametri *BG* ad lineam *ABG* continentem ipsum est sicut proportio diametri *EZ* ad lineam *DEZ* continentem ipsum, quod sic demonstratur.

Si non fuerit proportio amborum una, tunc sit proportio linee *GB* 10 ad lineam *AGB* sicut proportio *EZ* ad lineam *HU*. Et linea *HU* aut est longior aut brevior linea *DEZ*. Et ponam ipsam in primis breviorem, si fuerit possibile. Et dividam lineam *HU* in duo media super *T*. Et erigam super punctum *H* lineam equalem medietati linee *EZ*,

35 Et hoc quidem *P Zm* ex hoc quod *Ma* quod quidem *H* quod *T* / *ante* Iam *add MaR mg.* 5
35–46 Iam...figure *om. T*
39 eo *PR* illo *H*
41 est *om. Zm*
42 continet *HSMa* continent *P Zm*
44 ABG: A *S*
44–45 Et...declarare *om. S*
45 quod...declarare: propositum *H* / Et...figure *PRZm om. HS*
1 [v]: 6 *MaR*
2 continentem ipsum *tr. H*
3–39 Verbi... voluimus *tr. H* / voluimus *om. S*
3 Verbi gratia *om. TMa* / qui...circuli *om. T* qui sunt *R* / sunt *Ma*
4–5 sint... linea: dyameter primi sit linea BG secundi vero EZ *T*
5 ergo *om. H*
6 diametri *om. T* / lineam *om. T*
7 DEZ *PR Zm* EDZ *HT*
7–9 continentem... sit: quod si non sit igitur *T*
9 non fuerit *PZmMa* foret *H*
10 lineam[1] *om. T* / EZ: linea EZ *T* / lineam HU *tr. T.*
11 linea *om. T*
11–12 Et.... Et: si igitur est brevior igitur *T*
12 duo media: duas medietates *T* / super: in puncto *T*
13 medietati *om. T* medietate *R*

equal to it. This indeed is a contradiction and is impossible. Therefore, it has now been shown that the multiplication of the radius of any circle by 1/2 its circumference equals the surface area of the circle. And this is what we wished to demonstrate.

[Corollary:] And it is known from that which we have recounted that, when any arc at all of circle *ABG* is taken and two lines are drawn from its extremities to the center of the circle, the area of the sector which this arc and the two lines drawn from its extremities to the center contain is equal to the product of the radius of circle *ABG* and 1/2 the arc taken from it. And this is what we wished to show. And this is the form of the figure [Fig. 37].

[V.] THE RATIO OF THE DIAMETER OF ANY CIRCLE TO ITS CIRCUMFERENCE IS ONE [THAT IS, IS THE SAME FOR ALL CIRCLES].

For example, let there be two different circles, the circles *ABG*, *DEZ* [see Fig. 38]. Let their diameters be *BG*, the diameter of circle *ABG*, and *EZ*, the diameter of circle *DEZ*. I say, therefore, that

$$\text{diameter } BG/\text{circum } ABG = \text{diameter } EZ/\text{circum } DEZ,$$

which is demonstrated as follows:

If the two ratios were not equal, then let line BG/line $ABG = EZ/HU$, where line *HU* is longer or shorter than line *DEZ*. First I shall posit that it is shorter, if that is possible. And I shall bisect line *HU* at *T* and I shall erect perpendicularly upon line *HU* at point *H* a line equal to

35 et impossibile *om. Ar.*
36–37 medietatis...circuli: – ج ه – (*EG*)
37 linee...ipsum: محيط – ا ب ج – (*circumference ABG*)
38 *post* circuli *add. Ar.* – ا ب ج – (*ABG*)
39–45 Et....figure: وقد بان منه ان سطح نصف [الكرة فى نصف] القطر فى نصف اى قوس فرض يكون مساويا لمساحة القطلع الذى يحيط به تلك القوس و نصفا قطرين يمرّ ان بطر فيها
((The section in brackets is in the Arabic edition but, quite properly, is not in the Paris Arabic manuscript.)) (*And it is already known from this that the multiplication of the radius by the half of any assigned arc is equal to the area of the sector contained by this arc and two radii extending to the extremities of the arc.*)
5 EZ: – ه د – (*ED*)
5–8 Dico...demonstratur *om. Ar.*
9 Si... una: فان لم يكن كما ادعينا (*And so if it is not as we claimed*)
10, 13 EZ: – د ه – (*DE*)
12 si...possibile *om. Ar.*

stantem super lineam *HU* orthogonaliter, que sit linea *HK*. Et com-
15 plebo quadratum *KT*. Et quoniam linea *HK* est equalis medietati linee
EZ, et linea *HT* est brevior medietate linee *DZE*, erit quadratum *KT*
minus superficie circuli *DEZ*. Verum, proportio linee *KH* ad lineam
HT est sicut proportio medietatis linee *BG* ad medietatem linee *ABG*.
Et multiplicatio linee *KH* in lineam *HT* est superficies *KT*. Et multi-
20 plicatio medietatis linee *BG* in medietatem linee *ABG* est superficies
c. 2 circuli *ABG*. Ergo proportio superficiei circuli *ABG* | ad quadratum
KT est sicut proportio medietatis linee *BG* ad lineam *KH* duplicata.
Sed proportio medietatis linee *BG* ad lineam *KH* multiplicata est sicut
proportio linee *BG* ad duplum linee *KH* duplicata. Sed duplum linee
25 *KH* est equale linee *EZ*. Ergo proportio superficiei circuli *ABG* ad
quadratum *KT* est sicut proportio linee *BG* ad lineam *EZ* duplicata,
ac proportio superficiei circuli *ABG* ad superficiem circuli *DEZ* est
sicut proportio *BG* ad *EZ* duplicata, sicut declaravit Euclides. Ergo
proportio superficiei circuli *ABG* ad superficiem circuli *DEZ*, et
30 ad quadratum *KT*, est una. Ergo sunt equales. Sed quadratum *KT*
iam fuit minus superficie circuli *DEZ*, quod quidem est contrarium
et impossibile. Non est ergo linea *HU* brevior linea *DEZ*.

Et per huiusmodi dispositionem scitur quod linea *HU* non est
longior linea *DEZ*. Et cum linea *HU* non sit longior neque brevior
35 linea *DEZ*, tunc est equalis ei. Et proportio linee *BG* ad lineam *ABG*

14 linea *om. T*
15 quoniam: quia *T*
16 DZE: DEZ *H*
17 KH *ZmHRT* HK *P*
18 BG: AG *H*
19 KH: HK *H*
20 medietatis *om. T* | BG: KG *H*
21 *post* ABG[1] *add. H* Sit igitur proportio medietatis linee BG ad lineam KH multiplicata sicut proportio linee BG ad duplum linee KH duplicata (*istud totum om. PMaT hic; cf. lineas 23–24*) | Ergo: et ergo *T*
21–22 proportio... sicut: sicud est *T*
21 ABG[2]: ADG *H*
22 lineam *om. T* | duplicata: multiplicata *T*
23–24 Sed... duplicata *om. H hic sed cf. var. lineae 21*
23 medietatis *om. T*
24 linee[2] *om. T* | KH: HK *T*
24–25 Sed.... Ergo *om. T*
25 est: erit *H*
27 ac *PZm* aut *H* a *Ma*
28 sicut[2]... Euclides: per euclidem *T*
31 quidem *om. T*
32 Non... HU: Igitur linea HU non est *T* | ergo *om. H*
33–34 Et... DEZ *om. Ma*
33–35 Et.... ei: nec longior igitur est equalis *T*

15–16 Et...DZE *om. Ar.*
21–37 Et....voluimus: فنسبة سطح – ك ط –
الى دائرة – ا ب ج – كنسبة – ط ح – اعنى نصف
– د ه – الى نصف – ب ج – مثناة و هى نسبة
– د ه – الى – ب ج – مثناة و قد بين اقليدس
ان نسبة – د ه – الى – ب ج – مثناة كنسبة
دائرة – د ز ه – الى دائرة – ا ب ج – فنسبة سطح
– ك ط – الى – دائرة – ا ب ج – كنسبة دائرة –
د ه ز – اليها فسطح – ك ط – مساو لدائرة –

1/2 line EZ. This is line HK. And I shall complete square KT. And since line HK = 1/2 line EZ, and line $HT <$ 1/2 line DZE, square $KT <$ area of circle DEZ. Now line KH/line HT = 1/2 line BG/1/2 line ABG, and ($KH \cdot HT$) = area KT, and (1/2 line BG · 1/2 line ABG) = area circle ABG, and area circle ABG/square KT = (1/2 line BG/ line KH)^2. But (1/2 BG/ line KH)^2 = (line BG/2 line KH)^2, and 2 line KH = line EZ. Therefore, area circle ABG/square $KT = (BG/EZ)^2$, and area circle ABG/area circle $DEZ = (BG/EZ)^2$, as Euclid showed. Therefore, area circle ABG/area circle DEZ = area circle ABG/square KT. Therefore, area circle DEZ = square KT. But earlier [it was inferred from the assumption that] square $KT <$ area circle DEZ. This indeed is a contradiction and is impossible. Therefore, line HU is not less than line DEZ.

By a similar procedure it is [also] known that line HU is not longer than line DEZ. And since line HU is not longer and is not shorter than line DEZ, then it is equal to it; and line BG/line $ABG = ZE/HU$. And

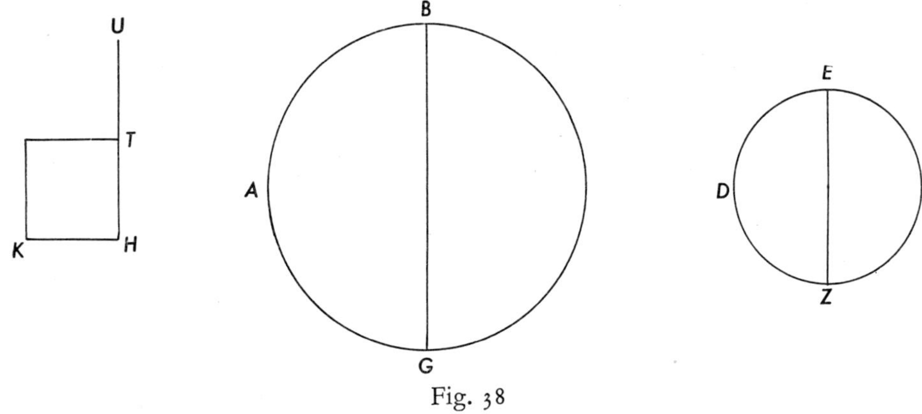

Fig. 38

Note: I have changed diameter EZ from a horizontal to a vertical orientation, in conformity with the drawing accompanying the Arabic text.

د ه ز – وكان اصغر منها هذا خلف فليس خطّ –
ح و – اقصر من محيط – د ه ز – و بمثل هذا
التدبير تبين انه ليس اطول منه فاذا نسبة –
د ه – الى محيط – د ه ز – كنسبة – ب ج – الى
محيط – ا ب ج – وكذلك فى كل دائرتين غيرهما
و ذلك ما اردناه

(And so KT^2/circle ABC = TH / ($\frac{1}{2}$ BG)^2 = ($\frac{1}{2}$ DE / $\frac{1}{2}BG$)^2 = (DE/BG)^2. And Euclid has already demonstrated that (DE/BG)^2 = circle DZE/circle ABG. And so KT^2 / circle ABG = circle DEZ / circle ABG. And so KT^2 = circle DEZ. But it was smaller than it. This is a contradiction. And so line HU is not shorter than circumference DEZ; and by a similar disposition it is demonstrated that it is a lot longer than it. And so DE / circum DEZ = BZ / circum ABG and it is thus for any two circles other than these. And this is what we wished.)

est sicut proportio *ZE* ad *HU*. Et linea *HU* est equalis linee *DEZ*. Iam ergo ostensum est quod proportio diametri omnis circuli ad lineam continentem ipsum est una. Et illud est quod demonstrare voluimus.

[VI.] CUM ERGO IAM MANIFESTUM SIT ILLUD QUOD NARRAVIMUS, TUNC OPORTET UT OSTENDAMUS PROPORTIONEM DIAMETRI CIRCULI AD LINEAM CONTINENTEM IPSUM.

5 Et operabimur in hoc per modum quo operatus est in eo Archimenides. Nam nullus illius scientie invenit aliquid usque ad hunc nostrum tempus preter ipsum in eo quod nobis apparuit. Et iste modus in inveniendo proportionem diametri ad lineam continentem, etsi non ostendat proportionem unius eorum ad alterum ita ut per eam racio-
10 cinetur secundum veritatem, tamen significat proportionem unius eorum ad alterum ad quemcunque finem voluerit inquisitor huius scientie de propinquitate, scilicet si voluerit inquisitor scire proportionem unius eorum ad alterum, verbi gratia, ut perveniat in propinquitate illius ad hoc, ut non sit inter ipsam et inter veritatem propor-
15 tionis unius eorum ad alterum, cum posita fuerit diametrus unum,
57r c. 1 nisi minus minuto, quod est pars sexagessima diametri, possit / illud. Et si voluerit pervenire in propinquitate illius ad hoc, ut non sit ei finis inter ipsam et inter veritatem proportionis unius eorum ad alterum nisi minus secundo, quod est pars sexagesima minuti, possit illud.
20 Et si voluerit ut perveniat in propinquitate illius ad quemcunque

36 ZE: linee ZE *H* / Et: sed *T* / linee DEZ *tr. T*

37 Iam...quod: ergo *T* / diametri omnis *tr. T*

38 Et...est *om. T*

38–39 demonstrare voluimus *HT tr. PZm* declarare voluimus *Ma*

1 [VI]: 7 *MaR*

1–24 Cum.... illud: Que igitur sit proportio dyametri circuli ad lineam continentem ipsum operabimur sicud Archimenides solus, ita quod non fallatur inquisitor in propinquitate veritatis proportionis ad alteram nisi minus minuto, quod est pars 60ª dyametri. Et si voluerit quod non medium nisi secundo servando (*! del. ?*), quod est pars 60ª minuti, et post illa ut perveniat ad quantum[cun]que finem voluerit computator raciocinari. *T*

1–173 Cum....quod² *om. S*

3 ad: et *H* /

4 ipsam *H* /

5 in hoc *om. H* / in eo *om. H*

6 ad hunc *om. H*

10 *de* significat *scr. P mg et Zm supra* i. ostendit / *post* significat *add. MaR* i. ostendit

11–12 huius...inquisitor *om. R* (*sed in PZmHMa*)

18 *de* proportionis *scr. P. mg. et Zm supra* vel mensure / *ante* proportionis *add. Ma* vel mensure

19 pars sexagesima *tr.* H

line HU = line DEZ. Therefore, it has now been demonstrated that the ratio of the diameter of every circle to its circumference is one. And this is what we wished to demonstrate.

[VI.] HENCE, SINCE WHAT WE HAVE RECOUNTED HAS NOW BECOME EVIDENT, THEN WE MUST SHOW [THAT IS, FIND] THE RATIO OF THE DIAMETER OF A CIRCLE TO ITS CIRCUMFERENCE.

And we shall proceed in this matter by the method which Archimedes used for it. For up to our time no one except him has discovered any knowledge of this, so far as we have seen. And this method of finding the ratio of the diameter to the circumference, although it does not reveal a true ratio that can be reckoned with, still does yield a ratio of the one to the other which is an approximation to any limit the investigator of this subject desires. That is, if the investigator wishes to know the ratio of the one to the other approximately so that, for example, between it and the true ratio there is less than a minute, i.e., a sexigesimal part of the diameter when the diameter is posited as one, that could be done. And if he wished to find an approximation of this to that so that less than a second, i.e., 1/60 of a minute, exists between it and the true ratio, that could be done. And if one wished to achieve an approximation of one to the other to any

1–4 Cum...ipsum: ثمّ لتبيّن نسبة القطر الى المحيط (*Then let us investigate the ratio of the diameter to the circumference*)

5 Et...in hoc *om. Ar.*

12–24 scilicet....illud *om. Ar.*

finem voluerit post illa duo, possit illud per illud quod narravit Archimenides. Et usi sunt hoc modo propinquitatis in omni computatione in qua cadunt radices surde, cum computator vult raciocinari per quantitatem eius. Et erit hoc ita. Incipiamus ergo declarare illud.

25 Lineemus ergo circulum *ATB*, cuius diameter sit *AB*, et ipsius centrum sit punctum *G* [Fig. 39]. Et protraham ex centro *G* lineam *GZ* continentem cum linea *GB* tertiam anguli recti. Et erigam super punctum *B* linee *GB* lineam *BZ* orthogonaliter. Manifestum est igitur quod arcus qui subtenditur angulo *BGZ* est medietas sexte circuli 30 *ATB* et quod linea *BZ* est medietas lateris exagoni continentis circulum *ATB*. Et dividam angulum *BGZ* in duo media cum linea *GE*. Et dividam angulum *BGE* in duo media per lineam *GU*. Et dividam angulum *BGU* in duo media per lineam *GD*. Et dividam angulum

24 *de* Et...ita *scr. P mg. et Zm mg.* (*et add. MaR post* eius): in alio, cum ergo hoc sit ita / *ante* Incipiamus *add. MaR mg.* 8
25 circulus *T* / sit *om. T*
25–26 et...punctum: centrum *T*
26 centrum sit *ZmHMa mg. P*
28 igitur *om. T*
29 sexte *PZm* seste *Ma* sexti *H*
31 BGZ: GBZ *T*
32 Et...GU *om. T* / BGE: BEG *H*
33 BGU *PZmMa* BGN *HT* (-N *pro* -U *hic et ubique in HT*)

27 GZ: – ج د – (*GD*) ((Note: The Arabic printed text has د (*D*) everywhere Gerard has Z; I am not noting any other place. Incidentally, the Paris Arabic MS has ح (*H*) for ج (*G*) and an ambiguous mark))

33 BGU: – ب ج – (*BG*) ((Note: This is – ب ح ر – in the Paris MS.)) / GD: – ج ز – (*GZ*) ((Note: Arabic text has ز (*Z*) wherever Gerard has D.))

desired limit beyond these two, that could be done by the method which Archimedes has recounted. And this method of approximation is used in every computation involving surd roots when a computator wishes to calculate with such a quantity. And it will be thus. Therefore, let us begin to show this.

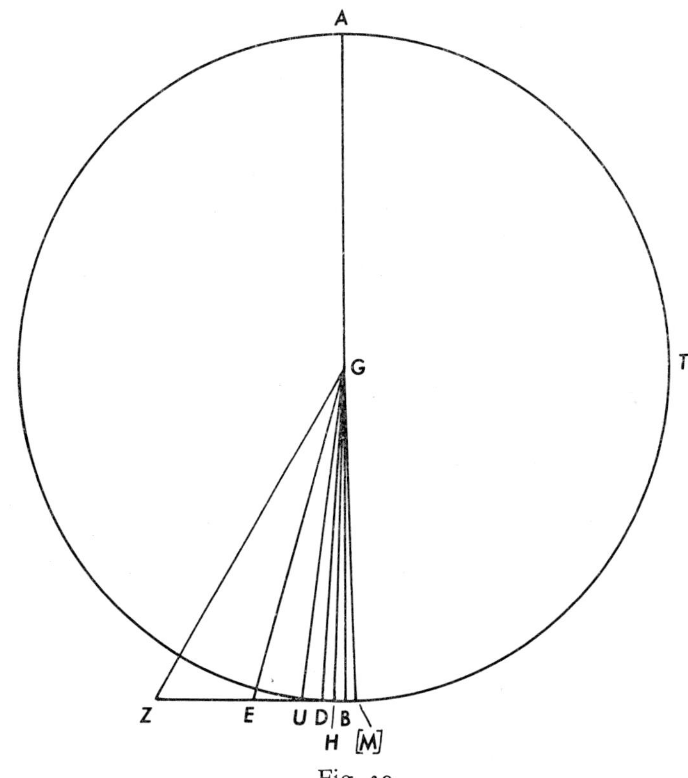

Fig. 39

Note: In MSS *P* and *H*, both halves of the proof are represented on one drawing, although that drawing is repeated. For Fig. 39 I have left off the inscribed figure covering the second half of the proof (see Fig. 40). Also, line *GM*, which is in the original Archimedean proof, is missing in this text and its drawing.

Let us draw circle *ATB*, whose diameter is *AB* and whose center is point *G* [see Fig. 39]. And I shall protract from center *G* line *GZ*, which contains with line *GB* a third of a right angle. And I shall erect line *BZ* perpendicularly on point *B* of line *GB*. It is evident, therefore, that the arc which is subtended by ∠ *BGZ* is 1/2 of 1/6 of circle *ATB* and that line *BZ* is 1/2 of a side of a hexagon containing circle *ATB*. Then I shall bisect ∠ *BGZ* by line *GE*, and ∠ *BGE* by line *GU*, and ∠ *BGU* by line

BGD in duo media per lineam *GH*. Manifestum est igitur quod arcus
35 qui subtenditur angulo *BGH* est pars centessima et nonagessima secunda circuli *ATB*, et quod linea *BH* est medietas lateris figure habentis nonaginta sex latera continentis circulum *ATB*.

Cum ergo hoc sit ita, tunc nos ponemus lineam *GZ* trecentum et sex propter facilitatem usus huius numeri in eo quod computatur.
40 Cum ergo fuerit linea *GZ* trecentum et sex, erit quadratum eius nonaginta tria millia et sexcentum et triginta sex. Et erit linea *BZ* centum et quinquaginta tria, quoniam angulus *BGZ* est tertia anguli recti et angulus *GBZ* est rectus. Et erit quadratum linee *BZ* viginti tria millia et quadringenta et novem. Et quadratum linee *GB* septuaginta millia
45 et ducenta et viginti septem. Ergo linea *GB* est plus ducentis et sexaginta quinque. Sed proportio duarum linearum *BG*, *GZ* agregatarum ad *BZ* est sicut proportio *GB* ad *BE*, propterea quod linea *GE* dividit angulum *BGZ* in duo media. Et due linee *BG*, *GZ* agregate sunt plus quingentis et septuaginta uno. Et linea *BZ* est centum et
50 quinquaginta tria. Ergo proportio *GB* ad *BE* est maior proportione quingentorum et septuaginta unius ad centum et quinquaginta tria. Ergo linea *GB* erit plus quingentis et septuaginta uno, cum fuerit *BE* centum et quinquaginta tria. Et quadratum *GB* est plus trecentis et viginti sex millibus et quadraginta uno. Et quadratum *BE* est viginti
55 tria millia et quadringenta et novem. Ergo quadratum *GE* est plus trecentis et quadraginta novem millibus et quadringentis et quinquaginta. Ergo linea *GE* est plus quingentis et nonaginta uno et octava
c. 2 / unius.

34 igitur *PZmHMa om. T* ergo *R*
36 est: erit *H*
38–39 Cum...sex: tunc *T*
38 trecenta *H*
39 huius *om. T*
40 fuerit: fuit *H* / Cum...eius: ponamus lineam GZ trecentum et 6 cuius linee quadratum necessario erit *T*
41 milia *H hic et ubique*
42 quoniam: quia *T*
43 BZ *ZmMaR* GZ *PH*
43–44 viginti...novem *del. m. rec. P et add. mg.* 23409
44 quadringenta *HMaT* quadreginta *R* quadriginta *P*
45 et[1] *om. H* / Ergo: igitur *T* / GB : BG *Zm* / *cum* ducentis *desinit T*
46 GZ: EZ *H*
51 ad: ac *H*
52 fuerit: fuerit positum *H*
53–54 trecentis...millibus: 300 et 26000 *H*
55 quadratum : 4 *H*
56 trecentis...millibus: 300 et 49000 *H*

38 Cum...ita *om. Ar.*
39 computatur: تتبين (*investigated*)
40 Cum...sex *om. Ar.*
42–43 et[2]...rectus: – ج ب د – (*GBD*)

GD, and $\angle BGD$ by line *GH*. It is evident, therefore, that the arc which is subtended by angle *BGH* is 1/192 of circle *ATB* and that line *BH* is 1/2 of a side of a figure having 96 sides which contains circle *ATB*.

Since, therefore, this is so, then let us assume line *GZ* to be 306 because of the facile utility of this number for computation. Therefore, since *GZ* is 306, $GZ^2 = 93{,}636$. And line $BZ = 153$, since $\angle BGZ = (1/3) \cdot 90°$ and $\angle GBZ = 90°$. And $BZ^2 = 23{,}409$, and $GB^2 = 70{,}227$. Hence, line $GB > 265$. But $(BG + GZ)/BZ = GB/BE$ because line *GE* bisects $\angle BGZ$. And $(BG + GZ) > 571$ and line $BZ = 153$. Hence, $(GB/BE) > (571/153)$. Hence, line $GB > 571$, when $BE = 153$. And $GB^2 > 326{,}041$. And $BE^2 = 23{,}409$. Hence, $GE^2 > 349{,}450$. Hence, line $GE > 591\frac{1}{8}$.

Et secundum exemplum quod narravimus declaratur quod pro-
60 portio linee *GB* ad *BU* est maior proportione mille et centum et sexaginta duorum et octave unius ad centum et quinquaginta tria. Et cum fuerit *BU* centum et quinquaginta tria, erit *GB* plus mille et centum et sexaginta duobus et octava unius. Et quadratum *GB* erit plus mille millibus et trecentis et quinquaginta millibus et quingentis et triginta
65 quattuor et quarta. Et quadratum *BU* erit viginti tria millia et quadringenta et novem. Et quadratum *GU* erit plus mille millibus et trecentis et septuaginta tribus millibus et nongentis et quadraginta tribus et quarta. Ergo linea *GU* est plus mille et centum et septuaginta duobus et octava unius.

70 Et secundum hoc exemplum quod narravimus declaratur quod proportio *GB* ad *BD* est maior proportione duorum millium et trecentorum et triginta quattuor et quarte unius ad centum et quinquaginta tria. Cum ergo fuerit linea *BD* centum et quinquaginta tria, erit *GB* plus duobus millibus et trecentis et triginta quattuor et quarta
75 unius. Et quadratum *GB* erit plus quinque mille millibus et quadringentis et quadraginta octo millibus et septingentis et viginti tribus. Et quadratum *BD* viginti tria millia et quadringenta et novem. Ergo quadratum *GD* erit plus quinquies mille millibus et quadringentis et septuaginta duobus millibus et centum et triginta duobus. Ergo linea
80 *GD* est plus duobus millibus et trecentis triginta novem et quarta unius.

Et secundum hoc exemplum quod narravimus declaratur quod proportio *GB* ad *BH* est maior proportione quattuor millium et sexcentorum et septuaginta trium et medietatis unius ad centum et quinqua-
85 ginta tria. Cum ergo fuerit linea *HB* centum et quinquaginta tria, erit linea *GB* plus quattuor millibus et sexcentis et septuaginta tribus et medietate unius. Et hec quidem est proportio lateris figure habentis nonaginta sex latera continentis circulum ad diametrum. Ergo pro-

63–64 mille...quinquaginta: 100000 (*!*) et 300 et 5 *H*
66–67 mille...millibus: 100000 (*!*) et 300 et 7300 *H*
75–76 quinque...millibus 5000000 et 400 48000 *H*
77 quadringenta et novem: 40 et 9 *H*
78–79 quadringentis...millibus: 400 72000 *H*
83–84 quattuor...sexcentorum: 4000 600 *H*
85 linea *om.* *H*
86 quattuor...sexcentis: 4000 600 *H*
87 *de* proportio *scr.* *P* *mg.* *et* *Zm* *supra* (*et* *add.* *Ma* *ante* proportio): vel mensura

And by using the example we have described, it is shown that $\frac{GB}{BU} > \frac{1,162\frac{1}{8}}{153}$. And when $BU = 153$, $GB > 1,162\frac{1}{8}$. And $GB^2 = 1,350,534\frac{1}{4}$. And $BU^2 = 23,409$. And $GU^2 > 1,373,943\frac{1}{4}$. Therefore, line $GU > 1,172\frac{1}{8}$.

And by using the example we have described it is shown that $\frac{GB}{BD} > \frac{2,334\frac{1}{4}}{153}$. When line $BD = 153$, therefore $GB > 2,334\frac{1}{4}$. And $GB^2 > 5,448,723$. And $BD^2 = 23,409$. Therefore, $GD^2 > 5,472,132$. Therefore, line $GD > 2,339\frac{1}{4}$.

And by using the example we have described it is shown that $\frac{GB}{BH} > \frac{4,673\frac{1}{2}}{153}$. When $HB = 153$, therefore $GB > 4,673\frac{1}{2}$. And this

59 quod narravimus: ذلك (*that*)

63–65 mille...quattuor: ١٣٥٣٤ (13,534) ((But it should be 1,350,534. The Paris Arabic manuscript appears to have 135,534, although there may be a dot between the two fives, and if so it is then correct.))

65 et quarta *om. Ar.*

66 GU: ـ ه ج ـ (*GE*)

67–68 et quarta *om. Ar.*

70, 82 quod narravimus: ذلك (*that*)

portio diametri ad omnia latera figure habentis nonaginta sex latera continentis circulum est maior proportione quattuor millium et sexcentorum et septuaginta trium et medietatis ad quattuordecim millia et sexcenta et octoginta octo. Iam ergo ostensum est quod proportio omnium laterum figure habentis nonaginta sex latera ad diametrum est minor tribus et septima unius.

Amplius protraham in circulo *ATB* cordam sexte, que sit linea *TB* [Fig. 40]. Et protraham *AT*. Et dividam angulum *TAB* in duo media per lineam *AN*. Et protraham cordam *NB* et dividam angulum *BAN* in duo media per lineam *AK*. Et protraham cordam *KB*. Et dividam angulum *KAB* in duo media cum linea *AL*. Et protraham cordam *LB* et dividam angulum *LAB* in duo media per lineam *AM*. Et protraham cordam *MB*. Ergo manifestum est quod corda *MB* est latus figure habentis nonaginta sex / latera quam continet circulus. Deinde ponam lineam *AB* mille et quingenta et sexaginta propter facilitatem usus huius numeri in eo quod volumus. Ergo erit corda *TB* septingenta et octoginta. Et erit quadratum *AB* duo mille millia et quadringenta et triginta tria millia et sexcenta. Et quadratum *BT* sexcentum et octo millia et quadringenta. Et erit quadratum *AT* mille millia et octingenta et vigintiquinque millia et ducenta. Ergo linea *TA* est minus mille et trecentis et quinquaginta uno. Sed proportio duarum linearum *TA*, *AB* coniunctarum ad *TB* est sicut proportio *AT* ad *TQ*. Et proportio *AT* ad *TQ* est sicut proportio *AN* ad *NB*. Et due linee *AT*, *AB* agregate sunt minus duobus millibus et nongentis et undecim. Et linea *TB* per hanc quantitatem est septingenta

91 ad: et *H*
97 cordam *PMaZm* lineam *H*
104 huius *om. H*
105 septingenta et octoginta: 70 et 80 *H*
105–106 duo...millia: 20000 et 400 et 3000 *H*
106 BT: TB *H*
107 sexcentum...millia: 6000 et 80000 *H*
108 octingenta et vigintiquinque millia: 800 et 25000 *H*
110 TA: TQ *H*
111 AT[1]...NB: AT et TQ est sicut proportio AT ad TB, TQ est sicut proportio ad NB *H*
111 *de* TQ...*NB scr. P mg. et Zm mg.* in alio, est sicut proportio AT ad TB
113, 115, 116 septingenta: septuaginta *H*

92–94 Iam...est: وهو (*And this is*)
97 AN: – ای – (*AY*) ((In Arabic text ی (*Y*) appears wherever Gerard has N; I note no more instances of it.))
104 in...volumus *om. Ar.*
111 TQ: – طع – ((Note: Arabic text has ع wherever Gerard has Q.)) / Et...est: وهو (*And this is*)
113 per...quantitatem *om. Ar.*

indeed is the ratio of a side of a figure containing the circle and having 96 sides to the diameter. Therefore, $\frac{\text{diameter}}{\text{perim poly 96 sides}} > \frac{4{,}673\frac{1}{2}}{14{,}688}$.

Therefore, it has now been demonstrated that $\frac{\text{perim poly 96 sides}}{\text{diameter}} < 3\frac{1}{7}$.

Further, I shall draw in circle ATB a chord subtending 1/6 [of the circle] and this chord is line TB [see Fig. 40]. I shall draw AT and bisect $\angle\ TAB$ by line AN. Then I shall draw chord NB and bisect $\angle\ BAN$

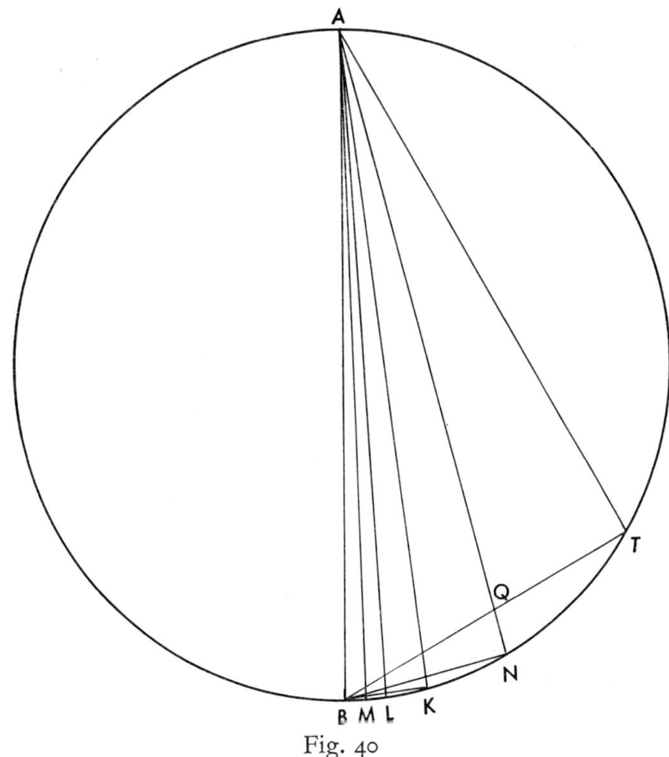

Fig. 40

by line AK. Then I shall draw chord KB and bisect $\angle\ KAB$ by line AL. Then I shall draw chord LB and bisect $\angle\ LAB$ by line AM. Then I shall draw chord MB. Therefore, it is evident that chord MB is the side of a figure having 96 sides which the circle contains. Then I shall assume line $AB = 1{,}560$ because of the facile utility of this number for what we wish. Therefore, chord $TB = 780$. And $AB^2 = 2{,}433{,}600$. And $BT^2 = 608{,}400$. And $AT^2 = 1{,}825{,}200$. Therefore, line $TA < 1{,}351$. But $(TA + AB)/TB = AT/TQ$. And $AT/TQ = AN/NB$, and $(AT + AB) < 2{,}911$, and, by this quantity, $TB = 780$. Therefore,

et octoginta. Ergo proportio AN ad NB est minor proportione duorum millium et nongentorum et undecim ad septingenta et octoginta. Cum ergo fuerit linea NB septingenta et octoginta, erit linea AN minus duobus millibus et nongentis et undecim. Ergo quadratum AN erit minus octo mille millibus et quadringentis et septuaginta tribus millibus et nongentis et viginti uno. Et quadratum NB erit sexcentum et octo millia et quadringenta. Ergo quadratum AB erit minus novem mille millibus et octoginta duobus millibus et trecentis et viginti uno. Ergo linea AB erit minus tribus millibus et tredecim et tribus quartis unius.

Et secundum exemplum quod narravimus declarabitur quod proportio AK ad KB est minor proportione quinque millium et nongentorum et viginti quattuor et trium quartarum unius ad septingenta et octoginta. Cum ergo fuerit linea KB septingenta et octoginta, erit linea AK minus quinque millibus et nongentis et viginti quattuor et tribus quartis unius. Et proportio quinque millium et nongentorum et viginti quattuor et trium quartarum unius ad septingenta et octoginta est sicut proportio mille et octingentorum et viginti trium ad ducenta et quadraginta. Ergo cum sit KB ducenta et quadraginta, erit AK minus mille et octingentis et viginti tribus. Et quadratum AK erit minus tribus mille millibus et trecentis et viginti tribus millibus et trecentis et viginti novem. Et quadratum KB erit quinquaginta septem millia et sexcenta. Ergo quadratum AB erit minus tribus mille millibus et trecentis et octoginta millibus et nongentis et viginti novem. Ergo linea AB erit minus mille et octingentis et triginta octo et novem partibus undecimis unius.

Et secundum exemplum illius quod narravimus declaratur quod proportio AL ad LB est minor proportione trium millium et sexcentorum et sexaginta unius et novem partium undecimarum unius / ad ducenta et quadraginta. Et proportio trium millium et sexcentorum

114 minor *ZmRH*, *corr. mg. P ex* sicut
116 Cum: si *H* / *de* NB *scr. P mg. et Zm mg.* in alio TB / *de* AN *scr. P mg et Zm mg.* in alio AN (*!*)
117–18 duobus.... minus *om. H*
117 *de* AN *scr. P mg. et Zm mg.* in alio AB
119 uno *om. H* / Et: ergo *H* / *de* NB *scr. P mg. et Zm supra* in alio TB
120 erit: est *H*
121 et[3] *om. H* / octoginta: octingenta *Zm*
126, 127, 130 septingenta: septuaginta *H*
127[1, 2], 130 octuaginta *H*
131 *de* proportio *scr. P mg. et Zm supra* vel mensura
137 octuaginta *H*
138 et octo *H*
141–42 sexcenta *H*
143 trium: AN *H*

$(AN/NB) < (2{,}911/780)$. Hence, when line $NB = 780$, line $AN < 2{,}911$. Therefore, $AN^2 < 8{,}473{,}921$. And $NB^2 = 608{,}400$. Therefore, $AB^2 < 9{,}082{,}321$. Therefore, line $AB < 3{,}013\frac{3}{4}$.

And by using the example which we have described it will be shown that $\frac{AK}{KB} < \frac{5{,}924\frac{3}{4}}{780}$. Therefore, when $KB = 780$, line $AK < 5{,}924\frac{3}{4}$. And $\frac{5{,}934\frac{3}{4}}{780} = \frac{1823}{240}$. Therefore, when $KB = 240$, $AK < 1{,}823$. And $AK^2 < 3{,}323{,}329$. And $KB^2 = 57{,}600$. Therefore, $AB^2 < 3{,}380{,}929$. Therefore, line $AB < 1{,}838\frac{9}{11}$.

And by using the example which we have described it is shown that $\frac{AL}{LB} < \frac{3{,}661\frac{9}{11}}{240}$. And $\frac{3{,}661\frac{9}{11}}{240} = \frac{1{,}007}{66}$. Therefore, when $LB = 66$,

114–15 Ergo...octoginta *om. Ar.*
124, 140, 152 quod narravimus: ذلك (*that*)
128, 133 AK: ((The MSS on which the printed text was based omitted AK, but it is in the Paris Arabic MS.))
132 Ergo...quadraginta *om. Ar.*
135–36 quinquaginta...sexcenta: ٧٧٦٠ (*7760*) ((This is erroneous; it is correctly given in Paris MS.))

et sexaginta unius et novem partium undecimarum unius ad ducenta
145 et quadraginta est sicut proportio mille et septem ad sexaginta sex. Cum ergo fuerit linea *LB* sexaginta sex, erit linea *AL* minus mille et septem. Ergo quadratum *AL* erit minus mille millibus et quattuordecim millibus et quadraginta novem. Et quadratum *LB* erit quattuor millia et trecenta et quinquaginta sex. Ergo quadratum *AB* erit minus
150 mille millibus et decem et octo millibus et quadringentis et quinque. Ergo linea *AB* erit minus mille et novem et sexta unius.

Et secundum exemplum quod narravimus declaratur quod proportio *AM* ad *MB* est minor proportione duorum millium et sedecim et sexte unius ad sexaginta sex. Cum ergo fuerit *MB* sexaginta sex, erit
155 *AM* minor duobus millibus et sedecim et sexta. Et quadratum *AM* erit minus quattuor mille millibus et sexaginta quattuor millibus et nongentis et viginti octo. Et quadratum *MB* erit quattuor millia et trecenta et quinquaginta sex. Ergo quadratum *AB* erit minus quattuor mille millibus et sexaginta novem millibus et ducentis et octoginta
160 quattuor. Ergo linea *AB* erit minus duobus millibus et decem et septem et quarta unius. Sed linea *MB* per hanc quantitatem est sexaginta sex. Et linea *MB* est latus figure habentis nonaginta sex latera que continetur a circulo. Ergo proportio diametri ad latera figure habentis nonaginta sex latera quam continet circulus est minor pro-
165 portione duum millium et decem et septem et quarte unius ad sex millia et trecenta et triginta sex. Iam ergo ostensum est quod proportio omnium laterum figure habentis nonaginta sex latera quam continet
58r c. 1 circulus ad diametrum est maior pro/portione trium et decem partium de septuaginta et una partibus ad unum. Et linea continens circulum
170 est longior omnibus lateribus figure habentis nonaginta sex latera quam continet circulus et brevior omnibus lateribus figure habentis nonaginta sex latera que continet circulum. Iam ergo manifestum est ex eo quod narravimus, quod proportio linee continentis circulum ad diametrum eius est maior proportione trium et decem partium de
175 septuaginta et una partibus ad unum et minor proportione trium et

148 LB: AL (AB?) *H*
149 *de* quinquaginta *scr. P mg. et Zm supra* in alio, non est hic
154 sexte *corr. ex* sexta in *PZmHMa*
159 octuaginta *H*
160 quatuor *Ma* duobus *PZm* in alio, quattuor *mg. P et Zm supra* in alio quatuor octuaginta *H*
161 Sed: et *H*
166 *de* triginta *scr. P mg et Zm supra* in alio, sexaginta
174 de: et *H*
175 et[2]: est *H*
175–76 *de* trium et septime *scr. P mg et Zm supra* in alio, decem partium de septuaginta

$AL < 1{,}007$. Therefore, $AL^2 < 1{,}014{,}049$. And $LB^2 = 4{,}356$. Therefore, $AB^2 < 1{,}018{,}405$. Therefore, line $AB < 1{,}009\frac{1}{6}$.

And by using the example which we have described it is shown that $\frac{AM}{MB} < \frac{2{,}016\frac{1}{6}}{66}$. Therefore, when $MB = 66$, $AM < 2{,}016\frac{1}{6}$. And $AM^2 < 4{,}064{,}928$. And $MB^2 = 4{,}356$. Therefore, $AB^2 < 4{,}069{,}284$. Therefore, line $AB < 2{,}017\frac{1}{4}$. But, by this quantity, line $MB = 66$. And line MB is the side of a polygon having 96 sides which is contained by the circle. Hence, $\frac{\text{diameter}}{\text{perim poly 96 sides}} < \frac{2{,}017\frac{1}{4}}{6{,}336}$. Hence, it has now been demonstrated that $\frac{\text{perim poly 96 sides}}{\text{diameter}} > \frac{3\frac{10}{71}}{1}$. And the circumference of the circle is longer than the perimeter of the polygon having 96 sides which is contained by the circle and shorter than the perimeter of the polygon having 96 sides which contains the circle. Hence, it has now become evident from our narrative that $\frac{3\frac{10}{17}}{1} < \frac{\text{circumference}}{\text{diameter}} < \frac{3\frac{1}{7}}{1}$.

147–48 mille...novem: ١٠٤٠٤٩ (104,049) ((Should be as Gerard and Paris Arabic MS have it: 1,014,049.))
154 MB...sex *om. Ar.*

septime ad unum. Et illud est quod declarare voluimus. Et hec est forma figure.

Et iam potest, qui querit, raciocinari proportionem linee circumflexe ad diametrum ut perveniat ex propinquitate numeri, cum quo
180 raciocinatur ad veritatem proportionis unius earum ad alteram ad quemcunque finem voluerit, secundum quod narravimus, propter hunc eundem modum quem fecit Archimenides.

[VII.] VOLO OSTENDERE QUOD, CUM ACCIPITUR SUPERFLUITAS MEDIETATIS OMNIUM LATERUM OMNIS TRIANGULI SUPER UNUMQUODQUE LATERUM EIUS, TUNC SI MULTIPLICATUR UNA TRIUM SUPERFLUITATUM IN ALI-
5 AM EARUM, DEINDE MULTIPLICATUR ILLUD QUOD AGREGATUR IN TERTIAM, POSTEA MULTIPLICATUR ID QUOD AGREGATUR IN MEDIETATEM OMNIUM LATERUM TRIANGULI, TUNC ILLUD QUOD AGREGATUR INDE EST EQUALE MULTIPLICATIONI EMBADI FIGURE IN SE.

10 Verbi gratia, sit triangulus *ABG* [Fig. 41]. Dico ergo quod quando accipitur superfluitas medietatis linearum *AB*, *BG*, *GA* coniunctarum super unamquamque linearum *AB*, *BG*, *GA*, deinde multiplicatur superfluitas medietatis linearum trium agregatarum super *AB* in superfluitatem medietatis earum super *BG*, postea multiplicatur illud

176 septime: septem (?) *H* / *post* unum *add. S* non igitur trium et septime ad unum sed minor / declarare: demonstrare *H*

176–77 Et[2]...figure *PZmMa om. HS*

178 Et *om. H* / Et...potest: sed *S* / post raciocinari *add. PS* eum *sed del. P* / proportionem: proportio *Ma*, *et ante eum habet Ma mg.* 9 *et del. PZm* per

179–82 ex....Archimenides: ad illum propinquum veritati sicud voluerit, ita quod non erit error nisi in minus minuto si voluerit et minus secundo si voluerit et minus tertio si voluerit (*cf. lineas* 13–19). Hanc probant sicud facit Archimenides ostendendo primo quod proportio omnium laterum superficiei habentis 96 latera continentis circulum ad dyametrum est minor tribus et septima unius, deinde quod proportio omnium laterum figure habentis 96 latera quam continet circulus ad dyametrum est maior proportione trium et 10 partium de 71 partibus ad unum. Igitur linea continens circulum est longior omnibus lateribus figure habentis 96 latera quam continet circulus et brevior omnibus lateribus figure habentis 96 latera que continet circulum et probat illud eodem modo sicud Archimenides. *S*

181–82 propter hunc: per *H*

182 Archimenides *ZmHMa* Archemides *P*

1 Volo...quod *om. S*

3 eius *om. H*

4 multiplicetur *H*

5 quod *om. H*

6 id *om. H*

8 inde *om. H*

14 earum *tr. H post* BG

And this is what we wished to show. And here is the form of the figure [Figs. 39 and 40].

And now anyone who wishes can calculate the ratio of a curvilinear line [i.e., an arc] to the diameter, arriving at an approximate number which can be reckoned by the same method that Archimedes used as close as is wished to the true ratio, just as we have explained it.

[VII.] I WISH TO DEMONSTRATE THAT, IF THE EXCESS OF ONE HALF OF THE PERIMETER OF EVERY TRIANGLE OVER EACH OF ITS SIDES IS TAKEN, AND IF ONE OF THE THREE EXCESSES IS MULTIPLIED BY THE SECOND AND THEN THAT PRODUCT IS MULTIPLIED BY THE THIRD EXCESS, AND FURTHER IF THE PRODUCT OF THE THREE EXCESSES IS MULTIPLIED BY ONE HALF OF THE PERIMETER OF THE TRIANGLE, THEN THE FINAL PRODUCT IS EQUAL TO THE AREA OF THE TRIANGLE SQUARED.

For example, let there be a $\triangle ABG$ [see Fig. 41]. I say, therefore, that, assuming the excess of $1/2\,(AB + BG + GA)$ over each of the sides AB, BG, GA, then $[1/2\,(AB+BG+GA) - AB]\cdot[1/2(AB+BG+GA)$

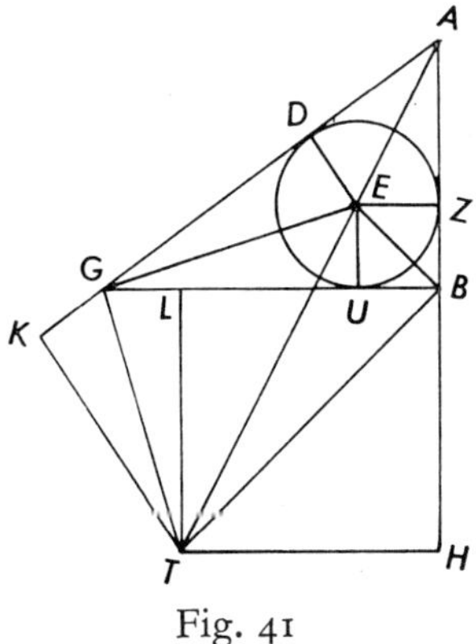

Fig. 41

176 declarare *om. Ar.*
176–77 Et...figure *om. Ar.*
178–82 Et....Archimenides: ومن الممكن ان يوصل بهذا الوجه بعينه الى اى غاية يراد من التدقيق فى هذا العمل
(*And it is possible with this same method to arrive at any limit of approximation in this operation.*)

1 Volo... quod *om. Ar.*
9 embadi: تكسير (*measure* or *magnitude*) ((See Arabic variant for Proposition I, line 21. The usual word translated by *embadum* is مساحة))
10–18 Dico....probatur *om. Ar.*

15 quod agregatur in superfluitatem medietatis earum super *GA*, deinde multiplicatur illud quod agregatur in medietatem linearum trium agregatarum, tunc illud quod agregatur inde est equale ei quod fit ex multiplicatione embadi trianguli *ABG* in se; quod sic probatur.

Revolvam in triangulo *ABG* maiorem circulum qui cadit in eo, 20 qui sit circulus *DZU*, et sit eius centrum *E*. Et protraham a centro lineas *ED*, *EU*, *EZ* ad puncta super que tangunt circulum latera trianguli. Et protraham lineam *AE*. Ostendam ergo quod *DA* est equalis *AZ* et *ZB* equalis *BU* et *UG* equalis *GD*, quoniam, quando linee contingentes circulum occurrunt sibi super punctum unum, tunc 25 ipse sunt equales, propterea quod angulus *EDA* est equalis angulo *EZA*, et unusquisque eorum est rectus. Et due linee *DE*, *EA* sunt equales duabus lineis *ZE*, *EA*. Ergo linea *DA* est equalis linee *AZ*. Et per huiusmodi modum scitur quod due linee *ZB*, *BU* sunt equales, et quod due linee *UG*, *GD* sunt equales. Et scitum est ex eo quod 30 narravimus, quod unaqueque duarum linearum *DA*, *AZ* est superfluitas medietatis linearum *AB*, *BG*, *GA* agregatarum super lineam *BG*, et quod unaqueque duarum linearum *ZB*, *BU* est superfluitas medietatis omnium laterum trianguli *ABG* super lineam *GA*, et quod unaqueque duarum linearum *DG*, *GU* est superfluitas medietatis 35 omnium laterum trianguli *ABG* super lineam *BA*. Deinde elongabimus lineam / *AE* usque ad *T* et elongabimus iterum lineam *AB* usque ad *H* et ponemus *AH* equalem medietati omnium laterum trianguli *ABG*. Declaratur ergo ex eo quod narravimus quod linea *HB* est equalis unicuique duarum linearum *DG*, *GU*. Et elongabimus *AG* 40 usque ad *K* et ponemus *AK* equalem *AH*. Ergo declaratur quod linea *GK* est equalis unicuique duarum linearum *ZB*, *BU*. Et protraham

c. 2

16 illud *S om. PZm H Ma*
16–17 in...inde *om. H*
18–124 quod....complere: habitam (*?* hic eam?) probat ex hoc quod multiplicatio superfluitatum medietatis omnium laterum trianguli super unumquodque laterum eius unius in alteram deinde eius quod aggregatur in tertiam est equale ei quod fit ex multiplicatione medietatis dyametri maioris circuli cadentis in tirangulo in se et eius quod aggregatur in medietatem omnium laterum trianguli. Ex hac propositione patet quod habito embado et duobus lateribus habetur tertium et similiter habitis tribus lateribus habetur embadus. *S*
20 DZU: DZM *H*
21 -U: -N *H hic et ubique*
22 *de* ostendam *scr. Pmg. et Zm supra* vel manifestum est
23–25 et[1]...equales *HR mg P et mg. Zm, et ante eum habent P et Zm* in alio
24 sibi *om. H*
29 sciendum *H*
35 *ante* ABG *del. P* A
40 K: N (*?*) *H* / AK: AN (*?*) *H*
41 GK: GN (*?*) *H*

$- BG] \cdot [1/2\,(AB + BG + GA) - GA] \cdot 1/2\,(AB + BG + GA) =$ (area $\triangle\ ABG)^2$, which is proved as follows.

I shall inscribe in $\triangle\ ABG$ the greatest circle which can fit in it, namely, the circle DZU. Let its center be E. I shall protract from the center lines ED, EU, EZ to the points of tangency of the sides of the triangle with the circle. I shall draw line AE. I shall show, therefore, that $DA = AZ$, $ZB = BU$, and $UG = GD$. For when lines tangent to a circle meet in a single point, they are equal, since $\angle\ EDA = \angle\ EZA$, both being right angles. But $DE + EA = ZE + EA$. Therefore, $DA = AZ$. In the same way it is known that $ZB = BU$ and that $UG = GD$. And it is known from what we have recounted that $DA = AZ = 1/2\,(AB + BG + GA) - BG$ and that $ZB = BU = 1/2\,(AB + BG + GA) - GA$, and that $DG = GU = 1/2\,(AB + BG + GA) - BA$. Then we shall extend line AE to T* and line AB to H and posit $AH = 1/2$ perimeter $\triangle\ ABG$. Therefore, it is shown from what we have recounted that $HB = DG = GU$. And we shall extend AG to K and posit $AK = AH$. Therefore, it is shown that line $GK =$ line $ZB =$ line BU. And I shall

* But the authors have not yet given us the conditions to determine T.

---

19 Revolvam: نرسم (*we draw*)
20 a centro *om. Ar.* ((Note: everywhere Gerard has Z, the Arabic text has د (*D*).))
21–22 super...trianguli: التماس (*of tangency*)
22 DA: – او – (*AU*) ((Gerard has D where Arabic text has و (*U*). I do not give the other instances.))
23 AZ: – ا د – (*AD*) ((Gerard has Z where Arabic text has د (*D*). I do not give the other instances.)) / BU: – ب ز – (*BZ*) ((Gerard has U where Arabic text has ز (*Z*). I note no other instances.))
23–29 quoniam....equales *om. Ar.*

31 AB, BG, GA *om. Ar.*
33, 34–35 medietatis...ABG: نصفه (*half of it*)
35–36 elongabimus...lineam[2] *om. Ar.*
36–106 usque[2]....se: الى ان يصير – ب ح – مثل – ج ز – و – ا ج – الى ان يصير – ج ك – مثل – ب ز – فيكون كل واحد من – ا ح – اك – مثل نصف جميع الاضلاع و نخرج من نقطتى – ح ك – عمودى – ح ط – ك ط – فيلتقيان ضرورة على نقطة واحدة من – ا ط – وهى نقطة – ط – مثلا و يكون – ط ح – ط ك – متساويين وان اردنا اخرجنا عمود – ح ط – و وصلنا – ط ك – و بينا انه ايضا عمود لتساوى ضلعى – ا ك

ex puncto *H* lineam *HT* super angulum rectum linee *AH* et protraham ex puncto *K* linea *KT* super angulum rectum linee *AK*. Ergo manifestum est quod linea *KT* est equalis linee *HT*. Et accipiam ex
45 linea *BG* equale *BH*, quod sit *BL*. Et protraham *TL*. Ergo manifestum est quod ipsa est perpendicularis super lineam *BG*. Propterea quod nos protraximus duas lineas *BT*, *TG*, ergo manifestum est quod augmentum quadrati *BT* super quadratum *TG* est equale augmento quadrati *BH* super quadratum *KG*. Sed *KG* est equalis *LG* et *BH* equalis *BL*.
50 Ergo augmentum quadrati *BT* super quadratum *TG* est equale augmento quadrati *BL* super quadratum *LG*. Propter illud ergo *TL* est perpendicularis super *BG*. Et *LT* equalis linee *TH*, propterea quod *BL* est equalis *BH* et *BT* est linea communis, et duo anguli *BLT*, *BHT* sunt recti. Et propter illud sunt duo anguli *LBT*, *TBH* equales
55 et duo anguli *LTB*, *BTH* equales. Et linea *BH* est continuata secundum rectitudinem cum linea *AB*, ergo duo anguli *UBZ*, *HBU* sunt equales duobus rectis. Et duo anguli *LBH* et *LTH* simul sunt equales duobus angulis rectis. Ergo angulus *LTH* est equalis angulo *ZBU*. Sed angulus *EBU* est medietas anguli *ZBU*. Et angulus *BTH* est
60 medietas anguli *LTH*. Ergo ipsi sunt equales. Et remanet ex triangulo

42 AH: AN (?) *H*
42–43 et...AK *om. H*
44 KT: HT *H* / HT: KT *H*
45 equalem *Zm*
46 est *om. Zm* / super lineam *bis H*
47–48 augmentum: angulum *H*
48 equale augmento: equalis angulo *H*
50 augmentum: angulum *H*
50–51 BT...quadrati *om. H*
51 illud: hoc *H*
52–53 propterea...anguli *om. H*
54–55 LBT...anguli *om. H*
56–57 cum...duobus *om. H*
59 Sed...ZBU *bis H*
60 remanet *Zm MaR* remamanet *P* remanent *H*

— عمود على — ب ج — وهو مساو — لط ح — لكون
— ب ح — مساويا — لب ل — و ب ط — مشتركا
و زاويتا — ح ل — قائمتين فتكون زاويتا — ل ب ط
— ح ب ط — متساويتين و نصل — ب ه — فزاويتا
— ز ب ه — د ب ه — متساويتان ولكون زاوية —
ل ب ح — مع زاوية — ل ط ح — كقائمتين تكون
زاوية — ز ب د — مساوية ازاوية — ل ط ح —

— ا ح — وكون — ا ط — مشتركا وتساوى زاويتى
— ح ا ط — ك ا ط — ونصل — ب ط — ط ج —
ونصل — ب ل — من — ب ج — مثل — ب ح —
ونصل — ط ل — فهو عمود على — ب ج — لان
الفضل بين مربعى خطّى — ب ط — ط ج — كالفضل
بين مربعى خطّى — ب ح — ج ك* — فلذلك — ط ل

ج ل — فالفضل بين مربعى خطّى — ب ط —
ط ج — كالفضل بين مربعى خطّى — ب ل —
ج ل

*Adds Paris and other MSS the following:
و — ب ح — مساو — ب ل — و — ج ك — مساو —

draw from point H line HT at a right angle to line AH and I shall draw from point K line KT at a right angle to line AK. Therefore, it is evident that $KT = HT$. And I shall take from line BG a line equal to BH, and this will be line BL. I shall draw TL. Hence it is evident that TL is perpendicular to line BG. For when we have drawn the two lines BT and TG, therefore it is evident that $BT^2 - TG^2 = BH^2 - KG^2$.* But $KG = LG$ and $BH = BL$. Hence, $BT^2 - TG^2 = BL^2 - LG^2$ [or, $BT^2 - BL^2 = TG^2 - LG^2$]. Accordingly TL is perpendicular to BG. And $LT = TH$ because $BL = BH$, line BT is common, and the two angles BLT and BHT are right angles. And because of this, $\angle LBT = \angle TBH$ and $\angle LTB = \angle BTH$. And line BH is a rectilinear continuation of line AB. Hence, $\angle UBZ + \angle HBU =$ 2 right angles. And $\angle LBH + \angle LTH =$ 2 right angles. Hence, $\angle LTH = \angle ZBU$. But $\angle EBU = 1/2 \angle ZBU$ and $\angle BTH = 1/2 \angle LTH$. Hence, $\angle EBU = \angle BTH$. And there

* For $BT^2 = BH^2 + HT^2$, and $TG^2 = KG^2 + TH^2$, by the Pythagorean theorem, and $KT = HT$. Therefore, $BT^2 - TG^2 = BH^2 - KG^2$.

ب ز – فى – زج – فى – ا د – واذا ضربنا هما
فى – ا ح – صار مربع – ه د – فى مربع – ا ح –

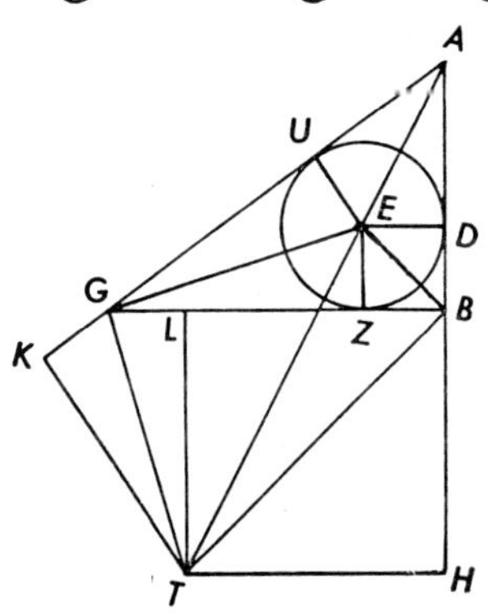

Note: The letters are transliterations of Arabic letters.

ونصفها لنصفها فزاوية – ه ب د – من مثلث –
ب د ه – مساوية لزاوية – ب ط ح – من مثلث
– ب ح ط – وزاويتا – ب د ه – ب ح ط –
قائمتان فمثلثا – ب ه د – ب ح ط – متشابهان
ونسبة – د ه – الى – د ب – كنسبة – ب ح –
الى – ح ط – و – د ب – مثل – ب ز – و – ب ح
– مثل – زج – فنسبة – ه د – الى – ز ب –
كنسبة – زج – الى – ح ط – وضرب – د ه – فى
– ح ط – مساو لضرب – ب ز – فى – زج –
وايضا نسبة مربع – ه د – الى ضرب – ه د –
فى – ح ط – اعنى الى ضرب – ب ز – فى – زج
– كنسبة – ه د – الى – ح ط – اعنى كنسبة –
ا د – الى – ا ح – فنسبة مربع – ه د – الى ضرب
– ب ز – فى – زج – كنسبة – ا د – الى – ا ح
– فضرب مربع – ه د – فى – ا ح – كضرب –

BTH angulus *TBH* equalis angulo *BEU* trianguli *BEU*. Ergo triangulus *EBU* est similis triangulo *TBH*. Ergo proportio *EU* ad *UB* est sicut proportio *BH* ad *HT*. Sed *ZB* est equalis *BU*. Et *EU* equalis *ZE*. Ergo proportio *EZ* ad *ZB* est sicut proportio *BH* ad *HT*. Ergo
65 quadratum* *EZ* in *HT* est equale quadrato *ZB* in *HB*. Sed proportio quadrati** *EZ* ad quadratum *EZ* in *HT* est sicut proportio *EZ* ad *HT*. Et proportio *EZ* ad *HT* est sicut proportio *AZ* ad *AH*. Ergo proportio *AZ* ad *AH* est sicut proportio quadrati *EZ* ad quadratum *EZ* in *HT*. Et quadratum *EZ* in *HT* est equale quadrato *ZB* in *HB*.
70 Ergo proportio quadrati *EZ* ad quadratum *ZB* in *BH* est sicut proportio *AZ* ad *AH*. Ergo illud quod fit ex multiplicatione quadrati *EZ* in lineam *AH* est equale ei quod fit ex multiplicatione quadrati *HB* in *BZ* per lineam *AZ*. At vero linee *AZ*, *ZB*, *BH* sunt super-
58v c. 1 fluitates medietatis linearum trianguli *AB*, / *BG*, *GA* agregatarum
75 super unamquamque linearum *AB*, *BG*, *GA*. Et linea *EZ* est medietas diametri circuli *DUZ*. Et linea *AH* est medietas linearum *AB*, *BG*, *GA* agregatarum.

Iam ergo manifestum est quod multiplicatio superfluitatum medietatis omnium laterum trianguli super unumquodque laterum eius unius
80 earum in alteram, deinde eius quod agregatur in tertiam, est equalis ei quod fit ex multiplicatione medietatis diametri maioris circuli cadentis in triangulo in se et eius quod agregatur in medietatem omnium

* Here used for "product." (See the Introduction, division 2, of this chapter.)

** Here used for "square of."

---

62 TBH: BTH *mg. H*
63 est[2] *om. Zm*
63–64 Sed...HT *om. H*
64 est *del. P?*, *om. Zm*
65 est equalis *bis H* / ZB: BZ *H*
67 *post* Et *del. P* ergo / AH: HA *H*
69 Et...HT *PZm* quod *H*
72 equalis *H*
73 *ante* At *scr. ZmP mg.*, *H tex.* in (pro *H*) alio, et quod agregatur
74 linearum trianguli *tr. Zm* / agregatarum *tr. H. post* trianguli
75 super...GA *om. H*
75–76 est....AH *om. H*
80 deinde eius *om. Zm* et *H* / equalis *P* equale *ZmHMa*

كضرب — ب ز — فى — زج — فى — ا د — فى —
ا ح — ولكون — ه د فى — ا ح — كتكسير المثلث
يكون مربع — ه د — فى مربع — ا ح — مربع
تكسير المثلث فاذا مربع تكسير المثلث مساوع
لضرب — ب ز — فى — زج — فى — ا د — فى —
ا ح — اعنى الفضول الثلاثه فى نصف جميع الاضلاع

(until BH = GZ, and AG until GK = BZ, and so AH as well as AK equals half of the sum of the sides. And from the two points H and K we draw two perpendiculars—HT and KT. And so they will necessarily meet in a single point of [*line*] *AT, namely, point T. And TH = TK.* [*Or*] *if we wish let us draw HT perpendi-*

remains of $\triangle BTH$, $\angle TBH$, which is equal to $\angle BEU$ of $\triangle BEU$. Therefore, $\triangle EBU$ is similar to $\triangle TBH$. Therefore, $EU/UB = BH/HT$. But $ZB = BU$ and $EU = ZE$. Therefore, $EZ/ZB = BH/HT$. Therefore, $(EZ \cdot HT) = (ZB \cdot HB)$. But $EZ^2/(EZ \cdot HT) = EZ/HT$ and $EZ/HT = AZ/AH$ [by similar triangles]. Therefore, $AZ/AH = EZ^2/(EZ \cdot HT)$. And $(EZ \cdot HT) = (ZB \cdot HB)$. Therefore, $EZ^2/(ZB \cdot BH) = AZ/AH$. Therefore, $(EZ^2 \cdot AH) = (HB \cdot BZ \cdot AZ)$. But lines AZ, ZB, BH are the excesses of $1/2\ (AB + BG + GA)$ over each of the sides, AB, BG, and GA. And line EZ is the radius of circle DUZ. And line $AH = 1/2\ (AB + BG + GA)$.

Therefore, it has now become evident that the multiplication of one of the excesses of 1/2 the perimeter of the triangle over each of its sides by the second excess, followed by the multiplication of this product by the third excess, is equal to the multiplication of the square of the radius of the greatest circle falling in the triangle by 1/2 the perimeter of the triangle.

cular and join TK and show that it also is perpendicular: For $AK = AH$, and AT is common, and $\angle HAT = \angle KAT$. And we join BT and TG, and from BG we cut $BL = BH$, and we draw TL. And it is perpendicular to BG because $BT^2 - TG^2 = BH^2 - GK^2$ [But $BH = BL$ and $GK = GL$; and so $BT^2 - TG^2 = BL^2 - GL^2$]. Accordingly, TL is perpendicular to BG. And $TL = TH$ because $BH = BL$ and BT is common and angles H and L are right angles. And so $\angle LBT = \angle HBT$. We draw BE. So $\angle ZBE = \angle DBE$. But since $\angle LBH + \angle LTH =$ 2 right angles, $\angle ZBD = \angle LTH$. And the half is equal to the half, i.e., $\angle EBD$ of $\triangle BDE$ equals $\angle BTH$ of $\triangle BHT$. And angles BDE** and BHT are right angles. And so triangles BED, BHT are similar. And [thus] $DE/DB = BH/HT$ and $DB = BZ$, $BH = ZG$. And so ED***$/ZB = ZG/HT$. And $DE \cdot HT$*

*The sentence in brackets is a translation of the material given in the asterisk, note, page 282. **Paris MS; "BZE" in published text. ***"HD" in published text; "ED" in other MSS.

laterum trianguli. Sed multiplicatio medietatis diametri maioris circuli cadentis in triangulo per medietatem omnium laterum trianguli est
85 mensuratio trianguli. Et multiplicatio eius iterum per medietatem diametri circuli est sicut multiplicatio medietatis diametri circuli in se, deinde quod agregatur in medietatem laterum trianguli. Deinde multiplicatio etiam eius quod agregatur in medietatem omnium laterum trianguli est equalis multiplicationi embadi superficiei trianguli in
90 se, propterea quod multiplicatio medietatis diametri maioris circuli cadentis in triangulo in medietatem omnium laterum trianguli est embadum trianguli, sicut ostendimus in illis que sunt premissa. Et multiplicatio *EZ* in se, deinde in *AH*, que est medietas omnium laterum figure *ABG*, est equalis multiplicationi *EZ* in *AH* et eius
95 quod agregatur in *EZ*. Ergo multiplicatio eius in *AH* et eius quod agregatur in se est equalis multiplicationi *BH* in *BZ*, deinde eius quod

83 *post* trianguli *add. mg. PZmMa et tex. HR* Sed multiplicatio quadrati EZ in AH est sicut multiplicatio AH in EZ et quod provenit in EZ. Ergo multiplicatio AH in EZ et quod provenit in EZ est sicut multiplicatio HB in BZ et eius (*om. H*) quod congregatur (agregatur *Ma*) in ZA. Ponamus* AH communem. Erit multiplicatio HB in BZ et quod congregatur in ZA* (*...* *om. H*) et quod (*om. H*) proveniet in AH equale (equalis *H*) multiplicationi ZE in AH et quod congregatur in EZ et totius quod proveniet in AH. Sed multiplicatio EZ in AH est embadum (embadi *MaR*) trianguli. Ergo erunt 3 (etiam *MaR*) quantitates: triangulus (trianguli *H* triangule *MaR*) EZ, et HA (AH, *Zm*). Ergo multiplicatio trianguli in EZ et eius quod congregatur (agregatur *MaR*) in AH est sicut multiplicatio AH in EZ et quod provenit in triangulum. Sed multiplicatio AH in EZ est triangulus (triangule *MaR*). Ergo multiplicatio trianguli in triangulum est equalis multiplicationi HB (in HB *Ma*) in BZ et eius quod provenit (pervenit *R*) in ZA et totius aggregati in AH. Et illud est quod demonstrare voluimus. *Et post* voluimus *add. H* in alio, iam ostensum est quod multiplicatio uniuscuiusque linearum AZ, ZB, BH in eo quod aggregatur ex multiplicatione unius earum in aliam deinde quod aggregatur in medietatem omnium laterum, est equalis ei quod aggregatur ex multiplicatione embadi trianguli in se et illud est quod demonstrare voluimus ad presens. (*Cf. var. lineae 90.*)

83–106 Sed.... voluimus *om. H hic et add. post* BH *in linea 119*

85 Et *om. H*

87 deinde[1]: inde *H*

88 etiam *om. H hic. cf. var.* 89

89 *post* trianguli *add. H* deinde etiam multiplicatio eius quod aggregatur / equalis *H* et equalis *P* e'quale *R* $^{u}_{q}$li *Ma*

90 *post* se *scr. P mg. et text. MaR* (*et Zm mg post* trianguli *in linea 83*) in (*om. MaR* alio (*om. MaR*), iam ergo ostensum est quod multiplicatio uniuscuiusque linearum AZ, ZB, BH in eo quod agregatur ex multiplicatione unius earum in aliam deinde quod agregatur in medietatem omnium laterum est equale ei quod agregatur ex multiplicatione embadi trianguli in se. Et illud est (*om. MaR*) quod demonstrare. (*Cf. var.* lineae *83*).

92 Et: est *H*

94 ABG: BG *H*

94–96 EZ... multiplicationi *bis H*

But the multiplication of the radius of the largest circle falling in the triangle by 1/2 the perimeter of the triangle is equal to the area of the triangle. And the multiplication of this product again by the radius of the circle equals the multiplication of the square of the radius by 1/2 the perimeter of the triangle. Then the further multiplication of this product by 1/2 the perimeter of the trianle equals to the area of the triangle squared, since the multiplication of the radius of the largest circle falling in the triangle by 1/2 the perimeter of the triangle is equal to the area of the triangle, as we demonstrate in those things which have been presented before. Therefore $(EZ^2 \cdot AH) = [(EZ \cdot AH) \cdot EZ]$. Therefore, $[(EZ \cdot AH) \cdot EZ] =$

$= BZ \cdot ZG$. *And also* $ED^2/ED \cdot HT = ED^2/BZ \cdot ZG = ED/HT$ *and* $ED/HT = AD/AH$. *And so* $ED^2/BZ \cdot ZG = AD/AH$. *Hence* $ED^2 \cdot AH = BZ \cdot ZG \cdot AD$. *And if we multiply them* ((i.e., each side of the equation)) *by* AH, *then* $ED^2 \cdot AH^2 = BZ \cdot ZG \cdot AD \cdot AH$. *But since* $ED \cdot AH$ = *area of the triangle, so* $ED^2 \cdot AH^2$ = *square of the area of the triangle. And thus the square of the area of the triangle is equal to the product* $BZ \cdot ZG \cdot AD \cdot AH$, *i.e.*, [*to the product of*] *the three excesses into half the sum of the sides.* ((The reader should compare this whole translation with the slightly different Latin text.))

agregatur in *AZ*. Ergo *EZ* iam multiplicata est in duos numeros, in *AH*, et fuit embadum, et in embadum, [i.e.,] mensurationem, et fuit equalis multiplicationi *BH* in *BZ*, deinde eius quod agregatur in *AZ*.
100 Ergo proportio linee *AH* ad embadum est sicut proportio embadi ad multiplicationem *AZ* in *ZB* deinde eius quod agregatur in *BH*. Ergo multiplicatio superfluitatum medietatis omnium laterum trianguli super unumquodque laterum trianguli unius earum in alteram et eius quod agregatur in tertiam deinde multiplicatio eius quod agrega-
105 tur in medietatem omnium laterum trianguli est equalis multiplicationi embadi trianguli in se. Et illud est quod declarare voluimus.

Et nobis quidem est possibile venire cum demonstratione super veritatem eius quod narravimus per modum alium. Cum factum sit, illud quod premissum est de narratione eius in hac figura manifestum.
110 Et illud est, quoniam proportio *EZ* ad *ZB* est sicut proportio *BH* ad *HT*, ergo quantitates *EZ*, *ZB*, *BH*, *HT* sunt proportionabiles. Ergo proportio prime ad quartam est proportio prime ad secundam multiplicata per proportionem prime ad tertiam, propterea quod quando
c. 2 ponitur quantitas secunda media inter pri/mam et quartam, fit pro-
115 portio prime ad secundam multiplicata per proportionem secunde ad quartam, existens proportio prime ad quartam; ergo proportio secunde ad quartam est proportio prime ad tertiam. Ergo proportio *EZ* ad *HT* est sicut proportio *EZ* ad *ZB* multiplicata per proportionem *EZ* ad *BH*. ⟨Et cum proportio *EZ* ad *HT* sit sicut proportio *AZ* ad *AH*
120 per similes triangulos, ergo proportio *AZ* ad *AH* est sicut proportio *EZ* ad *ZB* multiplicata per proportionem *EZ* ad *BH*, et ergo multiplicatio *AZ* in *ZB* et eius quod agregatur in *BH* est equalis multiplicationi *EZ* in se et eius quod agregatur in lineam *AH*. Et tunc possumus sicut prius demonstrationem complere.⟩

98 AH: OZH *Zm* / in *om.* *H* / mensurationi *H* / fuit: sunt *H*
99 multiplicationis *H*
101 in² *om.* *H*
103 in alteram *PZm* *om.* *Ma* in alterum *H*
106 declarare: demonstrare *H*
107 est possibile *tr.* *H*
110 Et: est quod *H*
111 proportionales *Zm*
113 per *PZmMa* in *B*
115 secundam multiplicata *scr. et del.* *H*
116 existens...quartam *scr. et del.* *H* / ergo: et *Zm*
118 per *om.* *H*
119 BH: HB *H* / *post* HB *add.* *H* *lineas 83–106*, *i.e.*, Sed...voluimus
119–24 ⟨Et....complere⟩ *supplevi*; *cf. Ar. Hic est lacuna in PZm*

107–108 Et...alium: وايضا بوجه اخر (*And also by another method*)
108–10 Cum....quoniam: بعد ان ثبت ان (*after it is evident that*)
111 ergo....proportionabilis *om.* *Ar.*

($BH \cdot BZ \cdot AZ$). Hence, EZ has now been multiplied by two numbers: (1) by AH, and this is the area of the triangle [i.e., $EZ \cdot AH =$ area of $\triangle$], and (2) by the area of the triangle itself, and this was equal to ($BH \cdot BZ \cdot AZ$) [i.e., ($EZ \cdot$ area of $\triangle$) $=$ ($BH \cdot BZ \cdot AZ$)]. Therefore, AH/area of $\triangle =$ area of $\triangle/(AZ \cdot ZB \cdot BH)$ [dividing (1) by (2)]. Therefore, the multiplication of one of the excesses of 1/2 the perimeter of the triangle over each of the sides of the triangle by the second excess, followed by the multiplication of this product by the third excess, and then followed once more by the multiplication of this product by 1/2 the perimeter of the triangle, is equal to the triangle squared. And this is what we wished to show.

We can demonstrate the truth of what we have recounted by another method. When this has been done, then what we have stated before in regard to this proposition is evident. And this is as follows: since $EZ/ZB = BH/HT$ [cf. line 64], hence the quantities EZ, ZB, BH, HT are proportional terms. Therefore,* the ratio of the first term to the fourth is equal to the ratio of the first to the second multiplied by the ratio of the first to the third, since, when the second term is placed as a mean between the first and the fourth, there results the ratio of the first to the second multiplied by the ratio of the second to the fourth [which compound ratio] equals the ratio of the first to the fourth. But the ratio of the second to the fourth is as the ratio of the first to the third. Therefore, $EZ/HT = (EZ/ZB) \cdot (EZ/BH)$.* And since $EZ/HT = AZ/AH$ by similar triangles, therefore $AZ/AH = (EZ/ZB) \cdot (EZ/BH)$. Hence, $(AZ \cdot ZB \cdot BH) = EZ^2 \cdot AH$. And then we complete the demonstration as before [in lines 78–106].

.... See Commentary, Proposition VII, lines 111–19

119–24 cum....complere〉: و – د ب – مثل
– ب ز – و – ب ح – مثل – زج – فنسبة –
ه د – الى – ح ط – اعنى نسبة – ا د – الى –
ا ح – مؤلفة من نسبة – ه د – الى – ب ز –
ومن نسبة – ه د – الى – زج – فضرب – ا د –
فى – ب ز – فى – زج – كضرب مربع – ه د –
فى – ا ح – ونتمم البرهان بالوجه المتقدّم
(*And* $DB = BZ$, $BH = ZG$; *and so* $ED/HT = AD/AH = (ED/BZ) \cdot (ED/ZG)$. *And so* $AD \cdot BZ \cdot ZG = ED^2 \cdot AH$. *And we complete the proof by the former method.*) ((See translation for Arabic variant, lines 36–106, for the rest of the proof. Although the texts are substantially the same, note that I have reconstructed Gerard's Latin text somewhat differently than the Arabic text since the first proof in the Gerard text is somewhat different from the current Arabic text; so my reconstruction must tie into the Gerard text rather than al-Ṭūsī's Arabic text.))

[VIII.] CUM PROTRAHUNTUR EX PUNCTO INTRA SPERAM QUATTUOR LINEE PERVENIENTES AD SUPERFICIEM SPERE, ET SUNT LINEE EQUALES, ET PUNCTA AD QUE PERVENIUNT LINEE NON SUNT IN SUPERFICIE UNA
5 RECTA, TUNC PUNCTUM QUOD EST INTRA SPERAM EST CENTRUM SPERE.

Verbi gratia, sit spera *ABGD* intra quam sit punctum *N* [Fig. 42]. Et protrahantur ex puncto *N* linee *NB*, *NE*, *ND*, *NG*, que sint equales. Et puncta *B*, *E*, *G*, *D* non sint in superficie una ⟨recta⟩.
10 Dico ergo quod punctum *N* est centrum spere; cuius hec est demonstratio.

Puncta *E*, *G*, *D* sunt in una superficie recta propter illud quod declaravit Euclides de hoc, quod omnis triangulus est in superficie una ⟨recta⟩ et propter illud puncta *B*, *E*, *G* sunt in superficie una
15 ⟨recta⟩. Revolvam ergo super puncta *E*, *G*, *D* circulum super quem sunt *E*, *G*, *D*, et super puncta *B*, *E*, *G* circulum *BEG*. Et protraham ex puncto *N* perpendicularem super superficiem circuli *BEG* cadentem super punctum *K*. Ergo manifestum est quod punctum *K* est centrum circuli *BEG* propterea quod si nos protraxerimus lineas *BK*,
20 *GK*, *EK* oportebit ut sint equales, quoniam linee *BN*, *EN*, *GN* sunt equales, et linea *NK* existit communis, et unaqueque linearum *BK*, *EK*, *GK* continet cum linea *KN* angulum rectum. Ergo circulus *BEG* est super superficiem spere *ABGD*. Et iam protracta est ex centro

1 [VIII]: 10 *mg. MaR* / ex: a *S*
2 provenientes *H*
3 linee *om. S* / ad que: atque *H*
4 proveniunt *H*
5 punctum: illud punctum *S* / quod est *om. S*
7–30 Verbi....eius *om. S*
7 *de* ABGD *scr. P mg. et Zm mg.* in alio, ABGDE
9 BEDG *H* / non *om. H*
9, 14, 15 ⟨recta⟩ *supplevi*
12 *post* Puncta *add. mg. PZm et text. HMaR* (*et del. H*) in (*om. MaR*, pro *H*) alio (*om. MaR*) B, E, G sunt (sint *MaR*) in superficie una recta secundum quod ostendit euclides (*add. H* puncta E, G) quod omnis* triangulus* (*···* *om. R hic*) est in una superficie recta (*om. MaR*, *add. R hic* omnis triangulus) et propter illud puncta E, G, D sunt in superficie una recta secundum quod ostendit euclides.
14 B, E, G: B, G, E *Zm*
15 ergo: igitur *H*
20–21 sunt equales *om. H*
21 communitas *H*
21–22 linearum...GK: linea BN est GN *H*
22 KN: NK *H*
23 super superficiem *ZmPMa* superficiem *R* superficies *H*

6, 7 N: ز (*Z*) ((Here and everywhere Gerard has N instead of ز (*Z*) of the printed Arabic text. I give no more instances of its variation.))
9 puncta...D *om. Ar.*
10–12 Dico...recta *om. Ar.*

[VIII.] WHEN FOUR LINES ARE DRAWN FROM A POINT WITHIN A SPHERE TO THE SURFACE OF THE SPHERE—AND THESE LINES ARE EQUAL AND THE POINTS AT WHICH THE LINES TERMINATE ARE NOT IN A SINGLE PLANE SURFACE—THEN THE POINT WHICH IS WITHIN THE SPHERE IS THE CENTER OF THE SPHERE.

For example, let there be a sphere *ABGD*, in which there is a point *N* [see Fig. 42]. And let equal lines *NB*, *NE*, *ND*, *NG* be drawn from point *N*. And points *B*, *E*, *G*, *D* are not in a single plane surface. I say, therefore, that point *N* is the center of the sphere.

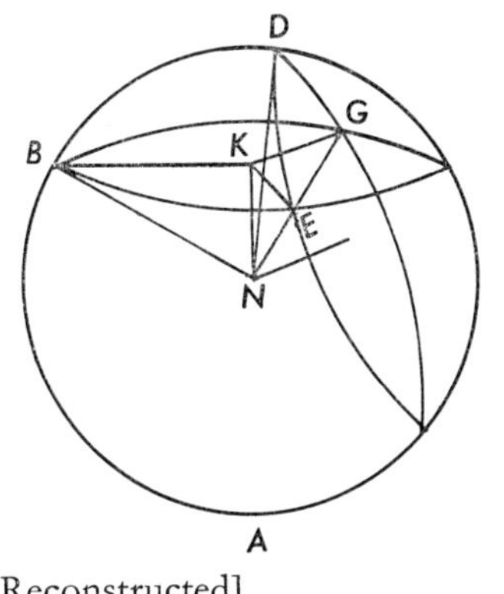

[Reconstructed]

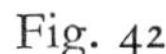
Fig. 42

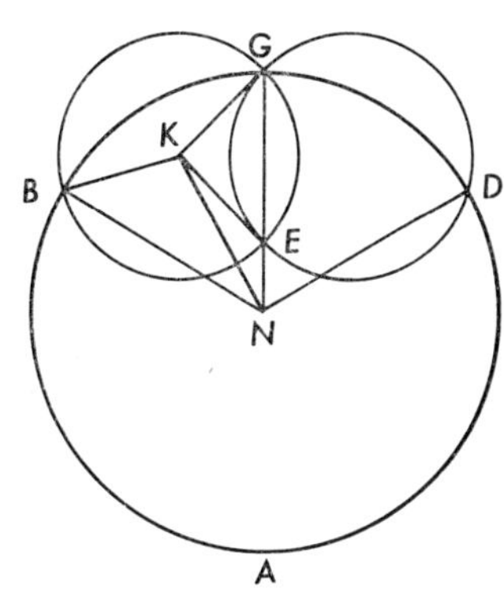

[As given in MS *P*]

Demonstration: Points *E*, *G*, *D* are in a single plane surface since Euclid has proved that every triangle is in a single plane surface. For this same reason, points *B*, *E*, *G* are in a single plane surface. Therefore, I shall describe circle *EGD* on points *E*, *G*, *D* and circle *BEG* on points *B*, *E*, *G*. And I shall protract a perpendicular from point *N* to the surface of circle *BEG*, falling on point *K*. Therefore, it has become evident that point *K* is the center of circle *BEG*. For if we draw lines *BK*, *GK*, *EK*, they are necessarily equal since lines *BN*, *EN*, *GN* are equal, line *NK* is common, and each of the lines *BK*, *EK*, *GK* contains a right angle with line *KN*. Therefore, circle *BEG* is on the surface of sphere *ABGD*. And a perpendicular on its surface has been drawn from its center, and this is

13 omnis triangulus: كل ثلث نقطة (*any three points*)
14 et...una *om. Ar.*
15–16 super...E, G. D: – ج د ه – (*GDE*)
17 *post* perpendicularem *add. Ar.* – زح – (*ZH*) ((Note: ح (*H*) replaces the *K* of Gerard's text throughout.))
17–18 cadentem...est[1]: فيقع (*And so it falls on*)
24 super...est *om. Ar.*

eius perpendicularis super superficiem eius, que est *KN*. Ergo centrum
25 spere est super lineam *KN*. Et similiter iam ostensum est quod [si sit]
59r c. 1 perpendicularis que protrahatur / ex puncto *N* et cadit super superficiem circuli *GDE*, tunc super ipsam est centrum circuli *EGD*. Et oportet propter illud ut super ipsam sit centrum spere. Ergo est punctum *N* centrum spere. Et illud est quod declarare voluimus. Et
30 hec est forma eius.

[IX.] CUM LINEA QUE PROTRAHITUR EX PUNCTO CAPITIS OMNIS PIRAMIDIS COLUMPNE AD CENTRUM BASIS EIUS EST PERPENDICULARIS SUPER BASIM IPSIUS, TUNC LINEE QUE PROTRAHUNTUR EX PUNCTO CAPITIS EIUS
5 AD CIRCULUM CONTINENTEM SUPERFICIEM BASIS EIUS SECUNDUM RECTITUDINEM SUNT EQUALES. ET MULTIPLICATIO UNIUS LINEARUM QUE PROTRAHUNTUR EX CAPITE EIUS AD CIRCULUM CONTINENTEM BASIM EIUS IN MEDIETATEM CIRCULI CONTINENTIS BASIM EIUS EST
10 EMBADUM SUPERFICIEI PIRAMIDIS COLUMPNE, SCILICET SUPERFICIEI EIUS QUE EST INTER PUNCTUM CAPITIS EIUS ET LINEAM CONTINENTEM BASIM EIUS.

Exempli causa, sit piramidis *ABGD* caput punctum *A* [Fig. 43]. Et circulus continens basim eius sit circulus *BGD*. Et centrum

24 eius[1, 2]: amborum *Zm* / in (?) centrum *H* / *ante* centrum *add. mg. P text. MaR* ex corolario et theorematis (tehorematis *Ma*) 1 Theodosii libri. ((This whole addition in *MaR* is enclosed in a box to indicate, I suppose, that it comes from a marginal note.))
25 est[1] *tr. H post* KN *in linea 25* / linea *H* / [si sit] *supplevi*
26 ex: a *Zm*
27 EGD: GED *H*
28 est *om. H*
29 N: M *H* / centrum spere *tr. Zm ante* punctum / declarare: demonstrare *H*
1 [IX]: 11 *mg. MaR*
2 *ante* omnis *add. PS* eius; *sed om. ZmMHa* / ad centrum: a centro *H*
3 est *om. H* / perpendiculariter *H* / super: supra *H*
5 continentem: contingentem *H*
13–77 Exempli.... voluimus *om. S*
13 capud *H*

---

25 *post* KN *add. Ar.* على ما تبين فى ثانى اشكال كتاب الاكر لثاوذوسيوس (*According to what is proved in Proposition II of the Book of the Spheres of Theodosius.*)
27–28 Et...spere: تمّر بمركز الكرة والعمود ان لا يتلاقيان الا عند – ز – (*passes through the center of the sphere and the perpendiculars meet only at Z*)
29 declarare *om. Ar.*
29–30 Et...eius *om. Ar.*
1–12 Cum...eous: كل مخروط مستدير قائم فسطح الخطّ الواصل بين رأسه واى نقطة فرضة

[the perpendicular] KN. Therefore, the center of the sphere is on line KN. And it has now been demonstrated in the same way that if a perpendicular is drawn from point N to fall on the suface of circle GDE, then the center of circle EGD is on it. And for the same reason [as above] the center of the sphere is on that perpendicular. [But point N is the common point of the two perpendiculars.] Therefore, point N is the center of the sphere. And this is what we wished to show. And here is its form. [Fig. 42].

[IX.] WHEN A LINE DRAWN FROM THE VERTEX OF ANY CONE TO THE CENTER OF ITS BASE IS PERPENDICULAR TO THE BASE, THEN THE STRAIGHT LINES DRAWN FROM ITS VERTEX TO THE CIRCUMFERENCE OF ITS BASE ARE EQUAL. AND THE MULTIPLICATION OF (1) ONE OF THE LINES DRAWN FROM ITS VERTEX TO THE CIRCUMFERENCE OF ITS BASE BY (2) ONE HALF THE CIRCUMFERENCE OF ITS BASE IS EQUAL TO THE AREA OF THE SURFACE OF THE CONE, THAT IS, ITS SURFACE BETWEEN THE VERTEX AND THE CIRCUMFERENCE OF ITS BASE.

For example, let there be a cone $ABGD$ with vertex A [see Fig. 43]. And let the circle containing its base be circle BGD with point E as

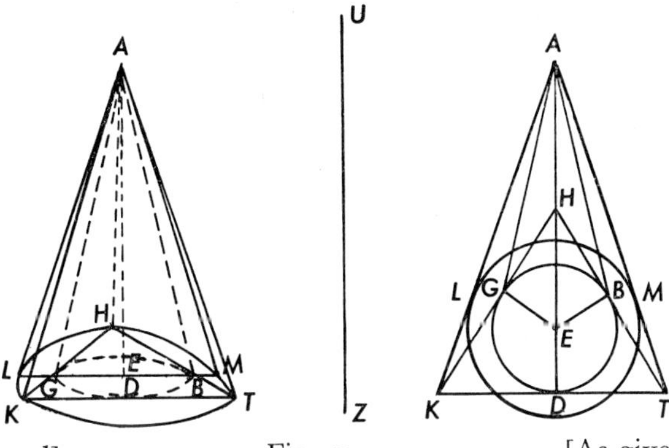

[Reconstructed] Fig. 43 [As given in MS P]

Note: The figure is slightly different in the Arabic text, but it is the same kind of drawing.

على محيط قاعدته فى نصف محيط قاعدته تساوى سطحه * المستدير

((*as in printed text, but one would expect سطح)) (*In the case of every right cone, the product of a line joining its vertex and any assigned point on the circumference of its base and one half of the circumference of the base equals the area of the cone.*)

14 circulus *om. Ar.*

15 eius sit punctum *E*. Et linea *AE* que protrahitur ex puncto *A* ad *E*, quod est centrum basis eius, sit perpendicularis super superficiem circuli *BGD*. Et propter illud linee que protrahuntur ex puncto *A* secundum rectitudinem ad lineam continentem circulum *BGD* erunt equales. Et protraham ex eo lineam unam, que sit linea *AB*. Dico 20 ergo quod multiplicatio linee *AB* in medietatem linee continentis circulum *BGD* est embadum superficiei piramidis *ABGD*, que elevatur ex circulo *BGD* ad punctum *A*; quod sic demonstratur.

Si non fuerit ita, tunc sit multiplicatio linee *AB* in quantitatem longiorem aut breviorem medietate circumferentie circuli *BGD* ipsum 25 embadum piramidis *ABGD*. Sit ergo in primis multiplicatio eius in quantitatem que sit longior medietate circuli *BGD* ipsum embadum piramidis *ABGD*. Et sit quantitas *UZ*. Et duplum *UZ* est longius circulo *BGD*. Ergo faciam super circulum *BGD* figuram habentem latera et angulos equales continentem ipsum. Et sint latera eius agre- 30 gata brevius duplo linee *UZ*, que sit figura *HTK*. Et protraham lineas *AH*, *AT*, *AK*. Et protraham iterum duas lineas *AG*, *AD*, que sint equales et equales linee *AB*. Ergo manifestum est quod linee *AB*, *AD*, *AG* cadunt orthogonaliter super lineas *HT*, *KT*, ⟨*KH*⟩ propterea quod axis piramidis qui egreditur ex puncto capitis eius ad cen- 35 trum circuli basis eius est perpendicularis super superficiem basis eius. c. 2 Tunc linee que protrahuntur ex punctis / *B*, *G*, *D* ad centrum erigun- tur super lineas *HT*, *TK*, *KH* orthogonaliter, quoniam sunt contingentes circulum. Et ostendam quod multiplicatio unius linearum *AB*, *AG*, *AD* in medietatem omnium linearum *HT*, *TK*, *HK* est embadum 40 superficiei corporis *ATHK*. Sed embadum superficiei corporis *AHTK* est maius embado superficiei piramidis columpne super quam est *ABGD*, quoniam ipsum continet illud. Et medietas omnium laterum figure *HTK* est brevior linea *UZ*. Et iam fuit multiplicatio linee *AB* in

17 ex: a *Zm*
27 U-: M- *H hic et ubique in hac propositione* / longior *H*
31 -K: -N *H hic et ubique in hac propositione*
33 cadunt: cadent *Zm* / ⟨HK⟩ *?Zm*
34 qui *P* que *H* quod *Ma*
37 *de* HT, TK *scr. P mg. et Zm mg.* in alio, HK, TK / KH: HK *Zm*
41 columpne *HMa* columne *PZm*
42 ABGD: BGD *H* / Et: et est *H*

---

15 punctum *om. Ar.*
15–16 Et linea...sit: ومحوره — ا ه — وهو (*And its axis is AE which is*)
17 circuli BGD: القاعدة حتى يكون المخروط قائما (*of the base, so that the cone is a right cone*)

its center. And let line *AE*, drawn from point *A* to *E* the center of its base, be perpendicular to the surface of circle *BGD*. Accordingly, the straight lines drawn from point *A* to the circumference of circle *BGD* will be equal. And I shall draw from it (i.e., *A*) one [such] line, namely line *AB*. I say, therefore, that the multiplication of line *AB* by 1/2 the circumference of circle *BGD* is equal to the surface area of cone *ABGD*, i.e., to the surface area extending from [the circumference of] circle *BGD* to point *A*

Demonstration: If it is not so, then let the multiplication of line *AB* by a quantity greater or less than 1/2 the circumference of circle *BGD* be equal to the area of cone *ABGD*. In the first place let the multiplication of it by a quantity greater than 1/2 the circumference *BGD* be equal to the area of cone *ABGD*. Let that [greater] quantity be *UZ*. And $2\,UZ >$ circum *BGD*. Therefore, I shall construct on circle *BGD* a regular polygon containing it. Let its perimeter be less than 2 *UZ*. This polygon is *HTK*. Then I shall draw lines *AH*, *AT*, and *AK*. And, further, I shall draw two lines, *AG* and *AD*, equal to each other and equal to line *AB*. It has become evident, therefore, that lines *AB*, *AD*, *AG* fall perpendicularly on lines *HT*, *KT*, and *KH*, since the axis of the cone from its vertex to the center of its base is perpendicular to the surface of its base. Then the lines drawn from points *B*, *G*, *D* to the center are perpendicular to lines *HT*, *TK*, and *KH*, since these lines are tangent to the circle. And I shall demonstrate that the multiplication of (1) one of the lines *AB*, *AG*, *AD* by (2) one half of the sum of lines *HT*, *TK*, *HK* is equal to the surface area of body *ATHK*. But the surface area of body *AHTK* is greater than the surface area of cone *ABGD*, since the one contains the other. Now 1/2 the perimeter of polygon *HTK* is less than line *UZ*. But the multiplication

17–19 Et....AB: ونصل – ا ب –
(*And we join AB*)
19–20 Dico...linee AB: فسطح – ا ب –
(*And so the product of AB...*)
21–22 ABGD....demonstratur *om. Ar.*
23–25 tunc....ABGD *om. Ar.*
25 eius: – ا ب – (*AB*)
26–27 BGD....ABGD *om. Ar.*
27–28 Et[2]...BGD[1] *om. Ar.*
30 *post* HTK *add. Ar.* ولتماس الدائرة على نقط – ب – ج – د –
(*And let it touch the circle at points B, G, D*)

33 ⟨KH⟩ is in the Arabic text: – ا ح –
34 axis.... eius: – ا ه – (*AE*)
35 basis eius: دائرة – ب ج د (*of circle BGD*)
36 B, G, D: التماس (*of tangency*)
37 lineas....KH: الاضلاع (*the sides*)
37–38 quoniam...circulum *om. Ar.*
38 Et ostendam quod: ولذلك يكون
(*And accordingly will...be*)
38–39 unius...AD: – ا ب – (*AB*)
39 linearum...HK: الاضلاع (*of the sides*)
40–42 ATHK....illud *om. Ar.*
43 HTK *om. Ar.*

lineam *UZ* ipsum embadum piramidis columpne. Et illud quod egre-
45 ditur ex conclusione est quod multiplicatio linee *AB* in id quod est brevius linea *UZ* est maior superficie piramidis columpne. Hoc est contrarium; ergo non est possibile ut multiplicatio linee *AB* in lineam que sit longior medietate circuli *BGD* sit embadum piramidis *ABGD*.

Amplius ponam lineam *UZ* breviorem medietate circuli *BGD*, si
50 fuerit illud possibile. Ergo multiplicatio linee *AB* in lineam *UZ* est embadum superficiei piramidis *ABGD*. Sed linea *UZ* est brevior medietate circumferentie circuli *BGD*. Ergo multiplicatio *AB* in medietatem circuli *BGD* est maior embado superficiei piramidis *ABGD*. Sit ergo illud ipsum embadum piramidis cuius basis est circulus *ML* et caput
55 eius punctum *A*. Cum ergo fiet in circulo *ML* figura habens latera et angulos equales non contingens circulum *BGD*, et protrahantur ex extremitatibus laterum huius figure linee ad punctum *A*, et protrahantur superficies triangulorum, erit embadum superficiei corporis cuius basis est figura habens latera facta in circulo *ML* et cuius caput
60 est punctum *A* minus embado superficiei piramidis cuius basis est circulus *ML* et cuius caput est *A*, quoniam piramis continet ipsum. Sed multiplicatio unius linearum que protrahuntur ex puncto *A* ad medietates laterum figure facte super circulum *ML* in medietatem omnium laterum figure facte in circulo *ML* est embadum superficiei
65 corporis cuius basis est figura facta in circulo *ML* et cuius caput est punctum *A*. Et linea que protrahitur ex puncto *A* ad medium uniuscuiusque laterum huius figure facte in circulo *ML* est longior *AB*. Et medietas omnium laterum figure facte est longior medietate circuli

45 cong!clusione *P* / AB: AD *H* / est *om. H*
46 est[2] *HMaZm om. P*
49–50 *de* ponam...possibile *scr. PZm mg.* (*et add. HMaR* ante* ponam): in (*om. HMaR*) alio (*om. HMaR*), ponam multiplicationem eius breviorem si est possibile. Ergo (*add. R* linea) erit. ((* *MaR* put the phrase in a box to indicate, I suppose, that it came from a marginal note.))
50 fuerit illud *tr. H* / UZ: ND *H*
51 pyramis *H*
52 circumferentie *mg. P*, *text ZmH om. Ma*/circumferentie circuli *tr. Zm*
53 pyramis *H*
55 in circulo *om. H* / figura *ZmH* figum *P* figuram *Ma*
56 protrahentur *Zm*
57 linee *P* linea *HMa* / ad *om. H*
57–58 protrahantur *P* protrahentur *ZmH* protrahuntur *Ma*
60 piramis *H*
62 puncto *om. H* / ad: et *H*
63–64 *de* super...laterum *scr. PZm mg.* (*et add. MaR* ante* super): in (*om. MaR*) alio (*om. MaR*), in circulo in medietatem (-m *om. MaR*) omnium laterum eius. ((**MaR* put the phrase in a box to indicate, I suppose, that it came from a marginal note.))
66 punctum *om. H* / protrahatur *H*

of line AB by line UZ was earlier [taken as] the area of the cone. And the inference from the conclusion is that the multiplication of line AB by a quantity less than UZ is greater than the surface of the cone. This is a contradiction. Hence, it is not possible that the multiplication of line AB by a line greater than 1/2 the circumference BGD is equal to the area of cone $ABGD$.

Further, I shall posit line UZ to be less than 1/2 the circumference BGD, if that is possible. Hence, the multiplication of line AB by line UZ is equal to the surface of cone $ABGD$. But line $UZ <$ 1/2 circum BGD. Hence, the multiplication of AB by 1/2 the circumference BGD is greater than the surface area of cone $ABGD$. Hence let this [product] be the area of a cone whose base is circle ML and whose vertex is point A. When a regular polygon is inscribed in circle ML—but not touching circle BGD—and when lines are drawn to point A from the extremities of the sides of this polygon and [thus] the surfaces of the triangles [formed by these lines and the sides of the polygon] are produced, then the surface area of the body whose base is the polygon inscribed in circle ML and whose vertex is point A is less than the surface area of the cone whose base is circle ML and whose vertex is point A, since the cone contains it. But the multiplication of (1) one of the lines drawn from point A to the middle points of the sides of the polygon inscribed in circle ML by (2) 1/2 the perimeter of the polygon inscribed in circle ML is equal to the surface area of the body whose base is the polygon inscribed in circle ML and whose vertex is point A. And the line drawn from point A to the middle of any one of the sides of the polygon inscribed in circle ML is greater than AB. Further, 1/2 the perimeter of the polygon is greater than 1/2 the circumference BGD.

44–48 Et...ABGD: فسطح المستدير اعظم مما هو محيط به هذا خلف (*And so the area of the cone is greater than that by which it is contained. This is contradictory.*)
49–50 BGD...possibile: و (*And*)
51 ABGD *om. Ar.*
51–52 linea....Ergo *om. Ar.*
53 circuli *om. Ar.* / est...piramidis: اعظم منه مساويا — — لسطح مخروط مستدير (*is equal to something greater than it, i.e., to the surface of a cone*)
57 extremitatibus...figure: زواياه (*its angles*)
57–58 et...triangulorum *om. Ar.*
59–60 cuius....A: الحادث (*new*)
61 et...ipsum *om. Ar.*
63 medietates: منتصف احد (*the middle of one*) / facte...ML: الذى لا يماس دائرة — ب ج د (*which does not touch circle BGD*)
64 omnium...ML: اضلاعه (*of its sides*)
65–66 corporis...A: ذلك المجسّم (*of that body*)
67 huius...ML *om. Ar.* / AB: خطّ — ا ب — (*line AB*)
68 omnium *om. Ar.* / facte *om. Ar.* / circuli: محيط دائرة (*of the circumference of the circle*)

BGD. Ergo multiplicatio linee *AB* in medietatem circumferentie *BGD*
70 est brevior multum embado superficiei piramidis cuius basis est circulus *ML*. Et iam fuit multiplicatio *AB* in medietatem circuli *BGD* ipsum embadum superficiei piramidis cuius basis est circulus *ML* et cuius caput est punctum *A*. Et superficies huius piramidis est maior superficie corporis habentis superficies. Hoc vero est contrarium. Ergo
59v c. 1 multiplicatio / linee *AB* in medietatem circumferentie circuli *BGD* est
76 embadum superficiei piramidis *ABGD*. Et illud est quod declarare voluimus.

[X.] OMNIS SUPERFICIES CUM ABSCIDIT QUAMLIBET PIRAMIDEM COLUMPNE, CUIUS BASIS EST CIRCULUS, ET EST EQUIDISTANS BASI EIUS, TUNC AMBARUM SECTIO COMMUNIS EST CIRCULUS. ET SI PROTRAHATUR EX CA-
5 PITE PIRAMIDIS LINEA AD CENTRUM BASIS EIUS, TUNC IPSA TRANSIT SUPER CENTRUM CIRCULI QUI EST SECTIO COMMUNIS.

Verbi gratia, sit piramis *ABGD*, cuius caput sit *A* et cuius basis sit circulus *BGD*, et ipsius centrum *E* [Fig. 44]. Et secet eam super-
10 ficies equidistans superficiei circuli *BGD*, et fiat sectio earum communis superficies *UTZ*, et protrahatur ex puncto *A* linea ad *E*, et

70 est[1] *H Zm supra om. PMa*
71 *de* circuli *scr. P mg.* (*et Zm supra in linea 52*) in alio, linee
72 ipsum: ipsum est *H*
73 cuius caput: caput eius *H*
75 medictatem *ZmHMa* medietate *P* / circuli *ZmP om. H* circulum *R* circulụ̄ *Ma*
76 declarare: demonstrare *H*

1 [X]: 12 *mg. MaR*
2 piramidis *H* / columpne *ZmHMa* columne *P*
3, 10 equedistans *H hic et ubique*
3 ambarum *ZmP* amborum *HMa*
5 linea: ex linea *H*
8–34 Verbi....eius *om. S*
11 protraham *H* / linea: lineam *H* / ad: at *H*

69–74 Ergo...superficies: فسطح المخروط المستدير الذى قاعدته – م ل – اصغر من سطح الجسم الذى فى داخله

(*And so the area of the cone whose base is ML is smaller than the area of the body which is inside it.*)

75 circumferentie *om. Ar.*
76 declarare *om. Ar.*
1–7 omnis....communis: كل مخروط مستدير قاعدته دائرة وقد فصله سطح مواز لقاعدته كان ذلك الفصل دائرة والمحور يمرّ بمركزها

(*The [common] section of any cone, whose base is a circle, and a plane parallel to the base is a circle, and the axis passes through its center.*)
8 ABGD *om. Ar.*
9–10 Et...BGD *om. Ar.*
10 earum *om. Ar.*
11 et[1]...E:–ا ه– والمحور (*And [its] axis AE*)

Therefore, the multiplication of line *AB* by 1/2 the circumference *BGD* is much less than the surface area of the cone whose base is circle *ML*. But the multiplication of *AB* by 1/2 the circumference *BGD* was [taken as equal to] the surface area of the cone whose base is circle *ML* and whose vertex is point *A*. And [yet] the surface of this cone is greater than the surface of the body having [triangular] surfaces. This indeed is a contradiction. Therefore, the multiplication of line *AB* by 1/2 the circumference of circle *BGD* is equal to the surface area of cone *ABGD*. And this is what we wished to show.

[X.] IN THE CASE OF EVERY [PLANE] SURFACE WHICH CUTS ANY CONE WHOSE BASE IS A CIRCLE—THE SURFACE BEING PARALLEL TO THAT BASE—THE COMMON SECTION [OF THE CUTTING SURFACE AND THE CONE] IS A CIRCLE. AND IF A LINE IS PROTRACTED FROM THE VERTEX OF THE CONE TO THE CENTER OF ITS BASE, THEN THAT LINE WILL PASS THROUGH THE CENTER OF THE CIRCLE WHICH IS THE COMMON SECTION.

For example, let there be a cone *ABGD*, whose vertex is *A* and whose base is circle *BGD* with center *E* [see Fig. 44]. And let a surface parallel to the surface of circle *BGD* cut the cone. Let their common section be *UTZ*. And let a line be protracted from point *A* to *E*, passing through

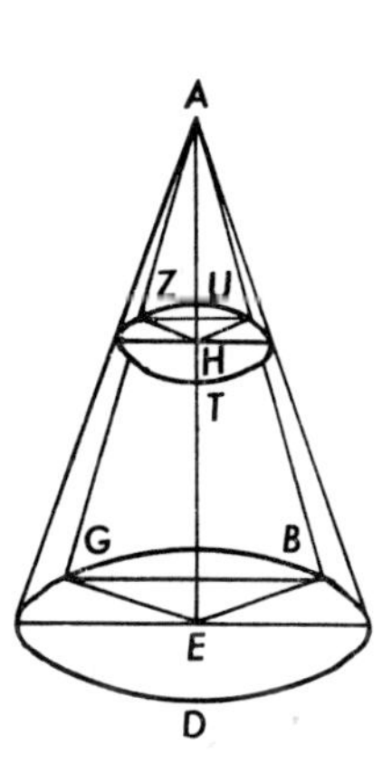

[Reconstructed]

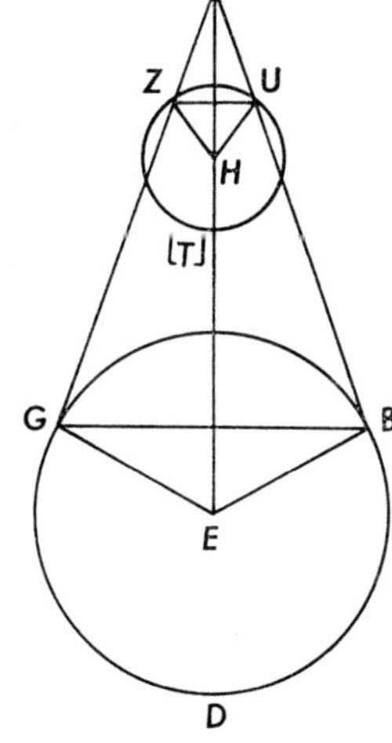

Fig. 44

[As given in Arabic texts and in MS *P*]

penetret linea superficiem *UTZ* super punctum *H*. Dico ergo quod linea *UTZ* continet circulum cuius centrum est punctum *H*, quod sic demonstratur.

15 Signabo super circulum *BGD* duo puncta *B*, *G*, et ponam arcum *BG* minorem semicirculo, et protraham duas lineas, *BE*, *EG*, et duas lineas *BA*, *AG*, et estimabo duas superficies duorum triangulorum *BAE*, *GAE* protractas. Ergo secat triangulus *BAE* superficiem *BGD* super lineam *BE* et secat superficiem *UTZ* super lineam *HU*. Ergo 20 due linee *HU*, *EB* sunt equidistantes, sicut narravit Euclides, et similiter due linee *EG*, *HZ* iterum sunt equidistantes. Et protraham iterum duas lineas *BG*, *UZ*. Ergo manifestum est quod utreque sunt equidistantes, propterea quod *BG* est sectio communis superficiei trianguli *BAG* et superficiei *BGD*, et linea *UZ* est sectio communis superficiei 25 trianguli *BAG* et superficiei *UTZ*. Ergo triangulus *BEG* habet latera equidistantia triangulo *UHZ*. Ergo anguli amborum sunt equales et latera utrorumque sunt proportionalia. Ergo proportio *BE* ad *EG* est sicut proportio *UH* ad *HZ*. Sed *BE* est equalis *GE*. Ergo due linee *UH*, *HZ* sunt equales. Et per huiusmodi dispositionem scitur quod 30 omnis linea egrediens ex puncto *H* ad quemlibet locum exitus circumferentie circuli *UTZ* est equalis unicuique duarum linearum *UH*, *HZ*. Ergo oportet ut sit linea *UTZ* circumferentia circuli et punctum *H* centrum circuli. Et illud est quod declarare voluimus. Et hec est forma eius.

12 superficies *H*
18 BAÇE *P* / protractis *H*
20–21 sicut... equidistantes *mg. P*, *Zm HMa*
25 superficies *H*
26 equedistancia *H*
28 Sed *om. H*
29 UH, HZ *PMaZm* UH et HZ *R* ZH, HN *H*
32 UTZ: NTZ et *H* / et: ad *H*
33–34 quod....eius: propositum *H*

---

12 UTZ: الفاصل (sectional)
12–14 Dico...demonstratur *om. Ar.*
15 circulum *om. Ar.*
16 duas lineas *om. Ar.*
16–17 et²...AG: – ب ا – ج ا – ب ج – (*BA*, *GA*, *BG*)
17–25 et....UTZ: فيمرّ مثلث – ا ب ه – بفصل – وح – من السطح الفاصل ومثلث – ا ه ج – بفصل – زح * – ومثلث – ا ب ج – بفصل – و ز – ويحدث مثلث ** – و زح ((*– زج – in the printed text.)) ((** مثل in the printed text.)) (*And triangle ABE cuts the [common] sectional surface in [common] sectional [line] UH and triangle AEG [cuts that surface] in [common] sectional [line] ZH and triangle ABG [cuts it] in [common] sectional [line] UZ and [so] triangle UZH is formed.*)
25–27 Ergo....proportionalia: وتكون اضلاعه موازية لاضلاع مثلث – ه ب ج كل لنطيره فيكونان متشابهين

surface *UTZ* at point *H*. I say, therefore, that line *UTZ* is the circumference of a circle whose center is point *H*.

Demonstration: I shall mark two points *B* and *G* on circle *BGD* and I shall assume that arc *BG* is less than a semicircle. Further, I shall draw two lines *BE* and *EG* and two [more] lines, *BA* and *AG*. And [thus] I shall consider the two surfaces of the two triangles, *BAE* and *GAE*, as drawn. Therefore, let △ *BAE* cut surface *BGD* in line *BE* and surface *UTZ* in line *HU*. Therefore, the two lines *HU*, *EB* are parallel, as Euclid explained. And similarly the two lines *EG* and *HZ* are also parallel. And I shall draw the two lines *BG*, *UZ*. Therefore, it is evident that they are parallel, since line *BG* is the common section of the surface of △ *BAG* and the surface of *BGD*, and line *UZ* is the common section of the surface of △ *BAG* and surface *UTZ*. Therefore, △ *BEG* has sides parallel to [those of] △ *UHZ*. Therefore, the angles of the one are equal to the angles of the other, and the sides of the one are proportional to the sides of the other. Therefore, $BE/EG = UH/HZ$. But $BE = GE$. Therefore, $UH = HZ$. And from a disposition of this kind it is known that every line going out from point *H* to any place on the circumference* of circle *UTZ* is equal to each of the two lines *UH*, *HZ*. Therefore, it is necessary that line *UTZ* is the circumference of a circle and point *H* is the center of the circle. And this is what we wished to show. This is its form [Fig. 44].

* See Commentary, Proposition X, lines 30–31.

(*And its sides are parallel to the sides of triangle EBG, each to the one facing it. And so they* ((the two triangles)) *are similar.*)
28 due linee *om. Ar.*
29 Et....quod: و (*And*)
30–31 quemlibet....HZ: محيط – و ز ط –
(*the circumference UZT*)
32 Ergo...circumferentia: فوزط (*And so UZT*)
33 declarare *om. Ar.*
33–34 Et....eius *om. Ar.*

[XI.] IN OMNI PORTIONE PIRAMIDIS COLUMPNE, CUIUS BASIS EST CIRCULUS ET CUIUS SUPERIUS EST CIRCULUS ET CUIUS BASIS SUPERFICIES EQUIDISTAT SUPERFICIEI SUPERIORIS EIUS ET LINEA QUE EGREDITUR EX CENTRO 5 BASIS EIUS AD CENTRUM SUPERFICIEI SUPERIORIS EIUS EST PERPENDICULARIS SUPER DUAS SUPERFICIES, TUNC SI PROTRAHANTUR IN BASI EIUS ET IN CIRCULO QUI EST IN SUPERIORE EIUS DUO DIAMETRI EQUIDISTANTES, ET CONTINUATUR QUOD EST INTER DUAS EXTREMINTATES c. 2 DUARUM DIAMETRO/RUM PER DUAS LINEAS, ERIT MUL- 11 TIPLICATIO UNIUS DUARUM LINEARUM IN MEDIETATEM CIRCUMFERENTIE CIRCULI BASIS EIUS ET IN MEDIETATEM LINEE CONTINENTIS SUPERIUS EIUS IPSUM EMBADUM QUOD ELEVATUR EX BASI EIUS ET PERVENIT AD 15 SUPERIUS PORTIONIS SECUNDUM RECTITUDINEM.

Verbi gratia, sit basis portionis piramidis columpne *UTZBGD* circulus *UTZ* [Fig. 45]. Et eius superius sit circulus *BGD*. Et sit linea *EH*, que continuat quod est inter duo centra, perpendicularis super superficiem *BGD* et super superficiem *TZU*. Et protrahantur in duo- 20 bus circulis *BGD*, *UTZ* due linee *BD*, *UZ* equidistantes. Et continuetur quod est inter duas extremitates earum cum duabus lineis *BU*,

1 [XI]: 13 *mg. MaR*
2 superius: superficies *H*
3 equedistat *H*
4 ex: a *S*
6 superficies: lineas *Zm*
7 protrahantur *ZmPMa* protrahatur *H* protrahuntur *S* / et *om. H*
10 duarum: duorum *Ma*
12 circumferentie *om. Zm* / circuli *om. S*
16–24 sit....quod: dico quod antecedente posito sicud patet in figura *S*
16 U-: N- *H hic et ubique*
20–21 continuentur *H*

1–15 In....rectitudinem: كل قطعة من مخروط مستدير قائم فيما بين دائرتين متوازيين فاذا اخرج فيهما قطر ان متوازيان ووصل بين اطرافهما بخطّين متقابلين كان سطح احد الخطّين فى نصفى محيطى الدائرتين مساويا لسطح القطعة المستدير (*In every segment of a right cone between two parallel circles when two parallel diameters are drawn in these circles and the extremities of them* ((the diameters)) *are joined by two opposite lines, then the multiplication of one of the two lines by the halves of the circumferences of the two circles is equal to the area of the segment of the cone.*)
16 UTZBGD: – ب ج و ط ز – (*BGUTZ*)
16, 17 circulus *om. Ar.*
17 eius superius: والاخرى التى تلى رأس المخروط (*And the other* [*circle*] *which is near the top of the cone.*)
17–19 Et²...TZU: و – ه ح – من المحور ما يقع بينهما وهوعمود على الدائرتين (*And EH is* [*that part*] *of the axis which falls between them and is perpendicular to the two circles.*)
19–20 in....UTZ *om. Ar.*
21 duabus lineis *om. Ar.*

[XI.] IN THE CASE OF EVERY SEGMENT OF A CONE* WHERE THE BASE IS A CIRCLE AND THE UPPER [PLANE SURFACE] IS A CIRCLE AND WHERE THE BASE SURFACE IS PARALLEL TO ITS UPPER SURFACE AND A LINE DRAWN FROM THE CENTER OF ITS BASE TO THE CENTER OF ITS UPPER SURFACE IS PERPENDICULAR TO BOTH SURFACES, THEN, IF A DIAMETER IN ITS BASE IS DRAWN PARALLEL TO A DIAMETER IN THE UPPER CIRCLE AND THE TWO LINES CONNECTING THE PAIRS OF EXTREMITIES OF THE TWO DIAMETERS ARE DRAWN, THE MULTIPLICATION OF ONE OF THE TWO LINES [OR SLANT HEIGHTS CONNECTING THE EXTREMITIES OF THE TWO DIAMETERS] BY THE SUM OF 1/2 THE CIRCUMFERENCE OF ITS BASE CIRCLE AND 1/2 THE CIRCUMFERENCE OF ITS UPPER CIRCLE IS EQUAL TO THE SURFACE AREA [THAT IS, LATERAL AREA] OF THE SEGMENT BETWEEN ITS BASE CIRCLE AND THE UPPER CIRCLE.

For example, let circle *UTZ* be the base of cone segment *UTZBGD* [see Fig. 45]. Let circle *BGD* be its upper surface. And let line *EH* which joins the two centers be perpendicular to surface *BGD* as well as to surface *TZU*. In the two circles *BGD* and *UTZ* let two parallel lines *BD* and *UZ* be drawn. Let their extremities be connected by two lines *BU*

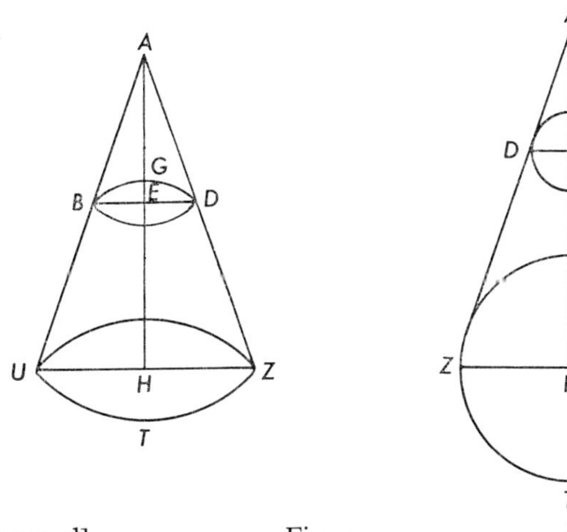

[Reconstructed] Fig. 45 [As given in MS *P*]

* I use the more general expression "segment of a cone" instead of "frustum of a cone" because the authors' use of this expression in Proposition XIII also includes a cone itself.

DZ que sunt equales, propterea quod linea *EH* iam secuit unamquamque duarum linearum *BD*, *UZ* in duo media, et est orthogonaliter erecta super unamquamque earum. Dico ergo quod multiplicatio
25 linee *BU* in medietatem circumferentie circuli *BGD* et in medietatem circuli *UTZ* est embadum superficiei portionis piramidis que elevatur ex circulo *UTZ* et pervenit ad circulum *BGD*, cuius hec est demonstratio.

Complebo piramidem *ATUZ*. Et notum est quod linea *EH* quando
30 extenditur secundum rectitudinem transibit super punctum *A*. Propter illud quod ostendimus quod linea que egreditur ex puncto *A* ad punctum *H* transit per punctum *E*, ergo linea *AH* egreditur ex capite piramidis ad centrum basis eius et cadit perpendicularis super basim. Ergo multiplicatio linee *AU* in medietatem circuli *UTZ* est embadum
35 superficiei piramidis *AUTZ*. Et multiplicatio linee *AB* in medietatem circuli *BGD* est embadum superficiei portionis piramidis super quam signatum est *ABGD*. Sed multiplicatio *AU* in medietatem circuli *UTZ* est sicut multiplicatio *AB* in medietatem circuli *UTZ* et *BU* in medietatem circuli *UTZ*. Verum multiplicatio *AB* in medietatem cir-
40 culi *UTZ* est sicut multiplicatio *AB* in medietatem circuli *BGD* et in superfluum medietatis circuli *UTZ* super medietatem circuli *BGD*. Verum multiplicatio *AB* per medietatem circuli *BGD* est embadum superficiei piramidis *ABGD*. Ergo remanet multiplicatio *AB* in superfluitatem medietatis circuli *UTZ* super medietatem circuli *BGD*,
45 et multiplicatio *BU* in medietatem circuli *UTZ*, et est embadum superficiei portionis piramidis super quam signate sunt *BGDUTZ*. Et illud quod fit ex multiplicatione linee *AB* in superfluitatem medietatis circuli *UTZ* super medietatem circuli *BGD* est equale multiplicationi linee *BU* in medietatem circuli *BGD*, propterea quod proportio linee
50 *AB* ad medietatem circuli *BGD* est sicut proportio *BU* ad superfluitatem medietatis circuli *UTZ* super medietatem circuli *BGD*.

27–96 cuius....eius *om.* *S*
30 extenditur *HRZm*(?) extendetur *P* exstenditur *Ma*
31–32 ex...egreditur *om.* *H*
32 transibit *P*
33 perpendiculariter (?) *H*
39 Verum *PZm* vero *H*
39–40 Verum...UTZ *om.* *Ma*
40 UTZ...in *PZm* BGD est in *H* BGD et *R*
41–43 UTZ....piramidis *Ma* *mg.* *P* *mg.* *Zm* *om.* *H*
43 ABGD *H* ABG *PZmMa* / AB: AD *H*
45 et[2] *om.* *H*
46 *ante* portionis *add.* *P* et *sed om.* *ZmHMa* / signati *H*
47–48 *post* circuli *inser.* *H* *lineas 52–70* ("Quod....circuli")
48 est equale: et equalis *H*
50 BGD: ABGD *H*
51 circuli[1]...medietatem *om.* *H*

and DZ. These two lines are equal since line EH bisects each of the two lines BD and UZ and is perpendicular to both of them. I say, therefore, that $BU \cdot (1/2 \text{ circum } BGD + 1/2 \text{ circum } UTZ) = \text{surf area cone seg } UTZBGD$.

Demonstration: I shall complete the cone $ATUZ$. And it is known that line EH, when extended rectilinearly, will pass through point A. Since we show that the line going out from point A to point H passes through point E, therefore line AH goes out from the vertex of the cone to the center of its base and is perpendicular to the base.

Therefore,

$$AU \cdot 1/2 \text{ circum } UTZ = \text{surf area cone } AUTZ.$$

And

$$AB \cdot 1/2 \text{ circum } BGD = \text{surf area cone } ABGD.$$

But

$$AU \cdot 1/2 \text{ circum } UTZ = [(AB \cdot 1/2 \text{ circum } UTZ) + (BU \cdot 1/2 \text{ circum } UTZ)].$$

Now.

$$AB \cdot 1/2 \text{ circum } UTZ = AB\,[1/2 \text{ circum } BGD + (1/2 \text{ circum } UTZ - 1/2 \text{ circum } BGD)]$$

and

$$AB \cdot 1/2 \text{ circum } BGD = \text{surf area cone } ABGD.$$

Therefore [1],

$$[AB \cdot (1/2 \text{ circum } UTZ - 1/2 \text{ circum } BGD) + (BU \cdot 1/2 \text{ circum } UTZ)] = \text{surf area cone seg } BGDUTZ.$$

And [so]

$$[AB \cdot (1/2 \text{ circum } UTZ - 1/2 \text{ circum } BGD)] = BU \cdot 1/2 \text{ circum } BGD,$$

since

$$\frac{\text{line } AB}{1/2 \text{ circum } BGD} = \frac{\text{line } BU}{(1/2 \text{ circum } UTZ - 1/2 \text{ circum } BGD)}.$$

22–24 que...earum *om. Ar.*
26–28 que...demonstratio *om. Ar.*
29 ATUZ: الى الرأس وهو – ا – (*up to the vertex A*)
29–30 Et....A: ونخرج – ح ه – الى – ا – وكذلك – و ب ، ز د – (*And we extend HE up to A and similarly UB, ZD [also up to A].*)
30–33 Propter....basim *om. Ar.*
34 Ergo: ومعلوم ان (*And it is known that*)
35 AUTZ: جميع (*of the whole*)
36–37 super...est *om. Ar.*
37 *post* ABGD *add. Ar.* وفضل الاول على الآخر هو السطح المستدير المحيط بالقطعة (*And the excess of the first over the second is the area of the segment of the cone.*)
37–43 Sed....ABGD *om. Ar.*
43 Ergo remanet: وذلك هو (*And that is*) ((The "that" refers to the "area of the segment of the cone"—noted in the variant given after ABGD in line 37.))
45–46 et...BGDUTZ *om. Ar.* ((But see variant for line 43.))
49–51 proportio....BGD: نسبة ا ب – الى – ب و – كنسبة نصف دائرة – ب ج د – الى فضل نصف دائرة – و ط ز * – على نصف دائرة – ب ج د – ((*corrected from و ط)) ($AB/BU = \frac{1}{2} BGD / [\frac{1}{2} UTZ - \frac{1}{2} BGD]$) ((Note: the rearrangement of the proportion in the Gerard translation.))

[Quod proportio linee *AB* ad medietatem circuli *BGD* sit sicut proportio linee *BU* ad superfluum medietatis circuli *UTZ* super medietatem circuli *BGD*, sic proba. Quia duo trianguli *ABE*, *AUH* 55 sunt similes, igitur proportio *AB* ad *BE* est sicut proportio *AU* ad *UH* et proportio *AB* ad *ED* est sicut proportio *AU* ad *HZ*. Ergo ex 24 quinti libri Euclidis est proportio *AB* ad *BD* sicut proportio *AU* ad *UZ*. Sed proportio *BD* ad lineam circumductam *BGD* est sicut proportio *UZ* ad lineam circumductam *UTZ*. Ergo secundum 60 equalitatem erit proportio *AB* ad lineam circumductam *BGD* sicut proportio *AU* ad lineam circumductam *UTZ*. Ergo proportio linee *AB* ad medietatem circumferentie circuli *BGD* sicut proportio *AU* ad medietatem circumferentie circuli *UTZ*, et hoc per impossibile probatur ex 24 quinti libri Euclidis. Sed cum he sint due quantitates, 65 scilicet linea *AU* et medietas circuli *UTZ*, et minuemus ex eis duas quantitates, scilicet lineam *AB* et medietatem circuli *BGD*, et est proportio diminute ad diminutam sicut totius ad totum, ergo est proportio reliqui, quod est *BU*, ad reliquum, quod est superfluum medietatis circuli *UTZ* super medietatem circuli *BGD*, sicut proportio totius 70 ad totum et sicut proportio *AB* ad medietatem circuli *BGD*. Et hoc est quod declarare voluimus.]

Ergo multiplicatio linee *AB* in superfluum medietatis circuli *UTZ* super medietatem circuli *BGD* est sicut multiplicatio *BU* in medietatem circuli / *BGD*. Ergo multiplicatio linee *BU* in medietatem cir- [60r c. 1] 75 culi *UTZ* et in medietatem circuli *BGD* est embadum superficiei portionis piramidis super quam sunt *BGD*, *UTZ*. Et illud est quod declarare voluimus.

52–54 Quod...BGD *om. Ma*
52–71 Quod....voluimus *mg. P et Zm om. hic sed add. fol. 77r in fine tractatus Ameti de proportione* (*Cf. var. linearum* 47–48, 54–71)
52 AB: AD *H*
53 ad: sicut *H*
54–71 sic....voluimus *add. mg. Ma et R in text.* (*Cf. var. linearum* 52–71)
54 sic proba *PZm* sic probat *Ma* quod sic probatur *H* / trianguli: anguli *H* / ABE, AUH: ABH, AUE *Zm* (*Zm tr.* H *et* B *ubique in lin. 52–56*)
56 ED *MaH* EB *P* / est *om. H*
56–57 HZ....ad *bis H*
58 UZ: TZN *H*
58–59 est sicut *tr. H*
63–64 probatur *HMa* proba *PZm*
65 et[2] *om. H*
67 diminutam (?)*P*, *ZmMa* diminutum *H*
70–71 Et...voluimus *om. H*
72 in *om. H*
76–77 declarare: demonstrare *H*

52–76 Quod....BGD, UTZ *om. Ar.*
76–77 declarare *om. Ar.*

*[Prove that $\frac{\text{line } AB}{1/2 \text{ circum } BGD} = \frac{\text{line } BU}{(1/2 \text{ circum } UTZ - 1/2 \text{ circum } BGD)}$ as follows: Since $\triangle\, ABE$ is similar to $\triangle\, AUH$, therefore $AB/BE = AU/UH$ and $AB/ED = AU/HZ$. Therefore, by V.24 of the book of Euclid, $AB/BD = AU/UZ$. But BD/circum $BGD = UZ$/circum UTZ. Therefore, by equality, AB/circum $BGD = AU$/circum UTZ. Hence, $\frac{\text{line } AB}{1/2 \text{ circum } BGD} = \frac{\text{line } AU}{1/2 \text{ circum } UTZ}$. This is proved by reduction to absurdity from V.24 of the book of Euclid. But since these are two quantities, i.e., line AU and 1/2 circumference UTZ, and we subtract from them two quantities, that is, line AB and 1/2 circumference BGD, and the ratio of remainder to remainder is as the ratio of whole to whole, therefore the ratio of [one] remainder, i.e., BU, to the [other] remainder, i.e., (1/2 circum UTZ — 1/2 circum BGD), is as the ratio of whole to whole, i.e., as the ratio of AB to 1/2 circumference BGD. And this is what we wished to show.]

Therefore [2],

$AB \cdot (1/2 \text{ circum } UTZ - 1/2 \text{ circum } BGD) = BU \cdot 1/2 \text{ circum } BGD$.

Hence [if formula 2 is substituted in formula 1],

$BU \cdot (1/2 \text{ circum } UTZ + 1/2 \text{ circum } BGD) = \text{surf area cone seg } BGDUTZ$.

And this is what we wished to show.

* See Commentary, Proposition XI, lines 52–71, for the material in brackets.

Iam ergo scitur ex eo quod narravimus quod si due linee *UB*, *BA* fuerint equales quocunque modo fuerit earum applicatio, scilicet se-
80 cundum rectitudinem aut non secundum rectitudinem, tunc multiplicatio unius earum in medietatem circuli *UTZ* et in circulum *BGD* est embadum superficiei corporis cuius caput est punctum *A* et cuius basis est superficies *UTZ*. Et hinc scitur quod si fuerint portiones plures piramidum columnarum composite ad invicem et fuerit super-
85 ficies superior portionis inferioris equalis basi superficiei portionis que est super eam et fuerit portio superior ex portionibus habens caput quod est piramis et fuerint bases portionum omnium equidistantes et fuerint linee que egrediuntur in omnibus portionibus ex basibus earum ad earum superiora secundum rectitudinem equales, tunc mul-
90 tiplicatio unius linearum que egrediuntur ex basibus portionum ad earum superiora in medietatem linee continentis basim portionis inferioris et in omnes lineas continentes omnes bases portionum que sunt super portionem inferiorem est embadum superficiei corporis compositi ex illis sectionibus, si sunt superficies sectionum continue secun-
95 dum rectitudinem aut sunt non secundum rectitudinem. Et illud est quod declarare voluimus. Et hec est forma eius.

78 *ante* Iam *mg. add. MaR* 14
79 quecunque *H* / scilicet *om. H*
80 aut...rectitudinem: circuli UTZ super medietatem circuli BG est embadum superficiei piramidis *H*
83 *de* hinc *scr. P mg. et Zm supra* in alio, propter hoc / hinc: propter hoc *Ma*
85 basi: basis *H* / superficiei: superioris *H*
86 *de* portionibus *scr. P mg. et Zm supra* (*et add. Ma* post* portionibus): in (*om. Ma*) alio (*om. Ma*), piramidis capitis. ((**Ma* adds the phrase in a box.))
87 piramidis *Zm* / *post* piramis *add. H* et basis piramidis capitis / fuerint *PZmMa* fuerit *R* similiter *H*
89 superiora: superficies *H*
91 basis *H*
96 declarare: demonstrare *H* / Et... eius *om. H*

---

83 superficies: دائرة (*circle*) / *post* scitur *add. Ar.* ايضا (*also*)
87 quod...piramis *om. Ar.*
90–91 que...superiora: تلك (*these*)
93 super... inferiorem: فوقها (*above it*)
95–96 Et.... eius *om. Ar.*

Hence it is now known from what we have recounted that, if two lines *UB*, *BA* are equal, regardless of whether the one is applied to the other as a continuous straight line or not, then the multiplication of either of them by the sum of 1/2 circumference *UTZ* and the circumference *BGD* is equal to the surface area of the body whose vertex is point *A* and whose base is the surface *UTZ*. And from this it is known that if there are several cone segments put together so that the upper surface of [each] lower segment is equal to the base surface of the segment which is [immediately] above it and if the top one of the segments is [itself] a cone and if the bases of all the segments are parallel and if the straight lines [i.e., slant heights] connecting each pair of lower and upper surfaces are equal, then the multiplication of (1) one of the slant heights by (2) [the sum of] 1/2 the circumference of the base of the lowest segment plus all the circumferences of the bases of the segments which are above the lowest segment is equal to the surface area of the body composed of all those segments, regardless of whether the surfaces are rectilinearly continuous or not.* And this is what we wished to show. And here is its form [Fig. 45].

* This simply means that the theorem holds even when corresponding line elements of the various segments are not in a continuous straight line, but are joined at some angle less than a straight line.

[XII.] CUM FUERIT CIRCULUS CUIUS DIAMETER SIT PROTRACTA, ET PROTRAHITUR EX CENTRO IPSIUS LINEA STANS SUPER DIAMETRUM ORTHOGONALITER ET PERVENIENS AD LINEAM CONTINENTEM ET SECATUR UNA 5 DUARUM MEDIETATUM CIRCULI IN DUO MEDIA, TUNC CUM DIVIDITUR UNA HARUM DUARUM QUARTARUM IN DIVISIONES EQUALES QUOTCUNQUE SINT, DEINDE PROTRAHITUR CORDA SECTIONIS CUIUS UNA EXTREMITAS EST PUNCTUM SUPER QUOD SECANT SE LINEA ERECTA 10 SUPER DIAMETRUM ET LINEA CONTINENS ET PRODUCITUR LINEA DIAMETRI IN PARTEM IN QUAM CONCURRUNT DONEC CONCURRUNT ET PROTRAHUNTUR IN CIRCULO CORDE EQUIDISTANTES LINEE DIAMETRI EX OMNIBUS PUNCTIS DIVISIONUM PER QUAS DIVISA EST QUAR- 15 TA CIRCULI, TUNC LINEA RECTA QUE EST INTER PUNCTUM SUPER QUOD EST CONCURSUS DUARUM LINEARUM PROTRACTARUM ET INTER CENTRUM CIRCULI EST EQUALIS MEDIETATI DIAMETRI ET CORDIS QUE PROTRACTE SUNT IN CIRCULO EQUIDISTANTIBUS DIAMETRO CON- 20 IUNCTIS.

Verbi gratia, sit circulus *ABG*, cuius diameter sit linea *AG* et cuius centrum sit punctum *D* [Fig. 46]. Et protrahatur ex eo linea *DB* erecta super lineam *AG* orthogonaliter et dividat arcum *ABG* c. 2 in duo media. Et / dividam quartam circuli super quam sunt *A*, *B* 25 in divisiones equales quot voluero et ponam eas divisiones *AZ*, *ZL*, *LB*. Et protraham cordam *BL* et faciam ipsam penetrare. Et elongabo iterum lineam *AG*, que est diameter, secundum rectitudinem donec

1 [XII]: 15 *mg. MaR*
1–76 Cum.... voluimus *om. S*
2 protrahatur *H*
3 supra *H*
3–4 proveniens *H*
6 dividatur *H* / harum duarum *tr. H* earum duarum *Zm*
9–10 *de* super... continens (*scr. P mg. et Zm supra* (*et add. HMa* ante* super): in (*om. HMa*) alio (*om. HMa*), sectionis medietatis diametri (diametris *Ma*) circuli (*add. Ma mg.* in alio) erecte (erecti *H*) cum linea continente. ((**Ma* puts the phrase in a box.))
12 in: a *H*
15 *de* recta *scr. P mg. Ma mg. et Zm supra* (*et add. Ma post* recta): in alio, divisa
17 protractarum *om. H*
18 corde *H*
25 quot *PZm* quod *H* quo *Ma* / eas: easdem *H*

[XII.] WHEN THERE IS A CIRCLE WHOSE DIAMETER IS DRAWN AND THERE IS DRAWN FROM ITS CENTER A LINE PERPENDICULAR TO THE DIAMETER AND TERMINATING AT THE CIRCUMFERENCE SO THAT ONE OF THE TWO HALVES OF THE CIRCLE IS BISECTED, AND THEN WHEN ONE OF THE TWO QUADRANTS IS DIVIDED INTO ANY NUMBER OF EQUAL PARTS AND THE CHORD OF THE SEGMENT, ONE OF WHOSE EXTREMITIES IS THE POINT OF INTERSECTION OF THE LINE ERECTED ON THE DIAMETER AND THE CIRCUMFERENCE, IS PRODUCED WHILE THE DIAMETER IS PRODUCED IN THE DIRECTION OF THEIR INTERSECTION UNTIL THE TWO LINES INTERSECT, AND THERE ARE DRAWN IN THE CIRCLE FROM THE POINTS AT WHICH THE QUADRANT ARC OF THE CIRCLE IS DIVIDED CHORDS PARALLEL TO THE DIAMETER, [IF ALL OF THIS IS DONE,] THEN THE STRAIGHT LINE BETWEEN THE POINT WHERE THE TWO EXTENDED LINES MEET AND THE CENTER OF THE CIRCLE IS EQUAL TO THE SUM OF THE RADIUS PLUS THE CHORDS DRAWN IN THE CIRCLE PARALLEL TO THE DIAMETER.

For example, let there be a circle *ABG* whose diameter is line *AG* and whose center is point *D* [see Fig. 46]. And from the center let line *DB* be drawn perpendicular to *AG*, thus bisecting arc *ABG*. And I shall divide the quadrant *AB* into as many equal parts as I wish, and I shall assume these parts to be *AZ*, *ZL*, *LB*. And I shall draw chord *BL* and make it continue. And I shall also extend line *AG*, the diameter,

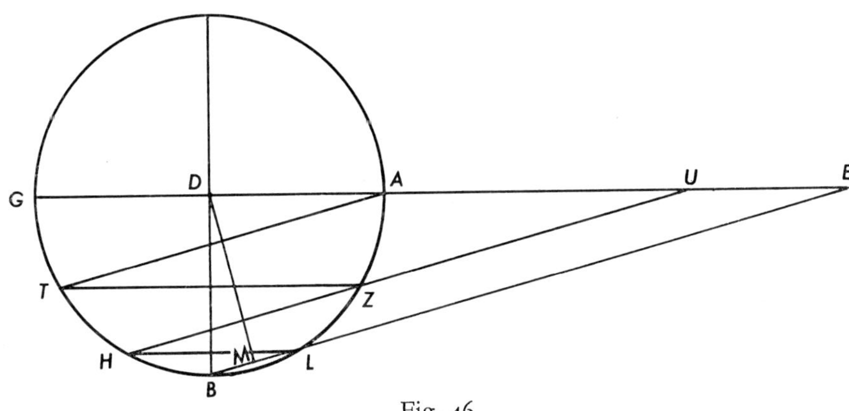

Fig. 46

1–20 Cum...coniunctis *om. Ar.*
23 lineam AG: القطر (*the diameter*)
23–24 et... media *om. Ar.*

concurrant super punctum *E*. Et protraham ex duobus punctis *Z*, *L* duas cordas *ZT*, *LH* equidistantes diametro *AG*. Dico ergo quod linea
30 *DE* est equalis medietati diametri et duabus cordis *ZT*, *LH* coniunctis, cuius hec est demonstratio.

Protraham lineam *TA* et protraham lineam *HZ* et faciam ipsam penetrare secundum rectitudinem donec occurrat linee *EG* super *U*. Et similiter faciam, si quarta circuli super quam sunt *A*, *B* fuerit divisa
35 in divisiones plures istis divisionibus. Linee ergo *TZ*, *HL* sunt equidistantes, quoniam taliter sunt protracte. Et linee *TA*, *HU*, *BE* sunt equidistantes propterea quod due divisiones *TH*, *HB* sunt equales duabus divisionibus *AZ*, *ZL*. Ergo quadratum *TAUZ* est equidistantium laterum. Ergo linea *TZ* est equalis *AU*. Et iterum quadratum
40 *HUEL* est equidistantium laterum. Ergo linea *HL* est equalis *UE*. Ergo tota linea *ED* est equalis duabus lineis *TZ*, *HL* et linee erecte que est medietas diametri coniunctis.

Si ergo nos protraxerimus in hac figura lineam ex centro et secuerit unam cordarum divisionum quarte circuli in duo media, sicut lineam
45 *DM*, tunc secatur linea *LB* super duo media super punctum *M* in duo media. Tunc iam scietur ex eo quod narravimus in hac figura quod multiplicatio medietatis corde *BL* in duas cordas equidistantes diametro et in medietatem diametri coniunctas est minor multiplicatione medietatis diametri in se et maior multiplicatione linee *DM* in se,
50 propterea quod triangulus *DMB* est similis triangulo *EDB* et est similis triangulo *EMD*. Ergo proportio linee *MB* ad *BD* est sicut proportio

28 Z, L: L Z *H*
32 *ante* TA *del. P* TH / HZ: HT *H*
33 occurrat: concurrant *H* / U: N *H hic et ubique*
34 si quarta: super quartam *H*
36 protracta *H* / linee *PZm* linea *HMa* / TA, HU: TA HN TA HN *H*
37 *de* due divisiones *scr. P mg. et Zm supra* (*ed add. Ma ante* due): in (*om. Ma*) alio (*om. RMa*) duo arcus
38 *de* duabus divisionibus *scr. P mg. et Zm supra* (*et add. Ma* ante* duabus): duobus arcubus.((**Ma* puts the phrase in a box.))
41 ED est *om. H*
44 *de* cordarum...quarte *scr. P mg. Zm.mg.* (*et add. HMa post* secuerit): in alio cordam (corda *Ma*) ex sectionibus (sactionibus! *Ma*) cordarum quarte / quarte: quarti *H* / sicut: super *H*
47, 48 in *om. H*

---

30 *post* diametri *add. Ar.* – ا ج – (*GA*)
31 cuius.... demonstratio *om. Ar.*
32 et...lineam *om. Ar.* / ipsam: – ز ح – (*HZ*)
33 secundum rectitudinem *om. Ar.*
34 quarta....B *om. Ar.*
35 istis divisionibus *om. Ar.* / TZ, HL: – ل ح – ز ط – ه ج – (*GE*, *TZ*, *HL*)
36 quoniam....protracte *om. Ar.*

rectilinearly until they [i.e., BL and AG] meet at point E. And I shall draw from the two points Z and L the two chords ZT and LH parallel to diameter AG. I say, therefore, that line DE = radius + ZT + LH.

Proof: I shall draw line TA and I shall draw line HZ, continuing the latter rectilinearly until it meets line EG at U. I shall proceed in a similar way if the quadrant AB is divided into more parts than these. Hence lines TZ and HL are parallel, since they are so drawn. And lines TA, HU, and BE are parallel, since $TH = AZ$ and $HB = ZL$. Therefore, the quadrilateral $TAUZ$ is a parallelogram. Therefore, line $TZ = AU$. And also quadrilateral $HUEL$ is a parallelogram. Therefore, line $HL = UE$. Therefore, the whole line $ED = TZ + HL$ + radius.

Hence in this figure we draw a line, e.g., line DM, from the center thus bisecting one of the chords of the quadrant, LB being the line bisected at point M. Then it will already be known, from what we have recounted concerning this figure, that the multiplication of (1) 1/2 chord BL by (2) the sum of the two chords parallel to the diameter plus the radius is less than the square of the radius and is greater than DM^2, because of the fact that the three triangles DMB, EDB, and EMD are similar. Therefore, $MB/BD = DB/BE$. Hence, $DB^2 = MB \cdot BE$, DB being the radius.

39–40 Et...laterum *om. Ar.*
41 tota linea *om. Ar.* / duabus lineis *om. Ar*
41–22 et....diametri: – ا د – (*AD*)
42 *post* coniunctis *add. Ar.* وذالك ما اردناه (*And that is what we wished.*)
43–47 in...quod: – د م – عمودا على وتر – ب ل – (*DM perpendicular to chord BL.*)
47 corde *om. Ar.*
47–48 in...coniunctas: – د ه – (*DE*)
50–65 et....se: لكون زاويتى – د م ب – ه ب د – قائمتين وزاوية – ب – مشتركة ونسبة – ب م – الى – م د – كنسبة – ب د – الى – د ه – فب م – د ه – مساو – ب د – فى – د م – و – ب د – فى – د م – اصغر من مربع – ب د – واعظم من مربع – م د – فاذا نصف – ب ل – فى نصف القطر وفى وترى – ط ز – ح ل – جميعا اصغر من مربع نصف القطر واعظم من مربع – د م – (*For the two angles DMB, EBD are right angles and angle B is common. And BM/MD = BD/DE, and so BM · DE = BD · DM, or [½ BL · DE = BD · DM], BD · DM < BD², and BD · DM > MD². And so [½ BL · radius · (TZ + HL)] < radius² and > DM².*)

DB ad *BE*. Et propter illud erit multiplicatio linee *DB*, que est medietas diametri, in se equalis multiplicationi linee *MB* in lineam *BE*. Verum linea *BE* est longior duabus cordis *ZT*, *LH* et medietate dia-
55 metri coniunctis, propterea quod iste coniuncte sunt *DE*, et linea *BE* est longior *DE*. Ergo multiplicatio linee *MB* in duas cordas *ZT*, *LH* et in medietatem diametri coniunctas est minor multiplicatione medietatis diametri in se. Et quoniam triangulus *DMB* est similis triangulo *EMD*, erit proportio *BM* ad *MD* sicut proportio *MD* ad *ME*.
60 Et similiter erit multiplicatio linee *BM* in lineam *ME* equalis multi-
60v c. 1 plicationi linee *MD* in se. Sed linea *ME* est minor duabus cordis *ZT*, *LH* et medietate / diametri coniunctis, propterea quod iste omnes sunt equales linee *DE*, et linea *DE* est longior *EM*. Ergo multiplicatio *MB* in duas cordas *ZT*, *LH* et in medietatem diametri coniunctas est
65 maior multiplicatione *DM* in se.

Iam ergo ostensum est quod in omni circulo in quo protrahitur ipsius diametrus deinde dividitur una duarum medietatum ipsius in duo media, postea dividitur una duarum quartarum in divisiones equales quotcunque fuerint et protrahuntur ex punctis divisionum omnium
70 corde in circulo equidistantes diametro, tunc multiplicatio medietatis corde unius sectionum quarte circuli in medietatem diametri et in omnes cordas que protracte sunt in circulo equidistantes diametro coniunctim est minor multiplicatione medietatis diametri in se et maior multiplicatione linee que egreditur ex centro et pervenit ad unam corda-
75 rum divisionum quarte circuli et dividit eam in duo media in se. Et illud est quod declarare voluimus.

52 *de* Et *scr. P mg. et Zm supra* in alio, similiter / Et *PZm* et similiter *H* similiter et *Ma* ((but *similiter* is in a box in Ma))
54 Verum *PZm om. H* Verum tamen *Ma*
54–55 LH...sunt *om. H*
56 LH: HL *H*
57–58 medietatis *Zm HR, mg. P mg. Ma*
59 BM: MB *Zm*
61 est minor *ZmMa bis P* est maior minori *H*
67, 68 dividatur *H*
70 corde: cordarum *H* / in: in cum (?) *H*
75 divisionem *H* / dividat *H*
76 declarare: demonstrare *H*

---

69 omnium *om. Ar.*
71 quarte circuli *om. Ar.*
72–73 que...coniunctim *om. Ar.*
75 quarte...media *om. Ar.*
76 quod...voluimus: المطلوب (*that which is sought*)

Now line $BE > (ZT + LH + BD)$, since $(ZT + LH + BD) = DE$ and $BE > DE$. Hence, line $MB \cdot (ZT + LH + BD) < BD^2$. And since $\triangle DMB$ is similar to $\triangle EMD$, $BM / MD = MD / ME$. And similarly $BM \cdot ME = MD^2$. But line $ME < (ZT + LH + BD)$, since

$$(ZT + LH + BD) = DE \text{ and } DE > EM.$$

Therefore,

$$MB \cdot (ZT + LH + BD) > DM^2.$$

Therefore it has now been demonstrated that in every circle where the diameter is drawn and one of the two halves of the circle is bisected and one of the two quadrants [thus formed] is then divided into any number of equal parts and from the [dividing] points of the parts are drawn chords in the circle parallel to the diameter, then the multiplication of one half of the chord of one of the segments of the quadrant by the sum of the radius plus all the chords drawn in the circle parallel to the diameter is less than the square of the radius and greater than the square of the line going out from the center which meets and bisects the chord of one of the parts of the quadrant. And this is what we wished to show.

[XIII.] CUM CECIDERIT IN MEDIETATE SPERE CORPUS QUOD CONTINEAT MEDIETAS SPERE ET FUERIT CORPUS COMPOSITUM EX PORTIONIBUS PIRAMIDUM COLUMPNARUM QUOTCUNQUE FUERINT, ET FUERIT SUPERFICIES SUPERIOR CUIUSQUE PORTIONIS EXISTENS BASIS PORTIONIS QUE EST SUPER EAM, ET FUERINT SUPERFICIES BASIUM PORTIONUM OMNIUM EQUIDISTANTES, ET FUERIT BASIS PORTIONIS INFERIORIS IPSA BASIS MEDIETATIS SPERE, ET FUERIT PORTIO SUPERIOR PIRAMIDIS PIRAMIS CAPITIS, ET PUNCTUM CAPITIS EIUS EST POLUS MEDIETATIS SPERE, ET FUERINT LINEE RECTE QUE EGREDIUNTUR EX OMNIBUS BASIBUS PORTIONUM AD ILLUD QUOD EST ALTIUS IN EIS SECUNDUM RECTITUDINEM EQUALES, ET CUM CECIDERIT IN CORPORE MEDIETAS SPERE QUAM CONTINEAT CORPUS, ET FUERIT SUPERFICIES BASIS HUIUS MEDIETATIS SPERE POSITA IN SUPERFICIE BASIS CORPORIS: TUNC EMBADUM SUPERFICIEI HUIUS CORPORIS ERIT MINUS DUPLO EMBADI SUPERFICIEI BASIS MEDIETATIS SPERE QUE CONTINET CORPUS ET MAIUS DUPLO EMBADI SUPERFICIEI BASIS MEDIETATIS SPERE QUAM CONTINET CORPUS.

Verbi gratia, sit medietas spere *ABGD* [Fig. 47]. Et circulus *ABG* sit circulus magnus, et eius superficies sit basis medietatis spere *ABGD*.

1 [XIII]: 16 *mg. Ma mg. R*
1–153 Cum....eius *om. S*
5 cuiusque: cuiuscunque *H*
7 equidistantes *PZm* extremitates *H*
13 altius *PZm* alteri *H* alterum *Ma*
18 erit: eius *H*
19 medietatis *HR om. PZmMa*
19–21 que...spere *om. H*
20–21 et...corpus *om. R*
22 ABGD: ABG *H*

6–7 et...equidistantes *tr. Ar. post* spere *in linea 11*
14 et cum: ثم (*then*)
17 corporis: النصف الاول (*of the first half*)
18–19, 20 embadi superficiei *om. Ar.*
19–20 que...corpus: الاولى (*of the first*)
21 quam...corpus: الثانية (*of the second*)
22 medietas *om. Ar.*
23 et...ABGD *om. Ar.*

[XIII.] WHEN THERE IS A BODY WHICH FALLS WITHIN A HEMISPHERE—AND WHICH [CONSEQUENTLY] THE HEMISPHERE CONTAINS—AND THE BODY IS COMPOSED OF ANY NUMBER OF SEGMENTS OF CONES* SUCH THAT THE UPPER [PLANE] SURFACE OF ANY SEGMENT IS THE BASE OF THE SEGMENT [IMMEDIATELY] ABOVE IT AND THE BASE SURFACES OF ALL THE SEGMENTS ARE PARALLEL, AND SUCH THAT THE BASE OF THE BOTTOM SEGMENT IS THE BASE OF THE HEMISPHERE, WHILE THE TOP SEGMENT IS ITSELF A CONE WITH ITS VERTEX A POLE OF THE HEMISPHERE, AND SUCH THAT THE SLANT HEIGHTS OF THE SEGMENTS ARE EQUAL, AND WHEN THERE IS INSCRIBED WITHIN THE BODY A HEMISPHERE WHICH THE BODY CONTAINS AND WHOSE BASE IS PLACED WITHIN THE SURFACE OF THE BASE OF THE BODY—[WHEN ALL OF THIS IS TRUE,] THEN THE SURFACE AREA OF THE BODY IS LESS THAN DOUBLE THE AREA OF THE BASE OF THE HEMISPHERE CONTAINING THE BODY AND MORE THAN DOUBLE THE AREA OF THE BASE OF THE HEMISPHERE WHICH THE BODY CONTAINS.

For example, let there be a hemisphere *ABGD* [see Fig. 47], and *ABG* a great circle of it, whose surface is the base of the hemisphere *ABGD*.

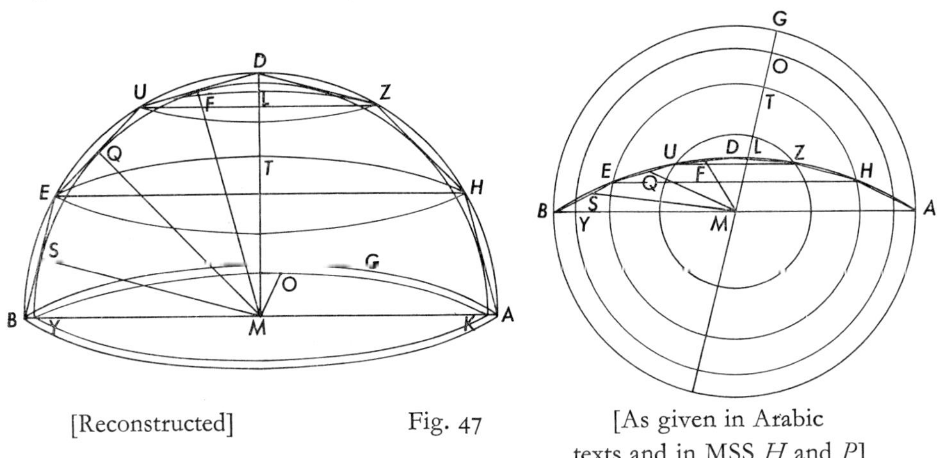

[Reconstructed] Fig. 47 [As given in Arabic texts and in MSS *H* and *P*]

Note: *AMB* is drawn off center in MS *P*, and lines *MF*, *MQ*, and *MS* are incorrectly drawn in MS *H*.

* As in Proposition XI, I have used here the expression "segment of a cone" to translate "portio piramidis columpne." Note that the authors use this to stand both for a frustum of a cone and for a small cone which stands on the uppermost frustum.

Et polus huius circuli magni sit punctum *D*. Et signabo in medietate
25 spere in primis corpus compositum ex portionibus quot voluero piramidum columpnarum secundum modum quem narravimus. Et ponam corpus in hac descriptione compositum ex tribus portionibus, que sint portiones *ABGH*, *EHTZ*, *ULZD*. Et basis corporis et basis medietatis spere *ABGD* est una. Et est superficies circuli *ABG*. Et caput
c. 2 / corporis est punctum *D*. Et est polus medietatis spere *ABGD*. Dico
31 ergo quod embadum superficiei corporis *ABGD* compositi ex embadis superficierum trium portionum piramidum, quarum una est superficies que elevatur ex circulo *ABG* secundum rectitudinem ad circulum *HTE* et superficies alia est illa que elevatur ex circulo *HTE*
35 secundum rectitudinem ad circulum *ULZ* et superficies tertia est que elevatur ex circulo *ULZ* secundum rectitudinem ad punctum *D*, est minus duplo embadi superficiei circuli *ABG*, cuius hec est demonstratio.

Protraham in spera *ABGD* medietatem circuli qui est ex circulis
40 magnis qui cadunt in spera transeuntis super polum qui est punctum *D*, qui sit arcus *ADB*. Et protraham lineam *AB*, que sit diameter spere, et dividam eam in duo media super punctum *M*. Et notum est quod punctum *M* est centrum spere. Et protraham duas lineas *HE*, *UZ*. Et notum est etiam quod utreque sunt equidistantes, et equidis-
45 tantes linee *AB*, propterea quod linee *AB*, *HE*, *ZU* sunt differentie communes super quas superficies circuli *ADB* secat superficies tres equidistantes, scilicet superficies circulorum *ABG*, *ETH*, *ULZ*. Et manifestum est quod linee *AB*, *HE*, *UZ* sunt corde circulorum *ABG*, *ETH*, *ULZ*, qui sunt bases portionum ex quibus componitur corpus
50 *ABGD*, propterea quod polum horum circulorum omnium est punctum *D* super quod transit medietas circuli *ADB*. Et protraham in

24 *de* circuli *scr. P mg. et Zm supra* in alio, spere / *ante* circuli *add. Ma* spere
26 quam *H*
27–28 que...portiones *ZmMa mg. P om. H*
28 ABGH...ULZD *scr. et del H* / EHTZ *corr. ex* EHLZ *in MSS*
30 *post* corporis *add. P et del.* eius
31 embadum *PZm* embado *H* embade *Ma*
34 *de* HTE[1] *scr. P mg.* in alio / HTE[1, 2]: ELH *Zm et Zm scr. supra* in alio, HTE / HTE[1]: pro alio habere (?) *H*
34–35 HTE[2]...ULZ *om. H*
34 HTE[2] *de scr. P mg.* in alio
35, 36 *de* ULZ *scr. P mg.* in alio / ULZ: UTZ *Zm et scr. Zm supra* in alio ULZ
35 U-: N- *H hic et ubique*
39 qui: quod *H*
42 notum: necessarium *H*
43 quod *bis H*
44 et equidistantes *om. H*
46 ADB: ABD *Zm*
48 corde: diametri *Zm*
49 corpus: totum corpus *H*

And the pole [of the axis] of this great circle is point *D*. And I shall first describe in the hemisphere a body composed of as many segments as I wish in the way that we have recounted. I shall posit as the body so described the body composed of three segments: the segments *ABGH*, *EHTZ*, and *ULZD*. And the base of this body is also the base of hemisphere *ABGD*, this base being the surface of circle *ABG*. And the vertex of the body is point *D*, which is also the pole of the hemisphere *ABGD*. I say, therefore, that the surface area of body *ABGD* which is composed of three segments of cones—i.e., (1) the surface extended rectilinearly [i.e., the lateral surface] between circle *ABG* and circle *HTE*, and (2) the surface extended rectilinearly between circle *HTE* and circle *ULZ*, and (3) the surface extended rectilinearly between circle *ULZ* and point *D*—is less than double the surface area of circle *ABG*.

Proof: I shall draw in [hemi]sphere *ABGD* a semicircle of one of the great circles of the sphere [and this is the semicircle] passing through polar point *D*. This semicircle is the arc *ADB*. And I shall draw line *AB* as a diameter of the sphere and I shall bisect it at point *M*. And it is known that *M* is the center of the sphere. And I shall draw the two lines *HE* and *UZ*. It is known also that they are both parallel [to each other] and parallel to line *AB*, since lines *AB*, *HE*, and *ZU* are the common sections of circle *ADB* and the three parallel surfaces of circles *ABG*, *ETH*, and *ULZ*. And it is evident that lines *AB*, *HE*, *UZ* are chords of circles *ABG*, *ETH*, and *ULZ* which are [themselves] the bases of the segments of which body *ABGD* is composed, since the pole [of the axis] of all these circles is point *D*, through which the semicircle *ADB* passes. And I shall draw

24–36 Et[2]....D: وليكون فيه مجسم على ما وصفنا مركب من ثلث قطع اولاها يرتفع من دائرة – ا ب ج – الى دائرة – ه ط ح – والثانية ترتفع منها الى دائرة – و ل ز – والثالثة ترتفع منها الى نقطة – د – نقول فالسطوح المستديرة المحيط بهذا الجسم جميعا

(And according to what we have described let there be [inscribed] in it a body compounded of three segments. The first extends from circle ABG to circle ETH and the second extends from it (ETH) to circle ULZ, and the third extends from it (ULZ) to point D. And so we say that the sum of the areas of the conic segments comprising the body)

37–38 cuius...demonstratio *om. Ar.*

39 *ante* spera *add. Ar.* نصف *(half)*

41 arcus *om. Ar.*

42 punctum *om. Ar.*

42–43 Et...spere *om. Ar.*

44 notum...quod *om. Ar.*

45 linee...ZU *om. Ar.*

47 equidistantes....superficies *om. Ar.*

47–48 ABG....corde: وهما قطر

(And they are diameters)

48 ABG *om. Ar.*

49–51 qui....ADB *om. Ar.*

51–52 in...portionibus *om. Ar.*

omnibus portionibus ex basibus earum ad ipsarum altiora lineas rectas, que sint linee *BE*, *EU*, *UD*. Et notum est quod ipse sunt equales, propterea quod ita posite sunt. Ergo medietatis circuli *ADB* iam
55 protracta est diametrus et est *AB*, et divisa est medietas circuli in duo media super *D*. Et divisus est arcus *DB* in divisiones equales, que sunt arcus *BE*, *EU*, *UD*, et protracte sunt ex duobus punctis *E*, *U* due corde equidistantes diametro, que sunt *UZ*, *EH*. Ergo multiplicatio medietatis unius cordarum *BE*, *EU*, *UD*, quecunque fuerit, in
60 duas lineas *UZ*, *HE* et in medietatem linee *AB* coniunctim est minor multiplicatione medietatis linee *AB* in se, propter illud cuius premisimus demonstrationem. Et iterum corpus *ABGD* est compositum ex portionibus piramidum columpnarum, et bases portionum omnium sunt equidistantes, et portio superior habet caput quod est piramis,
65 et linee recte que protrahuntur in omnibus portionibus ex basibus earum ad earum superiora secundum rectitudinem sunt equales. Ergo propter illud erit multiplicatio linee unius earum que protrahuntur ex basibus portionum ad superiora earum secundum rectitudinem in medietatem linee continentis basim portionis inferioris et in omnes
70 lineas continentes bases portionum que sunt super portionem inferiorem est embadum superficiei corporis, sicut o/stendimus in illis (61r c. 1) que sunt premissa. Ergo multiplicatio linee *BE* in duos circulos *ULZ*, *ETH* et in medietatem circuli *ABG* coniunctim est embadum superficiei corporis *ABGD*. Verum multiplicatio linee *BE* in duos circulos

53 sint: sunt *Zm*
60 et: etiam *H*
62 est *PZm* eorum *H* ex *Ma*
64 equidistantes: extremitates *H*
67 erit: est *H*
68 in *om.* *H*
69 basim *PZm mg.* *H* basis *Ma*
71–72 illis...premissa: premissis *H*
74 *add. Zm mg.* + *quod refert ad sequentes lineas in fol. 84r*:
Set multiplicatio linee EB in multiplicationen linee ZU in proportionem circumferentie circuli ULZ ad lineam ZU et in multiplicationem linee EH in proportionem circumferentie circuli ETH ad lineam EH et in multiplicationem medietatis diametri AB in proportionem medietatis circumferentie circuli ABG ad lineam MB que est medietas diametri est superficies corporis ABGD. Set multiplicatio EB in multiplicationem linee ZU in proportionem circumferentie circuli UZL ad lineam ZU est sicut multiplicatio EB in lineam ZU et eius quod provenit in proportionem circumferentie circuli UZL ad diametrum UZ. Ergo multiplicatio linee EB in lineas ZU, EH, MB et eius quod provenit in proportionem circumferentie circuli UZL ad lineam ZU est superficies corporis ABGD. Ergo et multiplicio medietatis linee EB in predicta omnia est medietas superficiei corporis ABGD. Et multiplicatio linee BS in lineas UZ, EH, MB est minus multiplicatione MB in se ipsam. Ergo multiplicatio linee BS in lineas ZU, EH, MB et eius quod provenit in proportionem circumferentie circuli UZL ad lineam ZU est minus

in all of the [surfaces of the] segments from their bases to those [bases immediately] above them straight lines, [i.e., slant heights] and these are the lines *BE*, *EU*, *UD*. And it is known that these lines are equal, since they were posited to be so. Hence a diameter of semicircle *ADB* has already been drawn and it is *AB*. And the semicircle [i.e., arc *ADB*] has been bisected at *D*. And arc *DB* has been divided into equal parts, namely, the arcs *BE*, *EU*, and *UD*. And from the two points *E* and *U* the two chords *UZ* and *EH* have been drawn parallel to the diameter. Therefore, the multiplication of one half of any one of the chords *BE*, *EU*, and *UD* by the sum $(UZ + HE + 1/2\ AB)$ is less than $(1/2\ AB)^2$, as we have demonstrated before [in Proposition XII]. Furthermore, body *ABGD* is composed of segments of cones in such a way that the bases of all the segments are parallel, the upper segment is a cone, and the straight lines drawn in all [the surfaces of] the segments from their bases to their upper [plane surfaces] rectilinearly [i.e., the straight lines constituting the slant heights] are equal. Therefore, as we demonstrated before [in Proposition XI], the multiplication of (1) one of these lines [i.e., slant heights] drawn from the bases of the segments to the upper [plane surfaces] rectilinearly by (2) the sum of one half the circumference of the base of the lowest segment plus all the circumferences of the bases of the segments above the lowest one is [equal to] the surface area of the body. Therefore,

$$BE \cdot (\text{circum } ULZ + \text{circum } ETH + 1/2 \text{ circum } ABG) = \text{surf area body } ABGD.$$

53 que...linee *om. Ar.* / notum...quod *om. Ar.*
54–58 Ergo...EH *om. Ar.*
59 cordarum...UD: منها (*of them*)
60 duas lineas *om. Ar.*
61–96 propter....spere: وايضا سطح واحد منها في نصف محيط دائرة – ا ب ج – وفي محيطي دائرة – ح ه ط – ز و ل – جميعا مثل السطح المحيط بالمجسّم لما مرّ وسطح واحد منها في نصف

– ا ب – و ي – ه ح – و ز – جميعا . ثم الحاصل فيما اذا ضرب فيه القطر حصل المحيط مساويا اسطح واحد منها في نصف محيط دائرة – ا ب ج – وفي محيطي دائرتي – ح ه ط – ز و ل – جميعا اعني السطح المحيط بالمجسّم وهو اقل من ضعف الحاصل من ضرب مربع نصف – ا ب – في ما اذا ضرب فيه القطر حصل المحيط ومربع نصف

75 *ULZ*, *ETH* et in medietatem circuli *ABG* est equalis ei quod fit ex multiplicatione linee *BE* in duas lineas *UZ*, *EH* et in medietatem linee *AB* coniunctim et multiplicationi eius quod agregatur inde in quantitatem in quam cum multiplicatur diameter est illud quod agregatur inde ipsa linea circumdans, propterea quod linee *UZ*, *EH*, *AB* sunt diametri 80 circulorum *ULZ*, *ETH*, *ABG*. Ergo multiplicatio linee *BE* in duas lineas *UZ*, *EH* et in medietatem linee *AB* coniunctim et multiplicatio eius quod agregatur in quantitatem in quam cum multiplicatur diameter est illud quod agregatur ipsa linea circumdans est embadum superficiei corporis *ABGD*. Sed multiplicatio medietatis linee *BE* in duas lineas 85 *UZ*, *EH* et in medietatem linee *AB* coniunctim et multiplicatio eius quod agregatur in quantitatem in quam cum multiplicatur diameter est illud quod agregatur ipsa linea circumdans est equalis medietati superficiei corporis *ABGD*. Et ipsa est minor multiplicatione medietatis linee *AB* in se et multiplicatione eius quod agregatur in quantitatem in quam 90 cum multiplicatur diameter est illud quod agregatur ipsa linea circumdans. Sed multiplicatio medietatis linee *AB* in se et multiplicatio eius quod agregatur in quantitatem in quam cum multiplicatur diameter est illud quod agregatur ipsa linea circumdans est embadum superficiei circuli *ABG*, propterea quod linea *AB* est eius diameter. Ergo super-95 ficies circuli *ABG* qui est basis corporis et medietatis spere que continet

multiplicatione linee MB in se et eius quod provenit in proportionem circumferentie circuli UZL ad lineam ZU. Set multiplicatio quadrati linee MB in proportionem circumferentie circuli UZL ad lineam ZU que est eadem proportioni circumferentie circuli ABG vel medietatis eius ad diametrum AB vel medietatem eius est superficies circuli ABG quia si posuero superficiem que provenit ex multiplicatione medietatis diametri AB in medietatem circumferentie circuli ABG que est equalis superficiei circuli ABG et quadravero diametri medietatem erit proportio illius superficiei equalis circulo ABG ad quadratum medietatis diametri AB sicut proportio medietatis circumferentie ad medietatem diametri. Cum ergo diviserimus superficiem illam equalem circulo per quadratum medietatis diametri proveniet proportio circuli ad quadratum que est proportio medietatis circumferentie ad medietatem diametri. Si ergo multiplicaverimus quadratum illud in illam proportionem proveniet circuli supreficies. Set multiplicatio quadrati in proportionem est maior multiplicatione medietatis linee EB in lineas ZU, HE, MB et eius quod provenit in proportionem quod est equale medietati superficiei corporis ABGD. Ergo duplum circuli est maius superficie corporis.

75 ei: etiam (?) *H*
77 inde *om.* *H*
78 inde *PZm* in *HMa*
81 et[1] *om.* *H*
84 BE: EB *H*
85 medietatem *ZmHMa* medietate *P*
90–91 *post* circumdans *add.* *H*, *Mamg.* *lineas 97–118* ("Quod.... forma")
92 cum *om.* *H*

Now

$$BE \cdot (\text{circum } ULZ + \text{circum } ETH + 1/2 \text{ circum } ABG) = BE \cdot (UZ + EH + 1/2\ AB) \cdot \pi^*,$$

since UZ, EH, and AB are the diameters of circles ULZ, ETH, and ABG. Therefore, $BE \cdot (UZ + EH + 1/2\ AB) \cdot \pi =$ surf area body $ABGD$. But $1/2\ BE \cdot (UZ + EH + 1/2\ AB) \cdot \pi = 1/2$ area body $ABGD$, and so $1/2\ BE \cdot (UZ + EH + 1/2\ AB) \cdot \pi < (AB/2)^2 \cdot \pi$. But $(AB/2)^2 \cdot \pi =$ area circle ABG, since line AB is its diameter. Therefore, circle ABG,

* I have used here the modern symbol π to stand for the phrase "the quantity which when multiplied by the diameter produces the circumference."

– ا ب – فيما اذا ضرب فيه القطر مساو لسطح الدائرة لأن ضرب نصف – ا ب – فيما اذا ضرب فيه القطر مثل المحيط سو نصف المحيط وضربه مرّة اخرى فى نصف – ا ب – هو سطح الدائرة فالسطح المحيط بالمجسّم اقل من ضعف * سطح دائرة – ا ب ج –

((* ضعف in Paris MS, نصف in printed text.)) (*And also the product of one of them and* $\frac{1}{2}$ *the circumference of circle* ABG *and the circumferences of circles* HET, ZUL *together is equal to the area of the surface of the body, according to what has passed [in Proposition* IX*]; and the product of one of them and* $\frac{1}{2}$ AB *and* $(EH + UZ)$, *and that whole product multiplied by* π, *is equal to the product of one of them and* $\frac{1}{2}$ *the circumference of circle* ABG *and the sum of the circumferences of circles* HET *and* ZUL, *i.e., to the area of the surface of the body. And this is less than* $2\pi\,(AB/2)^2$ and $\pi(AB/2)^2$ *is equal to the area of the circle, because* $(AB/2) \cdot \pi$ *is equal to* $\frac{1}{2}$ *the circumference of the circle and* $(AB/2) \cdot \pi \cdot (AB/2)$. *One half* AB *is the area of the circle. And so the area of the surface of the body is less than double the area of circle* ABG. ((As in the translation of Gerard's text, I have throughout rendered by the symbol π the phrase "that quantity which when multiplied by the diameter produces the circumference.))

corpus est plus medietate embadi corporis cadentis in medietate spere.

[Quod multiplicatio medietatis diametri in se et eius quod provenit in quantitatem in quam cum multiplicatur diameter provenit linea circumdans circuli sit equale superficiei circuli, ita ostenditur. Quo-
100 niam ponam *ET* equalem medietati circumferentie et *EZ* equalem medietati diametri (Fig. 48), et unam multiplicabo in alteram, erit ergo superficies *ZT* equalis superficiei circuli, et super *ZE* constituam quadratum, quod sit *ZL*, et ponam quod quantitas in quam cum multiplicatur diameter provenit circumferentia sit quantitas *RU*. Et quod
105 diameter cum multiplicatur in *RU* proveniet circumferentia, ergo circumferentia cum dividitur per diametrum provenit *RU*. Ergo *RU* est proportio circumferentie ad diametrum. Sed proportio totius ad totum est sicut medietatis ad medietatem et *ET* equatur medietati circumferentie et *EL* medietati diametri. Ergo proportio *TE* ad *EL*
110 est *RU*. Sed proportio *TE* ad *EL* est sicut proportio *TZ* ad *ZL*. Ergo proportio *TZ* ad *ZL* est *RU*. Ergo *TZ* cum dividetur per *RU* proveniet *ZL*. Ergo et *ZL* cum multiplicabitur in *RU* proveniet *ZT*. Sed *ZL* est quadratum medietatis diametri et *RU* est quantitas in quam cum multiplicatur diameter provenit circumferentia. Et *ZT* equatur superficei
115 circuli. Ergo multiplicatio medietatis diametri in se et eius quod provenit in quantitatem in quam cum multiplicatur diameter provenit circumferentia equatur superficiei circuli. Et hoc est quod voluimus, cuius hec est forma.]

96 medietate: medietati *H*
97–118 [Quod....forma] *mg. P*; *om. Zm?*; *cf. var. lin. 90–91*
97 Quod: sed quod *H*
98–99 linea...circuli[2] *corr. ex* linea circumdans est superficies circuli ABG in *mg. H et* superficies circuli *in mg. P et* vel circumferentia circuli sit equale superficiei circuli *supra mg. P et* vel superficies circuli *in Ma*
102 ZT: ET *H*
103 quod[1] *HMa om. P* / ZL: Z *H*
104, 106 proveniet *H*
107 ad[1] *PMa mg. H*
108 ET equatur: erit equatum *H*
109 medietati: medietate *H* / EL[2]: DL *H*
110 RU: NZ *H* / ad[1]: et *H*
111 RU[1]: NR *H*
112 Ergo *om. H*
113–114 et....diameter *om. H*
114 ZT: ZD *H* / equatum *H*
115–16 proveniet *H*
117 equatur: equalis *H* / *post* quod *add. H* demonstrare
118 cuius...forma *om. H* / *post* forma *add. Ma* vel circumferentia circuli sit equale superficiei circuli (*cf. var. lin. 98–99*).

97–118 [Quod....forma] *om. Ar.*

the base of the body and the hemisphere which contains the body, is equal to more than one half the area of the body falling within the hemisphere.

*[That the multiplication of the square of the radius by π is equal to the area of the circle is demonstrated as follows (see Fig. 48). Since I assume ET to be equal to one half the circumference and EZ equal to

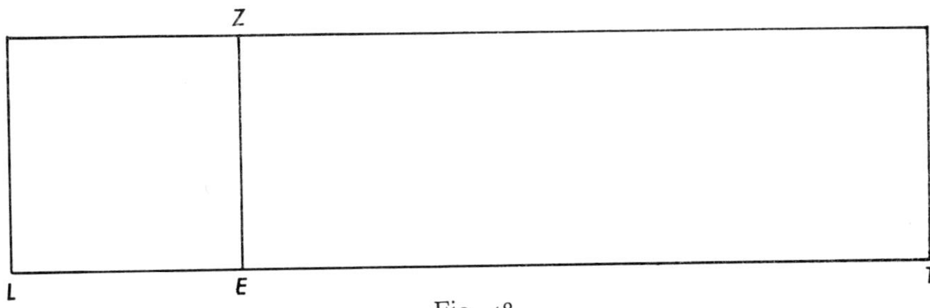

Fig. 48

Note: I have not reproduced line RU (which equals π).

the radius and I shall multiply one into the other, therefore surface ZT will be equal to the surface of the circle. And I shall construct a square on ZE, which square is ZL. I shall posit RU as the quantity which when multiplied by the diameter produces the circumference (i.e., as π). And since the multiplication of the diameter by RU will produce the circumference, therefore, when the circumference is divided by the diameter, RU is produced. Hence, RU = curcumference/diameter. But the ratio of the whole to the whole is as that of the half to the half, and ET is equal to half the circumference while EL is half the diameter. Hence, $TE/EL = RU$. But (TE/EL) = (area TZ/area ZL). Therefore, (area TZ/area TL) $= RU$. Hence, (area TZ/RU) = area ZL. Hence, (area $ZL \cdot RU$) = area ZT. But ZL is the square of the radius, $RU = \pi$, and ZT is the area of the circle. Hence, the multiplication of the square of the radius by π is equal to the area of the circle. And this is what we wished. This is its form (Fig. 48).]

* For the doubtful authenticity of lines 97–118, included here in brackets, see the Introduction, division 2, footnote 8, of this chapter.

Amplius describam in corpore *ABGD* medietatem spere quam con-
120 tinet corpus et sit superficies basis medietatis spere in superficie basis corporis, que est superficies circuli *ABG*, et est superficies circuli *OKY*. Et dividam lineas *BE*, *EU*, *UD* in duo media super puncta *S*, *Q*, *F* et protraham lineas *MS*, *MQ*, *MF*. Et notum est quod ipse sunt equales, propterea quod punctum *M* est centrum circuli *ABG* et corde
125 *BE*, *EU*, *UD* sunt equales. Et faciam in superficie huius circuli lineam *MO* non in superficie circuli *ADB*. Ergo puncta *S*, *Q*, *F*, *O* quattuor non sunt in superficie una. Et ad ea quidem omnia protracte sunt linee ex puncto *M*, que sunt linee *MS*, *MQ*, *MF*, *MO*, et sunt linee equales. Ergo punctum *M* est centrum spere quam continet corpus *ABGD*
130 et linea *MS* est medietas diametri eius. Et circulus *KOY* est basis medietatis spere. Ergo multiplicatio linee *MS* in se, deinde eius quod agregatur in quantitatem in quam cum multiplicatur diametrus est linea circumdans, est embadum circuli *KOY*. Sed multiplicatio medietatis linee *BE* in duas lineas *UZ*, *EH* et in medietatem linee *AB*
135 coniunctim est maior multiplicatione linee *MS* in se, propter illud

119 *de* describam *scr. P mg.* vel signabo *et Zm mg.* in alio, signabo / describam *P* signabo describam *HMa* / ABGD: ABG (?) *H*
119–20 continet *HMa* contineat *P Zm*
120 in superficie *PZm* in superficies *Ma* inferioris *H*
123 MF, MQ, MS *H*
125 EU: EM *H* / Et: quod *H*
126 non: que non sit *Zm* / punctum *H*
129 centrum: centrum circuli *Zm*
130 diametri: digitus *H*
131–35 deinde.... coniunctim *ZmHMa mg. P*
133 est *om. Zm.*
133–49 *de* Sed.... ABGD *scr. Zm in infer. mg. fol. 84r*: Set multiplicatio medietatis linee EB in lineas ZU, EH, MB coniunctas est maior multiplicatione linee MS in se, que est equalis MY que est medietas diametri circuli, scilicet KOY. Ponam proportionem medietatis circumferentie circuli KOY ad medietatem diametri ipsius KMY que est eadem que totius circumferentie ad totam diametrum. Ergo multiplicatio medietatis EB in lineas ZU, HE, MB coniunctas et eius quod provenit in proportionem circumferentie ad diametrum vel medietatis circumferentie ad medietatem diametri et hoc est medietas superficiei corporis ABGD quia proportio circumferentie ad diametrum suum est una est maius multiplicatione linee MY in se et eius quod provenit in proportionem circumferentie vel medietatis eius ad diametrum vel medietatem eius. Set multiplicatio linee MY in se et eius quod provenit in proportionem est superficies circuli KOY, ut probatum est. Ergo superficies circuli KOY est minus medietate superficiei corporis ABGD. Ergo duplum circuli est minus superficie totius corporis et hoc est quod demonstrare voluimus.
134 UZ: QZ *H*
135 maior: minor *Zm* / linee... se: medietas linee BE in duas lineas UZ, EH et in medietatem linee AB coniunctim *Zm*
135 propter illud *scr. et del.* (?) *H*

Now, further, I shall describe in body *ABGD* [Fig. 47] a hemisphere which the body contains and let the base of the hemisphere be inside the surface of the base of the body, i.e., inside the surface of circle *ABG*, and it (the base of the hemisphere) is the surface of circle *OKY*. And I shall bisect lines *BE*, *EU*, and *UD* at points *S*, *Q*, and *F*, and I shall draw lines *MS*, *MQ*, and *MF*. And it is known that these lines are equal, since point *M* is the center of circle *ABG* and the chords *BE*, *EU*, and *UD* are equal. And I shall produce in circle *OKY* line *OM*, which will not be in the surface of circle *ADB*. Therefore, the four points *S*, *Q*, *F*, and *O* are not in a single [plane] surface. And to these points the equal lines *MS*, *MQ*, *MF*, *MO* have been drawn from point *M*. Therefore, point *M* is the center of the sphere which body *ABGD* contains and line *MS* is its radius. And the circle *KOY* is the base of the hemisphere. Therefore, $MS^2 \cdot \pi =$ area circle *KOY*. But $BE \cdot (UZ + EH + 1/2\ AB) > MS^2$, as we have

121–22 et...OKY: يكون اصغر منها
(*will be smaller than it*)
123 Q: ع ((Note: here and everywhere))
124–25 propterea....equales: لانهـا اعمدة من المركز على اوتار متساوية و نرسم على مركز – م – وبعد – م س – فى سطح دائرة – ا ب ح – دائرة – ك ص ى –
(*For they are perpendiculars* [*drawn*] *from the center to equal chords and we describe circle KOY on center M with radius MS and within circle ABG.*) ((Note: Gerard represents ص by O, no doubt because he had already used S for س.))
126–53 Ergo....eius: ولأن خطوط – م س – م ع – م ف – م ص – الاربعة المتساوية التى ليست فى سطح واحد خرجت من نقطه – م – الى محيط الكرة الدخلة يكون – م – مركزا لها و – م س – نصف قطر لها ودائرة – ك س ى – قاعدة لها ومربع – م س – اضعر من سطح نصف – ب ه – فى نصف – ا ب – وفى – ه ح – و ز – جميعا فمربع – م س – فى المقدار الذى اذا ضرب فيه القطر حصل المحيط اعنى سطح دائرة – ك ص ى – اصغر من سطح نصف – ب ه – فى نصف – ا ب – وفى – ه ح – و ز – جميعا ثم الحاصل فى المقدار الـذى اذا ضرب فيـه القطر حصل المحيط اعنى نصف سطح المجسّم المحيـط بنصف الكرة الداخلـة فجميـع سطـح المجسّم اعظم من ضعف سطح دائرة – ك ص ى – وذلك ما اردناه

(*And because lines MS, MQ*, MF, and MO** are four equal lines which are not in one surface and are drawn from point M to the surface of the inside sphere, M is its center. And MS is half of its diameter; and circle KSY* ((!KOY?)) *is its base. And* $MS^2 < [\frac{1}{2} BE \cdot \frac{1}{2} AB \cdot (EH + UZ)]$. *And so* $MS^2 \cdot \pi =$ *area circle KOY, and* $(MS^2 \cdot \pi) < [\frac{1}{2} BE \cdot \frac{1}{2} AB \cdot (EH + UZ) \cdot \pi]$, *i.e.,* $(MS^2 \cdot \pi) < \frac{1}{2}$ *area of the body contained by the interior sphere. And so the area of the whole body* $< 2 \cdot$ *area circle KOY. Q.E.D.*) ((*Rendering ع by *Q*, as Gerard does)) ((**Rendering ص by *O*.))

c. 2 cuius demonstrationem premisi/mus. Ergo multiplicatio linee *MS* in se et multiplicatio eius quod agregatur in quantitatem in quam cum multiplicatur diameter est illud quod agregatur ipsa linea circumdans est equalis superficiei circuli *KOY*. Ergo superficies circuli *KOY* est
140 minor multiplicatione medietatis linee *BE* in duas lineas *UZ*, *EH* et in medietatem linee *AB* et multiplicatione eius quod agregatur in quantitatem in quam cum multiplicatur diameter est illud quod agregatur ipsa linea circumdans. Sed multiplicatio medietatis linee *BE* in duas lineas *UZ*, *EH* et in medietatem linee *AB* et multiplicatio eius
145 quod agregatur in quantitatem in quam cum multiplicatur diameter est illud quod agregatur ipsa linea circumdans est equalis medietati embadi superficiei corporis *ABGD*. Ergo embadum superficiei corporis *ABGD* est maius duplo embadi superficiei circuli *KOY*, que est basis medietatis spere quam continet corpus *ABGD*

150 Iam ergo ostensum est quod embadum superficiei corporis *ABGD* est minus duplo embadi basis medietatis spere que continet corpus et maius duplo embadi basis medietatis spere quam continet corpus *ABGD*. Et illud est quod declarare voluimus. Et hec est forma eius.

[XIV.] EMBADUM SUPERFICIEI OMNIS MEDIETATIS SPERE EST DUPLUM EMBADI SUPERFICIEI MAIORIS CIRCULI QUI CADIT IN EA.

Verbi gratia, sit medietas spere *BGAD*, et maior circulus qui cadit
5 in ea sit circulus *ABG*, et punctum *D* sit polus huis circuli [Fig. 49]. Dico ergo quod embadum superficiei medietatis spere *ABGD* est duplum embadi superficiei circuli *ABG*, quod sic probatur.

Si non fuerit duplum embadi circuli *ABG* equale superficiei medietatis spere *ABGD*, tunc sit duplum eius aut minus superficie medie-
10 tatis spere *ABGD* aut maius ea [, si fuerit possibile]. Sit ergo in primis duplum embadi circuli *ABG* minus embado superficiei medietatis spere *ABGD*, si fuerit illud possibile. Et sit duplum embadi circuli

136–37 cuius...se *om. H*
138 ipsa: illa *H*
139 KOY[1]: KOIB *H* / Ergo *tr. H post* circuli[2]
149 basis: embadum spere basis *H*
150 quod: quod cum *H*
152 et: est *H*
153 declarare: demonstrare *H* / Et...eius *om. H*
1 [XIV]: 17 *mg. MaR*
4–36 Verbi....est[1]: Ex quo infert *S*
4 BGAD: ABGD *Ar*
10 [si... possibile] *sic. PZmHMa*; *sed delendum est?*
12–13 si....ABG *om. H*

demonstrated earlier. Therefore, [since] $MS^2 \cdot \pi =$ area circle KOY, then area circle $KOY < 1/2\ BE \cdot (UZ + EH + 1/2\ AB) \cdot \pi$. But

$1/2\ BE \cdot (UZ + EH + 1/2\ AB) \cdot \pi = 1/2$ surf area body $ABGD$.

Therefore, the surface area of body $ABGD$ is greater than double the area of circle KOY, circle KOY being the base of the hemisphere which body $ABGD$ contains.

Therefore, it has now been demonstrated that the surface area of body $ABGD$ is less than double the area of the base of the hemisphere which contains the body and greater than double the area of the base of the sphere which body $ABGD$ contains. And this is what we wished to show. And this is its form [Fig. 47].

[XIV.] THE SURFACE AREA OF EVERY HEMISPHERE IS DOUBLE THE AREA OF THE GREATEST CIRCLE WHICH FALLS IN IT.

For example, let there be the hemisphere $BGAD$ and circle ABG the greatest circle falling in it, and let point D be the pole of this circle [see Fig. 49]. I say, therefore, that the surface area of hemisphere $ABGD$ is equal to double the area of circle ABG.

Proof: If double the area of circle ABG is not equal to the area of hemisphere $ABGD$, then it is less than the area of hemisphere $ABGD$ or greater than it. First, let double the area of circle ABG be less than the area of hemisphere $ABGD$, if that is possible. And let double the

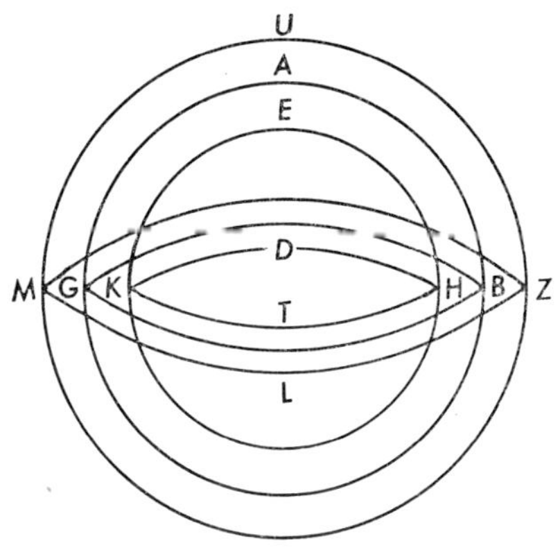

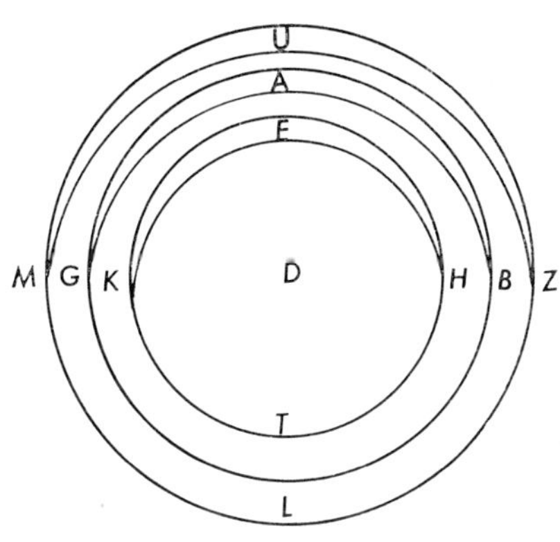

[Reconstructed] Fig. 49 [As given in MS P and in Arabic texts]

4–5 qui...ea: التى هو قاعدته (*which is its base*)
6–7 Dico...probatur *om. Ar.*
9–12 ABGD....possibile: فليكون اولا اصغر منه (*then let it at first be less than it*)
12–13 duplum...ABG *om. Ar.*

ABG equale superficiei medietatis spere minoris medietate spere *ABGD*, que sit medietas spere *EHTK*, Cum ergo fiet in medietate
15 spere *ABGD* corpus compositum ex portionibus piramidum columnarum, cuius basis sit superficies circuli *ABG* et cuius caput sit punc-
61v c. 1 tum *D*, et ponetur ut corpus non tangat / medietatem spere *EHTK*, tunc oportebit ex eis que premisimus ut embadum superficiei corporis *ABGD* sit minus duplo embadi superficei circuli *ABG*. Sed embadum
20 superficiei corporis *ABGD* est maius embado superficiei medietatis spere *EHTK*, quoniam continet ipsam. Ergo embadum superficiei medietatis spere *EHTK* est multo minus duplo embadi superficiei circuli *ABG*. Et iam fuit ei equalis. Hoc vero contrarium est et impossibile.

Et iterum sit duplum embadi superficiei circuli *ABG* maius embado
25 superficiei medietatis spere *ABGD*, si fuerit possibile illud. Et sit equale superficiei medietatis spere maioris medietate spere *ABGD*, que sit medietas spere *UZLM*. Cum ergo fiet in medietate spere *UZLM* corpus compositum ex portionibus piramidum columpnarum, cuius basis sit superficies circuli *UZLM* et cuius caput sit punctum *D*, et non
30 sit corpus tangens medietatem spere *ABGD*, tunc oportebit ex eo quod premisimus ut sit embadum superficiei corporis *UZLM* maius duplo embadi circuli *ABG*. Verum embadum superficiei medietatis spere *UZLM* est maius embado superficiei corporis *UZLM*. Ergo embadum medietatis spere *UZLM* est maius duplo embadi superficiei circuli
35 *ABG*. Sed iam fuit ei equale. Hoc vero est contrarium et impossibile.

Iam ergo ostensum est quod embadum superficiei omnis spere est quadruplum embadi superficiei maioris circuli cadentis in ea. Et illud est quod declarare voluimus. Et hec est forma eius.

13 equale *PZmMa* equalis *H*
18 *ante* tunc *del. H* cum ?
19 ABGD: ABG *H* / ABG: ABGḌ *P*
21 spere: sp̣ei *H*
23 ei: dg *H* / est: fuit *H*
24 Et *om. H*
25 ABGD: ẠABGD *P*
26 equale *ZmPMa* equalis *H* / medietate: medietatis *H*
27 U-: N- *H hic et ubique in hac propositione*
29 LZLM *H*
31 ut: quod *H*
33 maius: magis *H*
33–34 embado....maius *om. H hic sed cf. var. lin. 35*
34 superficiei *om. H*
35 *post* ABG *add. H* omisimus embadum superficiei medietatis spere NZLM est magis embado superficiei circuli AB (*!*) / Sed: et *H* / equale *corr. ex* equalis *in PHMa*
37–38 Et....eius *om. S*
38 Et...eius *om. H*

15 compositum...columnarum: كما وصفنا (*just as we have described*)
18–23 tunc....impossibile: كان سطحه اصغر من ضعف سطح دائرة – ا ب ج – و اعظم من سطح

area of circle *ABG* be equal to the area of a hemisphere smaller than hemisphere *ABGD*, namely, hemisphere *EHTK*. When, therefore, there is described in hemisphere *ABGD* a body composed of segments of cones, the base of which body is the surface of circle *ABG* and its vertex is point *D*, and it is posited that the body does not touch hemisphere *EHTK*, then from what we have proved before [in Proposition XIII] it will follow that the surface area of body *ABGD* is less than double the area of circle *ABG*. But the surface area of body *ABGD* is greater than the surface area of hemisphere *EHTK*, since the one contains the other. Therefore, the surface area of hemisphere *EHTK* is much less than double the area of circle *ABG*. But it was posited as equal to it. This indeed is a contradiction and is impossible.

Now again let double the area of circle *ABG* be greater than the surface area of hemisphere *ABGD*, if that is possible. Let it be equal to the area of a hemisphere greater than hemisphere *ABGD*, namely, hemisphere *UZLM*. When, therefore, there is inscribed in hemisphere *UZLM* a body composed of segments of cones, the base of which body is circle *UZLM* and its vertex is point *D* and the body does not touch hemisphere *ABGD*, then it will follow from what we have proved before that the surface area of body *UZLM* is greater than double the area of circle *ABG*. But the surface area of hemisphere *UZLM* is greater than the surface area of body *UZLM*. Therefore, the surface area of hemisphere *UZLM* is greater than double the area of circle *ABG*. But it was posited as equal to it. This indeed is a contradiction and is impossible.

Therefore, it has now been demonstrated that the surface area of any sphere is quadruple the area of the greatest circle falling in it. And this is what we wished to show. And this is its form [Fig. 49].

نصف كرة — ه ح ط ك ونصف سطح دائرة — ا ب ج — المساوى لسطح نصف كرة — ه ح ط ك — اعظم كثيرا منه هذا خلف
(Its surface was less than double the surface of circle ABG and greater than the surface of hemisphere EHTK. And double the area of circle ABG, which is equal to the area of the hemisphere EHTK, is much greater than it. [*But*] *this is a contradiction.*)

25 si....illud *om. Ar.*

26–27 maioris...spere[1] *om. Ar.*

28–29 compositum...D: كما وصفنا (*just as we have described*)

30–31 ex eo quod premisimus: لما سلف (*according to what went before.*) ((Note: this phrase is transferred to a position after ABG in line 32.))

32 Verum: و (*And*)

33 UZLM[2] *om. Ar. sed add.* لكونه محيطا به (*because the one contains the other*)

34 maius: اعظم كثيرا (*much greater*) / duplo *om. Ar.*

35 et impossibile *om. Ar. et hic add.* فاذا الحكم ثابت وذلك ما اردناه (*And so the rule is established. Q.E.D.*) ((Note: the Q.E.D. is given here instead of in lines 37–38 where it appears in the Latin text.))

[XV.] MULTIPLICATIO MEDIETATIS DIAMETRI OMNIS SPERE IN TERTIAM EMBADI SUPERFICIEI SUE EST EMBADUM MAGNITUDINIS SPERE.

Verbi gratia, sit spera *ABGD*, et medietas diametri eius sit linea
5 *SB* [Fig. 50]. Dico ergo quod multiplicatio linee *SB* in tertiam embadi superficiei spere *ABGD* est embadum magnitudinis spere *ABGD*, cuius hec est demonstratio.

Si non fuerit ita, tunc sit multiplicatio linee *SB* in tertiam embadi superficiei spere minoris aut maioris spera *ABGD* ipsum embadum
10 magnitudinis spere *ABGD*. Ponam ergo in primis multiplicationem linee *SB* in tertiam embadi superficiei spere maioris spera *ABGD* ipsum embadum magnitudinis spere *ABGD*. Et sit spera *ZULM*, cuius centrum et centrum spere *ABGD* sit unum. Ergo multiplicatio
c. 2 *SB* in tertiam emba/di spere *UZLM* est embadum magnitudinis spere
15 *ABGD*. Ergo cum fiet super speram *ABGD* corpus habens superficies continentes ipsum et non tangat speram *UZLM*, oportebit ex eo quod premisimus ut multiplicatio linee *SB* in tertiam embadi superficiei corporis quod continet speram *ABGD* sit maior embado spere *ABGD*. Sed multiplicatio linee *SB* in tertiam superficiei spere *UZLM*
20 est embadum magnitudinis spere *ABGD*. Ergo tertia embadi spere *UZLM* est minor tertia embadi superficiei corporis habentis superficies, et spera *UZLM* continet corpus. Hoc autem est contrarium.

1 [XV]: 18 *mg. Ma mg. R*
2–3 *post* embadum *add. S* et
4–38 Verbi.... voluimus *om. S*
7 cuius: secundum *H*
11 *ante* superficiei *scr. et del. H* et embadum
12–13 Et...unum: sit primum *H*
13 *ante* et *del. P* sit E
14 U-: N- *H hic et ubique in hac propositione*
16 tangat *HR* tangant *PZm* tagat *Ma*
17 *de* premisimus *scr. P. mg.* in conclusione 1 figure huius / superficiei: superficiem NZLM est embadum magnitudinis spere ABGD *H* (*Sed cf. var. lin. 19-20*)
19–20 UZLM.... ABGD *om. H*

---

5 SB: سف (*SF*)
5–7 Dico... demonstratio *om. Ar.*
8–15 Si.... ABGD:[1] فان لم يكن – س ف – فى ثلث سطح كرة – ا ب ج د – عظمها فليكن اولا اصغر من عظمها وليكن – س ب – فى ثلث سطح كرة اعظم من كرة – ا ب ج د – مساويا لعظم كرة – ا ب ج د – مثلا ككرة – و ز ل م – فليكن مركزاهما واحدا
(*And if the product of SF and $\frac{1}{3}$ the area of sphere ABGD is not equal to its volume, then at first let it be less than its volume. And let the product of SB and $\frac{1}{3}$ the area of a sphere larger than sphere ABGD be equal to the volume of sphere ABGD—and* [*this larger sphere is*] *for example UZLM. And let the centers of the two spheres be common.*)
15–16 habens.... ipsum: كما وصفنا
(*just as we have described*)
18–22 quod.... corpus: يساوى المجسّم ويكون اكبر من كرة – ا ب ج د – و يلزم منه ان يكون ثلث سطح المجسّم اعظم من ثلث كرة – و ز ل م – المحيط به

[XV.] THE MULTIPLICATION OF THE RADIUS OF EVERY SPHERE BY ONE THIRD OF ITS SURFACE AREA IS THE VOLUME OF THE SPHERE.

For example, let there be a sphere *ABGD* with radius *SB* [see Fig. 50]. I say, therefore, that the multiplication of line *SB* by one third of the surface area of sphere *ABGD* is the volume of sphere *ABGD*.

Proof: If this is not so, then let the multiplication of line *SB* by one third of the surface area of a sphere either larger than or smaller than sphere *ABGD* be [equal to] the volume of sphere *ABGD*. Hence, I shall posit first that the multiplication of line *SB* by one third of the surface area of a sphere larger than sphere *ABGD* is [equal to] the volume of sphere *ABGD*. Let this sphere be *ZULM* [*UZLM*], concentric with sphere

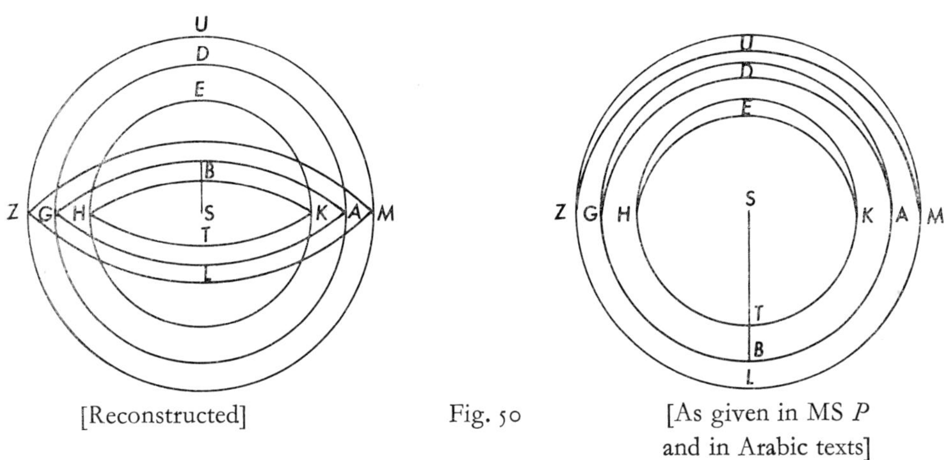

[Reconstructed] Fig. 50 [As given in MS *P* and in Arabic texts]

ABGD. Hence ($SB \cdot 1/3$ area sphere *UZLM*) $=$ vol sphere *ABGD*. Hence, when there is described about sphere *ABGD* a body having surfaces which bound it but which body does not touch sphere *UZLM*, it will follow from what we have proved before that the multiplication of line *SB* by one third of the surface area of the body which contains sphere *ABGD* is greater than the area of sphere *ABGD*. But the multiplication of line *SB* by one third of the area of sphere *UZLM* is [equal to] the volume of sphere *ABGD*. Therefore, one third of the area of sphere *UZLM* is less than one third of the surface area of the body having surfaces, while the sphere *UZLM* contains the body. This, however, is a contradiction.

(*equals the body and will be greater than sphere ABGD, and it follows from this that* $\frac{1}{3}$ *of the area of the body is greater than* $\frac{1}{3}$ *of* [*the area of*] *sphere UZLM containing it.*)

Et sit multiplicatio linee *SB* in tertiam embadi superficiei spere minoris spera *ABGD* ipsum embadum magnitudinis spere *ABGD*.
25 Et sit spera illa spera *EHTK*, cuius centrum et centrum spere *ABGD* sit unum. Ergo multiplicatio *SB* in tertiam embadi superficiei spere *EHTK* est embadum magnitudinis spere *ABGD*. Cum ergo fiet in spera *ABGD* corpus habens superficies continentes ipsum et non tangat speram *EHTK*, oportebit ex eo quod premisimus ut sit multi-
30 plicatio linee *SB* in tertiam embadi superficiei corporis habentis superficies quod continet spera *ABGD* minor embado spere *ABGD*. Sed multiplicatio linee *SB* in tertiam embadi superficiei spere *EHTK* est embadum magnitudinis spere *ABGD*. Ergo tertia embadi superficiei spere *EHTK* est maior tertia embadi superficiei corporis habentis
35 superficies, et corpus continet speram *EHTK*. Hoc vero est contrarium. Iam ergo declaratum est quod multiplicatio medietatis diametri spere in tertiam embadi superficiei eius est embadum magnitudinis eius. Et illud est quod declarare voluimus.

[XVI.] VOLO OSTENDERE QUOMODO PONANTUR INTER DUAS QUANTITATES DUE QUANTITATES ITA UT CONTINUENTUR QUANTITATES QUATTUOR SECUNDUM PROPORTIONEM UNAM.

5 Scientia enim illius valde utile est ei qui geometrie querit scientiam. Et hac eadem operatione extrahatur latus cubi, quod est quoniam quan-
62r c. 1 do illud quod est in cubo de unitatibus et partibus est no/tum et ponuntur inter numerum cubi et inter unum duo numeri continui secundum proportionem ⟨unam⟩, tunc ille qui sequitur unum ex duo-
10 bus numeris mediis est latus cubi.

25 spera[2] *om. H* / ETHK *H* / et: est *H*
27 fiet: fiat *H*
29 tangat *HMa* tangant *PZm* / *de* premisimus *scr. P mg.* in conclusione 2 figure huius
30–31 superficies: superficiem *H*
36 diametri *om. H*
37 embadi superficiei *tr. H*
38 quod...voluimus *om. H*
1 [XVI]: 19 *mg. Ma mg. R*
1–62 Volo....ostendere *om. S*
5 Scientia...scientiam *PZmMa mg. H* / Scientia enim: Nota quod scientia *H*
9 ⟨unam⟩ *supplevi*

23–38 Et....eius: ثم ليكن – س ب – فى ثلث سطح كرة اصغر من كرة – ا ب ج د – ككرة – ه ح ط ك – مساويا لعظم كرة – ا ب ج د – ونعمل فى كرة – ا ب ج د – مجسّما كما وصفنا بحيث لا يماس كرة – ه ح ط ك – و يجب مما مرّ ان – س ب – فى ثلث مساحة سطح المجسّم اصغر من مساحة كرة – ا ب ج د – فثلث سطح – ه ح ط ك – اعظم من ثلث سطح المحسّم المحيط به هذا حلف ، فاذا الحكم ثابت

And let the multiplication of line *SB* by one third of the surface area of a sphere less than sphere *ABGD* be [equal to] the volume of sphere *ABGD*. Let that [lesser] sphere be sphere *EHTK*, concentric with sphere *ABGD*. Therefore, the multiplication of *SB* by one third of the surface area of sphere *EHTK* is [equal to] the volume of sphere *ABGD*. When, therefore, there is inscribed in sphere *ABGD* a body having sufaces which bound it but which body does not touch sphere *EHTK*, it will follow from what we have proved before that the multiplication of line *SB* by one third of the surface area of the body having surfaces which the sphere *ABGD* contains is less than the area of sphere *ABGD*. But the multiplication of line *SB* by one third of the surface area of sphere *EHTK* is [equal to] the volume of sphere *ABGD*. Therefore, one third of the surface area of sphere *EHTK* is greater than one third the surface area of the body having surfaces, while the body contains sphere *EHTK*. This indeed is a contradiction. Therefore, it has now been shown that the multiplication of the radius of the sphere by one third of its surface area is [equal to] its volume. And this is what we wished to show.

[XVI.] I WISH TO DEMONSTRATE HOW TWO QUANTITIES ARE PLACED BETWEEN TWO QUANTITIES SO THAT THE FOUR QUANTITIES ARE IN CONTINUED PROPORTION.

For a knowledge of this [proposition] is very useful to anyone who seeks a knowledge of geometry. By this same operation is extracted the side of a cube; for when the cube is known in terms of units and parts, and between the number representing the cube and 1 are placed two [other] numbers in continued proportion [with the cube number and 1], then that number [representing that one] of the two mean numbers which follows 1 is the side of the cube.*

* See Commentary for a representation of this statement in modern notation.

(Then let the product of SB and $\frac{1}{3}$ the area of a sphere EHTK—a sphere smaller than sphere ABGD—be equal to the volume of sphere ABGD. And let us inscribe in sphere ABGD a body such as we have described before so that it does not touch sphere EHTK. And it is necessary from what has gone before [in the first part of the theorem] that the product of SB and $\frac{1}{3}$ the area of [the inscribed] body is less than the volume of sphere ABGD. And so $\frac{1}{3}$ the area of EHTK is greater than $\frac{1}{3}$ the area of the body containing it. [But] this is a contradiction. And so the rule is established.)

38 declarare *om. Ar.*

1–62 ((Note: al-Ṭūsī seems to have followed the original text more closely here than in most propositions.))

6 hac... extrahatur: به يعرف *(by means of this is known)*

7–8 illud... ponuntur: عرفنا مقدارين يقعان *(we know two quantities which fall...)*

Et hec quidem operatio quam narramus est viri ex antiquis qui dicitur Mileus*, cui est liber in geometria. Sint ita due quantitates inter quas volo ponere duas quantitates ita ut continuentur secundum proportionem unam quantitates M, N. Et sit quantitas M longior quan-
15 titate N. Revolvam autem circulum $ABGD$ [Fig. 51]. Et ponam diametrum eius et est AB equalem quantitati M. Et protraham in circulo $ABGD$ cordam AG equalem quantitati N. Et protraham ex extremitate diametri circuli $ABGD$, ex puncto B, lineam super rectos angulos, et producam lineam AG donec occurrat ei super punctum Z.
20 Et erigam super arcum AGB superficiem medietatis columne ita quod sint linee que protrahuntur in ea secundum rectitudinem ad arcum AGB perpendiculares super superficiem circuli $ABGD$. Et revolvam super lineam AB semicirculum, cuius superficies sit erecta ex superficie circuli $ABGD$ super angulos rectos super lineam AB et est arcus
25 AHE. Et figatur punctum A arcus AHE in loco suo sicut centrum et revolvatur arcus AHE super punctum A. Et sit superficies eius in revolutione sua erecta super superficiem circuli $ABGD$ super rectos angulos ut arcus AHE separet superficiem medietatis columne erectam super arcum AGB. Et figatur linea AB sicut meguar. Et revol-
30 vatur triangulus AZB super meguar AB donec occurrat linea AZ

* *Menelaus* in the Arabic text.

11 Et... quidem: Hec autem *H* / quam *H* qua *Ma* qui *P*
15 *ante* Et *scr. et del. H* cordam AG
16 equalem *ZmMa* equalis *H* equale *P*
19 concurrat *H*
23 *ante* sit *del. P* est
25 Et: quod *H*
26 revolvatur: revolvam *H* / arcum *H*
28 *de* separet *scr. P mg. et Zm supra* vel secet/separet: vel secet separet *Ma*
29 Et[1]: est *H*
30 super: supra *H*

12 Mileus: مانالاوس (*Menelaus*) / *post* geometria *add. Ar.* ونحن نصفه (*and we describe it*)
13–14 duas...quantitates: خطّى (*two lines*)
15 ABGD: – ا ب ج – (*ABG*) ((Note: here and throughout))
16–17 in....ABGD: فيها (*in it*)
17–18 extremitate...puncto *om. Ar.*
18–19 lineam...angulos: عمودا على – ا ب – (*a perpendicular on AB*)
19 punctum *om. Ar.*
24 super...AB *om. Ar.*
25 AHE[1]: ((Arabic has – ا ج ه – by mistake since in the rest of the passage it has – ا ح ه –))
26 punctum: مركز (*center*) / Et: بحيث (*so that*)
29 sicut meguar: كالمركز (*as a center*) ((This is no doubt a corrupt reading since the Arabic word محور for "axis" is better here.))
30 meguar: محور (*as an axis*) ((Here is the correct word for "axis."))

And the method which we describe is the method of one of the ancients called Menelaus, who wrote a *Book of Geometry*. Let *M* and *N* be the two quantities between which I wish to place two quantites so that they are [all] in continued proportion. And let quantity *M* be longer than quantity *N*. Then I shall describe circle *ABGD* [see Fig. 51]. I shall posit its diameter as *AB*, equal to quantity *M*. And I shall draw in circle *ABGD*

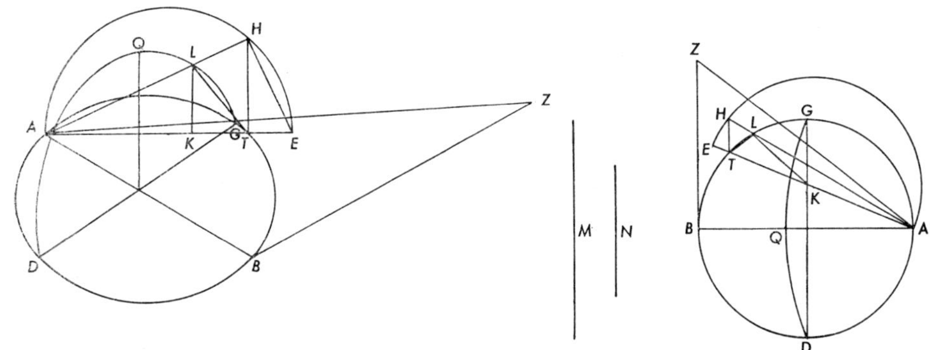

[Reconstructed] Fig. 51 [As given in MS *P*]

Note: The Arabic texts have a somewhat similar diagram, but like MS *H* the orientation is reversed to agree with the orientation of the Greek text of Eutocius and hence in my reconstruction I have followed this reversed orientation. Also, there are a number of errors in the drawing as given in MS *P*; the principal error is that *L* does not fall on semicircle *DQG*, and *LK* is not perpendicular to *ABGD*, as it should be.

the chord *AG*,* equal to quantity *N*. I shall draw a line at right angles to the diameter of circle *ABGD* from its extremity, point *B*, and I shall produce line *AG* until it meets the line from *B* at point *Z*. And I shall erect on arc *AGB* the surface of a half cylinder, in such a way that its rectilinear elements from arc *AGB* are perpendicular to the surface of circle *ABGD*. And I shall describe on line *AB* a semicircle whose surface is erected at right angles to the surface of circle *ABGD*. This semicircle is the arc *AHE*. And let point *A* of arc *AHE* be fixed in its place as a center and let arc *AHE* be rotated about point *A*, and its surface remains during its rotation perpendicular to the surface of circle *ABGD*, so that arc *AHE* intersects the surface of the half cylinder erected on arc *AGB*. Now let line *AB* be fixed as an axis, and let △ *AZB* be rotated about axis *AB* until line *AZ* makes a common section with the surface of the

* The chord *AG* is not actually drawn.

sectioni superficiei medietatis columpne et designet punctum *G*. Tunc ex linea *AZ* in revolutione sua medietatem circuli *GQD* erectam ex superficie circuli *ABGD* super angulos rectos, et signabo super locum in quo occurrit linea *AZ* sectioni superficiei medietatis columne punc-
35 tum *H*. Et figatur arcus *AHE* ex revolutione sua apud punctum *H*. Et protraham duas lineas *AH*, *AE*. Et signabo ubi occurrit linea *AH* arcui *GQD* punctum *L*. Et protraham ex puncto *L* perpendicularem super superficiem circuli *ABGD*, que sit *LK* propterea quod est differentia communis superficiei semicirculi *AHE* et superficiei semi-
40 circuli *GQD* et unaqueque harum duarum superficierum est erecta super superficiem circuli *ABGD* super angulos rectos. Ergo linea *LK* est perpendicularis. Et protraham lineam *LT*. Manifestum est igitur quia erigitur ex linea *AL* super angulos rectos, propterea quod multiplicatio linee *GK* cum linea *KD* est equalis multiplicationi
45 linee *LK* cum equali eius. Verum multiplicatio linee *GK* cum linea *KD*
c. 2 est equalis multiplicationi linee *TK* cum linea *KA*. Ergo multipli/catio linee *KT* cum linea *KA* est equalis multiplicationi linee *LK* cum equali eius. Ergo angulus *ALT* est rectus. Et iam ostensum est quod angulus *AHE* trianguli *AHE* est rectus, quoniam ipse est compositus super
50 medietatem circuli *AHE*, et quod angulus *ATH* trianguli *ATH* est rectus, quoniam *HT* est perpendicularis super superficiem circuli *ABG* et est una linearum que protrahuntur in medietate columne secundum rectitudinem ad arcum *AGB* in superficie circuli *ABGD*. Sed linea *AT* est in superficie circuli *ABGD*. Ergo angulus *ATH*, propter
55 illud quod diximus, est rectus. Ergo in unoquoque triangulorum *AHE ATH*, *ALT*, *AKL* est angulus rectus. Et angulo *HAE* communi in eis omnibus sunt trianguli similes. Ergo proportio *EA* ad *AH* est sicut proportio *AH* ad *AT* et sicut proportio *AT* ad *AL*. Sed linea *AE* est equalis quantitati *M* et linea *AL* est equalis quantitati N.

31 Tunc: et *H*
32 AZ: AT AZ *H* / GQD: GHD *H*
33 super locum *PZm* super circulum *Ma* lineam *H*
37 *post* L[2] *add. H* vel sectio (*cf. var. lin. 39*)
39 *de* differentia *scr. P mg. et Zm supra* vel sectio / differentia: vel sectio differentia *Ma* ((*Ma* puts *vel sectio* in a box))
40 *ante* superficierum *del. H* linearum
43 *post* igitur *del. P* et / erigitur: egreditur *H*
44 GK: GN *H* / KD *PZm* ND *H* GD *Ma*
45 LK: HZ *H* / GK: GN *H* / KD: ND *H*
46 TK: TN *H* / KA: NA *H*
47 KT: NT *H* / KA: NA *H* / LK: LN *H*
48 rectus: erectus *H* / est[2] *om. Zm*
51 HT *Zm Ar.* AT *PHMa*
53 superficiem *H*
55 triangulo *H* / AHE: AHT *H*
56 ALT, AKL: ANL *H* / HAE: AHE *H*
59 N *om. H*

half cylinder, and it marks point G.* Then by the rotation of line AZ I shall be erecting the semicircle GQD at right angles to the surface of circle $ABGD$. And I shall designate point H as the place where line AZ intersects the [common] section of the surface of the half cylinder [and arc AHE]. And let arc AHE stop rotating when it arrives at point H. Now I shall draw two lines AH, AE. I shall designate as point L the place where line AH intersects arc GQD. I shall draw from point L a perpendicular to the surface of circle $ABGD$, and this perpendicular is LK, since LK is the common intersection of the surface of semicircle AHE and the surface of semicircle GQD and each of these two surfaces is erected at right angles to the surface of circle $ABGD$. Hence, line LK is perpendicular. I shall also draw line LT. It is evident, therefore, that LT is erected at right angles to line AL; for $GK \cdot KD = LK^2$ [Euclid, VI.8, Por.], and $GK \cdot KD = TK \cdot KA$ [Euclid, III. 35]. Therefore, $KT \cdot KA = LK^2$. Hence, $\angle\ ALT$ is a right angle. And it now has been demonstrated that $\angle\ AHE$ of $\triangle\ AHE$ is a right angle, since it has been constructed upon semicircle AHE, and that $\angle\ ATH$ of $\triangle\ ATH$ is a right angle, since HT is perpendicular to the surface of circle ABG, and it is one of the straight line elements of the half cylinder protracted to arc ABG in the surface of circle $ABGD$; but line AT lies in the surface of circle $ABGD$, and therefore $\angle\ ATH$, because of what we have said, is a right angle. Therefore, in each of the triangles AHE, ATH, ALT, AKL, there is a right angle. And since $\angle\ HAE$ is common to all of them, the triangles are similar. Therefore, $EA/AH = AH/AT = AT/AL$. But line $EA = M$

* That is, it marks G as the beginning of the curve constituting the common intersection of line AZ and the surface of the half cylinder.

---

32 —Q—: ع ((Gerard again renders ʿain by Q.))

37–38 L²...LK: — ح — عمودا على سطح دائرة — ا ب ج — وهو خطّ — ح ط — ونخرج — ل ك — وهو عمود على سطح دائرة — ا ب ج — (*H a perpendicular to the surface of circle ABG and it is line HT and we draw LK and it is perpendicular to the surface of circle ABG*)

39 semiculi:[1] مثلث (*triangle*)

41–42 Ergo...perpendicularis *om. Ar.*

49 trianguli AHE *om. Ar.* / ipse...compositus: — ح ط — عمود (*HT is perpendicular*)

50 trianguli ATH *om. Ar.*

52–53 et...ABGD *om. Ar.*

54 ATH: — ا ل ط — (*ALT*)

56 AKL *om. Ar.*

56–57 Et....trianguli: و زاوية حادة مشتركة فهى (*And one angle is common; and so they are*)

60 Iam ergo ceciderunt inter duas quantitates *M*, *N* quantitates *AH*, *AT* et continuantur secundum proportionem unam. Et illud est quod voluimus ostendere.

[XVII.] Et quamvis demonstratio Milei in rebus quibus utitur in positione duarum quantitatum inter duas sit demonstratio certa erecta in mente, tamen est difficilis valde per inquisitionem et per eam est possibile ut ponamus inter duas quantitates duas quantitates ita ut 5 continuentur secundum proportionem. Nos autem querimus ut sit modus quo possibile sit nos consequi illud de quo narravimus facile.

Sint ergo due quantitates inter quas voluimus ponere duas quantitates ita ut continuentur secundum proportionem unam, due quantitates *A*, *B*. Et ponam quantitatem *GD* equalem quantitati *A* [Fig. 52]. 10 Et erigam ex linea *GD* lineam *DE* super rectum angulum [ex *GD*]. Et ponam lineam *DE* equalem quantitati *B*, et protraham lineam *GE* et extendam duas lineas *GD*, *ED* secundum rectitudinem, et non ponam eis utrisque finem determinatum. Et erigam super punctum *E* linee *GE* lineam orthogonaliter et protraham ipsam donec occurrat 15 linee que extenditur secundum rectitudinem cum linea *GD* super punctum *U*. Et erigam super punctum *G* lineam equidistantem linee *EU* donec occurrat linee que extensa est secundum rectitudinem cum linea *ED* super punctum *M* et est *MG*, et extendam lineam *MG* in partem alteram secundum rectitudinem usque ad punctum *P*, ut sit

1 [XVII]: 20 *mg. Ma mg. R*
1–60 Et....ostendere *om. S*
2 positione: portione *H* / certa: corporis *H*
3 *de* tamen est difficilis *scr. P mg. et Zm supra* (*et add. H post* mente): in alio tamen (*om. Zm*) non est difficilis
4 duas quantitates[2] *om. HMa*
5 querimus *om. H*
6 narramus *H*
9 GD: GB *H*
10 [ex GD] *delendum est?*
11 et...GE *om. Zm*
12 et...GD *ZmHMa om. P*
13 utrisque: nusquam (?) *H*
14 et...ipsam *om. Zm*
15–17 secundum....est *om. H*
19 usque *om. H*

---

62 ostendere *om. Ar.*
1–6 Et....facile: الأن الاشياء التى استعملها مانالاوس وان كان صحيحا فهى اما ان لايمكن ان يفعل واما يكون عسيرا جدا طلبنا لذلك وجها اسهل
((There appears to be some confusion of gender in this statement, but I have let it stand as in the printed text.)) (*And since the things which Menalaus used—even if they are true—are either not possible to execute or are very difficult, we have accordingly sought an easier method.*)
7–9 inter...quantitates *om. Ar.*
12 secundum rectitudinem *om. Ar.*
17–18 linee... linea *om. Ar.*
18–19 extendam.... P: ونخرجه (*And we extend it*)

and $AL = N$. So now, therefore, the quantities AH and AT have been placed between quantities M and N so that they are [all four] in continued proportion. And this is what we wished to demonstrate.

[XVII.] Although the demonstration of Menelaus concerning the placing of two quantities between two quantities is theoretically correct, still it is very difficult to follow it and by means of it [actually] to place two quantities between two quantities so that they are [all] in continued proportion. Accordingly we seek a method by which it is possible for us to obtain easily that which we have recounted.

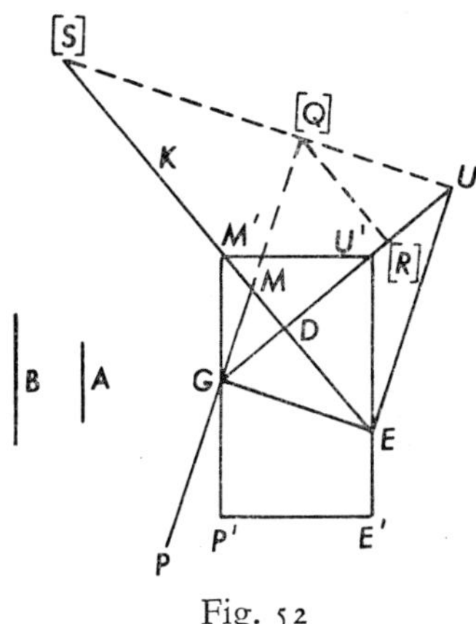

Fig. 52

Note: I have added the prime signs to M', U', E', and P' here and in the Latin text. In MS P the letters M, U, E, and P are used twice. Also, I have made EG perpendicular to MP and EU as the text demands, although they are not so drawn in MS P, and added the dotted lines and [Q], [R], and [S].

Let the quantities A and B be the two quantities between which we wished to place two quantities so that they are [all] in continued proportion. And I shall posit quantity GD equal to quantity A [see Fig. 52]. And I shall erect line DE perpendicular to line GD [at D.] And I shall posit line DE as equal to quantity B. And I shall draw line GE and extend the two lines GD and ED rectilinearly without positing any fixed length to them. I shall erect on point E of line GE a perpendicular, extending it until it meets the rectilinear extension of line GD at point U. And I shall erect at point G a line parallel to line EU, [extending it] until it meets the extension of ED at point M, and this line is MG. And I shall extend line MG in the other direction to point P so that line MP is equal to line EU.

20 linea *MP* equalis linee *EU*. Et estimabo quod posuimus *EU* moveri ex parte puncti *U* ad partem puncti *D*. Et sit extremitas eius que est apud punctum *U* inseparabilis in motu suo a linea *UD*. Et linea in motu suo non cesset transire super punctum *E* linee *GE*, ut quando 62v c. 1 mo/vetur *UE* sicut narravimus, tunc ubi est extremitas eius ex linea 25 *UG*, tunc linea *UE* in illa dispositione secundum rectitudinem sit extensum quod est inter punctum extremitatis eius et punctum *E'* sicut linea *UE*. Deinde signabo super lineam extensam secundum rectitudinem signum *K* et imaginabor quod linea *MP* moveatur ex parte puncti *M* ad partem puncti *K*. Et sit extremitas eius que est apud *M* 30 inseparabilis in motu suo a linea *DK*. Et linea *MP* in motu suo non cesset ire super punctum *G* linee *EG*, sicut narravimus de motu linee *UE*. Et imaginabor quod due linee *MP*, *UE* in motu suo sint equidistantes. Et imaginabor quod super extremitatem linee *UE* super punctum *E* sit linea erecta orthogonaliter super lineam *UE* sequens eam 35 in motu suo. Et non posita huic linee finem determinatum. Et hec linea non cesset abscidere lineam *MP* apud motum duarum linearum *UE*, *MP*. Cum ergo moventur due linee *UE*, *MP* et sunt in motibus suis equidistantes, adherent extremitates utrarumque duabus lineis *UD* *MK*, sicut narravimus. Et procul dubio linea erecta super lineam *UE* 40 orthogonaliter, que movetur cum ea et secat lineam *MP*, pervenit ad punctum *P'*. Quando ergo pervenit linea erecta super punctum *E* ad punctum *P'*, perveniunt illic due linee *U'E'*, *M'P'*. Et lineamus duas lineas *E'P'*, *U'M'*. Et scitum quidem est quod linea *E'P'* erigatur ex unaquaque duarum linearum *U'E'*, *M'P'* orthogonaliter, quoniam est 45 linea quam posuimus in principio esse erectam orthogonaliter ex linea

20 estimabo: continuabo *H* extimabo *Zm* / -U: -N *H hic et ubique in hac propositione*
22 in[1]: a *H* / *ante* UD *del. P* ED
23 cessat *H*
24 UE: PE *H* / ex: e *H*
25 UE: N^{E}_{G} *H*
26 punctum[1] *Ar.*, *corr. ex* motum puncti *in PZmHMa* / E': est *H*
27 UE: PM *Zm*
28 imaginabor *Zm* ymaginabor *H* iamginabo *P* immaginabor *Ma*
29 K: N *H*
30 DK: DN *H*
31 ire: nt *H* / linee[2]: nte *H*
32–33 equidistantes: extremitates *H*
33 ymaginabor *P* / linee UE: NE linee *H*
34 eam *om. Zm*
35–36 Et[2]...non *PZmMa* in alio, ut hec linea non *mg. P et supra Zm* pro alio ut hec linea non est ut linea non *H*
36 cessat *H*
37 UE, MP: MP NE *H*
38 equidistantes: extremitates *H* / utrarumque *PZm* utrorumque *Ma* utcunque *H*
39 MK: MN *H*
40 MP: PM *H*
43 est quod *tr. H* / erigatur: erigitur *H*
45 esse *om. H*

And I shall imagine that we have put *EU* in motion from point *U* in the direction of point *D*; and let the extremity at point *U* in the course of its movement be inseparable from line *UD*. And the line [*EU*] in the course of its motion continually passes through point *E* of line *GE*, so that when *UE* is moved as we have recounted, wherever the extremity of line *UE* is on line *UG*, in that position the straight line between that extremity [on line *UG*] and point *E′* is equal to line *UE*. Then I shall mark on line [*DE*] rectilinearly extended the point *K*, and I shall imagine that line *MP* is moved from the direction of point *M* toward the direction of point *K*. And let the extremity at *M* be in the course of its movement inseparable from line *DK*. And line *MP* in the course of its movement continues to pass through point *G* of line *EG*, just as we recounted concerning the motion of line *UE*. And I shall imagine that the two lines *MP* and *UE* in the course of their movements are [continually] parallel. And I shall imagine that the line erected perpendicular to line *UE* at its extremity *E* follows that line *UE* in its motion, and no fixed end is posited for this line. This line continues to cut line *MP* in the course of the movement of the two lines *UE* and *MP*. When, therefore, the two lines *UE* and *MP* are moved and remain parallel in the course of their movements, their extremities [continue to] adhere to the two lines *UD* and *MK*, as we have recounted. And without doubt the line erected perpendicular to *EU*, which is moved with and cuts line *MP*, arrives [some time] at point *P′*. When it arrives there, the two lines [*UE* and *MP*] have arrived at *U′E′*, *M′P′*. Let us draw the two lines *E′P′* and *U′M′*. And indeed it is known that line *E′P′* is perpendicular to each of the two lines *U′E′* and *M′P′*. since it is the line which we initially posited as being perpendicular to line

20 MP: – م ص – (*MṢ*) ((Note: Gerard renders ص by P))
25 UG: – و د – (*UD*)
26 punctum[1]: نقطة (*point*) *sed in Lat. MSS* motum puncti
27 sicut: من (*from*)
28 signum K: – ه د ك – خطّ (*line EDK*)
30 DK: – م ك – (*MK*)
45 in principio *om. Ar.*

UE et movetur cum ea donec pervenit ad punctum *P′*. Dico ergo quod due linee *DM′*, *DU′* sunt due quantitates que iam ceciderunt inter duas quantitates *GD*, *DE* et quod proportio *GD* ad *DM′* est sicut proportio *DM′* ad *DU′* et sicut proportio *DU′* ad *DE*, cuius 50 demonstratio est:

Quoniam due linee *U′E′*, *M′P′* sunt equidistantes et equales et duo anguli *U′E′P′* et *M′P′E′* sunt recti, tunc linea *U′M′* est equalis linee *E′P′* et unusquisque duorum angulorum *E′U′M′* et *P′M′U′* est rectus. At vero linea *M′D* est perpendicularis super lineam *U′D* et linea 55 *U′D* est perpendicularis super lineam *M′E*. Ergo proportio linee *GD* ad *DM′* est sicut proportio *DM′* ad *DU′* et sicut proportio *DU′* ad *DE*. Verum linea *GD* est equalis quantitati *A* et linea *DE* est equalis quantitati *B*; ergo due linee *DU′*, *DM′* iam ceciderunt inter duas quantitates *A*, *B* et continuantur secundum proportionem unam. Et 60 illud est quod voluimus ostendere.

c. 2 [XVIII.] / ET NOBIS QUIDEM POSSIBILE EST CUM HOC INGENIUM SIT INVENTUM UT DIVIDAMUS QUEMCUNQUE ANGULUM VOLUMUS IN TRES DIVISIONES EQUALES.

Sit itaque angulus *ABG* in primis minor recto. Et accipiam ex 5 duabus lineis *BA*, *BG* duas quantitates equales, que sint quantitates *BD*, *BE* [Fig. 53]. Et revolvam super centrum *B* et cum mensura longitudinis *BD* circulum *DEL*. Et extendam lineam *DB* usque ad *L*. Et protraham lineam *BZ* erectam super lineam *LD* orthogonaliter.

46 moventur *H*
48 *post* quod *add.* *Zm* ED ad GB et sicut / est *PZ* et est *HMa*
49 DU′: DE *H*
51 equidistantes: extremitates *H*
53 P′M′U′: PNM *H*
55 super lineam *om.* *H* / GD: DG *Zm*
55–57 *de* Ergo... DE[1] *scr.* *P mg. et om.* *Zm* (*et add.* *HMa** *ante* Erqo): ex corollario (capitulo *H*) 8e (8° *H*, octava *Ma*) sexti (sexto *H*, sesti *Ma*) euclidis (euclidi *Ma*). ((**Ma* puts this phrase in a box.))
55–56 GD ad *tr.* *H*
1 [XVIII]: 21 *mg.* *Ma*
1–46 Et.... figure *om.* *S*
1 *post* est *del.* *P* ut
3 voluerimus *H*
6 *de* BD, BE *scr.* *P mg. et Zm mg.* in alio, AE BD / BD, BE: AE, BD, DE *H* / cum *om.* *H*
7 DB: DA *H*

47 due... ceciderunt *om.* *Ar.*
54 U′D: – وج – (*UG*) ((This is also correct, since it is the same line))
57, 58 quantitati *om.* *Ar.*
60 ((After the conclusion of the proposition al-Ṭūsī adds a "mechanical" reconstruction of this method not in the Latin and presumably not in the original text of the Banū Mūsā. This has been given in German translation by Suter,

UE and moving with it until it [point G of line EG] arrived at point P'. I say, therefore, that the two lines DM', DU' are the two quantities which fall between the two [given] quantities GD, DE, and that $GD/DM' = DM'/DU' = DU'/DE$.

Proof: Since the two lines $U'E'$, $M'P'$ are parallel and equal and the two angles $U'E'P'$ and $M'P'E'$ are right angles, then line $U'M'$ = line $E'P'$ and each of the two angles $E'U'M'$ and $P'M'U'$ is a right angle. But line $M'D$ is perpendicular to line $U'D$ and line $U'D$ is perpendicular to line $M'E$. Therefore, $GD/DM' = DM'/DU' = DU'/DE$. But $GD = A$ and $DE = B$; therefore, the two lines DU', DM' now fall between the two quantities A and B and are in continued proportion with them. And this is what we wished to demonstrate.

[XVIII.] AND INDEED IT IS POSSIBLE, WHEN THIS [KIND OF] DEVICE* HAS BEEN FOUND, FOR US TO DIVIDE ANY ANGLE WE WISH INTO THREE EQUAL DIVISIONS.

And so let $\angle ABG$ at first be less than a right angle. And I shall take from the two lines BA and BG two equal quantities BD and BE [see Fig. 53]. And I shall describe circle DEL on center B with a radius BD. And I shall extend line DB up to L and erect line BZ perpendicularly on

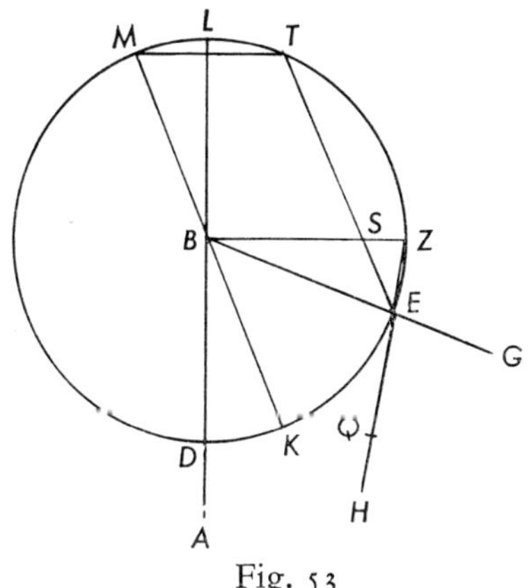

Fig. 53

* See Commentary and also Appendix VI.

Bibliotheca Mathematica, 3. Folge, vol. *3* (1902), pp. 269–70. Accordingly, I have omitted giving the Arabic here.))

2 ingenium: الحيلة (*artifice*) / sit inventum *om. Ar.*

5 BA: – ب ا ه – (*BAE*) / que.... quantitates *om. Ar.*

6 B: – ه – (*E*)

6–7 cum mensura longitudinis BD: ببعدهما (*And with their distances* [*as radii*])

Et lineabo lineam *EZ* et extendam ipsam usque ad *H*. Et non ponam
10 linee *ZH* finem determinatum. Et accipiam de linea *ZH* equale medietati diametri circuli, quod sit linea *ZQ*. Quando ergo ymaginamus quod linea *ZEH* movetur ad partem puncti *L* et punctum *Z* adherens est margini circuli in motu suo et linea *ZH* non cessat transire super punctum *E* circuli *DEL* et ymaginamus quod punctum *Z* non cessat
15 moveri donec fiat punctum *Q* super lineam *BZ*, oportet tunc ut sit arcus qui est inter locum ad quem pervenit punctum *Z* et inter punctum *L* tertia arcus *DE*; cuius demonstratio est:

Quod ego ponam locum ad quem pervenit punctum *Z* apud cursum puncti *Q* super lineam *BZ* apud punctum *T*. Et protraham lineam *TE*
20 secantem lineam *BZ* super punctum *S*. Ergo linea *TS* est equalis medietati diametri circuli, propterea quod est equalis linee *ZQ*. Et protraham ex *B* lineam equidistantem linee *TS*, que sit linea *MBK*. Et protraham lineam ex *T* ad *M*. Ergo linea *MT* et linea *TS* sunt equidistantes duabus lineis *MB*, *BS* et equales eis. Ergo linea *MT* est equidistans
25 linee *BS* et equalis ei. Sed linea *BS* est perpendicularis super diametrum *LD*. Ergo corda arcus *TM* erigitur ex diametro *LD* super duos angulos rectos. Ergo dividit diametrus *LD* cordam *MT* in duo media et dividit propter illud arcum *MT* in duo media super punctum *L*. Verum arcus *ML* est equalis arcui *DK*. Ergo arcus *DK* est equalis
30 medietati arcus *MT*. Sed arcus *MT* est equalis arcui *EK*, propterea quod linea *TE* equidistat linee *MK*. Ergo arcus *DK* est tertia arcus *DE*. Et similiter angulus *DBK* est tertia anguli *ABG*.

Et quoniam possibile est nobis per ingenium quod narravimus in

10 *ante* ZH *del. P* DH / equale *PZm* equalis *Ma* equalem *HR*
11 imaginabimur *Zm*
12 moventur *H*
13 margini: magini (*!*) *H*
14 imaginamur *Zm*
15 BZ: HZ *H*
16 ad: apud (*?*) *H*
16, 18 provenit *H*
21 *ante* Et *add. Ma* trinsecat super centrum ((*Ma puts this phrase in a box*))
21–22 *de* protraham ex B *scr. P mg. et Zm supra* in alio, transeat super centrum
24 est *om. H*
25 et: est *H*
26 super: supra *H*
29 arcus[1] *om. H*
30 MT[1]: ML *H*

---

10 linee ZH *om. Ar.*
11 ZQ: – زع – (*ZQ*) ((Again Gerard uses Q to transcribe ع))
12 ZEH: – زح – (*ZH*)
13 margini circuli: للمحيط (*to the circumference*)
17 *post* DE *add. Ar.* والزاوية التى يوترها هذه القوس ثلث زاوية – د ب ه – (*and the angle which this arc subtends,* [*i.e.,*] 1/3 *angle DBE*)

line LD. Further, I shall draw line EZ, extending it to H, but without assuming ZH to have any fixed length. And I shall cut from line ZH a line equal to the radius of the circle, namely, line ZQ. Therefore, when we imagine that line ZEH is moved in the direction of point L and that point Z [continually] adheres to the circumference in the course of its motion, and that line ZH continues to pass through point E of circle DEL, and we imagine that point Z continues to be moved until point Q falls on line BZ, then it is necessary for the arc between the point at which Z arrives and point L to be one third of arc DE.

Proof: For I posit point T as the place at which point Z arrives as point Q meets line BZ. And I shall draw line TE cutting line BZ at point S. Therefore, line TS is equal to the radius of the circle since it is equal to line ZQ. And I shall draw through B a line parallel to line TS, namely, line MBK. And I shall draw a line from T to M. Therefore, lines MT and TS are [respectively] parallel and equal to the two lines BS and MB. Therefore, line MT is parallel and equal to line BS. But line BS is perpendicular to the diameter LD. Therefore, the chord of arc TM forms two right angles with diameter LD. Therefore, diameter LD bisects chord MT and [therefore] it also bisects arc MT at point L. But arc ML = arc DK. Therefore, arc $DK = 1/2$ arc MT. But arc MT = arc EK, since line TE is parallel to line MK. Therefore, arc $DK = 1/3$ arc DE. Therefore, $\angle DBK = 1/3 \angle ABG$.

And since by means of the device which we have described in connection

18–19 apud....BZ *om. Ar.*
22 B: المركز (*the center*) / TS: – ط ه – (*TE*)
23–24 Ergo...eis: فطس – مساويا وموازيا – لم ب –
(*And so TS equals and is parallel to MB*)
25–26 diametrum *om. Ar*
26 corda arcus *om. Ar*
27–28 Ergo...L: ولذلك يكون منصفا بالقطر
(*And for this reason it will be divided into two by the diameter*)
29–31 Verum....MK: ويكون – م ل – مثل –
ل ط – و – د ك – مثل – م ل – و – م ط –
مساو – ط ه – فد ك – مثل نصف – ك ه –
(*And ML = LT, and DK = ML, and MT = TE. And so DK = ½ KE.*)
32 *post* ABG *add. Ar.* وذلك ما اردناه
(*And that is what we wished*)
33–46 Et...figure: ويحرك بالحيلة المذكورة – زح – على ان يتحرك – ز – على المحيط لا يفارقه ولا يزال يمرّ خطّ – زح – فى حركته على نقطة – ه – حتى تقع نقطة – ع – على خطّ – ب ز – ويتم المطلوب وان كانت الزاوية منفرجة نصفناها و ثلثنا النصف فيكون ثلثاه ثلث المنفرجة
(*And with the device we have mentioned ZH is moved so that Z is moved on the circumference inseparably and line ZH in its motion continues to intersect point E until point Q falls on line BZ and that which is sought is completed. And if it is an obtuse angle, we bisect it and trisect the half. And so two thirds of it will be one third of the obtuse* [*angle*].)

eis que premissa sunt et per ea que sunt ei similia ut moveamus lineam
35 *ZH* et ponamus extremitatem eius que est apud punctum *Z* revolvi
super marginem circuli inseparabilem ab ea et sit linea *ZH* in motu
63r c. 1 suo non transiens nisi super punctum *E* donec per/veniat punctum
Q per motum linee *ZH* super lineam *BZ*, ergo similiter est divisio
omnis anguli minoris recto in tres divisiones equales. Et per illud
40 possibile est nobis facile illud quod narravimus.

Et notum est quod si angulus quem dividere volumus in tres equales
divisiones est maior recto, dividemus ipsum in duas medietates; deinde
dividemus unam duarum medietatum in tres divisiones equales secundum quod narravimus. Manifestum est igitur quod iam tunc sci-
45 mus tertiam anguli qui est maior recto. Et illud est quod demonstrare
voluimus. Et hec est forma figure.

[XIX.] Et quoniam eius quod narravimus in positione duarum
quantitatum ita ut continuentur secundum proportionem unam iuvamentum est in cognitione lateris cubi, et significatio super illud non
est nisi ut innuatur ad illud tantum, sicut significant auctores geometrie
5 super radices surdas quibus non raciocinatur et sequitur computatorem
necessitas multociens ut raciocinetur per latus cubi, et cum veritate
mensure sue si est ex eis cum quibus raciocinatur, aut cum propinquitate si est ex eis cum quibus raciocinatur, sicut faciunt cum radicibus
surdis, tunc propter illud oportet nobis ut narremus cum propinqui-
10 tate latus cubi ut raciocinetur cum eo apud necessitatem. Et faciam in
illo modum quo in propinquitate rei ex veritate non est modus magis
ultimus; quod est quia narrabo modum qui significat super propin-
quitatem lateris cubi ex veritate mensure sue ad quemcunque finem

34 premisim'sa *P* / per: pt *H*
37 *post* perveniat *add.* *H* super
38 Q: K (?) *H*
41 voluimus *H*
42–43 est...dividemus *om.* *H*
46 Et...figure *om.* *H*
1 [XIX]: 22 *mg.* *Ma*
1–20 Et....ut: operacio ad extrahendum latus cubi secundum veritatem propinquam in radicibus surdis ubi numerus est omnino cubicus ita quod inter ipsum et veritatem non est nisi minus secundo si voluerit vel tertio si voluerit aut quantacunque parte vult est taliter *S*

1 positione *PZm* portione *H* porcionem *Ma*
3 super: tunc (?) *H*
4 ut: ut non (?) *H*
5 radices *PMa* vel, terminos *mg P et Zm supra* terminos radices *H* / rationatur (?) *H hic et ubique*
6 et *HMa* etiam *PZm*
8 *ante* si *del.* *P* qui est / *post* quibus *add.* *PZmMa* non; *sed om.* *Ar H* / *de* radicibus *scr.* *P mg. et Zm supra* vel terminis
9 *post* surdis *add.* *H* terminis
11 reei *H* / magis: maius *H*
12–13 propinquitate *H*

with the propostitions previously proved and by means of things which are similar to it it is possible for us to move line ZH so that point Z moves inseparably upon the circumference while line ZH in its motion continues to pass through point E until point Q arrives at line BZ, therefore in the same way every angle less than a right angle can be trisected. And by this [technique] we can easily do what we have recounted.

And it is known that if the angle we wish to trisect is greater than a right angle, we bisect it. Then let us trisect each of the two halves in the manner we have described. It is evident, therefore, that we now know the third of an angle greater than a right angle. And this is what we wished to demonstrate. And this is the form of the figure [Fig. 53].

[XIX.] And since that which we have recounted concerning the placing of two quantities so that they are in continued proportion is of assistance in the finding of the side of a cube—and its significance [here] is only that it leads to such an end, as remark the geometric authors in treating of surd roots with which there is no [exact] reckoning—and since a calculator must of necessity often reckon with the side of a cube, both with its true measure if it is such that he is reckoning with, or with an approximation if it is that with which he reckons as they do with surd roots, then accordingly we must describe a method of approximation of the side of a cube so that it can be reckoned with when necessary. In this matter I shall present a method which yields the best possible approximation of the true root. For I shall describe a method which gives an approximation of the true measure of the side of the cube to any limit the investigator

1 [XIX]: ((Note: This proposition is unnumbered in the published Arabic text))
1–62 Et....proportionaliter:

ينبغى لنا ان نصف
بعد ذلك تقريب ضلع المكعب لينطبق به عند
الحاجة ونعمل فى ذلك بالوجه الذى لا تقريب
البلغ منه . اعنى اذا اردنا ان تكون بينه وبين
الحقيقة مثلا اقلّ من دقيقه اومن ثانية قدرنا
عليه و العمل فيه ان صير المكعب الى اجزائها
ثوالث او سوادس او تواسع او غير ذلك ثم نطلب
مكعبا مساويا لذلك العدد ان كان والا طلبنا
اقرب مكعب اليه واذا وجدناه حفظنا ضلعه فان
كانت الاجزاء ثوالث فهو دقائق وان كانت سوادس
فهو ثوانى وعلى هذا القياس امر المسائل

(*We must describe after this* [*how to obtain*] *by approximation the side of a cube so that it will be reckoned with when necessary and*

vult querens, scilicet quando vult querens ut raciocinetur per latus
15 cubi cum propinquitate donec non sit inter illud quo raciocinatur et inter veritatem mensure eius, verbi gratia, nisi minus minuto, possit illud. Et si vult ut non sit inter illud et inter veritatem nisi minus secundo aut quacunque parte vult, possit illud per illud quod narrabo, si deus voluerit.

20 Et operatio quidem in illo est ut ponas illud quod est in cubo de unitatibus partium, eius tertia aut sexta aut nona aut ad quemcunque
c. 2 finem volumus ex / finibus computationis. Deinde inquiramus numerum cubi equalem numero partium quas habemus, scilicet secunda (!tertia?) aut sexta aut quecunque sint partes. Ergo inquisitio illius
25 est parva, cum nos non utamur fractionibus in eo quod querimus. Quod si non invenimus numerum cubi equalem numero partium quas habemus, accipimus numerum cubi propinquiorem numero partium que sunt nobiscum. Cum ergo inveniemus numerum cubi propinquiorem, servabimus ipsum. Deinde considerabimus partes que sunt
30 nobiscum. Nam si sunt tertia, tunc numerus harum partium que sunt in latere huius cubi est numerus minutorum que sunt in latere primi. Et si sunt partes sexta, tunc numerus partium que sunt in latere huius cubi est numerus secundorum, que sunt in latere cubi primi. Et secundum hoc exemplum quod est in latere huius cubi in secundis
35 faciemus in eis que sunt preter illud. Et iste quidem modus quem narravimus ex eis super quem non est necesse ponere demonstrationem neque ut addatur ex sermone in eo super illud quod diximus. Quod est quia omnes qui probabiliter dant computationem scient veritatem eius quod diximus quando legent librum nostrum.

40 Iam ergo declaratum est in eo quod narravimus de libro nostro isto super modum extrahendi mensuram linee continentis circulum ex

14 scilicet...querens *om. H* / *de* per *scr. P mg. et Zm supra* vel cum
21 *de* partium *scr. P mg. et Zm supra* in alio, partibus / eius *om. Zm* / nona: 3^a *H*
22 volueris *S* / inquiras *S*
23 habes *S* / scilicet: secundum (?) *H*
24 Ergo: vel, nam *mg. P et Zm supra* non rei *H*
25 non: iam *H* / eo: hoc *S*
26 inveniemus *H*
27 accipiamus *S*
28 Cum: quando *H*
30 sunt[1]: sint *S*
31 primi: primi cubi *S*
32 si *om. S* / partes: ille partes *S*
33 cubi primi *tr. S*
35–79 Et.... dei *om. S*
35 Et: est *H*
36 eis *P* eis est *ZmMa* ei est H
37 in eo: meo (?) *H*
38 omnes: omnis *H* / *ante* probabiliter *del. P* propter
40 isto *om. H*

desires. That is to say, if the investigator wishes to reckon with a side of a cube of such an approximation that the root he reckons with differs from the true measure by an amount that is only less than a minute, he can do this. And if he wishes that the difference between his approximation and the true measure is only less than a second, or than any other part he wishes, he can do this by the means which I shall describe, if God wills.

And the procedure is this*: Convert the cube from units into [sexagesimal] parts, i.e., to thirds, or to sixths, or to ninths, or to any desired limit of computation. Then we seek a cube number equal to the number of parts that we have, i.e., seconds (!thirds?), or sixths, or whatever parts we have. The search is a simple one since we do not use fractions in that which we are seeking; for if we do not find a cube number equal to the number of parts we have, we take the cube number nearest to the number of parts we have. When we find the nearest cube number we shall use it. Then we shall consider the parts we have. If they are thirds, then the number of parts in the side [or root] of this cube is the number of minutes. And if they are sixth parts, then the number of the parts in the side [or root] of this cube is [the number of] seconds. We shall proceed in regard to other parts according to the example of the side of the cube being in seconds. The method which we have described does not demand proof, nor is any commentary necessary beyond what we have said; for all those who calculate in a worthy manner will know the truth of what we have said when they read our book.

Therefore, it has now been shown, in what we have narrated in our book, how to find by the method that Archimedes used the measure of the

* See Commentary for a description of this procedure in modern notation.

[*we must*] *act in this matter with a method which yields the best possible approximation. That is to say, if we wish that between the approximation and the true* [*root*] *there is, for example, less than a minute or a second, we can do it and the procedure is this*: [*We*] *reduce the cube to its third parts, or to sixths, or to ninths, or to other parts. Then we seek the cube equal to the number, if this cube exists. If not, we seek the nearest cube to it. And when we have found it, we take its side* [*i.e., cube root*]. *And if the parts were thirds, then the side is in minutes; and if they were sixths, the side is in seconds. And* [*other*] *problems are treated according to this model.*)

diametro suo per modum quo operatur Archimenides et narravimus demonstrationem secundum quod multplicatio medietatis diametri circuli per medietatem linee continentis est embadum superficiei eius.
45 Et posuimus post illud modum communem quo scitur embadum omnis trianguli acutorum angulorum et orthogonii et ambligonii. Et isto modo quamvis iam usi sint multi hominum et sciverint ipsum, tamen ipsi etiam usi sunt eo, aut plures eorum, secundum modum credulitatis, preter quod sciverint demonstrationem super veritatem eius. Et po-
50 suimus post illud qualiter sciatur embadum superficiei piramidis columne et embadum superficiei sectoris piramidis columne. Et posuimus cum proportionibus id cuius necessitas sequitur nos in scientia embadi superficiei spere et embadi magnitudinis eius, quo fit iuvamentum in aliis ex scientia geometrie. Deinde posuimus post illud in eo qualiter
55 scitur embadum superficiei spere et ostendimus quia est equale quadruplo superficiei maioris circuli qui cadit in spera. Et narravimus qualiter scitur embadum magnitudinis spere et ostendimus quia est illud quod fit ex multiplicatione medietatis diametri spere in tertiam embadi superficiei eius. Et invenimus super illud quod narravimus de
60 illo cum demonstratione sufficientiam faciente geometrica. Et posuimus post illud qualiter ponantur due quantitates inter duas quantitates ita ut continuentur proportionaliter.

Et omne quod posuimus in libro nostro isto est ex eo quod nos docuimus nisi scientia extrahendi mensuram linee continentis circu-
65 lum ex diametro suo, nam est opus Archimenidis, et nisi scientia
63v c. 1 po/nendi duas quantitates inter duas quantitates ut continuentur secundum proportionem unam. Nam postquam posuimus nos in libro nostro isto opus quod operatus est Mileus in illo, posuimus cum eo illud quod fecimus nos in eo. Et posuimus iterum qualiter dividatur
70 angulus in tres divisiones equales.

42 Archimenides *Zm HR* Archiminides *P* Archeminades *Ma*
49 veritatem eius *tr. H*
50–51 piramidis... superficiei *PZm mg. H*
52 cum *om. H qui habet lacunam* / id: illud *H* / scientia: tertiam *H*
53 quo *PZm* que *H* qua *Ma* qui *R*
54 ex scientia geometrie *PZm* sex scienciam geometrie *Ma* ex sua geometria *H* / post *PZm om. H* postea *Ma*
57 magnitudinis *corr. ex* superficiei *in PZmH*; *scr. et del. Ma*
60–61 *de* posuimus *scr. P mg.* in alio, fecimus
61 *de* post illud *scr. P mg. et Zm supra* in alio, tunc
62 continuantur (?) *H*
63 in *om. H*
64 docuimus *PMa* in alio, fecimus *mg. P et Zm supra* determinavimus *H*
65 opus: operatio *H*
67 nos *om. H*
68 Mileus: Melius *H* Milleus *Zm*
70 angulus: triangulus *H*

circumference of a circle from its diameter. And we have described the demonstration according to which the multiplication of the radius by one half of the circumference is equal the area of its surface. And we have posed after that a common method by which is known the area of any triangle: acute, right, or obtuse. And although many people used this method and knew it, still even they—or [at least] many of them—used it credulously without knowing the demonstration of its truth. And then we posed how the surface of a cone is known and [as well] the surface area of a segment of a cone. And then we put forth with ratios that which we need for the knowledge of the surface area and volume of a sphere, which is of assistance for other matters of geometry. Then we posed after that how the surface area of a sphere is known and we demonstrated that it is equal to quadruple the area of a great circle of the sphere. And we described how the volume of a sphere is known and we demonstrated that it is equal to the product of the radius of the sphere and one third of its surface area. And we found that which we described by means of sufficient geometric demonstration. And then we posed how to place two quantities between two quantities so that they are [all] in continued proportion.

And everything which we have put in our book is our own teaching except the knowledge of finding the measure of the circumference of a circle from its diameter, for that is the work of Archimedes, and except the knowledge of placing two quantities between two quantities so that they are [all] in continued proportion. For although we have posed in our book in regard to the matter [of the two mean proportionals] the method that Menelaus fashioned, we put forth in addition our own method concerning it. And further we posed how to trisect an angle.

67–79 Nam....dei: فانه من عمل مانالاوس كما مر ذكره والحمد لله وحده تم الكتاب (*And it is from the work of Menelaus just as he has described it. Praise to God alone, the book is finished.*) ((al-Ṭūsī adds after the end of the text another proof of Proposition VII closer to that of Hero, but attributes it to one al-Khāzin.))

Et harum quidem rerum quas narravimus in nostro libro cognitio apud omnes qui querunt scientiam geometrie et computationis est magne quantitatis et animosa est earum operatio, et quod sequitur querentes hanc scientiam de necessitate earum. Nam scientia superficiei spere et magnitudinis eius que est ex eis proprie est ex eis quibus non vidi in illis quos testificati sumus ex illis qui sunt nostri temporis aliquem qui sciat modum quo computet ea convenientem veritati in illo qui vocet ut sciatur demonstratio super operationem eius. Completus est liber auxilio dei.*

* The Latinity of this presumably literal translation from the Arabic leaves much to be desired, and I am not at all sure my translation reflects the ultimate meaning of the Arabic authors. Unfortunately, this passage is missing in the Arabic version.

---

72 qui: quia *H*
73 est... operatio *om. H*
76 in: ex *H* / nostri: tante *H*
77 eam *H* / conveniente *Zm*
78 *post* eius *add. H* etc (?)
78–79 Completus... dei *om. Zm*
79 *post* liber *add. H* trium fratrum / *post* dei *add. Ma* Amen

And indeed the understanding of all these things we have recounted in our book is of great moment for all those who seek a knowledge of geometry and computation, and the use [of these things] is vital and they are necessary for those who seek this knowledge. For the knowledge of the surface and volume of a sphere which is one of the things [presented here] is properly a part of those things which no one of our time, as far as I have seen, knows how to compute by a method according with the truth [which is] in one who claims to know the demonstration of his method. This book has been completed with the help of God.

COMMENTARY

Proemium

4–34 "Propterea... Et." Notice that the Arabic text of al-Ṭūsī omits this first section of the Proemium. Latin manuscripts *S* and *T* omit much the same material, except that they both begin in line 35 and thus contain one line less of the Proemium. It is evident that the scribe of *S* had seen not only a truncated version (see division 2 of the Introduction) but also a fuller copy of the *Verba filiorum* which included the omitted material, for he notes that the *incipit* of the treatise is "Propterea quia vidimus" (see variant readings).

5 "magnitudinis." This is the word commonly used by Gerard to render what we call "volume."

44–46 "Euclides... latitudinem." Notice that when the Banū Mūsā cite Euclid's *Elements*, they do not give the specific number of the definition, axiom, or proposition, or even book. It is for this reason that, when in the marginal note to Proposition XI (lines 57, 64) the specific proposition number is cited, the genuineness of the note becomes suspect.

81 "quadratura." This is an unusual use of the word. Here it does not mean "quadrature" in the sense of "squaring" but rather it means "squareness," i.e., the unit-measure's character as a square. Cf. the Arabic reading in the Arabic variant readings.

Proposition I

1 "Figure laterate." The authors mean any "regular polygon." Generally this phrase is a shorthand way of indicating a regular polygon.

5–21 "Verbi.... voluimus." Compare the reading of *S* given in the variant readings. I translate this as follows: "It is proved by [1] resolving the figure containing the circle into triangles so that if it (the figure) is a triangle, [the resolution is] into 3 triangles, and by [2] drawing lines from the center of the circle to the points of contact of the circle with the figure—each line being a radius of the circle and falling perpendicularly on a side of the figure, i.e., the triangle, and by [3] using the following proposition: the area of any triangle arises from the multiplication of its altitude by one half of the base on which the altitude falls, [i.e., by using this proposition] as many times as there are triangles into which the figure is resolved."

18–21 "Et... corporis." This corollary concerns a regular polyhedron. As al-Ṭūsī suggests (see Arabic variant readings), this corollary can be proved by dividing the polyhedron into pyramids whose vertices are at the center of the sphere about which it is circumscribed and whose bases are the faces of the polyhedron, while the altitudes of these pyramids are radii of the sphere. Then the volume of each pyramid is $V = \frac{1}{3} a \cdot b$, where a is a radius and b is a face or base area. Thus the volume of the whole polyhedron becomes $V = \frac{1}{3} a \cdot B$, where B is the sum of all the b's of the individual pyramids.

Proposition II

5–20 "Verbi.... circuli." Manuscript *S* says only "He proves this by the preceding [proposition]." The proposition obviously assumes that an inscribed regular polygon must be less than the circle in which it is inscribed.

19–21 "Et... spere." By analogy the authors extend the proposition to a [regular] polyhedron inscribed in a sphere, so that $\frac{1}{3} r \cdot A <$ vol. sphere, where r is the radius and A is the surface area of the inscribed polyhedron. No proof is given.

Proposition III

1–36 "Si.... figure." This has as its basic purpose to demonstrate that when there is a circumference greater than (or less than) a given straight line, it is possible to find the perimeter of a regular inscribed (or circumscribed) polygon which is greater than (or less than) the given straight line. This is somewhat similar to the basic supposition specified by

Albert of Saxony, and tacitly assumed by the author of the Pseudo-Bradwardine text presented in Chapter V. This supposition is this: Given two unequal but continuous magnitudes, it is always possible to take away from the greater magnitude a third magnitude which is less than the greater of the two given magnitudes and greater than the lesser. Compare the Introduction to Chapter Five.

17–19 "Et... *EDZ*." One supposes that the authors would do this by the authority of Proposition XII.16 of the *Elements*, although that proposition is not cited. It should be pointed out here that Proposition XII.16 uses the corollary to X.1 in its proof, a corollary apparently missing in the Arabic versions of the *Elements*. See Chapter Three, page 60, footnote 1.

36 "voluimus." Here, al-Ṭūsī adds that the proposition is based on the assumption of the existence of a circle whose circumference is equal to a straight line (see Arabic variant readings and also page 63 above).

Proposition IV

1–38 Compare the proof of Leonardo Pisano in his *Practica geometrie*, edition of B. Boncompagni, *Scritti di Leonardo Pisano*, vol. 2 (Rome, 1862), p. 87, which from the form of the enunciation I would say originated with the Banū Mūsā text, although Leonardo does not spell out the proof in the same detail.

39–45 "Et.... declarare." This is the same corollary given in the Arabic text of the *De mensura circuli* and which, I suggested in Chapter Two, Section 2, appears to have been part of the Greek text in Hero's time. Cf. Hero of Alexandria, *Metrica*, I. 37, ed. of H. Schöne (Leipzig, 1903), p. 86.

Proposition V

1–39 "Verbi.... voluimus." Cf. *De curvis superficiebus*, Proposition III, as given below in Chapter Six. Compare also Pappus, *Collectio*, Book VIII. Proposition 22, ed. of F. Hultsch, vol. 3 (Berlin, 1878), p. 1105, Euclid XII. 2 is the basis of all three proofs (and the Banū Mūsā cite XII. 2, although not by number, in lines 25–28). It should be noticed that the Banū Mūsā make an interesting shift in the wording of the proposition. They abandon the earlier proportionality statement to the effect that the circumferences are to each other as the diameters, and they substitute instead that the ratio of any diameter to its circumference is a constant. This leads into the sixth proposition where such a constant ratio is determined.

Proposition VI

2–4 "proportionem... ipsum." As stated here and in Proposition V, the authors appear to be seeking the reciprocal of π. But actually, as the determination is detailed, we find that it is the ratio of the circumference to the diameter that is presented.

7–22 "Et.... Archimenides." Notice that the authors not only indicate the irrationality of the ratio of the diameter to the circumference but add that an approximation of the true ratio can be made which differs from the ratio by less than a minute (i.e., by less than a ratio of 1/60 of the diameter to 1), or by less than a second (i.e., by less than a ratio of 1/3600 of the diameter to 1), or by less than some other smaller fraction. Actually, of course, fractions used in this proposition are those of Archimedes and Eutocius and are not expressed as sexagesimal fractions. These remarks of the Banū Mūsā concerning approximation should be compared with the later statements of Proposition XIX. If the reader compares those figures used which go beyond the figures employed by Archimedes, he will see that they almost all agree with Eutocius' figures as given in his *Commentarius in dimensionem circuli*, ed. of J. L. Heiberg, *Archimedis opera omnia*, vol. *3* (Leipzig, 1915), pp. 232–58. This coincidence of the figures given by the Banū Mūsā with those of Eutocius would appear to be evidence in support of the view that Eutocius' commentary was known and used by the Islamic mathematicians. For the entirely different figures of Leonardo Pisano in his *Practica geometrie*, see *Scritti di Leonardo Pisano*, ed. of B. Boncompagni, vol. *2* (Rome, 1862), pp. 88–91.

28–31 "Manifestum... *ATB*." Compare Eutocius, *loc. cit.* I shall not specify the specific line numbers of the Eutocius text here and in the succeeding comments since the reader can easily follow them in the Heiberg text.

45–46 "Ergo... quinque." This agrees with the translation of the *De mensura circuli* possibly done by Plato of Tivoli (see lines 81–82 of that translation), where *EC* (equivalent to *GB* in the *Verba filiorum* and *EG* in the Greek text) is obviously assumed to be greater than 265, for it is said that "$EC/FC > 265/153$." In the Greek text of the *De mensura* EG/GZ is taken as equal to 265/153. This is repeated by Eutocius, who adds however that the square root of 70227 is 265 plus a very small fraction.

63–65 "Et... quarta." Where the Banū Mūsā have $1350534\frac{1}{4}$, Eutocius has $1350534 + \frac{1}{2} + \frac{1}{64}$. Note that the Florence Version of Proposition III (see Chapter Three, Section 3, line 76) gives the correct figure as an improper fraction.

66–68 "Et... quarta." Where the Banū Mūsā have $1373943\frac{1}{4}$, Eutocius has $1373943 + \frac{1}{2} + \frac{1}{64}$. Again note that the Florence Version (line 87) has the correct figure as an improper fraction.

73–76 "erit... tribus." Eutocius gives EG (equivalent to GB in the *Verba filiorum*) as equal to $2334\frac{1}{4}$ and $EG^2 = 5448723\frac{1}{16}$, while the Banū Mūsā simply say that $BG > 2334\frac{1}{4}$ and $GB^2 > 5448723$.

77–79 "Ergo... duobus[2]." Eutocius says that EK^2 (equivalent to GD^2 in the *Verba filiorum*) equals $5472132\frac{1}{16}$ while the Banū Mūsā say that $GD^2 > 5472132$.

87–88 "Et... diametrum." The authors have inadvertently reversed the order of the terms here, since GB/BH is equal to the ratio of the diameter to one side of the polygon having 96 sides. But in the next sentence the proper order is restored. Notice the absence here of the magnitude $H[M]$ which is the actual side of the polygon, although this side is specified in the Greek text (but there as LM). The Banū Mūsā did not bother to specify the side itself, since the ratio they do give, namely, GB/BH, is equivalent to the ratio of the diameter to the side, GB being half the diameter and BH half the side.

155–57 "Et... octo." Eutocius has AL^2 (equivalent to AM^2 in the *Verba filiorum*) equal to $4064928\frac{1}{36}$ while the Banū Mūsā say that $AM^2 < 4064928$.

158–60 "Ergo... quattuor." Eutocius has AG^2 (equivalent to AB^2 in the *Verba filiorum*) equal to $4069284\frac{1}{36}$ while the Banū Mūsā say that $AB^2 < 4069284$.

Proposition VII

1–124 "Volo.... complete." Some of the interesting history of this so-called "Hero's formula" is traced in Appendix IV. Suffice to say here, the proposition as given in this translation is the first treatment of the formula in Latin which also contains a proof, although the enunciation of the formula had been made by the *agrimensores* and in Plato of Tivoli's translation of Savasorda's *Liber embadorum*. Incidentally, F. Hultsch in his article on the history of Hero's formula (see Appendix IV) feels that Hero's proof is simpler and more elegant than the rather redundant proof of the Banū Mūsā.

83, 90 In the variant readings under these line numbers will be found two marginal notes that appear in P and Zm. Before considering their origin and significance, I shall first give them in translation: Note to line 83:

"(1) But $(EZ^2 \cdot AH) = [(AH \cdot EZ) \cdot EZ]$*. (2) Therefore, $[(AH \cdot EZ) \cdot EZ] = [(HB \cdot BZ) \cdot ZA]$. Let us multiply both products by AH.** (3) Then $[(HB \cdot BZ) \cdot ZA] \cdot AH = [(EZ \cdot AH) \cdot EZ] \cdot AH$. (4) But $(EZ \cdot AH)$ = area triangle. Therefore, [on the right side of equation (3)] there will be three quantities: Area triangle, EZ, and AH. (5) Hence $[(\text{Area } \triangle \cdot EZ) \cdot AH] = [(AH \cdot EZ) \cdot \text{Area } \triangle]$.+ But [as was said above in (4)] $(AH \cdot EZ)$ = area triangle. (6) Therefore, [substituting the equation of (4) twice in the equation of (3)] $(\text{Area } \triangle)^2 = [(HB \cdot BZ) \cdot ZA] \cdot AH$. And this is what we wished to demonstrate." Note to line 90: "Hence it has now been demonstrated that the multiplication of anyone of the lines AZ, ZB, BH by the product of the other two,++ and then by the semiperimeter, is equal to the area of the triangle multiplied by itself. And this is what we wished to demonstrate."

In note 7 of division 2 of the Introduction to this chapter I touched on the vexing question of the origin and proper position of these two marginal notes. But I must take up the matter in more detail. In manuscripts *P* and *Zm* the two notes appear separately in the margin. But in manuscript *H* the notes are joined and put into the text after "trianguli" in line 83. They there constitute a conclusion to the so-called first proof of Proposition VII. Accordingly, *H* transfers what are lines 83–106 in the text to a point after *BH* in line 119. There these lines become the conclusion of the second proof of Proposition VII. Actually, in *P* and *Zm* a lacuna follows line 119 and nothing more is given until the beginning of Proposition VIII.

Let us first consider certain aspects of *H*'s rearrangement of the text. *H* is certainly wrong in putting the two notes together, for the first note clearly is a complete conclusion to the argument, while the second note presents the last step in a similar concluding argument. And so if we put both of the notes in the text we not only have the last mathematical step of the argument twice (which is conceivable but not likely) but we also have the concluding phrase: "And this is

* I have used the parentheses and brackets, which are of course unnecessary in this modern notation, simply to indicate the successive multiplications given by the rhetorical expressions of the text.

** More literally: "Let us posit *AH* as common."

\+ Simply by rearranging the multipliers.

\+\+ This I believe to be the sense of the Latin, which is confusing.

what we wished to demonstrate" twice (and this is not probable). But suppose we accept for the moment that the first note of line 83 is the conclusion of the first proof, will *H*'s transference of lines 83–106 as a conclusion of the second proof stand-up? I think the answer must be in the negative. In the first place, *H* is the only copy to make this transference; for not only do the Latin manuscripts *P*, *Zm*, and *Ma* fail to have this arrangement but it is also absent in the Arabic text of al-Ṭūsī. And in the second place, lines 83–106 do not themselves form a sufficient conclusion mathematically to what has been given as the beginning of the second proof. That is to say, we must still add further steps before these lines become an acceptable conclusion. Let me explain this in more detail. As one can see from the text, the second "proof" starts off with the statement [1] $EZ/ZB = BH/HT$, which had been given earlier in line 64. The authors then show: [2] $EZ/HT = (EZ/ZB) \cdot (EZ/BH)$. Following this, the text breaks off in manuscripts *P* and *Zm*. But the Arabic text of al-Ṭūsī follows statement [2] by asserting, as the proof demands, the following:* [3] $EZ/HT = AZ/AH$ (compare line 67). It then concludes from [2] and [3] together: [4] $AZ/AH = EZ^2/(BH \cdot ZB)$, and its equivalent $(AZ \cdot BH \cdot ZB) = (EZ^2 \cdot AH)$. Then, as the Arabic text points out, we can complete the proof as in the first case (i.e., by using the procedure of lines 78–106). If, instead of doing what the Arabic text does, we follow *H* and add lines 83–90 after statement [2], it will be evident that statements [3] and [4], which are necessary parts of the proof, will be omitted since lines 83–106 only complete the proof after the derivation of [4] is assumed. Thus, no matter what we do about adding lines 83–106, we must add the further steps to complete the proof suggested by [1] and [2]. On the whole, it seems more probable that al-Ṭūsī is giving us what was essentially in the original Arabic text and that the "second proof" in the Arabic text was merely an alternate way to derive formula [4], i.e., to arrive at the same conclusion that was derived in a different way in the first proof and presented in lines 71–73.

But let us return to the question of the nature and origin of the two marginal notes. A careful study of the first note will show that it is an argument very much like that given in lines 78–106. In fact, the only differences are these: (a) while a number of the steps are common, the order of these steps differs in the two versions; and (b) in the note,

* I have changed al-Ṭūsī's lettering to the equivalent lettering used by Gerard.

the equation of step (2) is expanded to the basic equation of step (3), and thus ultimately to the equation of step (6), by the multiplication of each side of the equation by AH (a procedure that is also found in the text of al-Ṭūsī). But in lines 78–106 the same objective is achieved by dividing the equation $(EZ \cdot AH) = (\text{area } \triangle)$ by the equation $(EZ \cdot \text{area } \triangle) = (AZ \cdot ZB \cdot BH)$. The fact that al-Ṭūsī uses the procedure followed in the note seems to favor the conclusion that the note was added by Gerard from an alternate Arabic reading, a reading that was eventually taken up by al-Ṭūsī. On the other hand, two considerations throw some doubt on this conclusion. The first is that the introductory phrase *in alio* that precedes such alternate Arabic readings elsewhere in the text is missing in this case. Furthermore, it must be remembered that the second note (i.e., the note to line 90) is definitely labeled as an alternate reading, having the introductory phrase *in alio*, and this second note is an alternate reading for the last step of the argument that appears in both the first note and the text. Thus, if we accept the first note as an alternate Arabic reading, we would have to conclude that for the last step of the argument Gerard had before him *three* Arabic readings. But nowhere else in the text is there any evidence of three texts.

If the note to line 83 is not an alternate Arabic reading to lines 78–106 ,we are left with two other possibilities: (1) The note is from an Arabic text but applies to the conclusion of the second proof rather than the first. There is no positive evidence to support this conclusion. Furthermore, it is obvious that one would still need some additional material, namely, the assertion of step [3], even if he took this note as a conclusion for the second proof. I have already noted that no conclusion, other than the kind of statement found in al-Ṭūsī's text, was needed to complete the second proof. (2) The note was composed by Gerard of Cremona or some other Latin author with two objectives: to give a slightly different development of the conclusion from that found in the text and to reduce the argument to one involving only specific quantities rather than general statements. This seems to me to be a strong possibility.

111–19 "Ergo... *BH*." The reasoning in terms of specific quantities, as expressed in this sentence, is as follows: If (1) $EZ/ZB = BH/HT$, then (2) $EZ/HT = (EZ/ZB) \cdot (ZB/HT)$. From (1) we also know that (3) $ZB/HT = EZ/BH$. And so (4) $EZ/HT = (EZ/ZB) \cdot (EZ/HB)$. Q.E.D.

Proposition VIII

1–30 "Cum.... eius." Cf. the proof of Leonardo Pisano, *Scritti* (ed. of B. Boncompagni), vol. 2, (Rome, 1862), pp. 178–79.

25 "*KN*." At this point al-Ṭūsī cites Proposition II [of the first book] of the *Spherics* of Theodosius; actually, it is the corollary to that proposition that is pertinent.

Proposition IX

1–77 "Cum.... voluimus." For this proposition, cf. Archimedes, *De sphaera et cylindro*, Propositions 7–10, 14. Cf. also the proof of Leonardo Pisano, *Scritti*, vol. 2, pp. 179–80. It is obvious that Archimedes' Proposition 14, which is $A = \frac{1}{2} c \cdot r'$ where $r'^2 = s \cdot r$, reduces to the proposition of the Banū Mūsā, i.e., to $A = \frac{1}{2} c \cdot s$, s being the slant height, c and r being respectively the circumference and radius of the base circle. For if $A = \pi r'^2$, then $A = \pi \cdot s \cdot r$; and with $\pi = c/2r$, $A = \frac{1}{2} c \cdot s$. Cf. also Proposition I of the *De curvis superficiebus*, Chapter Six below. Finally, cf. Pappus *Math. Coll.*, VII, Proposition 165.

Proposition X

1–34 "Omnis.... eius." Cf. Proposition II of the short tract on the hyperbola edited in *Osiris*, vol *11* (1954), p. 370. Cf. also Leonardo Pisano, *Practica geometrie*, ed. cit. (pp. 180–81); this was drawn from the Banū Mūsā text.

30–31 "circumferentie... UTZ." Actually, we cannot call this "the circumference of circle *UTZ*" until the next sentence. It should read rather "line *UTZ*." After we have stated that all the lines from *H* to *UTZ* are equal to *UH* or *HZ*, then we can call *UTZ* the circumference of the circle with *H* as its center.

Proposition XI

1–96 "In.... eius." That this proposition follows from Proposition IX is shown much more economically by the use of modern notation. Assume that the segment of the cone has c_1 as the circumference of its lower base, c_2 as the circumference of the upper face, s as its slant height. Make a complete cone out of the segment. This cone has c_1 as the circumference of its base, $(p + s)$ as its slant height, and A_1 as its lateral surface area. In completing the larger cone, a smaller cone was added to the segment. That cone has c_2 as its base circumference, p as its slant height, and A_2 as its area. From Proposition IX,

$A_1 = \frac{1}{2} c_1 \cdot (p + s)$ and $A_2 = \frac{1}{2} c_2 \cdot p$.

Now [1]

$$A \text{ seg} = A_1 - A_2 = \tfrac{1}{2} c_1 (p + s) - \tfrac{1}{2} c_2 \cdot p$$
$$= (\tfrac{1}{2} c_1 \cdot p) + (\tfrac{1}{2} c_1 \cdot s) - (\tfrac{1}{2} c_2 \cdot p).$$

But $c_1/c_2 = (p + s)/p$ [by similar triangles].

Or [2] $c_1 \cdot p = (c_2 \cdot p) + (c_2 \cdot s)$. Substituting [2] in [1], we arrive at the required proposition, namely, A seg $= \frac{1}{2} s \cdot (c_1 + c_2)$. Cf. Leonardo Pisano, *Practica geometrie*, ed. cit., pp. 181–82; this was drawn directly from the Banū Mūsā text. Cf. Archimedes, *De sphaera et cylindro*, Proposition 16. Cf. also *De curvis superficiebus*, Proposition IV (see Chapter Six below). It is evident that the original proposition of Archimedes has been considerably altered by the author of the *De curvis superficiebus* and by the Banū Mūsā.

52–71 The material in brackets is added in the margin of *P*, *Zm*, and *Ma* and in the text of *H* and *R*. I have put it in brackets because I believe it to be an addition of Gerard's rather than a part of the text. For it is missing in the Arabic text and it cites a specific proposition of Euclid, a practice not followed in the Banū Mūsā.

Proposition XII

1–76 "Cum.... voluimus." Cf. Archimedes, *De sphaera et cylindro*, Propositions 21 and 22, and the *De curvis superficiebus*, Proposition V. Again we note that Leonardo Pisano drew from the Banū Mūsā: see Leonardo Pisano, *Practica geometrie*, ed. cit., p. 183.

43–76 "Si.... voluimus." This passage is an extension of, or corollary to, the enunciation which does not indicate the object of this additional proof.

Proposition XIII

1–153 "Cum.... eius." Cf. Archimedes. *De sphaera et cylindro*, Propositions 25 and 30. Also see Leonardo Pisano's *Practica geometrie*, ed. cit., 183–85.

77–79 "quantitatem... circumdans." This is the rhetorical expression used by the Banū Mūsā to designate what came to be designated as π. Notice that Leonardo Pisano, in taking over this proof, simply substitutes "3 1/7" whenever this expression appears in his proof (*Ibid.* 184–85).

97–118 "Quod.... forma." This is the section proving that $A = \pi r^2$. The more common formulation for the area of the circle in antiquity and the Middle Ages is $A = \frac{1}{2} c \cdot r$, as it appears in the *De mensura circuli* of Archimedes.

Proposition XIV

1–38 "Embadum.... eius." Cf. Archimedes, *De sphaera et cylindro*, Proposition 33, and the *De curvis superficiebus*, Proposition VI, Corollary. See also Leonardo Pisano, *Practica geometrie*, ed. cit., 185–86, and the anonymous *De ysoperimetris*, Appendix III, paragraph 8.

Proposition XV

1–38 "Multiplicatio.... voluimus. Cf. Archimedes, *De sphaera et cylindro*, Proposition 34, and the *De curvis superficiebus*, Proposition VIII. See Leonardo Pisano, *Practica geometrie*, ed. cit., 186–87 and the anonymous *De ysoperimetris*, Appendix III, paragraph 8.

Proposition XVI

6–10 "Et... cubi." The substance of this statement in modern notation is this: If a and 1 are the given quantities, a being equal to the cube, and $a/x = x/y = y/1$, x and y being the mean proportionals, then $y = \sqrt[3]{a}$ for $a \cdot y = x^2$, and $x = y^2$. Thus $a \cdot y = y^4$, or $a = y^3$.

12 "Mileus... geometria." The Banū Mūsā say that they took this proof from a *Liber in geometria* attributed to Mileus (i.e., Menelaus, as in the Arabic text*). While such a Greek text has not survived, we find this work mentioned in the *Fihrist* (cf. H. Suter, "Das Mathematiker-Verzeichniss im Fihrist des Ibn Abî Ja'ḳûb an-Nadîm," *Abhandlungen zur Geschichte der Mathematik*, 6. Heft [1892], p. 19). This work is also mentioned by al-Bīrūnī (Suter, *Bibliotheca Mathematica*, 3. Folge, vol. *11*, p. 69). Menelaus probably took it from Archytas (see next remarks).

12–62 "Sint.... ostendere." While attributed here to Menelaus, this is the solution which Eutocius in his commentary on the *De sphaera et cylindro* (ed. of Heiberg, *Arch. opera omnia*, vol. *3*, pp. 84.12–88.2) assigns to Archytas on the authority of Eudemus. Cf. the English translation of the passage by Ivor Thomas, *Selections Illustrating the History of Greek Mathematics*, vol. *1* (London, Cambridge, Mass., 1951), pp. 284–89. Thomas analyses the solution in modern notation which I have here adapted to the lettering found in the *Verba filiorum* (see Fig. 51 in the text): Take AB as the x axis, a perpendicular to A in the plane of $ADBG$ as the y axis, and a perpendicular to the plane of $ADBG$ at A as the z axis. AB, AG are the two given quantities, and we let $AB = a$, $AG = b$. The solution depends on finding point

* Obviously the Arabic form was without any diacritical mark for the nūn, so that Gerard read ميلاوس instead of منلاوس.

H, which is the intersection of three curves: (1) the cylinder $x^2 + y^2 = a \cdot x$, (2) the curve formed by the motion of semicircle AHE about A (a tore of inner diameter zero), $x^2 + y^2 + z^2 = a\sqrt{x^2 + y^2}$, and (3) the cone $x^2 + y^2 + z^2 = (a/b)^2 \cdot x^2$. Since H is the intersection, $AH = \sqrt{x^2 + y^2 + z^2}$, $AT = \sqrt{x^2 + y^2}$. From (2) it follows directly that $AH^2 = a \cdot AT$, or $a/AH = AH/AT$. From (1) and (3) it follows that $x^2 + y^2 + z^2 = (x^2 + y^2)^2/b^2$ and hence $\sqrt{x^2 + y^2 + z^2} = (x^2 + y^2)/b$, i.e., $AH = AT^2/b$, or $AH/AT = AT/b$. And so $a/AH = AH/AT = AT/b$, and thus AH and AT are the mean proportionals between a and b. As I have indicated in Appendix V, this proof was borrowed from the Banū Mūsā by Leonardo Pisano in his *Practica geometrie* and by Jordanus in his *De triangulis*. Of interest concerning the connection between Menelaus and Eutocius is the comment of J. L. Heiberg and E. Wiedemann (*Bibl. Math.*, 3. Folge, vol. *10*. 1910–11, p. 203). For the existence of the Eutocius commentary in Arabic manuscripts, see Appendix V, note 6.

Proposition XVII

1–60 "Et.... ostendere." This solution which the Banū Mūsā give as their own is very much like that given as Plato's solution by Eutocius (*Archimedis opera omnia*, vol. *3*, 56.13–58.14). Cf. Thomas' English translation, *op cit.* in previous comment, pp. 263-67, and T. Heath, *A History of Greek Mathematics*, vol. *1*, pp. 255–58. The reader will find Heath's analytical examination of this solution of interest.

For the additional mechanical device described in the Arabic text of al-Ṭūsī but not given by Gerard, see H. Suter, *Bibliotheca Mathematica*, *3*. Folge, vol. *3* (1902), pp. 269–70.

Proposition XVIII

1 "Ingenium." The reference is to a device or procedure similar to the one used in the second solution of the mean proportional problem, i.e., the procedure that allows us to move a line so that its extremity follows a curve while it passes through a certain point, the line being posited as undetermined in length. See also Appendix VI.

1–46 "Et.... figure." This solution of the problem is a mechanical one that serves to solve the problem as reduced to a *neusis* (i.e., to a "verging" problem). That is, it gives a mechanical procedure to find the crucial point S through which line ZH must pass in verging toward point E; for with the *neusis* solved, the trisection is solved. This mechanical procedure of moving ZH so that Z adheres continually to the circumference while ZH continually passes through point E permits Q (chosen on ZH so that ZQ is equal to the radius) to trace a conchoid which intersects BZ at S. This is a conchoid referred to a circular base, namely, the circumference of the circle. While this exact solution is not known in antiquity, a similar reduction of the trisection problem to a *neusis* is found in lemma eight of the *Liber assumptorum* (or *Lemmata*) found only in Arabic but attributed to Archimedes. For a discussion of both of these propositions and the subsequent history of the solution of the problem by the Banū Mūsā, see Appendix VI.

Proposition XIX

1–39 "Et.... nostrum." This method of approximating a cube number is a simple one. It involves the conversion of the number whose cube root is desired to an improper sexagesimal fraction such that the denominator is a perfect cube, and then finding the perfect cube number which is nearest to the numerator (i.e., $a^{\frac{1}{3}} = \sqrt[3]{a \cdot 60^n}/60^{\frac{n}{3}}$ where a is the number whose cube root we seek and n is some multiple of 3). As the authors suggest, if the original number is converted to "thirds" (i.e., $[a \cdot 60^3]/60^3$), then the root will be expressed in "minutes" (i.e., $a^{\frac{1}{3}} = \sqrt[3]{a \cdot 60^3}/60$); or if the number is converted to "sixths" (i.e., $[a \cdot 60^6]/60^6$), then the root will be expressed in "seconds" (i.e., $a^{\frac{1}{3}} = \sqrt[3]{a \cdot 60^6}/60^2$. If used seriously, this system requires an extensive table of cube numbers and roots.

Chapter five

Further Versions of the *De mensura circuli*

In Chapter Five I propose to include three more versions of Proposition I of the *De mensura circuli* of Archimedes. They could very well have been included in Chapter Three with the other emended versions of the *De mensura circuli* since they obviously resemble them in some respects. But I have put them together in this chapter because two of them, the Pseudo-Bradwardine Version and the Version of Albert of Saxony, do not directly use Proposition X.1 of Euclid's *Elements* as fundamental to their proofs, as do the emended versions of Chapter Three, while the third version, the so-called *Versio abbreviata*—even though it employs Proposition X.1—is fundamentally a shortened version of the Pseudo-Bradwardine text. Instead of using Proposition X.1, both the Pseudo-Bradwardine and Albert of Saxony versions assert the possibility (in the first part of the proof) of finding an inscribed regular polygon greater than a given surface (the right triangle with circumference and radius as the sides including the right angle) because of the continuous divisibility of the assumed excess of the circle over the triangle. That is, they do not start with a given regular polygon whose area differs from that of the circle by an amount greater than that by which the circle is said to exceed the triangle and then proceed on the authority of Proposition X.1 to reach a polygon whose area differs from that of the circle by an amount less than the assumed excess of the circle over the triangle, as was done in the emended versions of Chapter Three. Similarly, the proof developed in the first four propositions of the *Verba filiorum* given in Chapter Four does not directly use Proposition X.1 for asserting that we can obtain the desired polygon whose perimeter, itself less than the circumference, would be greater than a given line assumed to be less than the circumference of the circle, although in proving their assertion the Banū Mūsā appear to assume without specification Proposition XII.16, which uses the corollary to Proposition X.1 in its proof. In effect, all of the versions which do not directly use Proposition X.1 rest on the assumption made specific by Albert of Saxony that, given two unequal continuous (and comparable) magnitudes, it is always possible to take from the greater a third magnitude greater than the lesser and less than the greater of the two given magnitudes, a species of what Murdoch has called "betweenness" postulates.[1] The third magnitude is either a regular polygon greater than a triangle (as in the Pseudo-Bradwardine and Albert of Saxony versions) or a straight line (equal to the perimeter of such a regular polygon) greater than a given straight line and less than the circumference of the circle (as in the proof of the *Verba filiorum*). I do not mean to assert that either the author of the Pseudo-Bradwardine Version or Albert of Saxony directly followed the proof of the *Verba filiorum*, since the actual form of their proofs differs considerably from that of the *Verba filiorum*. But even granting the evident influences of the tradition of the emended versions of Chapter Three on the proofs of Pseudo-Bradwardine and Al-

[1] As Mr. Murdoch notes in his paper, "The Medieval Language of Proportions," in A. C. Crombie, ed., *Scientific Change* (New York, 1963), pp. 247–48 "betweenness" postulates assume the following: for all magnitudes a and b, when $a > b$, there is a third magnitude c such that $a > b > c$. In speaking of Albert's axiom, Mr. Murdoch adds, "This may be an odd way to state the property betweenness, but Albert is tailoring it to fit the Archimedean quadrature procedure of the *De mensura circuli*. By appeal to this, rather than to the traditional X.1, he has avoided the successive construction of inscribed and circumscribed polygons with increasing numbers of sides. This circumvention is effected because Albert has indeed packed a generalized X.1 (together with various assumptions validating the existence of additional magnitudes from consideration of the sums and differences of given magnitudes) into his postulate of continuity. Also notable is his grounding of his assumption upon the properties of divisibility of the geometric continuum, a move characteristic of the later medieval philosopher-mathematician." A "betweenness" proposition similar to Albert's postulate is found in Archimedes, *On Spiral Lines*, Proposition IV. This proposition could easily have been known to Albert of Saxony, since it was present in the form of the Moerbeke translation as Proposition II of a hybrid tract on quadrature inserted in the midst of Johannes de Muris' *De arte mensurandi* (MS Paris, BN lat. 7380, 59r): "2ª. Duabus lineis datis inequalibus, recta scilicet et circuli periferia, possibile est accipere rectam maiore datarum linearum minorem, minore autem maiorem." Cf. the Moerbeke translation in Vat. Ottob. lat. 1850, 12r. The hybrid tract is dated at Paris in 1340. See M. Clagett on Johannes de Muris in *Isis*, vol. *43* (1952), pp. 236–42. This tract will be discussed at length in Volume II.

bert of Saxony, it is not beyond possibility that these proofs were in some fashion influenced by the proof of the Banū Mūsā.

Other similarities are evident among the three versions of this chapter. They all depend on a basic proposition which asserts that a regular polygon is equal to a right triangle one of whose sides including the right angles is equal to the perimeter while the other is equal to the line drawn from the center of the polygon to the middle of one of the sides (Pseudo-Bradwardine Version, Proposition I; *Versio abbrevata*, Proposition I; Albert of Saxony Version, Conclusion V). Many of the versions in Chapter Three have a similar proposition, and sometimes they have such a proposition stated as a separate proposition (e.g., see the discussion of the equivalent proposition in the introduction to the Corpus Christi Version). But the versions of this chapter are the only ones to give the area of the polygon as equal to a right triangle; the other versions say the area is equal to the product of one half the perimeter and the line drawn from the center of the polygon to the middle of one of the sides, or something very similar to this. In short, only these versions of this chapter use the expression right triangle to stand for the area in question. I suppose that this was used by the author of the first of these versions as a kind of formal analogy with the statement of the principal quadrature proposition given in the *De mensura circuli* where the circle itself is said to be equal to a right triangle. I do not mean to press the point unduly, but the fact that all three of these versions use the same expression seems clearly to tie them together. It should also be noticed that the scholastic form of presentation is a further point of similarity in the three versions, although its use is much reduced in the *Versio abbreviata*.

1. The Pseudo-Bradwardine Version *(Versio Vaticana)*

Turning to the first of the three versions, the Pseudo-Bradwardine tract, we should observe first that the unique text of it is appended to the *Geometria* of Bradwardine in a manuscript that I judge to be of the fourteenth century (see Siglum below). It is as if Bradwardine—or, as is more likely, some other author—decided to expand in the form of an appendix the abbreviated treatment of the quadrature problem given in the body of the *Geometria* (see Chapter Two, Section 2, for that treatment). The title of our version, *Quadraturam circuli demonstrative concludere*, resembles the title of the chapter on quadrature found in the Gordanus tract (see Chapter Three, Section 4). The expression "to conclude demonstratively the quadrature of the circle" brings to mind the distinction made by Albert of Saxony in his *Questio de quadratura circuli* (Section 3 below, lines 77–83) between quadrature *ad intellectum* ("finding a square and demonstratively proving that it is equal to the circle") and quadrature *ad sensum* ("finding a square such that the sense reveals no difference between it and some circle"). The author goes on to assert that in order to reach the objective expressed in the title, he must first prove five other conclusions which will stand as premises for the proof of the main conclusion. As I have noted, the first conclusion equates a regular polygon to a right triangle with the perimeter and the line drawn from the center to the middle of one side as the sides including the right angle. As in the *Versio abbreviata*, the particular regular polygon used in the Pseudo-Bradwardine Version is a regular pentagon (while that used by Albert of Saxony is a square). The second proposition in the Pseudo-Bradwardine holds that the perimeter of any polygon circumscribed about a circle is longer than the circumference of the circle transformed into a straight line. The reader is referred to my introductions to the Cambridge and Corpus Christi versions in Chapter Three (Sections 1 and 5) for a discussion of similar statements in the various emended versions. Notice also that the direct conversion of curved line to a straight line (which indeed is implicitly assumed in all the proofs of Proposition I of the *De mensura circuli*) is also specifically made a part of the proof of Proposition III of the *Verba filiorum* (see line 16–17, and the Commentary, line 36). In the proof of Proposition II, Pseudo-Bradwardine makes use of Ptolemy's *Almagest*. Incidentally, the only other authority cited by our author is Euclid, whose *Elements* is cited three times (see the Commentary), twice in the first proposition and once in the last. The third and fourth propositions of the Pseudo-Bradwardine tract assert that when we have given a surface that is less than a given circle (Proposition III) or a surface greater than a given circle (Proposition IV) we can in the first case inscribe a regular polygon in the given circle which is greater than the given surface (Proposition III) or in the second case circumscribe a regular polygon about the given circle which is less than the proposed surface (Proposition IV). The proof of these propositions is based on the assumption that the latitude of excess between the given circle and the given surface is always divisible, which, as I have already noted earlier, is the feature distinguishing this version from the versions of Chapter Three. The fifth proposition merely holds that if a given surface is neither greater than nor less than a circle, it must be equal to the circle. This is said to be self-evident.

With these five propositions or conclusions assumed, the main quadrature proposition is then provable. For if the circle is said to be greater than the right triangle composed of circumference and radius at the right angle, a regular polygon which is greater than the triangle can be inscribed in the circle by Proposition III. But by Proposition I it is shown that such a polygon must be equal to a right triangle which is less than the proposed triangle composed of circumference and radius at the right angle. This then contradicts the inference drawn from the hypothesis that the circle is greater than the proposed right triangle; hence that hypothesis must be false. In precisely the same way, by the use of Propositions IV and I and II, the hypothesis that the circle is less than the proposed right triangle can be shown to be false. Thus, by Proposition V the main proposition follows.

My text has been constructed from the unique Vatican manuscript. Notice that the scribe of that manuscript often uses the spelling *pologonia* instead of *poligonia*, but I have everywhere used *poligonia*. Observe also the form *dactus* for *datus*. Here I have preserved the *dactus* readings (although a later hand has corrected *dacto* to *dato* in one place). Incidentally, it can be remarked that a later hand has often written over letters to clarify them. The scribe uses both *ci* and *ti* before vowels. I have used *ti* throughout. The diagrams are as in the manuscript, and the marginal folio references are, of course, to that manuscript.

Siglum of Manuscript

Aa = Vatican lat. 3102, 111v–112v, 14c.

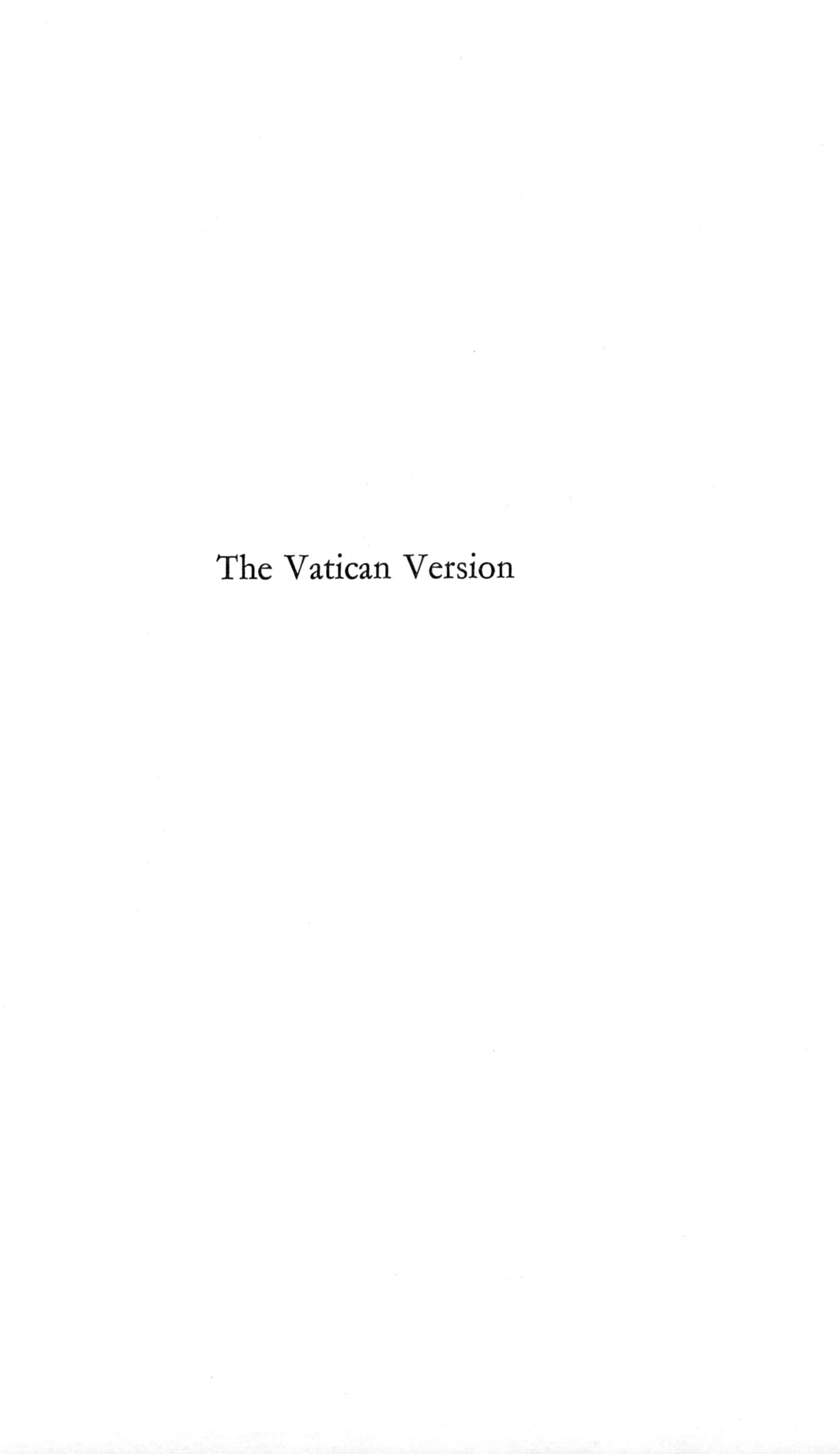

The Vatican Version

[Versio Vaticana]

111v / Quadraturam circuli demonstrative concludere. Pro evidenti demonstratione habenda quinque demonstrabiles conclusiones que premise erunt pro conclusione proposita. Quarum prima sit ista:

[I.] OMNIS FIGURA POLIGONIA EQUILATERA EST EQUALIS TRIANGULO ORTHOGONIO, CUIUS UNUM LATERUM RECTUM ANGULUM AMBIENTIUM ⟨EST PERPENDICULARIS⟩ A CENTRO POLIGONIE AD PUNCTUM MEDIUM UNIUS LATERIS POLIGONIE PROTRACTUM, ET ALTERUM EST EQUALE OMNIBUS LATERIBUS ILLIUS POLIGONIE SIMUL SUMPTIS.

Ista conclusio probatur, et ad probandum sit figura poligonia in forma, pentagonus *ABCDE*, a cuius centro *Z* ducatur *ZG* perpendicularis ad *G* punctum medium ipsius *DE* lateris pentagoni propositi [Fig. 54]. Protractis itaque duabus lineis *ZD* et *ZE* ad *E* et *D* angulos poligonie, patet quod inde causabitur triangulus *ZDE*, quem primo probo esse quintam partem pentagoni propositi, quia in dacto pentagono sunt precise quinque tanti trianguli [quorum cuiuslibet] quantitas est dactus triangulus *DEZ*. Igitur dactus triangulus est quinta pars pentagoni. Consequentia nota est. Antecedens apparet, nam protractis lineis ab ipso *Z* centro pentagoni ad omnes angulos ipsius patet quod in universo causabuntur quinque trianguli in predicto pentagono composito triangulo predicto, quorum 4 quilibet erit equalis dacto triangulo *ZDE*; quod probo. Omnes illi trianguli sunt equilateri cum

6–7 ⟨est perpendicularis⟩ *supplevi*: *cf.* *Versio abbreviata*, *2–3*
7 poligonie *corr. hic et ubique ex* pologonie
13 *ante* DE *del.* *Aa* GE
22 composito *corr. ex* compuncto (?)

[The Vatican Version]

To conclude demonstratively the quadrature of the circle. For an evident demonstration five demonstrable conclusions are needed—conclusions which will constitute premises for the [main] proposed conclusion. The first of these is this:

[I.] EVERY REGULAR POLYGON IS EQUAL TO A RIGHT TRIANGLE ONE OF WHOSE SIDES INCLUDING THE RIGHT ANGLE ⟨IS THE PERPENDICULAR⟩ DRAWN FROM THE CENTER OF THE POLYGON TO THE MIDDLE POINT OF ONE SIDE OF THE POLYGON AND THE OTHER IS EQUAL TO THE PERIMETER OF THE POLYGON.

Proof: For proving it, let there be a polygonal figure, the pentagon *ABCDE*, from whose center *Z* the perpendicular *ZG* is drawn to point *G*, the middle point of *DE*, a side of the proposed polygon [see Fig. 54].

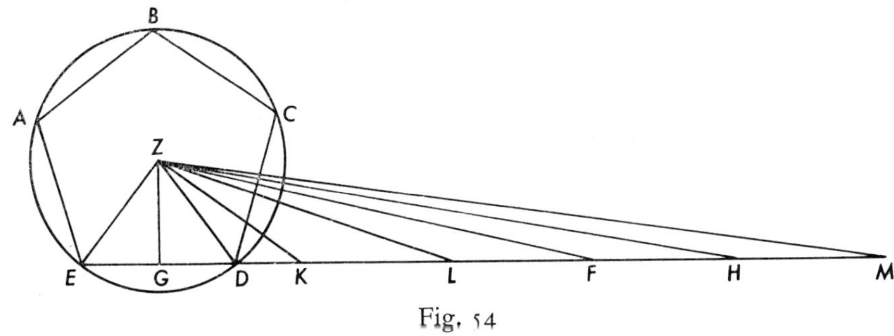

Fig. 54

When the two lines *ZD* and *ZE* have been drawn to angles *E* and *D* of the polygon, it is evident that △ *ZDE* will be formed. This triangle I first prove to be a fifth part of the proposed pentagon; for in the given pentagon there are exactly five such triangles each of which is equal to the given △ *DEZ*. Therefore, the given triangle is a fifth part of the pentagon. The consequence is known. The antecedent is apparent; for with lines drawn from center *Z* of the pentagon to all of its angles, it is evident that they will form *in toto* five triangles composing the aforesaid pentagon. Any one of the [other] four of these triangles will be equal to △ *ZDE*. This I prove. All of these triangles have sides equal [repectively]

triangulo *ZDE*; igitur equianguli; et igitur equales. Consequentie
25 note ex primo Euclidis; antecedens patet, quia latera eorum non sunt nisi latera dacti pentagoni, que per casum sunt equalia. Similiter linee a centro pentagoni protracte ad angulos eius sunt equalia, quoniam a centro ad circumferentiam circuli circumscribentis propositum pentagonum. Patet igitur quod omnes illi trianguli sunt equilateri cum
30 triangulo *ZDE*. Habito igitur quod triangulus *ZDE* est quinta pars pentagoni, protrahatur igitur medietas *GD* lateris pentagoni in continuum et directum quousque sit equalis omnibus lateribus pentagoni dacti simul sumptis, que sit *GM*, que dividatur in quinque partes equales 5 lateribus pentagoni, sic quod divisiones cadunt in punctis
35 *K*, *L*, *F*, *H*. Et tunc a punctis divisionum rectilinee ducantur ad *Z* centrum poligonie. Quibus stantibus patet quod in triangulo orthogonio *ZGM* causabuntur 5 trianguli, quos simul sumptos probo esse equales dicto pentagono. Nam quilibet istorum 5 triangulorum est equalis uni quinte. Ergo omnes 5 simul sunt equales pentagono. Con-
40 sequentia nota; antecedens probatur, scilicet, quod quilibet istorum triangulorum sit equalis uni quinte pentagoni, quia quilibet illorum est equalis triangulo *ZDE*. Igitur quilibet illorum est equalis uni quinte pentagoni, ut probatum est. Antecedens patet, quia quilibet illorum est equalis triangulo *ZGK*; nam triangulus *ZGK* est equalis
45 sibi et quilibet aliorum 4 est eidem equalis. Quod probo, quia omnes sunt super equales bases per casum et equalis altitudinis [quoniam] sunt inter lineas equidistantes, et hoc, si a centro *Z* duceretur equidistans linee *GM*. Igitur [per] 38 primi elementorum Euclidis omnes illi 4 trianguli sunt triangulo *ZGK* equales. Igitur cum ipse sit etiam
50 sibi ipsi equalis, quilibet illorum quinque triangulorum erit equalis triangulo *ZGK*. Tunc universaliter igitur quilibet illorum est equalis triangulo *ZGK*. Tunc universaliter igitur quilibet illorum est equalis *ZDE*. Tenet consequentia, quia triangulus *ZDE* est equalis triangulo *ZGK*, cum sint super equales bases et inter lineas equidistantes sicut
55 si de puncto produceretur equidistans a ipso *Z*. Sed cum triangulus *ZDE* sit quinta pars pentagoni, patet propositum, quod causat antecedens probandum. Habito igitur quod omnes illi quinque trianguli simul sumpti sunt equales pentagono dacto, arguitur sic: Omnes illi simul sumpti sunt idem quod *ZGM* triangulus orthogonius, cuius

30 *ante* Habito *del. Aa* probo igitur quod latera triangulorum   55 produceretur *corr. ex* producueatur (*?*)

to those of △ *ZDE*. Therefore, they are of equal angles; therefore, they are equal. The consequences are known from the first [book] of Euclid. The antecedent is evident because their sides are nothing more than the sides of the given pentagon, which are equal by construction. Similarly the lines drawn from the center of the pentagon to its angles are equal, being drawn from the center to the circumference of the circle circumscribing the proposed pentagon. It is evident, therefore, that all of these triangles have sides equal [respectively] to those of △ *ZDE*. Hence, with it granted that △ *ZDE* is a fifth part of the pentagon, *GD*, half of a side of the pentagon, is protracted continuously and directly until it equals the perimeter of the polygon. The [complete] protraction is *GM*, which we let be divided into five parts equal to the five sides of the pentagon, with *K*, *L*, *F*, and *H* as the points of division. Then from the points of division let straight lines be drawn to the center *Z* of the polygon. With all of this accepted, it is evident that five triangles will be formed in the right △ *ZGM*. I shall prove that these five triangles taken together are equal to the said pentagon, for any of these five triangles is equal to a fifth [of the pentagon]. Hence all five together are equal to the pentagon. The consequence is known; the antecedent is proved, namely, that any of these triangles is equal to one fifth of the pentagon since any of them is equal to △ *ZDE*. Therefore, any of them is equal to one fifth of the pentagon, as is proved. The antecedent is evident because any of them is equal to △ *ZGK*, for △ *ZGK* is equal to itself and any of the other four is equal to the same [quantity]. I prove this, for all of them are on equal bases, by reason of construction, and are of equal altitude, being between parallel lines if from *Z* a line were drawn parallel to line *GM*. Therefore, by I.38 of the *Elements* of Euclid, all four of these triangles are equal to △ *ZGK*. Therefore, since △ *ZGK* is equal to itself, any one of the five triangles will be equal to △ *ZGK*. So, universally, any one of them is equal to △ *ZGK* and, universally, any one of them is equal to *ZDE*. This consequence holds, since △*ZDE*=△*ZGK*, both triangles being on equal bases and between parallel lines if a line parallel [to *GK*] were drawn through point *Z*. But since △ *ZDE* is one fifth of the pentagon, the proposition is evident; for the antecedent causes it to be proved. Therefore, with it granted that these five triangles together are equal to the pentagon, it is argued as follows. All of these triangles taken together are [by construction] the same as the right △ *ZGM*, one of whose sides including the right angle, namely, *ZG*, is perpendicular at *G*, the middle [point]

60 unum laterum rectum angulum ambientium, scilicet *ZG*, est perpendicularis ad *G*, medium ipsius *DE* lateris poligonie descripte, et alterum, scilicet *GM*, trianguli predicti est equale omnibus lateribus poligonie simul sumptis, per casum. Sequitur quod figura poligonia equilatera est equalis triangulo orthogonio, cuius unum laterum etc., ut 65 dicit sumptio.

112r / Et sicut probatur de ista poligonia, ita potest conformiter de quibuscunque probari. Sequitur quod omnis figura poligonia equilatera est equalis triangulo orthogonio et hec erat presumptio probanda.

Secunda conclusio probanda erit ista:

70 [II.] LATERA CUIUSLIBET FIGURE POLIGONIE CIRCULO CIRCUMSCRIPTE SIMUL SUMPTA FACIUNT LINEAM LONGIOREM CIRCUMFERENTIA CIRCULI PREDICTI RECTIFICATA.

Probatur quia *AO* et *SA* medietates duorum laterum dacte poligo-75 nie incurventur ad circumferentiam circuli *SO*, punctis *O*, *S* manentibus fixis, cum protracta corda *SO* [Fig. 55]. Postquam arcus *SO* maior erit arcu inferiori ipsius circuli inscripti, quoniam eadem corda cordans duo arcus circulorum inequalium plus capit de minori quam

of side *DE* of the described polygon, and the other side of the aforesaid triangle, namely, *GM*, is equal to the perimeter of the polygon by construction. [For which reason] it follows that the regular polygon is equal to the right triangle, one of whose sides, etc., as the enunciation states.

And in the same way that it is proved for this polygon so it can be proved for any [other regular polygon]. [Thus] it follows that any regular polygon is equal to a right triangle, and this was the presumption to be proved.

The second conclusion to be proved will be this:

[II.] THE PERIMETER OF ANY POLYGON CIRCUMSCRIBED ABOUT A CIRCLE CONSTITUTES A LINE LONGER THAN THE CIRCUMFERENCE OF THE AFORESAID CIRCLE, [THE CIRCUMFERENCE] HAVING BEEN TRANSFORMED INTO A STRAIGHT LINE.

This is proved, because *AO* and *SA*, halves of two sides of the given polygon, are transformed into an arc of circle *S* [*A'*]*O*, points *S* and *O* remaining fixed and chord *SO* being drawn [between them, as in Fig. 55]. Inasmuch as arc *S*[*A'*] *O* will be greater than the inferior arc (*S*[*B*]*O*)—since a chord acting as a chord of two arcs of unequal circles intercepts more of the lesser circle than of the larger, as is evident in Book I* of the

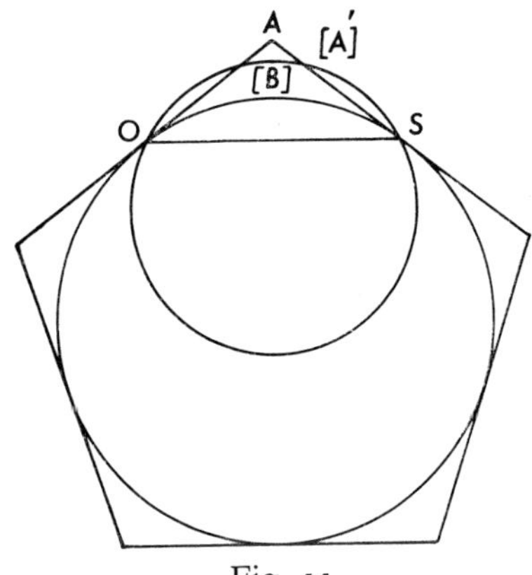

Fig. 55

Note: In the manuscript the smaller circle is not drawn so that *OS* is a chord of it as well as of the larger circle. Hence I have altered the drawing to conform with the text.

* This should, perhaps, refer rather to Book VI, Chapter 7, of the *Almagest* where the proposition is incidentally proved for a particular case (Ptolemy, *Syntaxis mathematica*, ed. J. L. Heiberg, vol. *1* [Leipzig, 1898], pp. 513–15. Cf. the Commentary below, line 79).

de maiori, ut patet in prima dictione almagesti Ptholomei, et cum
80 circumferentia, scilicet *OS*, sit quinta pars laterum pentagoni, et sic arcus *SO* circuli inscripti sit etiam quinta pars inscripti circuli, sequitur quod quinta pars laterum pentagoni sit maior quinta parte circumferentie circuli inscripti. Igitur omnia latera poligonie simul sumpta sunt maiora quam tota circumferentia circuli predicti. Igitur et cetera.

85 Tertia conclusio premittenda sit ista:

[III.] DACTA ALIQUA SUPERFICIE QUE SIT MINOR DACTO CIRCULO EST DABILIS ALIQUA FIGURA POLIGONIA INSCRIPTIBILIS EIDEM CIRCULO PROPOSITO MAIOR SUPERFICIE DATA.

90 Probatur quia ex quo ille circulus est maior dacta superficie et ipsam divisibiliter excedat, ergo infra latitudinem excessus locabilis est, et per consequens inscriptibilis est poligonia. Tenet consequentia, quia vel obstaret magnitudo excessus vel parvitas; non magnitudo, ut notum est, nec parvitas, ut probo, quia sicut magna poligonia est loca-
95 bilis intra magnum excessum ita intra quodcunque parvius potest locari minor, cum poligonia in infinitum sit divisibilis seu diminuibilis. Tunc universaliter intra illam latitudinem excessus locabilis et inscriptibilis est poligonia. Per consequens igitur illam sic locatam oporteat sic inscribi ipsi circulo et circumscribi dacte superficiei. Se-
100 quitur quod ipsa sic locata erit minor. Igitur conclusio valet.

Quarta conclusio:

[IV.] DACTA ALIQUA SUPERFICIE QUE SIT MAIOR CIRCULO DACTO DABILIS EST POLIGONIA CIRCUMSCRIPTIBILIS CIRCULO QUE SIT MAIOR CIRCULO DATO ET
105 MINOR SUPERFICIE PROPOSITA.

Inclusum sit minus includente et maius eo quod includit, proposito quod contingit aliquam poligoniam predicto circulo circumscribi minorem superficie dacta et maiorem circulo proposito. Patet quia ex quo dacta superficies est maior dacto circulo, ipsum tunc divisibiliter
110 excedit. Igitur eadem ratione qua supra intra latitudinem excessus potest colocari, et patet hec conclusio ut precedens.

81 inscripti[1] *corr. ex* circumscripti
85 *ante* conclusio *del. Aa* suppositio
88–89 maior superficie data *mg. Aa*
95 intra[1] *supra scr. Aa*
103–105 dabilis...proposita *mg. Aa* et maior superficie proposita *Aa*; (*sed corr. m.2* minor *ex* maior)
106 *ante* inclusum *del. Aa* con-

Almagest of Ptolemy—and since the arc $S[A']O$ is one fifth of the perimeter of the pentagon, while the arc $S[B]O$ is also one fifth of the inscribed circle, it follows that one fifth of the perimeter of the pentagon is greater than one fifth of the circumference of the inscribed circle. Therefore, the perimeter of the polygon is greater than the whole circumference of the aforesaid circle. Therefore, and so forth.

The third conclusion to be premised is this:

[III.] WITH SOME SURFACE GIVEN AS LESS THAN A GIVEN CIRCLE, THERE CAN BE GIVEN SOME [REGULAR] POLYGON INSCRIBABLE IN THAT PROPOSED CIRCLE WHICH WILL BE GREATER THAN THE GIVEN SURFACE.

Proof: From the fact that the circle is greater than the given surface and exceeds it by a divisible amount, therefore within the latitude of the excess [such] a polygon can be placed and consequently is inscribable. The consequence follows, for either the largeness or the smallness of the excess would prevent [the location of the polygon in the excess]. But not the largeness, as is known, nor the smallness, as I prove. For just as a large polygon can be placed within a large excess, so a smaller one can be placed within any smaller excess, since a polygon is divisible or diminishable to infinity. Then universally within that latitude of excess a polygon can be placed and inscribed. Hence, it ought to be so inscribed in that circle and circumscribed about the given surface. It follows that this polygon so placed will be less [than the circle]. Therefore, the conclusion is valid.

Fourth conclusion:

[IV.] WITH SOME SURFACE GIVEN AS GREATER THAN A GIVEN CIRCLE, THERE CAN BE GIVEN A POLYGON CIRCUMSCRIBABLE ABOUT THE CIRCLE WHICH IS GREATER THAN THE GIVEN CIRCLE AND LESS THAN THE PROPOSED SURFACE.

[The proof follows from this consideration:] something included is less than that which includes it and is greater than that which it includes, it having been proposed that some polygon can be circumscribed about the aforesaid circle which is less than the given surface and greater than the given circle. This is evident, for, since the given surface is greater than the given circle, it exceeds it by a divisible amount. Therefore, by the same argument as above, it (a polygon) can be placed within the latitude of excess. And this conclusion is evident in the same way as the preceding one.

Quinta conclusio:

[V.] SI ALIQUAM DACTAM SUPERFICIEM REPUGNAT ESSE MAIOREM ET ETIAM ESSE MINOREM CIRCULO
115 PROPOSITO, IPSA ERIT EI NECESSARIO EQUALIS.
Ista patet de se.

[VI.] HIIS PREMISSIS PROBO POSTEA QUADRATURAM CIRCULI CONCLUDERE.

Quia, sit circulus *Z* ad cuius centrum *Z* ducatur semidiameter *ZC*
120 [Fig. 56]. Similiter a centro *Z* protraham perpendicularem super semidiametrum *ZC* in continuum et directum usque erit equalis circumferentie circuli. Deinde a puncto *C* ducam lineam rectam ad punctum *G*, quod possum. Quibus sic protractis, patet quod inde causabitur triangulus *CGZ*. Tunc arguo sic: *CGZ* triangulus est equalis propo-
125 sito circulo *Z*. Igitur possibile est demonstrative probare propositum circulum esse quadrato equalem, et per consequens dactum circulum quadrare. Antecedens probo, quia oppositum implicat, ergo repugnat triangulum propositum esse maiorem dacto circulo et etiam implicat ipsum esse minorem ipso; ergo necessario erit sibi equalis. Conse-
130 quentia nota per quintam suppositionem. Antecedens probo quia quo ad utramque partem; primo quo ad primam, ut quod implicat ipsum esse maiorem. Quia, si est maior eo, ergo supra circulum circumscriptibilis est poligonia equilatera maior circulo proposito et minor triangulo dacto per quartam suppositionem. Sit igitur ita poligonia
135 circulo circumscripta, cuius unum latus *BCD* secundum medium eius punctum *C* contingens circulum in puncto contactus semidyametri cum circulo. Tunc arguo sic: Latera poligonie simul sumpta sunt maiora quam circumferentia dacti circuli inscripti rectificata, quare ipsa linea recta equalis circumferentie circuli propositi per secundam

126 circulum *mg. m.2* | *post* esse *del. m. 2* uni 135 BCD *corr. ex* GBD

Fifth conclusion:

[V.] IF IT IS CONTRADICTORY FOR SOME GIVEN SURFACE TO BE GREATER THAN OR LESS THAN A PROPOSED CIRCLE, IT WILL NECESSARILY BE EQUAL TO IT.

This is evident per se.

[VI.] WITH THESE [CONCLUSIONS] PREMISED, I SHALL THEN PROVE THAT THE QUADRATURE OF THE CIRCLE IS CONCLUDED.

For, let there be circle Z, to whose center Z radius ZC is drawn [see Fig. 56]. Also from center Z I shall protract continuously and directly a perpendicular to radius ZC until it will be equal to the circumference of

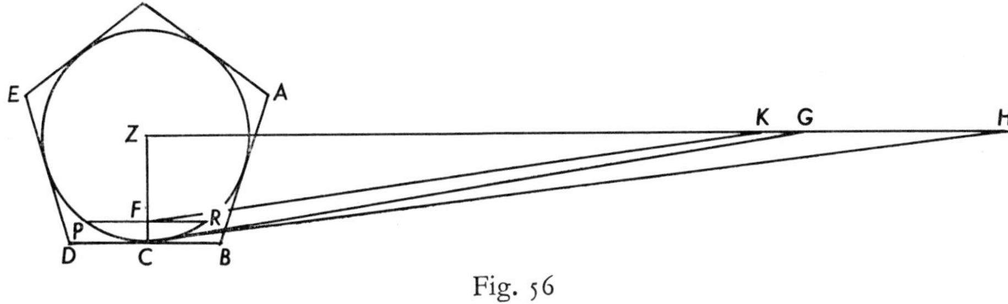

Fig. 56

the circle. Then from point C I shall draw a straight line to point G, which I can do. With these [lines] so drawn, it is evident that then $\triangle CGZ$ will be formed. Then I argue as follows: $\triangle CGZ$ is equal to the proposed circle Z; therefore, it is possible to prove demonstratively that a proposed circle is equal to a square and consequently [it is possible] to square a given circle. I prove the antecedent, for the opposite [to the antecedent] is contradictory. Hence it is contradictory that the proposed triangle is greater than the given circle and it is also impossible that it is less than it; therefore, it is by necessity equal to it. The consequence is known by the fifth supposition. The antecedent is proved in each of its parts. As for the first part, that is, that it is contradictory for the triangle to be greater: if it is greater than it, then there is circumscribable about the circle a regular polygon greater than the proposed circle and less than the given triangle, by the fourth supposition. Hence let there be such a polygon circumscribed about the circle, one side of which is BCD, tangent to the circle at its midpoint C, the point of contact of the radius with the circle. Then I argue thus: The perimeter of the polygon is greater than the circumference of the given circle transformed into a straight line. Hence it is greater than the straight line equal to the circumference of the proposed circle,

140 suppositionem. Protrahatur igitur *GZ* in continuum et directum quousque fiat equalis omnibus lateribus dacte poligonie, que sit *ZH*. Et tunc a puncto *C* ducam rectam *CH*. Arguo tunc sic: Triangulus *ZCH* est triangulus orthogonius, quia angulus *Z* est rectus per suppositum, cuius unum laterum rectum angulum ambientium, puta *ZH*, est equale 145 omnibus lateribus poligonie simul sumptis per suppositionem et reliquum, puta *ZC*, est perpendicularis a centro *Z* poligonie et circuli ad *C* medium punctum lateris poligonie per casum. Igitur triangulus *ZCH* est equalis poligonie dacte. Consequentia tenet per primam suppositionem. Sed cum dacta poligonia sit minor quam triangulus *ZCG* 150 primo positus, per quartam suppositionem igitur sequitur quod triangulus *ZCH* sibi equalis est minor eodem triangulo, scilicet *ZCG*, et per consequens totum est minus sua parte, quod implicat. Probo 112v etiam quod implicat dactum triangulum *ZCG* esse / minorem dacto circulo. Quia, si minor eo, igitur per tertiam suppositionem infra 155 dactum circulum inscriptibilis est poligonia equilatera minor dacto circulo et maior triangulo proposito. Sit igitur poligonia illa, cuius latus unum sit *RP* secans *ZC* semidiametrum in puncto F medio; quod non secat perpendiculariter non est cura, propter *Z* angulum esse rectum. Tunc notum est quod ex quo ista poligonia est inscripta 160 circulo proposito, latera eius simul sumpta erunt minora quam circumferentia circuli rectificata, et per consequens quam linea *ZG* per oppositum secunde suppositionis, quoniam si latera maioris sunt maiora per secundam suppositionem, tunc latera minoris erunt ipsa minora per oppositum. Sit igitur linea *ZK* equalis omnibus lateribus poligonie 165 simul sumptis, et tunc a puncto *F* ducam lineam *FK*. Tunc arguo sic: Triangulus *ZFK* est orthogonius, quia angulus *Z* est rectus per suppositum, cuius unum laterum angulum rectum ambientium est perpendicularis a centro poligonie *Z* ad *F* medium punctum *RP* lateris poligonie. Igitur triangulus *ZFK* est equalis dicte poligonie inscripte 170 circulo et cum illa sit maior triangulo *ZCG* proposito per tertiam conclusionem, sequitur quod triangulus *ZFK* sibi equalis erit maior eodem, scilicet triangulo *ZCG*, et per consequens pars maior toto, quod est iterum implicatur. Et cum iste implicationes sequuntur ex hoc, quod est triangulum *ZCG* propositum esse maiorem vel minorem 175 circulo *Z* proposito, sequitur quod non stat triangulum *ZCG* propositum esse maiorem vel minorem circulo *Z* proposito. Igitur per quin-

151 ZCH *corr. ex* ZCG

157 puncto *corr. ex.* punctum

by the second supposition. Hence let *GZ* be extended continuously and directly until it becomes equal to the perimeter of the given polygon, which continuation we let be *ZH*. And from *C* I then shall draw a straight line *CH*. Then I argue as follows: △ *ZCH* is a right triangle because ∠ *Z* is assumed to be a right angle. One of the sides of this triangle which include the right angle, namely, side *ZH*, is equal to the perimeter of the polygon, by supposition; the other side, namely, *ZC*, is the perpendicular drawn from *Z*, the center of the polygon and circle, to *C*, the middle point of the side of the polygon, by construction. Therefore, △ *ZCH* is equal to the given polygon. The consequence holds by the first supposition. But since the given polygon is less than △ *ZCG* as was first assumed, therefore by the fourth supposition it follows that △ *ZCH*, which is equal to the polygon, is less than that same △ *ZCG*, and consequently that the whole is less than the part, which is contradictory. I prove also that it is contradictory for the given △ *ZCG* to be less than the given circle. For, if less than it, then by the third supposition a regular polygon can be inscribed within the given circle which is less than the given circle and greater than the proposed triangle. Therefore, let that polygon be constructed with one side *RP* which cuts radius *ZC* in midpoint *F*. It is of no concern whether it cuts *ZC* perpendicularly, since *Z* is a right angle. Then from the fact that this polygon is inscribed in the proposed circle it is known that the perimeter is less than the circumference of a circle transformed into a straight line and consequently less than line *ZG*, by the contrary of the second supposition, since if the perimeter of a larger polygon is greater by the second supposition, then the perimeter of the lesser will be less *per oppositum*. Therefore, let line *ZK* be equal to the perimeter of the polygon. And then from point *F* I shall draw line *FK*. Then I argue as follows: △ *ZFK* is a right triangle because ∠ *Z* is assumed to be a right angle. One of the sides of this triangle which include the right angle is a perpendicular drawn from the center *Z* of the polygon to point *F*, the middle point of side *RP* of the polygon [, and the other is the perimeter of the inscribed polygon]. Hence △*ZFK* is equal to the said polygon inscribed in the circle. And since the polygon is greater than proposed △ *ZCG* by the third conclusion, it follows that △ *ZFK*, equal to the polygon, will be greater than △ *ZCG*, and consequently that a part is greater than the whole, which again is contradictory. And since these contradictions follow from this: that the proposed △ *ZCG* is greater than or less than the proposed circle *Z*, it follows that it is not possible for the proposed △ *ZCG* to be greater than or less than the proposed circle *Z*.

tam suppositionem necessario erit sibi equalis. Sed prima consequentia probatur, scilicet triangulus *ZCG* est equalis circulo *Z*. Igitur possibile est demonstrative probare dactum circulum *Z* esse equalem alicui 180 quadrato, et per consequens circulum propositum quadrare. Nam possibile est demonstrative probare dactum triangulum *ZCG* esse equalem uni dacto quadrato, ex scientia 14 propositionis secundi elementorum Euclidis. Igitur cum dactus triangulus sit equalis dacto circulo *Z*, sequitur quod etiam demonstrative erit dactum circulum esse 185 eidem quadrato equalem. Et sic patet propositum nostrum quod a primo proposuimus.

Explicit Geometria Braduardi[ni] et Quadratura circuli.

Therefore, by the fifth supposition, it will by necessity be equal to it. But the first consequence is [now] proved, namely, that △ *ZCG* is equal to circle *Z*. Therefore, it is possible to prove demonstratively that the given circle *Z* is equal to some square and consequently [it is possible] to square a proposed circle. For it is possible to prove demonstratively that the given △ *ZCG* is equal to one given square from knowledge of Proposition II.14 of the *Elements* of Euclid. Therefore, since the given triangle is equal to the given circle *Z*, it follows demonstratively that the given circle is also equal to the same square. And thus is evident the proposition which we proposed at the beginning.

[Here] ends the *Geometry* of Bradwardine and the *Quadrature of the Circle*.

COMMENTARY

25 "ex primo Euclidis." Presumably the author refers to Proposition I.4 of the *Elements*. For I.4, see the Commentary to the Naples Version (Chapter Three, Section 2), line 36.

48 "38 primi." For I.38 of the *Elements*, see the Commentary to the Munich Version (Chapter Three, Section 6), line 226.

79 "prima dictione almagesti." As I indicated in the text, the reference seems to be rather to the seventh chapter of Book VI. It may be that we should not attempt to translate *dictio* as "book" but merely as "assertion." However, if this second translation is used, I do not know the precise connotation of *prima*; i.e., I am not sure what "as evident in the first assertion of the *Almagest*" would mean. This proposition for which the *Almagest* is cited as authority can also be found in Jordanus, *De triangulis*, Book III, Proposition 3 (ed. of M. Curtze, Thorn, 1887, p. 20), and in Bradwardine's *De continuo*, Proposition 19 (MS Thorn R 4° 2, p. 161), as well as in his *Geometria speculativa*, Tract I, Chap. 4, conclusion 3 (ed. of Paris, 1495, 7v-8r).

90–97 "quia... diminuibilis." The expression here of what is at base a simple idea of the divisibility of the excess as a continuous magnitude seems to be most inept and clumsy.

106 "Inclusum... includit." This might better have been set out as a supposition earlier and used in connection with the proof of Proposition II.

128 "implicat." *Implicare* in the medieval texts often stands for *implicare contradictionem*.

182–83 "14... Euclidis." For II.14 of the *Elements*, see the Commentary to the Cambridge Version (Chapter III, Section 1), lines 82–83.

187 "Explicit... circuli." This is the colophon for the whole manuscript and thus includes the *Geometria* of Bradwardine as well as the *Quadratura circuli*.

2. The Abbreviated Version of the Pseudo-Bradwardine Text

An abbreviated version of the Pseudo-Bradwardine *Quadrature of the Circle*—given and discussed in Section 1 of this chapter—is found in a unique fifteenth-century manuscript of Florence (see the Siglum below).[1] While taking over the main lines of proof found in the longer treatise, the author of the abbreviated version has made interesting alterations. Although he substantially follows the proofs of the longer tract, he omits most of the scholastic elaboration, e.g., omitting the constant reiteration of consequence, antecedent, and the like. This is true for his first and third propositions which are drawn from Propositions I and VI of the Pseudo-Bradwardine tract. The author of the abbreviated version in addition omits entirely Propositions II and V of the longer text, while he makes a single proposition—(Proposition II)—of Propositions III and IV of the Pseudo-Bradwardine work. Furthermore, the proof that he gives for his second proposition is completely different from the proofs of the Pseudo-Brad-

[1] This quadrature proof is a part of an interesting collection of miscellaneous geometrical problems that occupies folios 46r-55r. It is designated on f. 46r as *Bachon Alardus in 10 Euclidis*. This appears to be a reference to what I have called the third version of Adelard of Bath's rendering of the *Elements* of Euclid, where Adelard's name is given in the form of *Alardus Batoniensis*. There is perhaps a further confusion here with Roger Bacon, who cites this version of the Elements, using the form *Alardus Batoniensis* (see M. Clagett, "The Medieval Latin Translations from the Arabic of the Elements of Euclid," *Isis*, vol. 44, 1953, p. 23, n. 18). The quadrature proof itself does not appear to have any connection with the third version of Adelard and I assume it was added by a fourteenth century (or perhaps fifteenth century) author who was collecting and expounding miscellaneous geometrical problems. For another reference to Alardus, see Appendix I, Section 1, Sigla.

wardine Propositions III and IV. It depends not on the statement that a polygon can be inscribed or circumscribed in the excess between triangle and circle because of the infinite divisibility of that excess but rather on Proposition X.1 of the *Elements* of Euclid. In short, he follows the same procedure that is used in the various emended versions of the *De mensura circuli* given in Chapter Three above. Actually, the letters he employs on Figure 58 are those of the Gerard of Cremona translation. He also depends on the tradition of Gerard of Cremona's translation of the *De mensura circuli* for the enunciation of his Proposition III. While the longer text simply says that the objective is to conclude the quadrature of the circle demonstratively, the Florence author substitutes Gerard of Cremona's wording of Proposition I of the *De mensura circuli*. Hence it is quite evident that the author of the abbreviated version conflates the Pseudo-Bradwardine text with the popular tradition of the *De mensura circuli*.

My text of the *Versio abbreviata* follows the unique manuscript closely. I have generally followed the scribe in his variant spellings. However, where the scribe has sometimes used *orthogonius* and sometimes *ortogonius*, I have used *orthogonius* throughout. The scribe or author sometimes uses the adjectival form *pentagona* (to stand for *figura pentagona*) and sometimes the noun *pentagonus*. The figures are as in the manuscript and the marginal folio numbers are to the unique manuscript.

Siglum of Manuscript

Ea = Florence, Bibl. Naz. Conv. Soppr. J.IX. 26, 49v–50v, late 15c.

[Versio abbreviata]

49v / [I.] OMNIS FIGURA POLIGONIA EQUILATERA ET EQUIANGULA EST EQUALIS TRIANGULO ORTHOGONIO, CUIUS UNUM LATERUM RECTUM ANGULUM CONTINENTIUM EST PERPENDICULARIS A CENTRO POLIGONIE AD PUNCTUM MEDIUM UNIUS LATERIS IPSIUS POLIGONIE PRO-5 TRACTUM ET ALTERUM EST EQUALE OMNIBUS LATERIBUS ILLIUS POLIGONIE SIMUL SUMPTIS.

Ista probatur data figura poligonia, ut pentagona *ABDEC* [Fig. 57], a cuius centro *Z* ducatur *ZG* perpendicularis ad latus *DE*, scilicet 10 super medium eius in puncto *G*. Fiet igitur triangulus *DZE*, qui est quinta pars illius pentagoni, ut faciliter deducitur: Cum omnia latera sint equalia et linee a centro procedentes ad 5 angulos pentagoni sint equales, sequitur quod illi trianguli sint equianguli et equilateri et equales igitur. Deinde protrahatur linea *GE* quam libuerit in directum, 15 de qua resecetur primo *EK* equalis *GE* et producatur linea *ZK*. Patet igitur quod triangulus *ZEK* est equalis triangulo *ZDG*, cum sint super equales bases inter duas equedistantes aut equalis altitudinis. Igitur sequitur quod totus triangulus *ZGK* sit equalis toti triangulo *ZDE*, quia ab eodem et eidem dentis et additis equalibus restat idem 20 et equalis. Deinde producam lineam *ZL* capientem de basi ad libitum productam portionem equalem basi *GK*, et consimiliter faciam producendo lineam *ZF*, *ZH*, et *ZM*. Ex quo patet quod ego constituo 5 triangulos, quorum quilibet est equalis alteri et per consequens triangulo *ZGK*, qui est quinta pars pentagoni, et per consequens omnes

7 poligonie *corr. ex* polligonie   11 deducitur *corr. ex* deducentur

[The Abbreviated Version]

[I.] EVERY REGULAR POLYGON IS EQUAL TO A RIGHT TRIANGLE ONE OF WHOSE SIDES INCLUDING THE RIGHT ANGLE IS THE PERPENDICULAR DRAWN FROM THE CENTER OF THE POLYGON TO THE MIDDLE POINT OF ONE SIDE OF THE POLYGON AND THE OTHER IS EQUAL TO THE PERIMETER OF THE POLYGON.

This is proved, assuming as the given polygon a pentagon *ABDEC* [see Fig. 57], from whose center *Z* [line] *ZG* is drawn perpendicular to side *DE*, i.e., at its middle point *G*. Hence let △ *DZE* be formed, which

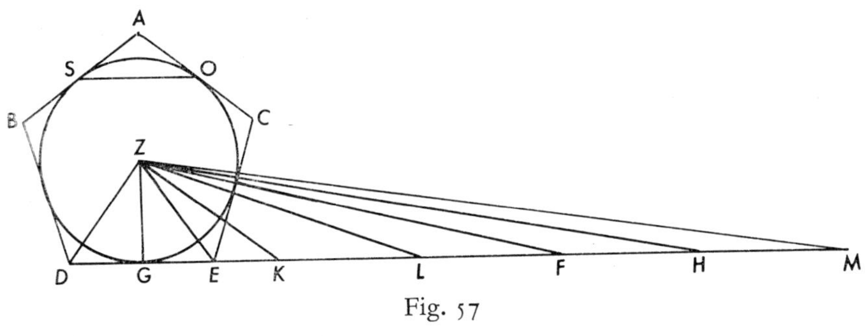

Fig. 57

is a fifth part of the pentagon, as is easily deduced: Since all of the sides are equal and since the lines drawn from the center to the five angles of the pentagon are equal, it follows that these triangles are equiangular and equilateral, and hence equal. Then let line *GE* be protracted rectilinearly as far as is desired. From this protraction in the first place will be cut *EK* equal to *GE*, and let line *ZK* be drawn. It is evident, therefore, that △ *ZEK* = △ *ZDG*, since they are on equal bases and are between parallel lines, i.e., are of equal altitude. Therefore, it follows that the whole △ *ZGK* is equal to the whole △ *ZDE*, for if equals are subtracted from or are added to equals, the results are the same and equal. Then I shall produce line *ZL*, having cut off from the protracted base a segment (*KL*) equal to base *GK*. And similarly I shall draw lines *ZF*, *ZH*, and *ZM*. From this it is evident that I form five triangles, any one of which is equal to another and consequently to △ *ZGK*, which is a fifth part of the pen-

25 simul equantur pentagono dato. Sed linea *GM* est equalis omnibus lateribus pentagoni illius simul sumptis. Quare patet propositum.

[II.] ITEM SECUNDO SUPPONITUR QUOD DATA ALIQUA SUPERFICIE BREVIORE DATO CIRCULO EST DANDA ALIQUA FIGURA POLIGONIA INSCRIPTIBILIS EIDEM CIRCULO 30 MAIOR SUPERFICIE DATA ET BREVIOR CIRCULO PROPOSITO, ET SIMILITER DATA ALIQUA SUPERFICIE MAIORE 50r / CIRCULO DATO CONTINGIT ALIQUAM POLIGONIAM EIDEM CIRCULO CONSCRIBI BREVIOREM SUPERFICIE DATA ET MAIOREM CIRCULO PREDICTO.

35 Probatur, sit enim circulus *ABGD* [Fig. 58] et sit primo data figura maior ipse circulus. Faciam autem in circulo quadratum *AG*. Et secabo arcum *AB* per medium in puncto *F* et copulabo *AF*, *BF*, et similiter faciam per totum circulum in arcubus similibus. Iam ergo separatum est ex residuis circuli portionibus plus medietate ipsarum, et est *AFB* 40 triangulus et sibi similes. Cum ergo fecerimus ita per illud quod sequet, remanebunt portiones que erunt minores quantitate eius quod circulus addit super datam superficiem. Et sic figura poligonia quam continet circulus erit tunc maior eadem data superficie, quod est propositum.

39 AFB: AFL (?) *Ea*

42 poligonia *corr. ex* poligoniam

tagon. And consequently all of them taken together equal the given pentagon. But line *GM* is equal to the perimeter of the pentagon. Therefore, that which is proposed is evident.

[II.] ALSO IT IS SUPPOSED IN THE SECOND PLACE THAT WITH SOME SURFACE GIVEN AS SMALLER THAN A GIVEN CIRCLE, THERE CAN BE GIVEN SOME [REGULAR] POLYGON INSCRIBED IN THE SAME CIRCLE WHICH IS GREATER THAN THE GIVEN SURFACE AND LESS THAN THE PROPOSED CIRCLE. AND SIMILARLY WITH SOME SURFACE GIVEN AS GREATER THAN A GIVEN CIRCLE, THERE CAN BE CIRCUMSCRIBED ABOUT THE SAME CIRCLE SOME [REGULAR] POLYGON WHICH IS LESS THAN THE GIVEN SURFACE AND GREATER THAN THE AFORESAID CIRCLE.

Proof: Let there be a circle *ABGD* [see Fig. 58], and in the first place let the circle be greater than the given figure. Moreover, I shall construct

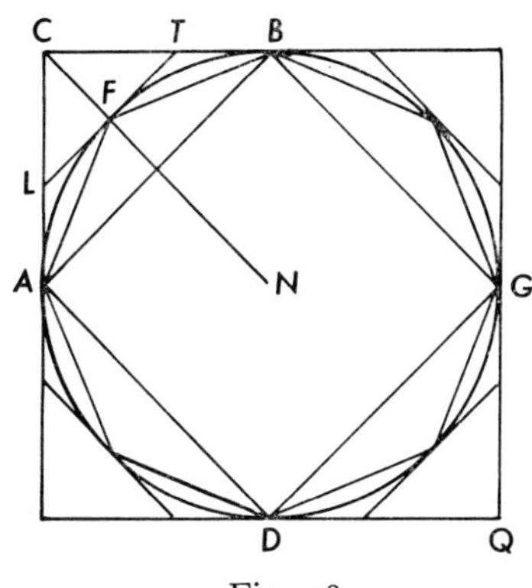

Fig. 58

square *AG* in the circle; and I shall bisect arc *AB* at point *F*, joining *AF* and *BF*. And I shall proceed in the same way in the similar arcs throughout the whole circle. Therefore, there now has been taken away from the segments left over [between the perimeter of the square and the circumference of the circle] more than half of these segments. That [portion thus taken away from the segments] consists in △ *AFB* and those [triangles] similar to it. When we have proceeded successively in this manner, there will [sometime] remain segments which will be less than the quantity by which the circle exceeds the given surface. And so the polygon which the circle contains will then be greater than the given surface, which is that which has been proposed.

Si autem circulus sit brevior, circumscribatur sibi quadratum *CQ*
45 contingens circulum in punctis *A*, *B*, *G*, *D* et dividam arcum *AB* ut prius per medium in puncto *F* et consimiliter dividantur omnes arcus similes illius circuli. Et ducatur a puncto *F* sectionis linea *LT*, et sic fiat in similibus sectionibus, sic quod tales linee contingant circulum in punctis sectionum. Deinde ab angulo *C* quadrati circumscripti pro-
50 ducam ad centrum lineam *CN*, que utique transibit per punctum *F* cum dividat arcum *AB* in duo media, et per consequens est perpendicularis in puncto *F* super lineam *LT*. Iam ergo ex quadrato *CQ* est separatum plus medietate eius et est talis separatio circulus predictus. Item cum linea *LT* iam sit divisa in duo media supra *F* et linea *CF*
55 sit perpendicularis super eam, et similter est de omnibus similibus, et quoniam linee *LC* et *CT* sunt maius *LT*, igitur earum medietas est maior medietate ipsius. Igitur linea *CT* est maior linea *FT*, et per consequens linea *TB*, quia *FT* et *TB* sunt equales. Ergo triangulus *FCT* est maior medietate figure *FCB*. Igitur a fortiori est multo maior
60 medietate figure que continetur *FC* et *CB* et arcu *FB*. Et similiter dicam de triangulo *CFL* et omnibus similibus quod est multo maior
50v medietate figure *CFA* que continetur / duabus lineis *CF* et *CA* et arcu *FA*. Ergo totus triangulus *LTC* est multo maior medietate totius figure *AFBC*, et similiter sunt trianguli sibi similes plus medietate
65 portionum aliarum sibi similium. Cum ergo fecerimus sic successive, remanebunt portiones supra circulum que cum agregabuntur erunt minus augmento date superficiei supra circulum, quod erat propositum secundum.

His suppositis sit ista conclusio:

70 [III.] OMNIS CIRCULUS ORTHOGONIO TRIANGULO EST EQUALIS, CUIUS UNUM DUORUM LATERUM RECTUM ANGULUM CONTINENTIUM MEDIETATI DYAMETRI CIRCULI EQUATUR ET ALTERUM IPSORUM LINEE CIRCULUM CONTINENTI.

75 Sit enim circulus *Z*, cuius centrum *Z* [Fig. 59], a quo ducatur semidiameter *ZD*, et protraham ab eodem centro perpendiculariter super lineam *ZD* lineam *ZG* equalem circumferentie ipsius circuli. Ducatur itaque linea *DG*. Dico ergo triangulum *DZG* esse equalem dato circulo et per consequens circulum posse quadrari. Consequentia nota

45 dividam *corr. ex* dividat
47 ducatur *corr. ex* ducantur
51 est *corr. ex* et
56 medietas *corr. ex* meditatas
76 ZD *corr. Ea ex* ZGD / protraham *corr. ex* protrahatur

If, moreover, the circle is less [than the given surface], let there be circumscribed about it square *CQ*, which is tangent to the circle in points *A*, *B*, *G*, and *D*. And I shall bisect arc *AB* as before, in point *F*. And in the same way let all the similar arcs of the whole circle be bisected. And let line *LT* be drawn at the division point *F*—and do the same thing at the [other] similar divisions—so that such lines are tangent to the circle at the points of division. Then from angle *C* of the circumscribed square I shall draw line *CN* to the center, and it will intersect point *F* as it bisects arc *AB*. And consequently it is perpendicular to line *LT* at point *F*. Therefore, there has now been taken away from square *CQ* more than its half by the aforesaid circle. Also, since line *LT* is already bisected at point *F* with *CF* perpendicular to *LT*—and all of the other similar [corners of the square] are treated in the same way—and since *LC* and *CT* are greater than *LT*, therefore their half is greater than its half. Therefore, line *CT* > line *FT*, and consequently line *CT* > line *TB* since *FT* = *TB*. Therefore, △ *FCT* is greater than 1/2 △ *FCB*. Hence a fortiori it is much greater than half the figure contained by [straight lines] *FC* and *CB* and arc *FB*. And I shall say in the same way regarding △ *CFL*, and all the similar figures, that it is much greater than half the figure *CFA* contained by the two [straight] lines *CF* and *CA* and arc *FA*. Therefore, the whole △ *LTC* is much greater than half the whole figure *AFBC*; and in the same way the triangles similar [to △ *LTC*] are greater than half of the other figures similar [to figure *AFBC*]. Therefore, when we have proceeded in this way successively, there will [sometime] remain figures beyond the circle which when added together will be less than the quantity by which the given surface exceeds the circle; which was the second thing proposed.

With these [two conclusions] supposed, there follows this conclusion:

[III.] EVERY CIRCLE IS EQUAL TO A RIGHT TRIANGLE ONE OF WHOSE TWO SIDES INCLUDING THE RIGHT ANGLE IS EQUAL TO THE RADIUS OF THE CIRCLE AND THE OTHER OF THESE TO THE CIRCUMFERENCE OF THE CIRCLE.

Let there be a circle *Z*, whose center is *Z* [see Fig. 59]. From *Z* let the radius *ZD* be drawn. And from that same center I shall draw as a perpendicular to line *ZD* a line *ZG* equal to the circumference of the circle. And so let *DG* be drawn. I say, therefore, that △ *DZG* is equal to the given circle and consequently that the circle can be squared. The consequence is

80 et antecedens probatur; quia, si non, sit ergo primo maior. Ergo super dicto circulo circumscriptibilis est figura poligonia minor dato triangulo et maior ipso circulo per secundam suppositionem. Sit ergo ista poligonia figura pentagona, gratia exempli, cuius unum laterum sit *BDF*, cuius medius punctus sit *D*, scilicet in contactu istius figure
85 cum circulo, et ducatur semidyameter *ZD*. Tunc sic latera ipsius pentagoni maiora sunt circumferentia circuli igitur maiora linea *ZG*. Sint ergo equalia linee *ZH*. Ergo triangulus orthogonius *DZH* est equalis illi pentagono. Sed ille pentagonus est brevior triangulo *DZG*. Igitur triangulus *DZH* brevior est triangulo *DZG*, totum sua parte, quod
90 est impossibile.

Si autem dicatur quod iste traingulus *DZG* sit minor dato circulo, igitur per eandem secundam suppositionem eidem circulo inscriptibilis est figura poligonia que erit maior dicto triangulo et brevior ipso circulo, que iterum sit pentagona, cuius unum laterum sit *NO*,
95 cuius semidyameter sit *ZR*. Tunc sic latera istius pentagoni sunt minora circumferentia circuli et per consequens linea *ZG*. Sint ergo equalia linee *ZK*. Et ducatur *RK*. Igitur triangulus orthogonius *RZK* est equalis dato pentagono ex prima suppositione. Igitur ipse triangulus *RZK* est maior triangulo orthogonio *DZG*, cum pentagonus iste
100 sit maior eodem. Igitur pars est maior suo toto, quod est impossibile ut prius, et cetera.

81 circumscriptibilis *corr. ex* inscriptibilis

known; and the antecedent is proved. For if not, let it in the first place be greater. Therefore, one can circumscribe about the said circle a polygon less than the given triangle and greater than that circle, by the second supposition. Therefore, let this polygon be a pentagon, for example, one of whose sides *BDF* has a midpoint *D*, evidentally the point of tangency of the polygon and the circle; and let the radius *ZD* be drawn. And so

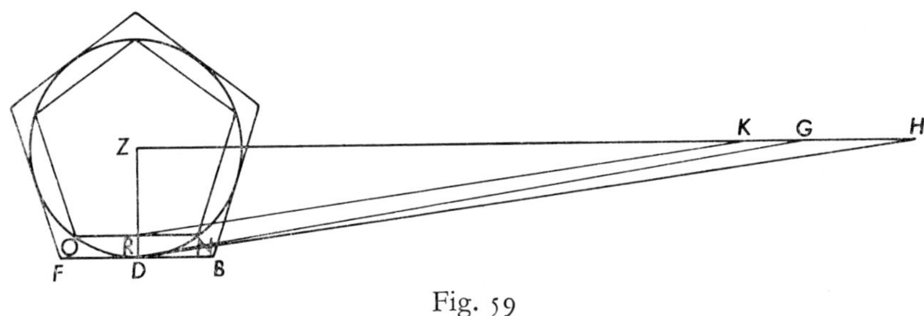

Fig. 59

then the perimeter of the pentagon is greater than the circumference of the circle; hence it (the perimeter) is greater than line *ZG*. Therefore, let it be equal to line *ZH*. Therefore, right △ *DZH* is equal to the pentagon But that pentagon is less than △ *DZG*. Therefore, △ *DZH* < △ *DZG*, the whole is less than its part, which is impossible.

If, moreover, it is said that this △ *DZG* is less than the given circle, then by the second supposition there can be inscribed in the same circle a [regular] polygon which is greater than the said triangle and less than that circle, which polygon again is a pentagon with one side *NO* and radius [i.e., line from the center to the midpoint of one of its sides] *ZR*. Then in this case the perimeter of the pentagon is less than the circumference of the circle and so less than line *ZG*. Hence let it be equal to line *ZK*. And let *RK* be drawn. Therefore, right △ *RZK* is equal to the given pentagon, by the first supposition. Therefore, this △ *RZK* is greater than right △ *DZG*, since the pentagon is greater than it. Therefore, the part is greater than its whole, which as before is impossible, and so on.

COMMENTARY

19–20 "quia... equalis." This is a compendious statement of Axioms 2 and 3 of Book I of the *Elements* of Euclid.

27–68 "Item... secundum." Notice that in this proposition the polygon to be inscribed in the circle is not a pentagon, as in the first and third propositions (and in the equivalent propositions of the Pseudo-Bradwardine Version), but rather is some regular polygon with $2n$ sides where n is some even number. This is of course the polygon employed in all of the versions of the *De mensura circuli* given in Chapter Three.

79 "Consequentia nota." This simply means that the over-all implication is sound, that is, the conclusion (or consequent, using medieval terminology) does indeed follow if the antecedent is true. It is, of course, the antecedent, namely, that triangle DZG is equal to the circle, which is the object of this proposition.

3. Albert of Saxony's *Question on the Quadrature of the Circle*

Of the three treatments of the quadrature problem in this chapter, Albert of Saxony's *Questio* is by far the most detailed and interesting one. I would suppose that it dates from the time of Albert's teaching career at the University of Paris and thus before he became first rector of the University of Vienna in 1366.[1] It seems probable that this *questio* was one of a series of geometrical questions; at least it appears in one of its two copies with another geometrical question:[2] Utrum dyameter alicuius quadrati sit

[1] M. Clagett, *The Science of Mechanics in the Middle Ages* (Madison, 1959). p. 641.

[2] The text of this additional question was published by H. Suter in the *Zeitschrift für Mathematik und Physik*, vol. 32 (1887), Hist.-lit. Abtheilung, pp. 43–54.

commensurabilis costae eiusdem, a question composed either by Albert or his contemporary and colleague at Paris, Nicole Oresme.[3]

As a scholastic *questio*, Albert's small tract has all the common features of the fourteenth-century disputative form: affirmative arguments (*rationes quod sic*), negative arguments (*rationes quod non*), distinctions (*distinctiones*) and conclusions (*conclusiones*) which constitute a determination of the question, and finally comments on the initial arguments (*rationes principales*). The scholastic technique is also illustrated by the close attention of the author to the logical form of argument, where the common logical terms are employed: *consequentia* (implication), *antecedens* (antecedent part of the *consequentia*), *consequent* (the conclusion of the *consequentia*), major and minor. As in the case of the Munich Version given in Chapter Three, Albert outlines the logical structure of his proofs before proceeding to the detailed proofs.

One of the most interesting parts of the introductory arguments—or "principal reasons," as they are often called—is the citation of Antiphon and Bryson via Aristotle's *Physica* and *Elenchi*. I have discussed this citation in the Commentary. I can merely note here that it is by no means simple to decide what Antiphon and Bryson had concluded concerning quadrature. In viewing the principal reasons, the reader must remember that such initial arguments are often rather faulty and loose, being commonplace arguments that are ordinarily not taken very seriously by the author but which for tradition's sake he must consider.

Incidentally, it is something of a breach of scholastic form for Albert to present the affirmative reasons prior to the negative in view of the fact that he will ultimately decide the question in the affirmative. It is far more common for an author to present negative arguments first if he is to decide the question in the affirmative, and the affirmative arguments first if he is to decide it in the negative. The affirmative arguments initially presented are three: (1) Since there can be a square less than the circle and one greater than the circle, there can be one equal to the circle. (2) If such a square equal to the circle cannot be found, we would then have a case of the passage from "greater" to "lesser" without going through "equal," which is impossible. (3) The relationship of "squaring" to a circle is like that of "cubing" to a sphere; but a sphere can be "cubed" since the content of a spherical vase can be poured into a cubical vase.

[3] V. P. Zoubov, "Quelques observations sur l'auteur du traité anonyme 'Utrum dyamter alicuius quadrati sit commensurabilis costae ejusdem,'" *Isis*, vol. 50 (1959), pp. 130–34.

Following upon the brief statement of these affirmative arguments those in opposition are then presented: (1) If a circle can be squared, a square can be converted into a circle; but no art of converting a square into a circle has come down to us. (2) It is impossible to take a given [inscribed] square and, by pushing out the sides, to make a circle equal to that square, since the circle is considered to be the most capacious of all isoperimetric figures. (3) If we assume from Euclid and Campanus that the equality of figures has to be proved by the equality of angles, it will be impossible to find a square equal to a circle for it is impossible to find equality between a semicircle and half of a square since it is impossible to find equality of angles between a semicircle and half a square.

With the preliminary affirmative and negative arguments presented, Albert then goes on to his own views which are in the form of a series of distinctions and conclusions leading up to the ultimate proof of the quadrature of the circle. The first distinction notes that there are five ways in which the expression "quadrature of a circle" can be understood. The first way rests on the ultimate semantic identification of the word quadrature with "dividing into four" and merely means dividing the circle into four quadrants by two orthogonally intersecting diameters; this common usage Albert ascribes to Campanus and it reminds us of the reference to quadrature in Aristophanes' *Birds* (see below, Commentary, lines 49–53). The second way of understanding quadrature is as a procedure of fashioning a squarelike figure out of a circle by rearranging its parts, perhaps as in the accompanying figure ◻. The expression is also used in a third way to mean the finding of a square whose perimeter is equal to the circumference of a circle but whose area is not equal to the area of the circle; this usage he once more ascribes to Campanus. The fourth way of understanding quadrature—labeled later as erroneous—is as the finding of a square whose perimeter is equal to the circumference of a circle and which at the same time is equal in area to the circle. Finally, the fifth meaning assigned to quadrature is simply the finding of a square equal to a circle.

In the second of his distinctions Albert notes that quadrature of circle in the third and fifth meanings can be considered "with respect to sense," or "with respect to intellect" (i.e., demonstratively), or "with respect to both sense and intellect." "With respect to sense" in the third meaning of quadrature merely means that a square can be found whose perimeter does not sensibly differ from the circumference of the circle; while in connection with the fifth meaning it means that a square can be found whose area does not sensibly differ from that of the circle. "With respect to intellect" in connection with the third way of understanding quadrature means that one can demonstratively prove the equality of the perimeter of a square and the circumference of the circle, and in connection with the fifth way it means that one can demonstratively prove the one area is equal to the other.

From these distinctions Albert of Saxony then goes on to his conclusions. (1) Quadrature understood in the first way as quartering a circle is, of course, possible and need not be demonstrated. (2) As understood in the second way, quadrature is possible only so long as we do not take quadrature in this sense to mean that the "squarelike" figure is equal to the circle. (3) Quadrature in its third connotation is possible both "with respect to sense" and "with respect to intellect." Campanus, Albert says, has shown that this is so "with respect to sense" (see Appendix I) when he made the assumption that the circumference contains the diamter $3\frac{1}{7}$ times. As for quadrature in the third connotation, "with respect to intellect," our author believes that it is in such a connotation that Aristotle understands quadrature when he says in his *Categories* (Chapter 7, 7b): "quadrature of a circle is knowable but not yet known." Albert believes that Aristotle said this because no one had yet demonstrated that the circumference is $3\frac{1}{7}$ the diameter, nor in fact that any straight line is equal to the circumference. "Nevertheless," Albert says, "it [i.e., quadrature in the third sense] is demonstrable to the intellect, although such is difficult." (4) Quadrature in the fourth sense—where not only is the perimeter of the square equal to the circumference of the circle but also the area of the same square is equal to the area of the circle—is quite properly said to be impossible by Albert. The reason given is that the circle is the "most capacious" of isoperimetric figures.

Albert has now cleared the way to talk of quadrature in the fifth sense and thus to prove demonstratively that there is a square equal to a circle. To do this, he first proves in the fifth conclusion that every regular polygon is equal to a right triangle one of whose two sides including the right angle is equal to the perimeter of the polygon while the other is equal to the line drawn perpendicularly from the center of the polygon to the midpoint of one of its sides. The wording of this conclusion is obviously drawn from the same tradition as that of the Pseudo-Bradwardine Version. It is easily proved by resolving the regular polygon into triangles, as was done in other emended versions of the *De mensura circuli* (e.g., see the remarks concerning Propositions IIIA and IIIB in the introduction to the Corpus Christi Version in Chapter III, Section 5).

It is the sixth conclusion of Albert's *Questio* that is the first proposition of the *De mensura circuli.* Albert has taken the wording of this proposition from one of the versions based on Gerard of Cremona's translation. Its wording is closest to that found in the Naples Version (see Chapter Three, Section 2). It might be argued that Albert did not know the original translation of Gerard in its entirety, since if he had he would in all probability have mentioned Proposition III when talking about the problem of determining the ratio of the circumference to the diameter. In initiating his proof of the sixth conclusion, Albert first makes three suppositions. The first is in two parts: (1) one figure inscribed in another is less than it, and (2) the perimeter of the inscribed figure is less than that of the figure in which it is inscribed. The second part is similar to the third postulate of the Cambridge Version and in a general way to Propositions I and II of the Corpus Christi Version and to the second conclusion of the Pseudo-Bradwardine Version. The first part is like the statement of line 106 in the Pseudo-Bradwardine Version. The second of Albert's suppositions states that in the case of two right triangles the one is greater whose two sides including the right angle are greater, or one of whose two sides is greater while the other side is equal. The third supposition holds that with two unequal, continuous magnitudes given, it is possible to take from the greater a quantity greater than the lesser. This is said to follow from the fact that any excess by which one quantity exceeds another is divisible. In discussing the Pseudo-Bradwardine tract, I have already indicated the importance of Albert of Saxony's supposition for proving the main proposition, particularly when compared with the proof based directly on Proposition X.1 of the *Elements* of Euclid.[4]

Albert's proof of the sixth conclusion essentially shows that if the given right triangle is assumed to be less than the circle, by Supposition III there can be inscribed in the circle a regular polygon which is greater than the triangle. This polygon, however, is equal to a right triangle which is then

[4] Professor Murdoch, in his forthcoming text of Bradwardine's *De continuo*, is planning to discuss in considerable detail the relationship of this postulate to Proposition X.1 and to Book V, definition 4, of the *Elements*, as well as to Archimedes' Lemma. In all of the latter, "exceeding" or "falling short" is emphasized rather than the "betweenness" notion of Albert's postulate. Murdoch finds that the similarity between Albert's postulate and X.1 lies in their dependence on a system of magnitudes as continuous and their inconsistency with a system of magnitudes as discontinuous, while Book V, definition 4, and Archimedes' Lemma are consistent with a system of magnitudes as discontinuous. Incidentally, as I have noted in the Introduction to this chapter, note 1, Archimedes in his *On Spiral Lines*,

both greater than the given right triangle (by Supposition III) and less than it (by Conclusion V and Suppositions I and II), an obvious contradiction. If the given right triangle is said to be greater than the circle, a similar contradiction ensues. Hence the given right triangle must be equal to the circle. Then in the seventh conclusion the right triangle is converted to a square by reference to Proposition II.14 of the *Elements*.

Following the proof of the seventh conclusion, Albert discusses the initial arguments with which he opened the question. In the course of the discussion of the first affirmative argument Albert notes that perhaps Bryson intended by quadrature the finding of not only a square equal in area to a circle but one whose perimeter is equal to the circumference of the circle. Such, however, does not appear to have been the intention of Bryson (see the Commentary below, lines 3–4). Albert adds that the argument of Bryson (*ratio Brissonis*) does prove quadrature but not in the sense understood by Bryson. By the argument of Bryson, he means, I take it, that since there is a square greater than the circle and one less than the circle, there must be one equal to it. This Albert believes to be true enough in this case; but such an argument cannot be universally true, i.e., in the passage from "less" to "greater" it is possible not to pass through "equal." A case in point concerns the angle of a circular segment, the angle formed by an arc and a chord. Albert assumes that there can be a rectilinear acute angle less than the angle of the circular segment and a right angle greater than the angle of the segment without there being any rectilinear angle equal to the angle of the segment. Albert also believes that the second affirmative argument arrives at the correct conclusion, but once again he notes here the case of the angle of a segment as an example of the possibility of transition from "less" to "greater" without arriving at equal. Here he directly follows and quotes Campanus' commentary on Proposition III.15 (Greek text, III.16) of the *Elements* (see the Commentary below,

Proposition IV, has a proposition similar to Albert's axiom, "Given two unequal lines, viz., a straight line and the circumference of a circle, it is possible to find a straight line less than the greater of the given lines and greater than the less." The proof of this is based on Archimedes' Lemma (Vat. Ottob. lat. 1850, 12r): "Duabus lineis datis inequalibus, recta scilicet et circuli periferia, possibile est accipere rectam maiore quidem datarum linearum minorem, minore autem maiorem. Quotiens enim quo excedit maior linea minorem ipsi compositus excedet rectam, et in tot equalia divisa recta una decisio minor erit excessu. Si quidem igitur et periferia maior recta una decisione apposita ad rectam, minore quidem datarum palam quod maior erit, maiore autem minor; etenim que apponitur, minor erit excessu."

line 45). Albert disposes of the initial negative arguments easily and his refutation demands no special comment.

My text is based on the two manuscripts indicated in the Sigla below, but I have generally preferred *Z* since *Fa* includes alterations arising from the fact that its author, Wigand Durnheimer, has incorporated the *Questio* of Albert into an extended geometrical tract of his own (for this tract, see Chapter Three, Section 4, Introduction). Wigand accordingly felt free to make additions or deletions. For example, he deleted lines 1–47 and lines 264–318 of Albert's text—the lines from 264–318 that refer to Proposition VII and the initial *rationes*. He is thus left with Albert's *distinctiones* and the first six *conclusiones*. It will be obvious to the reader who makes the comparison that my text diverges only slightly from the very good text already published by H. Suter on the basis of *Z*.[5] The drawings, in general, are reproduced from *Z*, although Fig. 61 has been emended by me to accord with the intention of the text. The marginal folio numbers are from *Z*.

Sigla of manuscripts

Z = Bern, Bürgerbibliothek A.50, 169r–172r, 15c.
Fa = Vienna, Nationalbibliothek cod. 5257, 64v–67r, 1390.

[5] H. Suter, "Der Tractatus 'De quadratura circuli' des Albertus de Saxonia," *Zeitschrift für Mathematik und Physik*, vol. *29* (1884), Hist.-lit. Abtheilung, pp. 81–101.

The Question of Albert of Saxony on the Quadrature of the Circle

169r

Questio Alberti de Saxonia de quadratura circuli

Queritur utrum quadrare circulum sit possibile. Arguitur primo quod sic, auctoritate Antifontis et Brissonis, qui ut dicit philosophus primo elencorum et primo physicorum circulum quadrare sunt conati. 5 Probatur etiam ratione, quia cuicunque est dare quadratum maius et quadratum minus, sibi est dare quadratum equale; sed circulo est dare quadratum maius, scilicet quadratum circulo circumscriptum, et quadratum minus, scilicet quadratum circulo inscriptum; igitur etiam est dare quadratum circulo equale. 2º sic: si non esset dare quadratum 10 circulo equale, sequeretur quod fieret transitus de maiore ad minus, sive de extremo ad extremum transeundo per omnia media, et tamen nunquam perveniretur ad equale vel ad medium. Sed hoc est falsum. Igitur probo consequentiam, quia sit unum quadratum circulo inscriptum et incipiat hoc continue augeri uniformiter donec fiat maius cir- 15 culo. Si igitur fuit aliquando circulo equale, habetur propositum; si non, transitus factus est de minore ad maius respectu istius circuli et nunquam perventum est ad equale. 3º sic: sicut se habet spera ad cubicari ita circulus ad quadrari. Sed spera potest cubicari, ut patet si aquam replentem vas spericum infundamus ad vas quadraticum sive 20 cubicum.

Oppositum arguitur sic: Si circulus potest quadrari, quadratum posset circulari. Tenet consequentia, cum non videatur esse maior ratio de uno quam de alio. Falsitas consequentis videtur tenere ex eo quod non videtur esse tradita aliqua ars quadratum circulandi. 2º sic: circu- 25 lum quadrare est quadratum equale circulo invenire; sed hoc invenire est impossibile, quia si latera quadrati extendantur equaliter a centro, superficies circularis erit multo capacior quam erat ante ipsum qua-

1–47 Queritur....quesito *om. Fa et scr.* area omnis circuli equalis est thetragonismo sub medietate circumferentie et medietate dyametri contento. Ad declarationem maioris huius conclusionis distinguo de quadratura circuli et pono aliquas conclusiones ex quibus patebit conclusio.

The Question of Albert of Saxony on the Quadrature of the Circle

It is sought whether it is possible to square a circle. It is argued affirmatively: in the first place by the authority of Antiphon and Bryson, who, as the Philosopher says in the first [book] of the *Elenchi* and in the first [book] of the *Physics*, attempted to square the circle. It is also proved by argument. For if there can be given a square greater than something and a square less than it, there can also be given a square equal to it. But there can be given a square greater than the circle—that is, a square circumscribed about the circle—and also a lesser square—that is, a square inscribed in the circle. Therefore, there also can be given a square equal to the circle.

Second affirmative argument: If there could not be given a square equal to a circle, it would follow that there would take place passage from "greater" to "lesser," or from extreme to extreme, through all the means without ever arriving at "equal" or "middle." But this is false. Therefore, I prove the consequence. For let there be one square inscribed in a circle and let this square begin to be continually and uniformly increased until it becomes larger than the circle. If, therefore, it was at some time equal to the circle, we have the proposition; if not, then passage has been made from "lesser" to "greater" with respect to that circle without ever arriving at "equal."

Third affirmative argument: As a sphere is related to cubing so a circle is related to squaring. But a sphere can be cubed, as is obvious if we pour the water filling up a spherical vase into a squared or cubic vase.

On the opposite side it is argued as follows: If a circle can be squared, a square could be "circled." The consequence holds, for there appears to be no greater reason for the one than for the other. The falsity of the consequent seems to follow because there does not seem to have been transmitted any art of "circling" the square.

Second negative argument: To square a circle is to find a square equal to a circle. But to find this is impossible, for if the sides of a square are extended equally from the center, the circular surface will be more ca-

dratum, quod patet ex libello de corporibus ysoperimetris ubi dicitur, spera omnium corporum ysoperimetrorum esse maximum. Item con-
30 fertur ex alio, quia non potest inveniri quadratum cuius medietas sit equalis medietati circuli, igitur nec potest reperiri quadratum quod sit equale circulo. Tenet consequentia, quia quorum dimidia sunt inequalia et ipsa. Antecedens probatur, quia omnis figura rectilinea sub angulis continetur, quibus impossibile est angulos semicirculi esse
35 equales, igitur nullius circuli medietas medietati quadrati potest esse equalis. Tenet consequentia, quia per equalitatem angulorum Euclides et Campanus probant equalitatem figurarum, sicut patet in 4ª propositione et in commento eiusdem primi Euclidis, que incipit: "omnium duorum et cetera." Antecedens quo ad primam eius partem patet,
40 scilicet quod omnis figura rectilinea sub angulis continetur, quia si non sub angulis sed sub angulo posset contineri, sequeretur quod due linee recte possent claudere superficiem, quod est contra ultimam petitionem primi Euclidis. Sed quo ad secundam partem patet, scilicet quod anguli rectilinei et anguli semicirculi non possunt esse equales;
169v nam hoc demonstratum est / super 15 3 huius, scilicet Euclidis.

46 In ista questione primo distinguendum est de quadratura circuli; 2º ponende sunt conclusiones et respondendum est quesito.

Quantum ad primum, sciendum [est] quod quadraturam circuli possumus intelligere quinque modis. Uno modo quadrare circulum est
50 ipsum duabus dyametris orthogonaliter se secantibus in centro in quatuor partes equaliter dividere, et isto modo loquitur Campanus in theorica sua et multi alii doctorum de quadratura circuli quando circulum iubent quadrari. Secundo modo per quadraturam circuli possumus intelligere, ipsum in quadratum vel in figuram aliqualiter simi-
55 lem quadrato reducere per partium decisionem et earum situs transpositionem, et isto modo rudes loquuntur et intelligunt de quadratura circuli. Tertio modo per quadraturam circuli possumus intelligere inventionem alicuius quadrati, non tamen quod sit equale circulo, sed latera eius simul iuncta sunt equalia circumferentie circuli in rectum
60 extense, et isto modo Campanus quadravit circulum. Quarto modo possumus intelligere per quadrare circulum invenire quadratum equale circulo, cuius latera simul iuncta cum hoc sint equalia circumferentie circuli in rectum extense. Quinto modo possumus intelligere per quadraturam circuli invenire unum quadratum equale circulo.

52 doctores *Fa*

pacious than this square was before. This is evident from the *Booklet on Isoperimetric Bodies* where it is said that the sphere is the maximum of all isoperimetric bodies. It also follows from another [argument]: Since a square cannot be found whose half is equal to half of a circle, therefore a square cannot be found which equals a circle. The consequence holds, since those whose halves are unequal are themselves unequal. The antecedent is proved, for every rectilinear figure is contained by angles to which the angles of a semicircle cannot possibly be equal; therefore, the half of no circle can be equal to the half of a square. The consequence holds, because Euclid and Campanus prove the equality of figures by the equality of angles, as in proposition I.4 of Euclid and its commentary, which begins: "Of any two, etc." The first part of the antecedent is evident, namely, that every rectilinear figure is contained by angles; for if it were not contained by angles but rather by an angle, it would follow that two straight lines could enclose a surface, which is against the last axiom of [Book] I of Euclid. But the second part is evident, namely, that rectilinear angles and the angles of a semicircle cannot be equal; for this has been demonstrated in III.15 of this book, i.e., [the *Elements*] of Euclid.

In this question we must first make some distinctions regarding quadrature of a circle. Then, secondly, conclusions must be drawn and response made to that which was sought.

As for the first, it should be known that we can understand "quadrature of a circle" in five ways. In one way, to square a circle is to divide it into four equal parts by two diameters which intersect orthogonally in the center. Campanus speaks in this way in his *Theory* [*of the Planets*], and [so also do] many other scholars [writing] on quadrature of a circle when they direct one to square (that is, *quarter*) a circle. In a second way we can understand by quadrature of a circle the reduction of a circle to a square or to some figure in some fashion similar to a square by cutting off parts and by transposing them. It is in this way that the ignorant speak of quadrature of a circle. In a third way we can understand by quadrature of a circle the finding of some square which is not equal to the circle but whose sides joined together are equal to the circumference extended in a straight line. And in this way Campanus squared the circle. In a fourth way we can understand by squaring a circle the finding of a square equal to a circle and whose sides joined together are also equal to the circumference of the circle extended in a straight line. In a fifth way we can understand by quadrature of a circle the finding of a square equal to the circle.

65 Secunda distinctio est ista, quod quadratura circuli 3° et 5° modo dicta, quedam est ad sensum, quedam ad intellectum, et quedam ad utrumque. Quadrare circulum tertio modo ad intellectum est probare aliquod quadratum cuius latera simul iuncta sunt equalia circumferentie circuli in rectum extense, et isto modo non est dubium quin circulus 70 sic quadrari possit, quia non est dubium quin aliqua linea recta sit equalis circumferentie in rectum extense, que si in 4^or^ partes equales dividatur, et partes ad invicem ad angulos rectos iungantur, constituunt unum quadratum. Quadrare autem circulum tertio modo ad sensum est invenire unum quadratum cuius latera simul iuncta con- 75 stituunt unam lineam, que si obiceretur visui una cum circumferentia circuli in rectum extensa, visus nequaquam posset ponere differentiam inter illas, sed iudicaret unam esse tantam quantam aliam. Sic similiter quadratura circuli 5° modo dicta, quedam est ad intellectum, ut invenire unum quadratum et hoc demonstrative probare esse equale circulo; 80 et quedam est ad sensum, ut facere et invenire unum quadratum inter quod et aliquem circulum sensus nequit ponere differentiam, nec considerare que istarum superficierum sit maioris capacitatis, ymmo iudicat unam alteri esse equalem. Quadrare autem circulum 3° modo ad utrumque, scilicet tam ad sensum quam ad intellectum, est aliquod 85 quadratum invenire et demonstrative probare quod illius latera simul iuncta sunt equalia circumferentie alicuius circuli in rectum extense, et cum hoc quod inter latera illius quadrati simul iuncta et circumferentiam circuli in rectum extensam sensus non posset ponere differentiam; sic conformiter dicamus de quadratura circuli 5° modo dicta 90 ad utrumque, scilicet tam ad sensum quam ad intellectum.

Quantum ad secundum principale sit prima conclusio ista: Quadratura circuli primo modo dicta est possibilis. Hoc patet cuilibet intuenti; quare non oportet ipsam modo demonstrare.

Secunda conclusio est quod loquendo de quadratura circuli secundo 170r modo dicta, scilicet quod iste partes circum/ferentie circumiacte con- 96 stituant quadratum equale circulo, dico quod est impossibilis nec scibilis nec demonstrabilis. Probatur quia, postquam partes circumferentie sic sunt circumiacte, non resultat figura quadrata, cum non omnes eius anguli sunt recti, ymmo nullus eius angulus est rectus. Et ex alia,

90 *post* intellectum *add. Fa* ita quod est invenire unum quadratum et hoc demonstrative probatur esse equale circulo atque inter illud et inter aliquem circulum sensus non potest ponere differentiam nec considerare que istarum superficierum sit maioris capacitatis.

The second distinction is this: Quadrature spoken of in the third and fifth ways is sometimes [considered] with respect to sense and sometimes with respect to intellect, and sometimes with respect to both. To square a circle in the third way with respect to intellect is to prove that there is some square whose sides joined together are equal to the circumference of the circle extended in a straight line. And in this way there is no doubt that a circle can be squared because there is no doubt that there is some straight line equal to the circumference extended in a straight line. If this line is divided into four equal parts and the parts are mutually joined at right angles, they form a square. Now to square a circle in the third way with respect to sense is to find a square whose sides joined together form a straight line so that if one visually compares it with the circumference of the circle extended in a straight line he cannot see any difference between them but would judge one to be as long as the other. In the same way quadrature of a circle is spoken of in the fifth way sometimes with respect to intellect, as in finding a square and demonstratively proving that it is equal to the circle, and sometimes with respect to sense, as in constructing and finding a square such that the sense reveals no difference between it and some circle, and one would not consider one of these to be more capacious than the other; rather he would judge one to be equal to the other. To square a circle in the third way with respect to both [i.e., to sense and intellect] is to find some square and to prove demonstratively that its sides when joined together are equal to the circumference of some circle extended in a straight line, and furthermore the senses could not determine any difference between the sides of the square joined together and the circumference o- the circle extended into a straight line. We can speak in a similar way concerning quadrature of a circle in the fifth way with respect to both, i.e., to sense as well as to intellect.

As for the second principal [part of our determination], let the first conclusion be this: Quadrature of a circle spoken of in the first way is possible. This is evident to anyone who is attentive; therefore, it is not necessary to demonstrate this now.

The second conclusion is that in speaking of quadrature of a circle understood in the second way, namely, that the parts circumjacent to the circumference form a square equal to the circle, I assert that this is impossible; neither it is knowable nor demonstrable. This is proved: For after the parts are thus made circumjacent to the circumference, a squared figure does not result since not all of its angles are right angles, but in fact no angle of it is a right angle. And also because this figure is not equal

100 quia ista figura non est equalis circulo ex eo quod est circulo inscripta. Si tamen intelligitur quod circumferentia potest sic dividi in quatuor partes equales que per earum transpostitionem constituunt talem figuram aliqualiter similem quadrato, non est dubium quin hoc sit possibile, sicut possibile est quadrare circulum primo modo.

105 Tertia conclusio: Quadratura circuli tertio modo dicta est possibilis et ad sensum et ad intellectum. Probatur primo quod ad sensum per Campanum qui sic quadravit circulum, asserans secundum assertionem multorum philosophorum, circumferentiam circuli continere dyametrum ter et septimam eius partem, et cum dyameter sit linea recta, 110 patet quod si tres dyametros cum septima parte eiusdem dyametri ad invicem iungamus, constituent unam lineam rectam equalem circumferentie circuli in rectum extense, que si visui obiceretur una cum circumferentia illius circuli in rectum extensa, visus nequit ponere differentiam inter eas, sed penitus iudicaret eas esse equales; etiam 115 adhuc si una alteri supponeretur; et ista linea tunc divideretur in 4^or^ partes equales, et si fieret unum quadratum ex illis, esset quadrare circulum 3° modo; et hoc est possibile, igitur et cetera. Quod autem quadratura circuli 3° modo dicta sit possibilis ad intellectum patet auctoritate Aristotilis (!) in predicamentis, ubi dicit, "quadratura 120 circuli, si est scibilis, nondum scita." Hoc autem non intellexit de quadratura circuli primo modo dicta, quia illa est scita, nec de quadratura 2° modo dicta, quia illa est impossibilis, nec 4° modo dicta, quia ista etiam est impossibilis, ut statim probabitur, nec 5° modo dicta, quia ista est scita, ut patebit in una conclusione ponendarum; 125 ergo intellexit de quadratura 3° modo dicta. Ista enim, quamvis sit scibilis forte, nondum est scita, ex eo quod forte nondum per artem est inventum nec demonstratum ad intellectum circumferentiam circuli habere se in proportione tripla sexquiseptima ad dyametrum, nec aliquam lineam rectam equalem circumferentie; nichilominus est de-130 monstrabile ad intellectum quamvis difficile. Et ideo quadratura ipsius circuli Campani est ad sensum non ad intellectum.

Quarta conclusio: Quadratura circuli quarto modo dicta est impossibilis, scilicet aliquod quadratum esse equale circulo, cuius latera simul iuncta sint equalia circumferentie circuli in rectum extense. Hec

124 in...ponendarum *om. Fa* 131 *ante* Campani *add. Z* ipsius *sed om. Fa*

to the circle since it is inscribed in the circle. If, however, it is understood [by this] that the circumference can be so divided into four equal parts which when they are transposed form a figure in some way similar to a square, there is no doubt that this is possible, just as it is possible to square a circle in the first way.

Third conclusion: Quadrature of a circle spoken of in the third way is possible with respect both to sense and to intellect. It is proved in the first place with respect to sense by Campanus, who squared a circle in this way, asserting as did many of the philosophers that the circumference of the circle contains the diameter three times and its seventh part. And since the diameter is a straight line, it is evident that if we join together three diameters with a seventh part of the diameter, they will form a straight line equal to the circumference extended in a straight line. If such a straight line is visually compared with the circumference of the circle extended in a straight line, one cannot see any difference between them but would judge them to be completely equal; even if one were superposed on the other, this would be so. And if this line were divided into four equal parts and if a square were formed from these parts, this would be squaring a circle in the third way. And this is possible; therefore, et cetera. That, moreover, quadrature of a circle spoken of in the third way is possible with respect to intellect is evident by the authority of Aristotle in the *Categories* where he says, "quadrature of a circle, if it is knowable, is not yet known." But he has not understood quadrature of a circle in the first way because that is known, nor in the second way because that is impossible, nor in the fourth way because this too is impossible, as I shall immediately prove, nor in the fifth way because this is known, as will be evident in one of the conclusions to be posited. Therefore, [by exclusion,] he understood quadrature spoken of in the third way. For this way, although perhaps knowable, "is not yet known" since perhaps it has (had?) not yet been found by art nor demonstrated to the intellect that the circumference of a circle is related in a $3\frac{1}{7}$ ratio to the diameter, nor that any straight line is equal to a circumference. Nevertheless, it is demonstrable to the intellect, although it is difficult. And, therefore, Campanus' quadrature of this circle is with respect to sense not to intellect.

The fourth conclusion: Quadrature of a circle spoken of in the fourth way is impossible, i.e., that some square is equal to a circle [and that at the same time] its sides joined together are equal to the circumference of a circle extended in a straight line. This [conclusion] is evident from the fact

135 [conclusio] patet ex eo quod figura circularis inter omnes alias est capacissima.

Breviter de quadratura circuli nec primo modo nec 2° nec 3° nec 4° modo principaliter intendo, sed de quadratura circuli 5° modo dicta principaliter intendo, scilicet demonstrative probare aliquod quadra-140 tum esse equale circulo.

Quinta conclusio ad probandum quadraturam circuli hic intentam sit ista: Omnis figura rectilinea equiangula et equilatera est equalis triangulo orthogono, cuius alterum laterum rectum angulum continentium est equale linee recte quam omnia latera illius figure simul 145 iuncta constituunt et reliquum laterum angulum rectum constituentium equale linee a centro eiusdem figure ad aliquod suorum laterum perpendiculariter ducte.

170v Verbi gratia, sit figura rectilinea equilatera *ABCD* et equian/gula, cuius centrum sit *E*, ducaturque linea a centro *E* [Fig. 60] usque ad 150 aliquod latus illius figure perpendiculariter, scilicet ad latus *AB*, tangens latus *AB* in puncto *F*; sitque triangulus orthogonus *EFG*, cuius alterum laterum rectum angulum constituentium, scilicet *EG*, sit equale omnibus lateribus illius figure simul iunctis, et reliquum laterum rectum angulum constituentium, scilicet *EF*, sit equale linee, vel sit 155 ipsamet linea perpendiculariter ducta a centro *E* ad latus *AB* tangens ipsum in puncto *F*. Tunc dico quod figura *ABCD* est equalis triangulo *EFG*. Hoc probatur sic: Illa tota sunt equalia quorum partes similis denominationis unius sunt equales partibus similis denominationis alterius. Sed modo ita est quod partes similis denominationis 160 figure equiangule et equilatere *ABCD* sunt equales partibus similis denominationis trianguli *EFG*; igitur et cetera. Maior est nota per unam propositionem quinti Euclidis, que sic dicit, "Si fuerint quotlibet quantitates aliarum totidem equemultiplices aut singule singulis equales, necesse est quemadmodum una earum ad sui comparem to-165 tum quoque ex hiis aggregatum ad omnes illas pariter acceptas simili-

135 alias: ysoperimetras sibi *Fa*
136 *post* capacissima *add. Fa* ut patuit de ysoperimetris propositione ultima
138 hic intendo *Fa*
148 *supra mg.* 170*v scr. Z* Albertus de Saxonia de quadratura circuli
163 aliarum *Fa* aliarumque *Z*
164 ad sui comparem *Fa, om. Z., cf. Adelard II, V.1*
165 quoque *Fa* -que *Z, cf. Adel. II, V.1*

that a circular figure is the most capacious of all [isoperimetric figures].

In brief, I do not principally intend quadrature of a circle spoken of in the first, second, third, or fourth way, but rather in the fifth way: that is, to prove demonstratively that some square is equal to a circle.

The fifth conclusion—for proving quadrature of a circle as here intended—is this: Every regular polygon is equal to a right triangle, one of whose two sides containing the right angle is equal to a straight line compounded of all the sides of the polygon joined together, and the other side of those comprising the right angle is equal to a line drawn perpendicularly from the center of the same figure to one of its sides.

For exemple, let the regular polygon be *ABCD* with center *E* [see Fig. 60]. And let a line be drawn perpendicularly from the center *E* to one side of the figure, namely, to side *AB*, touching side *AB* in point *F*.

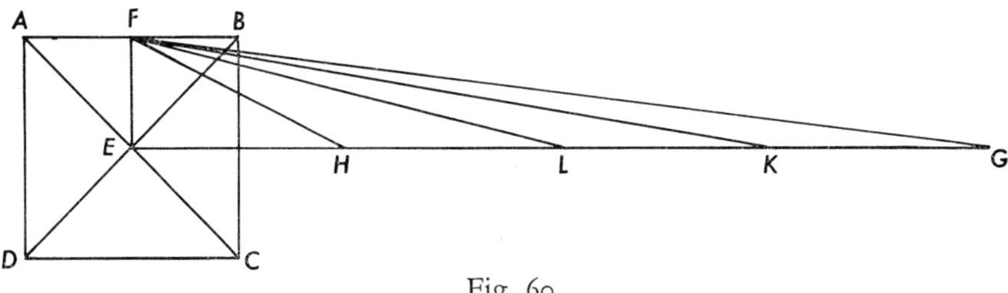

Fig. 60

Note: I have transposed letters *C* and *D* to conform with the text.

And let the right triangle be *EFG* with one of the two sides comprising the right angle, i.e., *EG*, equal to all of the sides of the figure joined together and with the other of the sides comprising the right angle, that is, *EF*, equal to the line, or being the very line, perpendicularly drawn from the center *E* to side *AB*, touching it in point *F*. Then I say that figure *ABCD* is equal to △ *EFG*. This is proved as follows: Those wholes are equal where the parts of similar denomination of the one are equal to the parts of the same denomination of the other. But now it is so that the parts of similar denomination of the regular polygon *ABCD* are equal to the parts of the same denomination of △ *EFG*; therefore, et cetera. The major is known by the [first] proposition of [Book] V of [the *Elements* of] Euclid: "If there is any number of quantites which are equal multiples of just as many other [quantities], i.e., the quantities are equal in multitude, it is necessary that just as one of them is related to its comparable term so the whole aggregate of these [quantities of one set] is related to the aggregate of the other quantities [of the second set]." But the minor is proved. For

ter se habere." Sed minor probatur: ducam enim a centro *E* ad quemlibet angulum figure predicte *ABCD* lineam rectam et erunt quatuor trianguli equales, scilicet *EBA* et *EBC* et *ECD* et *EDA*, qui ex eo sunt equales quod cuiuslibet latera unius sunt equalia lateribus alterius,
170 et quilibet istorum triangulorum est quarta pars figure equilatere et equiangule *ABCD*. Rursus dividam latus *EG* trianguli *EFG* in 4or partes equales, quelibet illarum partium erit equalis uni lateri figure *ABCD*, quarum una partium lateris *EG* sit *EH*, altera *HL*, 3a *LK*, 4a *KG*. Postea ducam a puncto *F* ad punctum *H* linee *EG* lineam
175 *FH*, et ad punctum *L* lineam *FL*, et ad punctum *K* lineam *FK*, et erit triangulus *EFG* resolutus in 4or triangulos equales, ex eo quod omnes illi trianguli 4or cadunt super equales bases et altitudo omnium eorum est linea *EF*; quare per primam 6i Euclidis erunt equales. Et similiter quilibet triangulorum *EFG* trianguli est equalis cuilibet trian-
180 gulorum figure *ABCD*, ex eo quod basis cuiuslibet trianguli trianguli *EFG* est equalis basi cuiuslibet trianguli figure *ABCD* et una est altitudo omnium, scilicet linea *EF*; quare per primam 6i Euclidis sequitur eos esse equales. Prima enim 6i Euclidis dicit sic: "Si duarum superficierum equedistantium laterum sive triangulorum fuerit alti-
185 tudo una, tanta erit alterutra earum ad alteram, quanta sua basis ad basim alterius." Erunt igitur 4or partes seu 4or quarte trianguli *EFG* equales quatuor quartis figure equiangule et equilatere *ABCD*. Quare sequitur, cum partes similis denominationis figure equilatere et equiangule *ABCD* sint equales totidem partibus similis denominationis
190 trianguli *EFG*, sequitur totam superficiem figure *ABCD* esse equalem toti superficiei trianguli *EFG*, quod fuit probandum. Et idem esset si talis figura equilatera et equiangula esset pentagona vel exagona vel quotcunque angulorum vel quotcunque laterum equalium.

Sexta conclusio: Omnis circulus est equalis triangulo orthogono,
195 cuius alterum laterum rectum angulum continentium est equale cir-
171r cumferentie in rectum extense et reliquum latus rectum angulum / continentium est equale semidyametro eiusdem circuli.

Ad quam conclusionem probandam suppono primo omnem figuram alteri inscriptam illi cui inscribitur minorem esse, eiusque latera
200 simul iuncta lateribus illius cui inscribitur breviora. 2o suppono omnium triangulorum orthogonorum illum maiorem esse cuius ambo

178 Euclidis *Fa*, *om.* *Z*
196 *supra mg.* *171r scr.* *Z* De quadratura circuli
197 *post* circuli *add.* *Fa* Hoc est Archimenidis in libello de mensura circuli

I shall draw a straight line from center E to each angle of the aforesaid figure $ABCD$ and four equal triangles will be formed, namely, EBA, EBC, ECD, and EDA. These triangles are equal because the sides of any one are equal to the sides of any other one, and each of these triangles is one fourth of the regular polygon $ABCD$. Further, I shall divide side EG of $\triangle EFG$ into four equal parts. Each of these parts will be equal to a side of figure $ABCD$. Let one part of side EG be EH, the second HL, the third LK, and the fourth KG. Afterwards I shall draw line FH from point F to point H in line EG, and line FL to point L, and line FK to point K. And [so] $\triangle EFG$ will be resolved into four equal triangles because all four triangles fall on equal bases and line EF is the altitude of all of them. Hence by VI.1 of [the *Elements* of] of Euclid they will be equal. And similarly any of the triangles making up $\triangle EFG$ is equal to any of the triangles making up figure $ABCD$, because the base of any triangle of $\triangle EFG$ is equal to the base of any triangle of figure $ABCD$, and there is a single altitude for all the triangles, namely, line EF. Hence by VI.1 of Euclid it follows that they are equal. For VI.1 of Euclid says: "If the altitude of two parallelograms or triangles is one, then they will be to one another as their bases." Therefore, the four parts, or four fourths, of $\triangle EFG$ will be equal to the four fourths of the regular polygon $ABCD$. Therefore, it follows, since the parts of one denomination of the regular polygon $ABCD$ are equal to just as many parts of the same denomination of $\triangle EFG$, that the whole surface of figure $ABCD$ is equal to the whole surface of $\triangle EFG$; which was to be proved. And it would be the same if the regular polygon were a pentagon, or a hexagon, or a regular polygon of any number of sides and angles.

The sixth conclusion: Every circle is equal to a right triangle one of whose sides containing the right angle is equal to the circumference extended in a straight line and the other side of those sides containing the right angle is equal to the radius of the same circle.

For the proof of this conclusion I posit in the first place that every figure inscribed in another figure is less than the figure in which it is inscribed, and that its sides when joined together are less than the sides of the figure in which it is inscribed. In the second place, I posit that of any [pair of] right triangles the one is greater whose two sides including the right angle are greater than those same sides of the other triangle, or one of whose [two sides containing the right angle] is equal to one side of the other triangle and the second side is greater than the corresponding side of the other triangle. This is illustrated by triangles EBC, EDF, and

latera angulum rectum constituentia sunt maiora reliquis lateribus reliqui trianguli orthogoni angulum rectum constituentibus, aut alterum reliquo unius et alterum reliquo ipsum respiciente equale, sicut
205 patet in hiis triangulis *EBC* et *EDF* et *EDA*. [Fig. 61] 3° suppono propositis duabus quantitatibus continuis a maiore maiorem minore posse resecare. Ista patet ex eo quod quilibet excessus quo una quantitas aliam excedit est divisibilis. Hiis suppositis probo conclusionem sic:

210 Esto enim circulus *AB*, cuius centrum sit *E*; sitque triangulus orthogonus *EBC*, cuius latus *EC* sit equale circumferentie circuli *AB* in rectum extense et reliquum latus rectum angulum constituentium, scilicet *EB*, sit semidyameter seu equale semidyametro circuli *AB*. Tunc dico triangulum *EBC* esse equalem circulo *AB* [Fig. 61]. Pro-
215 batur sic: quia triangulus *EBC* nec est maior nec est minor circulo *AB*, igitur sibi equalis. Tenet consequentia, quia omnes quantitates quarum una non est maior nec minor alia sunt equales. Sed probo antecedens et primo quod triangulus *EBC* non sit minor circulo *AB*. Nam si sic, tunc per tertiam suppositionem a circulo *AB* figura maior triangulo
220 *EBC* posset resecari. Sed hoc est falsum; nam sit illa figura poligonia equilatera et equiangula circulo inscripta *FGHIKLMN*, a cuius centro *A* ducatur linea *AO* perpendiculariter super latus *FG* [Fig. 62]. Tunc per conclusionem quintam iam ante demonstratam illa figura poligonia et octogonia erit equalis triangulo orthogono, cuius alterum
225 laterum rectum angulum continentium est equale omnibus lateribus illius figure simul iunctis et reliquum est equale linee perpendiculari *AO*. Sed latera predicte figure simul iuncta circumferentia circuli in

203 trianguli *Fa* anguli *Z*
205 EDF et EDA *Z* EDC et EBA *Fa*
212–13 angulum... scilicet *Fa*, *om. Z*

EDA [in Figure 61]. Third, I suppose that with two continuous quantities proposed, a magnitude greater than the "lesser" can be cut from the "greater." This is evident because any excess by which any quantity exceeds another is divisible. With these things posited, I prove the conclusion as follows:

Let there be a circle *AB*, whose center is *E*; and let there be a right △ *EBC*, whose side *EC* is equal to the circumference of circle *AB*

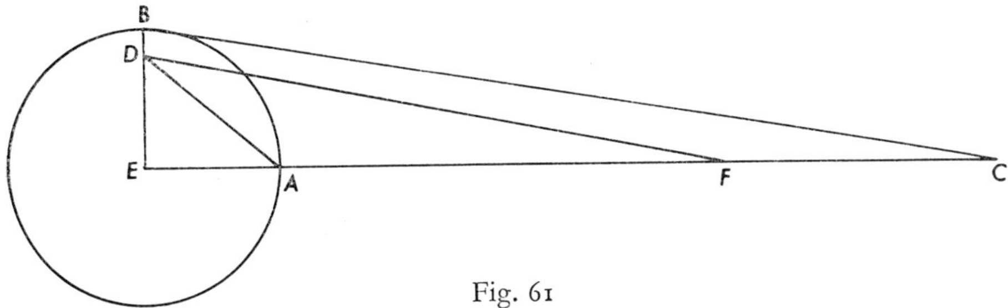

Fig. 61

Note: The drawing has been altered to agree with the text. *F* has been added and the position of *A* changed from the left side of the circle to the right side.

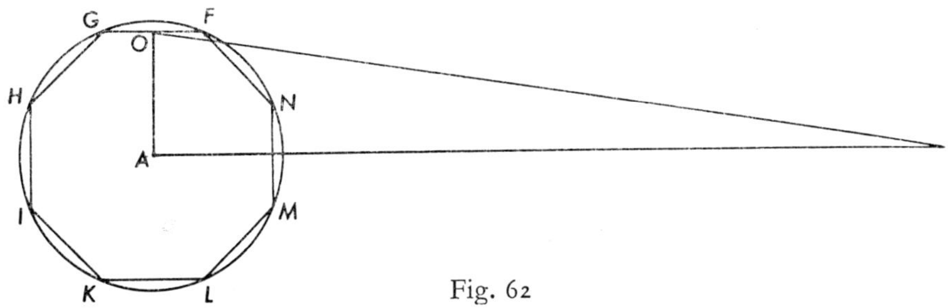

Fig. 62

extended in a straight line and whose other side containing the right angle, namely, *EB*, is the radius, or is equal to the radius, of circle *AB* [see Fig. 61]. Then I say that △ *EBC* is equal to circle *AB*. This is proved as follows: Since △ *EBC* is neither greater than nor less than circle *AB*, therefore it is equal to it. The consequence holds, for any [two] quantities, one of which is neither greater than nor less than the other, are equal. But I prove the antecedent, and in the first place, that △ *EBC* is not less than circle *AB*. For if so, then by the third supposition a figure greater than △ *EBC* could be cut from circle *AB*. But this is false. For let there be inscribed in the circle regular polygon *FGHIKLMN*, from whose center *A* is drawn line *AO* perpendicular to side *FG* [see Fig. 62]. Then by the fifth conclusion, demonstrated earlier, that regular polygon—an octagon—

rectum extensa sunt breviora, ut patet per secundam partem prime suppositionis, et linea *AO* est brevior semidyametro. Igitur triangulus 230 orthogonus, cuius duo latera rectum angulum constituentia sunt eedem linee, est minor triangulo orthogono *EBC* per secundam suppositionem, quem scilicet triangulum *EBC* velles esse equalem circulo *AB*. Quare etiam sequitur figuram illam poligoniam a circulo resecatam non esse maiorem triangulo *EBC*. Quare sequitur quod circulus 235 *AB* non est maior triangulo *EBC*, quod fuit probandum.

Non potest dici quod triangulus *EBC* sit maior circulo *AB*, quia sic a triangulo *EBC* posset una quantitas maior circulo *AB* resecari per tertiam suppositionem. Sed hoc est falsum, quia si sic, tunc resecetur ab eo una quantitas seu una figura poligonia, que sit *APQRSTVX* 240 [Fig. 63], que si est maior circulo *AB* potest sibi, vel equali sibi, esse circumscripta, que tamen si est circumscripta circulo *AB* probo quod impossibile est ipsam esse minorem triangulo *EBC*. Nam si esset minor triangulo *EBC* et resecata ab eo, sequeretur triangulum orthogonum, cuius alterum laterum rectum angulum constituentium est 245 equale omnibus lateribus figure circumscripte simul iunctis et reliquum

232 EBC *Fa* ABC *Z*
234 *post* EBC *add. Fa* quare sequitur quod a circulo AB non potes quantitatem maiorem resecare triangulo EBC

will be equal to a right triangle one of whose sides containing the right angle is equal to all of the sides of the polygon taken together and the other side is equal to the perpendicular *AO*. But the sides of the aforesaid polygon joined together are less than the circumference of the circle extended in a straight line, as is obvious by the second part of the first supposition. Further, line *AO* is less than the radius. Therefore, the right triangle whose two sides containing the right angle are these same lines is less than right △ *EBC* [Fig. 61] by the second supposition—△ *EBC* being the triangle you wish to be equal to circle *AB*. Therefore, it also

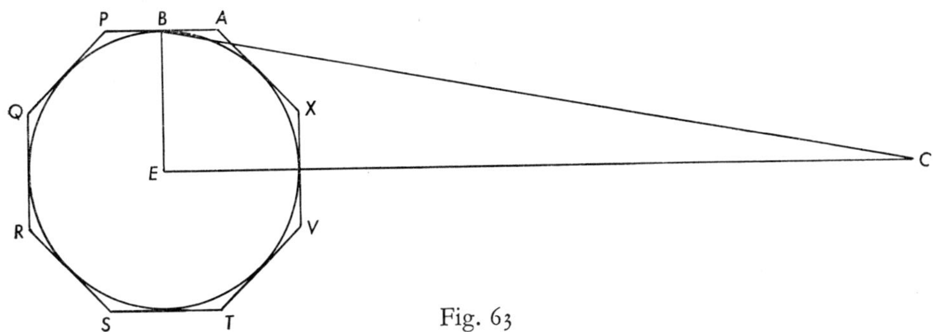

Fig. 63

Note: Rather than triangle *EBC* taken before in Fig. 61 (as formed from the circumference and radius of the circle), one would expect another triangle equal to the circumscribed octagon in the same manner as in Fig. 62.

follows that the polygonal figure cut from the circle is not greater than △ *EBC*. Therefore, it follows that circle *AB* is not greater than △ *EBC*, which was to be proved.

It cannot be said that △ *EBC* is greater than circle *AB*, because then there could be cut from △ *EBC* a quantity greater than circle *AB*, by the third supposition. But this is false. For if so, then let such a quantity—or a polygon—be cut from it; let this polygon be *APQRSTVX* [see Fig. 63]. Since it is greater than circle *AB*, then it can be circumscribed about circle *AB*, or about something equal to circle *AB*. But if it is circumscribed about circle *AB*, I prove that it is impossible for it to be less than △ *EBC*. For if it were less than △ *EBC* and it were cut from *EBC*, it would follow that the right triangle one of whose sides including the right angle is equal to the perimeter of the circumscribed figure and the other of the sides equal to the line drawn perpendicularly from the center of the figure to one of its sides is less than △ *EBC*, which is false. The consequence holds by the fifth conclusion demonstrated earlier. For

laterum equale linee ducte a centro eiusdem figure perpendiculariter
171v ad unum suorum laterum, esse / minorem triangulo *EBC*, quod est
falsum. Tenet consequentia per quintam conclusionem prius demonstratam, [ex] eo quod iste triangulus est equalis illi figure. Falsitas
250 consequentis patet, quia omnia latera figure circumscripte sunt longiora circumferentia circuli inscripti in rectum extensa, ut patet per secundam partem prime suppositionis, que quidem latera omnia figure circumscripte simul iuncta unum de lateribus trianguli equalis illi figure poligonie angulum rectum continentibus constituunt, et reliquum
255 de eisdem lateribus angulum rectum constituentibus est equale semidyametro. Igitur per secundam partem secunde suppositionis triangulus talis orthogonus qui est equalis figure illi poligonie circumscripte circulo est maior triangulo *EBC*. Quare etiam figura illa poligonia circulo sic circumscripta est maior triangulo *EBC*. Quare sequitur
260 quod non est sibi minor, nec ab eo resecata, et eodem modo arguerem de quacunque alia figura poligonia maiori ipso circulo. Sic igitur probatum est quod triangulus *EBC* non est minor nec maior circulo *AB*. Quare sequitur quod sit equalis, quod fuit probandum.

7[a] conclusio: Quadratura circuli quinto modo dicta est scibilis et
265 scita, demonstrabilis et demonstrata.

Probatur: demonstratum est circulo equalem triangulum assignare, et demonstratum est quadratum equale triangulo invenire, igitur demonstratum et scitum est quadratum equale circulo invenire. Tenet consequentia, quia quecunque sunt equalia inter se, quicquid est equale
270 uni est equale alteri. Sed antecedens quo ad primam eius partem patet ex dictis et quo ad secundam eius partem patet ex ultima 2[i] Euclidis, que dicit: "Dato trigono equum quadratum invenire."

Ad rationes, et primo ad rationes quod sic. Ille enim probant intentum; verumptamen ponentes illas rationes ad alium intellectum inten-
275 debant de quadratura circuli; et ad illum intellectum Aristoteles improbavit Brissonem et Antifontem. Brisso forte intendebat quod esset dare quadratum equale circulo, cuius omnia latera simul iuncta essent equalia circumferentie in rectum extense, et hoc est impossibile, sicut dictum est in una conclusione, quamvis etiam ratio Brissonis bene
280 probet quadraturam circuli quamvis non ad intellectum Brissonis, cum

247 *supra mg. 171v scr. Z* De quadratura circuli
249 [ex] *supplevi*
251 circumferentia *Fa* circumferentie *Z*
260 eo *corr. ex* ea
264–318 *7a*.... cetera *om. Fa*
266 *post* traiangulum *del. Z* triangulum
273 quod sic *corr. ex* ad oppositum *in Z*

this triangle is equal to that polygon. The falsity of the consequent is obvious. For the perimeter of the circumscribed figure is greater than the circumference of the inscribed circle extended in a straight line, as is shown by the second part of the first supposition. Now the perimeter of the circumscribed figure constitutes one of the sides including the right angle of the triangle equal to the polygon, and the other side of those sides containing the right angle is equal to the radius. Therefore, by the second part of the second supposition, such a right triangle, which is equal to the polygon circumscribed about the circle, is greater than △ *EBC*. Hence, the polygon so circumscribed about the circle is also greater than △ *EBC*. Therefore, it follows that it is not less than it, nor is it less than any part cut from △ *EBC*. And in the same way I would argue concerning any other polygon greater than that circle. Hence it has thus been proved that △ *EBC* is neither less than nor greater than circle *AB*. Therefore, it follows that it is equal to it, which was to be proved.

Seventh conclusion: Quadrature of a circle spoken of in the fifth way is knowable and known, demonstrable and demonstrated.

Proof: [When] it has been demonstrated that a triangle equal to a circle can be designated and it has been demonstrated that a square equal to a triangle can be found, therefore it has been demonstrated and made known that a square equal to a circle can be found. The consequence holds, for wherever there are quantities mutually equal, whatever is equal to the one is also equal to the other. But the first part of the antecedent is evident from what has been said [in the sixth conclusion] and its second part is evident from the last [proposition] of [Book] II of Euclid, which says: "To find a square equal to a given triangle."

Now for the [initial] arguments—first those in the affirmative: For they prove what is intended. Still, those positing these arguments were intending quadrature of a circle in another meaning—in the meaning understood in Aristotle's refutation of Bryson and Antiphon. Bryson perhaps intended that a square equal to a circle can be given where the perimeter of the square is equal to the circumference extended in a straight line; and this is impossible, as was said in one conclusion, although the argument of Bryson might well prove the quadrature of the circle but not in terms of the meaning of Bryson, since it holds and is true that the major of that argument is not universally true, for there can be truly given a rectilinear angle greater than an angle of a segment*—as in the case of a

* A mixed angle formed by an arc of a circle and a chord (in this case a diameter of the circle).

hoc tamen stat et verum est quod maior illius rationis non est universaliter vera, quia angulo portionis datur bene angulus rectilineus maior, sicut est angulus rectus, et angulus rectilineus minor; non tamen datur angulo portionis angulus rectilineus equalis. Et igitur, quamvis verum 285 sit quod ratio prima concludit, tamen maior illius rationis non est universaliter vera. Similiter dico de 2ª: quamvis verum sit quod ipsa concludit, tamen illud quod reputatur esse falsum in ea est valde possibile; potest enim fieri transitus de minore ad maius transeundo quodlibet punctum inter maius et minus et nunquam pervenire ad 290 equale. Verbi gratia, aliquis enim angulus rectilineus est minor angulo portionis et aliquis est maior eodem, sicut est angulus rectus; et possibile est quod angulus rectilineus acutus continue augeatur donec fiat equalis angulo recto et tamen nunquam fit equalis angulo portionis. Quod patet si aliqua linea recta una cum dyametro alicuius circuli 172r constitueret angulum acutum et continue moveretur / versus lineam 296 contigentem circulum donec ipsum cooperiret; et istud satis clare ponit Campanus et deducit super 15 propositionem tertii Euclidis. 3ª ratio probat verum, sicut est illud quod alie rationes intendunt concludere.

300 Postea ad rationes que vidantur esse contra conclusionem principaliter intentam. Ad primam quando dicitur, "si circulus posset quadrari, quadratum posset circulari," concedo ad istum intellectum, quod possibile est invenire circulum equalem quadrato, sicut est possible invenire quadratum equale circulo, et hoc ad intellectum positum in 305 questione. Et quando dicitur, "si latera alicuius quadrati extendantur equaliter a centro, non erit superficies equalis superficiei prius extente," certe verum est. Nec hoc oportet ad quadraturam circuli; ymmo dictum est quod impossible est aliquod |quadratum esse equale circulo, cuius omnia latera simul iuncta sint equalia circumferentie in rectum 310 extense. Ad aliam quando dicebatur, "impossibile est invenire quadratum cuius medietas sit equalis medietati circuli," hoc nego. Et quando dicebatur, "data aliqua medietate alicuius quadrati anguli istius figure non sunt equales aliquibus angulis semicirculi," [hoc] admitto. Et quando dicitur, "igitur nec medietas quadrati medietati cir-315 culi est equalis," nego consequentiam. Et quando dicitur, "Euclides et Campanus per equalitatem angulorum probant equalitatem figurarum," bene volo. Ex hoc tamen non sequitur quod inequalitatem angulorum sequatur inequalitas figurarum, et cetera.

283 non *corr. ex* $\overline{\text{non}}$ *in Z* 295 acutum *corr. ex* rectum *in Z*

right angle, and there can [also] be given a rectilinear angle less than such an angle of a segment; yet there cannot be given a rectilinear angle equal to an angle of a segment. And, therefore, although the first argument follows, yet the major in that argument is not universally true. I answer in the same way regarding the second argument. Although the conclusion is true, yet that which is reputed to be false in it is quite possible. For there can be passage from "lesser" to "greater" by going through every point between "lesser" and "greater" without ever arriving at "equal." For example, some rectilinear angle is less than an angle of a segment and some is greater than that angle of a segment, as for example a right angle, and it is possible that an acute rectilinear angle be continually increased until it becomes equal to a right angle and yet it never becomes equal to the angle of a segment. This is evident in the case where a straight line comprises an acute angle with the diameter of some circle and this line is continually rotated toward a line tangent to the circle until it coincides with that line. And this Campanus clearly posits and deduces [in the commentary] on III.15 of Euclid. The third argument proves to be true according to the intention of the other arguments.

Now for the arguments that might seem to be against the conclusion as principally intended. To the first argument when it is said: "If a circle could be squared, a square could be 'circled'," I concede that in this meaning [of quadrature] it is possible to find a circle equal to a square just as it is possible to find a square equal to a circle—and this with respect to the meaning posited in the question. And when it is said: "If the sides of some square are extended equally from the center, there will not be a surface equal to the surface before its extension," this is certainly true. But this is not necessary for quadrature of a circle. Nay, it has been said that it is impossible that some square is equal to a circle where all the sides of the square taken together are equal to the circumference extended in a straight line. To the argument where it is was said: "It is impossible to find a square where half of it is equal to half a circle," I deny this. And when it was said: "With half of some square given, the angles of this figure are not equal to the angles of the semicircle," I admit this. And when it is said: "therefore half of the square is not equal to half the circle," I deny the consequence. And when it is said: "Euclid and Campanus prove the equality of figures by the equality of angles," I agree completely. From this, however, it does not follow that from the inequality of the angles follows the inequality of the figures, and so on.

COMMENTARY

3–4 "auctoritate... conati." Aristotle in his *Physics* I.2, 185a, lines 14–17, makes the following reference to Antiphon: "The exponent of any science is not called upon to solve every kind of difficulty that may be raised, but only such as arise through false deductions from the principles of the science: with others than these he need not concern himself. For example, it is for the geometer to expose the quadrature by means of segments, but it is not the business of the geometer to refute the argument of Antiphon." (English translation of T. L. Heath, *Mathematics in Aristotle* [Oxford, 1949], p. 94.) A similar reference is made to both Antiphon (a contemporary of Socrates) and his junior contemporary Bryson in the second chapter of the *Sophisitici elenchi*, 11.171b, line 34, to 172a, line 7: "The eristic disputant stands somewhat in the same relation to the dialectician as the pseudographer to the geometer; for in his paralogisms he starts from the same principles as dialectics uses, just as the pseudographer apes the geometer. But the pseudographer is not eristic, because he deduces his fallacies from the principles and conclusions which fall under the particular art (geometry); the other, who uses the principles coming under dialectic, will clearly be eristic with reference to other subjects. Thus, for example, the squaring of the circle by means of lunules is not eristic, but the quadrature of Bryson is eristic; the reasoning used in the former cannot be applied to any subject other than geometry alone, whereas Bryson's argument is directed to the mass of people who do not know what is possible and what impossible in each department, for it will fit any. And the same is true of Antiphon's quadrature." (English translation of Heath, *ibid.*, p. 47.) For Bryson, the reader should compare the references in the *Posterior Analytics*, I.9, 75b, line 37, to 76a, line 3, and *Soph. elen.*, 11. 171b, lines 12–18. From the commentaries of Themistius (*In physicam*, ed. of Schenkl, p. 4.2., *et seq.*) and Simplicius (*In physicam*, ed. of H. Diels, p. 54, line 20, to p. 55, line 24) the procedure of Antiphon can be reconstructed. He apparently started with a square (Simplicius) or equilateral triangle (Themistius) inscribed in the circle.

Then in the segments between the inscribed figure and the circle he inscribed isosceles triangles, producing a polygon of twice as many sides. This same procedure was used successively and, according to Simplicius, "Antiphon thought that in this way the area of the circle would be used up and we would sometimes have a polygon inscribed in the circle the sides of which would, owing to their smallness, coincide with the circumference of the circle. And, as we can make a square equal to any polygon ... we shall be in a position to make a square equal to a circle." Simplicius further tells us that according to Alexander the principle that is violated by Antiphon is that a circle touches a straight line in one point (alone), or, according to Eudemus, that magnitudes are divisible without limit. Be that as it may, it seems apparent that Antiphon had an intuition of the circumference as the limit of the perimeter, and our texts have shown how the Antiphon procedure was modified and taken up as a part of the exhaustion process, supposedly invented by Eudoxus and used by Euclid and Archimedes.

When we turn to Bryson's quadrature we are faced with a more difficult task of interpretation. Heath (*op. cit.*, p. 48) has summarized the suggestions of the various commentators on Bryson's method. "The comment by the Pseudo-Alexander (?Michael Ephesius) on *Soph. El.* c. 11 is scarcely worth consideration. According to this, Bryson inscribed a square in a circle and circumscribed another about it, while he also took a square intermediate between them (the commentator does not say how constructed); then he argued that, as the intermediate square is less than the outer square and greater than the inner, while the circle is also less than the outer and greater than the inner square, and, as *things which are greater and less than the same things respectively are equal*, it follows that the circle is equal to the intermediate square. Upon this the commentator observes that not only is the assumed axiom applicable to other things as well as geometrical magnitudes, but it is also false, because 8 and 9 are both less than 10 and greater than 7 and yet are not equal. Evidently the commentator was completely at sea. According to Philoponus, however, Alexander made a better suggestion, namely, that the circle is greater than *every* inscribed rectilinear figure and less than *every* circumscribed figure; again, a *rectilinear* figure drawn between the inscribed and circumscribed figures is less than the latter and greater than the former; but things greater and less than the same thing are equal to one another; therefore the circle is equal to the intermediate rectilinear figure. Themistius, too, observes that Bryson said that the circle is greater than *all* inscribed rectilinear figures and less than *all* circumscribed figures." Cf. the medieval translation of Themistius' *Paraphrasis* of the *Posterior Analytics*, ed. of J. R. O'Donnell, *Mediaeval Studies*, vol. *20* (1958), p. 265: "Et propter id non convenit alicui ut aestimet quod quadratura qua Brisso quadrat circulum sit demonstratio geometrica. Quod est quia utitur in ea propositione, cujus receptio est necessaria, et quamvis ipsa sit certa, vera, verumtamen communis. Et est quia dicit quod res quae sunt majores et minores unis et eisdem rebus, et minores, sunt aequales ad invicem. Haec enim propositio non verificatur in magnitudine tantum, sed verificatur et in numero et in tempore et in rebus aliis multis. Et iste sermo, quem addidit Brisso ad hoc, et putavit quod jam quadravit circulum, non est ex eis; quod sit necessarium in hoc nostro sermone: [Verumtamen rememoror ejus diligenti scientiam et desideranti eam.] Inquit Brisso: circulus est major omnibus figuris polygoniis, (136r) quae describuntur intra ipsum, et minor omnibus figuris multorum angulorum, quae describuntur super ipsum deforis. Et similiter est dispositio figurae multorum angulorum descriptae in eo quod est inter figuras descriptas intra circulum et super ipsum deforis. Necesse est ergo ut sit circulus et haec figura polygonia major rebus unis et eisdem et minor rebus unis et eisdem. Oportet ergo inde ut sint haec duo aequalia propter propositionem susceptam quae dicta est." Heath then goes on to suggest that Bryson conceived of compressing circumscribed and inscribed figures into one so that they coincide both with one another and with the curvilinear figure in question. However, I wonder if Albert's suggestion to the effect that if there is a square larger and one lesser there must be one equal to the circle is perhaps not close to the intent of Bryson's argument. As to just what kind of mean the intermediate square is, this is another matter, and one argued at length. Is it an arithmetic mean or a geometric mean or some other mean between an inscribed and circumscribed square? It would seem to me that the tracts given below in Appendix I are the heirs of the Bryson tradition. Actually, the remark of Proclus quoted by Philoponus seems pertinent to this interpretation of Bryson: "The circle is greater than every inscribed rectilinear figure and less than every circumscribed rectilinear figure, and where, in relation to a thing, there exists a greater and a less, there exists also an equal to it." As Heath points out (*op. cit.*, p. 49), the conclusion is "that there *exists* a rectilinear figure equal to the circle, though Bryson did not attempt to construct it." Cf. O.

Becker, "Eudoxus-Studien, II," *Quellen und Studien zur Geschichte der Mathematik, Astronomie und Physik*, Abt. B. Studien, vol. *2* (1933), p. 369, *et seq.* The reader should compare the interpretation of Albert to that of Thomas Aquinas (*Opera omnia*, vol. *1*, *In libros posteriorum analyticorum* [Rome, 1882], p. 304, Text no. 23; Lect. 17): "'Est enim sic demonstrare' etc.; probat propositum, scilicet quod non sufficiat ex veris et immediatis aliquid demonstrare, quia sic contingeret aliquid demonstrare, sicut Bryso demonstravit tetragonismum, idest quadraturam circuli, ostendens aliquod quadratum esse circulo aequale per aliqua principia communia, hoc modo: In quocunque genere est invenire aliquod maius et minus alicui, in eodem est invenire et illi aequale; in genere autem quadratorum est invenire aliquod quadratum minus circulo, quod scilicet scribitur intra circulum, et aliquod maius circulo, intra quod circulus describitur; ergo est invenire aliquod quadratum circulo aequale. Haec quidem probatio est secundum commune: aequale enim, et maius, et minus, excedunt genus quadranguli et circuli. Unde patet quod huiusmodi rationes demonstrant secundum aliquod commune, quia medium alteri inest, quam ei de quo fit demonstratio; et ideo huiusmodi rationes conveniunt aliis, et non conveniunt istis, de quibus dantur." Consult also the remarks of Campanus on the passage from "less" to "greater" without going through "equal" given below in this Commentary, line 45, although that comment is more pertinent to the second affirmative argument given in lines 9–17. Murdoch points out that Albert has given two forms of the assumption "where there is a greater and a lesser, there must be an equal." The first form—a static one—which posits the existence of an "equal to" from the existence of a "greater" and "lesser" is present in lines 5–9, while the second, "kinematic" form which emphasizes the passage through equality in going from "greater" to "lesser" is contained in lines 9–17.

28–29 "libello... maximum." Cf. the Commentary to the Munich Version (Chapter Three, Section 6), lines 179–81.

37–39 "4a... cetera." For Proposition I.4 of the *Elements* in the Adelard II Version, see the Commentary to the Naples Version (Chapter Three, Section 2), line 36. For the Campanus text, see the edition of Basel, 1546, p. 9.

42–43 "ultimam... Euclidis." For the last axiom in the Adelard II Version of the *Elements*, see my "The Medieval Latin Translations from the Arabic of the *Elements* of Euclid," *Isis*, vol. *44* (1953), p. 31: "Item duas lineas rectas superficiem non (*or* nullam) concludere."

45 "15... Euclidis." Proposition III.15 of the Adelard II Version of the *Elements* (III.16 of the Greek text) runs (with the addition of Campanus from the Campanus edition [Basel, 1546], pp. 66–67): "Si ab altero terminorum diametri cuiuslibet circuli orthogonaliter linea recta ducatur, extra circulum eam cadere necesse est. Atque inter illam et circulum, aliam lineam rectam capi impossibile est. Angulum autem ab illa et circumferentia contentum, omnium acutorum angulorum esse angustissimum. Angulum vero intrinsecum a diametro et circunferentia contentum, omnium angulorum acutorum esse amplissimum necesse est.... CAMPANI additio. Ex hoc notandum, quod non valet ista argumentatio, hoc transit a minori ad maius et per omnia media: ergo per aequale. Nec ista. Contingit reperire maius hoc, et minus eodem: ergo contingit reperire aequale, hoc autem sic patet. Sit circulus *a b* super centrum *c*, cuius diameter *a c b*, et ducatur ab eius termino *a* linea *a d* orthogonaliter, eritque contingens circulum per correlarium huius. Describatur iterum super punctum *a* secundum quantitatem diametri *a b*, circulus *b e d*, et imaginetur linea *a b* moveri super punctum *a* per circunferentiam arcus *b e d*, ita quod punctum *b* numeret omnia puncta arcus *b e d*, quousque perveniat ad lineam *a d*, et cooperiat ipsam. Et quia angulus *b a d* est rectus: erit ut non fit sumere aliquem

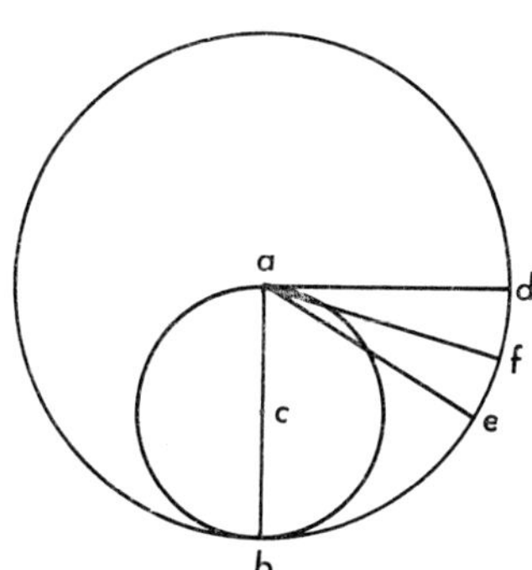

angulum acutum cui aequalem non fecerit linea *a b* cum diametro *a c b* minoris circuli, quia transivit ad anglum rectum, dinumerans situm omnium angulorum acutorum, quorem manifestum est quosdam esse minores angulo semicirculi contento a semicircunferentia *a b* et diametro *a c b*, et angulum rectum manifestum est esse maiorem eodem. Dico quod nullus in transitu ab acutis minoribus ad rectum maiorem intermedius: fuit ei aequalis. Si enim fuerit aliquis, sit ut illum fecerit linea *a b*, cum punctus *b* fuit in puncto *e* arcus *b e d*. Quia ergo angulus *e a b* est aequalis angulo semicirculi praedicto, angulus autem semicirculi est amplissimus omnium acutorum per ultimam partem huius, erit angulus *e a b* amplissimus omnium acutorum. Dividatur ergo angulus *e a d*, sicut proposuit 9 primi, per aequalia, ducta linea *a f*, eritque (per 9 conceptionem) angulus *f a b* amplior angulo *e a b*, quare erit aliquid amplius amplissimo: quod est impossibile. Vel sic. Cum angulus *e a b* sit aequalis angulo semicirculi sicut ponitur, at angulus semicirculi cum angulo contingentiae est aequalis uni recto, similiter quoque angulus *e a b* cum angulo *e a d* est aequalis uni recto, erit angulus *e a d* aequalis angulo contingentiae: et quia angulus contingentiae est angustissimus omnium acutorum per 3 partem huius: erit similiter angulus *e a d*, ei aequalis, angustissimus omnium acutorum, sed angulus *e a f* est eo angustior per conceptionem: erit ergo aliquid angustius angustissimo: quod est impossibile. Non ergo erit angulus rectilineus aequalis angulo semicirculi. Et quia transitur a minori ad maius et non per aequale: item quia est reperire minorem eo et maiorem: patet instantia contra utranque argumentationem praedictam...." (A few commas have been deleted from the text as given in the edition.) Cf. the similar argument of Campanus in his addition to III.30 (*ed. cit.*, p. 78). For an interesting historical account of these "mixed" angles composed of a curve and straight line (i.e., the so-called "angle of a semicircle," "the angle of a segment," "the angle of contingence [= horn angle])," see T. Heath, *Euclid, The Elements*, vol. 2 (Annapolis, 1947), pp. 39–43. For Murdoch's comments on this passage, see his paper, "The Medieval Language of Proportions," in A. C. Crombie, ed., *Scientific Change* (New York, 1963), pp. 248–49.

49–53 "Uno... quadrari." As Heath reports (*History of Greek Mathematics*, vol. 1 [Oxford, 1921], pp. 220–21), the recognition of the similarity of the words for "quadrature" and "quartering" is an ancient one and is at the base of the remark made by Aristophanes in the *Birds* (1005) which has Meton the astronomer bringing a rule and compass to make a construction "in order that your circle may become a square." Heath says, "This is a play on words, because what Meton really does is to divide a circle into four quadrants by two diameters at right angles to one other;... the word τετράγωνος really means 'with four right angles (at the centre),' and not 'square,' but the word conveys a laughing allusion to the problem of squaring all the same." For Campanus' use of the word *quadratura* in the sense of "quartering," see his *Theorica Planetarum* (Oxford, Bodl. Auct. F.3.13, 118r, c.2). Cf. *ibid.*, 127r, c.2, where he says: "Et quadrabo istos circulos duabus diametris ortogonaliter se secantibus super centrum eorum."

94–104 "Secunda.... modo." I am not sure what is intended here. In the introduction to this text I have suggested that a figure such as this ☐ is implied. My authority for this is lines 101–103 where the author speaks of dividing the circumference into four parts and rearranging the parts to form a figure "in some way similar to a square."

106–117 "Probatur... cetera." See the tract of Campanus in Appendix I.

119–20 "Aristotilis... scita." See the *Categories*, Chapter 7, 7b. (See the Commentary to Appendix I, Section 3, lines 2–3.)

161–66 "Maior... habere." Proposition V.1 is quoted here directly from the Adelard II Version of the *Elements*, no doubt from Campanus' commentary (see the edition of Basel, 1546, p. 114).

178, 182, 183 "primam sexti." For Proposition VI.1 of the *Elements* in the Adelard II Version, see the Commentary to Chapter Three, Section 1, the Cambridge Version, line 70. Albert's quotation is exactly as in the Adelard text except that after *duarum* (line 183) of Albert's text, the Adelard text adds *rectilinearum*.

271 "ultima secundi." For II.14 of the *Elements* in the Adelard II Version, see the Commentary to the Cambridge Version, lines 82–83. Note that in line 272 Albert uses *invenire* while the Adelard text had *describere*.

276 "Brissonem et Antifontem." See above in this Commentary, lines 3–4.

282–98 "quia... verum." See above in this Commentary, line 45.

Chapter six

Archimedes' *De sphaera et cylindro* in the Latin West before 1269

1. A Latin Fragment of the *De sphaera et cylindro*

I pointed out in Chapter Four that the first extensive knowledge of the conclusions of Archimedes' *On the Sphere and the Cylinder* passed to the Latin West with the translation by Gerard Cremona of the so-called *Verba filiorum* of the Arabic mathematicians, the Banū Mūsā. But the enunciations and proofs of the Archimedean propositions found in the *Verba filiorum*—although representative of the spirit of Archimedes' work—were in no sense direct translations of *On the Sphere and the Cylinder*. It seems evident that the Latin Schoolmen possessed no complete translation of this work until the general translation of Archimedes' works by William Moerbeke in 1269. But, while this is true, it should be noted that several years ago I discovered and published in *Isis* (vol. *43*, [1952], pp. 36–38) a fragment of what appears to be an earlier translation of a very small part of *On the Sphere and the Cylinder*. I published it on the basis of a singly Digby manuscript (*S* in the sigla below). The fragment appeared there among the abridgements of several works translated by Gerard of Cremona, e.g., the *De mensura circuli* of Archimedes, the *Verba filiorum*, the *De diversitate aspectus* of Alkindi. I suspected at that time that this series of abridgements had been made from some codex of Gerard translations like that of Paris, BN lat. 9335. It is of interest that a collection of Gerard's translations appearing in a manuscript of Madrid, Biblioteca Nacional 10010, contains some of the same works and includes our fragment on folio 84r. This Madrid manuscript is, like the part of the Digby manuscript including the fragment, from the fourteenth century. It is particularly worth noting that our fragment of the *De sphaera* in Digby 168 is immediately followed on the same folio by the enunciations of the propositions of the *De mensura circuli* in the Gerard of Cremona translation. It is not unreasonable to suppose, therefore, that our fragment also stems from some translation by Gerard of Cremona. That it was, at least, translated from the Arabic rather than from the Greek might be inferred from the use of terminology common to translations from the Arabic rather than from the Greek: *piramis columpna* instead of *conus*; *columpna* instead of *cylindrus*; *medietas diametri* instead of *que ex centro*. However, such an argument is by no means conclusive. Against Gerard's role as translator of the *De sphaera* we can note that such a translation is not mentioned in the vita composed by his students.[1] Still, other works that we know to be translations of Gerard are also missing from the list of translations in the vita.[2]

It will be noticed on examining the six propositions of the fragment given below that only the enunciations are presented. These enunciations are found in the introductions to the two books of Archimedes' *De sphaera*. Thus Propositions 1, 2, 3, and 5 are found in the introduction to Book I in the same order as they are given in the fragment.[3] In the introduction to Book II, Propositions 1, 2, 3, and 6 are stated.[4] Only Proposition 4 is not given exactly as in the introductions. Proposition 4 of the fragment describes a pyramid, constructed on one of the parallelogram faces of a *serratile*, as being two thirds of the volume of the *serratile*. (A *serratile* is a triangular prism and here it is presumably turned over on its side so that the side becomes the base.) But Archimedes in the introduction to Book I of *On the Sphere and Cylinder* gives the regular formula for the volume of a pyramid as one third the volume of a prism of the same base and altitude.[5] And he gives this theorem between the propositions which are numbered 3 and 5 in the fragment, and hence we ought to conclude that Proposition 4 of the fragment suffered alteration either in translation or copying, the scribe or translator changing "one third" to "two thirds" and "prism" to "serratile." Our final conclusion, then, must be that in spite of the altered form of the fourth proposition, the source of the fragment is the introductory material to the *De sphaera*.

My new text of this fragment is based primarily on the Madrid manu-

[1] For the vita, see Chapter Two, Section 2, note 3.

[2] G. Sarton, *Introduction to the History of Science*, vol. *2* (Baltimore, 1931), p. 338.

[3] Archimedes, *Opera omnia*, ed. of J. L. Heiberg, vol *1* (Leipzig, 1910), pp. 2–4.

[4] *Ibid.*, p. 168.

[5] *Ibid.*, p. 4.

script with variant readings from the Digby manuscript, which is unfortunately heavily stained on the page including the fragment. The propositions are numbered only in the Digby manuscript. The marginal folio number is to the Madrid manuscript.

Sigla of Manuscripts

Zm = Madrid, Bibl. Nac. 10010, 84r, 14c.
S = Oxford, Bodl. Libr., Digby 168, 122r (old page 121r), 14c.

Propositiones Archimenidis

84r 1. Omnis spere superficies que continet ipsam est quadruplum (!) maioris circuli cadentis in illa spera.

2. Omnis portionis spere superficies est equalis circulo, ex cuius 5 centro egreditur linea ad lineam continentem ipsum equalis linee recte que egreditur ex puncto capitis portionis ad lineam continentem circulum qui est basis illius portionis.

3. Omnis columpna, cuius basis est equalis maiori circulo cadenti in spera aliqua et est altitudo eius equalis diametro illius spere, 10 est semel et semis equalis illi spere, et iterum superficies illius columpne et due bases eius aggregate sunt semel et semis equalis superficiei spere.

4. Omnis piramis, cuius basis est figura rectilinea, autem ipsa et serratile semel sunt super basem unam et est utriusque altitudo una, 15 tunc illa piramis est due tertie serratilis.

5. Omnis piramis columpne, cuius basis est basis columpne et altitudo est equalis altitudini eius, est tertia eius.

6. Omnis sector spere est equalis piramidi columpne, cuius basis est circulus equalis superficiei portionis que est in sectore spere et 20 altitudo est equalis medietati diametri spere.

1 Propositiones *S om. Zm*
9 dyametro *S*
10 *post* equalis *S scr. et del.* est
13 autem *corr. ex* aut *in S et Zm*
20 dyameter *S*

Propositions of Archimedes

1. The surface containing every sphere is quadruple the greatest circle falling in the sphere.

2. The surface of every segment of a sphere is equal to the circle whose radius is equal to the straight line extending from the vertex of the segment to the circumference of the circle which is the base of the segment.

3. Every cylinder whose base is equal to the greatest circle falling in some sphere and whose altitude is equal to the diameter of the sphere is equal to three halves of the sphere, and further the [lateral] surface of that cylinder together with its bases is equal to three halves the surface of the sphere.

4. Every pyramid whose base is a rectilinear figure and which is [elevated] upon the same base as a *serratile** and has the same altitude as the *serratile* is two thirds of the *serratile*.

5. Every cone whose base is the base of a cylinder and whose altitude is equal to the altitude of the cylinder is one third of the cylinder.

6. Every sector of a sphere is equal to a cone whose base circle is equal to the surface of the segment which is in the sector of the sphere and whose altitude is equal to the radius of the sphere.

* A triangular prism, but here standing on its side as a base; i.e., here the pyramid and *serratile* are apparently both constructed on one of the parallelograms of the *serratile*. See the Commentary, lines 13–15.

COMMENTARY

2–3 "Omnis...spera." Cf. the translation by William Moerbeke of this same statement (Note: I include here for convenience the whole passage from the introduction to Book I of the *De sphaera*, MS Vat. Ottob. lat. 1850, 23v): "Primum quidem quod omnis spere superficies quadrupla est maximi circuli eorum qui in spera. Deinde autem quod superficiei omnis portionis spere equalis est circulus, cuius que ex centro est equalis recte ei que a vertice portionis ducitur ad periferiam circuli qui est basis portionis. Ad hoc autem quod omnis cylindrus habens basem quidem equalem maximo circulo eorum qui in spera, altitudinem autem equalem diametro spere, et ipse est emiolius spere, et superficies ipsius superficiei spere.... Quod omnis pyramis est tertia pars prismatis habentis basem eandem pyramidi et altitudinem equalem, et quod omnis conus est tertia pars cylindri habentis basem eandem cono et altitudinem equalem...." Cf. the different wording of this first proposition in the *De curvis superficiebus*, Section 2 below, Proposition VI, lines 43–44. Cf. *Verba filiorum*, Proposition XIV, lines 36–37, and also *De ysoperimetris*, MS Oxford, Bodl. Auct. F.5.28, 106v (see Appendix III, passage 8).

8–12 "Omnis...spere." See the previous comment for the Moerbeke translation of this same statement. Cf. *De curvis superficiebus* (Section 2 below, Proposition VIII, lines 1–5).

13–15 "Omnis...serratilis." Cf. Commentary, lines 2–3, above, for the Moerbeke rendering of this statement. The Adelard-Campanus version of the *Elements* (ed. of Basel [1546], p. 346) defines the *corpus serratile* as follows: "Corpus serratile dicitur, quod quinque superficiebus, quarum tres parallelogrammae sunt, duae vero triangulae, continentur." Apparently in changing from prism to *serratile* the translator has limited the proposition to a pyramid with a parallelogram as a base. If, however, the pyramid and *serratile* were constructed on a triangular face, then the pyramid would be one third of the *serratile*, as in the Archimedean proposition.

16–17 "Omnis...eius." Cf. Commentary, lines 2–3, above, for the Moerbeke rendering.

18–20 "Omnis...spere." Moerbeke's translation of these lines from Book II of the *De sphaera* runs (MS. cit., 31r): "...omnis sector solidus [spere] est equalis cono basem quidem habenti circulum equalem superficiei portionis eius que in sectore spere, altitudinem autem equalem diametro ei que ex centro spere." Cf. the *De curvis superficiebus*, Section 2, below, Proposition IX, lines 1–3, and the Commentary to that proposition.

2. The *Liber de curvis superficiebus* of Johannes de Tinemue

Much more significant for the spread of knowledge of the *De sphaera et cylindro* of Archimedes than the brief fragment described in the first section of this chapter was the so-called *Liber de curvis superficiebus* of a mathematician whose name seems to have been Johannes de Tinemue. This treatise was, next to the *De mensura circuli*, the most popular Archimedean work circulating in the Latin West during the thirteenth and fourteenth centuries. This is shown by the fact that at least twelve manuscripts of it are still extant. Furthermore, it was widely quoted by geometers of this period, as we shall see, and although it is probably of Greek origin—as I shall argue—it circulated in the thirteenth century with the tracts of Arabic origin. While I have spoken of it as an "Archimedean" work, I do not mean to imply that it was a work composed by the great Syracusan mathematician but only that it was inspired throughout by the techniques and conclusions found in Book I of the *De sphaera et cylindro* and the *De mensura circuli*.

Before taking up the difficult problem of the origins of this tract and the identity of its author, I should first remark that there was some variation in the title and in the author's name. Among the geometers the work was ordinarily known as *De curvis superficiebus Archimenidis*, and indeed this title appears in some of the manuscripts.[1] Manuscript *B*, our most trustworthy manuscript of the thirteenth century, has what I believe to be the original title given in the hand of the scribe of the main text:

[1] See below, notes 20 through 25, and see the various forms of the title given in the variant readings.

Liber magistri Johannis de Tinemue de curvis superficiebus.[2] This exact title is repeated twice in manuscript *J*, also of the thirteenth century.[3] An alternate title is found in the colophon of the best manuscripts and the title of manuscript *F*: *Commentum Johannis de Tin9 (or Tin̄) in demonstrationes Archimenidis* (although in *F* the name is spelled out as "Ioannis de Chinemue"). Finally, we can remark that a Latin author of the first half of the thirteenth century, Gerard of Brussels, cited our tract under the title of *De pyramidibus* (presumably abbreviated for *De pyramidibus rotundis*),[4] i.e., *On Cones*, no doubt because of the prominence in the treatise of propositions involving cones.

It will be evident from the titles already mentioned that there was some variation among the manuscripts in the name of the author. MS *B*, in the same hand as is used in the rest of the text, has in the lower lefthand cornor of folio 111r the form "Johannes de Tinemue," as I have indicated above. MS *J* also has the same form. The late manuscript *F* apparently altered the place name slightly to "Chinemue," and incidentally the same altered form appears in another work on quadrature in that manuscript (see Appendix I). MS *A* gives the different form "Tinnenie." In the colophons of the various manuscripts the following forms are found: "Tiṅ," "Tin̄," "Thin̄," "Tl̄n," "Thin9."[5] In the colophons of two manuscripts (*A* and *E*) the name "Johannes" is dropped entirely and the name of a Latin scholar is substituted: Gervasius de Essexta (*A*) or Gervasius de Assassia (*E*). But since MS *A*—which I believe to be the beginning of this tradition—has the name of Johannes in the title, it seems clear to me that the Latin scholar Gervasius took the work as it existed in the first tradition of *B* and *C* and made some slight changes to produce the second tradition. One of those changes was to substitute his own name for that of Johannes de Tinemue in the colophon.

This brings us to the problem of when the treatise was originally composed. There seems to be little doubt that it existed in some form in

[2] It should be pointed out that at the beginning of the tract MS *B* has *De curvis superficiebus Archimenides* in what I believe to be a hand later than that in which the tract is written. The title as indicated above is in the left-hand lower corner of the first folio of the work (111v) and is in the same hand as the rest of the text.

[3] See variant readings.

[4] Gerard of Brussels, *Liber de motu*, II. 2 (ed. of M. Clagett, *Osiris*, vol. *12*, 1956, p. 132, lines 7–8): "Patet ergo per primam *de piramidibus* quod triangulus LNP equatur curve superficiei rotunde piramidis *OAE*." (Cf. the variant readings on p. 114, line 74, p. 121, lines 8–10, of the Gerard text, except that the second reference ought also to be to the fourth rather than to the fifth proposition of the *De curvis superficiebus*).

[5] See the variant forms of the colophon as given in the variant readings.

antiquity or the early Byzantine period. This I believe to be confirmed by the fact that Hero of Alexandria quotes a conclusion equivalent to its third proposition in his *Mechanics*[6] and that the author of the *Stereometry*, which is also a part of the Heronian corpus, quotes what is equivalent to its tenth proposition.[7] Furthermore, an examination of the content of the Latin treatise shows that the author had more detailed knowledge of the *De sphaera et cylindro* of Archimedes than would have been possible for a Latin author of the early thirteenth century, when the Latin version first became known.[8] The most probable conclusion is that the tract was translated. Against my earlier opinion[9] and the opinion of Heiberg,[10] I am

[6] Hero in the *Mechanics*, I. 4 (ed. of L. Nix, in *Heronis Alexandrini opera... omnia*, vol. *2* [Leipzig, 1900], p. 11), argues that since the diameter of one circle is twice that of another, the semicircumference of the one is twice the semicircumference of the other. He adds, "Archimedes has already proved this." Archimedes does not directly prove this, either in the *Measurement of the Circle* or in the *Sphere and the Cylinder* (although to be sure the determination of π as a constant could serve as a basis of proof). But the text of the *De curvis superficiebus* includes this as its third proposition.

[7] *Stereometrica*, ed. of J. L. Heiberg in *Heronis Alexandrini opera...omnia*, vol. *5* (Leipzig, 1914), p. 8, lines 6–7; cf. *scholia*, p. 229. In both references it is said that Archimedes proves the equality of eleven cubes of the diameter to twenty-one spheres. But nowhere does Archimedes directly prove this in the extant genuine works. I would suppose, therefore, that this is an indication that the *De curvis superficiebus* (which does include this proposition) was circulating and that even then it was associated in some fashion with Archimedes.

[8] For example, in the course of Proposition VII (lines 179–80), the commentator says that the author (i.e., Archimedes) does not treat the case where the number of sides of half of the regular polygon inscribed in a circle and rotating about the diameter of the circle to make a solid composed of conical segments is an odd number. He offers this omission as an excuse for not investigating it himself. I submit that this is a clear indication that Johannes de Tinemue was following along the text of Archimedes, particularly since the enunciation of Proposition VII seems to come almost directly from the text of *De sphaera et cylindro* and was not available in Latin before 1269 in any of the possible sources of Archmidean material; and the *De curvis superficiebus* was quoted by Gerard of Brussels in his *Liber de motu*, which has to be dated before 1260, and probably should be dated a good deal earlier—see Clagett (cited in note 4 above), pp. 104–107. I suppose it is not impossible to argue that the author of the *De curvis superficiebus* was a Latin Schoolman who was familiar with Greek texts and constructed his Latin text as a kind of paraphrase and commentary on Greek materials, but I think this quite unlikely. At least I know of no other such composition at this time. The only points favorable to an independent Latin composition are that the citations of Euclid seem to be in the form preferred by Latin authors and that Proposition XII. 10 of the Greek text is cited as XII. 9 in the manner of the Adelard and Arabic versions. (To be sure it may well be that these versions ultimately follow a Greek order employed by Hero.)

[9] "The *De curvis superficiebus Archimenidis*," *Osiris*, vol. *11* (1954), pp. 297–99.

[10] J. L. Heiberg, *Archimedis opera omnia*, 2 ed., vol. *3* (Leipzig, 1915), p. XCVIII.

now convinced that the translation was made from the Greek rather than from the Arabic. The Latin text retains no trace of the phrasing so characteristic of works translated from the Arabic (as, for example, the use of *iam* with the perfect tense to translate *qad* with the perfect). On the other hand, lines 18–22 of the first proposition appear to be a translation from the Greek, for it would be surprising if the proper name Pallas (line 19) would survive transmission in unmutilated form from Greek through Arabic to Latin. Further evidence is perhaps provided by the use of the word *discolos* in line 18, which is the accusative plural of *discolus* or, more properly, *dyscolus*, a rare Latin word used to transliterate the Greek word δύσκολος. It is possible that this word was used by the translator of our tract simply because he found the same word used in the Greek text.[11] The continuous use of the term *falsigraphus*,[12] which is commonly used to render the Greek ψευδογράφος, could be another indication of a tract that is Greek in origin. And finally the *explicit* of the treatise including the word *tiphis* (transliterating the Greek τῖφυς) seems to add further weight to the Greek origin of the tract. Nothing can be argued from the geometrical terms, for the terms employed in this text are those employed in translations from both Arabic and Greek.

Assuming that the *De curvis superficiebus* was in its origin a Greek tract, we must then face the problem of identifying the author. Unfortunately, I have not been able to find any trace of a Johannes de Tinemue in antique or Byzantine sources. If the quotation of the third proposition in Hero's *Mechanics* was in the original text of Hero's work and was indeed a quotation of this work, then the treatise cannot be any later than the first century A.D.[13] But even if the quotation in Hero's *Mechanics* is an interpolation in the Arabic text (which I strongly doubt, since it is apparently in all of the Arabic manuscripts), or somewhat more likely an interpolation in the Greek text prior to its translation into Arabic,[14] the antique or Byzantine

[11] I suppose one could argue that since the Latin word *dyscolus* was used in the Vulgate translation of the New Testament (I. Petr. 2. 18) it might have been used by a Latin author who was, in preparing the *De curvis superficiebus*, making an original composition rather than doing a translation, but I have already argued against the text being an original composition.

[12] See Chapter Three, Section 5, Corpus Christi Version. See also Commentary, Section 5, Chapter Three, lines 218–19.

[13] See note 6 above. On the basis of astronomical evidence Hero can rather confidently be dated in the first century. See O. Neugebauer, "Über eine Methode zur Distanzbestimmung Alexandria-Rom bei Heron," *Det Kgl. Danske Videnskabernes Selskab. Historisk-filologiske Meddelelser*, vol. *26*, 2 (1938), p. 23, and his note in *Isis*, vol. *39* (1948), pp. 243–44.

[14] On the question of interpolations in the text of the *Mechanics*, see Nix's intro-

origin of the tract seems to follow from the fact that the tenth proposition was quoted in the *Stereometry*. The *Stereometry*, while probably not by Hero, certainly existed in late antique or early Byzantine times.

I have remarked on the fact that two of the manuscripts include the name of a Latin Schoolman, Gervasius of Essex, who apparently was responsible for that revision of the text which I have called Tradition II. I have not been able to identify any such person from Essex, but there was a Johannes Gervasius of Exter (*Exonia*) who was chancellor of York and who was consecrated Bishop of Winton in the Roman Curia in 1261.[15] A scribal confusion between *Essexta* and *Exonia* is possible. The fact that Johannes Gervasius was in Italy in 1261 and died in Viterbo in 1268 is perhaps an indication that he was a member of that remarkable scientific group which included Witelo, Campanus de Novara, William of Moerbeke, and possibly Aquinas.[16] If so, this would seem to establish his interest in scientific matters. He might well have been the reviser of the *Liber de curvis superficiebus*.

The *De curvis superficiebus* is a treatise of ten propositions with several corollaries dealing for the most part with the surfaces and volumes of cones, cylinders, and spheres. The meaning of the title is evident: On Figures Bounded by Curved Surfaces. In brief, the main objectives of the tract are those of Book I of the *De sphaera et cylindro* of Archimedes. Curtze, Heiberg, and Björnbo tended to identify the two works,[17] and Heiberg published the enunciations of the propositions without proofs. But none of these authors seems to have realized that the *De curvis superficiebus*, while something of a commentary on the *De sphaera et cylindro*, is an independent treatise rather than a fragment or paraphrase of Archimedes' work. Incidentally, they all miss the important point partially recognized in the catalogue of Libri manuscripts[18] that the form and techniques of the *De mensura circuli* influenced the author of the *De curvis*

duction (*op. cit.* in note 6 above), pp. XXVIII–XXXIII.

15 Th. Tanner, *Bibliotheca Britannico-Hibernica* (London, 1748), p. 313.

16 For a brief account of this group, see M. Grabmann, *Guglielmo Moerbeke* (Rome, 1946), pp. 56–62.

17 M. Curtze, "Über eine Handschrift etc.," *Zeitschrift für Mathematik und Physik*, vol. *28* (1883). Hist.–lit. Abt., pp. 1–13. A. A. Björnbo, "Handschriftenbeschreibung etc.," *Avhandlungen zur Geschichte der Mathematischen Wissenschaften, 26.* Heft (1912), p. 128. Cf. *Bibliotheca mathematica*, Dritte Folge, vol. *9* (1908), p. 152. Heiberg's comments, with the text of the enunciations of the *De curvis superficiebus*, are found in the work cited in note 10 above, *loc. cit.*

18 *Catalogue of the Extraordinary Collection of Splendid Manuscripts... Formed by M. Guglielmo Libri*, London, 1859, 147 (referring to MS No. 665).

superficiebus. For example, the expression of the enunciation of Proposition I of the *De curvis superficiebus* was influenced by the statement of Proposition I of the *De mensura circuli*. Incidentally, the author several times cites the *De mensura circuli* of Archimedes, sometimes with that title and sometimes as *De quadratura circuli*. Ordinarily it is the first proposition of the *De mensura circuli* that is cited, but the second proposition is referred to twice in the course of Proposition X of the *De curvis superficiebus* (see lines 4, 13), and the third proposition is noted in the course of Proposition III (see lines 34–35). Euclid's *Elements* is the only other work referred to by the author, and he cites it about twenty times, although to be sure many of the citations are to the same propositions, namely, Propositions XII.9 and XII.11. He also borrows from the tradition of the *Elements* the use of terms to label the various parts of a proof: *exemplum* (for the Greek *ekthesis* or setting out), *dispositio* (for the Greek *kataskue* or additional construction needed for the proof), *ratio* (for the Greek *apodeixis* or proof). He also uses the term *improbatio* to label a formal refutation and *dissolutio* for a proof *per impossibile*. I have shown elsewhere how these terms were used by Greek, Latin, and Arabic authors in connection with the *Elements*.[19]

Not only was this work known to Gerard of Brussels by the middle of the thirteenth century, as I have indicated above (note 4), but Roger Bacon speaks of it in his *Communia mathematica*.[20] In the fourteenth century the *De curvis superficiebus* was cited by a number of authors: Thomas Bradwardine,[21]

[19] See my "King Alfred and the *Elements* of Euclid," *Isis*, vol. *45* (1954), pp. 272–73. The terms used in the text of the *De curvis superficiebus* are particularly like those found in the introduction to the version of the *Elements* which I have designated as Adelard III but which is probably to be dated close to 1200. The definitions found in the Adelard III introduction which are pertinent to our present discussion are the following (*ibid*, p. 274): "Exemplum est alicuius figure suppositio. Dispositio est aliarum figurarum ad exemplum applicatio. Ratio est principiorum vel suarum conclusionum inductio, i.e., argumentum.... Instantie dissolutio est cum falsigraphus insistit non sic vel aliter accidere quam geometer affirmat."

[20] For citation by Bacon, see Chapter IV, division 1, note 3, where our tract is called *de Curvis Superficiebus Archimenidis*.

[21] Thomas of Bradwardine, *Tractatus de proportionibus*, chap. IV (ed. of H. Lamar Crosby, Jr., pp. 124, 126): "Quorumlibet duorum circulorum circumferentiae suis diametris sunt proportionales. (Et hoc est quinta conclusio *De curvis superficiebus*.)... Cuiuslibet spherae superficies aequalis est quandrangulo qui sub lineis aequalibus diametri spherae et circumferentiae maximi circuli continetur. (Et hoc est octava Archimenidis *De curvis superficiebus*.)" Bradwardine makes a similar reference to the so-called fifth conclusion in his *De continuo* (MS Thorn, 4° 2, p. 164). The propositions cited by Bradwardine are actually the third and sixth propositions of the *De curvis superficiebus*. It is obvious that Bradwardine used a text of the tradition of *D* which

Nicole Oresme,[22] Francischus de Ferraria,[23] an anonymous author of a commentary on the *Liber de ponderibus*,[24] and the anonymous author of the *Liber de inquisicione capacitatis figurarum*.[25] Furthermore, sometime prior to 1328 (see note 21) a Latin author added the two propositions given in Section 3 of this chapter, while another Latin author, probably in the fourteenth century, paraphrased the proofs of the whole tract and added three additional propositions (Section 4 of this chapter), and still a third author (perhaps originally Gervasius) supplied an additional proof to Proposition VII (Section 5 of this chapter).

The text of the *Liber de curvis superficiebus* which I have presented here differs in only a few places from my previous effort,[26] although I have considerably expanded the variant readings. In general, I have preferred the readings of the first tradition. Manuscripts *B* and *C* of that tradition have been employed throughout the text. Manuscript *B* is not the archetype but must be quite a faithful copy. It is only rarely that I have felt it necessary to correct its reading by an appeal to Manuscript *C* or to the second tradition. Manuscript *J*, also of the thirteenth century, may well have been the link between the two traditions, as it seems to have been in the case of the *Liber de motu* of Gerard of Brussels, which I have edited

contained the two extra marginal propositions. Since the *De proportionibus* was composed in 1328, it is evident that the tradition represented by *D* is prior to that date.

[22] Nicole Oresme, *De configurationibus qualitatum*, MS Bibl. Nat. lat. 7371, 227v: "Nunc autem est ita quod proportio circumferentiarum in quantitate est sicut proportio semidiametrorum circulorum quorum sunt circumferentie, ut patet [per] quintam conclusionem Archimenidis *De curvis superficiebus*." Oresme has also used a manuscript of the tradition represented by *D*.

[23] Francischus de Ferraria, *Questio de proportionibus*, MS Oxford, Bodl. Canon. Misc. 226, 58v: "Deinde suppono unam suppositionem Euclidis *De curvis superficiebus*...et est: Qu[or]um angulorum et circulorum circumferentie suis dyametris sunt proportionabiles, igitur qualis est proportio dyametri *a* ad suam circumferentiam talis est proportio dyametri *b* ad suam circumferentiam." Francischus has apparently changed *Archimenidis* to *Euclidis*.

[24] In E. Moody and M. Clagett, *The Medieval Science of Weights* (Madison, 1952), p. 302: "...per quintam Archimedis *De curvis superficiebus*, eo quod eadem est proportio diametrorum vel semidiametrorum vel circumferentiarum." Once more we should note that the author of this tract has used a text in the tradition of manuscript *D*, where the third proposition is numbered as the fifth.

[25] *De inquisicione capacitatis figurarum*, ed. of M. Curtze in *Abhandlungen zur Geschichte der Mathematik*, *8*. Heft (1898), p. 39: "Qui cubus per 10^{am} Archimenisis de curvis superficiebus habet se ad sphaeram dati circuli, sicut 21 ad 11....ergo per sextam Archimenidis de curvis superficiebus ipsum parallelogrammum *acdb* est aequale embado sphaerae circuli dati.... [p. 50] et per secundam Archimenidis provenit tota curva superficies columpnae."

[26] Published in the article cited in note 9 above.

elsewhere.[27] For the second tradition I have used only manuscript *A*, its best manuscript, throughout the whole text. Most of the remaining manuscripts of both traditions specified in the Sigla have been collated through line 47 of the first proposition, enough to indicate such groupings as can be detected. Manuscript D_1 was copied from *D* and not very carefully; I have presented here its variant readings only through line 22 of the first proposition (in addition to reporting its variant readings for the colophon). I see no reason to include any more of its careless misreadings. I have also given the variant readings for the poor manuscript *E* only through line 22 of Proposition I. Manuscript *M* is a paraphrase of the text and I have treated it separately in Section 4 of this chapter. Some of the detailed characteristics of each of the manuscript have been included in the Sigla list below. Following my usual custom I have numbered the propositions successively with Roman numerals.[28] The figures are those found in *B*.[29] The marginal folio numbers refer to manuscript *B*.

Sigla of Manuscripts

Tradition I

B = Oxford, Bodleian Library, Auct. F.5.28, 111r–116r, 13c. On the whole this is the most reliable manuscript. Most of its orthographic peculiarities are reflected in the text here published. However, we can here note that it almost always has *ci* before a vowel rather than *ti*, as

[27] See note 4 for my text of the *Liber de motu*. I was able to use the Berlin Manuscript (here designated as *J* but there designated as *B*) in the preparation of that text. As I noted in the introduction to that text, the Berlin manuscript seems to have occupied a middle position between the Oxford and Naples manuscripts (here designated as *B* and *A* but there designated as *O* and *N*); see p. 109 of that text. The identity of *J* and Libri MS 665 is argued on pp. 103–104.

[28] Actually, manuscripts *A* and *D* number the propositions with Arabic numerals in the margin: 1ª, 2ª, and so forth (although *A* omits numbers for Propositions VIII, IX, and X). *B* uses Roman numerals but divides the propositions into two books: I through VI for the first book, and I through IIII for the second book. *J* makes the same division but uses a hybrid system of Roman and Arabic numerals (see variant readings). The only other manuscript to reflect this division into books is *A*, although to be sure it numbers the propositions successively so that Proposition VII is 7ª rather than I.

[29] Inside the figure for Proposition VII in MS *B* we find written in the same hand as that used in the text *mors Hugonis*, i.e., "the death of Hugo." One is tempted to think that reference is to Hugo, the twelfth-century author of a *Practica geometrie*, but this is merely a guess. Or it may be a comment of the scribe of *B*, discouraged by the length and complexity of Proposition VII. If this is so, then the scribe's name could have been Hugo.

in *circumferencia*. I have in my text employed the *ti* form. *B* starts off with the spelling *pyramis* (i.e., with *py* for all forms of this word) but for the major part of the work employs *piramis*. I have used *piramis* throughout. *B* also vacillates between *basem* and *basim*, between *demissa* and *dimissa*, generally preferring the latter of each of these pairs of terms. The form *corellarium* is used, which I have rejected by following other manuscripts. *B* divides the propositions into two books, numbering the propositions of the first book I through VI, and those of the second book I through IV. But this numbering, added as rubrics, may be in a later hand. There are a number of marginal notes in *B* which are entirely explanatory or constitute a geometric elaboration of the text. These have not been included in the variant readings since they appear to be the notes of the scribe of *B* rather than of the author of the text. They are of exactly the same sort as are found in all other works in the codex.

C = Oxford, Bodleian Library, Digby 174, 174v–178r, 13c. This manuscript is very close to *B* and, in fact, may be a copy of that manuscript. Its orthographic peculiarities are evident in the variant readings, but a few of them can be noted here: *teorema* for *theorema* in *B*, *ypothenussalium* for *ypothenusalium* in *B*, *demissa* for *dimissa*, ordinarily *basim* but occasionally *basem*, occasionally *ortogonio* for *orthogonio*. One unusual feature of this copy is the judicious use of commas, a practice very rare indeed among medieval manuscripts, which generally have periods, colons, and dashes used without much attention to rule or discrimination. There are a few interesting notes in a sixteenth-century hand which I have not included in my variant readings. There is a note dated: 1551, 6 May, on folio 175v.

J = Berlin, Deutsche Staatsbibliothek (MS now at Marburg, Westdeutsche Bibliothek), Quarto 150, 90r–94v, 13c. A Gothic hand very much like that of *B*. A copy close to the tradition of *B* and *C*, but not quite as accurate as those manuscripts. Within that tradition *H* and *F* resemble *J* quite closely. I have collected it with the text through Proposition I, line 47, and in the explicit. The forms *piramis*, *ortogonio*, and sometimes *hypothenusa* (for *ypothenusa*) are used. In numbering the propositions and dividing them into two books, it resembles *B*, but interestingly enough it omits lines 26–28 of Proposition I in the same manner as C, although these lines are present in *B* and even in Tradition II. This manuscript is identical with Libri MS 665 described in the Libri sale catalogue of London, 1859, pp. 145–48 (see notes 18 and 27 above).

D = Florence, Biblioteca Nazionale, Conv. Soppr. J.V.30, 1r–4v, 14c. A good copy of this tradition, being quite close to *B* and *C*. It employs *hypothenusse* for *ypothenuse* in *B*, that is, *ss* is found instead of *s* in the various forms of this word in those instances when the word is written out and not abbreviated. We also find in *D* the peculiar form *correlarium*, resembling the *corellarium* of *B* in using *e* as the second vowel. In *D* the letter *y* replaces *i* in the various forms of *diameter*, as, *dyametros*. *D* includes two extra propositions on the lower margins of folios 1r and 1v. These additional propositions appear to be in the same hand (only very much larger in form). They have been edited and are discussed in Section 3 of this chapter. The propositions in this copy are numbered in the margin: 1ª, 2ª, etc. Collation of *D* with the text is through line 47 of Proposition I.

D_1 = Paris, Bibliothèque Nationale, MS lat. 11247, 2r–25v, late 15c or 16c. Copied directly from *D* but includes many misreadings. The propositions are unnumbered and no figure after the first is given. I have included variant readings only through line 22 of the first proposition. Its orthographic peculiarities for the most part duplicate those of *D*.

H = Basel, Univ. Bibl. F.II.33, 151r–153r, 14c. This is another good copy, very close to *B* and *C* in its readings. It was perhaps copied from *J*. I have given variant readings through the first 47 lines of proposition I. The propositions are unnumbered in *H*.

F = Vienna, Nationalbibliothek, cod. 5303, 11r–18v, 15–16c. *F* is of the tradition of *B* and *C*. It has close similarities with *J* and *H*. Orthographic distinctions: *pyramis* for *piramis*, *hypothenuse* for *ypothenuse*. The propositions are unnumbered in *F*. *F* has been collated with the text through line 47 of Proposition I.

I = Dresden, Sächs. Landesbibliothek, Db. 86, 188r–194v, early 14c. The manuscript is badly stained (and often is not readable) but where it is free from staining it is a very good copy and quite close to *B* (e.g., even in the spelling *corellario*, cf. Proposition I, line 104). Occasional orthographic peculiarities: *polligonie* for *poligonie*, *habitum* for *ambitum*. The propositions are unnumbered in *I*. *I* has been collated with the text through line 47 of Proposition I.

M = Florence, Bibl. Naz., Conv. Soppr. J.V.18, 92v–96v (lower pagination). This text consists of a paraphrase but the enunciations appear to be drawn from Tradition I. See Section 4 of this chapter.

TRADITION II

A = Naples, Biblioteca Nazionale, VIII.C.22, 57r–60r, 13c. This is the best of three manuscripts of the second tradition, the tradition perhaps inaugurated by Gervasius of Essexta. Notice that in addition to the substitution of his own name for that of Johannes in the colophon, Gervasius eliminated the parenthetical paragraph included in lines 18–22 of Proposition I. The scribe uses the *ci* form as does *B*; he also often employs *maiore* for *maiori*, *basem* for *basim*, *sexcupla* for *sextupla*, *theoreuma* for *theorema*. The propositions are numbered through the seventh: 1ª, 2ª, and so forth.

E = London, British Museum, Harleian 625, 137r–139v, 14c (?). This is a very free copy of the second tradition. I have included variant readings only through line 22 of the first proposition. Note the form Gervasius de Assassia in the colophon. The propositions are unnumbered.

G = Cambridge, University Library, Mm.III.11, 196r–198v, 15c. Corresponds closely to *A* and *E* but cuts off in the middle of Proposition VII. Hence it does not have Gervasius' name. I have included variant readings through line 47 of Proposition I. The propositions are unnumbered.

Incipit Liber Magistri Johannis de Tinemue de curvis superficiebus [Archimenidis]

Liber I

111r I. / CUIUSLIBET ROTUNDE PIRAMIDIS CURVA SUPERFICIES EST EQUALIS TRIANGULO ORTHOGONIO, CUIUS UNUM LATERUM RECTUM ANGULUM CONTINENTIUM EQUATUR YPOTHENUSE PIRAMIDIS, RELIQUUM CIRCUM-
5 FERENTIE BASIS.

Sit *ACNP* circulus circa centrum *D* [Fig. 64.]. Et sit *DE* linea cathetus perpendicularis ad *AN*, *CP* diametros, *EA* vero sit ypothenusa dimissa ab *E* in *A*. Stante igitur *DE* immota, circumvolvatur *DEA* triangulus orthogonius ad perficiendam piramidem rotundam
10 per *QPONMCB* donec redeat ad punctum *A* unde ceperat. Sit item

Title Incipit....I *mg.* *BJ** *om.* *IGMD*$_1$ Incipit liber Johannis de tinennie de curvis superficiebus. Liber I *A* De curvis superficiebus Archimenides *manu recentiori* (?) *B* Archimedes (*C manu recentiori*; Archimenides *H*) de curvis superficiebus *CH* Arcimenidis (*ff. 1r, 2r, 3r* Archimenidis, *4r*) de curvis superficiebus (? *2v*) *mg.* *D* In nomine domini nostri Iesu christi. Incipit commentum Ioannis de Chinemue in demonstrationes Archmiedis (!) *F* De curvis superficiebus *E* liber de curvis superficiebus *M*/[Archimenidis] *supplevi*; *sed cf. var. supra*

1 I *B* I^a *J* 1^a *AD* ((Note: the propositions are numbered only in MSS BJAD; see the Introduction to this chapter, note 28.))

2 est equalis *tr.* *AEG*

3 continentium: continens *D* continens non *D*$_1$

4 *ante* reliquum *add.* *E* et

6 circulus *om.* *AEG* / centrum: diametrum *AEG* / Et *om.* *G* / sit^2 *om.* *H*

7 cathetis *D*$_1$ / AN, CP: AC NP *DD*$_1$ / EA...sit: et vero EA *E* / vero: vero figure (?)*H* vero linea *JF*

8 dimissa *BIF* demissa *JACHGDD*$_1$ protrahata *E* / immota: immoto *F* inmota *IJD*$_1$

9 orthonogius: orthogonius intellectualiter *JFH* / perficiendam: faciendum *H*

9–10 piramidem...per: rotundam superficiem piramidem *A* rotundam pyramidem *EG*

10 *ante* Q- *add.* *A* A- / QP-: APP- *D*$_1$ / -NM-: -MN- *F* / donec: circumferentiam donec *HFJ* / A *om.* *BF* / inceperat *J*

**J* has the title twice on folio 90r, and following the title as given in the upper margin we read *de curvis liber primus*.

Here Begins the Book of Master Johannes of Tinemue on the Curved Surfaces [of Archimedes]

Book I

I. THE LATERAL SURFACE OF ANY [RIGHT CIRCULAR] CONE IS EQUAL TO A RIGHT TRIANGLE, ONE OF WHOSE TWO SIDES CONTAINING THE RIGHT ANGLE IS EQUAL TO THE SLANT HEIGHT OF THE CONE, WHILE THE OTHER [IS EQUAL] TO THE CIRCUMFERENCE OF THE BASE.

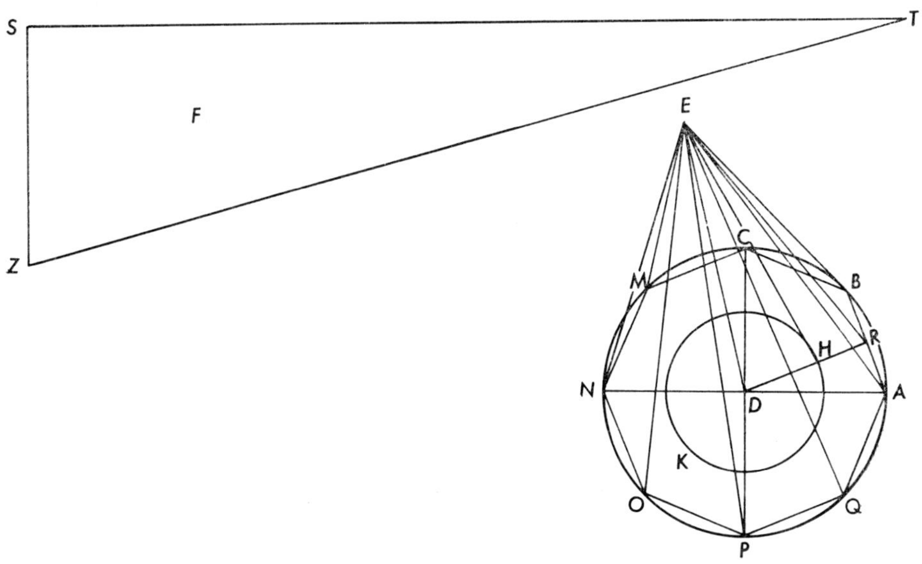

Fig. 64

Let *ACNP* be a circle with center *D* [see Fig. 64]. And let line *DE* be a line perpendicular to diameters *AN* and *CP*, while *EA* is a slant height drawn from *E* to *A*. Therefore, with *DE* as an axis, let right △ *DEA* be rotated through *QPONMCB* until it returns to point *A* from which it began, thereby forming a cone. Also, let slant height *EA* be

EA ypothenusa equalis linee *SZ* et tota circumferentia *ACNP* equalis *ST*. Et sit *S* angulus rectus.

Dico itaque quod curva superficies rotunde piramidis *DEA* est equalis triangulo orthogonio *F*.

15 Rationis causa. Presentis demonstrationis ypothesis, et tota sequentium theorematum series, lineam rectam curve et superficiem rectam curve esse equalem sibi postulat admitti.

Unde discolos et obiurgantes non cogit ad scientiam, cum et ipsi sapere non audeant, cum etiam laborent ut nesciant, cum et a pallade 20 ultro vultus detegentes oculos avertant. Hec igitur hiis, qui rerum subtilium fugas, quantitatum miracula, proportionum nexus, nature deposita rimantur, proposuit philiosophus.

Triangulus itaque *F* aut est equalis curve superficiei *DEA* rotunde piramidis aut curve superficiei rotunde piramidis site in minori basi 25 quam *ACNP* circulus, que piramis sit equalis altitudinis piramidi *DEA*, id est, cuius cathetus sit *DE*, aut curve superficiei rotunde

11 *SZ*: scilicet Z DD_1
11–12 et...ST *om.* *AG* et ST equalis circumferentie basis pyramidis *E*
11 equalis[2]: equatur *I* sit equalis *F*
12 *post* rectus *add.* *E* et protrahatur linea TZ ita quod totus ille triangulus sit *F*
13 itaque: ergo DD_1
14 orthogonio *om.* *EF*
15 Rationis causa: ratio *AGE* / demonstrationis *om.* *HF* / *ante* ypothesis *add.* $AGDD_1$ et / hypothenussis D_1 / ypothesis et tota: omnis *E*
15–16 subsequentium *A*
15–16 sequentium...series: series et theorematum sequentium *E* sequencium cathetum series *J*
16 teorematum *C* / series: superficies *scr. et del. H et add. H mg* series
16–17 lineam...curve *CHFJ* lineam curvam recte et superficiem rectam curve *B* lineam rectam curvam curve et superficiem rectam curve *D* lineam rectam curvam curve et superficiem rectam curve D_1 lineam rectam curve et curvam recte et superficiem rectam curve *I* superficiem curvam recte ut lineam curvam recte (recte *om. A suprascr. G*) *AG* superficiem recte curvam et lineam curve rectam *E*
17 esse equalem *tr.* *HJF* / equalem sibi *tr.* *D* / admitto *H*
18–22 Unde....philosophus *om.* *AEG*
18 discelos *C* / cum: et cum *I*
19 etiam: ter D_1/ut: nec (?) *H* / cum *om.* *HF* / et: illic (?)*D* autem D_1
20 ultro vultus: vultum ultro *HF* ultimum multum D_1 / detegentes: detegente *CDJ* detegentum *HF* deteguntem D_1 / Hec: hoc D_1
21 subtilians fiiguris *I* / fugas: fugeras *H* fugas *J* fugeras *mg. J*
22 rimantur: rimanantur *D* nominantur D_1 / *ante* proposuit *add.* *JHF* physice filiis
23 itaque *om.* *H* / est *om.* *AG* / curve superficiei *tr.* *E* / DEA: dicte *AG*
24 *post* aut *scr. et del. C* maioris aut minoris aut / curve superficiei rotunde *om.* *AG* *I* (?) / aut...piramidis *om.* *HF*
25 quam: quam sit *AG* / circulus *om.* *AG* / equalis *om.* *J* / pyramidis *D*
26 aut: aut est equalis *AG*
26–28 aut...DE *om.* *CJ*

equal to line SZ and the whole circumference $ACNP$ be equal to ST; and let S be a right angle.

And so I say that the lateral surface of cone DEA is equal to right △ F.

A supposition necessary for the proof: The hypothesis of the present demonstration as well as the whole series of subsequent theorems postulates that a straight line can be supposed equal to a curved line and a plane surface to a curved surface.

Whence he does not urge to knowledge those who are objectors and are difficult, since they do not venture to learn—for they even work in order not to know—and since, on uncovering their faces, they avert their eyes away from Pallas. Therefore, the philosopher has proposed these things for those who examine the flights of subtle things, the miracles of quantities the connections of proportions, the things committed to nature.

And so △ F either is equal to (1) the lateral surface of cone DEA or to (2) the lateral surface of a cone situated on a base less than circle $ACNP$ but whose altitude is equal to that of cone DEA, namely, to perpendicular DE, or to (3) the lateral surface of a cone situated on a base larger than

piramidis site in maiori basi quam sit *ACNP*, que piramis sit equalis altitudinis piramidi *DEA*, id est, cuius cathetus sit *DE*. Cum enim triangulum *F* alicui curve superficiei rotunde piramidis quivis bene
30 sanus admittat, aliter quam diximus nec ratio veritatis indagatrix capit, nec humanus animus sibi consonus aliter esse posse animadvertit. Fere enim ex immediatis trimembris illa connectitur divisio.

Sit itaque, quod mentiatur falsigraphus, *F* non esse equalem curve superficiei rotunde piramidis *DAE*, sed minoris. Esto ergo *F* equalis
35 curve superficiei rotunde piramidis site in minori basi quam *ACNP* circulus. Et basis illius piramidis sit circulus *HK* circa centrum *D*. Et sit cathetus illius piramidis *DE*.

Dispositio. Inscribatur circulo *ACNP* figura poligonia equalium laterum et angulorum et sit *ACNP* ita quod non contingat circulum
40 *HK*. Deinde ab *E* dimittantur ypothenuse singule ad singulos angulos figure poligonie *ABCMNOPQ*. Dimittatur etiam ab *E* linea *ER* perpendicularis ad *AB*. Sit item *EH* ypothenusa dimissa ab *E* ad *H* ubi *DR* linea secat circulum *HK*. Est itaque *DEH* triangulus orthogonius qui intellectualiter circumvolvatur per circumferentiam *HK*
45 donec redeat ad *H* unde ceperat ad perficiendam piramidem rotundam *DEH*, cuius piramidis curve superficiei sit secundum falsigraphum *F* triangulus equalis.

Dissolutio. *ER* est perpendicularis ad *AB*. Ergo quod fit ex ductu *ER* in *RB* est duplum trianguli *ERB* per [41] theorema primi Elemen-

27 base *H* / ACNP: ACNP circulus *HF* / piramis: pyramidis *H*
27–34 site....piramidis *om. G*
27–32 quam....divisio: et cetera *A*
28 cuius *om. HF supra scr. J*
30 admittit *H*
31 consonus: bene constitutus *F* / posse animadvertit: advertit *HF* / Fere *CHDF* Tunc *B*
32 illa *om. D*
33 itaque: ergo *A*
33–34 curve superficiei *tr. D*
34 rotunde *om. JHF* / DAE: DEA *AGD* / ergo: igitur *HF*
35 superficiei: superficiei alicuius *J* / rotunde piramidis *tr. JAG* / minore *AG* / quam *BHF* quam sit *ACIG*
36 circulus:[1] circulus sit *HF* / Et: et sit *HF* / illius: istius *H* alicuius *J* / sit *om. CHFD supra scr. J*
36–37 sit....piramidis *om. I*
38 ACNP *om. AG*
39 et sit ACNP *om. AG* / sit *om. D* / ita quod: in qua *HF*
39 contingant *D* / circulum: minorem circulum AG
40 *post* HK *add. H* per X II[ndi] (?) *et add. F* per *cum lacuna* / ab *E om. AG* / demittantur *HFGD mg. J* / angulos *om. D*
41 demittatur *ACHFGDJ* / linea *om. ADG*
42 demissa *CHGFDJ* / ad: in *AG*
45 redeat: veniat *HF* / piramidem rotundam *tr. HF*
46 DEH: HDE *AG* / curve superficiei *tr. AG*
49–50 in....ER *om. B*
49 [41] *supra scr. C manu recentiori* / theorema: 12 *A*

ACNP but whose altitude is equal to that of cone *DEA*, namely, to perpendicular *DE*. For, since anyone who is quite sound admits that △ *F* is equal to the surface of some cone, no searcher after truth can understand it in any way other than as we have stated it; nor can a sound human mind think that it can be otherwise. For this [logical] division arises almost immediately from the three terms.

And so let it be, as the pseudographer falsely says, that *F* is not equal to the lateral surface of cone *DAE* but [to the surface] of one that is less. Therefore, let *F* be equal to the lateral surface of a cone situated on a base less than circle *ACNP*. And let the base of this cone be circle *HK* with center *D*. And let the altitude of this cone be *DE*.

Disposition*: Let there be inscribed in circle *ACNP* a regular polygon, *ACNP*, in such a way that it does not touch circle *HK*. Then from *E* let slant heights be drawn to each of the angles of polygon *ABCMNOPQ*. Also, from *E* let *ER* be drawn perpendicular to *AB*. Also, let *EH* be the slant height drawn from *E* to *H* where line *DR* cuts circle *HK*. And so right △ *DEH*, which we mentally rotate through circumference *HK* until it returns to *H* from where it began, forms cone *DEH*, to whose lateral surface—according to the pseudographer—△ *F* is equal.

Dissolution**: *ER* is perpendicular to *AB*. Therefore, $ER \cdot RB = 2 \triangle ERB$, by I.41 of the *Elements*. Similarly, $ER \cdot RA = 2 \triangle ERA$.

* That is, "further construction"; see page 444 for the use of this term.

** That is, "the argument by reduction to absurdity"; see page 444.

50 torum. Similiter id quod fit ex ductu *ER* in *RA* est duplum trianguli *ERA*. Ergo quod fit ex ductu *ER* in *RA* et *RB* est duplum trianguli *EAB*. Ergo per primam secundi id quod fit ex ductu *ER* in *AB* duplum est trianguli *EAB*. Eadem ratione quod fit ex ductu *ER* in *CB* duplum est trianguli *ECB*. Eadem quoque ratione si bene numeraveris, 55 invenies id quod fit ex ductu *ER* in *AB*, *BC*, *CM*, *MN*, *NO*, *OP*, *PQ*, *QA*, latera poligonii, esse duplum omnium triangulorum ypothenusalium, scilicet, *EAB*, *ECB*, *ECM*, et reliquorum ypothenusalium. Sed *R* angulus est rectus. Ergo *EA* maior est quam *ER*; et circumferentia *ACNP* maior est ambitu poligonii, scilicet, *AB*, *BC*, *CM*, 60 *MN*, *NO*, *OP*, *PQ*, *QA*, lineis rectis. Ergo maius est id quod fit ex ductu *EA* in circumferentiam quam id quod fit ex ductu *ER* in ambitum poligonii. Sed id quod fit ex ductu *EA* in totam circumferentiam est duplum trianguli *F*, quia *ST* est equalis circumferentie *ACNP* et *EA* est equalis *SZ*. Et id quod fit ex *ER* in ambitum poli- 65 gonii est duplum omnium triangulorum ypothenusalium pariter ac- 111v ceptorum. Ergo maius est duplum trianguli *F* / quam duplum omnium triangulorum ypothenusalium pariter acceptorum. Ergo subduplum subduplo maius, scilicet triangulus *F* omnibus triangulis ypothenusalibus pariter sumptis. Sed trianguli ypothenusales pariter 70 sumpti maiores sunt curva superficie *DEH* rotunde piramidis incluse. Ergo triangulus *F* est maior curva superficie rotunde piramidis *DEH*. Non ergo est ei equalis, contra hoc quod predixit falsigraphus. Relinquitur ergo quod non possit esse *F* equalis curve superficiei rotunde piramidis site in minori basi quam *ACNP* circulus, que piramis sit 75 equalis altitudinis piramidis *DAE*, id est cathetus sit *DE*.

Sit item, secundum falsigraphum, *F* equalis curve superficiei rotunde piramidis site in maiori basi quam *ACNP* circulus, cuius pira-

51 RA: ER *A* / RB: RB coniunctim *A* / est duplum *tr.* *AC*
52–54 EAB....trinaguli *om.* *C*
52–53 Ergo...EAB *om.* *A*
53 quod: id quod *A*
54 duplum est *tr.* *A* / quoque: que *A*
55 ductu *om.* *A*
56 latera poligonii *AC om.* *BI*
56, 57 ypothenussalium *C*
57 scilicet *om.* *A*
57 et...ypothenusalium: et cetera *A*
58 EA...ER: ER est minor EA *CD* / maior: minor *A*
60 NO...QA: et ceteris *A*
61 in[1]: in totam *A*
62 id *om.* *A*
64 id *om.* *A* / fit *om.* *A* / ambitum: habitum *I*
66–67 Ergo...acceptorum *om.* *A*
68 maius *om.* *A* / F *om.* *A*
69 sumptis: acceptis *A*
69–70 pariter sumpti *om.* *A*
72 contra: quod est contra *A* / predixit: dicit *A*
75 id est *BCDI* cuius *A* et *F*
76 item: tunc *A*
77 maiore *A*

Hence, $ER(RA + RB) = 2 \triangle EAB$. Therefore, by II.1 [of the *Elements*], $ER \cdot AB = 2 \triangle EAB$. By the same reasoning, $ER \cdot CB = 2 \triangle ECB$. By the same argument repeated a number of times, $ER \cdot (AB + BC + CM + MN + NO + OP + PQ + QA) = 2 \cdot (\triangle EAB + \triangle ECB + \triangle ECM + \triangle ENM + \triangle ENO + \triangle EPO + \triangle EPQ + \triangle EQA)$, that is, the product of ER and all of the sides of the polygon is equal to double all of the face triangles. But $\angle R$ is a right angle. Hence $EA > ER$, and circumference $ACNP$ is greater than the perimeter of the polygon, that is, the circumference is greater than the sum of the straight lines AB, BC, CM, MN, NO, OP, PQ, QA. Hence, ($EA \cdot$ circumference) $>$ ($ER \cdot$ perimeter of polygon). But ($EA \cdot$ circumference) $= 2 \triangle F$, because ST is equal to circumference $ACNP$ and $EA = SZ$. Furthermore ($ER \cdot$ perimeter of polygon) $= 2 \cdot$ (the sum of all the face triangles). Therefore, $2 \triangle F > 2 \cdot$ (the sum of all the face triangles). Therefore, half of the one is greater than half of the other, that is, $\triangle F >$ (the sum of all the face triangles). But the sum of all the face triangles is greater than the lateral surface of the included cone DEH. Therefore, $\triangle F$ is greater than the lateral surface of cone DEH. Therefore, it is not equal to it, and this is contradictory to that which the pseudographer stated earlier. It remains, therefore, that F could not be equal to the lateral surface of a cone situated on a base less than circle $ACNP$ but with an altitude equal to that of cone DAE, that is, with altitude DE.

Following the pseudographer once more, let F be equal to the lateral surface of a cone situated on a base greater than circle $ACNP$ but whose

midis cathetus sit *DE*. Brevitatis tamen causa sit *DEH* piramis rotunda proposita, cuius ypothenusa *EH* sit equalis *SZ*. Et circumferentia
80 *KH* sit equalis *ST*. Dicatque falsigraphus *F* non esse equalem curve superficiei piramidis *DEH*, sed alterius piramidis eiusdem altitudinis site in maiori basi. Et sit basis illius piramidis circulus *ACNP*. Piramis vero fundata super *ACNP* sit *DEA*, cuius scilicet *DEA* curve superficiei sit, secundum falsigraphum, *F* equalis.

85 Improbatio. In hoc triangulo *DEH*, *D* est rectus; maneat enim tota prior linearum dispositio. Ergo *H* est acutus. Ergo *EHR* est obtusus; ergo maior quam *R*. Ergo *ER* est maius *EH*; et omnia latera poligonii pariter sumpta sunt maiora quam circumferentia *HK*. Ergo maius est id quod fit ex *ER* in ambitum poligonii quam id quod fit
90 ex *EH* in *HK*, scilicet, *SZ* in *ST*. Sed quod fit ex *SZ* in *ST* est duplum trianguli *F*. Et quod fit ex *ER* in ambitum poligonii duplum est ad omnes triangulos ypothenusales. Ergo maiores sunt omnes trianguli ypothenusales quam *F*, subduplum subduplo sicut et duplum duplo maius. Sed *F*, secundum falsigraphum, est equalis curve superficiei
95 piramidis *DEA*. Ergo omnes trianguli ypothenusales sunt maiores curva superficie *DEA* rotunde piramidis. Superficies inclusa maior sit superficie includente, quod est impossibile. Relinquitur ergo quod curva superficies *DAE* non sit equalis *F*.

Cum ergo *F* non sit equalis curve superficiei alicuius rotunde pira-
100 midis site in maiori basi vel in minori quam in proposite piramidis basi, que sit eiusdem altitudinis proposite piramidi, relinquitur *F* esse equalem curve superficiei proposite piramidis, quod proposui. Hoc itaque theorema satis elegans, satis eleganter propositum, satis eleganti venustatur corollario.

105 [Corollarium:] Ex hoc manifestum quod proportio curve superficiei

79 EH sit *tr.* *A*
80 KH: HK *A* / phalsigraphus *I*
81 eiusdem altitudinis *om.* *A*
82 maiore base *A*
85 D: D angulus *A* / maneat *B* manet *AC*
88 pariter: insimul *A*
89 ex: ex ductu *A*
89–91 quam....poligonii *om.* *A*
91 duplum est *tr.* *A*
93 F: F triangulus *A* / et *om.* *A*
94 maius *om.* *A* / F: F triangulus *A*, *et tr.* *A post* falsigraphum
96 DEA....piramidis: piramidis DEA *A*
96–97 Superficies...superficie: Et sit superficies inclusa maior *A*
99 alicuius *om.* *A*
100 base *A*
100–101 in[3]...basi *A* piramis proposita *BC*
101 eiusdem: equalis *A*
102 proposui: proposuimus *A*
103 theorema *BI* theoreuma *AH* teorema *C* / satis[2]...satis[3] *om.* *A*
104 corellario *BI* corrollario *F* correlario (?) *D* corollarium *J*
105 quod: est quod *A*

altitude is DE. For the sake of brevity, however, let cone DEH be the proposed cone whose slant height EH is equal to SZ. And let circumference KH be equal to ST. And let the pseudographer say that F is not equal to the lateral surface of cone DEH, but [to that] of another cone of the same altitude but situated on a greater base. Let the base of that cone be circle $ACNP$, while the cone based on $ACNP$ we let be DEA. It is the lateral surface of this cone DEA that the pseudographer supposes F equals.

Refutation: In this $\triangle\ DEH$, $\angle\ D$ is a right angle, for the whole arrangement of lines remains as before. Therefore, $\angle\ H$ is an acute angle. Therefore, $\angle\ EHR$ is obtuse and thus greater than $\angle\ R$. Therefore, $ER > EH$ and the perimeter of the polygon is greater than circumference HK. Therefore, ($ER \cdot$ perimeter of polygon) $>$ ($EH \cdot HK$), and hence ($ER \cdot$ perimeter of polygon) $>$ ($SZ \cdot ST$). But ($SZ \cdot ST$) $= 2\ \triangle\ F$, and ($ER \cdot$ perimeter of polygon) $= 2 \cdot$ (the sum of all the face triangles). Hence the sum of all the face triangles is greater than F, the halves being related as their doubles. But F according to the pseudographer is equal to the lateral surface of cone DEA. Therefore, the sum of all the face triangles is greater than the lateral surface of cone DEA, the included surface being greater than the including surface, which is impossible. It remains, therefore, that the lateral surface of DEA is not equal to F.

Since, therefore, F is not equal to the lateral surface of some cone of the same altitude but situated on a base greater than or less than the base of the proposed cone, it remains that F is equal to the lateral surface of the proposed cone, which is the proposition I have put forth. And so this theorem, sufficiently elegant and having been proposed with sufficient elegance, made beautiful by a sufficiently elegant corollary.

[Corollary:] From this it is evident that the ratio of the lateral surface

rotunde piramidis ad suam basim est sicut ypothenuse sue ad semidiametrum basis sue. Archimenides enim in quadratura circuli ostendit circulum esse equalem triangulo orthogonio, cuius unum laterum rectum angulum continentium equatur circumferentie circuli, reliquum
110 vero semidiametro. Cum itaque ex ductu circumferentie in semidiametrum fiat superficies dupla circuli et eiusdem circumferentie in ypothenusam fiat superficies dupla curve superficiei piramidis, erit proportio producti ad productum que producentis ad producens. Ergo que est proportio dupli curve superficiei piramidis ad duplum sue basis,
115 eadem est ypothenuse ad semidiametrum. Sed duplorum et subduplorum eadem est proportio. Ergo proportio curve superficiei piramidis ad suam basim que ypothenuse ad semidiametrum basis.

II. CUIUSLIBET COLUMPNE ROTUNDE CURVA SUPERFICIES EQUALIS EST TETRAGONO QUI CONTINETUR SUB LINEIS EQUALIBUS AXI COLUMPNE ET CIRCUMFERENTIE BASIS.

5 Hoc eodem genere demonstrationis quo et precedens theorema astruitur. Sive enim dicatur quadrangulum equale curve superficiei maioris columpne eiusdem altitudinis, sive minoris columpne quam sit illa cuius axis et circumferentia basis sunt equalia lineis continentibus quadrangulum, probabitur contrarium.

10 Describatur enim intra basim maiorem poligonium equalium laterum minorem circulum non contingentium. Et erigantur a singulis angulis perpendiculares basi et equales axi, que omnes cadunt in curvam superficiem maioris columpne; alioquin due linee super punctum unum eidem linee orthogonaliter insistent, propter lineam que suo
15 circuitu describit superficiem columpnalem Si igitur capita linearum perpendicularium rectis lineis coniungantur, fient quadranguli rectanguli interiorem columpnam minime contingentes, qui omnes pariter accepti erunt maiores curva superficie minoris columpne incluse et
112r minores mai/oris, et equales ei quod fit ex ductu perpendicularium

106 basim: piramidem *A*
107 Archimedes *F*
111 circuli *C* trianguli circuli *B* ad circulum *A*
112 curve...primidis *A* superficie piramidis *BI* pramidis superficie *C*
116 proportio: que est *A*
117 basem *A* / que: eadem est *A* / *post* basis *add. A* et e contrario

1 II *B om. C* II[a] *J* 2[a] *AD*
5 theoreuma *A*
8 sunt: sint *A*
10 basem *A*
14 eidem...insistent: orthogonaliter insistunt uni et eidem linee *A* / insisterunt (?) *C*
16–17 rectianguli *B*

of a cone to its base is the same as that of its slant height to the radius of its base. For Archimedes in the *Quadrature of the Circle* shows that a circle is equal to a right triangle one of whose sides containing the right angle is equal to the circumference of the circle while the other [is equal] to the radius. And so, since from the product of the circumference and radius arises a surface double that of the circle and from the product of the same circumference and the slant height arises a surface double that of the cone, the ratio of product to product will be as the ratio of multiplying factors to multiplying factors. Therefore, [after the elimination of common factors] the ratio of double the lateral surface of a cone to double its base is as the ratio of the slant height to the radius [of the base]. But the ratio of doubles is the same as that of their halves. Therefore, the ratio of the lateral surface of a cone to its base is as that of the slant height to the radius of the base.

II. THE LATERAL SURFACE OF ANY [RIGHT] CYLINDER IS EQUAL TO THE RECTANGLE CONTAINED BY LINES EQUAL [RESPECTIVELY] TO THE AXIS OF THE CYLINDER AND THE CIRCUMFERENCE OF THE BASE.

This is provided with the same kind of demonstration as the preceding theorem. For if it be said that the rectangle is equal to the lateral surface of a cylinder with the same axis but which is either on a greater or lesser base than that whose axis and base circumference are equal to the lines containing the rectangle, the contrary will be proved.

For within such a larger base let there be described a regular polygon whose sides do not touch the smaller circle [which is the base of the proposed cylinder]. And let there be erected from each of the angles [of the polygon] a line perpendicular to the base and equal to the axis, all of which lines fall in the lateral surface of the greater cylinder. Moreover, two [such] lines [opposite to one another] will be perpendicular to the same line [each] at one [terminal] point. When this line [with its perpendiculars] is rotated, it will generate a cylindrical surface. If, therefore, the terminal points [of each adjacent pair] of perpendicular lines are joined by straight lines, there will be produced [a set of] rectangles which do not touch the interior cylinder at all. All of these rectangles taken together will be greater than the lateral surface of the lesser, included cylinder and will be less than [the surface of] the greater [cylinder], and they will be equal to the product of [one of] the perpendiculars, or the axis, and the

20 sive axis in totum ambitum poligonii, qui maior est circumferentia basis minoris et minor maioris. Observanti ergo ordinem precedentis demonstrationis facile patebit propositum.

[Corollarium:] Ex hoc igitur manifestum quod proportio curve superficiei columpne ad curvam ⟨superficiem⟩ sue ⟨rotunde⟩ piramidis 25 est tanquam proportio axis columpne ad medietatem ypothenuse piramidis. Erit etiam proportio curve superficiei columpne ad suam basim sicut axis columpne ad quartam partem diametri basis.

Quia ex ductu circumferentie circuli in medietatem ypothenuse fit curva superficies piramidis per proximam et ex ductu eiusdem circum- 30 ferentie in axem columpne fit curva superficies columpne per istam, ergo cum productorum et producentium eadem sit proportio, erit proportio curve superficiei columpne ad curvam piramidis tanquam axis columpne ad medietatem ypothenuse piramidis.

Amplius, ex ductu circumferentie circuli in semidiametri medietatem 35 provenit area circuli, et ex ductu eiusdem circumferentie in axem columpne provenit curva superficies columpne; ergo cum productorum et producentium eadem sit proportio, erit proportio curve superficiei columpne ad suam basim que est axis columpne ad quartam partem diametri basis. Et sic duplex patet corollarium suo iunctum 40 theoremati.

III. QUORUMLIBET DUORUM CIRCULORUM CIRCUMFERENTIE SUIS DIAMETRIS SUNT PROPORTIONALES.

Describantur duo circuli *ABC*, *EFG* circa centra *D*, *H* [Fig. 65]. Dico quod proportio diametri *CDA* ad diametrum *GHE* que circum- 5 ferentie *ABC* ad circumferentiam *EFG*. Si enim ita non fuerit, erit proportio *CA* diametri ad *GE* diametrum que circumferentie *ABC* ad minorem vel maiorem circumferentiam quam *EFG*.

Et sit primo ad minorem, scilicet circumferentiam *IK*. Describatur ergo intra circulum *EFG* poligonium equalium laterum et angulorum

20 sive: sue *A*
21 minoris: maioris minoris *A*
23 igitur: etiam *A* / quod: est quod *A*
26 basem *A*
28 Quia: quoniam *A*
31 *ante* proportio *del. A* productio
33 piramidis *om. B*
34 circuli: basis *A* / semidiametri medietatem: medietatem sui semidiametri *A*
39–40 Et...theoremati: Sic ergo patet duplex corollarium premissi theoreumatis *A*
39 duplex: dupliciter*B* / corellarium *B*
40 teorematí *C*
1 III *B om. C* III[a] *J* 3[a] *AD* / circulorum circumferentie *tr. A*
4 quod *A* quoniam *BC* / diametri CDA *tr. A* / que: est que *A*
5 ita *AB supra scr. C*
7 minorem vel maiorem: maiorem vel minorem *A* / EFG: sit EFG *A*
8 scilicet *tr. A ante* IK

perimeter of the polygon which is [itself] greater than the circumference of the lesser base and less than that of the greater. Therefore, that which was proposed will be easily evident to one observing the order of the preceding demonstration.

[Corollary:] From this it is evident, therefore, that the ratio of the lateral surface of a [right] cylinder to the lateral surface of its cone [constructed on the same base] is as the ratio of the axis of the cylinder to one half the slant height of the cone. Furthermore, the ratio of the lateral surface of the cylinder to its base will be as that of the axis of the cylinder to one foutrh the diameter of the base.

Since from the product of the circumference of the [base] circle and one half the slant height arises the lateral surface of a cone, by the preceding [theorem], and from the product of the same circumference and the axis of the cylinder arises the lateral surface of the cylinder, by this [theorem], therefore, the products and the factors producing the products being of the same ratio, the ratio of the lateral surface of the cylinder to the lateral surface of the cone will be as that of the axis of the cylinder to one half the slant height of the cone.

Further, from the product of the circumference of the [base] circle and one half of its radius arises the area of the circle, and from the product of the same circumference and the axis of the cylinder arises the lateral surface of the cylinder. Therefore, since the ratio of the products is the same as that of the factors producing the products, the ratio of the lateral surface of the cylinder to its base will be as that of the axis of the cylinder to one fourth the diameter of its base. And thus the two-part corollary joined to its theorem is evident.

III. THE CIRCUMFERENCES OF ANY TWO CIRCLES ARE PROPORTIONAL TO THEIR DIAMETERS.

Let the two circles ABC, EFG be described about centers D and H [see Fig. 65]. I say that the ratio of diameter CDA to diameter GHE is as that of circumference ABC to circumference EFG. For if it were not thus, the ratio of diameter CA to diameter GE will be as the ratio of circumference ABC to a circumference which is either greater than or less than EFG.

Now at first let [the ratio of diameters] be as that [of ABC] to a lesser circumference, namely, IK. Therefore, let there be described within circle EFG a regular polygon whose sides do not touch circle IK at all. And

10 minime contingens *IK* circulum. Et describatur simile poligonium intra circulum *ABC* et sit *IK* circa centrum *H*. Ratio. Age, que est proportio *CA* diametri ad *GE* eadem est *ABC* circumferentie ad *IK* circumferentiam ex ypothesi falsigraphi; et que est *CA* ad *EG* eadem est ambitus poligonii ad ambitum poligonii, per primam duodecimi. Ergo 15 que est *ABC* circumferentie ad *IK* circumferentiam eadem est ambitus poligonii ad ambitum poligonii. Ergo permutatim que est proportio *ABC* circumferentie ad ambitum poligonii *ABC* eadem est circumferentie *IK* ad ambitum poligonii *EFG*. Sed *ABC* est maior ambitu sui poligonii. Ergo *IK* circumferentia maior est ambitu poligonii *EFG*, 20 quod est falsum. Relinquitur ergo quod non sit proportio *ABC* circumferentie ad minorem circumferentiam quam *EFG* que diametri *CA* ad diametrum *EG*.

Sit itaque proportio *ABC* ad maiorem circumferentiam quam *EFG* que diametri ad diametrum; sitque *MS*. Ergo e contrario proportio 25 *EG* diametri ad *AC* que *MS* ad *ABC*. Est itaque proportio *EG* ad *AC* que *EFG* circumferentie ad aliquam circumferentiam; sitque illa *H*. Que est proportio *EG* ad *AC* eadem est tam *MS* circumferentie ad *ABC* circumferentiam quam *EFG* circumferentie ad *H*. Ergo que est proportio *MS* ad *ABC* eadem est *EFG* ad *H*. Ergo permutatim 30 que est *MS* ad *EFG* eadem est *ABC* ad *H*. Sed *MS* est maior *EFG*. Ergo *ABC* est maior *H*. Est itaque proportio *EG* diametri ad *AC*

11 Ratio *AC om. B*
12 CA: EA *A*
14–16 Ergo...poligonii *om. A*
17 ad *A* in *BC* / poligonii ABC: sui poligonii *A*
18 ABC: EFG *AC*
19 *post* poligonii *add. A* vel ABC
20 est falsum *tr. A* / ergo *om. C*
21 que: que est *A*
23 itaque: ergo *A*
27 *post* H *add. A* ratio
28 circumferentie *om. A*
29 est proportio *om. B* / proportio *om. C*, *sed supra scr. C in manu recentiori*

let there be described a similar polygon within circle *ABC*; let the center of *IK* be *H*. Proceed with the demonstration as follows: the ratio of diameter *CA* to diameter *GE* is as that of circumference *ABC* to circumference *IK*, by the hypothesis of the pseudographer, and *CA* is to *EG* as the perimeter of [one] polygon is to the perimeter of the [other] polygon, by XII.1 [of the *Elements*]. Therefore, circumference *ABC* is to circumference *IK* as the perimeter of the one polygon [within *ABC*] is to the perimeter of the polygon [within *EFG*]. Therefore, permutatively, circumference *ABC* is to the perimeter of the polygon within *ABC* as circumference *IK* is to the perimeter of the polygon within *EFG*. But [circumference] *ABC* is greater than the perimeter of its [inscribed] polygon. Therefore, circumference *IK* is greater than the perimeter of

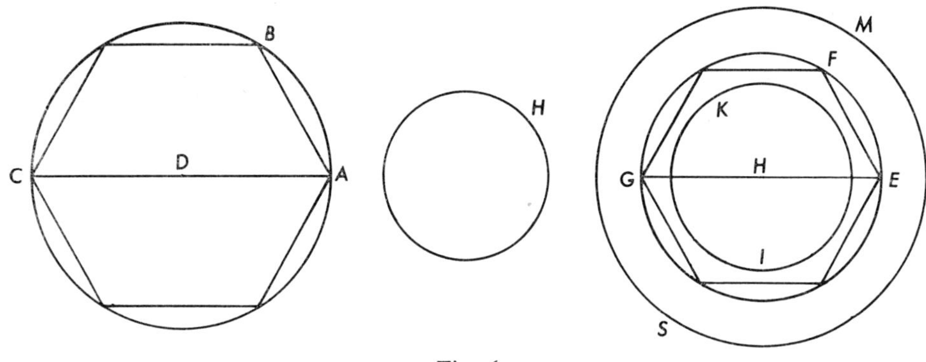

Fig. 65

polygon *EFG*, which is false. It remains, therefore, that the ratio of circumference *ABC* to a circumference less than *EFG* is not as that of diameter *CA* to diameter *EG*.

And so let the ratio of *ABC* to a circumference greater than *EFG* be as diameter to diameter. And this [greater circumference] we let be *MS*. Therefore, by inversion the ratio of diameter *EG* to [diameter] *AC* is as that of [circumference] *MS* to [circumference] *ABC*. And so the ratio of *EG* to *AC* is that of circumference *EFG* to some circumference, and that [latter] circumference we let be *H*. [Thus] the ratio of *EG* to *AC* is as the ratio of circumference *MS* to circumference *ABC* and as the ratio of circumference *EFG* to [circumference] *H*. Therefore, [by the equality of ratios] the ratio of *MS* to *ABC* is as that of *EFG* to *H*. Therefore, by the alternation [of ratios] *MS* is to *EFG* as *ABC* is to *H*. But [circumference] *MS* is greater than [circumference] *EFG*. Therefore, [circumference] *ABC* is greater than [circumference] *H*. And so the ratio of diameter *EG*

que est *EFG* ad circumferentiam *H* minorem *ABC*, quod prius improbatum est. Relinquitur ergo propositum inconcussum.

Idem aliter posset demonstrari per tertiam Archimenidis de men-
35 sura circuli. Sed non est adeo sufficienter.

[Corollarium.] Ex hoc liquet theoremate quod id quod fit ex ductu cuiuslibet circumferentie in diametrum alterius circuli est equale ei quod fit ex ductu secunde circumferentie in diametrum alterius circuli.

IV. QUARUMLIBET DUARUM PIRAMIDUM ROTUNDARUM INEQUALIUM ET SIMILIUM CURVE SUPERFICIES HABENT DIFFERENTIAM EQUALEM EI QUOD FIT EX DUCTU DIFFERENTIE YPOTHENUSARUM IN DIMIDIAS
5 CIRCUMFERENTIAS SUARUM BASIUM.

Esto exemplum *OEC* piramis rotunda [Fig. 66], cuius basis circulus *EQB*, cathetus *OC*, ypothenusa *CE*, curva superficies *R*, et minor piramis *DIM* similis piramidi *OEC*, cuius basis *ANM*, cathetus *DI*, ypothenusa *IM*, curva superficies *L*. Dico ergo quod differentia aug-

32 EFG: EFG circumferentie *A*
33 inconcussum *om. A*
35 non *om. A* / est *om. C*
36 [corollarium] *mg. C*
36–38 Ex...circuli *om. F*
38 circuli: circuli scilicet prioris *A*
1 IV: IIII *B om. C* IIII[a] *J* 4[a] *AD*
2 curve *om. A*
7 EQB: ECB *A* / OC: OE *A*

to [diameter] *AC* is as that of [circumference] *EFG* to circumference *H* where *H* is less than [circumference] *ABC*, which [proportionality] was refuted earlier. Therefore, that which was proposed remains firm.

The same thing can be demonstrated in another way by [using Proposition] III of Archimedes' *On the Measurement of the Circle*, but not so adequately.

[Corollary:] From this theorem it is clear that the surface arising from the product of any circumference and the diameter of another circle is equal to that which arises from the product of the second circumference and the diameter of the first circle.

IV. THE DIFFERENCE BETWEEN THE LATERAL SURFACES OF ANY TWO UNEQUAL BUT SIMILAR CONES IS EQUAL TO THAT WHICH ARISES FROM THE MULTIPLICATION OF THE DIFFERENCE BETWEEN THE SLANT HEIGHTS BY HALF [THE SUM OF] THE CIRCUMFERENCES OF THEIR BASES.

For example, let *OEC* be a cone [see Fig. 66], whose base circle is *EQB*, altitude is *OC*, slant height is *CE*, and whose lateral surface is *R*. And [let there also be] a lesser cone *DIM* similar to cone *OEC*. Its base we let be *ANM*, its altitude *DI*, its slant height *IM*, and its lateral surface

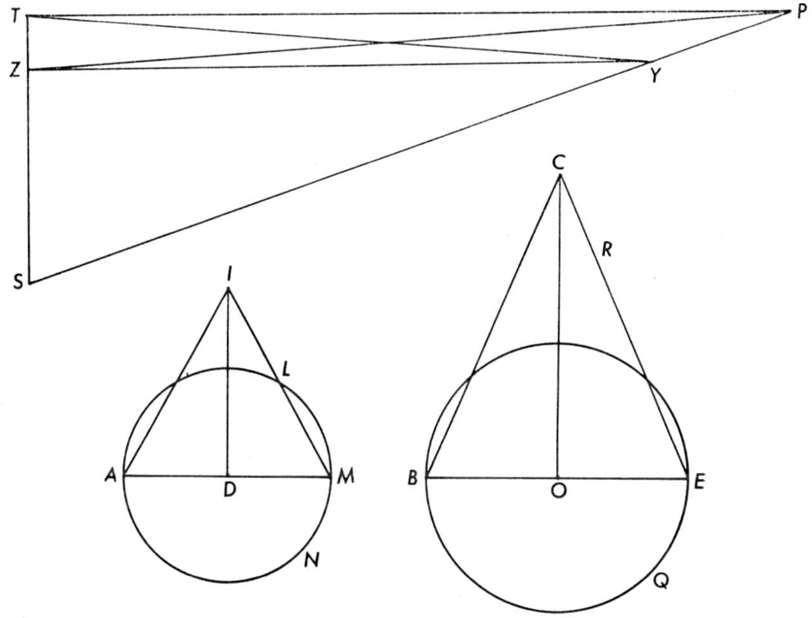

Fig. 66

Note: I have added the prime sign to O′ here and in the text.

10 menti *R* superficiei curve ad superficiem *L* est id quod fit ex ductu
112v augmenti *CE* ypothenuse super *IM* in medietates / circumferentiarum, *QBE*, *ANM*.

Sit enim trigonus orthogonius *STP*, cuius latus *ST* sit equale *CE* et *TP* sit equalis circumferentie *EQB*. Est enim trigonus *STP* equalis
15 *R*, per primam huius. Resecetur, ab *ST*, *ZS* equalis *IM*, *ZY* equidistante demissa a *Z*.

Ratio. Age. Tam *Z* quam *T* est rectus, et *S* est communis, et *Y*, *P* sunt equales propter *ZY*, *PT* equidistantes. Ergo trigonus *STP* est similis trigono *SZY*. Ergo proportio *ST* ad *TP* que *SZ* ad *ZY*. Item
20 que est proportio diametri *EB* ad diametrum *MA* eadem est circumferentie *EQB* ad circumferentiam *ANM*, per proximam. Et que est *EB* ad *AM* eadem est *CE* ypothenuse ad *IM* ypothenusam, per diffinitonem similium piramidum. Ergo proportio *CE* ad *IM* que *EQB* ad *ANM*. Sed *ST*, *CE*; et *EQB*, *TP*; et *IM*, *SZ* sunt equales. Ergo
25 proportio *ST* ad *SZ* que *TP* ad *ANM*. Sed que est *ST* ad *SZ* eadem est *TP* ad *ZY*, propter *STP*, *SYZ* triangulos similes. Ergo *ANM*, *ZY* sunt equales, cum *TP* ad illa eadem sit proportio. Sit ergo trigonus *SYZ* equalis *L*, per primam huius. Patet ergo quod differentia *R* ad *L* est superficies *ZYPT*. Ducatur ergo linea *ZP* a *Z* in *P* et *YT* a *Y* in *T*.
30 Ratio. *ZTY*, *YPZ* sunt trianguli super *ZY* basim inter *ZY*, *TP* equidistantes. Ergo sunt equales. Ergo id quod est ex *ZT* in *ZY* est duplum ad utrumque, et sic ad *ZPY*. Ergo id quod fit ex *TZ* in medietatem *ZY* est equale *ZPY*. Similiter id quod fit ex *TZ* in *TP* est duplum *TPZ*. Ergo quod fit ex *ZT* in medietatem *TP* est equale *TPZ*. Et sic
35 *ZTPY* superficies, que est differentia *R* ad *L*, fit ex ductu differentie *CE* ad *IM*, que est linea *ZT*, in medietatem circumferentie *EQB* et medietatem *ANM*, mediantibus *TP*, *ZY* quod proposuimus.

V. SI IN CIRCULO DESCRIPTI POLIGONII EQUILATERI ET EQUIANGULI MEDIETAS AD TERMINOS DIAMETRI TERMINATA, DIAMETRO STANTI, CIRCUMDUCATUR, E-

12 QBE: scilicet ABC *A*
14 EQB: EOB *A*
16 Z: T *A*
19 que: que est *A* / ZY: ZP *A*
24 TP: SP *A* / TP *corr.* *C* *ex* SP / IM, SZ: IMS, ZY *A* / equales: equalia *A* equa *C*
26 SYZ: SZY *A* / triangulos similes *tr.* *A*
27 Sit: Est *AC* / SYZ: STY *A*
28 primam: proximam *A*
29 ZYPT *B* YZTP *A* ZYTP *C* ((with signs to reverse *T* and *P* in C))
30 sunt *om.* *C* / basem *A* / TP: et TP lineas *A*
31 id quod *AC* quod *B* / ZT in ZY *B* TZ in TY *A* TZ in ZY *C*
32 id *om.* *C*
34 TPZ[1] *A* ZTP *BC* / TPZ[2]: ZTP *A*
1 V *B* *om.* *C* V[a] *J* 5[a] *AD*
3 stanti: circumstanti *A*

L. I say, therefore, that

$(\text{lat surf } R - \text{lat surf } L) = (CE - IM) \cdot (1/2 \text{ circum } QBE + 1/2 \text{ circum } ANM)$.

For let there be a right $\triangle\ STP$, whose side ST is equal to CE and TP is equal to circumference EQB, for $\triangle\ STP = R$, by the first [proposition] of this [work]. Let line ZS, equal to line IM, be cut from ST, with line ZY drawn from Z parallel [to TP].

Now proceed with the proof: Both $\angle\ Z$ and $\angle\ T$ are right angles, $\angle\ S$ is common, and $\angle\ Y = \angle\ P$ since ZY and PT are parallel. Therefore, $\triangle\ STP$ is similar to $\triangle\ SZY$. Therefore, $ST/TP = SZ/ZY$. Also, diameter EB is to diameter MA as circumference EQB is to circumference ANM, by the preceding [proposition]. And EB is to AM as the slant height CE is to the slant height IM, by the definition of similar cones. Therefore, $CE/IM = EQB/ANM$. But $ST = CE$, $EQB = TP$, and $IM = SZ$. Therefore, $ST/SZ = TP/ANM$. But $ST/SZ = TP/ZY$, since $\triangle\ STP$ and $\triangle\ SYZ$ are similar. Therefore, $ANM = ZY$, since TP has the same ratio to [each of] them. Therefore, let $\triangle\ SYZ = L$, by the first [proposition] of this [work]. It is evident, therefore, that $(R - L) =$ surface $ZYPT$. Then let line ZP be drawn from Z to P and YT from Y to T.

[Further] argument. ZYT and YPZ are triangles [constructed] on base ZY and between parallel lines ZY and TP. Therefore the triangles are equal. Therefore $(ZT \cdot ZY) = 2 \cdot (\text{each triangle}) = 2\ \triangle\ ZPY$. Therefore $(TZ \cdot 1/2\ ZY) = \triangle\ ZPY$. Similarly $(TZ \cdot TP) = 2\ \triangle\ TPZ$. Therefore $(ZT \cdot \frac{1}{2}\ TP) = \triangle\ TPZ$. And thus

$\text{surf } ZTPY = (R - L) = (CE - IM) \cdot (\frac{1}{2} \text{ circum } EQB + \frac{1}{2} \text{ circum } ANM)$,

since $ZTPY = ZT \cdot (\frac{1}{2}ZY + \frac{1}{2}TP)$, and $(CE - IM) = ZT$. This is what we proposed.

V. IF HALF OF A REGULAR POLYGON—TERMINATED AT THE ENDS OF A DIAMETER AND DESCRIBED IN A CIRCLE—IS ROTATED WHILE THE DIAMETER REMAINS FIXED, THE

RUNT CONICE SUPERFICIES TOTIUS CORPORIS EQUALES
5 EI QUOD FIT EX DUCTU LATERIS CIRCUMDUCTI IN OMNES CIRCUMFERENTIAS DESCRIPTAS AB ANGULIS POLIGONII, SIVE EI QUOD FIT EX DUCTU CIRCUMFERENTIE CIRCULI CONTINGENTIS POLIGONIUM IN LINEAM QUE CUM DIAMETRO EIUSDEM CIRCULI ET LATERE POLIGO-
10 NII IN EODEM CIRCULO CONSTITUIT TRIANGULUM ORTHOGONIUM.

Esto exemplum circulus *AHID* et poligonium equilaterum ei inscriptum [Fig. 67].

Dispostio. A singulis angulis poligonii ad oppositos ducantur linee,
15 et sint *NM*, *FG*, *HD*, *LE*, *BC*, secantes diametrum *AI* perpendiculariter. Ne tamen minus diligens lector scrupulum, quo progrediens pedem offendat, possit reperire, sic probetur illas perpendiculariter secare *AI*.

Ratio. Age. *ADC*, *AHB* arcus sunt equales. Ergo *CIY*, *BIY* anguli
20 cadentes in illos sunt equales. Item *YBI*, *YCI* in *BI*, *CI* arcus equales sunt equales. Ergo cum in *BIY* triangulo, *B*, *I*, *Y* anguli sunt equales duobus rectis, similiter et in triangulo *YIC*, *Y*, *I*, *C* anguli sunt equales duobus rectis, gemini *Y* sunt equales, et sic uterque rectus. Ergo *BC* est perpendicularis ad *AI*. Simili ratione *NM* perpendicularis ad *AI*.

12 ei: illi *A*
14 angulis *om.* *A*
15 NM: MN *A*
19 AHB *om.* *C*

CONICAL SURFACES OF THE WHOLE BODY [FORMED] WILL BE EQUAL [IN SUM] TO THAT WHICH ARISES FROM THE PRODUCT OF A SIDE OF THE ROTATED [POLYGON] AND THE SUM OF THE CIRCUMFERENCES DESCRIBED BY THE ANGLES OF THE POLYGON, OR [IT IS EQUAL] TO THAT WHICH ARISES FROM THE PRODUCT OF THE CIRCUMFERENCE OF THE CIRCLE TOUCHING THE POLYGON AND THE LINE WHICH FORMS A RIGHT TRIANGLE ALONG WITH THE DIAMETER OF THE SAME CIRCLE AND A SIDE OF THE POLYGON IN THE SAME CIRCLE.

For example, let there be circle *AHID* and a regular polygon inscribed in it [see Fig. 67].

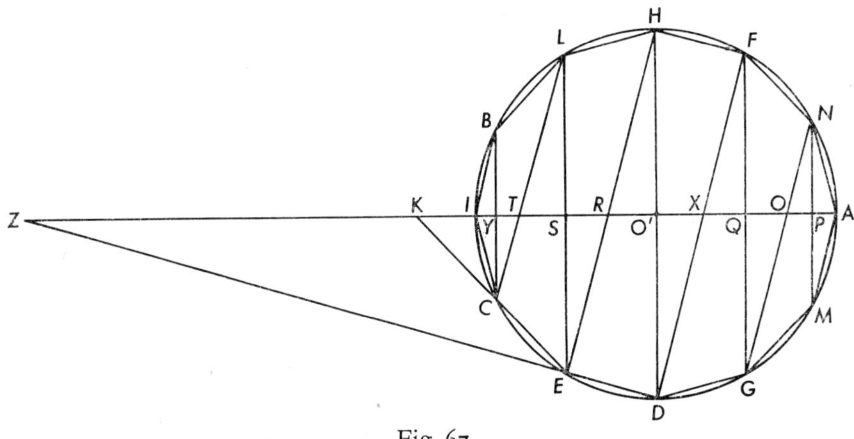

Fig. 67

Disposition: From the individual angles of the polygon let lines be drawn to the opposite angles, namely, lines *NM*, *FG*, *HD*, *LE*, and *BC*, all these lines cutting diameter *AI* perpendicularly. Lest, however, the less attentive reader be able to find an obstacle against which he strikes his foot as he goes forth, let it be proved as follows that these lines cut *AI* perpendicularly.

Proceed with the proof: Arcs *ADC* and *AHB* are equal. Therefore, angles *CIY* and *BIY* falling in those arcs are equal. Also, [angles] *YBI* and *YCI* in equal arcs *BI* and *CI* are equal. Therefore, since angles *B*, *I*, *Y* in △ *BIY* are equal to two right angles and similarly angles *Y*, *I*, *C* in △ *YIC* are equal to two right angles, the pair of angles at *Y* are equal, and thus each is equal to a right angle. Therefore, *BC* is perpendicular to *AI*. By a similar argument *NM* is perpendicular to *AI*. Also, *LC* intersects

25 Item *LC* secat *LE*, *BC*, et facit *L*, *C* angulos cadentes in *LB*, *CE* equos arcus equales et coalternos. Ergo *LE*, *BC* sunt equidistantes, et eadem ratione, *LE*, *HD*, et *FG*, et *NM*.

Protrahuntur deinde oblique secantes diametrum *AI* linee *NG*, *FD*, *HE*, *LC*. Protrahatur deinde *AI* in occursum *EC* et *ED* extra 30 circulum, singulis angulis et punctis rationi necessariis per equalia notatis. Erunt ergo *NM*, *FG*, *HD*, *LE*, *BC* diametri circulorum quos describunt anguli semipoligonii *ADI* circumducti. Dico itaque quod conica superficies corporis poligonii fit ex ductu unius lateris poligonii, scilicet *DE*, in omnes circumferentias [circulorum] quorum 35 diametri sunt *MN*, *FG*, *HD*, *LE*, *BC*; et etiam est equalis ei quod fit ex *HE* in *ADIH* circumferentiam.

Rationis causa, superficiem piramidis *YIC* describit ypothenusa *IC*. Et est axis illius piramidis *IY*, diameter basis linea *BC*. Ergo per primam huius superficies curva piramidis *YIC*, quam describit *IC* cir- 113r cumvoluta, est superficies que fit / ex ductu *IC* in medietatem circum- 41 ferentie *BC* diametri. Item in *SEK*, *YCK* triangulis, et *Y* et *S* est rectus et *K* communis. Ergo *SEK*, *YCK* trianguli sunt similes. Item superficies curva quam continent *LB*, *CE* et circumferentie diametrorum *LE*, *BC*, scilicet quam describit linea *CE* circumvoluta, est differentia 45 duarum piramidum inequalium et similium, quarum minoris axis est *YK*, ypothenusa *KC*, diameter basis *BC*; maioris piramidis axis *KS*, ypothenusa *KE*, diameter basis *LE*. Ergo per proximam illa superficies, quam describit *CE* circumvoluta, fit ex ductu *CE* in circumferentie *BC* medietatem et medietatem circumferentie *LE*. Ergo curva 50 superficies piramidis *YIC*, quam describit *IC*, et curva superficies quam describit *CE*, pariter sumpte, fiunt ex ductu *CE* in circumferentiam *BC* et medietatem circumferentie *LE*, quoniam *IC*, *CE* latera poligonii sunt equalia. Item curva superficies quam describit *DE* circumvoluta est differentia piramidum similium et inequalium *O'DZ*, 55 *SEZ*. Ergo illa superficies fit ex ductu *ED* in medietatem circumferen-

25 LE *AC* IC (*?*) *B* / LB, CE: LC LB *A*
26 equos...equales: arcus equales equos *A*
29 Protrahantur *B*
31 ergo: igitur *A*
32 itaque *om. A*
35 ei: illi *A*
36 ADIH: AHI *A*
39 superficies curva *tr. A*
41 triangulis *A om. BC*
42 K: K est *A*
42–43 superficies curva *tr. A*
43 circumferentie diametrorum *A* circumferentia diametrum *B* circumferentia diametrorum *C*
46 piramidis *A om. BC*
47 LE: LEI *C*, *sed supra scr. C* LE
48–51 circumvoluta...CE[1] *AB mg. C*
52 LE *om. A* / *post* LE *del. C* et medietatem circumferentie DH
55 ED: DE *C*

LE and *BC* and makes equal alternate angles *L* and *C* falling in equal arcs *LB* and *CE*. Therefore, *LE* and *BC* are parallel, and by the same argument *LE*, *HD*, *FG*, and *NM* are parallel.

Then let the oblique lines *NG*, *FD*, *HE*, and *LC*, all cutting the diameter *AI*, be drawn. Then let *AI* be extended to meet *EC* and *ED* outside of the circle, with the separate angles and points necessary for the argument marked off by equal lines. Therefore, *NM*, *FG*, *HD*, *LE* and *BC* will be the diameters of the circles which the angles of the half polygon *ADI* describe in rotation. And so I say that the conical surface of the polygonal body arises from the product of one side of the polygon, namely, *DE*, and the sum of the circumferences [of the circles] whose diameters are *MN*, *FG*, *HD*, *LE*, and *BC*; and also that it is equal to the product of *HE* and circumference *ADIH*.

For the sake of the argument, [let us say that] hypotenuse *IC* describes the surface of cone *YIC*. And the axis of this cone is *IY*; the diameter of its base is line *BC*. Therefore, by the first [proposition] of this [work], the lateral surface of cone *YIC*, which *IC* describes in rotation, is the surface arising from the product of *IC* and one half the circumference of which *BC* is the diameter. Also in triangles *SEK*, and *YCK*, *Y* and *S* are right angles and $\angle K$ is common. Therefore, triangles *SEK* and *YCK* are similar. Further, the lateral surface contained by *LB*, *CE* and the circumferences of diameters *LE* and *BC*, i.e., the lateral surface which line *CE* describes in its revolution, is [equal to] the difference between the two unequal but similar cones, the smaller of which is that one with axis *YK*, slant height *KC*, and base diameter *BC*, and the larger of which is that one with axis *KS*, slant height *KE*, and base diameter *LE*. Therefore, by the preceding [proposition], that surface which *CE* describes in its revolution is equal to $CE \cdot (1/2 \text{ circum } BC + 1/2 \text{ circum } LE)$. Therefore, the lateral surface *YIC* described by *IC* and the lateral surface described by *CE* taken together are equal to the product of (1) *CE* and (2) the sum of circumference *BC* and one half the circumference *LE*, since *IC* and *CE*, as sides of the [regular] polygon, are equal. Further, the lateral surface described by *DE* in revolution is [equal to] the difference of the similar and unequal cones *O′DZ* and *SEZ*. Therefore, that surface is equal to the product of (1) *ED* and (2) the sum of 1/2 circumference

tie *LE* et medietatem circumferentie *DH*. Et sic superficies curva quam describit *IC* et quam describit *CE* et quam describit *DE* sunt equales ei quod fit ex ductu *DE* in circumferentias *BC* et *LE* et medietatem circumferentie *HD*. Simili ratione invenies curvam super-
60 ficiem quam describit *AM* et quam describit *MG* et quam describit *GD*, scilicet medietatem totius conice superficiei poligonii corporis, esse id quod fit ex *GD*, sive *DE* illi equali, in circumferentiam *NM* et *FG* et medietatem circumferentie *HD*. Ergo si sufficienter enumeres, reperies totam conicam superficiem fieri ex uno latere poligonii,
65 scilicet *DE* in omnes circumferentias quas describunt anguli poligonii, quarum diametri sunt *NM*, *FG*, *HD*, *LE*, et *BC*. Et sic prior pars constat propositi.

Rusus, *Y* undique est rectus, et *C*, *B* anguli cadentes in *IB*, *IC* equos arcus sunt equales. Ergo et *T*, *I* anguli sunt equales. Ergo *BYI*,
70 *TYC* sunt trianguli similes. Simili quoque ratione *TYC*, *TSL*, *SRE*, *RO′H*, *O′XD*, *XQF*, *QOG*, *OPN*, *PAM* sunt similes. Ergo que est proportio *IY* ad *BY* eadem est *TY*, *YC*; et *ST*, *SL*; et *RS*, *SE*; et *RO′*, *O′H*; et *XO′*, *O′D*; et *XQ*, *QF*; et *OQ*, *QG*; et *OP*, *PN*; et *AP*, *PM*. Ergo que est proportio *IY* ad *BY* eadem est totius *AI* ad
75 omnes rectas lineas, *BC*, *LE*, *HD*, *FG*, *NM*. Sed que est *AI* ad *NM* et *FG* et *HD* et *LE* et *BC*, eadem est *ADIH* circumferentie ad omnes circumferentias quas describunt anguli poligonii, quarum diametri sunt *NM*, *FG*, *HD*, *LE*, et *BC*, per tertiam huius et penultimam quinti. Ergo que est proportio *IY* ad *BY* eadem est *ADIH* ad
80 quinque circumferentias diametrorum *NM*, *FG*, *HD*, *LE*, *BC* [,quas describunt anguli poligonii. Sed *HED* angulus est rectus cadens in semicirculum; ergo est equalis *Y*. Et *H*, *B* sunt equales, ut qui cadunt in *DE*, *IC* arcus equales. Et sic *BYI*, *DEH* trianguli sunt similes. Ergo que est proportio *IY* ad *BY* eadem est *DE* ad *EH*. Ergo que
85 est *DE* ad *EH* eadem est *ADIH* circumferentie ad quinque circumferentias]. Ergo quod fit ex ductu primi in ultimum, scilicet *ED* in

56 *post* LE *del. C* et medietatem circumferentie LE
62 NM: AM *AC*
64 reperies: invenies *A*
65 scilicet...poligonii *om. A*
67 constat *tr. A ante* prior *in linea 66*
68 IB: LB *C*
69 equos arcus *tr. A*
69 anguli *A om. BC*
70 TYC: TIC *A* / trianguli *A om. BC* / quoque: que *A* / TSL: TIL *A*
71 *RO′H*: HRO *A* / OPN: ONP *A*
72 IY: TY *A*
73 O′H: CH *A* / OQ: CQ *A*
78 et[1] *om. A*
79 IY: QY *C* / ADIH: totius ADIH *A*
80–86 [quas....cricumferentias] *supplevi, sed cf. mg. B ubi scr. post* IY ad BY *in linea 79* DE ad EH cum trianguli DEH, BIY sunt similes

LE and 1/2 circumference DH. And thus the lateral surfaces described by IC, CE, and DE are [together] equal to the product of (1) DE and (2) the sum of circumference BC, circumference LE, and one half the circumference HD. By a similar argument you will find that the lateral surfaces described by AM, MG, and CD, i.e., 1/2 the whole conical surface of the polygonal body, is equal to the product of (1) GD, or its equal DE, and (2) the sum of circumference NM, circumference FG, and 1/2 circumference HD. Therefore, if you make the complete enumeration, you will find that the whole conical surface arises from the product of (1) one side of the polygon and (2) the sum of the circumferences described by the angles of the polygon, i.e., it is equal to $DE \cdot$ (circum of NM + circum of FG + circum of HD + circum of LE + circum of BC). And so the first part of the proposition is evident.

Again, every angle at Y is a right angle and angles C and B falling in equal arcs IB and IC are equal. Therefore, angles T and I are equal. Therefore, $\triangle BYI$ is similar to $\triangle TYC$. By a similar argument, triangles TYC, TSL, SRE, $RO'H$, $O'XD$, XQF, QOG, OPN, and PAM are similar. Therefore, $IY/BY = TY/YC = ST/SL = RS/SE = RO'/O'H = XO'/O'D = XQ/QF = OQ/QG = OP/PN = AP/PM$. Therefore, $IY/BY = AI/(BC + LE + HD + FG + NM)$ [by Euclid V.1]. But $AI/(NM + FG + HD + LE + BC)$ = circum $ADIH$/(circum of NM + circum of FG + circum of HD + circum of LE + circum of BC), by [Proposition] III of this [work] and the penultimate [proposition] of [Book] V [of the *Elements*]. Hence IY/BY = circum $ADIH$/(circum of NM + circum of FG + circum of HD + circum of LE + circum of BC). [But $\angle HED$ is a right angle, falling as it does in a semicircle. Hence $\angle HED = \angle Y$. And $\angle H = \angle B$, as they fall in equal arcs DE and IC. Thus triangles BYI and DEH are similar. Hence $IY/BY = DE/EH$. Hence DE/EH = circum $ADIH$/the 5 circumferences.*] Hence ($ED \cdot$ the 5 circumferences) = ($EH \cdot$ circum $ADIH$). But ($ED \cdot$ the 5 circumferen-

* That is, (circum of NM + circum of FG + circum of HD + circum of LE + circum of BC).

omnes circumferentias, est equale ei quod fit ex *EH* in *ADIH*. Sed ex *DE* in quinque circumferentias fit tota conica superficies corporis poligonii, ut preostensum est. Ergo tota superficies corporis poligonii
90 equatur ei quod fit ex *EH* in *ADIH* circumferentiam. Et sic utraque propositi pars plene constat

Eodem modo poterit probari propositum si fuerint latera semipoligonii numero imparia. Descripta tamen superficies a medio latere columpnalis erit. Esto enim figura X equalium angulorum inscripta
95 circulo *ADIH* et equalium laterum [Fig. 68]. Ductis lineis a singulis angulis ad oppositos angulos, sicut prius, perpendiculariter ad *AI*, et [duce] alias oblique secantes *AI*, sicut in dispositione proxima, deinde ab *O* centro ducatur *OD* perpendicularis ad *GE*, secans *GE* in duo equa, et *OH* perpendicularis ad *FL*, secans *FL* in duo equa.
100 Deinde ab *O* in *F* et in *E* ducantur due linee *OF*, *OE*.

Ratio. Age. *HF*, *GD* arcus sunt equales, quia sunt medietates *GE*, *FL* equalium. Et *FA* et *GA* sunt equales. Ergo *AH*, *AD* sunt equales arcus. Sed utraque est quarta totius circumferentie. Ergo *O* gemini centrales cadentes in illos sunt recti. Ergo *HOD* est una recta
105 linea. Similiter cum *FAE* sit semicirculus, erit *FOE* una linea recta. Item cum *HD* sit perpendicularis ad *GE* et *FL* ex dispositione, erunt *GE*, *FL* equidistantes. Et sunt equales ex ypothesi. Ergo *FGEL* est parallelogramum rectangulum. Cum itaque *FG* et *LE* sunt equales et equidistantes, similiter *FL* et *GE*, et stanti *M'P*, circumvolvatur
110 superficies *FE*, et fiet columpna, cuius basis erit circulus quem describit *LE*; axis vero *M'P*. Fit autem curva superficies illius columpne

87 ei: illi *A* / ADIH: ADIH circumferentiam *A* / Set *A*
90 EH: HE *A*
91 propositi...constat: pars constat propositi *A*
92 Eodem: eodem autem *A*
94 columpnalis erit *tr.* *A*
98–99 ad...perpendicularis *om.* *A*
99 secans *om.* *B*
101 arcus...equales: sunt equales arcus *A*
102 equales: equalia *A*
104 cadentes...recti: sunt equales cadentes in illos arcus quare recta *A*
105 recta *A* *om.* *BC*
106 GE: G *A*
107 Et *A* Sed et *BC*
108 et *A* *om.* *BC*
109 et[3] *om.* *A*
110 erit: est *A*
111 curva superficies: illa superficies curva *A*

ces) is equal to the whole conical surface of the polygonal body, as was just demonstrated. Therefore, the whole surface of the polygonal body is equal to the product of EH and circumference $ADIH$. And thus each part of the proposition is fully evident.

By the same method the proposition could be proved if the number of the sides of the half polygon was an odd number. However, the surface described by the middle side will be cylindrical. For let there be described in circle $ADIH$ a regular polygon of ten sides [see Fig. 68]. Having drawn lines connecting each angle with the angle opposite to it, as before, these lines being perpendicular to AI, [draw] other [lines] cutting AI obliquely, as in the previous construction. Then from center O let

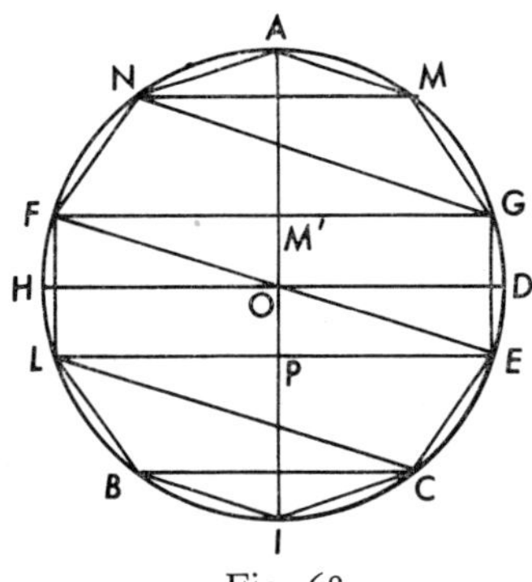

Fig. 68
Note: I have added the prime sign to M'.

OD be drawn perpendicular to GE, bisecting GE, and OH perpendicular to FL, bisecting FL. Then from O let there be drawn two lines OF and OE to F and E [respectively].

Proceed with the argument: HF and GD are equal arcs because they are halves of equal arcs GE and FL, and FA and GA are equal. Therefore AH and AD are equal arcs. But each is a quadrant of the whole circumference. Therefore, both the central angles O falling in those arc are right angles. Therefore, HOD is a straight line. Similarly, since FAE is a semicircle, FOE will be a straight line. Also, since HD is perpendicular to GE and to FL by construction, GE and FL are parallel, and they are equal by hypothesis. Therefore $FGEL$ is a rectangle. And so, since FG is equal and parallel to LE, and similarly FL to GE, with $M'P$ fixed, surface FE is rotated and a cylinder is produced whose base circle will be the circle described by LE and whose axis is $M'P$. Moreover, the lateral surface of this cylinder is equal to the product of (1) the axis

ex ductu axis *M'P*, sive *FL*, in circumferentiam diametri *LE*, per secundam huius. Ratiocinando ergo ut superius, proba quod conica superficies poligonii corporis fit ex *FL* in circumferentias quatuor
115 diametrorum, *NM*, *FG*, *LE*, *BC*. Similiter etiam ut supra probetur conicas illas superficies fieri ex ductu *FG* in *ADIH* circumferentiam. Et sic omni modo constat propositum.

113v /VI. CUIUSLIBET SPERE SUPERFICIES EST EQUALIS QUADRANGULO RECTANGULO QUI SUB LINEIS EQUALIBUS DIAMETRO SPERE ET CIRCUMFERENTIE MAXIMI CIRCULI CONTINETUR.

5 Esto exemplum *ACBD* circumferentia, diameter cuius sit *AB* [Fig. 69]. Stanti ergo *AB*, circumducatur *ACB*, et fiat spera. Dico quod id quod fit ex ductu diametri in *ACBD* circumferentiam est equale superficiei spere *ACBD*.

Sin autem, sit equalis superficiei minoris spere vel maioris; et primo
10 minoris spere, scilicet spere quam describit *SNH* semicirculus circumvolutus circa *O* centrum, constituitur.

Dispositio. Inscribatur circulo *ACBD* poligonium equalium laterum et angulorum, circulum *SNH* minime contingentium. Deinde ab *O* centro ducatur *OI* perpendicularis ad *AF* et secans *AF* in duo
15 equa et alia a *B* in *F*.

Ratio. Age. *AB* diameter est maior *BF*. Ergo id quod fit ex *AB*

114 corporis *om.* *A*
114–15 quatuor diametrorum *tr.* *A*
115 etiam: et *A*
116 ADIH / AHDI *ABC*
117 omni modo: omnino *A*
1 VI *B* *om.* *C* VI[a] / 6[a] *AD* / est equalis *tr.* *A*
2 rectangulo *om.* *A*
5 Esto *A* *om.* *BC* / cuius sit: eius *A*
7 ACBD *A* / ABCD *BC*
8 spere *om.* *A*
13 SNH minime *B* SH non *A* SH minime *C*

M'P, or [its equal] *FL*, and (2) the circumference of diameter *LE*, by the second [proposition] of this [work]. Therefore, by reasoning as above, prove that the conical surface of the polygonal body is equal to the product of (1) *FL* and (2) the circumferences of the four diameters *NM*, *FG*, *LE*, and *BC*. Also let it be proved in the same way as above that those conical surfaces [making up the surface of the whole polygonal body] are equal to the product of *FG* and circumference *ADIH*. And so this proposition is evident in every way.

VI. THE SURFACE OF ANY SPHERE IS EQUAL TO THE RECTANGLE CONTAINED BY LINES EQUAL TO THE DIAMETER OF THE SPHERE AND THE CIRCUMFERENCE OF THE GREATEST CIRCLE [OF THE SPHERE].

For example, let *ACBD* be a circumference whose diameter is *AB* [see Fig. 69]. With *AB* fixed, let *ACB* be rotated, producing a sphere.

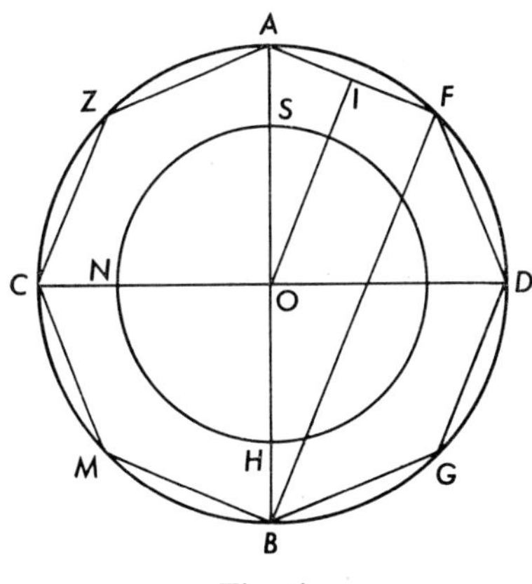

Fig. 69

I say that the product of (1) the diameter and (2) circumference *ACBD* is equal to the surface of sphere *ACBD*.

But if not, let it be equal to the surface of sphere that is less or greater than [*ACBD*]. In the first place let it be [equal to the surface] of a lesser sphere, i.e., of the sphere which semicircle *SNH* describes in rotation about center *O*.

Disposition: Let there be inscribed in circle *ACBD* a regular polygon whose sides do not touch circle *SNH* at all. Then let *OI* be drawn from center *O* perpendicular to *AF* and bisecting *AF*. Let another line be drawn from *B* to *F*.

Proceed with the proof: Diameter *AB* is greater than *BF*. Therefore,

in *ACBD* circumferentiam maius est quam id quod fit ex *BF* in *ACBD*. Sed id quod fit ex *BF* in *ACBD* est equale conicis superficiebus corporis poligonii per proximam. Et conice superficies corporis poligonii sunt maiores superficie spere *SH* interioris. Ergo id quod fit ex ductu *AB* in *ACBD* circumferentiam maius est superficie spere *SH*; contra preconcessa. Relinquitur ergo id quod fit ex ductu *AB* in *ACBD* non est equale superficiei minoris spere quam *ACBD*.

Sit itaque si fieri potest equale superficiei spere maioris. Brevitatis tamen causa, prior observetur figura, et sit *SH* spera proposita et diameter *SH*. Dico quod fit ex *HS* diametro in *HS* circumferentiam non est equale superficiei maioris spere quam *SH*. Sin autem, sit equalis superficiei spere *ACBD*.

Ratio. Age. *F* cadens in semicirculum *ADB* est rectus. Similiter et *I* rectus ex dispostitione; et *A* communis. Ergo *AIO*, *AFB* trianguli sunt similes. Ergo que est proportio *AF* ad *AI* eadem est *BF* ad *OI*. Sed *AF* est dupla ad *AI*. Ergo *BF* est dupla *OI*. Sed *OI* est maior quam *OS*, que subdupla est ad *HS* diametrum. Ergo *BF* est maior *HS*; et *ACBD* circumferentia maior *SNH* circumferentia. Ergo id quod fit ex *BF* in *ACBD* est maius eo quod fit ex *HS* in *SNH* circumferentiam. Sed quod fit ex *BF* in *ACBD* est equale conicis superficiebus corporis poligonii per proximam. Et quod fit ex *HS* diametro in *SNH* circumferentiam est equale superficiei spere *ACBD* secundum falsigraphum. Ergo conice superficies corporis poligonii sunt maiores superficie *ACBD* spere, inclusum includente, quod est impossibile. Relinquitur propositum.

[Corollaria:] Ex hoc ergo manifestum quoniam superficies spere est quadrupla maximo circulo eiusdem spere, et equalis curve superficiei columpne, cuius tam axis quam diameter basis equatur diametro

18 BF: FB *C* | ACBD[2]: ADBC *A* | equales *A*
19–20 corporis poligonii *tr. A*
20 spere SH *tr. A*
21–23 circumferentiam...ACBD[1] *AB mg. C*
22 ergo: ergo quod *A*
23 ACBD[1]: ABCD *C* | minoris spere *tr. A* | ACBD[2]: ABCD *C*
24 itaque: ergo *A*
24 equale...spere *A om. BC*
25 prior: prius posita *A* | sit *A om. BC*
26 HS diametro: ductu HS diametri *A*
29 ADB: ABD *A*
29–30 Similiter et I: Et similieter I est *A*
30 *post* A *add. A* est | AIO, AFB: AIC AFD *A*
32 Ergo...OI[1] *om. A*
33 que: quia *A*
35 ex[1]: ex ductu *A* | ACBD: ACBD circumferentiam *A*
36 BF: ductu *A* | ACBD: ACBD circumferentiam *A*
38 HS: HH *B*
41 propositum: ergo propositum *A*
42 ergo *om. A* | manifestum: manifestum est *A*

$(AB \cdot \text{circum } ACBD) > (BF \cdot \text{circum } ACBD)$. But $(BF \cdot \text{circum } ACBD)$ is equal to the conical surfaces of the polygonal body, by the preceding [proposition]. Further, the conical surfaces of the polygonal body are greater than the surface of the interior sphere SH. Therefore, $(AB \cdot \text{circum } ACBD) >$ surface of sphere SH, which is contradictory to what was conceded earlier. It remains, therefore, that the product of AB and $ACBD$ is not equal to the surface of a sphere less than $ACBD$.

And so, if possible, let it be made equal to the surface of a sphere greater [than sphere $ACBD$]. For the sake of brevity, however, let the prior figure [Fig. 69] be consulted, and let sphere SH be the proposed sphere with diameter SH. I say that the product of diameter HS and the circumference of HS is not equal to the surface of a sphere greater than SH. But if this is not [equal to the surface of sphere SH], let it be equal to the surface of sphere $ACBD$.

Proceed with the proof: F, falling in semicircle ADB, is a right angle. And similarly I is a right angle by construction, and $\angle A$ is common. Therefore, triangles AIO and AFB are similar. Therefore, $AF/AI = BF/OI$. But $AF = 2\ AI$. Therefore, $BF = 2\ OI$. But OI is greater than OS, which is one half diameter HS. Therefore, BF is greater than HS, and circumference $ACBD$ is greater than circumference SNH. Therefore, $(BF \cdot \text{circum } ACBD) > (HS \cdot \text{circum } SNH)$. But $(BF \cdot \text{circum } ACBD)$ is equal to the conical surfaces of the polygonal body, by the preceding [proposition]. And the product of diameter HS and circumference SNH is equal to the surface of sphere $ACBD$, according to the pseudographer. Therefore, the conical surfaces of the polygonal body are greater than the surface of sphere $ACBD$, i.e., the "included" [is greater] than the "including." which is impossible. [Hence] the proposition remains [as the only true possibility].

[Corollaries:] From this, therefore, it is evident that the surface of a sphere is quadruple [the area of] the greatest circle of the same sphere, and [it is further evident that] it is equal to the lateral surface of a cylinder

45 spere. Tota quoque huius columpne superficies superficiei spere sexquialtera est.

Ut enim in quadratura circuli ostensum est quod fit ex ductu quarte partis diametri in circumferentiam est equale circulo. Ergo quod fit ex diametro in circumferentiam, scilicet superficies spere, est quad-
50 rupla ad circulum. Item ex ductu diametri in circumferentiam maximi circuli fit superficies spere, et ex ductu axis in circumferentiam basis fit curva superficies columpne. Si ergo et axis et diameter basis columpne sit equalis diametro spere, erit superficies spere superficiei illius columpne equalis. Ex hoc liquet quod curva superficies columpne cum
55 duobus circulis qui sunt extremitates columpne est sexquialtera ad superficiem spere. Et hoc erat probandum.

Liber secundus

[Descriptio:] Omne corpus erectum super basim supereminenti superficiei equalem et similem et equidistantem sub basi et altitudine dicitur contineri.

5 VII. OMNE SOLIDUM CONICARUM SUPERFICIERUM INSCRIPTIBILE ET CIRCUMSCRIPTIBILE SPERE EQUUM EST PIRAMIDI CUIUS BASIS SIT EQUALIS SUPERFICIEI SOLIDI ET ALTITUDO SEMIDIAMETRO SPERE INSCRIPTE SOLIDO.

Esto solidum conicarum superficierum *CAQB* inscriptibile et cir-
10 cumscriptibile spere [Fig. 70]. Dico quod corpus *ACBQ* est equale piramidi cuius basis est equalis conicis superficiebus *ACBQ* corporis propositi et altitudo equalis semidiametro spere inscripte solido *ACBQ*

Rationis causa. Protrahantur a centro *O* ad tria latera superficiei poligonie, que circumvoluta facit solidum, tres perpendiculares. Et
114r sint *OI*, *OT*, *OZ*. / Protrahantur item ad quatuor angulos *C*, *F*, *N*,

46 est *om. C*
48 *ante* diametri *del. B* circumferentie / diametri: circumferentie diametri *C*
49 diametro: ductu diametri *A*
52 et[1] *om. B*
54 equalis *tr. A post* spere / equale *C*
56 erat probandum *A* est *BC*, *sed post* est *add. C manu recentiori* quod ostendere voluimus / *in mg. habet B manu recentiori* Explicit primus liber
1 Liber secundus *mg. B mg. A* secundus liber Archimenidis de curvis superficiebus *mg. B manu recentiori* secundus *mg. J* (*f. 93r*)
2 [Descriptio] *mg. H om. alii MSS* diffinitio *mg. D*
5 VII: 7[a] *AD* I *B* I[a] *J* / solidum *BC supra scr. A* corpus *in textu A*
10 ACBQ *corr. ex* ACQB *in BCA*
11 ACBQ *C* ACQB *BA*
13 Protrahatur *C*
14 poligonii *A*
15 OT, OZ: OZ, OT *C*

whose axis and base diameter are each equal to the diameter of the sphere. Further, the whole surface of this cylinder is three halves the surface of the sphere.

For in the *Quadrature of the Circle* it was demonstrated that the product of (1) one quarter of the diameter and (2) the circumference is equal to the circle. Therefore, that which arises from the product of the diameter and the circumference—that is, the surface of the sphere—is quadruple the circle. Also [if] from the product of the diameter and the circumference of the greatest circle arises the surface of a sphere, and if from the product of the axis and the base circumference the lateral surface of a cylinder is produced, then, when the axis and the base diameter of the cylinder are each equal to the diameter of the sphere, the surface of the sphere will be equal to the [lateral] surface of the cylinder. From this it is evident that the lateral surface of the cylinder together with the two circles which are the bases of the cylinder is three halves the surface of the sphere. And this is what was to be proved.

Book II

[Description:] Every body erected on a base which is parallel, equal, and similar to a surface rising directly above [the base] is said to be contained by the base and the altitude.

VII. EVERY SOLID CONSISTING OF CONICAL SURFACES AND INSCRIBABLE OR CIRCUMSCRIBABLE IN A SPHERE IS EQUAL TO A CONE WHOSE BASE IS EQUAL TO THE SURFACE OF THE SOLID AND WHOSE ALTITUDE [IS EQUAL] TO THE RADIUS OF THE SPHERE INSCRIBED IN THE SOLID.

Let there be a solid *CAQB* (*ACBQ*) having conical surfaces, which solid is inscribable and circumscribable in a sphere [see Fig. 70]. I say that body *ACBQ* is equal to a cone whose base is equal to the conical surfaces of the proposed body *ACBQ* and whose altitude is equal to the radius of the sphere inscribed in solid *ACBQ*.

[Construction] needed for the proof: Let there be drawn three perpendiculars from center *O* to three of the sides of the polygonal surface whose rotation makes the solid. Let these perpendiculars be *OI*, *OT*, and *OZ*.

16 *A* a centro quatuor linee *OC*, *OF*, *ON*, *OA*. Protrahantur itaque *AN*, *NF* in occursus linee *OC*. Ab *F* item ducatur perpendicularis *FP* ad *OC*, et ab *N* perpendicularis *NV* ad *OC*.

Ratio. Age. Triangulus *OFC*, stanti *OC*, circumductus facit du-
20 plicem piramidem *OFC*, equalem piramidibus *OPF*, *PFC*. Sed columpna que fit ex ductu *PC* in circulum quem describit *FP* circumvoluta est tripla ad piramidem *PFC* per IX duodecimi Elementorum. Et columpna que fit ex ductu *PO* axis in circulum quem describit *FP* est tripla ad piramidem *FPO* per eandem. Quod ex ductu axis in
25 basim fiat columpna sumat demonstrator pro rato per precendentem descriptionem. Quod ergo fit ex ductu totius *OC* in eundem circulum triplum est duplicis piramidis *OFC* per primam secundi. Sed tam *I* quam *P* est rectus et *C* communis. Ergo *OIC*, *CPF* trianguli sunt similes. Ergo que est porportio linee *OC* ad *OI* eadem est *CF* ad *FP*.
30 Et que est proportio *CF* ad *FP* eadem est curve superficiei piramidis *FCP*, quam describit linea *FC*, ad circulum quem describit *FP* per corollarium prime huius. Ergo quod fit ex ductu *OC* linee in circulum descriptum ab *FP*, quod, scilicet, est aliqua columpna per descriptionem precedentem, est equale ei quod fit ex *OI* linea in curvam super-
35 ficiem piramidis *FPC*, quam describit linea *FC*. Sed quod fit ex ductu *OC* in circulum quem describit *FP* circumvoluta est columpna tripla duplici piramidi *OPF*, *PFC*, per IX duodecimi Elementorum et

16 itaque: que *C*
17 *ante* NF *del.* *C* AF
17, 21, 23 FP: SP *A*
21 circumvoluta *om.* *B*
22 IX: XI *A*
23 PO: PC *A*
24 FPO: SPC *A*
25 sumat *B* sumit *A* summit *C*
28 CPF: PFC *A*
29, 30 CF ad FP: OF ad SP *A*
30 Et: Sed *A* / proportio *om.* *A*
31 FCP: FOP *A* / FP: SP *A*
32 corellarium *B*
33 FP: SP *A*
34 ei: illi *A* / ex: ex ductu *A*
33–34 descriptionem precedentem *tr.* *A*
34 linee *A*
36 FP: SP *A*
37 IX: XI *A*

Also let the four lines OC, OF, ON, and OA be drawn from the center to the four angles C, F, N, and A. And then let AN and NF be extended until they meet line OC [extended]. Also, from F let FP be drawn perpendicular to OC, and from N, NV perpendicular to OC.

Proceed with the proof: With OC fixed, $\triangle\, OFC$ by rotation produces the double cone OFC, equal to cones OPF and PFC. But the cylinder which arises from the product of PC and the circle described by FP in rotation is three times the cone PFC, by [Proposition] XII.9* of the *Elements* And the cone which arises from the product of axis PO and the circle described by FP is three times the cone FPO, by the same [proposition]. The demonstrator assumes that the cylinder arises from the product of

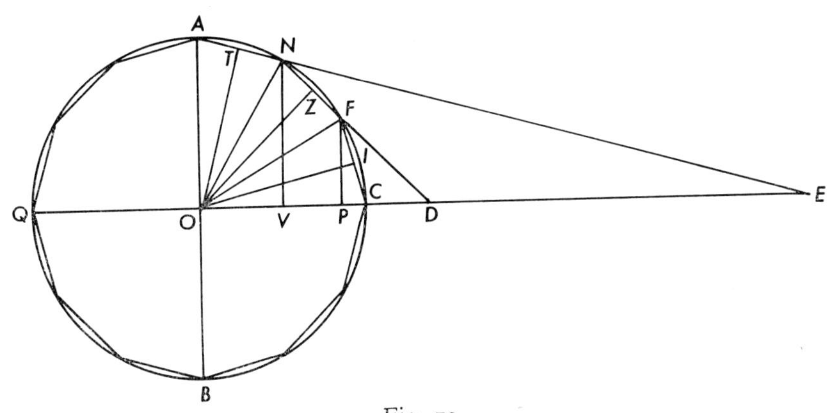

Fig. 70

the axis and the base, as is confirmed by the preceding description [given just prior to Proposition VII]. Therefore, the product of the whole OC and the same circle is three times the double cone OFC, by [Proposition] II.1 [of the *Elements*]. But both $\angle\, I$ and $\angle\, P$ are right angles, and $\angle\, C$ is common. Therefore, triangles OIC and CPF are similar. Therefore, $OC/OI = CF/FP$. But the ratio of CF to FP is the same as that of the lateral surface of cone FCP described by line FC to the circle described by FP, by the corollary of [Proposition] I of this [work.] Therefore, the product of line OC and the circle described by FP, which product evidently is equivalent to some cylinder by the preceding description, is equal to the product of line OI and the lateral surface of cone FPC described by line FC. But the product of OC and the circle described by FP in rotation is [equal to] a cylinder which is [in volume] three times the double cone OPF, PFC, by [Proposition] XII.9 of the *Elements* and by the preceding

* Greek text XII.10 throughout this proposition.

precedentem descriptionem. Ergo solidum quod fit ex ductu linee *OI* in curvam superficiem quam describit *FC* in *FCP* piramidem est
40 equale illi columpne, et ita est triplum ad duplicem piramidem *OPF*, *PFC*. Sed illius eiusdem columpne, que fit ex *OI* in curvam superficiem que sit equalis curve superficiei piramidis *FPC*, quam superficiem describit *FC* circumvoluta, subtripla est piramis cuius altitudo *OI*, basis vero equalis curve superficiei quam describit *FC*, per IX
45 duodecimi Elementorum. Ergo duplex piramis *OFP*, *FPC* est equalis piramidi cuius altitudo *OI*, basis vero circulus equalis curve superficiei quam describit *FC*. Sit ergo illa piramis rotunda piramis *M*.

Amplius, ex ductu *OV* axis in circulum descriptum ab *NV* fit columpna tripla piramidi *NVO* et ex ductu *VD* in eundem circulum
50 fit columpna tripla piramidi *NDV*. Ergo per primam secundi quod fit ex *OD* in eundem circulum est triplum duplici piramidi *DNV*, *NVO*. Item tam *V* quam *Z* est recus et *D* est communis. Ergo *NVD*, *DZO* trianguli sunt similes. Ergo que est proportio *OD* ad *ZO* eadem est *DN* ad *NV*. Sed que est *DN* ad *NV* eadem est curve superficiei
55 piramidis *DNV*, quam describit *DN* circumvoluta, ad circulum cuius semidiameter est *NV*, per corollarium prime huius. Ergo que est proportio linee *OD* ad *OZ* eadem est curve superficiei quam describit *ND* ad circulum quem describit *NV*. Ergo quod fit ex ductu linee *OD* in circulum descriptum ab *NV*, primi, scilicet, in ultimum, equum
60 est ei quod fit ex ductu linee *OZ* in curvam superficiem quam describit *ND*, scilicet unius medii in reliquum. Sed ex ductu *OD* in circulum quem describit *NV* fit columpna, per precedentem descriptionem, tripla duplici piramidi *NVO*, *NVD*, per IX duodecimi Elementorum. Ergo quod fit ex ductu linee *OZ* in curvam superficiem piramidis
65 *NDV*, quam describit linea *ND*, est equale columpne triple ad duplicem piramidem *NVO*, *NVD*. Ergo piramis cuius altitudo est *OZ*, basis vero circulus equalis curve superficiei piramidis *NVD*, quam

38–39 linee OI *tr. A*
39 FCP: FOP *A*
42 superficiei *om. A* / FPC: SPC *A*
43 piramidis *A*
45 FPC: SPC *A*
46 equalis *om. A*
52 NVD: NOD *A*
53 est *om. C*
54 NV[1]: NO *C*
55 DN: linea DN *A*
56 semi- *supra scr. B* / corellarium *B* / Ergo *om. A*
59 OD *om. C*
60 linee OZ *tr. A*
61 scilicet *A om. BC*
63 tripla: est tripla *A* / piramidi *om. A* / IX: XI *A*
65 equale columpne *B* equalis *A* equalis columpne *C*
66 est *A om. BC*
67 equalis *A om. BC*

description. Therefore, the solid arising from the product of line OI and the lateral surface of cone FCP described by FC is equal to that cylinder, and thus is triple the double cone OPF, PFC. But the cone whose altitude is OI and whose base is equal to the lateral surface which FC describes is one third that same cylinder which arises from the product of OI and the lateral surface of cone FPC described by FC in rotation; this is by [Proposition] XII.9 of the *Elements*. Therefore, the double cone OFP, FPC is equal to the cone whose altitude is OI and whose base circle is equal to the lateral surface described by FC. Therefore, let this cone be cone M.

Further, from the product of axis OV and the circle described by NV arises a cylinder which is three times the cone NVO, and from the product of VD and the same circle arises a cylinder which is three times the cone NDV. Therefore, by [Proposition] II.1 [of the *Elements*], the product of OD and the same circle is three times the double cone DNV, NVO. Also, V as well as Z is a right angle, and $\angle D$ is common. Therefore, triangles NVD and DZO are similar. Therefore, $OD/ZO = DN/NV$. But as DN is to NV, so the lateral surface of cone DNV described by DN in rotation is to the circle whose radius is NV, by the corollary of the first [proposition] of this [work]. Therefore, the ratio of OD to OZ is the same as that of the lateral surface described by ND to the circle described by NV. Therefore, the product of line OD and the circle described by NV, i.e., the product of the first and the last term, is equal to the product of line OZ and the lateral surface described by ND, i.e., the product of the two middle terms. But the product of OD and the circle described by NV equals, by the preceding description, a cylinder which is three times the double cone NVO, NVD, by [Proposition] XII.9 of the *Elements*. Therefore, the product of line OZ and the lateral surface of cone NDV described by line ND is equal to a cylinder which is three times the double cone NVO, NVD. Therefore, the cone whose altitude is OZ and whose base is a circle equal to the lateral surface of cone NVD described by ND is

describit *ND*, est equalis duplici piramidi *NVO*, *NVD*, cum earum utraque sit ad idem subtripla. Item tam *P* quam *Z* est rectus et *D* communis. Ergo *DFP*, *DOZ* trianguli sunt similes. Ergo proportio *OD* ad *OZ* que *FD* ad *FP*. Sed que *FD* ad *FP* eadem est curve superficiei piramidis *FDP*, quam describit *FD*, ad circulum cuius semidiameter est linea *FP* per corollarium prime huius. Ergo proportio *OD* ad *OZ* que curve superficiei quam describit *FD* ad circulum quem describit *FP*. Ergo quod fit ex ductu linee *OD* in circulum quem describit *FP* est equale ei quod fit ex ductu *OZ* in curvam superficiem piramidis *FDP*, quam superficiem describit linea *DF*. Sed ex ductu *OD* in circulum quem describit *FP* est columpna tripla duplici piramidi *FDP*, *FOP*. Et quod fit ex ductu *OZ* in circulum equalem curve superficiei piramidis *FDP*, quam superficiem describit *FD*, est equale columpne triple ad piramidem cuius *OZ* altitudo, basis vero circulus equalis curve superficiei piramidis *FDP*, quam superficiem describit *FD*, per precedentem descriptionem et IX duodecimi Elementorum. Ergo piramis cuius altitudo *OZ*, basis vero circulus equalis curve superficiei quam describit *FD* in piramidem *FDP* est equalis duplici piramidi *FDP*, *FPO*; subtriplum subtriplo equale sicut triplum / triplo. Sed piramis cuius altitudo *OZ*, basis vero circulus equalis curve superficiei piramidis *NVD*, quam superficiem describit linea *ND* circumvoluta, *VD* stanti, preostensa est esse equalis duplici piramidi *NVO*, *NVD*. Ergo piramis cuius altitudo *OZ*, basis vero circulus equalis curve superficiei quam describit *NF* circumvoluta, *PV* stanti, est equalis differentie duplicium piramidum *DVN*, *NVO* et *DPF*, *FPO*, scilicet corpori quod describit triangulus *NFO*, dependens a linea *ND*, circumvolutus, *DO* stanti. Illa piramis sit *S*.

Regula enim hec firmissima est: Quarumlibet duarum pirimidum inequalium [eiusdem altitudinis] differentia equatur piramidi eiusdem altitudinis cuius basis equatur differentie basium illarum piramidum. Quod sic probatur. Sit piramis *DBO*, cuius basis circulus *AB*, cathetus *OD*, ypothenusa *DB*, et minor piramis *ODE* eiusdem altitudinis,

68 NVO, NVD: NVD, NVO *C*
69 D: D est *A*
72–74 cuius...circulum *om. A*
73 corellarium *B*
77 piramidis *A* piramidum *BC*
78 OD *om. A* / columpna: conica *A*
79 equale *C*
81 columpne: conice superficiei *A* / OZ altitudo *tr. A*
83 IX: XI *A*
86 equale: est equale *A*
93 FPO: PF *A*
94 Illa piramis *om. A*
95–96 piramidum inequalium *tr. A*
96 [eiusdem altitudinis] *supplevi*
98 AB: IB *A*

equal to the double cone NVO, NVD, since each of them is one third of the same quantity. Also, P as well as Z is a right angle, and $\angle D$ is common. Therefore, triangles DFP and DOZ are similar. Therefore, $OD/OZ = FD/FP$. But the ratio of FD to FP is the same as that of the lateral surface of cone FDP described by FD to the circle whose radius is line FP, by the corollary of the first [proposition] of this [work]. Therefore,

$$\frac{OD}{OZ} = \frac{\text{lateral surface described by } FD}{\text{circle described by } FP}.$$

Therefore, the product of line OD and the circle described by FP is equal to the product of OZ and the lateral surface of cone FDP described by line DF. But from the product of OD and the circle described by FP arises a cylinder which is three times the double cone FDP, FOP. And the product of OZ and the circle equal to the lateral surface of cone FDP described by FD is equal to a cylinder which is three times the cone whose altitude is OZ and whose base is equal to the lateral surface of cone FDP described by FD, by the preceding description and [Proposition] XII.9 of the *Elements*. Therefore, the cone whose altitude is OZ and whose base is a circle equal to the lateral surface which FD describes in cone FDP is equal to the double cone FDP, FPO, for one third is to one third as triple is to triple. But the cone whose altitude is OZ and whose base is a circle equal to the lateral surface of cone NVD described by line ND in rotation, with VD fixed, has previously been shown to be equal to the double cone NVO, NVD. Therefore, the cone whose altitude is OZ and whose base is a circle equal to the lateral surface described by NF in rotation, with PV fixed, is equal to the difference between the double cones DVN, NVO and DPF, FPO, i.e., to the body described by $\triangle NFO$ in rotation, where line ND is revolved with DO remaining fixed. Let this cone be S.

For the following rule is firmly established: The difference [in volume] between two unequal [right circular] cones [of the same altitude] is equal to a [third right circular] cone of the same altitude whose base is equal to the difference [in area] between the bases of those cones. Proof: Let there be cone DBO, whose base circle is AB, altitude OD, and slant

100 cuius basis circulus *GE*, ypothenusa *DE* [Fig. 71]. Differentia *AB* circuli ad *GE* circulum vocetur *K*. Sitque circulus *M* equalis *K*, et piramis *MS* fundata super *M* circulum, cuius altitudo equalis *OD*.

Ratio. Age. Que est proportio *DBO* piramidis ad *DOE* piramidem eadem est *AB* circuli basis ad *GE* circulum basim, per XI duodecimi. 105 Ergo que est proportio *DBO* ad augmentum *DBO* super *DOE* eadem est *AB* circuli ad *K* superficiem per eversam proportionem. Ergo a pari que est *DBO* ad augmentum suum super *DOE* eadem est *AB* circuli ad *M* circulum. Sed que est *AB* circuli basis ad *M* circulum basim eadem est *DBO* piramidis ad *MST* piramidem, cum sint eius- 110 dem altitudinis per XI duodecimi. Ergo proportio *DBO* ad augmentum suum supra *DOE* que *DBO* ad *MST* piramidem. Ergo augmentum *DBO* piramidis super *DOE* est equale *MST* piramidi, quod proposuimus. Principali ergo proposito insistamus.

Amplius, in *AOE*, *TOE* tam *O* quam *T* angulus est rectus, et *E* 115 est communis, quod ut in anteproxima propostitione probari potest. Ergo *OAE*, *OTE* trianguli sunt similes. Ergo proportio *OE* ad *OT* est que *EA* ad *OA*. Et que est *AE* ad *AO* eadem est curve superficiei *AOE* ad circulum quem describit *AO*. Ergo que est proportio *OE* ad *OT* eadem est curve superficiei piramidis *AOE*, quam superficiem 120 describit *AE*, ad circulum quem describit *AO*. Ergo solidum quod fit ex ductu *OE* in circulum quem describit *AO* est equale solido

100 DE: OD *C*
102 M: AN *A* / equalis: est equalis *A*
103 DBO: ABO *A*
104, 106, 107, 108 AB: IB *A*
104 basem *A* / XI *A* IX *BC*
107 DBO: proportio DBO *A*
109 basem *AC* / est *om.* *C*
111 supra: super *C* / DBO: DBC *A*
114–15 in...communis *A* AOC angulus est rectus *BC*
115 potest *A* potest et E est communis *BC*
117 EA: AE *A* / OA: AO *A* / est[2] *B* *om.* *AC*
118–20 Ergo...AO *om.* *A*
118 est *om.* *B*

height DB [see Fig. 71]. And [let there be] a smaller cone ODE of the same altitude, whose base is circle GE, and whose slant height is DE. The difference [in area] between circles AB and GE we let be designated as K. Then let there be a circle M equal to K, and erected upon circle M a cone MS, whose altitude is equal to OD.

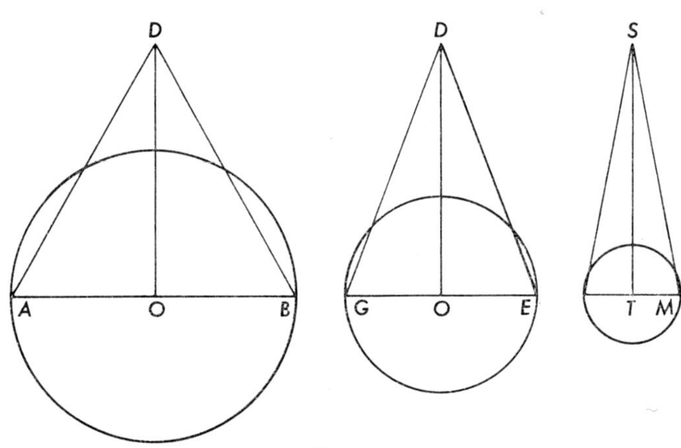

Fig. 71

Now proceed with the proof: $\dfrac{\text{cone } DBO}{\text{cone } DOE} = \dfrac{\text{base circle } AB}{\text{base circle } GE}$ by XII.11 [of the *Elements*]. Therefore, $\dfrac{\text{cone } DBO}{\text{cone } DBO - \text{cone } DOE} = \dfrac{\text{circle } AB}{\text{surface } K}$, by the conversion of ratios. Therefore, by equality,

$$\frac{\text{cone } DBO}{\text{cone } DBO - \text{cone } DOE} = \frac{\text{circle } AB}{\text{circle } M}.$$

But base circle AB/base circle M = cone DBO/ cone MST, since they are of the same altitude; this is by XII.11 [of the *Elements*]. Therefore,

$$\frac{\text{cone } DBO}{\text{cone } DBO - \text{cone } DOE} = \frac{\text{cone } DBO}{\text{cone } MST}.$$

Therefore, cone DBO — cone DOE = cone MST, which we proposed. Therefore, let us press on to the principal proposition [see Fig. 70, above].

Further, in [triangles] AOE and TOE, O as well as T is a right angle, and $\angle E$ is common, which can be proved as in the proposition before the preceding one [i.e., as in Proposition V]. Therefore, triangles OAE and OTE are similar. Therefore, $OE/OT = EA/OA$. And

$$\frac{AE}{AO} = \frac{\text{lat surf cone } AOE}{\text{circle described by } AO}.$$

quod fit ex ductu *OT* in curvam superficiem piramidis *AOE*, quam superficiem describit *AE*. Sed quod fit ex *OE* in circulum quem describit *AO* est columpna tripla ad piramidem *AOE*. Et id quod fit
125 ex ductu *OT* in curvam superficiem piramidis *AOE*, quam superficiem describit linea *AE*, est equale columpne triple ad piramidem, cuius altitudo *OT*, basis vero circulus equalis curve superficiei piramidis *AOE*, quam superficiem describit linea *AE*. Ergo piramis *AOE* est equalis piramidi cuius altitudo *OT*, basis vero circulus equalis curve
130 superficiei piramidis *AOE*; subtriplum subtriplo sicut triplum triplo.

Deinde simili ratione sicut superius respiciantur *OTE*, *NVE* trianguli similes, cum tam *V* quam *T* sit rectus et *E* communis. Ergo proportio *OE* ad *OT* que *EN* ad *NV*. Sed que *EN* ad *NV* eadem est curve superficiei piramidis *NVE*, quam superficiem describit linea
135 *NE*, ad circulum cuius semidiameter est *NV*, per corollarium prime huius. Ergo que est proportio *OE* ad *OT* eadem est curve superficiei piramidis *NVE*, quam superficiem describit *NE*, ad circulum quem describit *NV*. Ergo quod fit ex *OE* in circulum quem describit *NV* est equale illi quod fit ex ductu *OT* in curvam superficiem quam de-
140 scribit *NE*. Sed ex *OE* in circulum quem describit *NV* fit columpna tripla duplicis piramidis *NVE*, *NVO*, per precedentem descriptionem et per IX duodecimi. Et quod fit ex *OT* in curvam superficiem quam describit *NE* est columpna tripla piramidis, cuius altitudo *OT*, basis vero circulus equalis curve superficiei quam describit *NE*, per eandem
145 descriptionem et per IX duodecimi. Ergo duplex piramis *NVO*, *NVE* est equalis piramidi cuius altitudo *OT*, basis vero circulus equalis curve superficiei quam describit *NE*. Sed piramis *AOE* preostensa est esse equalis piramidi cuius altitudo *OT*, basis vero circulus equalis curve superficiei quam describit *AE*. Ergo piramis cuius altitudo *OT*, basis
150 vero circulus equalis curve superficiei quam describit *AN* circumvoluta, *OV* stanti, est equalis corpori quod est differentia piramidis *AOE* ad duplicem piramidem *NVE*, *NVO*, scilicet corpus quod describit triangulus *NOA*, dependens a linea *AE*, circumvolutus, *OE*

123, 125², 134, 137 superficiem *om.* *A*
132 cum: quorum *A* / sit: angulus est *A* / E: E est *A*
134 NVE *B* NVB *A* NVH (?) *C*
135 cuius semidiameter *A* *tr.* *BC* / corellarium *B*
141 duplicis piramidis *tr.* *A*
142 IX: XI *A* / fit *om.* *A*
143 *post* NE *add.* *B* per eandem descriptionem et per IX XII^mi / piramidis: duplcis piramidis *A* / OT: est OT *A*
144 equalis *om.* *A*
145 IX: XI *A*
147–50 NE.... describit *om.* *A*
152 NVO *om.* *A*

Therefore. $\frac{OE}{OT} = \frac{\text{lat surf cone } AOE \text{ described by } AE}{\text{circle described by } AO}$. Therefore,

$(OE \cdot \text{circle described by } AO) = (OT \cdot \text{lat surf cone } AOE \text{ described by } AE)$. But the product of OE and the circle described by AO is a cylinder which is three times cone AOE; and the product of OT and the lateral surface of cone AOE described by line AE is equal to a cylinder which is three times the cone whose altitude is OT and whose base circle is equal to the lateral surface of cone AOE described by line AE. Therefore, cone AOE is equal to a cone whose altitude is OT and whose base circle is equal to the lateral surface of cone AOE, since 1/3 is to 1/3 as 3 is to 3.

Then by an argument similar to that above, triangles OTE and NVE are regarded as similar since V as well as T is a right angle and $\angle E$ is common. Therefore, $OE/OT = EN/NV$. But

$$\frac{EN}{NV} = \frac{\text{lat surf cone } NVE \text{ described by line } NE}{\text{circle of radius } NV},$$

by the corollary of the first [proposition] of this [work]. Therefore,

$$\frac{OE}{OT} = \frac{\text{lat surf cone } NVE \text{ described by line } NE}{\text{circle described by } NV}.$$ Therefore,

$(OE \cdot \text{circle described by } NV) = (OT \cdot \text{lat surf described by } NE)$. But the product of OE and the circle described by NV is a cylinder which is three times the double cone NVE, NVO, by the preceding description and by [Proposition] XII. 9 [of the *Elements*]. And the product of OT and the lateral surface described by NE is a cylinder which is three times a cone whose altitude is OT and whose base circle is equal to the lateral surface described by NE, by that same description and by XII.9 [of the *Elements*]. Therefore, the double cone NVO, NVE is equal to a cone whose altitude is OT and whose base circle is equal to the lateral surface described by NE. But cone AOE was earlier shown to be equal to the cone whose altitude is OT and whose base circle is equal to the lateral surface described by AE. Therefore, the cone whose altitude is OT and whose base circle is equal to the lateral surface described by AN in rotation, with OV remaining fixed, is equal to the body constituted by the difference [in volume] between cone AOE and the double cone NVE, NVO, i.e., the body which triangle NOA describes when AE is revolved around OE as an axis, by the rule set out earlier. Let

stanti, per premissam regulam. Illa piramis que est equalis huic differentie / sit *G*. Ergo si memineris priorum *G*, *S*, *M* piramides sunt equales medietati corporis poligonii, et earum tres bases sunt equales curve superficiei medietatis poligonii, et earum altitudo, *OT* sive *OI* sive *OZ*, que sunt equales medietati diametri spere que inscribitur poligonio.

Simili demonstrationis progressu probetur alia medietas corporis poligonii, scilicet *AQB*, esse equalis tribus piramidibus quarum altitudo sit equalis semidiametro spere inscripte solido, bases vero equales curve superficiei conice medietatis poligonii. Sit deinde piramis *R* fundata super basim circulum *LK* equalem sex basibus sex piramidum que probate sunt esse equales corpori poligonio. Sit etiam *R* equalis altitudinis illis sex piramidibus. Erit ergo piramis *R* illis sex piramidibus equalis, quia que est proportio piramidis ad piramidem eadem est basis ad basim, cum sint eiusdem altitudinis, per XI duodecimi. Erit ergo piramis *R* mediantibus sex piramidibus equalis corpori solido poligonio, eiusque basis circulus *LK* equalis conicis superficiebus corporis, et altitudo *R* equalis *OT* linee que est equalis semdiametro spere inscripte solido. Relinquitur ergo ratum quod longe diuque venati sumus.

Si tamen poligonium equilaterum et equiangulorum circulo inscribatur, cuius medietas sit laterum numero imparium, et stanti diametro circumvoluatur et faciat solidum poligonium, dubium esse potest utrum illud solidum sit equale piramidi cuius altitudo sit equalis semidiametro spere inscripte solido et basis sit circulus equalis conice superficiei solidi. Quod quia auctor non proposuit, et nos illud investigare omittimus et diligenti relinquimus posteritati*.

154 stante *C*
155 S, M *tr.* *A*
156–57 et...poligonii *om.* *A*
158 que *A* quod *BC*
160 demonstrationis...probetur: rationis progressu et demonstratione consimili probatur *A*
161 esse *om.* *A*
164 basem *A*
165 poligonii *A*
166–67 Erit...equalis *om.* *A*
167 eadem: ea *A*
168 basem *AC*
171 corporis *om.* *A*
172–73 longe diuque: diu longeque *A*
175 sit laterum *tr.* *A* / stante *C*
179-80 auctor...posteritati: non propositum ab auctore nos investigare omittimus et posteritatis diligentie relinquemus *A*

* See Section 5 below for the proof by the scribe of Manuscript *E* of the theorem for the case where the half-polygon has an odd number of sides.

the cone equal to this difference be G. Therefore, if you recall the earlier statements, cones G, M, and S are equal [in sum] to one half of the polygonal body, and their three bases are equal [in sum] to the lateral surface of half that polygonal [body], while their altitude, OT or OI or OZ, is equal to the radius of the sphere inscribed in the polygonal [body].

By a similar line of demonstration, let it be proved that the other half of the polygonal body, i.e., AQB, is equal to three cones each of whose altitudes is equal to the radius of the sphere inscribed in the solid and whose bases are equal [in sum] to the conical curved surface of the half polygonal [body]. Then let there be a cone R* erected on base circle LK equal to the six bases of the six cones which have been proved to be equal [in sum] to the polygonal body. Also let R be of altitude equal to that of [each of] the six cones. Therefore, cone R will be equal to these six cones, for the ratio of cone to cone is the same as that of base to base when the cones are of the same altitude, by XII.11 [of the *Elements*]. Therefore, cone R by the mediacy of the six cones will be equal to the solid polygonal body, and its (R's) base circle LK is equal to the conical surfaces of the body, and R's altitude is equal to line OT, which is [itself] equal to the radius of the sphere inscribed in the solid. Therefore, what we sought a long time ago remains as established.

If, however, half of the regular polygon inscribed in the circle has an odd number of sides, it can be doubted whether the solid formed by the polygon's rotation about a fixed diameter is equal to a cone whose altitude is equal to the radius of the sphere inscribed in the solid and whose base is a circle equal to the conical surface of the solid. But since the author has not proposed it, we shall refrain from investigating it, leaving it to a diligent posterity.

* R is not actually constructed on the figure.

VIII. OMNIS COLUMPNA CUIUS ALTITUDO DIAMETRO SPERE ET BASIS MAXIMO CIRCULO IN SPERA FUERINT EQUALES SEXQUIALTERA EST SPERE, SICUT ET TOTA SUPERFICIES COLUMPNE SUPERFICIEI SPERE SEXQUIALTERA EST.

Ne quis veritati demonstratoris oblatret et demonstrationis tenorem interrumpat, priusquam propositum aggrediamur, quoddam elementum ad propositum perutile proponatur, et probetur quod tale est. Si fuerit proportio primi ad secundum que tertii ad quartum, fueritque id quod fit ex ductu quinti in primum maius eo quod fit ex ductu sexti in secundum, erit quoque id quod fit ex ductu quinti in tertium maius eo quod fit ex ductu sexti in quartum.

Exemplum. Sit *A* ad *B* ut *C* ad *D* et fiat ex ductu *E* in *A* quantitas *G* et ex *F* in *B* quantitas *H*; sitque *G* maior quam *H* [Fig. 72]. Et item ex *E* in *C* fiat quantitas *M* et ex *F* in *D* quantitas *S*. Dico quod *M* est maior *S*.

Dispositio. Ducatur *E* in *B* et fiat *R*, et *E* in *D* et fiat *N*. Ratio. Age. Ex *E* in *A* fit *G*; ex *E* in *B* fit *R*; ergo cum productorum et producentium eadem sit proportio, erit *G* ad *R* ut *A* ad *B*. Item ex *E* in *B* fit *R*; ex *F* in *B* fit *H*, ergo que est proportio *E* ad *F* eadem est *R* ad *H*. Rursum, ex *E* in *C* fit *M*; ex *E* in *D* fit *N*, ergo proportio *C* ad *D* est que *M* ad *N*. Item ex *E* in *D* fit *N*; ex *F* in *D* fit *S*, ergo que est proportio *E* ad *F* eadem est *N* ad *S*. Sic ergo collectis premissis, erit proportio *G* ad *R* que *M* ad *N* et *R* ad *H* que *N* ad *S*. Ergo per equam proportionem que est proportio *G* ad *H* eadem est *M* ad *S*. Sed *G* est maior *H*. Ergo *M* est maior *S*. Sic quod cum iam constet, principali attingamus proposito.

Esto exemplum *QM* columpna cuius altitudo linea *QM* est equalis

1 VIII: 8ª *D om. AC* II *B* 2ª *J* secunda conclusio *mg. B manu recentiori* / *mg. B manu recentiori*: glossum ex magna causa quod (quia *?*) includitur notabilis conclusio...
2 in spera *A om. BC*
6–7 tenorem interrumpat *A tr. BC*
7 quoddam *om. A*
8 perutile: utile *A*
13 fiat *B* fit *AC*
14 item: tunc *A*
15 fiat *AC om. B* / et *om. A*
17 E²: C *A*
19 producentium et productorum *A* / sit *om. A*
22 est *om. A*
23 eadem: ea *A*
24 R²: H *C* / N²: M *A*
26 S *om. C* / Sic *om. AC*
27 attingamur (*?*) *C*
28 est *om. A*

VIII. EVERY CYLINDER WHOSE ALTITUDE IS EQUAL TO THE DIAMETER OF A SPHERE AND WHOSE BASE IS EQUAL TO THE GREATEST CIRCLE IN THE SPHERE IS THREE HALVES THE SPHERE, JUST AS THE TOTAL SURFACE OF THE CYLINDER IS THREE HALVES THE SURFACE OF THE SPHERE.

Lest anyone carp at the soundness of the demonstrator and interrupt the course of the demonstration, let us—before we undertake that which we have proposed—posit and prove a certain principle useful for [proving] the proposition. [The principle is this:] If the ratio of the first [term] to the second is as that of the third to the fourth, and if the product of the fifth and the first is greater than the product of the sixth and the second, so also will the product of the fifth and the third be greater than the product of the sixth and the fourth.

Exemplification: Let $A/B = C/D$. Let $E \cdot A = G$, and $F \cdot B = H$, and let $G > H$ [see Fig. 72]. And also let $E \cdot C = M$, and $F \cdot D = S$. I say that $M > S$.

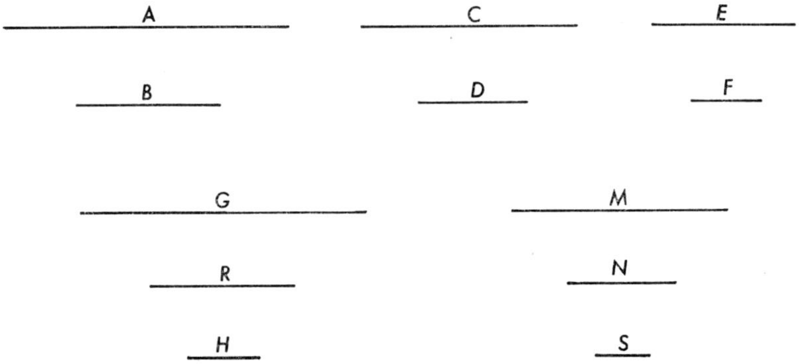

A C E
B D F
G M
R N
H S

Fig. 72

Disposition: Let $E \cdot B = R$, and $E \cdot D = N$. Now proceed with the proof: With $E \cdot A = G$, and $E \cdot B = R$, therefore, since the ratio of the products and that of the quantities producing the products are the same, $G/R = A/B$. Also, with $E \cdot B = R$, and $F \cdot B = H$, therefore $E/F = R/H$; again with $E \cdot C = M$, and $E \cdot D = N$, therefore $C/D = M/N$. Also, with $E \cdot D = N$, and $F \cdot D = S$, therefore $E/F = N/S$. Therefore, with all of these statements set forth together, $G/R = M/N$, and $R/H = N/S$. Therefore, by the equality of ratios, $G/H = M/S$. But $G > H$; therefore $M > S$. Since this is now evident, let us take up the principal proposition.

For example, let there be a cylinder QM whose altitude, line QM, is

AE diametro spere *F* et basis eius circulus *RS* maximo circulo *F*
30 spere, scilicet circulo *ADEC*, sit equalis [Fig. 73]. Dico quod columpna *QM* est sexquialtera spere *F*. Sin autem, sit *QM* sexquialtera maioris vel minoris, et primo *K* spere minoris.

Dispositio. Inscribatur *ACED* circulo poligonium equalium laterum et numero parium, circulum *H* minime contingentium. Deinde
35 a *G* in *E* ducatur linea recta *EG*, et ab *O* centro utriusque circuli ducatur *OI* perpendicularis ad *AG*. Stanti ergo *AE*, circumvolvatur poligonium et fiat corpus conicarum superficierum. Et circumvolvatur semicirculus *ADE* fiatque spera *F*, et *HYT* semicirculus fiatque spera *K*. Medietas autem corporis conicarum superficierum dicatur *P*.

40 Ratio. Age. Quod fit ex ductu *AO* in circumferentiam *ADEC* est
115v duplum circuli *ADEC*, per / primam Archimenidis. Et quod fit ex *EG* in eandem circumferentiam est duplum *P* per quintam huius. Ergo que est proportio *AO* ad *EG* eadem est circuli *ADEC* ad superficiem *P*. Sed *AO* est maior *OI*, cum *I* sit rectus, et *AE* maior *EG*,
45 cum *G* cadens in semicirculum sit rectus. Ergo maius est id quod fit ex ductu *AE* in *AO* quam ex *OI* in *EG*. Sit ergo *OA* primum, *EG* secundum, circulus *ADEC* tertium, *P* quartum, *AE* quintum, *OI* sextum. Ergo per id quod paulo ante principale propositum probavimus maius est id quod fit ex *AE* quinto in *ADEC* circulum tertium,
50 scilicet columpna *QM*, que dicta est esse sexquialtera spere *K*, eo quod fit ex ductu *OI* sexti in *P* superficiem quartum. Sed quod fit ex ductu *OI* in superficiem *P* est triplum semisolidi, per proximam. Ibi enim probatum est quod tres piramides quarum altitudo est *OI*, bases vero sunt equales conice superficiei semipoligonii, sunt equales semipoli-
55 gonio, et ad illas tres triplum est quod fit ex ductu *OI* in *P* superficiem conicam semipoligonii. Ergo quod fit ex *OI* in superficiem *P* sexquialterum est totius solidi. Ergo collectis prioribus columpna *QM*, que est sexquialtera ad speram *K*, est maior quam sexquialtera totius solidi conicarum superficierum. Ergo *K* subsexquialterum est maius
60 quam solidum, pars suo toto, quod est impossibile. Relinquitur ergo

30 ADEC: ADOE *A* / quod: quoniam *A*
32 primo *tr.* *A* *post* minoris[2]
33 ACED: ACDE *A*
39 corporis *tr.* *A* *post* superficierum
42 *ante* huius *del.* *B* IIII[ti]; *C habet* IIII[m]
43 AO: AD *A*
44 AE: AE est *A*
45 id *om.* *A*
48–49 principale...probavimus *AC* premissimus *B*
49 id *om.* *A* / circulum *om.* *A*
54 sunt[1] *om.* *A*
55 tres *A* tres est *BC* / triplum est *tr.* *A* / quod: cum quod *B* / OI: M *A*
58 maior: maius *A* / sexquialterum *A*

equal to diameter AE of sphere F and whose base circle RS is equal to the greatest circle of sphere F, i.e., to circle $ADEC$ [see Fig. 73]. I say that cylinder QM is 3/2 sphere F. For if not, let QM be 3/2 [a sphere] greater or less [than F]. In the first place [let it be 3/2] a lesser sphere K.

Disposition: Let there be inscribed in circle $ACED$ a regular polygon of an even number of sides, which sides do not touch circle H at all. Then let straight line EG be drawn from G to E, and let line OI be drawn from center O of each circle perpendicular to AG. With AE as an axis, let the polygon be rotated, forming a body with conical surfaces. And let the semicircle ADE be rotated, forming sphere F; and let semicircle HYT be rotated, forming sphere K. Moreover, let half of the body with conical surfaces be designated as P.

Proceed with the proof: ($AO \cdot$ circum $ADEC$) = 2 circle $ADEC$, by the first [proposition of the *Measurement of the Circle*] of Archimedes. And ($EG \cdot$ circum $ADEC$) = (2 · surface of P), by the fifth [proposition] of this [work]. Therefore, AO/EG = circle $ADEC$/surface of P. But $AO > OI$, since $\angle I$ is a right angle; and $AE > EG$, since $\angle G$, falling in a semicircle, is a right angle. Therefore, $(AE \cdot AO) > (OI \cdot EG)$. Therefore, let OA be the first term, EG the second, circle $ADEC$ the third, [the surface of] P the fourth, AE the fifth, and OI the sixth. Hence, by that principle which we proved just before [undertaking] the principal proposition, the product of the fifth term AE and the third term circle $ADEC$, which product is equal to cylinder QM, itself said to be equal to 3/2 sphere K, is greater than the product of the sixth term OI and the fourth term surface of P. But the product of OI and the surface of P is three times the semisolid, by the previous [proposition]. For there it was proved that three cones each of whose altitudes is OI and whose bases are [in sum] equal to the conical surfaces of half the polygonal body are equal to the [volume of the] half of the polygonal body; and ($OI \cdot$ surface of P) = (3 · the three cones). Therefore, ($OI \cdot$ surface of P) = (3/2 the whole solid). Therefore, with all the prior statements brought together, [it is evident that] cylinder QM, which is 3/2 sphere K, is greater than 3/2 the whole solid with conical surfaces. Therefore, [sphere] K, which is 2/3 [cylinder QM], is greater than the solid, [that is,] the part is greater than

quod columpna *QM* non est sexquialtera spere minoris quam *F*.
Dico etiam quod neque maioris. Sed ne tempus nugis teramus prior maneat figura. Et esto linea *QM* equalis linee *HT*, diametro spere *K*, et circulus *RS* equalis *HYT* maximo circulo spere *K*. Dico itaque
65 quod columpna *QM* est sexquialtera spere *K*, falsigraphus immo spere *F* maioris. Inconcussis omnibus maneat dispostitio que et prius, addito eo quod *AO* secetur in *Z*, ita quod proportio *AO* ad *HO* sit tanquam *HO* ad *ZO*.

Ratio. Age. Que est proportio *EG* ad *AO* eadem est superficiei *P*
70 ad circulum *ADEC*, sicut prius probavimus. Et que est *AO* ad *ZO*, id est, *AO* semidiametri ad *HO* semidiametrum, duplicata, eadem est *ADEC* circuli ad *HYT* circulum, per primam duodecimi. Ergo per equam proportionem, que est proportio linee *EG* ad lineam *ZO* eadem est superficiei *P* ad *HYT* circulum. Sed quod fit ex *OI* in *EG* suum
75 duplum maius est eo quod fit ex *TH* in *OZ*. Ergo per regulam in initio huius propositionis probatam, quod fit ex ductu *OI* in superficiem *P* est maius eo quod fit ex ductu *TH* diametri in *HYT* circulum. Sed quod fit ex ductu *OI* in superficiem *P* est sexquialterum totius solidi conicarum superficierum, per proximam. Et quod fit ex *HT*
80 in circulum *HYT* est columpna *QM*, quam dixit sexquialteram spere *F* esse falsigraphus. Ergo maius est sexquialterum solidi conicarum superficierum quam columpna *QM* sexquialtera spere *F*. Ergo subsexquialterum subsexquialtero maius, solidum spera *F*, pars suo toto, quod est impossibile. Relinquitur ergo columpnam non esse sexqui-
85 alteram maiori spere vel minori quam illi cuius diameter axi columpne et maximus circulus basi columpne equatur. Sic ergo proposito non sumus defraudati.

62 teramus: teneamus *A*
63 equalis linee *A* linea equalis *BC*
64 RS: KS *C* / circulo *tr. A ante* HYT
65 columpna QM *A* QM conica *BC* / est sexquialtera *tr. A*
75 fit *om. A*
75–76 in[2]...propositionis: superius *A*
77–78 est....P *om. A*
80 circulum HYT *tr. A*
80–81 sexquialteram...falisgraphus: falsigraphus esse sexquialteram spere F *A*
83 maius: maius est *A* / spera: spere *A*
84 columpnam: columpnam QM *C*
85 maiori spere *tr. A*
86–87 Sic...defraudate: Et sic constat propositum *A*

the whole, which is impossible. It remains, therefore, that cylinder QM is not 3/2 of a sphere less than F.

I say also that neither is it [three halves] of one greater [than F]. But in order not to waste time in idle speech, let us keep the prior figure [Fig. 73]. Let line QM be equal to line HT, the diameter of sphere K, and circle RS to HYT, the greatest circle of sphere K. And so I say that cylinder QM is 3/2 sphere K, while the pseudographer [says] rather that cylinder QM is 3/2 the larger sphere F. With all things unchanged, the disposition [of quantites and magnitudes] remains as before, it having been added that AO is cut at Z so that $AO/HO = HO/ZO$.

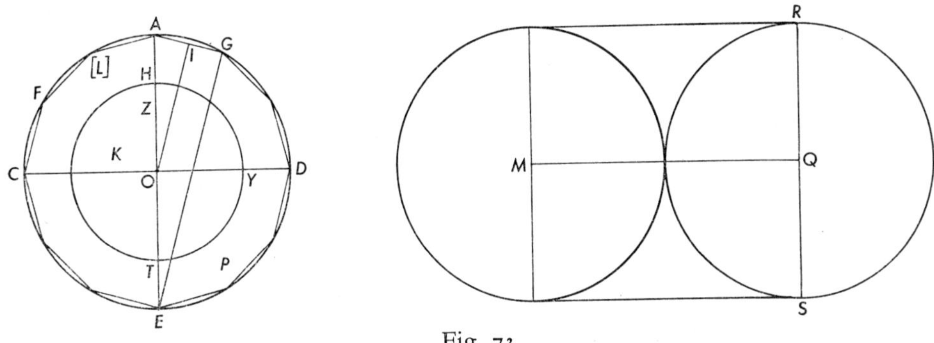

Fig. 73

Note: The figure of the cylinder MQ is rotaded so that altitude MQ stands as a vertical.

Proceed with the proof: EG/AO = surface of P/circle $ADEC$, as we proved earlier. And $AO/ZO = (AO/OH)^2$ = circle $ADEC$/circle HYT, by [Proposition] XII.1 [of the *Elements*]. Hence by the equality of ratios EG/ZO = surface of P/circle HYT. But with $EG = 2\ OI$, $(OI \cdot EG) > (TH \cdot OZ)$. Therefore, by the rule proved in the beginning of this proposition, ($OI \cdot$ surface of P) > (diameter $TH \cdot$ circle HYT). However, ($OI \cdot$ surface of P) = (3/2 the whole solid with conical surfaces), by the preceding [proposition]; and ($HT \cdot$ circle HYT) = cylinder QM, which the pseudographer has said to be 3/2 sphere F. Therefore, 3/2 the solid with conical surfaces is greater than cylinder QM, i.e., greater than 3/2 sphere F. Therefore, 2/3 (3/2 the solid) > 2/3 (3/2 sphere F), and so the solid is greater than sphere F, i.e., the part is greater than the whole, which is impossible. It remains, therefore, that a cylinder is not 3/2 a sphere larger or smaller than the sphere whose diameter is equal to the axis of the cylinder and whose greatest circle is equal to the base of the cylinder. And so, therefore, we have not been deceived in our proposition.

IX. OMNIS SPERA EST EQUALIS ROTUNDE PIRAMIDI CUIUS BASIS EQUATUR SUPERFICIEI SPERE ET ALTITUDO SEMIDIAMETRO SPERE.

Esto exemplum *A* piramis rotunda, cuius basis circulus *NM* sit
5 equalis superficiei spere *C* et *AB* altitudo *A* sit equalis *DC* semidiametro *C* spere [Fig. 74]. Dico quod spera *C* est equalis piramidi *A*.

Dispositio. Sumatur enim piramis *E*, cuius altitudo *EI* sit equalis *AB* et eius basis equalis maximo circulo spere *C*. Sumatur etiam columpna cuius basis sit equalis maximo circulo spere *C* et *GH* eius
10 axis sit duplus ad *DC*.

Ratio. Age. *GH* columpna est duplo altior piramide *E* et sita est in equali basi cum *E*. Ergo columpna *GH* est sextupla ad *E* piramidem. Si enim essent eiusdem altitudinis columpna esset tripla ad *E* per IX duodecimi. Sed columpna *GH* sita est in basi *FG* circulo
15 equali maximo circulo spere *C* et axis *GH* est equalis diametro spere *C* ex dispositione. Ergo columpna *GH* est sexquialtera ad speram *C*, per proximam. Sed columpna *GH* erat sextupla ad *E*. Ergo spera *C* est quadrupla ad *E*. Item ex ductu diametri *C* spere in circumferentiam
116r maximi circuli *C* spere, i.e., basis *E* pi/ramidis, sit curva superficies
20 spere *C*, per sextam huius, scilicet *MN* circulus. Et quod fit ex ductu diametri spere *C* in circumferentiam maximi circuli spere *C* est quadruplum ad circulum maximum spere *C*, per primam Archimenidis de mensura circuli. Ergo a duplici pari circulus *MN* est quadruplus basi piramidis *E*. Sed *E* et *A* piramides sunt eiusdem altitudinis. Ergo *A*

1 IX *om.* *AC* 9ª *D* III *B* 3ª *J* / est equalis rotunde *A* rotunde est equalis *BC*
5 DC *om.* *A*
7 enim *om.* *A*
14 IX: XI *A* / FG: FH *A*
18, 19 C spere *tr.* *A*
21 C[1] *om.* *A*
22 circulum maximum *tr.* *A*
22–23 per...circuli *A* *om.* *BC*

IX. EVERY SPHERE IS EQUAL TO A CONE WHOSE BASE IS EQUAL TO THE SURFACE OF THE SPHERE AND WHOSE ALTITUDE IS EQUAL TO THE RADIUS OF THE SPHERE.

For example, let there be a cone A, whose base circle NM is equal to the surface of sphere C and whose altitude AB is equal to radius DC of sphere C [see Fig. 74]. I say that sphere C is equal to cone A.

Disposition: Let there be posited cone E, whose altitude EI is equal to AB and whose base is equal to the greatest circle of sphere C. Also let a cylinder be posited whose base is equal to the greatest circle of sphere C and whose axis GH is 2 DC.

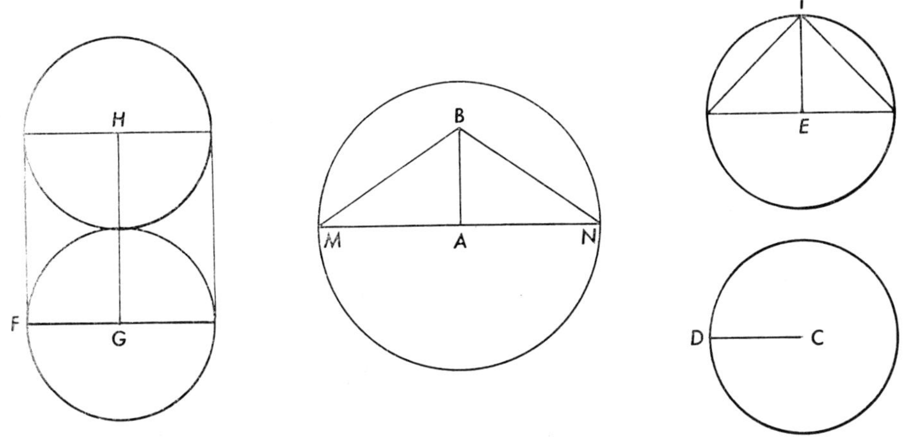

Fig. 74

Proceed with the proof: Cylinder GH is twice the height of cone E and is situated on a base equal to that of E. Hence cylinder $GH = 6$ cone E, for if the altitudes were the same, the cylinder would be three times E, by XII.9 [of the *Elements*]. But cylinder GH is situated on base circle FG equal to the greatest circle of sphere C, and axis GH is equal to the diameter of sphere C by construction. Hence cylinder $GH = 3/2$ sphere C, by the preceding [proposition]. But cylinder GH was [equal to] 6 E. Hence sphere $C = 4\ E$. Also, the product of the diameter of sphere C and the circumference of the greatest circle of sphere C or of the base of cone E is equal to the surface of sphere C, by the sixth [proposition] of this [work], i.e., [this product is equal] to circle MN. And the product of the diameter of sphere C and the circumference of the greatest circle of sphere C is four times the greatest circle of sphere C, by the first [proposition] of Archimedes' *Measurement of the Circle*. Therefore, by equality twice, circle $MN = 4$ base of cone E. But cones E and A are of the

25 piramis est quadrupla *E* piramidi per XII. Sed *C* spera fuerat quadrupla ad *E*. Ergo *A* piramis est equalis *C* spere. Ad quod astruendum aspiravimus.

X. CUIUSLIBET SPERE PROPORTIO AD CUBUM SUI DIAMETRI EST TANQUAM PROPORTIO UNDECIM AD VIGINTI UNUM.

Hoc respondet secunde propostitioni Archimenidis de quadratura
5 curculi, et ex ea fidem sumit. Sed cum illa processerit per fere, et hec similiter, unde nec hec nec illa est vera; sed vulgariter per integra procedit.

Esto exemplum *D* spera et cubus fundatus super *EH*, quadratum linee *OM* equalis *AB* diametro *D* spere [Fig. 75]. Dico quod que
10 proportio cubi *EH* ad speram *D* eadem est XXI ad XI. Fiat enim columpna *OM*, cuius tam axis quam diameter basis equatur *AB*.

Ratio. Age. Proportio *GF* quadrati ad *O* circulum est sicut XIIII ad XI, per secundam Archimenidis de quadratura circuli. Sed ex ductu *OM* altitudinis in *GF* quadratum fit cubus, et ex *OM* in *O* circulum
15 sit columpna. Ergo cum productorum et producentium eadem sit proportio, erit proportio cubi ad columpnam tanquam *GF* basis ad *O* circulum. Ergo a pari proportio cubi ad columpnam sicut XIIII ad XI. Sed proportio columpne ad speram est sciut XXI ad XIIII, cum columpna sit sexquialtera ad speram, per anteproximam. Sit ergo
20 cubus primum, columpna secundum, XIIII tertium, XI quartum, spera quintum, XXI sextum. Ergo per XXIIII quinti Elementorum in proportione equalitatis erunt quantus cubus ad speram tantus numerus XXI ad numerum XI.

Sicque tiphis noster portum tenet in quem iam dudum vela suc-

25 C spera *tr. A*
1 X *om. AC* 10ª *D* IIII *B* 4ª *J*
2 proportio *om. A*
4 Hec *A*
5 ex...sumit *om. A* / summit *C* / processerit *A* precesserit *BC*
8 EH: GH *A*
9 D spere *tr. A* / que *om. A*
10 cubi EH: est cubi CH *A* / eadem est: que *A*
11 basis *om. A*
12 quadrati *A* quadranguli *BC*
12–13 XIIII ad XI: 11 ad 14 *A*
20 secundum: quasi secundum *C*
21 XXIIII quinti *B* 23ᵃᵐ 12ᵐⁱ *A* XXIIIIª primi *C*
23 numerum *om. A*
24–28 Sicque... Archimenidis *om. H*
24 typhis *F* / tiphis noster: timba nostra *E* classis nostra, i.e. navis *mg. D* clasis nostra, i.e. navis D_1 / quem: quam *E* / iam *om. JF*
24–25 succinxerat *AEFDD*$_1$ succingerat *B* succingserat *C* succingcerat *J*

same altitude. Therefore, cone $A = 4$ cone E, by [the eleventh proposition of Book] XII [of the *Elements*]. But sphere C was $4\ E$. Therefore, cone $A =$ sphere C. And this is the porism to which we aspired.

X. THE RATIO OF ANY SPHERE TO THE CUBE OF ITS DIAMETER IS AS THE RATIO OF XI TO XXI.

This corresponds to the second proposition of Archimedes' *On the Quadrature of the circle*, and it assumes its credence from that [proposition]. But since that [proposition] proceeded only by approximation, as also does this one, neither this nor that one is true, but proceeds only in a common way by means of integers.

For example, let there be a sphere D and a cube based on EH which is the square of line OM, OM being equal to diameter AB of sphere D [see Fig. 75]. I say that cube EH/sphere $D = 21/11$. For let there be constructed

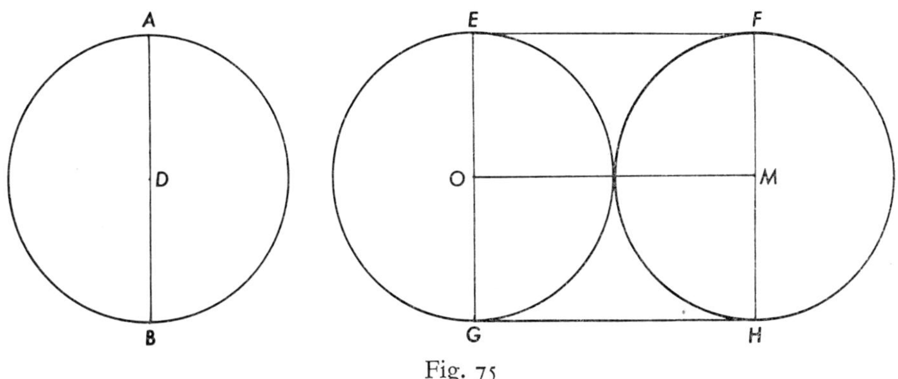

Fig. 75

Note: The figure of the cylinder OM is rotated so that altitude OM rotates as a vertical.

a cylinder OM, whose axis as well as its base diameter is equal to AB.

Proceed with the proof: Square GF/circle $O = 14/11$, by the second [proposition] of Archimedes' *Quadrature of the Circle*. But (altitude $OM \cdot$ square GF) $=$ cube EH, and ($OM \cdot$ circle O) $=$ cylinder OM. Therefore, since the ratios of the products and the quantities producing the products are the same, cube [EH]/cylinder [OM] $=$ square GF/circle O. Therefore, by equality, cube [EH]/cylinder [OM] $= 14/11$. However, cylinder [OM]/sphere [D] $= 21/14$, since cylinder [OM] $= 3/2$ sphere [D], by the next to the last proposition [i.e., Proposition VIII]. Hence let the cube be the first term, the cylinder the second, 14 the third, 11 the fourth, the sphere the fifth, and 21 the sixth. Therefore, by [Proposition] V.24 of the *Elements*, cube/sphere $= 21/11$.

25 cinxerat. Iamque cum bibulis hereat harenis anchora Archimenidis remigii, Johannes navigationis grates ageret summo creatori.

Explicit commentum Johannis de Tinemue in demonstrationes Archimenidis.

25 bibulis *om.* D_1 *cum lacuna* / hereat: habeat D_1 / *post* anchora *add.* *BJ* simul / Archimenidis *AEJF* Archimenides *BCID*

25–28 anchora.... Archimenidis *BCID* anchora remigii Archimenidis. Explicit commentum Gervasii de Essexta (Assassia *E*) *AE*

26 Johannes: et Johannes *J* / grates ageret: grate agit *I* grates agit DD_1 / summo *tr.* *J* *ante* grates

27–28 Explicit... Archimenidis *om.* *JF*

27 Tinemue *expandi ex* Tin' *B* Tin̄ *C* Thin̄ *D* Thin⁹ D_1 Tl̄n *I* ((Note: My expansion is based on the form given in the lower right margin of the first folio, 111r, of *B*.))

28 *post* Archimenidis *add.* *mg.* *B* *manu recentiori* super duos libros. hic finis. *et mg. superius B habet manu recentiori* Archimenides. Secundus liber. finis.

And so our Tiphys reaches the port toward which he set sail but a short time ago. And now when the anchor of Archimedes the rower sits fast in the thirsty sands, let Johannes extend the thanks of navigation to the Supreme Creator.

Here ends the commentary of Johannes de Tinemue on the demonstrations of Archimedes.

COMMENTARY

Proposition I

1–5 "Cuiuslibet...basis." The form of this enunciation was obviously suggested by the first proposition of the *De mensura circuli*. This proposition must be distinguished from, but compared to, those given by Archimedes in the *De sphaera et cylindro* (Proposition 14 of Book I) and the Banū Mūsā in the *Verba filiorum* (Proposition IX). For the citation of these and the comparable propositions of Pappus and Leonardo Pisano, see Commentary, *Verba filiorum*, Proposition IX, page 63. The proof is for a right circular cone.

15–17 "Rationis...admitti." This supposition is comparable to the first postulate of the Cambridge Version of the *De mensura circuli* (see Chapter Three, Section 1, above). These lines, together with lines 18–22, form a rather intrusive comment, which perhaps suggests that Johannes was commenting on a primitive form of this work rather than on the *De sphaera et cylindro*.

23–102 "Triangulus....proposui." We can represent the spirit of the proof in an abbreviated form (see Fig. 64), using modern symbols as follows.*

Proof:

(1) To prove: $S = F$, where S is the lateral surface of a right circular cone of slant height l and base circumference c, and F is the area of right triangle such that $F = \frac{1}{2}(l \cdot c)$.

(2) Either $F = S$ or $F \neq S$. If $F \neq S$, then $F > S$, or $F < S$.

* The following table relates the symbols used in this commentary to the quantities given in the figure and text:

$l = SZ =$ slant height EB
$c = ST =$ circum of circle $ACNP$
$c_1 =$ circum of circle HK
$t =$ side of polygon $ACNP$
$h =$ altitude of face triangle EAB

(3) Suppose first that $F < S$, and by the basic postulate $F = S_1$, where S_1 is the lateral surface of some cone whose altitude is the same as that of the cone of surface S, but whose base is less than that of cone S. Let c_1 be the circumference of the base of S_1. Hence, $c_1 < c$.

(4) Now inscribe in c a regular polygon, one of whose sides is t and whose perimeter is p, such that $c > p > c_1$ and p does not touch c_1 at any place. (Cf. Proposition III of the *Verba filiorum*.) Then construct a pyramid with a polygon of perimeter p as the base and with an identical altitude as that of cone of surface S. Let the surface of this pyramid be S_2. Let the altitude of one of its triangular faces be h.

(5) Then $S_2 = \frac{1}{2}(h \cdot p)$, since the area of any one triangle is $\frac{1}{2}(h \cdot t)$ and the sum of all t's is p.

(6) Hence $S_2 < F$, since $F = \frac{1}{2}(l \cdot c)$ and $l > h$ and $c > p$.

(7) Therefore $S_2 < S_1$, from (3) and (6).

(8) But S_2 was constructed as including S_1; hence, in actuality $S_2 > S_1$.

(9) Since (7) is the contradiction of the fact of (8), the supposition from which (7) was derived, namely, that $F < S$, is false.

(10) Then suppose, if it is possible, that $F > S$ and thus that $F = S_1$ where S_1 is the surface of a cone of the same altitude as cone S but whose circumference c_1 of its base is greater than c.

(11) Now circumscribe about c a regular polygon of side t and perimeter p such that $c_1 > p > c$ and p does not touch c_1. Erect on p a pyramid of surface S_2 with the same altitude as the cone of surface S, and, as before, let h be the altitude of one of its triangular faces.

(12) Then $S_2 = \frac{1}{2}(h \cdot p)$. (Cf. step 5.)

(13) Hence $S_2 > F$, since $h = l$ and $p > c$.

(14) Therefore $S_2 > S_1$, since $F = S_1$ by hypothesis.

(15) But S_2 was constructed, in fact, within S_1, and so $S_2 < S_1$.

(16) Since (14) is contradictory to the fact of (15), the supposition from which (14) was derived, namely, $F > S$, is false.

(17) Hence, since $F \ngtr S$ and $F \nless S$, then $F = S$, or $S = \frac{1}{2}(l \cdot c)$. Q.E.D.

49 "per [41] theorema primi." See Commentary, Cambridge Version of the *De mensura circuli*, Chapter Three, Section 1, line 33.

52 "primam secundi." See Commentary, Corpus Christi Version, Chapter Five, Section 5, lines 66–67.

105–117 "Ex....basis." This corollary is identical to Proposition 15 of Book I of Archimedes' *De sphaera et cylindro* (ed. of Heiberg, p. 69). Incidentally, the actual wording of Johannes' quotation of Proposition I differs from that of the Gerard of Cremona translation. I would suspect that, if this were an original Latin composition rather than a translation, Gerard's translation would have been quoted rather exactly, as is usually the case with later Latin geometrical treatises. In short, the somewhat different form of the quotation from Proposition I of the *De mensura circuli* may be another indication that we are faced here with a translation rather than an original composition.

Proposition II

1–22 "Cuiuslibet....propositum." Cf. the comparable but different theorem of Archimedes' *De sphaera et cylindro*, Book I, Proposition 13 (ed. of Heiberg, pp. 53–62). The proof is similar to that for the first proposition and so is merely sketched by the author. Again I use modern symbols to present the tenor of the proof.

The given rectangle of adjacent sides equal to the axis of the cylinder and the circumference of its base must be either equal to the lateral surface S of the cylinder or not equal to it. If it is not equal to S, then it must be equal to the surface S_1 of some other cylinder with the same axis but with a smaller or larger base. Suppose first that the quadrangle is equal to the lateral surface S_1 of a cylinder of smaller base. Then inscribe within the base of the cylinder of surface S a regular polygon whose perimeter is greater than the circumference of the base of the cylinder of S_1 and whose sides in no place touch that circumference. Then construct a prismatic figure within S the base of which figure is the regular polygon previously inscribed. The lateral surface of this prism can be shown to be at the same time smaller and larger than the surface S_1, an obvious contradiction negating the assumption that the surface S_1 of a cylinder with a base smaller than that of the cylinder of surface S is the desired surface to which the rectangle must be equated. In the second half of the proof the surface of a similar prism circumscribed around S is shown to be simultaneously larger and smaller than the surface S_1 of a cylinder with the same axis as that of the cylinder of surface S but whose base is larger. From this contradiction it is apparent that the desired surface S_1 cannot be of a cylinder of larger base. Since then S_1 cannot be the surface of a cylinder of either larger or smaller base, it must be the surface of a cylinder of the same base as that of which S is the surface. In short, S_1 must be identical with S, and so the given rectangle is equal to S. Q.E.D.

23–40 "Ex....theoremati." Both corollaries are proved by the application of simple proportions.

Proposition III

1–35 "Quorumlibet....sufficienter." For a discussion of the comparable theorems given by Pappus and the authors of the *Verba filiorum*, see the Commentary to Proposition V of the *Verba filiorum* in Chapter Four above. This proposition is patently an auxiliary theorem to be used later; it does not concern itself with a "curved surface," as do the other theorems of this treatise. Johannes' proof (see Fig. 65) of this theorem follows:*

(1) Let d_1, d_2 be the diameters of circles whose circumferences are c_1, c_2.

(2) Then either $d_1/d_2 = c_1/c_2$ or $d_1/d_2 \neq c_1/c_2$. And if $d_1/d_2 \neq c_1/c_2$, then $d_1/d_2 = c_1/c_3$ where c_3 is either greater or less than c_2.

(3) Suppose first that $d_1/d_2 = c_1/c_3$ where c_3 is less than c_2.

(4) Now describe a regular polygon within c_2 such that its perimeter p_2 is less than c_2 but greater than c_3 (Cf. Proposition I, step 4). Also describe a similar polygon within c_1 whose perimeter is p_1.

(5) Then $c_1/p_1 = c_3/p_2$ because (a) $d_1/d_2 = c_1/c_3$ by hypothesis, and (b) $d_1/d_2 = p_1/p_2$ by Euclid, XII.1. (See Commentary, line 14, below.)

(6) But $c_1 > p_1$ and thus $c_3 > p_2$. (Cf. *De Sphaera*, Book I, *post* postulate 5.)

(7) However $p_2 > c_3$ by construction.

* The following table relates the symbols used in this commentary to the quantities given in the figure and text:

$d_1 = CA$	c_3 = circum of circle MS (2nd half of proof)
$d_2 = GE$	c_4 = circum of circle H
c_1 = circum of circle CBA	p_1 = perim of polygon BAC
c_2 = circum of circle FEG	p_2 = perim of polygon FEG
c_3 = circum of circle IK (1st half of proof)	

(8) Since (7) is in fact true and contradicts (6), the inference of (6) must be false and thus the supposition from which it was drawn, namely, that $c_3 < c_2$, must be false.

(9) Now suppose that $d_1/d_2 = c_1/c_3$ where $c_3 > c_2$.

(10) Then let $d_2/d_1 = c_2/c_4$ where $c_4 \neq c_1$.

(11) Then $c_3/c_2 = c_1/c_4$ because $c_3/c_1 = c_2/c_4$, this being true by eliminating d_1 and d_2 from the equations of (9) and (10).

(12) Hence $c_1 > c_4$ since from (9) $c_3 > c_2$.

(13) Therefore $d_2/d_1 = c_2/c_4$ where $c_4 < c_1$.

(14) But (13) can be shown to be false by steps similar to those of (3) to (8). Hence if (13) is false, the supposition from which it was derived is false, namely, that $c_3 > c_2$.

(15) Since $c_3 \not> c_2$ and $c_3 \not< c_2$, then $c_3 = c_2$ and $d_1/d_2 = c_1/c_2$. Q.E.D.

The author concludes the proof with the statement that the proposition can be proved in another way by the application of the third proposition of the *De mensura circuli* of Archimedes.

14 "primam duodecimi." Although the author of the *Liber de curvis* no doubt precedes the Adelard II translation in point of time, in order to keep uniformity of citation (and since the proposition as given in Adelard II is basically the same as in the Greek text) I shall quote the Adelard II Version (Brit. Museum, Add. 34018, 63r): "Omnium duarum superficierum multiangularum inter duos circulos descriptarum est proportio alterius ad alteram tanquam proportio quadratorum que ex diametris circulorum eas circumscribentium proveniunt." Now if polygon$_1$/polygon$_2 = d_1^2/d_2^2$ [by Proposition XII.1] and if polygon$_1$/polygon$_2 = p_1^2/p_2^2$, then it is obvious that $d_1/d_2 = p_1/p_2$, as stated in step (5) of the argument given in the previous comment.

Proposition IV

1–37 "Quarumlibet....proposuimus." Again (see Fig. 66) let us restate Johannes' argument.* In the fourth proposition Johannes will prove that $S_2 - S_1 = \frac{1}{2}(l_2 - l_1) \cdot (c_2 + c_1)$ where S_2 and S_1 are the lateral surfaces of similar cones and l_2 and l_1 are their slant heights and

* The following table relates the symbols used in this commentary to the quantities given in the figure and text:

$l_1 = IM = SZ$ $\qquad$ $c_1 = ZY =$ circum of circle MNA

$l_2 = CE = ST$ $\qquad$ $c_2 = TP =$ circum of circle EQB

c_2 and c_1 are the circumferences of their bases. Cf. Archimedes, *De sphaera et cylindro*, Book I, Proposition 16, and *Verba filiorum*, Proposition XI; both are cited in the commentary to Proposition XI of the *Verba filiorum* in Chapter IV above. The proof follows:

(1) Let there be a right traingle $F_2 = \frac{1}{2}(l_2 \cdot c_2)$, and thus, by Proposition I of the *De curvis*, we can say that $S_2 = F_2$.

(2) From l_2 subtract the length l_1 and erect a line x parallel to c_2. There will be formed then a triangle $F_1 = l_1 \cdot x$ similer to triangle F_2.

(3) Then $l_2/l_1 = c_2/x$ since F_1 is similar to F_2.

(4) Hence $x = c_1$ since (a) $l_2/l_1 = d_2/d_1$ by definition of similar cones, and (b) $d_2/d_1 = c_2/c_1$ by Proposition III of the *De curvis*, and hence (c) $l_2/l_1 = c_2/c_1$.

(5) Hence $F_1 = S_1$ by Proposition I of the *De curvis.*

(6) Thus $S_2 - S_1 = F_2 - F_1$.

(7) But $F_2 - F_1$ is a quadrangular area whose base is $l_2 - l_1$ and whose erect sides are c_2, c_1. This area can be divided into two triangles with bases respectively c_2 and c_1 and with the same altitude $l_2 - l_1$.

(8) $F_2 - F_1 = \frac{1}{2}(l_2 - l_1) \cdot (c_2 + c_1) = S_2 - S_1$, by the addition of the triangles formed in (7). Q.E.D.

Proposition V

1–117 "Si....propositum." Cf. *De sphaera et cylindro*, Book I, Proposition 24 (ed. of Heiberg, 95–97). The fifth proposition is crucial for Johannes' ultimate determination of the surface area of the sphere. This proposition finds the surface area of a solid composed of conoid segments generated by the revolution of a regular polygon inscribed in a circle around one of the diameters of the circle. Although the author does not stipulate in the statement of the proposition the number of sides of the original polygon whose revolution describes the solid in question, like Archimedes he assumes in his proof that the number of sides of this polygon is divisible by four.

To understand what Johannes is proving and the nature of his proof,* let us take an equilateral polygon of $4\,n$ sides, each side of magnitude t (see Fig. 67). Let this polygon be inscribed in a circle

* The following table relates the symbols used in this commentary to the quantities given in the figure and text:

$c = ADIH$	$h_1 = YI$	$r_1 = BY$
$d = AI$	$h_2 = TY$	$r_2 = LS$
$f = EH$	$h_3 = ST$	$t = DE$

of circumference c and diameter d. Then let the circle with the inscribed polygon be rotated about a diameter of the circle. By this rotation the circle describes a sphere and the polygon describes a figure with conical surfaces (cf. *De sphaera*, Book I, Proposition 24).

The first part of Johannes' proposition states that the surface of this solid S will be given as follows: $S = t\,(c_1 + c_2 + \ldots + c_{n-1})$ where $c_1, c_2, \ldots, c_{n-1}$ are the circumferences of the circles described by the vertices of the angles of the polygon in the course of its rotation; t is the side of the polygon.

The second part of the proposition affirms further that the same surface will also be given by this relationship: $S = c \cdot f$, where f is one of the sides of the right triangles whose hypotenuse is the diameter of the circle, d, and whose other side is a side t of the polygon; c is the circumference of the circle.

The first part of the proposition is easily proved by adding up the conical surfaces $s_1, s_2, \ldots, s_n$ described by n sides of the polygon as it rotates. These surfaces are, of course, equivalent to the total surface S. The proof follows:

(1) $s_1 = \frac{1}{2}\, t \cdot c_1$ by *De curvis*, Proposition I.

(2) $s_2 = \frac{1}{2}\, t\,(c_1 + c_2)$ by *De curvis*, Proposition IV.

$s_3 = \frac{1}{2}\, t\,(c_2 + c_3)$ by *De curvis*, Proposition IV.

....

$s_{n-1} = \frac{1}{2}\, t\,(c_{n-2} + c_{n-1})$ by *De curvis*, Proposition IV.

$s_n = \frac{1}{2}\, t \cdot c_{n-1}$ by *De curvis*, Proposition I.

(3) Hence $S = t\,(c_1 + c_2 + \ldots + c_{n-1})$ by addition. Q.E.D.

The second part of the proposition, namely, $S = c \cdot f$, is proved as follows (referring to Fig. 67):

(1) $h_1/r_1 = h_2/r_1 = h_3/r_2 = \ldots h_{2n-3}/r_{n-1} = h_{2n-2}/r_{n-1} = t/f$ by the proportional sides of similar triangles.

(2) Now $h_1 + h_2 + h_3 + h_4 + \ldots + h_{2n-3} + h_{2n-2} = d$ and $r_1 + r_1 + r_2 + r_2 + \ldots + r_{n-1} + r_{n-1} = d_1 + d_2 + \ldots + d_{n-1}$.

(3) Thus $d/(d_1 + d_2 + \ldots + d_{n-1}) = t/f$ from (1) and (2) together.

(4) But $d/(d_1 + d_2 + \ldots + d_{n-1}) = c/(c_1 + c_2 + \ldots + c_{n-1})$ by *De curvis*, Proposition III.

(5) Hence $t/f = c/(c_1 + c_2 + \ldots + c_{n-1})$.

(6) But $S = t\,(c_1 + c_2 + \ldots + c_{n-1})$ from the first part of this proposition.

(7) Hence $S = c \cdot f$. Q.E.D.

An additional proof for a solid generated by the revolution of a semipolygon of odd-numbered sides follows the same form of the principal proof. It needs no further analysis.

78–79 "penultimam quinti." Although, as I have said, it is improbable that this work is an original Latin composition and hence the author had no access to the Adelard II version of the *Elements*, we can cite the Adelard wording of Proposition V.24 (Ms. cit., f. 21r) for the sake of uniformity of citation throughout this volume and since the Adelard reading of this proposition is equivalent to the reading of the Greek text: "Si fuerit proportio primi ad secundam tanquam tertii ad quartum, proportio vero quinti ad secundum tanquam sexti ad quartum, erit proportio primi et quinti pariter acceptorum ad secundum tanquam sexti et tertii pariter acceptorum ad quartum."

Proposition VI

1–41 "cuislibet....propositum." The comparative propositions of Archimedes and the Banū Mūsā are noted below in the Commentary to lines 42–56. Johannes is now prepared to treat the surface area of a sphere which he finds in this proposition to be equal to a rectangle, one of whose two adjacent sides is equal to the diameter of the sphere and the other to the circumference of the sphere. Restating Johannes' proposition symbolically, it reads $S = d \cdot c$ where S is the surface of the sphere, d and c the diameter and circumference of a great circle of the sphere. Proof* (see Fig. 69):

(1) Either $d \cdot c = S$, or $d \cdot c \neq S$.

(2) If $d \cdot c = S$, then $d \cdot c = S_1$ where $S_1 < S$ or $S_1 > S$ (by the basic postulate).

(3) Let us assume first that $d \cdot c = S_1$ where $S_1 < S$. Let us further suppose that S_1 is the surface of a sphere concentric with S. We designate the circumferences of great circles of these spheres c_1 and c. Of course $c > c_1$.

(4) Now inscribe within c an equilateral polygon of side t and of perimeter p such that $c > p > c_1$ and p does not touch c_1. Then let c revolve about d to produce S and similarly c_1 to produce

* The following table relates the symbols used in this commentary to the quantities given in the figure and text:

$p = AF + FD + \ldots + ZA$	$c = ADBC$
$t = AF$	$c_1 = SNH$
$d = AB$	$q = FB$

S_1 and p to produce the surface S_2 composed of a series of conical surfaces.

(5) Then $d \cdot c > q \cdot c$ since $d > q$, d being opposite a right angle.

(6) Hence $S_1 > S_2$ since (a) $d \cdot c = S_1$ from (3) and (b) $q \cdot c = S_2$, by *De curvis*, Proposition V.

(7) But, in fact, $S_2 > S_1$ by construction.

(8) Since (6) is contrary to the fact of (7), then (6) is false, and the assumption from which it follows is false, namely, that $S_1 < S$.

(9) By a similar series of steps Johannes derives a contradiction from $S_1 > S$.

(10) Hence, if $S_1 \ngtr S$ and $S_1 \nless S$, $S_1 = S$, and thus $d \cdot c = S$. Q.E.D.

42–56 "Ex....probandum." Three corollaries to this proposition are given and proved from it easily. The first states that the surface of the sphere is equal to four times the area of a great circle of the sphere. (For the statement of this by Archimedes and the Banū Mūsā, see the Commentary to Proposition XIV of the *Verba filiorum* in Chapter IV above.) From the first proposition of the *De mensura circuli*, Johannes asserts that the area of a circle is equal to one quarter the product of the diameter and the circumference. From his own Proposition VI he gives the surface of a sphere as the product of the diameter and circumference; hence the surface of the sphere is four times the area of a great circle.

The second corollary states that the surface of a sphere is equal to the lateral surface of a cylinder whose axis and base diameter are equal to the diameter of the sphere. This corollary is immediately evident from Proposition II of the *De curvis* taken together with Proposition VI. Similarly, if we add the area of the two end circles of such a cylinder to find its total surface, we shall have $S = \frac{3}{2} d \cdot c$, as the last corollary asserts.

Description and Proposition VII

2–4 "Omne...contineri." Although this so-called "description" is phrased in a very general way, it is used in Proposition VII only to justify the conclusion that the volume of a right cylinder is equal to the product of the base and altitude.

5–173 "Omne....sumus." Having in the first six propositions determined the areas of certain "curved" surfaces including finally that of a sphere, Johannes is now ready to go on to the problem of the volumes of the solids bounded by the curved surface areas already discussed, with the primary objective of determining the volume of a sphere. Before going to the question of the volume of a sphere, Johannes introduces this crucial Proposition VII, which seeks the volume of the solid whose area has been found in Proposition V, i.e., the solid generated by the the rotation of a regular polygon inscribed in a circle around the diameter of the circle. This is virtually the same theorem as that given by Archimedes, *De sphaera et cylindro*, Book I, Proposition 26 (ed. of Heiberg, pp. 101–105). Archimedes' proof, however, is different from the one presented here by Johannes. Johannes again implies in this proposition, although he does not state it, that the number of the sides of the polygon is divisible by four.

Proposition VII states, and Johannes ventures to prove, that $V = P = \frac{1}{3}S \cdot r_i$ where V is the volume of the solid formed by the rotation of a regular polygon of $4n$ sides around the diameter of the circle in which it is inscribed, P is the volume of an equivalent cone, S is the surface of the solid of volume V, and r_i is the radius of a sphere inscribed in the solid of volume V. Proof* (see Fig. 70):

(1) Construction: Take a quarter of the circle circumscribing the polygon of $4n$ sides. Inscribed in that quarter there will be n sides (Johannes takes a figure which has 3 sides; I have generalized the proof for a figure with n sides), which we designate $t_1, t_2, \ldots, t_n$. If we connect the midpoints of these sides with the center of the circle, we shall have equal lines $r_{i1}, r_{i2}, \ldots, r_{in}$, each of them equal to the radius of a circle inscribed within the polygon. Then if we connect the center with the extremities of the sides of the polygon, we shall have equal lines $r_{c1}, r_{c2}, \ldots, r_{cn}$, each equal to the radius of the circle in which the polygon is inscribed. Now t_1 intersects r_{c1}; but extend t_2 by the line e_1 and extend r_{c1} by f_1. Similarly, extend t_3 until it joins the extension of r_{c1}. Call these extensions e_2 and f_2 respectively. Make similar extensions for the remaining sides. If we rotate this quarter of a circle with its appended lines, the quarter itself forms a hemisphere, and the quarter of the regular polygon forms a solid of volume $V/2$. Each side of the polygon

* The following table relates the symbols used in this commentary to the quantities given in the figure and text:

$r_{c1} = OC$	$e_1 = FD$	$r_{i2} = OZ$	$t_3 = AN$
$r_{c2} = OF$	$e_2 = NE$	$r_{i3} = OT$	$f_1 = DC$
$r_{c3} = ON$	$r_1 = FP$	$t_1 = FC$	$f_2 = DE$
$r_{c4} = OA$	$r_{i1} = OI$	$t_2 = NF$	

describes conical surfaces $s_1, s_2, \ldots, s_n$. The circles forming the bases of these conical surfaces are of areas $a_1; a_1, a_2; \ldots a_{n-1}, a_n$; the radius of a is r_1, that of a_2 is r_2, etc. Furthermore, $V/2$ is composed of a series of special volumes, $v_1 + v_2 + \ldots + v_n$; v_1 is the volume of the double cone formed by the rotation around r_{c1} of the triangle of sides t_1, r_{c1}, and r_{c2}; v_2 is the volume generated by the revolution of the triangle of sides t_2, r_{c2}, r_{c3}. And similarly for $v_3 \ldots v_n$.

The following auxiliary volumes should be noted:

(a) v_{e1} formed by the rotation of triangle with sides e_1, $r_{c1} + f_1$, and r_{c2}.

(b) v_{e1t2} formed by the rotation of the triangle with sides $e_1 + t_2$, $r_{c1} + f_1$, and r_{c3}.

(c) v_{e2} formed by the rotation of the triangle of sides e_2, $r_{c1} + f_1 + f_2$, and r_{c3}.

(d) v_{e2t3} formed by the rotation of the triangle of sides $t_3 + e_2$, $r_{c1} + f_1 + f_2, r_{c4}$.

(2) Now $v_1 = \frac{1}{3}r_{i1} \cdot s_1$ because (a) $t_1/r_1 = r_{c1}/r_{i1}$ by similar triangles; and (b) $t_1/r_1 = s_1/a_1$ by *De curvis*, Corollary to Proposition I; and thus (c) $r_{c1} \cdot a_1 = r_{i1} \cdot s_1$; but (d) $r_{c1} \cdot a_1 = 3\ v_1$ by Euclid, Proposition XII.9 (Gr. XII.10).

(3) And $v_2 = \frac{1}{3}r_{i2} \cdot s_2$ because (a) $v_2 = v_{e1t2} - v_{e1}$ and (b) $v_{e1t2} = \frac{1}{3}r_{i2} \cdot (s_e + s_2)$ by steps like (2) (note: s_e is the surface generated by e_1 in its revolution); and (c) $v_{e1} = \frac{1}{3}r_{i2} \cdot s_e$.

(4) Similarly $v_3 = \frac{1}{3}r_{i3} \cdot s_3$.

....

$$v = \tfrac{1}{3}r_{in} \cdot s_n.$$

(5) Hence $V = \frac{1}{3}r_i \cdot S$ because (a) $r_{i1} = r_{i2} = r_{i3} = \ldots = r_{in} = r_i$; (b) $V/2 = v_1 + v_2 + \ldots + v_n$; and (c) $S/2 = s_1 + s_2 + \ldots + s_n$. Q.E.D.

22, 37, 44–45, 63, 83, 142, 145 "IX duodecimi." This is Proposition XII.10 in the Greek text. Notice that it is XII.9 in both the Adelard and Arabic texts. For the possible bearing of this fact upon the question of whether the *Liber de curvis superficiebus* is an original Latin composition or translation, see note 8 of the Introduction to this section. The text of the proposition runs in the Adelard II Version as follows (*ms. cit.*, 65v): "Omnis columna rotunda pyramidi sue tripla esse probatur."

50 "primam secundi." See the Commentary, the Corpus Christi Version, Chapter Three, Section 5, lines 66–67.

95–113 "Regula....insistamus." Here Johannes digresses to prove by reference to Euclid that the difference between two unequal cones of the same altitude is equal to a cone of the same altitude whose base is equal to the difference between the bases of the original cones.

104, 110, 168 "XI duodecimi." Proposition XII.11 in the Adelard II Version of the *Elements* runs (*ms. cit.*, 66r): "Omnes duas rotundas pyramides sive columnas eque altas suis basibus proportionales esse necesse est."

174–80 "Si....posteritati." The significance of this passage for the question of whether this tract is an original composition or transaction is discussed in note 8 of the Introduction to this section.

Proposition VIII

1–5 "Omnis....est." The eighth proposition of the *De curvis* is the familiar corollary of Archimedes' *De sphaera et cylindro*, Book I, Proposition 34 (ed. of Heiberg, pp. 131–33), asserting that a cylinder with an altitude equal to the diameter of a sphere has a volume 3/2 that of the sphere. It was also known to the Latin readers of the anonymous *De ysoperimetris*. (See Appendix III, paragraph 8.) It should be noted that with Johannes this is the fundamental theorem regarding the volume of a sphere, while with Archimedes this proposition is derived from an earlier more fundamental proposition. Hence there is no direct correspondence between the proofs of Johannes and Archimedes.

6–27 "Ne....proposito." Johannes initiates the proof with an *elementum* on proportions to the effect that if there are six terms, a through f, such that $a/b = c/d$ and $e \cdot a > f \cdot b$, then $e \cdot c > f \cdot d$.

28–87 "Esto....defraudati." The proof (see Fig. 73) can be summarized as follows*: $C = \frac{3}{2} V$, where V is the volume of a sphere whose diameter is d and C is the volume of a cylinder whose altitude and base diameter both are equal to d.

(1) Either $C = \frac{3}{2} V$ or $C \neq \frac{3}{2} V$.

(2) If $C \neq \frac{3}{2} V$, then there is some sphere V_1 such that $C = \frac{3}{2} V_1$. And either $V_1 < V$ or $V_1 > V$.

(3) Assume first that $V_1 < V$, with V and V_1 being concentric.

(4) Construction: Inscribe within c, the circumference of a great

* The following table relates the symbols used in this commentary to the quantities given in the figure and text:

$c = ADEC$	$h = OI$	$r = AO$
$c_1 = HYT$	$q = GE$	$d = AE$
$p = AG + \ldots + [L]A$		

circle of V, a regular polygon of perimeter p such that $c > p > c_1$ where c_1 is the circumference of a great circle of sphere V_1 and where p does not touch c_1. Construct h and q as in Proposition VI. Let V_2 be the volume of the solid formed by the rotation of the polygon around the diameter d. Let S be the surface of that solid. P is the surface of half that solid. The radius of sphere V is r. The area of great circle of radius r we designate as A.

(5) Then $r/q = A/P$ because (a) $r \cdot c = 2\,A$ by *De mensura circuli*, Proposition I; and (b) $q \cdot c = 2\,P$ by *De curvis*, Proposition IV.

(6) And $r \cdot d > h \cdot q$ since $r > h$ and $d > q$.

(7) Hence $d \cdot A > h \cdot P$ since in (5) and (6) we have six terms fulfilling the conditions of the *elementum* advanced at the beginning of the proposition.

(8) Thus $C > \frac{3}{2}\,V_2$ because (a) $d \cdot A = C$ and (b) $h \cdot P = \frac{3}{2}\,V_2$.

(9) But $C > \frac{3}{2}\,V_1$ since $V_2 > V_1$ by construction.

(10) Thus the fact of (9) contradicts the assumption of $C = \frac{3}{2}\,V_1$ where $V_1 < V$; and hence that assumption must be false.

(11) Assuming the other possibility, that $C = \frac{3}{2}\,V_1$ where $V_1 > V$, Johannes shows by similar steps that a contradiction follows.

(12) Hence, if $V_1 \nless V$ and $V_1 \ngtr V$, $V_1 = V$ and $C = \frac{3}{2}\,V$. Q.E.D.

The proof for the surfaces was given in Proposition VI.

72 "primam duodecimi." See the Commentary to Proposition III, line 14.

Proposition IX

1–27 "omnis....aspiravimus." The ninth proposition equates the volume of a sphere to a cone whose base is equal to the surface of the sphere and whose altitude is equal to the radius of the sphere, that is, $V = \frac{1}{3}\,d \cdot c \cdot r$. This formulation is equivalent to the more familiar modern form using π, namely, $V = \frac{4}{3}\pi\,r_3$. Johannes' proof is simple. The relationship between a cylinder and cone is known and also that between a cylinder and a sphere. Hence the relationship between a cone and a sphere immediately follows. Cf. the statements of this proposition by Archimedes, the Banū Mūsā, Leonardo Pisano, and the author of the *De ysoperimetris* in the Commentary to Proposition XV of the *Verba filiorum* (Chapter Four above).

14 "IX duodecimi." See the Commentary to Proposition VII, line 22.

25 "XII." See the Commentary to Proposition VII, line 104.

Proposition X

1–23 "Cuislibet.... XI." The tenth and the last proposition of the *De curvis* has no parallel proposition in the *De sphaera et cylindro*. But it represents, as Johannes says, a proposition similar to that of the second proposition of the *De mensura circuli*. Proposition II of the *De mensura circuli* showed that, if we assume a value of π as $3\frac{1}{7}$, the ratio of the area of a circle to the square on its diameter is as 11 to 14. Making a similar approximation for π, this last proposition of the *De curvis* asserts that the ratio of the volume of a sphere to cube of its diameter is as 11 to 21. Johannes carefully points out at the beginning of the proof that both he and Archimedes are proceeding by approximation, and actually neither Archimedes' nor his own proposition is true (*vera*). Inverting his ratios, we can represent his proof as follows:

To prove: $V/d^3 = 11/21$

(1) $C/d^3 = 11/14$, C being the volume of a cylinder of axis d and base diameter d, because (a) $C/d^3 = A/d^2$, A being the area of a circle of diameter d; (b) $A/d^2 = 11/14$, by *De mensura circuli*, Proposition I.

(2) $C/V = 3/2$, by the *De curvis*, Proposition VIII.

(3) Hence $V/d^3 = 11/21$. Q.E.D.

21 "XXIIII quinti." Proposition V.24 runs in the Adelard II Version (*ms. cit.*, 21r): "Si fuerit proportio primi ad secundum tanquam tertii ad quartum, proportio vero quinti ad secundum tanquam sexti ad quartum, erit proportio primi et quinti pariter acceptorum ad secundum tanquam sexti et tertii pariter acceptorum ad quartum."

3. Two Propositions Interposed in the Text of the *Liber de curvis superficiebus*

In the fourteenth-century manuscript *D* (Florence, Bibl. Naz., Conv. Soppr. J.V.30) of the *Liber de curvis superficiebus* there occur on the margin of folio 1r and 1v two additional (or interposed) propositions concerned, respectively, with the difference between the lateral surfaces of two cones and with the difference between the lateral surfaces of two cylinders. I have not found these additional propositions in any other medieval manuscript, but they are present in manuscript D_1 (Paris, Bibl. Nat., Fonds latin 11247), which is a careless copy made from D in the sixteenth century. In D_1 the first additional proposition is placed in the body of the text after Proposition I, with a marginal note stating that "it was in the margin" (*erat in margine*), which means, of course, that the additional proposition was in the margin of manuscript D from which D_1 was copied. Similarly, the second additional proposition was placed by the scribe of D_1 in the text of the *Liber de curvis superficiebus* after Proposition III. Again, the scribe of D_1 noted that *hoc erat in margine*.

It seems evident that these two additional propositions had some currency from the early fourteenth century in spite of our not being able to find further manuscripts including them. For, as I indicated in Section 2 of this chapter (see notes 21, 22, and 24), from at least the year 1328, geometers tended to cite Proposition III of the *Liber de curvis superficiebus* as the fifth proposition. The most plausible way to account for this is to assume that there existed a commonly known text (the source of D) in which the two additional propositions had been added to the body of the text before the regular Proposition III.

The first additional proposition depends directly on Proposition I of the *Liber de curvis superficiebus*. It holds that the difference between the lateral surfaces of two unequal cones having equal slant heights is equal to the lateral surface of a third cone with the same slant height but with a base circumference (or diameter) equal to the difference between the base circumferences (or diameters) of the first two cones. By Proposition I of the *Liber de curvis superficiebus*, each of the surfaces of the two unequal cones is equated to a right triangle in which the right angle is contained by a line equal to the slant height and a line equal to the circumference of the base. The triangular difference between the two right triangles is then shown to be equal to a third right triangle which is equal to the lateral surface of the third cone. This proposition should be compared with Proposition IV of the *Liber de curvis superficiebus*.

The second additional proposition depends on Proposition II of the *Liber de curvis superficiebus*. It states that the difference between the lateral surfaces of two unequal cylinders having the same altitude is equal to the lateral surface of a third cylinder with the same altitude but having a base circumference (or diameter) equal to the difference between the base circumferences (or diameters) of the first two cylinders. By Propositions II of the *Liber de curvis superficiebus*, each of the surfaces of the two cylinders is equated to a rectangle with sides equal respectively to the altitude of the cylinder and the circumference of its base. The rectangular difference between these two rectangles then can be shown to be equal to the lateral surface of the third cylinder. It is at the end of this proposition that we find the term "interposed" (*interposite*) applied to the first proposition (and by implication to the second proposition as well).

I have already commented upon the characteristics of the two manuscripts D and D_1 in the Sigla of Section 2 of this chapter. I have followed D exclusively in my text, noting, however, the variant readings of D_1. The marginal references are to D. The figures are as in D except that I have rotated the quadrilateral $DEMN$ in Fig. 75 through 90 degrees.

[Interposed Propositions]

[Propositiones interposite]

1r /[I.] OMNIUM DUARUM ROTUNDARUM PYRAMIDUM INEQUALIUM, QUARUM YPOTHENUSE FUERINT EQUALES, SUARUM CURVARUM SUPERFICIERUM DIFFERENTIA EST EQUALIS CURVE SUPERFICIEI ROTUNDE PYRAMIDIS, 5 CUIUS YPOTHENUSA EST EQUALIS YPOTHENUSE RELIQUARUM PYRAMIDUM, CIRCUMFERENTIA VERO BASIS EQUALIS DIFFERENTIE CIRCUMFERENTIARUM BASIUM RELIQUARUM, VEL DIAMETER BASIS EQUALIS DIFFERENTIE DIAMETRORUM BASIUM RELIQUARUM.

10 Sint due pyramides inequales rotunde, quarum ypothenuse sint equales. Maior quidem *DHE*, eius basis circulus *GE*, ypothenusa vero *DE* [Fig. 76]. Minor vero *DCK*, cuius basis circulus *FK*, ypothenusa *DK*. Et sit circumferentia circuli *AL* equalis differentie circumferentiarum duorum circulorum *GE* et *FK*. Et super circulum *AL* erigatur 15 pyramis *DBL*, cuius ypothenusa *DL* ponatur equalis ypothenuse *DE*: fiatque, per hanc primam huius, triangulus orthogonius *DEN* equalis curve superficiei pyramidis *DHE*, ita quod *DE* latus trianguli sit equalis *DE* ypothenuse, et *DN* equale circumferentie circuli *GE*, et angulus *D* rectus. Et ex lateri *DN* resecetur *DL*, equalis circumfe- 20 rentie circuli *FK*. Et quia *DE*, *DK* sunt equales et angulus *D* rectus,

16 triangulis D_1 19 resetur D_1

[Interposed Propositions]

[I.] IN THE CASE OF TWO UNEQUAL CONES WHOSE SLANT HEIGHTS ARE EQUAL, THE DIFFERENCE BETWEEN THEIR LATERAL SURFACES IS EQUAL TO THE LATERAL SURFACE OF A [THIRD] CONE WHOSE SLANT HEIGHT IS EQUAL TO THE SLANT HEIGHT OF THE OTHER CONES AND WHOSE BASE CIRCUMFERENCE IS EQUAL TO THE DIFFERENCE BETWEEN THE BASE CIRCUMFERENCES OF THE OTHER CONES, OR WHOSE BASE DIAMETER IS EQUAL TO THE DIFFERENCE BETWEEN THE BASE DIAMETERS OF THE OTHER CONES.

Let there be two unequal cones whose slant heights are equal. The greater one we let be *DHE*, its base circle *GE*, and its slant height *DE* [see Fig. 76]. The smaller cone we let be *DCK*, its base circle *FK*, and

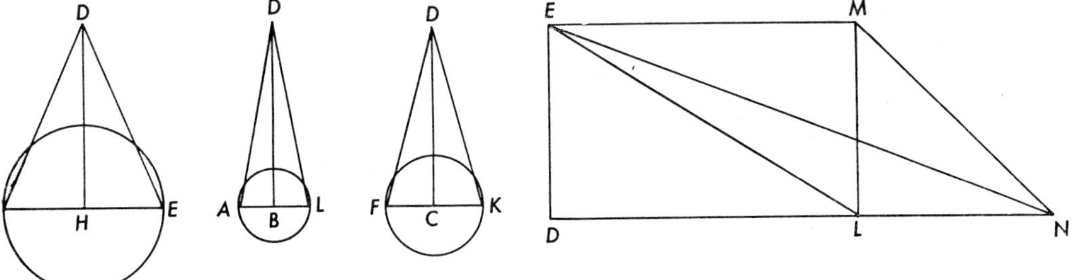

Fig. 76

its slant height *DK*. And let the circumference of circle *AL* be equal to the difference between the circumferences of the two circles *GE* and *FK*. And on circle *AL* let cone *DBL* be erected; its slant height *DL* is posited as equal to slant height *DE*. Then by the first [proposition] of this [work], let right △ *DEN* be constructed as equal to the lateral surface of cone *DHE*, so that side *DE* of the triangle is equal to slant height *DE* and [side] *DN* is equal to the circumference of circle *GE*, *D* being a right angle. And from side *DN* let *DL*, equal to the circumference of circle *FK*, be cut. And because [side] *DE* and [slant height] *DK* are equal and

ergo, per primam huius, triangulus *DEL* equalis curve superficiei pyramidis *DCK*. Ergo differentia curve superficiei pyramidis *DHE* et pyramidis *DCK* erit triangulus *ELN*. Et differentia circumferentiarum circulorum *GE* et *FK* erit linea *LN*. Dico tunc curvam superficiem pyramidis *DBL* esse equalem triangulo *LEN*. 25

Quod sic probatur: A puncto *L* educatur *LM* perpendicularis super lineam *DN* et equalis ypothenuse *DL*, et trahatur linea *MN*. Eritque triangulus *MLN*, per hanc primam, equalis curve superficiei pyramidis *DBL*. Sic triangulus *MLN*, per 37 et 33 primi, et per 28 primi Euclidis, est equalis triangulo *ELN*. Quare et curva superficies *DBL* erit etiam equalis triangulo *ELN*, quod est propositum. 30

Patet etiam, per 3 huius, quod sicut se habet circumferentia *GE* ad circumferentiam *AL*, ita se habet diameter *GE* ad diametrum *AL*. Et iterum, per eandem, sicut se habet circumferentia *FK* ad circumferentiam *AL*, ita se habet diameter *FK* ad diametrum *AL*. Ergo, per penultimam quinti, sicut se habent circumferentie *GE*, *FK* simul ad circumferentiam *AL*, ita se habent diametri *GE*, *FK* simul ad diametrum *AL*. Ergo, per conversam proportionalitatem, sicut se habet circumferentia *AL* ad circumferentias *GE*, *FK* simul, ita se habet diameter *AL* ad diametros *GE*, *FK* simul. Sed circumferentia *AL* est equalis differentie circumferentiarum *GE*, *FK*. Ergo diameter *AL* est equalis differentie diametrorum *GE*, *FK*, quod fuit ultimo propositum. 35 40

1r / [II.] OMNIUM DUARUM ROTUNDARUM COLUMNARUM INEQUALIUM, EQUE ALTARUM, SUARUM SUPERFICIERUM CURVARUM DIFFERENTIA EST EQUALIS CURVE SUPERFICIEI ROTUNDE COLUMNE, EQUE ALTE, CUIUS BASIS CIRCUMFERENTIA DIFFERENTIE CIRCUMFERENTIARUM RELIQUARUM BASIUM EST EQUALIS, VEL CUIUS BASIS DIAMETER DIFFERENTIE DIAMETRORUM RELIQUARUM BASIUM EST EQUALIS. 5

Sint predicte columne *AB*, *CD*; et fac tetragonum *GL* [Fig. 77], per 2 huius, equalem curve superficiei columne *AB*, et tetragonum *GK* 10

21 superficiem (?) D_1
24 circuli D_1 / LN: in D_1 / tunc curvam: curva tunc D_1
28 MLN: MRA D_1
29 Sic: dt D_1 / MLN: MIN D_1 / primi et: et primam D_1
33, 35 diameter: diametrum D_1
36 quinti *om.* D_1 *et habet lacunam* / GE: EG D_1
39, 40 simul: similiter D_1
40 diameter: diametri D_1
42 est equalis: et circulis D_1
4 eque *om.* D_1
7 diametri D_1

D is a right angle, therefore by the first [proposition] of this [work] $\triangle$ DEL is equal to the lateral surface of cone DCK. Therefore, $\triangle$ ELN will be [equal to] the difference between the lateral surfaces of cone DHE and cone DCK, and line LN will be [equal to] the difference between the circumferences of circle GE and circle FK. Then I say that the lateral surface of cone DBL is equal to $\triangle$ LEN.

This is proved as follows: From point L let line LM be erected as a perpendicular to line DN and equal to slant height DL; and let line MN be drawn. And $\triangle$ MLN, by this first [proposition], will be equal to the lateral surface of cone DBL. So $\triangle$ $MNL = \triangle ELN$, by [Propositions] I.37, I.33, and I.28 of Euclid. Therefore, the lateral surface of DBL will also be equal to $\triangle$ ELN, which has been proposed.

It is also evident, by the third [proposition] of this [work], that

$$\frac{\text{circum } GE}{\text{circum } AL} = \frac{\text{diam } GE}{\text{diam } AL},$$

and further that $\frac{\text{circum } FK}{\text{circum } AL} = \frac{\text{diam } FK}{\text{diam } AL}$. Therefore, by the penultimate [proposition] of [Book] V [of the *Elements*],

$$\frac{\text{circum } GE + \text{circum } FK}{\text{circum } AL} = \frac{\text{diam } GE + \text{diam } FK}{\text{diam } AL}.$$

Therefore, by inverse proportionality,

$$\frac{\text{circum } AL}{\text{circum } GE + \text{circum } FK} = \frac{\text{diam } AL}{\text{diam } GE + \text{diam } FK}.$$

But circum AL = circum GE − circum FK. Therefore, diam AL = diam GE − diam FK,* which was the last thing proposed.

[II.] IN THE CASE OF TWO UNEQUAL CYLINDERS WHOSE ALTITUDES ARE EQUAL, THE DIFFERENCE BETWEEN THEIR LATERAL SURFACES IS EQUAL TO THE LATERAL SURFACE OF A [THIRD] CYLINDER OF THE SAME ALTITUDE WHOSE BASE CIRCUMFERENCE IS EQUAL TO THE DIFFERENCE BETWEEN THE BASE CIRCUMFERENCES OF THE OTHER [CYLINDERS], OR WHOSE BASE DIAMETER IS EQUAL TO THE DIFFERENCE BETWEEN THE BASE DIAMETERS OF THE OTHER [CYLINDERS].

Let the aforesaid cylinders be AB and CD [see Fig. 77], and by the second [proposition] of this [work] make the rectangle GL equal to the lateral surface of cylinder AB and the rectangle GK equal to the lateral

* See the Commentary, lines 35–43.

equalem curve superficiei columne *CD*. Et proba tetragonum *HL* esse equalem curve superficiei columne *EF*, que ponitur eque alta columne *AB*, et circumferentia *EF* equalis differentie circumferentiarum *AB* et *CD*, qui quidem modus demonstrandi similis demon-
15 strationi alterius interposite [propositionis].

11 HL: HI D_1 12 eque: equi D_1

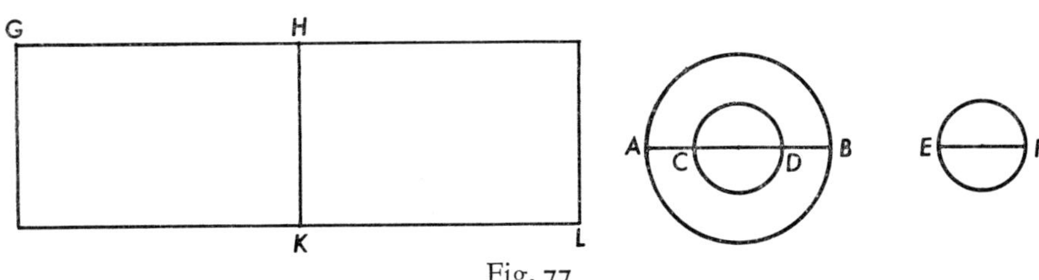

Fig. 77

surface of cylinder CD. And prove that rectangle HL is equal to the lateral surface of cylinder EF, which is posited as being equally high as cylinder AB, and circumference EF is equal to the difference between circumferences AB and CD, which method of demonstration is similar to the demonstration of the other interposed [proposition].

COMMENTARY

Proposition I

29 "per 37...28." Proposition I.37 in the Adelard II version of the *Elements* runs (Brit. Mus. Addit. 34018, 6r): "Equales sunt sibi cuncti trianguli qui super eandem basim atque inter duas lineas equidistantes fuerint constituti." Proposition I.33 (*ibid.*, 5v): "Si summitatibus duarum linearum equidistantium et equalis quantitatis alie due linee applicentur, ispe* etiam equales et equidistantes erunt." Proposition I.28 (*ibid.*, 5r): "Si linea una duabus [lineis rectis] supervenerit, fuertitque angulus eius intrinsecus** angulo intrinseco sibi opposito equalis, aut duo anguli intrinseci ex une parte duobus [angulis] rectis equales, ille due linee erunt equidistantes." Words in brackets in the Campanus edition of Basel, 1546.

35–43 "Ergo....propositum." Actually, what the author tacitly assumes at this point in the proof is that if Proposition V.24 of the *Elements*, using *componendo*, is accepted, then an equivalent proposition using *separando* is true, that is, if $(a + b)/c = (d + e)/f$, then $(a - b)/c = (d - e)/f$. Our author's argument, in effect, is that if $c = a - b$, then $f = d - e$, a, b, and c being the circumferences GE, FK, and AL, and d, e, and f being the diameters GE, FK, and AL. The transformation of

* ipse etiam *corr. m. rec. ex* utrobique ** *corr. ex* extrinsecus

V.24 assumed here is Simson's first corollary to Proposition V.24 (see T. L. Heath, *Euclid, The Elements* (vol. 2, p. 184).

36 "penultimam quinti," See above, this chapter, Commentary to Section 2, the *Liber de curvis superficiebus*, Proposition V, lines 78–79.

4. Three Propositions Added to the *Liber de curvis superficiebus*

In speaking of the popularity of the *Liber de curvis superficiebus* in Section 2 of this chapter, I noted that a reworking of the text appears in manuscript *M* (Florence, Bibl. Naz., Conv. Soppr. J.V.18, 92r–96v, 14c). After giving Proposition I of the Cambridge Version of the *De mensura circuli* (see Chapter Three, Section 1), the Latin author of this text then abridges the ten propositions of the *Liber de curvis superficiebus* and finally gives three further propositions on curved surfaces. Unfortunately, *M* is a very poor copy and needs much editorial emendation, both as to figures and text, in order to make mathematical sense. I can only hope that a better manuscript of this version will be found, but until such time I shall limit myself to editing and reconstructing the three additional propositions. The extensive revision needed for these three propositions will, I am sure, show the wisdom of my temporarily setting aside the preparation of the text of the abridgment of the ten propositions from the *Liber de curvis superficiebus*. The three additional propositions on curved surfaces I have numbered III (XI), IV (XII), and V (XIII). The numbers III, IV, and V indicate that these are three further propositions beyond the two propositions given in Section 3 of this chapter, while the numbers XI, XII, and XIII show that these propositions follow directly upon the ten propositions of the *Liber de curvis superficiebus* as given in manuscript *M*. I see no reason to doubt that these three additional propositions were, like the abridgement of the *Liber de curvis superficiebus* here given, done by a Latin author. For our author specifically distinguishes these additional propositions from those of the *Liber de curvis superficiebus* by referring the latter to Archimedes ("Arch."). Furthermore, if we assume that the author of these additional propositions was also responsible for the inclusion of Proposition I of the *De mensura circuli* in the Cambridge Version, then their Latin origin is assured, since that version was shown in Chapter Three above to be of Latin origin. Incidentally, the addition of Proposition I from the *De mensura circuli* as an introductory proposition makes considerable mathematical sense, since this proposition is frequently employed in the *Liber de curvis superficiebus* and the additional propositions.

Proposition III (XI), it should be noted, applies the so-called exhaustion method to a problem involving the surface of a segment of a sphere. It is evident that the author did not have access to Propositions 42 and 43 of Book I of the *De sphaera et cylindro* of Archimedes, for if he had had such access, he would not have had to apply the exhaustion procedure but only to apply simple proportions, as Archimedes did in proving Proposition 3 of Book II. It will be noted that Archimedes' Proposition II.3 is our author's corollary, except that Archimedes presented it as a problem:[1] "To cut a given sphere by a plane so that the surfaces of the segments may have to one another a given ratio." The whole proof of Proposition III (XI) was no doubt somewhat confused in its original form and the scribe has increased that confusion. But I am reasonably sure that my restored text represents the tenor of the original proof. Proposition IV (XII) is auxiliary to the last of the three propositions, while the last proposition itself tells us how to determine the volume of a truncated cone.

My text of the three propositions is freely corrected from *M* (although all of the original readings are, of course, included in the variant readings). There are certain peculiarities of the text that should be noted. The author often in a shorthand way refers to the "segment of the sphere," when he means "the surface of the segment of the sphere." I have added "surface" in my translation. Similarly, the author uses "poligonium" when he means the solid formed by the rotation of the polygon. The scribe of *M* uses both *ti* and *ci* before a vowel. I have always used *ti*, for example, *circumferentia*. The diagrams (Figs. 78 and 79) for Proposition III (XI) I have drawn from the text since they were missing in *M*. The only changes I have made in Fig. 80 are to put the circles in perspective, to draw a circle about *DEF*, and to relocate *K* slightly to indicate that it falls on the inner circle. The marginal references are to manuscript *M*, which has been described in Section 2 above.

[1] In the Moerbeke translation (MS Vat. Ottob. lat. 1850, 32r), Proposition II. 3, runs: "Tertium problema erat hoc: datam speram plano secare ita ut portionum superficies ad invicem habeant proportionem eandem date."

[Propositiones addite Libro de curvis superficiebus]

95v c. 1 / [III (XI).] ⟨SI⟩ CIRCULUS SPERAM PER INEQUALIA SECET, DIAMETRO SPERE PER CENTRUM EIUS TRANSEUNTE, ERIT PROPORTIO EADEM, SET ALTERNATIM SUMPTA, INTER PORTIONES DIAMETRI ET DIAMETRUM ET INTER 5 SUPERFICIES PORTIONUM SPERE ET DIVIDENTEM CIRCULUM.

Fiat ergo spera *ABGD*, diameter *AG*, circulus *M*, cuius diameter est *BD* [Fig. 78]. Dico ergo quod que est proportio totalis diametri ad illam partem que est *EG* ea est curve superficiei *ABD* ad circulum 10 secantem.

Si non, ergo ea est proportio minoris spere portionis vel maioris

4 portiones *correxi ex* proportiones / diametrum *corr. ex* diameterum 5 portionum *corr. ex* proportionum

[Propositions Added to the Book on Curved Surfaces]

[III (XI).] IF A CIRCLE CUTS A SPHERE INTO UNEQUAL [SEGMENTS] AND THE DIAMETER OF THE SPHERE PASSES THROUGH THE CENTER OF THE CIRCLE, THE INVERSE RATIO [OF EACH] OF THE SEGMENTS OF THE DIAMETER TO THE DIAMETER IS THE SAME AS THE RATIO [OF THAT ONE] OF THE SURFACES OF THE SEGMENTS OF THE SPHERE [WHOSE AXIS IS THE OTHER SEGMENT OF THE DIAMETER] TO THE [AREA OF THE] CUTTING CIRCLE.

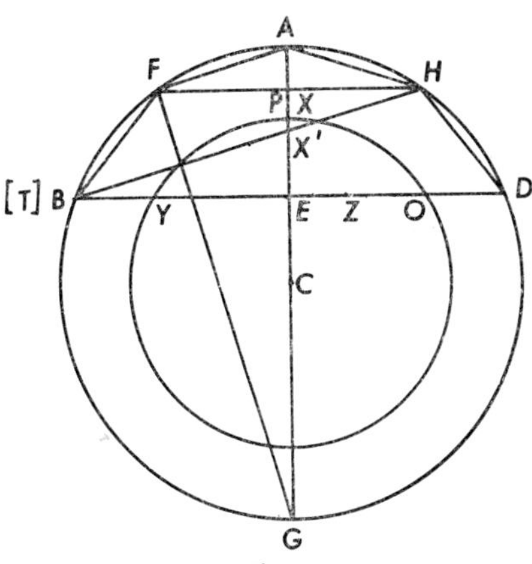

Fig. 78
Note: I have added the prime sign to X'.

Therefore, let sphere $ABDG$ be constructed [see Fig. 78] with diameter AG, and let the [cutting] circle be M with diameter BD [and center E]. I say, therefore, that the ratio of the whole diameter to its segment EG is the same as that of the curved surface [of sphere segment] ABD to the [area of the] cutting circle.

If not, the ratio is as that of [the surface of a] segment of a smaller or larger sphere to that circle. If it is of a [segment of a] smaller [sphere], let it be inscribed in it [i.e., in the given sphere]. Let the smaller sphere

ad illum circulum. Si minoris, inscribatur ei et sit *YXO*. Deinde per
c. 2 XII Euclidis inscribere poligonium maiori, scilicet quod non / tangat
interiorem. Deinde duobus lateribus poligonii subtrade (!) basim *FH*
15 et a *B* ad *H* duc lineam rectam et aliam ab *F* ad *G*. Age ergo ita V[a]
Arch[imenidis] probatur, et id quod fit ex ductu *AF* in circumferentiam *FH* et circumferentiam *BE* est equale curve superficiei quam
describunt *AF* et *BF*.

Item *AFG* et *AFP* sunt similes. Ergo que est proportio lateris *FG*
20 ad *FA* ea est *FP* ad *AP*. A simili que est *FG* ad *FA* ea est *PH* ad
PX' et ea est a simili *BE* ad *X'E*. Ergo per coniunctum que est
unius antecedentis ad unum consequens ea est omnium antecedentium
ad omnia consequentia. Ergo que est *FG* ad *FA* ea est *FPH* et *BE* ad
totalem lineam *AE*. Set que est linee ad lineam ea est circumferentie
25 ad circumferentiam, ut probabitur inferius. Ergo que est proportio
FG ad *FA* ea est circumferentie *FH* et circumferentie *BE* ad circumferentiam *AE*. Ergo quod fit ex *FG* in circumferentiam *AE* est
equale curve quam describunt *AF* et *BF*.

Item *AG* et *BD* secant se ad rectos in circulo. Ergo que est pro-
30 portio *EG* ad *BE* ea est *BE* ad *AE* et ea est *AE* ad aliam; sit illa
ZE. Et sit *TE* equalis *BE*. Age ergo quod fit ex ductu *BE* in *ZE* est
equum quadrato *AE*. Ergo que est proportio *GE* ad *BE* ea est *BE* ad
EA et ea est *EA* ad *EZ*. Ergo eversim que est *AE* ad *BE* ea est *ZE*
ad *EA*. Sint ergo *GE* cum *EA* primum, *BEZ* secundum, *BE* tertium,
35 *EA* quartum. Ergo id quod fit ex *GA* in *EA* est equum ei quod
fit ex *BZ* in *BE*. Set *EB* est equale *TE*. Ergo quod fit ex *ZT* in *BE*
est equum ei quod fit ex *GA* in *EA*. Et quod fit ex *GA* in circumferentiam *AE* est equum ei quod fit *ZT* in circumferentiam *BE*. Set

12 YXO *corr. ex* YZO
14 *ante* basim *add. M* poligonii *quod delevi*
15 ad[2] *corr. ex* a
17 BE *corr. ex* BL (?)
18 describunt *corr. ex* describitur
20 ea[2] *corr. ex* eam
21 BE ad X'E *corr. ex* BL ad XL
22 unius *corr. ex* minus (?)
23 BE *corr. ex* BL
24 AE *corr. ex* AL
26 *post* FA *scr. M* in circumferentiam ea est PH ad PZ FA *sed del. M* in circumferentiam *et delevi* ea est PH ad PZ FA / BE *corr. ex* BL
31 ZE[1] *corr. ex* AZE / ZE[2] *corr. ex* XE
32 *post* AE *iter. M lineas 29–32* (Ergo.... AE) *per errorem* ergo que est proportio EG ad BE, ea est BE ad AE, et etiam AE ad aliam; sit ille XE (!) est equum quadrato AE. Ergo que est et sit TE equalis BE. Age ergo que (!) fit ex ductu BE in XE (!) est equum quadrato AE. *Sed delevi totum.*
33 *post* BE *add. M* ad; *delevi*
34 BEZ *corr. ex* BET
36 quod *corr. ex* que
37 Et *corr. ex* est equum ei / GA *corr. ex* BX in BE *quod est sublineatun in M*

segment be YXO. Then by the [14th proposition of the] twelfth [book] of Euclid*: [the problem is] to inscribe a polygonal body [i.e., a solid] in the greater [sphere] so that it does not touch the interior [sphere]. Then from two sides of the polygon draw base FH, then draw a straight line from B to H and another one from F to G. Proceed, therefore, as in the proof of the fifth proposition [of the *Book on Curved Surfaces*] of Archimedes. And so, $AF \cdot$(circum FH** + circum BE***) = lat surf described by AF and BF.

Also, triangles AFG and AFP are similar. Therefore $FG/FA = FP/AP$. Similarly, $FG/FA = PH/PX' = BE/X'E$. Therefore, by the composition of ratios, the ratio of one antecedent to one consequent is as the ratio of all the antecedents [together] to all the consequents [together]. Therefore, $\frac{FG}{FA} = \frac{(FP + PH + BE)}{(AP + PX' + X'E)} = \frac{FH + BE}{AE}$. But [since] line is to line as circumference to circumference, as will be proved below, therefore $\frac{FG}{FA} = \frac{(\text{circum } FH + \text{circum } BE)}{\text{circum } AE}$.

Therefore, ($FG \cdot$ circum AE) = lat surf described by AF and BF.

Also, AG and BD intersect each other at right angles in the circle. Therefore, $EG/BE = BE/AE = AE/ZE$ [ZE being so constructed to complete the proportion]. And let $TE = BE$. Therefore, $BE \cdot ZE = AE^2$. Therefore, [since] $GE/BE = BE/EA = EA/EZ$, by the inversion of ratios, $AE/BE = ZE/EA$. Therefore, [by composition of ratios,] $\frac{GE + EA}{BE + EZ} = \frac{BE}{EA}$. Therefore, $GA \cdot EA = BZ \cdot BE$. But $EB = TE$. Hence $ZT \cdot BE = GA \cdot EA$. And ($GA \cdot$ circum AE) = ($ZT \cdot$ circum BE). But circum BE = 1/2 circum BD, because diameter is to diameter as circum-

* Proposition XII. 14, Adelard text, is XII. 17 in the Greek text.

** Where FH is considered as a diameter.

*** That is, 1/2 circum BD, where BD is a diameter equal to twice BE considered as a diameter.

circumferentia *BE* est subdupla ad circumferentiam *BD*, quia que
40 diametri ad diametrum et circumferentie ad circumferentiam. Ergo quod fit ex *AG* in circumferentiam *AE* equum est ei quod fit ex *ZT* in medietatem circumferentie *BD*. Ergo maius est quod fit ex *AG* in circumferentiam *AE* quam quod fit ex *FG* ⟨in circumferentiam *AE*⟩. Ergo quod fit ex *AG* in circumferentiam *AE* est maius quam
45 curva *AF* et *FB*. Ergo quod fit ex *ZT* in medietatem circumferentie *BD* est maius illa eadem curva. Set id quod fit ex *ZT* in medietatem circumferentie *BD* se habet ad circulum *BD* ut se habet *ZT* ad *BE* lineam, per primam sexti Euclidis; et per primam de mensura circuli sic circulus diametri *BD* fit ex linea *BE* in dimidiam circumferentiam
50 ipsius circuli, ut patet. At sicut *GE* ad *EA* ita est *BE* ad *EZ*. Ergo eversim sicut *AE* est ad *EG* ita *ZE* ad *EB*. Ergo coniunctim sicut est *AEG* ad *EG* ita est *ZET* ad *EB*. Ergo sic se habet id quod fit
96r c. 1 ex *TZ* / in dimidiam circumferentiam ipsius circuli *BD* ad circulum diametri *BD* ita se habet *AEG* ad *EG*. Set superficies portionis spere
55 *YXO* ita se habet ad circulum *BD* sicud habet ⟨*AG* ad *EG* per⟩ ipothesim. Ergo eodem modo se habent ad illum circulum portio spere *XYO* et id quod fit ex *TZ* in dimidiam circumferentiam *BD*. Ergo id quod fit ex *TZ* in dimidiam circumferentiam *BD* est equale illi portioni spere. Set positum est quod illud est maius curva poligonii.
60 Ergo illa curva spere est maior curva poligonii; set includit illum. Ergo inclusum est maius includente. Relinquit ergo quod non sit proportio portionis minoris spere ad circulum sicut diametri ad illam partem.

Ergo maioris; et sit idem centrum; et sit proportio portionis spere
65 maioris ad circulum *BED* sicut *AG* ad *EG*; et sit illa portio *Y'OX* [Fig. 79]. Dividatur iterum arcus *ABD* in arcus quos in medio contingentes linee convenient infra *Y'OX*, ut sit una superficies inclusa lineis

39 circumferentia *corr. ex* circumferentiam / BE *corr. ex* B
41 ZT *corr. ex* FG ergo quod fit ex AG et
42 *post* BD *habet M lacunam et postea* AG maior (?) FG *quod delevi*
42 medietatem circumferentie *corr. ex* circumferentiam
47 habet[2] *corr. ex* habent / BE *corr. ex* BD
49 *post* fit *add. M* linea *quod delevi*
50 EZ *corr. ex* EX
53 ad circulum *bis M*
54 AEG ad EG *corr. ex* AET ad G
55 YXO *corr. ex* XEX
57 XYO *corr. ex* XYE (?)
57, 58 TZ *corr. ex* TX
59 positum *corr. ex* posterius
60 maior *corr. ex* maiori
62 *post* portionis *add. M* diametri *quod delevi*
64 *ante* Ergo *add. et. del. M* Set arcus BAD in arcus quos in medio contingentes linee convenient infra YOX ut sit una.
65 portio *corr. ex* proportio

ference is to circumference. Hence $(AG \cdot \text{circum } AE) = (ZT \cdot 1/2 \text{ circum } BD)$. Hence $(AG \cdot \text{circum } AE) > (FG \cdot \text{circum } AE)$. Hence $(AG \cdot \text{circum } AE) >$ surf described by AF and FB. Therefore, $(ZT \cdot 1/2 \text{ circum } BD) >$ surf described by AF and FB. But $\frac{ZT \cdot 1/2 \text{ circum } BD}{\text{circle } BD} = \frac{ZT}{BE}$, by [Proposition] VI.1 of Euclid. And by the first [proposition] *On the Measurement of the Circle* [of Archimedes], the circle of diameter BD thus arises from the product of the line BE and one half the circumference of that circle, as is evident. But $GE/EA = BE/EZ$. Hence, by inversion, $AE/EG = ZE/EB$. Hence, by composition of ratios,

$$\frac{AE + EG}{EG} = \frac{ZE + ET}{EB}$$

[, EB being equal to ET]. Hence, $(TZ \cdot 1/2 \text{ circum } BD)/\text{circle } BD = (AE + EG)/EG$. But the surface of sphere segment YXO is related to circle BD as AG is to EG, by hypothesis. Therefore, in the same way,

$$\frac{\text{Surf segment of sphere } YXO}{\text{circle BD}} = \frac{TZ \cdot 1/2 \text{ circum } BD}{\text{circle BD}}$$

Therefore, $(TZ \cdot 1/2 \text{ circum } BD) =$ (surf segment of sphere XYO). But it was posited that it was greater than the curved surface of the [rotated] polygon. Therefore, that curved surface of the sphere is greater than the curved surface of the [rotated] polygon. But the latter includes it and so the "included" is greater than the "including," [which is impossible]. Therefore, it remains that the ratio of the [surface of the] segment of a lesser sphere to the circle is not as the diameter to its segment.

Therefore, [let us try the segment of a] greater [sphere]. Let it have the same center [as the given sphere]. And let the ratio [of the surface] of a segment of greater sphere to circle BED be as AG to EG. And let the segment be $Y'OX$ [see Fig. 79]. Again let arc ABD be divided into arcs at whose middle points tangent lines within $Y'OX$ meet so that a single

FH, ⟨*HK*,⟩ *KL*, *LM*, *MF*. Eductis hinc modo (?) eius superficiei medietas circumducti describit solidum circa portionem spere; et con-
70 tingit *HK* in puncto *Y*, et protrahantur linee *YC*, *FL*, *HL*. Quoniam ergo illud poligonium circumscribitur, erunt trianguli *YCK* et *KPH* similes, quia angulus *Y* est rectus propter angulum contingentie et angulus *P* rectus et angulus *K* communis. Ergo tertius tertio. Ergo que est proportio linee *CY* ad lineam *YK* eadem est proportio *HP* ad
75 *KP*. A simili que est proportio *CY* ad *YK* eadem *PL* ad *PN*, propter triangulos similes; et eadem sunt ad ea. Ergo que est *CY* ad *YK* eadem est *HL* et *FE* ⟨ad *KE*⟩. Ergo quod fit ex ductu *YC* in circumferentiam *KE* est equum ei quod fit ex ductu *YK* in circumferentiam *HL* et circumferentiam *FE*. Ergo quod fit ex ductu dupli *CY* in eandem
80 circumferentiam est equum ei quod fit ex ductu dupli *YK* in easdem circumferentias. Set duplum *CY* est *AG* diameter, scilicet duplum *YK* est *HK*. Ergo illud quod fit ex ductu *GA* in circumferentiam *KE* est equum ei quod fit ex ductu *KH* in circumferentiam *HL* et *FE*. Set ultimum est equale illis duabus superficiebus solidi, ut in
85 quinta Arch[imenidis] probatur. Ergo quod fit ex ductu *AG* in circumferentiam *KE* est equum illis duabus superficiebus. Set id quod fit ex ductu *AG* in circumferentiam *KE* est maius eo quod fit ex *GA* in circumferentiam *AE*. Set illud est equum ei quod fit ex ductu *TZ* in circumferentiam *BE*, ut ostensum est in priori parte huius proposi-
90 tionis, eadem facta dispositione de *T* et *Z*. Ergo quod fit ex ductu *AG* in circumferentiam *KE* est maius eo quod fit ex ductu *TZ* in circumferentiam *BE*. Set id quod fit ex *TZ* in circumferentiam *BE* est equum portioni superficiei sperice *Y'OX*. Set probatur illud omne equum predictis duabus superficiebus. Ergo ille due superficies sunt maiores
c. 2 illa portione. Ergo inclu/sum est maius includente. Et sic patet pro-
96 positum.

71 erunt trianguli *corr. ex* ert anguli
74 CY *corr. ex* CX
79 FE *corr. ex* BE
81 CY *corr. ex* EX
84 FE *corr. ex* FD / duabus *corr. ex* duobus
86 *post* equum *add. et del. M* ei quod fit ex ductu HK
95 portione *corr. ex* proportione

surface is included by lines FH, HK, KL, LM, and MF. Now with these lines so drawn the half of this surface when rotated describes a solid about the segment of the sphere; and it [the sphere] touches HK in point Y. And let lines YC, FL, and HL be drawn. Therefore, since that polygon is circumscribed, triangles YCK and KPH will be similar because $\angle Y$ is a right angle, being an angle of tangency, and $\angle P$ is a right angle, and $\angle K$ is common. Therefore, the third [angle is equal]

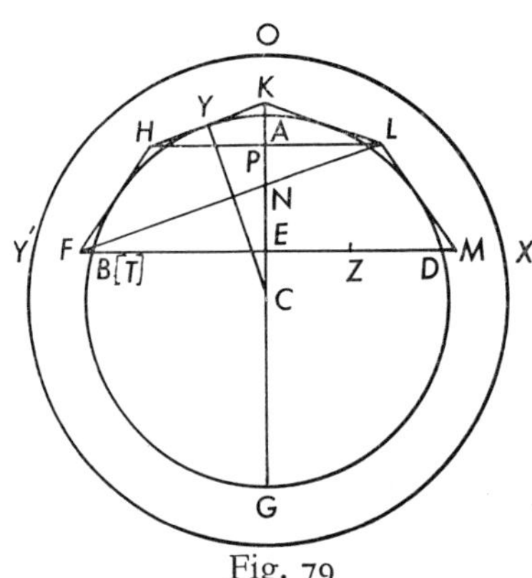

Fig. 79

Note: I have added the prime sign to Y'.

to the third [angle]. Hence $CY/YK = HP/KP$. Similarly, $CY/YK = PL/PN$, because of similar triangles. Therefore, by equality [and composition], $CY/YK = (HL + FE)/KE$ [and hence

$$\frac{CY}{YK} = \frac{\text{circum } HL + \text{circum } FE}{\text{circum } KE}\Bigg].$$

Therefore, $(YC \cdot \text{circum } KE) = YK \cdot (\text{circum } HL + \text{circum } FE)$. Therefore, $(2\ CY \cdot \text{circum } KE) = 2\ YK \cdot (\text{circum } HL + \text{circum } FE)$. But $2\ CY =$ diameter AG, and $2\ YK = HK$. Therefore, $(GA \cdot \text{circum } KE) = KH \cdot (\text{circum } HL + \text{circum } FE)$. But $KH \cdot (\text{circum } HL + \text{circum } FE)$ is equal to the two surfaces of the solid, as is proved in the fifth [proposition] of [the *Book on Curved Surfaces* of] Archimedes. Therefore, $(AG \cdot \text{circum } KE)$ is equal to those two surfaces. But $(AG \cdot \text{circum } KE) > (GA \cdot \text{circum } AE)$, [since circum $KE >$ circum AE]. But when T and Z are placed as before, $(GA \cdot \text{circum } AE) = (TZ \cdot \text{circum } BE)$, as was shown in the earlier part of this proposition. Therefore, $(AG \cdot \text{circum } KE) > (TZ \cdot \text{circum } BE)$. But $(TZ \cdot \text{circum } BE) =$ surf spherical segment $Y'OX$; but it is proved that $(AG \cdot \text{circum } EK) =$ surf described by KH and HF.

Similiter proportio alterius portionis ad eundem circulum non erit maior vel minor quam est proportio *AG* ad *AE*. Relinquitur ergo quod eadem proportio superficiei portionis sperice *BAD* ad superficiem alterius sicut proportio *AE* ad *EG*. Quoniam que est proportio *BAD* ad circulum *BED* eadem est *AG* ad *EG*, at similiter que est proportio circuli *BED* ad superficiem sperice portionis *BGD* ea est *AE* ad *AG*; ergo per eversam proportionalitatem que est portionis sperice ad portionem spericam, ea est *AE* ad *EG*.

[IV (XII).] SI ROTUNDE PIRAMIDIS AXIS IN TERTIAM PARTEM BASIS DUCATUR, PRODUCETUR SOLIDUM EQUALE PIRAMIDI.

Quia quod fit ex ductu axis in totalem basim piramidis est triplum piramidis, ut probatur in XII Euclidis, ergo quod fit ex ductu eiusdem ⟨in⟩ tertiam partem illius est equale piramidi.

[V (XIII).] IN OMNI CURTA PIRAMIDE SI A CENTRO BASIS AD YPOTHENUSAM DUCATUR PERPENDICULARIS, QUOD FIT EX EADEM PERPENDICULARI IN TERTIAM SUPERFICIEI PIRAMIDIS CUM QUOD EX AXE IN TERTIAM CIRCULI SUPERIORIS EQUATUR IPSI PIRAMIDI.

Sit ergo circulus *BYC*, cuius centrum *E*, diameter *BC*, et sit basis piramidis *BAC*, et sit *DLF* equidistans *BC* [Fig. 80]. Et circulus illius sit *MK* circa centrum *E*, axi tam sit eadem. Sumatur [*CEY*] sector circuli basis, tertia pars scilicet. Et deinde ad ypothenusam *AC* ducatur a centro perpendicularis *EP*. Est igitur *EK* equalis *LF* et sector *KEM* tertia pars circuli *KM*. Inde sic, id quod fit ex ductu *AE* in illum sectorem *CEY* est equale hec totali piramidi. Set quod fit ex *AL* in *KME* est equale hec piramidi parti *DLF*. Ergo quod fit ex ductu relique partis eius linee in eundem sectorem cum eo quod fit ex ductu

97, 99, 102, 103 portionis *corr. ex* proportionis
Prop. IV (XII)
4 triplum *corr. ex* duplum
Prop V (XIII)
1 curta *corr. ex* curva
3–4 superficiei *corr. ex* superficie
6 BC *corr. ex* BI
11 *post* ductu *add. et del. M* relique
13 DLF *corr. ex* DEF

Hence those two, surfaces [described by KH and HF] are together greater than [the surface of] that segment [$Y'OX$] Therefore, the "included" is greater than the "including", [which is impossible]. And so that which was proposed is evident.

Similarly, the ratio of the [surface of the] remaining segment to the [area of the] same circle will be neither greater nor less than the ratio of AG to AE. It remains, therefore, that

$$\frac{\text{surf spherical segment } BAD}{\text{surf spherical segment } BGD} = \frac{AE}{EG}, \text{ for}$$

$$\frac{\text{surf segment } BAD}{\text{circle } BED} = \frac{AG}{EG}, \text{ and similarly, } \frac{\text{circle } BED}{\text{surf segment } BGD} = \frac{AE}{AG}.$$

Hence, by inverse proportionality [and the elimination of the product of AG and circle BED], the [surface of the one] spherical segment is to the [surface of the other] spherical segment as AE is to EG.

[IV (XII).] IF THE AXIS OF A CONE IS MULTIPLIED BY ONE THIRD OF ITS BASE, A SOLID EQUAL TO THE CONE WILL BE PRODUCED.

Since the product of the axis and the whole base is three times the cone, as is proved in [Proposition 9* of Book] XII of Euclid's [*Elements*], therefore the product of the same [axis] and one third of its base is equal to the cone.

[V (XIII).] IN EVERY TRUNCATED CONE THE VOLUME IS EQUAL TO THE SUM OF (*a*.) THE PRODUCT OF (1) A PERPENDICULAR DRAWN FROM THE CENTER OF ITS [LOWER] BASE TO A SLANT HEIGHT [OF THE TRUNCATED CONE] AND (2) ONE THIRD OF THE [LATERAL] SURFACE OF THE [TRUNCATED] CONE AND (*b*.) THE PRODUCT OF THE AXIS AND ONE-THIRD PART OF THE UPPER [BASE] CIRCLE.

Let there be a circle BYC, whose center is E and whose diameter is BC [see Fig. 80]. And let it be the [lower] base of cone BAC, and let DLF be parallel to BC. And let there be a circle MK of center E within that [base circle which is equal to the circle of diameter DLF], and whose axis is the same. Let there be taken a sector [CEY] which is one third of the base circle. And then let $\perp EP$ be drawn from the center to the slant height AC. Hence $EK = LF$, and sector $KEM = 1/3$ circle KM. Then, ($AE \cdot$ sector CEY) = whole cone [BAC]. But ($AL \cdot$ sector KME) = partial cone DLF [A]. Hence the product of the difference in altitudes

* XII. 9 in the Adelard II text; XII. 10 in the Greek text.

15 totalis linee in reliquam partem, in *KCMY*, est equale relique parti eius totalis piramidis, scilicet curte piramidi. Ergo quod fit ex ductu *LE* in *EKM* cum eo quod fit ex *AE* in *KCMY* est equale curte piramidi.

Item anguli *L* et *E* sunt recti et *A* est communis; ergo trianguli 20 *ALF* et *AEC* sunt similes. Ergo que proportio *LF* ad *FA* ea est *EC* ad *CA*. Set que ypothenuse ad semidiametrum basis ea est curve piramidis ad circulum basis, per primam Arch[imenidis]. Item *P* angulus est rectus et *A* communis. Ergo *APE* et *CEA* sunt similes. Ergo que est proportio *CE* ad *CA* ea est *EP* ad *AE*. Set que est *EC* 25 ad *AC* eadem est curve totalis piramidis ad circulum basis. Set que totius ad totum ea est tertie ad tertiam. Ergo tertia pars superficiei piramidis *ADF* ad sectorem *EMK* sicut *AC* ad *EC*. Ergo..., i.e., superficiei pramidis *CF* ad superficiem residuam, i.e., *YCKM*, sicut *AC* ad *EC* et etiam sicut *AE* ad *EP*, propter similitudinem triangu-30 lorum. Ergo quod fit ⟨ex *EP*⟩ in tertiam superficiei *CF* est equum ei quod ex *AE* in superficiem *KCMY*. Set illud est equale curte piramidi cum eo quod *EL* in sectorem *MEK*, hoc est, per tertiam partem 96v c. 1 superioris circuli. Set quia ⟨*EP*⟩ est perpendicularis / ad *APC*, relinquitur ergo propositum.

15 *post* equale *add. et del. M* curte piramidi
16 totalis *corr. ex* c'te
17 KCMY *corr. ex* KC et MY
20 FA *corr. ex* LA
21 ypothenuse *corr. ex* altitudinis
23 APE *corr. ex* APA
25 AC *corr. ex* AP
27 ... *hic habet M lacunam*
29–30 *ante* triangulorum *del. M* c'te
30 superficiei *corr. M ex* superficiem
33 perpendicularis *corr. ex* perpendiculares / APC *corr. ex* IPO
34 *post* propositum *habet M* est *quod delevi*

[i.e., LE] and the same sector [i.e., KEM] plus the product of the whole altitude [i.e., AE] and the difference between the sectors, i.e., $KCMY$, is equal to the difference between the cones, i.e., to the truncated cone. Hence $(LE \cdot \text{sector } EKM) + (AE \cdot \text{sector } KCMY) =$ the truncated cone.

Also, angles L and E are right angles and $\angle A$ is common. Therefore, triangles ALF and AEC are similar. Therefore, $LF/FA = EC/CA$. But the slant height is to the radius of the base as the lateral surface of a cone is to the [area of the] base circle, by the first [proposition of the *Book on Curved Surfaces*] of Archimedes. Also, $\angle P$ is a right angle and $\angle A$ is common. Therefore, [triangles] APE and CEA are similar.

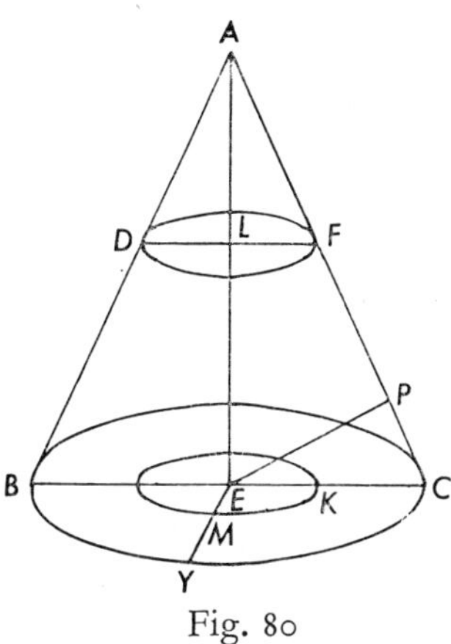

Fig. 80

Hence $CE/CA = EP/AE$. But EC is to AC as the lateral surface of the whole cone is to the base circle. But the ratio of whole to whole is as the ratio of their third parts. Therefore, $\dfrac{1/3 \text{ lat surf cone } ADF}{\text{sector } EMK} = \dfrac{AC}{EC}$.

Hence [the ratio of the difference between the surfaces of the two cones,] i.e., of the surface of the [truncated] cone of [slant height] CF, is to the difference between the surfaces [of the circular sectors], i.e., $YCKM$, as AC to EC, and also as AE to EP, because of the similarity of triangles. Therefore, $(EP \cdot 1/3 \text{ surf of trunc cone } CF) = (AE \cdot \text{surf } KCMY)$. But the product of AE and surface $KCMY$ is equal to the truncated cone together with* the product of EL and sector MEK, i.e., the third part of the upper circle. But since EP is a perpendicular to APC, therefore the proposition remains.

* Should be "minus" rather than "together with"; see the Commentary, Proposition V, step (7).

COMMENTARY

Proposition III (XI)

1–104 "⟨Si⟩....*EG*." It will perhaps be helpful to recapitulate the confusing argument so that it might be more easily followed:

Suppose that *Sps* is the surface of a segment of a sphere generated by the rotation of circular segment *BAD* around its axis *AEG*. Suppose that *Cd* is the area of the circle which cuts the sphere into unequal parts. *BD* is the diameter of that circle. It is perpendicular to the diameter of the sphere, namely, *AEG*. Then Proposition III holds that $Sps/Cd = AG/EG$.

Proof (see Fig. 78):

(1) [Either $Sps/Cd = AG/EG$ or $Sps/Cd \neq AG/EG$. If $Sps/Cd \neq AG/EG$, then there is some surface *Sx* of a segment of a sphere concentric with a given sphere such that $Sx/Cd = AG/EG$ and] $Sx < Sps$, or $Sx > Sps$.

(2) Suppose that $Sx = Syox < Sps$.

(3) $AF \cdot$ (circum FH + circum BE) = Spoly *ABD*, where Spoly *ABD* is the surface formed by the rotation of the two sides of a regular polygon inscribed within segment *BAD*. This follows from the procedures used in Proposition V of the *De curvis superficiebus* (q.v.).

(4) $\triangle\ AFG$ is similar to $\triangle\ AFP$, since both have a right angle and a common angle. Hence, $FG/FA = FP/AP$. By similar procedures, $FG/FA = FP/AP = PH/PX' = BE/X'E$.

(5) $FG/FA = (FPH + BE)/AE =$ (circum FPH + circum BE)/circum AE, by addition and [from Proposition III, *De curvis superficiebus*].

(6) Hence, $FG \cdot$ circum AE = Spoly *ABD*, from (3) and (5) together.

(7) $EG/BE = BE/AE$, since *BE* is perpendicular to *AG* within a semicircle.

(8) Then $BE/AE = AE/EZ$ where *EZ* is constructed to make the proportion hold true. Thus $BE \cdot EZ = AE^2$. By conversion, $AE/BE = EZ/EA$, and $(GE + AE)/BEZ = BE/EA$, by

composition. Or $GA \cdot EA = BZ \cdot BE$. But $BE = ET$ (i.e., B and T appear to be identical). Hence, $ZT \cdot BE = GA \cdot EA$.

(9) Therefore, $ZT \cdot$ circum BE = $GA \cdot$ circum AE, since BE/EA = circum BE/circum EA [by *De curvis superficiebus*, Proposition III]. But circum $BE = \frac{1}{2}$ circum BD. Hence, $GA \cdot$ circum $AE = ZT \cdot \frac{1}{2}$ circum BD.

(10) $GA \cdot$ circum $AE >$ Spoly ABD, since Spoly ABD = $FG \cdot$ circum AE from step (6), and $GA > FG$, GA being opposite a right angle.

(11) Therefore, $ZT \cdot \frac{1}{2}$ circum $BD >$ Spoly ABD.

(12) But $(ZT \cdot \frac{1}{2}$ circum $BD)/(BE \cdot \frac{1}{2}$ circum $BD) = ZT/BE$.

(13) But $ZT/BE = AG/EG$, since from (8) $GE/EA = BE/ZE$, and conversely $AE/EG = ZE/EB$, and by addition $AEG/EG = ZET/EB$.

(14) Therefore, Syox $= ZT \cdot \frac{1}{2}$ circum BD, since by hypothesis Syox$/Cd = AG/EG$, and $Cd = BE \cdot \frac{1}{2}$ circum BD, by Proposition I of the *De mensura circuli* and by substituting known and equivalent quantities in the equation of (12).

(15) Therefore, Syox $>$ Spoly ABD, from (11) and (14) simultaneously.

(16) But in actuality Spoly ABD includes Syox and hence must be greater. This contradicts the deduction of (15) and hence the assumption on which it is based, namely, that Syox $<$ Sps.

Part II of Proof (see Fig. 79).

(17) Since Syox $\nless$ Sps, let us suppose that Syox $>$ Sps.

(18) As before, Syox has the same center as Sps, but is now circumscribed. Polygon $FHKLM$ is circumscribed about BAD, so that its sides are tangent to BAD but do not touch the outer circle (by *Elements* XII.14 = Greek XII.17). The rotation of $FHKLM$ forms Spoly FKM. Note: side HK is tangent to BAD at Y and the perpendicular YC is constructed. HPL and FNL are drawn as indicated. Z and T are marked, as in the first part of the proof.

(19) $\triangle CYK$ is similar to $\triangle HPK$, since $\angle P = \angle Y$, both being right angles, and $\angle K$ is common. CYK is also similar to KPL, PLN, and FNE.

(20) $CY/YK = HP/KP = PL/PN = FE/NE$, because of similar triangles.

(21) $CY/YK = (HL + FE)/KE$, by addition of proportions.

(22) $CY \cdot$ circum $KE = YK \cdot$ circum $(HL + FE)$, by *De curvis superficiebus*, Proposition III. Thus $2\ CY \cdot$ circum $KE = 2\ YK \cdot$ circum $(HL + FE)$.

(23) Therefore, $AG \cdot$ circum $KE = KH \cdot$ circum $(HL + FE)$, since $2\ CY = AG$, and $2\ YK = HK$.

(24) Hence, $AG \cdot$ circum $KE =$ Spoly FKM, since by *De curvis*, Proposition V, Spoly $FKM = KH \cdot$ circum $(HL + FE)$.

(25) $AG \cdot$ circum $KE > AG \cdot$ circum AE [since $KE > AE$], and $AG \cdot$ circum $AE = ZT \cdot$ circum BE, as shown in the first part.

(26) Hence, $AG \cdot$ circum $KE > ZT \cdot$ circum BE.

(27) But $ZT \cdot$ circum $BE =$ Syox, as in first part [by the assumption that Syox/circle $CD = AG/EG$].

(28) Therefore, $AG \cdot$ circum $KE >$ Syox, or Spoly $FKM >$ Syox. But this is impossible since Syox includes Spoly FKM. Hence Syox $\ngtr$ Sps. Since Syox $\nless$ Sps and Syox $\ngtr$ Sps, then Syox $=$ Sps, and the proposition follows.

In the same way it can be shown that when Sps is the surface of the segment BGD, then Sps/Cd $= AG/AE$. Thus, by eliminating Cd, surface seg BAD/surface seg $BGD = AE/EG$, as the corollary tells us. Cf. Archimedes, *De sphaera et cylindro*, Book II, Proposition 3.

13 "XII Euclidis." Proposition XII.14 (= Gr. XII.17) runs in the Adelard-Campanus version (ed. Basel, 1546): "Duabus sphaeris unum centrum habentibus propositis, intra maiorem earum solidum multarum basium superficiem minoris sphaerae minime tangentium figuraliter constituere...."

25 "ut probabitur inferius." It is not "proved below," but rather follows from Proposition III of the *Liber de curvis superficiebus*.

48 "primam sexti." See the Commentary, Chapter Three, Section 3, the Cambridge Version, line 70.

Proposition IV (XII)

5 "XII Euclidis." See above, the Commentary to Section 2, Proposition VII, line 22.

Proposition V (XIII)

1–34 "In....propositum." This proof is recapitulated as follows:

To prove that the volume of a truncated [right circular] cone is equal to the sum of (a) the product of the perpendicular from the center of the base to the slant height multiplied by one third of the surface of the truncated cone and (b) the product of the altitude multiplied by one third of the area of the superior circle, or, referring to Fig. 80, trunc cone $BDFC = [EP \cdot (\frac{1}{3}$ surf $BDFC) + LE \cdot (\frac{1}{3}$ circle $DF)]$.

(1) $CEY \cdot AE =$ cone ABC, and $KME \cdot AL =$ cone ADF, assuming $EK = LF$, $KEM = \frac{1}{3}$ circle $KM = \frac{1}{3}$ circle DF [and $CEY = \frac{1}{3}$ circle BYC, and applying Proposition IV above].

(2) Trunc cone $= (LE \cdot EKM) + (AE \cdot KMYC)$, since trunc cone $=$ cone $ABC -$ cone ADF, or $[(AL + LE) \cdot (KEM + MYKC)] - (AL \cdot EKM) =$ trunc cone.

(3) $\triangle\ ALF$ is similar to $\triangle\ AEC$, since L and E are right angles, and A is common. Hence $LF/AF = EC/AC$.

(4) $\triangle\ APE$ is similar to $\triangle\ CEA$, since P is a right angle and A is common. Hence $AE/EP = AC/EC = AF/LF$.

(5) Surf ADF/circle $KM = AF/LF$ and surf ABC/circle $BYC = AC/EC$, by Corollary of Proposition I of the *De curvis superficiebus*. Hence, $[\frac{1}{3}$ surf $ADF]/EKM = [\frac{1}{3}$ surf $ABC]/CEY$, EKM and CEY each being one third of their circles.

(6) Hence, by the subtraction of proportions, $[\frac{1}{3}$ surf trunc cone $BDFC]/YMKC = AE/EP$.

(7) But $AE \cdot YMKC =$ trunc cone $BDFC - (LE \cdot EKM)$, from (2).

(8) Hence, $[EP \cdot \frac{1}{3}$ surf $BDFC] + (LE \cdot EKM) =$ vol trunc cone $BDFC$. Q.E.D.

5. An Anonymous Comment on Proposition Seven of the *Liber de curvis superficiebus*

It will be recalled that after proving the seventh proposition of the *Liber de curvis superficiebus* for the case of a solid produced by the rotation of an inscribed regular polygon whose half has an even number of sides (that is, a polygon of $4n$ sides), Johannes de Tinemue remarks that there is some doubt as to whether the proposition holds for a solid produced by the rotation of a half polygon with an odd number of sides, but that since the author (i.e., Archimedes) had not taken the matter up, he, Johannes, would leave its treatment to a diligent posterity. At the end of the text of the *Liber de curvis superficiebus* in manuscript E (British Museum, Harleian

625, 139v, 14c), a Latin author has taken up Johannes' challenge and a proof is there presented for the case where the half polygon has an odd number of sides. Of course, the middle side in such a case does not describe a conical surface as to the other sides. It describes rather a cylindrical surface. One could say, accordingly, that such a case does not fall under the proposition as stated, for that proposition is stated in terms of a body with "conical surfaces," and such a body with exclusively conical surfaces would be formed only by a half polygon having an even number of sides. Hence both Johannes' remarks and the further remarks of the Latin commentator of manuscript *E* are to some extent inappropriate unless we extend the definition of the body under investigation.

In his additional comment the Latin author accepts from Proposition VII that the volumes of each of the solids formed by the rotation of triangles *OFC*, *OFN*, *OAP*, and *OAQ* around diameter *CQ* is equal to a cone whose altitude is the line drawn from the center of the circle to the middle point of each side of the polygon and whose base is the surface described by that side of the polygon. There remains for proof, then, only the case of the volume of the solid formed by the rotation of the middle triangle *ONA* about diameter *CQ*. The author easily shows that it too is equal to a cone whose altitude is equal to *OT* and whose base is the cylindrical surface described by side *NA*. Since the altitude is the same in each of the various cones to which the various parts of the total solid are equal, it is evident that the whole solid will be equal to a cone whose altitude is the same as that of each of the individual cones (that is, which is equal to the radius of a sphere inscribed in the solid) and whose base is the sum of all the bases of the individual cones (and that sum is comprised of all the surfaces described by the sides of the half polygon in revolution, or, in short, to the whole exterior surface of the solid). Hence the proposition applies equally well to a solid formed by a half polygon with an odd number of sides as it does to a solid formed by a half polygon with an even number of sides.

The Latin commentator concludes with a short proof that the proposition applies to the solid formed by the rotation of a polygon where a pair of opposite sides are bisected by the diameter about which the rotation is performed. The only thing that is necessary to prove in connection with this case is obvious, namely, that the solids formed by the rotation of the sides which the diameter bisects are equal to cones each of which has as its altitude a line drawn from the center of the circle to the middle point of the rotated side and has as its base a circle equal to that described by the side in rotation. All of the other partial solids are covered by the proof in Proposition VII. Hence Proposition VII applies to such a solid. The commentator then errs in stating that if we rotate a regular polygon of this last type about a diameter which connects two opposite angles of the polygon, the solid formed in this rotation will be equal to the solid formed when we rotate the same polygon about the diameter which bisects two opposite sides. For in fact, the surfaces of these solids will not be equal, and so even though the same formula of Proposition VII applies to both cases, their volumes cannot be equal. MS *E* has been described in Section 2 above.

[Commentum in septimam propositionem Libri de curvis superficiebus]

139v / Quia iste commentator Archimenidis non probat conclusionem septimam indistincte sed in illo casu solummodo quo corpus conicarum superficierum inscriptibile spere et circumscriptibile habeat superficies numero pares, cum tamen indistincte loquatur conclusio, 5 ideo ne illius conclusionis probatio claudicet uno pede probabo conclusionem in illo casu quo superficies prenominati corporis sunt impares numero, ut sic posteritatis diligencia cui ipse commentator in fine conclusionis septime hoc probandum relinquit nubem obfuscantis dubietatis extenuet et lucem certitudinis imprimat illustrantis.

10 Sit igitur corpus conicarum superficierum *CNAQB* numero imparium [Fig. 81] inscriptibile et circumscriptibile spere. Tunc ex modo arguendi in septima conclusione patet quod piramis cuius altitudo est *OI* et basis equalis curve superficiei quam describit linea *FC* circumvoluta cum linea *FO*, stanti linea *CO*, est equalis corpori incluso infra 15 *CFO* lineas circumvolutas, stanti eidem *CO*. Et per idem piramis cuius altitudo est *ZO* et basis equalis superficiei quam describit linea *FN*

Comment on Proposition VII of the Book on Curved Surfaces

Because this commentator on Archimedes does not prove the seventh conclusion generally but only in the case where the body with conical surfaces which is inscribable and circumscribable in a sphere has an even number of surfaces—and yet the conclusion is stated in a general fashion—therefore in order that the proof of the conclusion does not limp along on one foot I shall prove the conclusion for the case where the surfaces of the above designated solid are odd in number. Thus the "diligence of

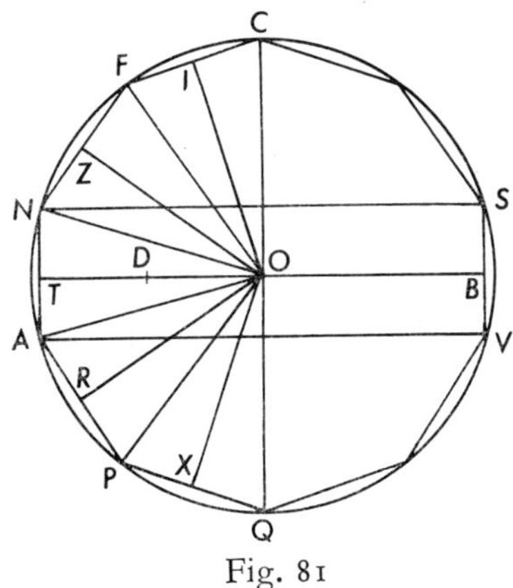

Fig. 81

posterity" to which the commentator in the end of the seventh conclusion leaves the proof may thin out the cloud of obscuring doubt and impress [on it] the light of illuminating certitude.

Therefore, let the body which has an odd number of conical surfaces and which is inscribable and circumscribable in a sphere be *CNAQB* [see Fig. 81]. Then by the manner of arguing followed in the seventh conclusion it is evident that the cone whose altitude is *OI* and whose base is equal to the lateral surface described by the revolution of *FC* when it is revolved along with *FO*, and *CO* is the axis, is equal to the body included within lines *CF* and *FO* when they are revolved about *CO* as an axis. And by the same argument the cone whose altitude is *ZO*, and whose

in circumvolucione poligonii est equalis corpori incluso infra *FNO* lineas circumvolutas per circumvolucionem poligonii. Et eadem ratione arguitur de piramide cuius altitudo est *OR* et basis equalis superficiei descripte ab *AP* et etiam de piramide cuius altitudo est *OX* et basis equalis superficiei descripte a *PQ* circumvoluta. Igitur non restat aliud probare nisi quod piramis cuius altitudo est *OT* et basis equalis curve superficiei quam describit linea *AN* in circumvolucione poligonii sit equalis corpori incluso infra *ANO* lineas circumvolutas in circumvolucionem poligonii.

Hoc autem sic probatur: Linea *AN* est equedistans linee *COQ*, eo quod *CN* arcus est equalis arcui *AQ*. Igitur ex circumvolucione linee *AN* in circumvolucione poligonii fit columpna *AS* cuius axis est equalis linee *AN* et basis est circulus cuius diameter est linea *AV*, seu linea *NS* sui equalis vel linea *TB* utrique equalis. Cum igitur linea *TOB* dividat tam *AN* quam *SV* in duo equalia, patet quod columpna *TS* sit medietas columpne *AS*, posito quod *TB* sit diameter basis columpne *TS*.

Arguatur tunc sic: Sicut *NT* linea, que est altitudo columpne *TS*, se habet ad quartam partem diametri basis, que sit *DO*, illa curva superficies columpne *TS* quam describit *TN* circumvoluta se habet ad superficiem basis cuius diameter est linea *TB*, per secundam partem correlarii secunde huius. Sed ex ductu *TN* primi in basem columpne quartum resultat illa columpna *TS*. Igitur ex ductu *DO* secundi in curvam superficiem columpne tertii resultat eadem columpna, per 15 sexti vel 19am septimi Euclidis. Tunc sic: ex ductu *OD* linee in curvam superficiem columpne *TS* fit columpna *TS*. Igitur per 15 quinti ex ductu *OT* semidiametri in eandem curvam superficiem columpne *TS* fit columpna dupla ad *TS*. Igitur piramis cuius altitudo est linea *TO* et basis equalis curve superficiei quam describit linea *TN* est tertia pars duplicis columpne ad *TS*. Illa igitur piramis est sexquitertia ad columpnam *TS*, sed corpus quod includit triangulus *TNO* in circumvolucione poligonii est sexquitertia ad columpnam *TS*, eo quod ille triangulus circumvolutus includit totam columpnam excepta piramide *NOS* eiusdem basis et eiusdem altitudinis que est subtripla ad columpnam per 9am duodecimi Euclidis. Igitur piramis cuius altitudo est linea *TO* et basis eqalis curve superficiei columpne *TS*, quam describit linea *TN* circumvoluta, est equalis corpori incluso a trian-

20 OX *corr. ex.* AX

base is equal to the surface described by line *FN* when the polygon is rotated, is equal to the body included within △ *FNO* as it is rotated with the rotation of the polygon. And by the same reasoning it is argued concerning the cone whose altitude is *OR* and whose base is equal to the surface described by *AP* and also concerning the cone whose altitude is *OX* and whose base is equal to the surface described by *PQ* in revolution. Therefore, nothing remains to be proved except that the cone whose altitude is *OT* and whose base is equal to the lateral surface described by *AN* in the rotation of the polygon is equal to the body included within △ *ANO* as it revolves with the rotation of the polygon.

This is proved as follows: Line *AN* is parallel to line *COQ* because arc *CN* is equal to arc *AQ*. Therefore, by the revolution of line *AN* as the polygon rotates a cylinder *AS* is produced whose axis is equal to line *AN* and whose base is the circle whose diameter is line *AV* or line *NS* equal to it or line *TB* equal to each of them. Therefore, since line *TOB* bisects both *AN* and *SV*, it is evident that cylinder *TS* is equal to one half of cylinder *AS*, it having been posited that *TB* is the diameter of the base of cylinder *TS*.

Then let it be argued as follows: Line *NT*, which is the altitude of cylinder *TS*, is related to one fourth of the diameter of the base, which is *DO*, as the lateral surface of cylinder *TS*, described by *TN* in revolution, is related to the base circle whose diameter is line *TB*, by the second part of the corollary of the second [proposition] of this [work]. But the product of the first term, *TN*, and the fourth term, the base of the cylinder, is equal to cylinder *TS*. Therefore, the product of the second term, *DO*, and the third term, the lateral surface of the cylinder, is equal to the same cylinder, by VI.15 (VI.16, *Greek text*) or VII.19 of [the *Elements* of] Euclid. Then as follows: The product of line *OD* and the lateral surface of cylinder *TS* is equal to cylinder *TS*. Therefore, by V.15 [of the *Elements*], the product of radius *OT* and the same lateral surface of cylinder *TS* is equal to a cylinder double *TS*. Therefore, the cone whose altitude is line *TO* and whose base is equal to the lateral surface described by line *TN* is one third the cylinder which is double *TS*. Therefore, that cone is four thirds cylinder *TS*; but the body which △ *TNO* includes in its revolution is four thirds cylinder *TS*, because that triangle when it has revolved includes all of the cylinder except the cone *NOS* having the same base and altitude—and cone *NOS* is one third the cylinder by XII.9 (XII.10, *Greek text*) of Euclid. Thus the cone whose altitude is line *TO* and base is equal to the lateral surface of cylinder *TS*, described by line *TN* in revolution, is equal

gulo *TNO* circumvoluto et eadem ratione piramis cuius altitudo est
55 linea *TO* et basis equalis curve superficiei quam describit linea *TA* circumvoluta est equalis corpori incluso a triangulo *TAO* circumvoluto. Igitur piramis cuius altitudo est linea *TO* et basis equalis toti superficiei curve quam describit totalis linea *NA* circumvoluta est equalis corpori incluso a toto triangulo *ANO* circumvoluto. Sic igitur
60 liquet protracta (?) probatio quam commentator in conclusione septima aggredi recusavit.

Item licet medietas poligonii que circumvoluta facit corpus conicarum superficierum circumscriptibile et inscriptibile spere non tangat extremitates diametri—immo diameter secet duo eius latera ut unum in
65 duo equalia—adhuc nichilominus in illo casu verificabitur conclusio. Nam statim ad oculum liquet quod in illo casu linea *DE* circumvoluta describit superficiem planam circularem [Fig. 82] et per consequens *ODE* triangulus circumvolutus describit piramidem cuius altitudo est linea *OE* et basis circulus cuius diameter est linea *DF*. Cum igitur in
70 aliis partibus poligonii stat probatio facta in commento septime conclusionis, igitur piramis cuius altitudo est linea *EO* et basis equalis omnibus superficiebus exterioribus corporis descripti per semipoligonium *GHE* est equalis illi corpori sic descripto. Et ex hoc liquet quod si poligonium equilaterum et equiangulum circulo inscribatur,
75 quod sit *CHK* gratia exempli, sive medietas eius tangens extrema diametri, scilicet *HEK*, circumvolvatur, stanti diametro *HK*, sive alia

to the body included by △ *TNO* when it has been revolved. And by the same argument the cone whose altitude is line *TO* and whose base is equal to the lateral surface described by line *TA* in revolution is equal to the body included by △ *TAO* when it has revolved. Therefore, the cone whose altitude is line *TO* and whose base is equal to the whole lateral surface described by the whole line *NA* in revolution is equal to the body included by the whole △ *ANO* when it has revolved. Therefore, the protracted proof, which the commentator refused to take up in the seventh conclusion, is clear.

Also, even though the half polygon which when rotated produces a body with conical surfaces which is circumscribable and inscribable in a sphere

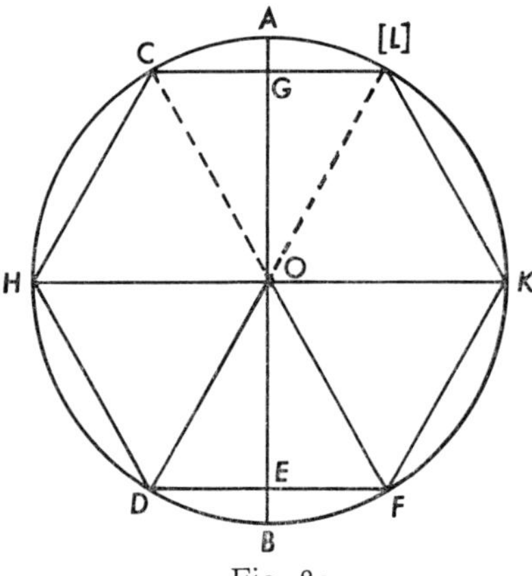

Fig. 82
Note: I have added the dotted lines.

does not touch the extremities of the diameter [about which the polygon rotates]—and in fact the diameter bisects two of its sides—still the conclusion will be verified in this instance. For it is immediately clear to the eye that in this case line *DE* when rotated describes a plane circular surface [see Fig. 82]. Consequently, △ *ODE* when rotated [about *AB*] describes a cone whose altitude is line *OE* and whose base circle has as its diameter line *DF*. Since the proof produced in the comment to the seventh consion is valid for the other parts of the polygon, therefore the cone whose altitude is line *EO* and whose base is equal to all of the exterior surfaces of the body described by the half polygon *GHE* is equal to the body so described. And from this it is evident that, in the case of the regular polygon *CHK* inscribed in the circle, if the half polygon which touches the extremes of the diameter, i.e., [half polygon] *HEK*, is rotated about *HK* as an axis, the body described by that half will be equal to the body which

eius medietas, scilicet *GHE*, cuius media puncta oppositorum laterum tangunt diametrum *AB*, circumvolvatur, stanti diametro *AB*, corpus ab una medietate descriptum erit equale corpori quod ab alia
80 medietate describitur. Hoc enim patet, eo quod utrumque corpus uni et eidem piramidi est equale; patet etiam quod eorum superficies sunt equales.

is described by another half in which the middle points of opposite sides touch the diameter, namely, [half polygon] GHE, when it rotates about diameter AB. This is evident because each body is equal to one and the same cone; it is also evident that their surfaces are equal.

COMMENTARY

7 "posteritatis diligencia." The reader will recall that manuscript E is of the second tradition of the manuscripts of the *Liber de curvis superficiebus* and hence it is this reading which is quoted rather than the reading "diligenti...posteritati" given in the first tradition (see Section 2, above, Proposition VII, line 180, and the variant reading for lines 179–80).

41 "15 sexti vel 19^am^ septimi." Proposition VI.15 [Greek VI.16] in the Adelard II Version of the *Elements* runs (Brit. Mus. Addit. 34018, 23v): "Si fuerint quatuor linee proportionales, quod sub prima et ultima rectangulum continetur equum erit ei quod sub duabus reliquis...." Proposition VII.19 (*ibid.*, 29r): "Si fuerint quatuor numeri proportionales, quod ex ductu primi in ultimum producetur equum erit ei quod ex ductu secundi in tertium...." Actually VII.19 does not properly apply, since in this comment the Latin commentator is concerned with geometric magnitudes rather than with numbers.

42–43 "15 quinti." See Commentary, Chapter Three, Section 5, the Corpus Christi Version, line 278.

51 "9^am^ duodecimi." See above, Commentary to Section 2, Proposition VII, line 22.

73–82 "Et...equales." Here the commentator makes a mistake—since the two solids do not equal the same cone and their surfaces are not equal. He was thrown off by the fact that Proposition VII can be used to determine the volume of each of the solids.

Chapter seven

The Arabo-Latin Tradition of Archimedes in Retrospect

The first phase of our study of Archimedes in the Middle Ages is completed with the presentation, in Chapter Six, of the various remanents of Archimedes' *De sphaera et cylindro* introduced in the Latin West prior to Moerbeke's translations in 1269 and of the later fragments dependent on these remanents. For the most part, the texts which have been presented in this volume stem either directly or indirectly from works in the Arabic tradition of Archimedes. The only exception is the text of the *Liber de curvis superficiebus*, which was probably translated from the Greek rather than from the Arabic. But even that work circulated with the various tracts of the Arabo-Latin tradition and thus belongs more properly to a study of that tradition than to the study of Moerbeke's translations from the Greek which will occupy us in the second volume. The main results of our study of the Arabo-Latin tradition may be summarized briefly.

1. Two of Archimedes' works played the central role in the introduction of the Arabic tradition of Archimedes to the Latin West: *De mensura circuli* and *De sphaera et cylindro*. Of the first, the Arabo-Latin tradition, particularly in the translation of Gerard of Cremona (see Chapter Two), presented a complete picture, at least of the form of that treatise that was current in late antiquity and among the Arabs. Of the second of these works, it was primarily the propositions of the first book that circulated in the Latin West. Principally responsible for introducing these propositions with their Archimedean-type proofs were the *Verba filiorum* of the Banū Mūsā (Chapter Four) and the *Liber de curvis superficiebus* of Johannes de Tinemue (Chapter Six). With the appearance of these various translations, the Latin geometer learned how to prove theorems concerned with the area of a

circle in terms of its circumference and radius (and also in terms of π and the radius), the relation of the square of the diameter of a circle to the area of the circle, the calculation of π as lying between $3\frac{10}{71}$ and $3\frac{1}{7}$, the lateral area and volume of a right circular cone, the surface area and volume of a right cylinder, and the area and volume of a sphere. Incidentally, some of the conclusions of the *De sphaera et cylindro* circulated from the early thirteenth century in the *De ysoperimetris*, which was translated from the Greek but which often formed a part of the mathematical codexes in the Arabo-Latin tradition (see Appendix III, paragraphs 7 and 8).

2. In addition to these Archimedean propositions, further geometrical theorems and techniques were introduced into the Latin West by the spread of the *Verba filiorum* and the *Liber de curvis superficiebus*. The former gave the Latin Schoolmen their first contact with the problem of the finding of two mean proportionals between two given quantities, transmitting an elegant proof that appears to go back to Archytas of Taras and a mechanical proof associated by Greek mathematicians with Plato (see Chapter Four and Appendix V). This work also presented to the medieval geometer the problem of trisecting an angle with a solution by reduction to a *neusis*, or verging problem, of the same sort as is found in the *Liber assumptorum* (or *Lemmata*) ascribed by the Arabs to Archimedes (see Chapter Four and Appendix VI). And while the Latin mathematicians had early access to the theorem known as Hero's formula for the area of a triangle in terms of its sides, the *Verba filiorum* gave them their first proof of that theorem (see Chapter Four and Appendix IV). The additional contributions of the *Liber de curvis superficiebus* beyond its Archimedean propositions were more modest: a proof of the theorem stating that the circumferences of any two circles are proportional to their diameters and proof of a theorem like that of Proposition II of the *De mensura circuli* but applied instead to a sphere, namely, that the cube of the diameter of a sphere is to the volume of the sphere as 21 is to 11 (see Chapter Six).

3. Perhaps even more important than their introduction of the conclusions reached in the two basic Archimedean treatises, the various Latin translations provided the medieval geometers with a knowledge of the so-called method of exhaustion developed by the Greeks primarily to treat problems involving the areas and volumes of figures bounded by curved lines and curved surfaces. With the Archimedean models before them, certain of the medieval mathematicians went on to apply it to problems not found in the translations. For exemple, the exhaustion procedure was used by a Latin author, who reworked the *Liber de curvis*

superficiebus, to prove a theorem relative to the area of a spherical segment (see Chapter Six, Section 4)—a theorem equivalent to the third proposition of Book II of Archimedes *De sphaera et cylindro*; this proposition, however, Archimedes had not proved by exhaustion. The exhaustion technique was even employed in kinematic problems by Gerard of Brussels in a manner suggestive of its Archimedean source (see Chapter One, note 17).

4. The texts introducing the conclusions and methods of the two Archimedean treatises became themselves the object of much paraphrasing and commentary, as is evident in the seven* different versions of the *De mensura circuli* presented in Chapter Three, the three versions of the same work in Chapter Five, and in the emendations and paraphrases of the *Liber de curvis superficiebus* in Chapter Six. Some reworking of the inherited material was also present in the works of the two famous mathematicians of the thirteenth century, Jordanus de Nemore and Leonardo Pisano, where, for example, another solution of the problem of finding two mean proportionals is added to the solutions transmitted by the *Verba filiorum* (see Appendix V). Incidentally, there is also found associated with the name of Jordanus a somewhat different proof of Hero's theorem on the area of a triangle in terms of its sides—a proof much closer to the original proof given by Hero (see Appendix IV). Jordanus also suggests another solution of the *neusis* problem to which the Banū Mūsā had reduced the trisection of an angle. His suggested solution is based on a proposition cited from a contemporary *Perspective* and appears to relate to a solution given by Alhazen in his *Optics* that depends on conic sections (see Appendix VI). As these commentaries and emended versions spread, the medieval geometers gained more facility and confidence in their own abilities. Thus a mathematician of the fourteenth century was able to correct in detail the calculations for the determination of π that he found in an earlier commentary (see Chapter Three, Section 3). Incidentally, this commentary with its corrections, while somewhat reminiscent of Eutocius' commentary on the *Measurement of the Circle*, is quite original in its methods of calculation. One is also struck by the originality of certain of the emended versions of the *De mensura circuli*, particularly that of Albert of Saxony in the fourteenth century (Chapter Five, Section 3).

5. In addition to the evidence of the popularity of the Archimedean tracts represented by the various commentaries and reworkings mentioned above, further evidence is present in the extensive citation of the transla-

* Counting as separate versions the two proofs of Proposition I found in the Florence codex treated in Section 3 of Chapter Three.

tions in the course of the thirteenth and fourteenth centuries, as, for example, in the works of Jordanus, Leonardo Pisano, Roger Bacon, Campanus, Thomas Bradwardine, Franciscus de Ferraria, Nicole Oresme, Albert of Saxony, Wigandus Durnheimer, and a number of anonymous authors (Chapter Two, Section 2; Chapter Three, Section 4; Chapter Four, Section 1; Chapter Five, Section 3; Chapter Six, Section 2).

6. At the time that the various Archimedean tracts were being received, cited, and commented upon, Euclid's *Elements* was becoming the dominant influence on medieval mathematics. It is not surprising, therefore, to find that the mathematicians commenting on or reworking the Archimedean tracts took great pains to introduce numerous propositions from the *Elements* to serve as an underpinning to the Archimedean proofs. To take only one—although no doubt the most important—example, in the seven versions of Proposition I of the *De mensura circuli* presented in Chapter Three and in one of the three versions of that same proposition given in Chapter Five, Proposition X.1 of Euclid's *Elements* was cited as the basic proposition supporting the convergent procedure that is essential in the exhaustion method. It was not until the fourteenth century, when mathematical studies had become more completely intertwined with philosophical studies, that Aristotle was also cited as an authority (see Chapter Five, Section 3).

7. A striking feature of the various emendations and reworkings of the *De mensura circuli* given in Chapters Three and Five is their self-conscious concern with the logical features of the proofs. In using the exhaustion method which involves *reductio ad absurdum*, the various authors often specifically and carefully point out where the contradiction lies instead of leaving it to the inference of the reader, as was done in the original Archimedean texts. Further, as we come into the fourteenth century, we find the philosophical influence becoming more apparent in the form and organization of the tracts. With this philosophical and scholastic influence comes a modification of the Hellenic mathematical form of presentation and a consequent introduction of the form used in medieval logical tracts and *questiones*, a form which employs *rationes*, *distinctiones*, *consequentiae*, *antecedentes*, *consequentes*, and the like. The extreme case of the changed form is Albert of Saxony's treatment of the problem of the quadrature of the circle (Chapter Five, Section 3), but it exists to some extent in the other tracts of Chapter Five, as well as in the Gordanus and Munich versions of Chapter Three.

8. Also reflective of the penetration of the philosophical and scholastic

point of view is the distinction made by Albert of Saxony between the understanding of the quadrature problem "with respect to sense" and its understanding "with respect to intellect." To find a square equal to a circle "with respect to sense" means that we must find a square which does not sensibly differ from the circle; but "with respect to intellect" it means that we must demonstratively prove that there is a square equal to the circle. Incidentally, the introduction of physical arguments in the course of a demonstration is occasionally noticeable, as in the Corpus Christi Version of the *De mensura circuli* (Chapter Three, Section 5), where the postulate to the effect that a curved line can be equal to a straight line is supported by reference to the bending into curved form of a hair or silk thread, or as in Albert of Saxony's tract where he reports a common argument in support of quadrature which holds that the contents of a spherical vase can be poured into a cubical vase.

9. A further indication of the increasing juncture of philosophical and mathematical exposition is evident in the abandonment of Proposition X.1 of the *Elements* as the acknowledged basis of the proof of Proposition I of the *De mensura circuli* in favor of a postulate derived from the continuous divisibility of a continuum. The best expression of this occurs in what we may call Albert of Saxony's postulate (Chapter Five, Section 3): With two continuous quantities proposed, a magnitude greater than the "lesser" can be cut from the "greater," a postulate only implicit in the treatment of the quadrature of the circle by the Banū Mūsā (Chapter Four). Similarly reflective of philosophical penetration into mathematics is the discussion of the "if a greater and a lesser, then an equal" postulates (e.g., see Albert of Saxony's text in Chapter Five, Section 3, and particularly the comments to that text, lines 3–4, 45).

10. Much more remote and indirect is the influence of the Arabic tradition of Archimedes' physical works on Latin science. The *Equilibrium of Planes* was not known until its translation from the Greek in 1269. However, it is quite evident that the statical works associated with the name of Jordanus that predate 1269 owe something in their general mathematical-deductive form to the Hellenistic tradition of statical works inaugurated by Archimedes. In the conversion of material beam segments to weightless lines, or of weights to geometrical areas, the medieval student of statics is at least the heir of Archimedes. And more specifically, when Jordanus and his followers replace a material segment of a balance beam by a weightless line together with a weight hung from the midpoint of that line, we can detect the ultimate influence of Archimedes' concept of

center of gravity and the more direct influence of Gerard of Cremona's translation of Thābit ibn Qurra's *Liber karastonis* (see Chapter One, p. 9) and of the short anonymous tract *De canonio* translated in the early thirteenth century from the Greek.

11. In addition to his general but indirect influence on statics, we can also note the very modest influence of Archimedes' *On Floating Bodies.* The genuine tract was not translated until 1269 by William Moerbeke. But certain of its conclusions (including the celebrated "Principle of Archimedes") are reflected in the Pseudo-Archimedean *De insidentibus in humidum*, a work of the thirteenth century probably translated from the Arabic, or at least dependent on Arabic sources (see Chapter I, p. 8).

In concluding our summary of the Arabo-Latin tradition and its influence on medieval mathematical work, it is of importance to reiterate that this tradition was not simply displaced by the corpus of translations made from the Greek by William of Moerbeke, the corpus that will be examined in volume II. The continuing citation and copying of these treatises translated from the Arabic takes us down into the Renaissance, when they were finally displaced by a renewed interest in the Greek text.

APPENDIXES

Appendix I

Some Non-Archimedian Treatments of Quadrature

1. The Theorem of Jordanus

In this first appendix I have included several theorems and tracts in which the quadrature problem is taken up in a manner other than that of Archimedes. I am purposely omitting the treatises already discussed by Paul Tannery (see Chapter Two, page 15, note 2). My first text is of a theorem drawn from the thirteenth-century *Liber de triangulis* of Jordanus de Nemore. While my text is very close to that already published by M. Curtze on the basis of manuscript *I* (see the Sigla below), some changes have been made as the result of comparing this text with other manuscripts. In order to bring the text into conformity with the other texts of this volume, I have altered the line numbers, capitalized the letters designating magnitudes, and changed Curtze's punctuation. Incidentally, Curtze mistakes somewhat the intent of the proposition when he summarizes it as follows:[1] "Interessant ist die Art, in welcher Jordanus versucht ein dem Kreise gleiches Quadrat zu finden. Er benutzt dazu (No. 16) eine Figur, welche zwischen dem Kreise und dem umgeschriebenen Quadrate als mittlere Proportionale gefunden werden soll. Ist diese Figur ein Kreis, so ist das diesem umgeschriebene Quadrat das gesuchte; ist sie kein Kreis, sondern eine geradlinige Figur, so kann man sie als Quadrat darstellen. Dann zeigt er, dass man immer ein Rechteck finden kann—die eine Seite ist die Seite des gegebenen die andere des gefundenen Quadrates—welches gleich dem Kreis ist. Wie er die mittlere Proportionale finden will, sagt er nicht, weil dies

[1] See the text of Curtze (cited in the Sigla under *I*), p. XIII.

eben die Quadratur des Kreises voraussetzen würde, die er eben sucht." But it is clear that Jordanus in this proposition is not finding "a figure that is the mean proportional between the circle and a circumscribed square," but rather is finding a figure which is the third continuously proportional term after the circumscribed square and the circle considered as the first two terms. In the first part of the proof that figure is posited as being a circle, while in the second it is a rectilinear figure. In both cases it is shown that the desired square equal to the proposed circle is, like the original circle, a mean proportional between the circumscribed square and the third proportional term. But it is evident that we are not told how to construct the third proportional surface, whether it is a circle or a rectilinear figure, but only that there exists such a third proportional term. We are accordingly at a loss to construct the desired square equal to the proposed circle.[2]

It will be evident to the reader that this theorem of Jordanus has an intimate relationship with the first of the two quadrature proofs found in

[2] There is a very curious, corrupt, and erroneous treatment of the problem of constructing a square equal to a proposed circle in a Cambridge University Library manuscript, Ee.III.61, 176r–v, which appears to be an effort to meet the implied challenge of the Jordanus theorem. Here the author pretends to construct, merely by using compass and straight line, a second circle that constitutes the third proportional term after the square circumscribed about the first proposed circle and the proposed circle itself. This construction is, of course, impossible. After so constructing the third proportional, he then further circumscribes a square about the circle constructed as the third proportional term. Then, as in the Jordanus proof, this new circumscribed square is shown to be equivalent to the originally proposed circle. The text of this treatment is so corrupt (even though perfectly legible) that I have hesitated to attempt a reconstruction without another manuscript. It begins: "Proposito circulo quadratum equale describere. Falsigraphus insistat...," and ends: "...Ergo *ABCD* circulus et quadratum *HI* sunt equales, cum utrumque sit medium proportionale ad *FG* quadratum et *QRST* circulum, quod fuit propositum. Explicit." It should be noted that this text is to be dated after the appearance of Grosseteste's commentary on the *Posterior Analytics*, which it cites (f. 176r: "licet secundum lincolniensem primo posteriorum..."). A somewhat similar erroneous proposition by one Fredricus is found in the fourteenth-century MS Basel, Bibl. Univ., F.II.33, 137r. It begins: "Quadratura circuli nichil est nisi inventio embadi quadrati equalis embadi totius circuli. Sed quod circulus fiat quadratum impossible est per 15 3[ti] Euclidis quia linea curva nunquam fit recta neque angulus incidentie fit angulus contingentie licet fiat maior vel minor et transeat per equale. Faciam ergo circulum *ABGD*...." It ends: "...proportio dupla, quod declarare volumus. Hec est quadratura circuli secundum Fredricum ex geometricis extracta principiis." His basic error is in thinking that a circle is the mean between an inscribed and circumscribed square which he has constructed ("medium inter quadrata duo, scilicet exterius et interius, quod erit equale embadi circuli..."). Notice that in the introductory part which I have quoted Fredricus raises but ignores the same objection that Albert of Saxony had raised regarding horn angles (see Chapter Five, Section 3, Albert of Saxony's

a thirteenth-century Bodleian manuscript (Digby 174) and published here in the next section of this appendix. Accordingly, the question arises as to whether Jordanus' theorem was the source of the Digby proof or whether Jordanus' theorem stems from that proof. In the succeeding section I shall argue the first of these alternatives.

The only orthographic point worth noticing is that I have followed *I(Ed)* in using the very common medieval misspelling of *paralellogramum*. The figures are those given in *I(Ed)*, *Oa*, and *Sl*. The marginal folio references are to manuscript *I*.

Sigla of Manuscripts

I = Dresden, Sächs. Landesbibliothek, Db 86, 58v–59r, early 14c. I have used the transcription of M. Curtze, *Jordani Nemorarii Geometria vel de triangulis libri IV*, in *Mitteilungen des Coppernicus-Vereins für Wissenschaft und Kunst, VI.* Heft (Thorn, 1887), pp. 36–37.

Ca = Oxford, Corpus Christi College 251, 84v, 13c. The proposition stands alone in this manuscript.

Sl = British Museum, Sloane 285, 89v, 14c.

E = British Museum, Harleian 625, 128v, 14c.

Oa = Paris, BN lat. 7434, 85r, 14c.

[See also a somewhat extended copy of this proposition in the collection of quadrature proofs given in Glasgow University Library MS BE 8–y.18, 210v–211r. It is there designated as *Quadratura circuli secundum magistrum alardum in maiori commento*. As I noted above in Chapter V, p. 388n, *Alardus* was the form of Adelard of Bath's name used in the so-called Adelard III version of the *Elements*. I shall comment further on this extended copy of Jordanus in an appendix to Volume II.]

Questio de quadratura circuli, Commentary to line 45). Note also that Fredricus refers to a demonstration of the Commentator on the first book of the *Physics* where he speaks of lunes ("demonstratio commentatoris in primo physicorum de hiis decisionibus que sunt per lunulas"). See Averroes, *Commentaries on the Physics* (ed. Venice, 1562, 11v, c.1): "...qui existemavit se quadrare circulum, cum quadravit lunares figuras, quae sunt portiones circulorum, oportet Geometria contradicere ei: quia peccavit in eis, que sunt post principia; portiones enim cadentes in circulo, licet possint facere quadrata aequalia eis, tamen non sequitur ex hoc facere quadrata aequalia circulo: nam illae portiones non consumunt circulum per divisionem." The Latin translation of this passage may have been the source of the mistaken identification of lunes with segments of circles in various versions of the *De mensura çirculi* (see page 62 above). Finally, Fredricus speaks of the difficulty of finding the quadrature by means of numbers (Per numeros difficile est invenire...").

Additional Manuscripts, Not Used for Text

Basel, Univ. Bibl. F.II.33, 146r–150v, 14c. Variant readings of this manuscript are given by Curtze in his edition.

Bruges, Stadsbibliotheek 530, 1v–8v, 14c. This manuscript does not contain this proposition; in fact, it contains no proposition of the *De triangulis* after IV.11.

Paris, BN lat. 7378A, 29r–36r, 14c.

Florenze, Bibl. Naz. Conv. Soppr. J.X.40, 47r–66r, 15c.

Florence, Bibl. Naz. Conv. Soppr. J.I.32, 124r–135v. This manuscript dates from the end of the thirteenth century, but the last proposition given is IV.10.

Jordanus, *On Triangles*, Book IV, Proposition 16

[Propositio 16ª libri quarti Jordani de triangulis]

58v / Proposito circulo equale quadratum constituere

Esto exemplum *A* circulus [Fig. 83]. Disposicio: adiciatur et alius circulus *B*, quorum utrique circumscribatur quadratum, et diameter, eritque quadratum circumscriptum quasi quadratum diametri ipsius
5 circuli. Ergo secundum secundam duodecimi eadem est proporcio *A*, *B* circulorum et *DE*, *FG* quadratorum. Ergo permutatim eadem est *DE*, *A* et *FG*, *B*. Statuatur itaque *C* quasi tercia superficies proporcionalis post *DE* et *A*. Erit autem vel circulus vel aliusmodi, ut rectilinea, superficies. Sit primo circulus *HK* quadrato circumscripto.
10 Itaque eadem est proporcio *DE*, *A* et *A*, *C*. Sed item secundum secundam duodecimi eadem est *DE*, *A* et *HK*, *C*; ergo tam *A* quam *HK* est medium proporcionale inter *DE* et *C*. Ergo *A* circulus et *HK* quadratum sunt equalia, quod proposuimus.

Amplius sit *C* alia figura quam circulus. Ea itaque redigatur in
15 quadratum secundum ultimam secundi, angulis signatis per *R*, *S*, *Y*, *X*. Itaque, cum *DE* sit maior extremitas, maior est quam *RY*, ergo et latus latere maius. Resecetur ergo de *MD*, *MT* ad equalitatem *RX* et exibit *MN* paralellogramum ex *ME*, *MT* descriptum. Ergo *MN*

1 equalem *Oa* / constituere: describere *Ca*
2 Esto exemplum: sit *E* / exemplum: exemplo *Ca* / Disposicio *om. Oa* proponitur cui *E* / adiciatur *ESlOa* et adiciatur *Ca* adiiciatur *I(Ed)* / et *om. E*
3 circulus: qui sit *E* / et: et probabitur *E* / diametrus *E*
4 quasi: quia *Ca* q *Oa* / quadratum² *om. Ca*
4–5 ipsius circuli *om. E*
5 Ergo: igitur *E* / est *om. Sl*
6 circulorum: circulo *Sl* / mutatim *SlOa*
7 DE, A *I(Ed) Sl* A, DE *Ca* DE ad A *EOa* / FG, B: FG ad B *EOa* / C: E *Oa* / quasi: a *Ca*
8 post: p⁰ *Oa* / Erit autem: aut erit *Ca* Erit autem *EOa* / vel²: et *Sl*
10 DE, A: DE ad A *EOa* / A, C: A ad C *E* A ad E *Oa* / item *I(Ed)*, *om. Ca* tunc *SlOa* et per *E*
11 DE, A: DE ad A *EOa* / HK, C: HK ad C *E* HK ad E *Oa*
14 C: E *Oa* / Ea itaque: que *E* / redigatur: redigatum *Sl* redicatur *Oa*
15 angulis...per: quod sit *E* / R: K (?) *Ca* / Y: V *Ca*
16 Itaque cum *tr. E* / RY: KY *Ca* RI *Oa*
17 et *om. CaOa* / resecentur *Sl* / de *EaSl*, *om. EOa* / MT: in MT (?) *E* T *Oa* / RX: KX *Ca*
18 exhibit *Sl* / MT: in MT *E*
18–21 *de* Ergo...proporcionalis *scr. I(Ed) mg.* Cuius demonstracio est. Proporcio DE ad RY est MD ad RX sive proporcio (*sed del.* ?) MT duplicata. Sed que est proporcio MD at MT ea est DE ad CY (MN ?) duplicata; quare MN est media (medium ?) proporcio (proporcionale ?).
18 MN: MN paralellogrammum *E*

[Jordanus, *On Triangles,* Book IV, Proposition 16*]

To Form a Square Equal to a Given Circle

For example, let the circle be A [see Fig. 83].

Disposition: Let another circle B with its diameter be added; let a square be described about each of those circles. And the circumscribed square [in each case] will be as a square of the diameter of the circle. Hence, by [Proposition] XII.2 [of the *Elements*], circle A/circle B = square DE/square FG. Therefore, by permutation, $DE/A = FG/B$. Let there be formed a third surface C, which is a [third] proportional

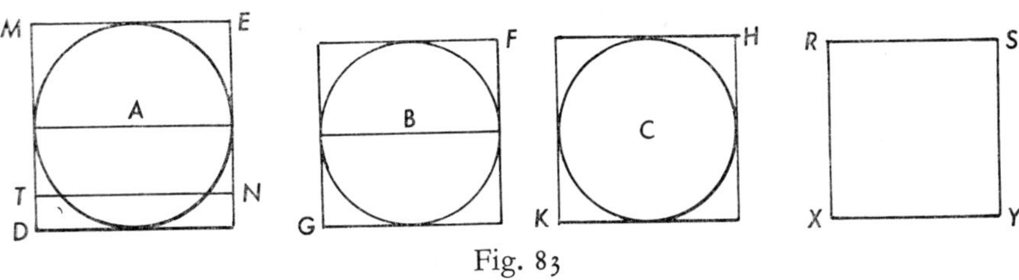

Fig. 83

[term] following DE and A. Now C will either be a circle or a surface of another kind, like a rectilinear surface. In the first place, let it be a circle which is circumscribed by square HK. And so, $DE/A = A/C$, but also, by [Proposition] XII.2 [of the *Elements*], $DE/A = HK/C$. Therefore, HK as well as A is a mean proportional between DE and C. Therefore, circle A and square HK are equal, which we proposed.

Next, let C be some [rectilinear] figure other than a circle. Then let it be converted into a square by the last [proposition] of [Book] II [of the *Elements*], with its angles designated as R, S, Y, and X. And so, since DE is the larger extreme [among the three proportional terms], DE is greater than square RY, and a side [of DE] is greater than a side [of RY]. Therefore, let MT, equal to RX, be cut from MD. Then a parallelogram MN—contained by ME and MT—is described. Therefore, MN is the mean

* See the discussion of the number of this proposition in the Commentary below.

est medium proporcionale inter *DE*, *RY*, scilicet quadrata suorum
20 laterum, cum quilibet tetragonus inter quadrata suorum laterum medio
loco est proporcionalis. Sed *A* circulus fuit medium proporcionale
inter eadem. Ergo *A* circulus et *MN* paralellogramum sunt equalia.
59r / Quadretur ergo *MN* secundum ultimam secundi, et erit eius quadratum equale *A* circulo proposito, quod proposuimus.

19 proporcionaliter *SlOa* / RY: RV *Ca* KY *Oa* / scilicet *I(Ed)*, *om. SlE* seu *Ca*
20–21 cum...proporcionalis *om. Ca*
20 cum *E*, *om. ICaSl* ⟨nam⟩ *Ed* quia *Oa* / laterum[2] *E*, *om. ICaSlOa* / *post* laterum[2] *add. E* sit
21 loco est *om. E* / *post* proporcionalis *add. E* ut equalia patet ex prima sexti
22 eadem: DE et RY *E* / A *om. Oa* / et *om. Oa* / sunt *om. Oa* / equalia: equales *Ca*
23 ergo: igitur *Oa*
24 *post* proposuimus *add. Oa* et cetera

proportional between *DE* and *RY*, which are the squares of its sides,* since a rectangle is the mean proportional between the squares of its sides. But circle *A* was the mean proportional between them [i.e., between square *DE* and *C* (or square *RY*)]. Therefore, circle *A* and parallelogram *MN* are equal. Therefore, let *MN* be converted to a square by the last [proposition] of [Book] II [of the *Elements*], and this square will be equal to the given circle *A*, which we proposed.

COMMENTARY

Title It should be remarked that the division into books which is adopted by Curtze is taken (and modified) from the designation of a later owner of manuscript *I* (*ed. cit. in siglis*, p. XI). The only other manuscript which I have examined that reflects some division into books is *E*, where we find on f. 128v (containing our proposition) the designation "L. III"; but in this copy the theorems are unnumbered. In manuscript *Sl* the proposition which I am here publishing is numbered "61."

5,11 "secundam duodecimi." Proposition XII.2 in the Adelard II Version of the *Elements* runs (Brit. Museum Addit. 34018, 63v): "Omnium duorum circulorum proportio est alterius ad alterum tanquam proportio quadrati sue diametri ad quadratum diametri alterius."

15 "ultimam secundi." See Chapter Three, Section 1, the Cambridge Version of the *De mensura circuli*, Commentary to lines 82–83.

* A note in *I*(*Ed*) tells us: "The proof of this follows: square DE/square RY = line MD^2/line RX^2 = line MD^2/line MT^2. But MD/MT = square DE/paral MN. Therefore, square DE/square RY = square DE^2/paral MN [or, square DE/paral MN= paral MN/square RY]. Therefore, *MN* is the mean proportional." (See variant for lines 18–21 for the Latin text of this marginal comment.)

2. Two Anonymous Quadrature Proofs of the Thirteenth Century

The two proofs included in this section are taken from the thirteenth-century Digby manuscript designated as *C* in the Sigla below. They follow directly on the proof of a third proposition which reads: "Si circulo inscribatur quadratum et eidem circulo describatur aliquod quadratum, octogonium inscriptum eidem circulo erit inter illa duo quadrata medio loco proportionale." The hand in which these three proofs are written is different from that of the preceding items in this codex, but it too dates from the thirteenth century. The same three items are also included in the late fifteenth-century codex *F*, the proof for the proposition on the octagon being on folio 18v, that for the first quadrature proof on folio 19r, and the second quadrature proof on folio 21v (following the text of Gerard of Cremona's translation of the *De mensura circuli*). That the three pieces were copied by the scribe of *F* from manuscript *C* is a possibility. I have already indicated in Chapter Six, Section 2, that the text of the *De curvis superficiebus* found in *F* was at least of the tradition of *C*. The fact that the three extra items were all in *F* as well as in *C* further establishes the close relationship between these two manuscripts. A telling point in arriving at the conclusion that the three extra items were copied by *F* from *C* (or the tradition of *C*) is the fact that in just the places where the scribe of *C* was careless in the formation of his letters the scribe of *F* has been thrown off the track, thus producing a variant reading. For example, in line 33 of the second quadrature proof, *C* has *huisset* to stand for *habuisset*, having neglected to use an abbreviation sign for the omitted letters. Furthermore, the scribe has so written *hui* that it looks like *bea*. Thus, when the scribe of *F* copied this word, he produced the nonsense word *beasset*.

A word must be said about the title accompanying the two quadrature proofs in the Digby manuscript: *De circulo quadrando*. It is written on the top of folio 137r in a later fifteenth-century or sixteenth-century hand, although the first of the two proofs starts on the last line of 136v. In manuscript *F*, folio 18v, at the beginning of the proposition on the octagon, the scribe has written *Quadratura circuli*. And then later (folio 21v), when giving the second proof for quadrature, he writes: *Demonstratio Ioannis de Chinemue de quadratura circuli*. No such title appears in manuscript *C*, his source. Hence we must suppose that the title was a free invention by the scribe of *F* arising from the association of this fragment in *C* with the *Liber de curvis superficiebus*, which was composed by Johannes.

I raised the question in the Introduction to the first section of this appendix of the relationship of the first quadrature proof of the Digby manuscript with the theorem on quadrature given by Jordanus. It is my opinion that the Digby proof was composed after Jordanus' theorem. This is deduced from the fact that the Digby author repeats, in a way similar to that of Jordanus, proofs for two possibilities, namely, that the third proportional is either a circle or that it is a rectilinear figure. Following this the author then brings up a third possibility not covered by the other two, that the third proportional is a surface contained by a straight line and a curved line, suggesting a circular segment as an example. I would suppose that had Jordanus been using the Digby proof he would have made some reference to the third possibility.

The most interesting point to notice about the second of the two quadrature proofs is that the author recognized that the chief difficulty with his proof lay in finding which ratio between lines is equivalent to the ratio which the circumscribed square has to the circle.

My text of the two proofs is taken almost exclusively from manuscript *C*, as are the diagrams. The only change necessary in the drawing is the alteration of a letter of the square constructed on *HI* (Fig. 85). Both manuscripts have *K* marking one of the lower angles on their drawings but have *N* in the text. I have changed the *K* to *N* to agree with the text. The marginal folio references are to manuscript *C*.

Sigla of Manuscripts

C = Oxford, Bodleian, Library, Digby 174, 136v–137r, 13c.
F = Vienna, Nationalbibliothek cod. 5303, 19r, 21v, 15–16c.

[De circulo quadrando]

136v / [1.] Sit circulus *O* inscriptus quadrato *A* [Fig. 84]. Que est proportio
137r *A* ad *O* eadem est *O* / ad aliam quantitatem rectilineam vel curvilineam. Et sit primo *A* ad *O* que *O* ⟨ad⟩ *M* circulum. Circulo igitur *M* circumscribatur *S* quadratum. Ratio. Age. Que est proportio *A* ad *S* eadem est *O* ad *M*, per XII. Ergo que est *A* ad *O* eadem est *S* ad *M*. Sed que est *A* ad *O* eadem est *O* ad *M*. Ergo que est *O* ad *M* ea est *S* ad *M*. Ergo *S* quadratum *O* circulo est equale. Sit deinde *A* ad *O* sicut *O* ad rectilineam superficiem. Quecunque superficies illa sit nisi sit quadratum, redigatur in quadratum. Sit igitur *A* ad *O* sicut *O* ad *S* quadratum. Ergo *O* circulus est medio loco proportionalis inter *A* et *S* quadrata. Sed quod fit ex *S* latere in *A* latus est medio loco proportionale inter eadem. Ergo superficies ex *A* linea in *S* lineam est equale *O* circulo. Redigatur ergo illud parallelogramum ⟨in⟩ quadratum. Et erit illud quadratum similiter equale *O* circulo et sic quadratus et circulus propositus. Impediet tamen quis demonstrationis progressum dicens et que est proportio *A* quadrati ad *O* circulum eadem est *O* ad superficiem contentam linea recta et linea curva qualis est portio circuli *ACB*, et sic impeditur propositum.

[2.] Sit circulus *O*, *AC* quadrato inscriptus [Fig. 85]. Sitque *AC* ad *O* sicut linea *FG* ad *M*. Nullius enim quantitatis ad aliam aliqua potest esse proportio que non posset in lineis inveniri, licet inveniri

4 que O: quomodo *F* / ⟨ad⟩ *supplevi*
5 est proportio *tr. F*
8 circulo *F* circuiẹlo *C* / equalis *F* / sicut *F* sic *C*
12 proportionalis *F*
14 ⟨in⟩ *supplevi*
15 quadratus et circulus: circulus quadratus est *F*
16 demonstratoris *F*
17 et *om. F*
18 portio *F* proportio *C*
20 *ante* Sit *scr. F* Demonstratio Ioannis de Chinemue de quadratura circuli / Sitque: Sit itaque *F*
22 non...lineis: in lineis non valeat *F* / posset: ps^et *(?) C* valeat *F*

On the Circle to be Squared

[1.] Let there be a circle O inscribed in square A [see Fig. 84]. The ratio of A to O is the same as the ratio of O to another rectilinear (actually, rectiplanar) or curvilinear (actually, curviplanar) quantity. And in the first place let the ratio of A to O be as that of O to circle M. Therefore, let square S be circumscribed about circle M. Proceed with the proof. $A/S = O/M$, by [Proposition 2* of Book] XII [of the *Elements*]. Therefore, $A/O = S/M$. But $A/O = O/M$. Hence, $O/M = S/M$. Hence, square S is equal to circle O. Then let the ratio of A to O be as that of O to a rectilinear (actually, rectiplanar) surface. If that surface is anything but a square, let it be converted into a square. Hence, $A/O = O/\text{square } S$.

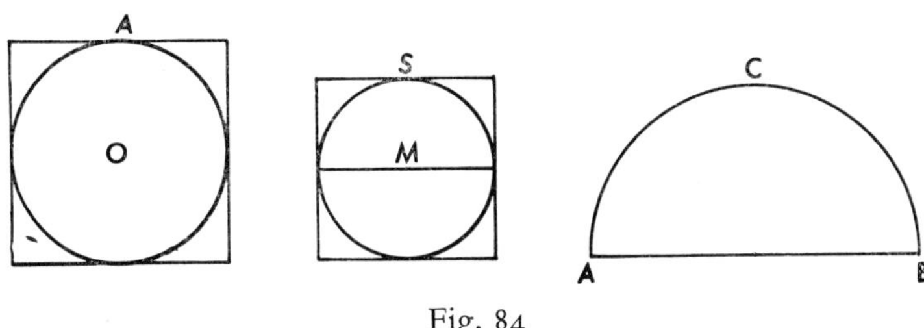

Fig. 84

Hence, circle O is the mean proportional between squares A and S. But the product of a side of S and a side of A is the mean proportional between those same [squares]. Hence, the surface arising from the product of line A and line S is equal to circle O. Therefore, let that parallelogram be converted into a square. And that square will in the same way be equal to circle O, and thus both the square and the proposed circle [are equal to O]. However some one might impede the course of the demonstration by saying that the ratio of square A to circle O is the same as that of O to a surface contained by a straight line and curved line, as for example the circular segment ACB, and so the proposition is impeded.

[2.] Let circle O be inscribed in square AC [see Fig. 85]. And let $AC/O = \text{line } FG/\text{line } M$. For there is no ratio of one quantity to another that cannot be found as existing between [some] lines, although we might not know how to find these lines. And so let DE be the mean proportional be-

* See line 5 of the Commentary to Appendix I, Section 1, the Jordanus theorem.

a nobis non sciantur linee ille. Sit itaque *DE* medio loco proportionale inter *FG* et *M*, et quadrentur *FG*, *DE*. Sit item linea *AB* ad *HI* sicut
25 *FG* ad *DE*, et quadretur *HI*. Ratio. Que est proportio *FG* ad *M* eadem est *FG* ad *DE* duplicata. Ergo a pari que est *FG* ad *M* ea est *FS* quadrati ad *DR* quadratum. Sed que est *FG* ad *M* ea est *AC* ad *O*. Ergo proportio *FS* ad *DR* sicut *AC* ad *O*. Sed que est *FG* ad *DE* eadem est *AB* ad *HI*. Ergo cum eas surgant superficies similes, pro-
30 portio *FS* ad *DR* est sicut proportio *AC* ad *HN*. Ergo a pari que est *AC* ad *O* eadem est *AC* ad *HN*. Ergo circulus *O*, *HN* quadratum sunt equa. Sic igitur liquet quod aliquis circulus alicui quadrato est equalis. Siquem itaque tam arguto naturaliter habuisset acumen quod sciret invenire duas lineas quarum una se haberet ad aliam ea proporti-
35 one qua quadratum ad circulum ei inscriptum, ille nostro iam dicto usus artificio cuilibet circulo proposito statim quadratum equum inveniret.

23 linee ille *tr.* *F*
25 HI *C* $\overset{h}{\text{bi}}$ *F*
26 est[1]: est proportio *F*
27 FS: fl *F* / DR: da *F*
28 FS: $\overset{s}{\text{fg}}$ *F* / DR: k *F*
30 FS: fl *F* / DR: dk *F*
31 O: ạo *F* / circulus O *tr.* *F*
33 habuisset: huisset *C* beasset *F* / acumen: acumie *vel* accumē *C* acuīe *F*
34 se haberet *tr.* *F*

tween *FG* and *M*, and let *FG* and *DE* be squared. Also let line *AB*/line *HI* = *FG*/*DE*, and let *HI* be squared. Proof: *FG*/*M* = FG^2/DE^2. Therefore, by equality, *FG*/*M* = square *FS*/square *DR*. But *FG*/*M* = *AC*/*O*. Therefore, square *FS*/square *DR* = *AC*/*O*. But *FG*/*DE* = *AB*/*HI*. Therefore, since similar surfaces are erected upon them [i.e., upon lines

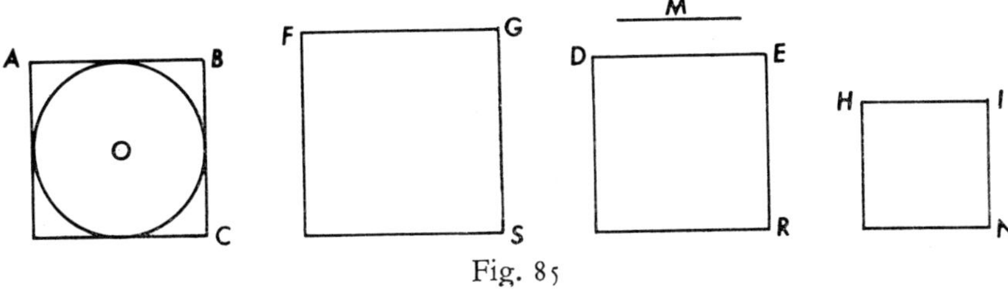

Fig. 85

FG, *DE* and *AB*, *HI*], square *FS*/square *DR* = square *AC*/square *HN*. Therefore, by equality, square *AC*/*O* = square *AC*/square *HN*. Therefore, circle *O* and square *HN* are equal. Hence it is clear that some circle is equal to some square. And thus, with it so argued, if anyone naturally possessed the acumen to know how to find two lines related in the same ratio as a square to the circle inscribed in it, he would be able by using the method which we have just described to find immediately a square equal to the proposed circle.

3. The *Quadratura circuli* Attributed to Campanus

Among the various non-Archimedean tracts on quadrature, we can single out a short work of seven conclusions which is attributed in six of a total of seven manuscripts to the thirteenth-century mathematician, Campanus de Novara.[1] In one manuscript (*Ec*, see variant readings, after line 121) and one edition (Ed_2, variant readings, line 1) the tract is said to have been composed at Rome. I am somewhat reluctant to concede that this elementary piece, exhibiting such immaturity, could be the work of the same man who composed the standard commentary on the *Elements* of Euclid. The

[1] MSS *Q*, *Eb*, *Ec*, *Mi*, *Er*, *Bd* (see Sigla and variant readings after line 121). I do not know of any manuscript supporting the attribution to one Guilhelmus Anglus by John Bale, *Index of British and Other Writers*, edited by R. L. Poole with the help of Mary Bateson (1902), p. 114.

same doubt had occurred to the scribe of our Milan manuscript (*Mi*) for, after giving the regular colophon of the work with its attribution of the tract to Campanus, we read in smaller letters the following: "Campanus never composed [this] for he would have forgotten his commentary [had he done so]."[2] Be that as it may, it was accepted as being by Campanus in the fourteenth century by Albert of Saxony.[3] On the other hand, in the edition of 1495, its authorship is designated as "a certain archbishop of the Franciscan order."[4] When one compares variant readings, it becomes evident that this edition is related to the manuscript tradition represented by manuscript *Ga*. But in that manuscript our tract begins: "Franco, *scolasticus* of Liège, has written this book on quadrature of the circle to Archbishop Hermann." One wonders whether or not some later scribe (or the editor of the 1495 edition) somehow altered this beginning, or one similar to it, to produce the attribution to "a certain archbishop of the Franciscan order." Perhaps this scribe or editor, knowing that Franco of Liège had produced an entirely different work on quadrature (see Chapter Two, Section 1, note 2), thought that this work was the reply of the archbishop to Franco. If this is what happened, then that scribe or editor was guilty of a historical blunder in making the "certain archbishop" a Franciscan, since any contemporary of Franco's would have lived toward the middle of the eleventh century. I think that the attribution to Franco himself that appears in manuscript *Ga* is doubtful, not only because all other manuscripts ascribe the treatise to Campanus but also because of the existence of an entirely different tract on quadrature by Franco. True, the geometrical knowledge presented in the tract under consideration is almost as primitive as that of the quadrature tract that goes under the name of Franco, but the syllo-

[2]See the variant readings, line 121. Cf. the gloss of *Ga* ("nondum demonstratum") and the skeptical poem of *Ec*. Incidentally, M. Chasles, H. Suter, and M. Cantor in more recent times have also expressed this same doubt: M. Chasles *Aperçu historique sur l'origine et le développement des méthodes en géométrie*, 3d ed. (Paris, 1889), p. 515; H. Suter, "Der Tractatus 'De quadratura circuli' des Albertus de Saxonia," *Zeitschrift für Mathematik und Physik*, vol. *29* (1884), Hist.-lit. Abt., p. 95; M. Cantor, *Vorlesungen über Geschichte der Mathematik*, 2d ed., vol. *2* (Leipzig, 1899–1900), p. 101. Cf. the notes by A. Sturm and G. Eneström in the *Bibliotheca mathematica* Dritte Folge, vol. *3* (1908), pp. 152–53.

[3] See Chapter Five, Section 3, lines 57–60.

[4] See variant readings, line 1. Apparently this attribution was repeated in a later edition of Paris, 1530, which I have not seen (cf. Eneström and Suter, *loc. cit.*, in note 2, above); it appears to have been a reprint of the 1495 edition. However, in what I take to be the second edition (Ed_2) published by L. Gaurico, the tract is ascribed to Campanus (see the variant readings, line 1) and apparently also in the edition by Johannes Caesarius Juliacensis (see Sigla).

gistic structure of conclusions V–VII seems to put the treatise at a somewhat later date.

It may also be of interest in the dating of our tract to examine the short poem that accompanies it in all but one of the copies that I have examined. It runs as follows (see the variant readings, after line 121):

> Rem novam mirabilem, quadraturam circuli
> Velud inscruptabilem, apud doctos populi
> Olim scibilem, puri cernunt oculi
> Vere demonstrabilem, nunc in fine seculi.

This we can translate as follows:

> Quadrature of the circle—a new marvelous thing
> Once knowable among scholars as uninvestigable
> But now in the end of the age clear eyes discern it
> As truly demonstrable.

A number of questions arise concerning this poem. Was it written by the author of the tract? Supposing the answer to be yes because of its presence in so many copies, are the words *in fine seculi* to be translated in a general way as "in the end of the age"? Or are they to be rendered more specifically as "in the end of the century"? If the latter, which century is implied? It would seem to me that the only possible centuries are the twelfth and thirteenth. I reject the eleventh century because of the syllogistic form of the last three conclusions. Supposing for the moment that the thirteenth century is meant, then we must ask the further question of how literally "end" is to be understood. If it is taken to mean 1299, then the author could not have been Campanus, who died sometime between 1296 and 1298.[5] If "end" simply means "toward the end," however, then it is possible for Campanus to have composed the work. But in actuality, as I have said earlier, there is some doubt that an excellent geometer of Campanus' stature would have composed such a trite work. In fact, it is hard to conceive how any geometer of the late thirteenth century could have used the expression *quadratura circuli* in the peculiar way it is used in our tract, in view of the fairly wide circulation of Archimedes' *De mensura circuli* in the thirteenth century. Hence, we might well conclude that the twelfth century is a better possibility for the time of composition of our tract. (I am, of course, not at all sure that the expression *in fine seculi* does not simply mean "in the end of the age," having eschatological implications.)

[5] Giovanni Vacca in the *Enciclopedia italiana* (under "Campano da Novara") says: "Morì poco dopo aver fatto testamento il 9 settembre 1296; risulta defunto in un documento del 28 luglio 1298."

The elementary and trivial character of the treatise will be immediately evident to the reader. Its final objective is given in the "seventh" conclusion, where it is concluded that every circle is equal to a square whose side the diameter exceeds by three half-parts, where the diameter is divided into seven parts. It is evident that the expression "equal to" in this proposition means "equal with respect to perimeters," that is, if we assume a circumference of 22 parts with a diameter of 7 parts (or π equal to $3\frac{1}{7}$), the square whose perimeter is equal to the circumference of the circle will be the one whose side is $5\frac{1}{2}$ parts. The author does not seem to realize that he is using an approximation when he makes the simultaneous division of the circumference into 22 parts and the diameter into 7, for throughout he uses the phrase "precisely exceeds," singularly inappropriate for any calculation involving π.

The proof of the main propositions (V–VII) depends on four preliminary conclusions. The first shows how to divide a circle (and hence its circumference) into four equal parts by two orthogonally intersecting diameters. The second then goes on to state that a straight line can be made equivalent to the circumference by assuming that the circumference consists of 22 parts and the diameter of 7 parts. The third conclusion purports to show how a straight line can be divided into four equal parts. But in fact the demonstration is concerned with constructing a straight line four times as long as a given line. In the fourth conclusion a square is constructed from four equal straight lines.

With the four first conclusions "proved" and considered as auxiliary propositions, the fifth, sixth, and seventh are presented as the major premise, the minor premise, and the conclusion of a syllogism described by the author as being of "the first mode of the first figure." The fifth conclusion holds that a plane figure contained by a single circularly drawn line whose diameter exceeds a quarter part of the figure's perimeter by three half-parts is equal to (but actually its perimeter is equal to) a square (actually, to the perimeter of that square) whose side the diameter exceeds by three half-parts. The sixth proposition then states that every circle is such a figure contained by a single circularly drawn line. Finally, in the last proposition it is concluded, as we have noted, that every circle is equivalent (in respect to perimeters) to a square whose side the diameter exceeds by three half-parts, i.e., whose side is $5\frac{1}{2}$ parts when the diameter is 7 parts and the circumference 22 parts.

It should be remarked that our text follows manuscript *Q* in numbering the conclusions I through VII. But there is some doubt as to whether conclusions VI and VII were so numbered in the original tract, since it is evident that conclusions V–VII constitute a kind of integrated syllogistic argument and since the other manuscripts and the first edition do not number the sixth and seventh conclusions (except *Bd*, which numbers the sixth, and perhaps *Mi*, which has an unreadable marginal statement that might be interpreted as 7ª *conclusio*). Furthermore, it will be noticed that in the proemium, line 7, the fifth conclusion is spoken of as the one "principally intended." On the other hand, the textual reference in line 109 to the "sixth (6ª)" proposition, while absent in manuscripts *Er*, *Ga*, and *Ed*, is present not only in *Q* and *Bd*, which number the sixth proposition, but also in *Eb*, *Ec*, and *Mi* even though those manuscripts appear to have abandoned the numbering for that proposition.[6]

I have collated all seven manuscripts with the first edition. Unfortunately, no copy goes back of the fourteenth century, all of the manuscripts clustering in a period of less than a century in length. It is difficult to set the manuscripts precisely in relationship with each other, but a few observations can be made. There appear to be two main traditions, the one represented by *Q*, *Eb*, *Ec*, *Bd*, *Er*, *Mi*, and the other by *Ga*, *Ed*. We have already discussed how that second tradition bears indications of authorship that diverge from the main attribution to Campanus of the first tradition. Within the first tradition *Q* quite often diverges from other manuscripts and sometimes in a most intelligent way, and so on occasion I have followed *Q* against the other manuscripts. Within the group of the remaining manuscripts we often find *Ec* (perhaps the earliest of the fourteenth-century manuscripts) pairing with *Bd*; but neither is always satisfactory nor can constitute the sole basis of the text. We also quite commonly find *Ec* paired with *Eb*, and both grouped with *Er*. In lieu of finding earlier and more satisfactory manuscripts, I have had recourse in setting up the text to the usual procedure of insuring linguistic and stylistic uniformity together with mathematical consistency. In view of what has been said in this introduction concerning the authorship and also of the fact that the title is drawn from *Bd* alone, I have bracketed it. There is little variation among the manuscripts in respect to the figures accompanying the text.[7] The marginal references are to manuscript *Q*.

[6] See the variant readings for lines 84, 98, 102, 108–9, 109, 110.

[7] Fig. 86(*a*) is missing only in *Mi*. Fig. 86(*b*) is in all copies. Fig. 87 has been taken from *Ga* and *Ed*. *Eb*, *Ec*, *Bd*, *Er*, and *Mi* have instead a row of six intersecting circles in the manner of the three circles in Fig. 88, thus indicating a line divided into seven parts (but *Ec* and *Bd* have them opposite Proposition V and *Bd* has on folio 91r the circumference divided into 22 parts by a series of circles in much the same man-

Sigla of Manuscripts and Editions

Q = Paris, BN lat. 7378A, 18r–v, 14c. The numbering of the conclusions is given in the margin. The spelling is generally that adopted in the text, except that for the most part *ci* rather than *ti* is used before a vowel, e.g., *circumferencia*.

Eb = Erfurt, Stadtbibliothek, Amplon. F. 178, 138–139v, item No. 15, middle of 14c. Similar to *Ec*. Propositions are unnumbered.

Ec = Erfurt, Stadtbibliothek, Amplon. Q. 361, 79v–80r, item No. 23, first half 14c. A good text, but occasional omissions and the substitution of *separantur* for *superantur*. The propositions are unnumbered. This manuscript perhaps was the source of Ed_2.

Bd = Oxford, Bodleian Library, Digby 147, 89r–91r, 14c. Similar to *Ec*. Has *diameter* throughout. Uses *ci* before a vowel. Numbered through Proposition VI.

Mi = Milan, Bibl. Ambros. H. 144 Inf., 145r–146r, 15c. For the numbering procedure, see the variant readings. Spelling peculiarities: *seccare allius*, *contangens*, *inteligatur*, *tolerent*, *transendit*.

Er = Erfurt, Stadtbibliothek, Amplon. Q. 385, 51r–53r, item No. 3; text ends on 52v, diagrams are on 52v and 53r, late 14c. Occasionally careless with case endings. The propositions are numbered in the margin through the fourth conclusion. The scribe writes *separantur* for *superantur*.

Ga = Cambridge University Library, Ee.III.61, 176v–177v, 15c. Of the same tradition as *Ed*. Substitutes an attribution to Franco of Liège for the customary proemium with its reference to the *Predicaments* of Aristotle. Uses *ci* instead of *ti* before a vowel, e.g., *circumferencia*. Uses *diameter* instead of *dyameter* found in most of the other manuscripts.

Ed = Edition with Thomas Bradwardine, *Geometria speculativa* (Paris, 1495), 20v–21v. *Ed* is somewhat prolix, tending to add phrases. The conclusions are numbered through the fifth, in the text. Spelling peculi-

ner). Fig. 88 is the same in all copies. The four parts of Fig. 89 are completely given only in *Ga* and *Er*. *Eb*, *Ec*, *Bd*, and *Mi* have (*a*), (*b*), and (*d*) and *Q* has just (*d*), while *Ed* has none, substituting an erroneous figure. Fig. 90 is missing in *Ga* and *Q*; only the circle is drawn in *Eb*, *Ec*, and *Mi* (and then the diameter of the circle in *Ec* is divided into seven parts, in *Mi* it is erroneously divided into six parts, and in *Eb* into eight parts). In *Er* the figure is as I have given it except that the text of the fifth conclusion is written within the square and that of the sixth and seventh within the bottom of the circle. In *Bd* the figure is as I have given it. In *Ed* the square and the circle are separated. Incidentally, the figures (not the conclusions) are numbered in *Ga*.

arities: *sciencia*, *tercia*, occasionally *interseccetur*, *diameter* (instead of *dyameter* as in MSS), *premituntur*, *phisicam*, *capud*.

Ed_2 = L. Gaurico, *Tetragonismus, id est circuli quadratura, per Campanum, Archimedem Syracusanum atque Boetium...adinventa* (Venice, 1503). I have compared this edition with the manuscripts and the first edition. At times it agrees with the earlier edition, but also occasionally with the manuscripts against the edition. I think that Gaurico had both the edition and some manuscript (perhaps *Ec*) before him in preparing this edition. I have included variant readings of Ed_2 only for the beginning and for the numbering of the sixth proposition. After each proposition, Ed_2 includes an added comment by Gauricus.

[Before this volume went to Press, further manuscripts of the Campanus *De quadratura circuli* were discovered: (1) Paris, BN lat. 16089, 188r–189r, 13–14c; (2) Cambridge University Library, MS 1572; (3) Glasgow University Library, MS BE 8-y.18, 208r–209r; (4) Naples, Bibl. Naz. VIII.C.19, 235r–37r; (5) Munich, Bay. Staatsbibl. 4377, 136v–37. I have had an opportunity to examine only the Glasgow University Library manuscript. Its title runs: "Incipit quadratura circuli secundum magistrum campanum cum commento eidem asscripto." It is not a particularly faithful copy.]

Other Editions, Not Used for Text

Johannes Caesarius Juliacensis, *De quadratura circuli demonstratio ex Campano*, 1507. Cited by H. Suter, *op. cit.*, p. 95, in note 2 above.

Edited with Thomas Bradwardine, *Geometria speculativa* (Paris, 1503). Apparently this is a reprint of the first edition of 1495, since here it is also attributed to "quodam archiepiscopo ordinis fratum minorum," according to Suter, *ibid.*

[Incipit tractatus de quadratura circuli secundum Campanum]*

18r / Aristotiles (!) in eo qui de cathegoriis libro inscribitur dicit quadratura quidem circuli scibilis est, scientia autem eius nundum est, et in plerisque locis reprehendit multos et magnos qui hoc demonstrare
5 conantes turpiter erraverunt. Et ideo hic quadratura circuli demonstratur. Et primo premittuntur quatuor conclusiones et probantur, secundo ex hiis inducitur quinta principaliter intenta inferius scilicet ibi: "Omnis figura plana etc."

* For reasons of typography, the title has been set as two lines, although in the system of line numbers here adopted it is considered as a single line.

---

1 [Incipit...Campanum] *Bd*, *om.* *QEbEc ErMi* Franco scolasticus Loadicensis (*!*) ad Hermanum Archiepiscopum scripsit hunc librum de quadratura circuli *Ga* Tractatus de quadratura circuli editus a quodam archiepiscopo ordinis fratrum minorum. Prohemium *Ed* Campani viri clarissimi tetragonismus idest circuli quadratura rome edite cum additionibus Gaurici Ed_2

2–6 Aristotiles...primo: Et hoc ad probandum *Ga*

2–8 Aristotiles...etc.: Ad demonstrandum igitur circuli quadraturam campanus noster primo quatuor conclusiones et quidem facillimas secundo autem ex his inducitur quinta que simul cum sexta totam de circuli tetragonismo demonstrationem manifestissime concludit Ed_2

2 Aristotiles: Aristoteles *EcEd* / cathegoriis *QEbBdEd* kategoriis *Ec* kathegoriis *Er* categoriis *Mi* / libro inscribitur *EbErEd tr. Q* libro dicitur *Mi* scribitur libro *Ec* / dicit: ostendit hoc *Ec* dicit hoc quod *Bd*

3 quidem circuli *tr. Q* / autem: enim *Q* vero *Mi* / nundum *QEbMi* nondum *EcBdErEd* / est: est inventa *mg. QEd*

3, 4 et *om. Eb*

4 in: im *EbEd* / plerisque: quampluribus *Mi* / et magnos *om. Mi*

5–6 conantes...premittuntur: a magistro Compano (*!*) et primo preintelligatur *Er*

5 turpiter: enormiter *Ed* / Et ideo hic *EbEcBdMi* ideo hic *Q* hic vero *Ed*

6 premituntur *MiEd* / conclusiones: contradictiones *Er*

6–7 et²...inducitur *om. Q* et probantur secundo *Er* et probantur ex hiis inducitur *Mi*

6–7 probantur secundo *om. Ga*

7 *post* inducitur *add. Ed* et concluditur / inducitur...scilicet: inducit conclusionem principaliter intentam *Ga* / intenta: intelecta *Mi* intellecta *Eb* / inferius... etc. *om. BdEd* / inferius scilicet ibi *Q* inferius ibi *EbMi* ex hiis infertur ibi *Er* inferius *Ec*

8 plana *QEr om. GaEc* plana unica *EbMi*

[Here Begins the Treatise of Campanus on the Quadrature of the Circle]

Aristotle in the book which is entitled *On Categories* says, "Quadrature of the circle is knowable, but not yet known." And in many places he censures a great many who, in attempting its demonstration, have erred in an unsightly way. Therefore, quadrature of the circle is here demonstrated. In the first place, four conclusions are premised and proved. Then, secondly, from these conclusions is inferred a fifth conclusion below, the one principally intended: "Every plane figure, etc."

[I.] Prima conclusio: LINEAM ORBICULARITER DUCTAM
10 BINA DYAMETRO IN QUATUOR EQUALIA SECARE.

Dyameter enim est linea recta ab extremo in extremum per centrum ducta dividens figuras in partes equales [Fig. 86*a*]. Ergo si sint due dyametri sese intersecantes in centro ad angulos rectos, dividunt figuram in quatuor partes equales [Fig. 86*b*]. Et nota quod "dyameter"
15 dicitur a "dya," quod est "duo," et "metros," quod est "mensura," quasi "duorum mensura," scilicet duarum medietatum.

[II.] Secunda conclusio: LINEE ORBICULARITER DUCTE LINEAM RECTAM EQUALITER DARE.

Iuxta mathematicam scientiam et phisicam veritatem circulus divi-
20 ditur in 22 partes equales, et remota una, scilicet 22ª parte, tertia remanentis, scilicet septem, est dyameter circuli. Tripletur igitur dyameter et addatur 7ª dyametri ordinenturque huius partes in rectum et

9 Prima conclusio *mg. Q mg. Bd Ed om. Ga Eb* Est ergo prima conclusio ista *Ec* Prima propositio *Mi* Suppositio prima *mg. Er*
10 equalia *om. Er* / secare: seccare contingit incipit comentum *Mi* / *post* secare *add. Er* hec suppositio sic declaratur
11 enim *om. EbGaEd* / in: ad *BdGa*
12 figuras: figuram *QGaEd* / partes: duas partes *Ed* / equales: equaliter ut patet in illa prima figura *Ga* / *post* equales *add. Ed* ut patet hic in prima figura / Ergo...due: si vero due sunt *Ed* / si *om. Eb* si vero *Ga* / sint: sunt *QEd* / sint *tr. Ga post* sese *in linea 13* / duo *Er*
13 secantes *Ga* / in centro *om. Er* / rectos: equales *GaEd*
13–14 dividunt figuram: dividitur figura *Q* divident *EbEc*
14–15 equales....a: ut hic patet per secumdam figuram dicitur autem diameter ad *Ed*
14–15 Et...dicitur: ut patet dicitur enim diameter *Ga*
15 dya: dia *BdGaEd* / quod est² *om. EcGa*
16 quasi: quia *Ga* / quasi...mensura *om. Bd* / mensura: mensuratur *Er* / scilicet *QEcEd* sive *EbMiEr* / scilicet...medietatem *om. Ga*
17 Secunda conclusio *mg. Q mg. Bd Ed om EbEcGa* secunda propositio *Mi* Secunda *mg. Er* / rectam *om. Mi*
18 equaliter: equalem *ErGaEd* / dare: dare contingit *Mi* / *post* dare *add. Ec* super est possibile *et Eb* supple est *et Bd* suple est possibile
19 mathematicam *QEcEr* mathematicorum *GaEd* mathematicarum *EbBdMi* / phisicam *EcBdErGaEd* physicorum *Q* physicam *EbMi*
19–20 dividitur *tr. Mi post* partes / dividitur *om. Eb* divisus *Er et tr. post* equales (*Note*: *in mg. Ga*: supponit quod probandum est, ibi assumit conclusionem)
20 22 *om. Er* / equales *om. ErEd* / et: quarum *Ed* / remota una *tr. Ed* / tertia: tercia pars summe *Ed* tertia pars *Er*
20–21 tertia remanentis *tr. Er* / remanentis: remanens *Q*
21 scilicet: cum *Bd* / septem: 7 *Ga* septinarius sive septem *Ed* 7ª *QEbEcMi* septem illa *Er* (*Note*: septinarius et septem *tr. Ed post* circuli) / triplicetur *Bd*
22 addetur *Ec* / 7ª: septima pars *Er* / dyametri: diametri pars *Ed* / ordinenturque *QGaEd* et ordinentur *ErMi* ordinentur *EbEcBd* / huius partes *QGa tr. Ed* huiusmodi partes *EcBdMi* iste partes *Er* / rectum: recto *Ed*

[I.] The first conclusion: TO DIVIDE A CIRCULARLY DRAWN LINE INTO FOUR EQUAL PARTS BY A PAIR OF DIAMETERS.

For a diameter is a straight line drawn through the center from one extreme to the other extreme which divides figures into equal parts [see Fig. 86*a*]. Therefore, if two diameters intersect each other in the center at right angles, they divide the figure into four equal parts [see Fig. 86*b*]. And it is noted that "diameter" is named from "dya," which is "two,"

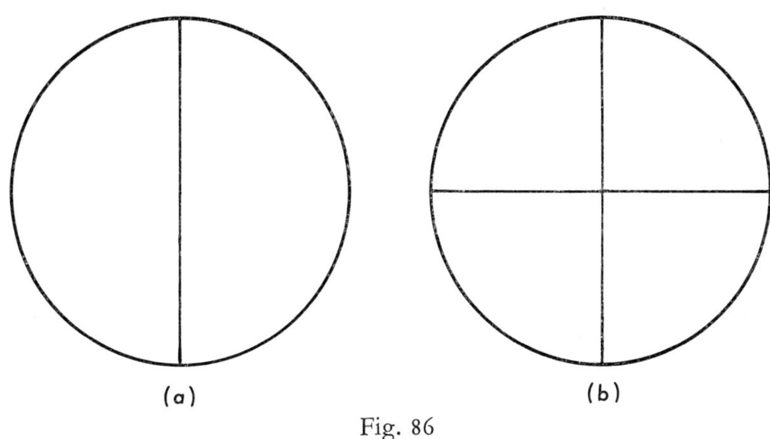

Fig. 86

and "metros," which is "measure," and so it is equivalent to "the measure of two things," that is, of two halves.

[II.] The second conclusion: TO GIVE A STRAIGHT LINE EQUAL TO A CIRCULARLY DRAWN LINE.

Using mathematical knowledge and physical truth, a circle is divided into 22 equal parts, and with one part subtracted, that is, the 22nd part, a third of the remainder, namely, 7, is the diameter of the circle. Therefore, let the diameter be tripled and let there be added [to the result] a seventh of the diameter, and let these parts be ordered in a straight line. We shall

habebitur linea recta equalis circulari, ut in figura apparet [Fig. 87].

[III.] Tertia conclusio: LINEAM RECTAM IN QUATUOR
25 EQUALIA SECARE.

Fiat circulus unus [Fig. 88]. Deinde circino non restricto nec ampliato ponatur pes circini in circumferentia et circumducatur ut secundus circulus constituatur qui in duobus locis intersecet primum et intersecetur ab eo, transiens per centrum primi. Deinde ducatur linea
30 recta per ambo centra ab extremo in extremum utriusque circuli et ubi terminabitur hec linea in circumferentia secundi circuli ponatur pes

23 circulari: circulari linee *ErEd* linee circulari *Ga* / *ante* ut *add. Er* cuius iste sint dyameter (?) / in...apparet *Q* in figura patet manifeste *EbEcBd* patet in presenti figura manifeste *Mi* patet in figura *GaEr* hic liquidum est videre *Ed* (*Note*: patet in figure *tr. Ga post* rectum *et hic add.* satis patet)

24 Tertia conclusio *mg.Q mg.Bd Ed* (*sed* Tercia *in Ed*) *om. GaEbEc*; Quarta. Tertia propositio huius libri *Mi* tertia *Er* (*Note: opposite the third conclusion in Ga is the marginal comment*: immo hic lineam quadruplicit)

26 *ante* Fiat *add. Ed* quod sic patet *et Ga* hec conclusio probatur sic / *post* unus *add. Er* cum circino / Deinde: de *Q*

26–27 *post* ampliato *add. Ed* sed stante uniformiter ut prius

27 circini: circuli *Q* circini fixus *Er* / ducatur *GaEd* / ut: et *Ed*

28 qui: quod *Q* / intersecet *EbEcGaEd* intersecat *QBd* interseccet *Mi* intersecabit *Er* / primum: primum circulum *Er* / et *om. Eb*

29 intersecatur *Er* intersecabitur *Bd* / *post* eo *add. Er* in eisdem locis / transiens: transietque *Er* / primi: primi circuli *Er* / Deinde: post hec *Er* de hinc *EbEd* / ducatur: deducatur *Er*

30 centra *om. Mi* per *Eb* / ab...circuli: utriusque circuli ista quod in utriusque circumferencia terminetur *Er* / in: ad *Bd*

31 ubi: non *Ec* / terminatur *MiGa* / hec *om. Q* illa *Er* / in *om. Q*

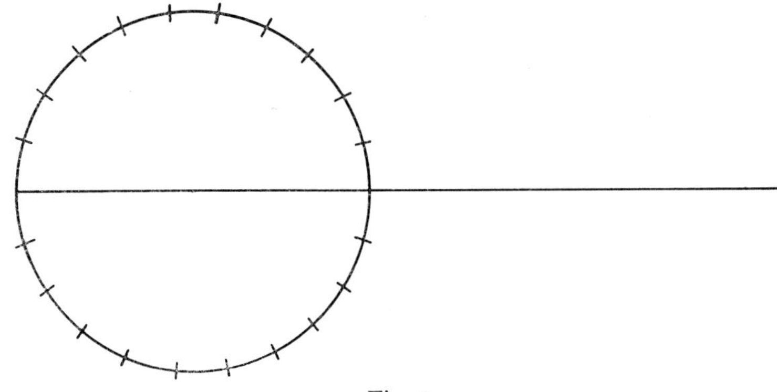

Fig. 87

have [accordingly] a straight line equal to a circular line, as is apparent in the figure [Fig. 87].

[III.] The third conclusion: TO CUT A STRAIGHT LINE INTO FOUR EQUAL PARTS.

Let one circle be drawn [see Fig. 88]. Then with the compass neither closed nor opened further, let a foot of the compass be placed in the circumference [and the other foot] revolved so that a second circle is

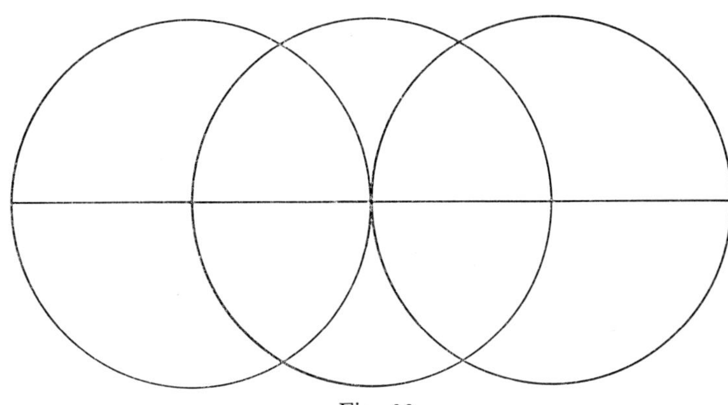

Fig. 88

formed which intersects the first, and is intersected by it, in two places. This second circle passes through the center of the first one. Then let a straight line be drawn through both centers from one extreme to the other of each circle, and where this line is terminated in the circumference of the second circle let one foot of the compass be placed with the same arrangement as before. And let [the other foot of the compass] be revolved so that a third circle is formed which intersects the second circle, and is

circini sub dispositione priori, et circumducatur ut tertius circulus constituatur qui in duobus locis intersecet secundum et intersecetur ab eo, contingens primum in centro secundi. Trahatur predicta linea
35 recta usque ad circumferentiam tertii circuli, ut in predicta figura [Fig. 88] patet. Predicta igitur linea recta transiens per tria centra ab extremo primi circuli usque ad extremum tertii dividitur in quatuor partes equales, nam quelibet partes due predicte linee sunt in eodem circulo a centro ad circumferentiam ducte; ergo sunt equales. Et
40 quoniam quecunque uni et eidem sunt equalia, inter se sunt equalia. Sequitur quod quelibet pars linee in uno predictorum crirculorum contenta est equalis cuilibet parti linee in alio circulo cicumscripte. Hoc idem aliter probatur sic: Fiat circulus unus. Deinde pede circini non diversificati posito in circumferentia eiusdem circuli, reliquus pes fiat
45 eiusdem circini non variati protendatur extra circulum supradictum ibique fixo centro ducatur ut secundus circulus constituatur contingens primum in puncto. Posito in puncto pede circini non mutati ducatur

32–33 circumducatur...constituatur: fiat tercius circulus *Bd*
32 ducatur *GaEd*
33 qui *om. Er* | intersecetur: tersecetur *Er* secetur *Ga* intersecabitur *Bd*
34 contangens *Mi* | in centro: et centrum *Ed* | secundi: secundi circuli *Ga* | trahaturque *EcBdGaEd* | predicta linea *tr. EcBd*
35 recta *om. BdQMi*
35 tertii circuli *tr. Er*
35–36 in...patet *EbQMi* in figura patet *Ec* patet in figura *ErBd* patet in 4ª figura *Ga* patet in figura presenti *Ed*
36 Producta *Ed* | igitur: ergo *EbEcBdEr* | recta *om. QEr* | transiens *iter. Mi* | per ...centra *om. Q*
37 primi circuli *tr. BdEr* | usque *om. Ga* | tertii *om. Er* | *post* tertii *add. Er* circuli | quatuor *om. Bd*
38 quilibet *Er* | partes due *tr. BdGaEd*
39 ducta *Er* | ergo: igitur *Q* | sunt *om. QEbEcMi* | Et *om. Q*
40 eidem: eidem numero *Ga* | sunt[1]: fuerint *Bd* | equalia[1]: equales *Q* | inter *QEbEcBd* ipsa inter *ErEd* illa inter *Mi* et inter *Ga*
40–41 sunt[2]....quod: similiter quare *Ga*
41 Sequitur *om. Ed* | quod: ergo quod *Mi* ergo *Ed* | in *om. Eb* | uno...circulorum: predictis circulis *Bd* | circulorum *om. Er*
42 parti: alii parti *Ed* | allio *Mi* | circumscripte: circumscripto *Q* contente *Ed*
43 Hoc...sic *EbQMi* Hoc idem probatur sic *EcBd* Hic probatur aliter *Er* Item potest fieri alio modo *Ed* ad idem aliter *Ga* | circulus unus *tr. QEr* | unus *om. Er* | Deinde: ut deinde *Er*
44 diversificato *Bd* | reliquus *om. EcBd* | pes: autem pes *Ed*
45 eiusdem: ipsius *GaEd* | variati: diversificati *Ec* | supradictum: predictum *Bd*
46 centro: centro circuli *Ec* | producatur *Bd* | ut: ubi *Ec* | contangens *Mi*
47 Posito in puncto *EbEcMi* posito *Q* posito in puncto contingencie *Ga* positoque in puncto contingencie *Ed*
47–48 Posito...ut: sue circumferentie et in eodem puncto figatur pes circini non variati et *Er*
47 pede *om. Eb* pedis *Bd* | mutati *QEd* variati vel mutati *EcBd* variati nec mutati *Mi*

intersected by it, in two places. This third circle touches the first circle in the center of the second. Let the aforesaid straight line be extended up to the circumference of the third circle, as is evident in the aforesaid figure [Fig. 88]. Therefore, the aforesaid straight line, passing through the three centers from the extreme of the first circle to the extreme of the third circle, is divided into four equal parts, for any two parts of the aforesaid line are radii of the same circle; therefore, they are equal. And since all are equal to one and the same quantity, they are mutually equal. It follows, therefore, that any part of a line contained in one of the aforesaid circles is equal to any part of the line described in another circle.

This same thing is proved in another way as follows. Let one circle be drawn. Then with one foot of a compass [whose span] remains unchanged placed in the circumference of the same circle, the other foot of the same unchanged compass is extended outside the above-mentioned circle. And with the center fixed there [outside of the circle], let it [the other foot] be revolved so that a second circle is formed which is tangent to the first in a point. Then with a foot of the same compass, still unchanged, placed

alius pes circini, ut tertius circulus constituatur intersecans duo predictos circulos, transiens per eorum centra. Tunc trahatur linea recta
50 per eorum centra tria que secatur in quatuor partes equales, ut in figura [Fig. 88] manifestum est.

[IV.] Quarta conclusio: EX QUATUOR RECTIS LINEIS EQUALIBUS QUADRATUM EQUILATERUM COLLOCARE.

Hoc quidem manifestum est, et nichilominus potest demonstrari sic:
55 Fiant due linee recte sese in capite contingentes, ex quarum contactu constituatur unus angulus rectus. Deinde ponatur pes primus in contactu ipsarum linearum et reliquus pes in capite alterius linearum predictarum, ducaturque ad caput alterius linearum [Fig. 89*a*]; nec circulus actu compleatur sed completus intelligatur. Deinde ponatur pes
60 circini non variati in capite alterius linearum predictarum versus circumferentiam, que scilicet due linee supradicte sunt due semidyametri circuli prelibati. Alter vero pes ponatur in centro predicti circuli, et ducatur constituens circulum intersecantem predictum et se per illum

48 allius *Mi*
48–51 circini....est: et cetera ut prius *Ga*
48 intersecans *correxi ex* interseccans *Mi et* invicem secans *Ed et* intersecantes *Q*
48–49 intersecans...circulos *om. Ec* / intersecans...transiens *om. Bd*
48–50 intersecans...tria: que transiet per duorum circulorum centra et intersecabit eos. Hoc facto trahatur linea recta per eorum centra *Er*
48 duo *ErMi* duos *QEd*
49 Tunc *om. Bd*
49–50 Tunc...centra *om. Mi*
50 eorum....tria: tria centra *EcBdEd* / seccantur *Mi* secabitur *Er*
50–51 in²...est *Q* patet in figura *Eb* patet in presenti figura *Ec* patet *Bd* patet in figura presenti *Mi* prius *Er* manifestum est nam quelibet due partes etc. ut supra quod patet in hac figura *Ed*
52 Quarta conclusio *mg.Q mg.Bd Ed om. EbEcGa* quarta propositio *Mi* quarta *mg.Er* / rectis lineis *tr. QEr*
53 equilaterum *om. MiEd* / collocare: constituere *GaEd* colocare comentum *Mi*
54 et nichilominus potest *EbEcBdEd* sed potest nichilominus *Q* et potest *Mi* potest tamen *Er* nihilominus potest *Ga* / sic *om. Er*
55 Fiant: sint *Ed* / contingentes *tr. Q post* sese / contangentes *Mi* / ex quarum: et quatuor *Eb*
56 constituatur...primus: primus fixus *Er* / unus *om. Ga* unicus *Ed* / primus: primus circini *Q* circini *GaEd*
57 et reliquus pes *EbEcBdEr* et reliquus *QMi* reliquus vero pes *GaEd*
57–58 alterius...nec: eius ducaturque sic ut predictarum linearum angulus in circulo includatur et ducatur pes usque ad caput alterius linearum ut *Er*
58–59 ducaturque...compleatur: nec compleatur circulus *EcBd*
58 -que: quod *Eb* / ad: usque ad *Ga* / capud *Ed* / linearum: linee *GaEd* / *ante* nec *add. Ga* ut patet in 5ª figura
59 actu *om. Ed* / completus intelligatur *tr. Bd* / inteligatur *EbMi* / *post* intelligatur *add. Ed* sicut patet in hac figura *et EcBd* sicut / Deinde: de hinc *Mi*
60 predictarum *om. BdGa* / versus: huius *Er*
61 supradicte: simul dicte *Er* / sint *Eb* / due²: duo *Bd*
62 circuli prelibati *tr. EcBd* / Alter vero: reliquus *Ga*
63 predictum: circulum predictum *Ec*
63–64 et...in *om. Ga*

at the point [of tangency], let the other foot of the compass be revolved so that a third circle is formed which intersects the two aforesaid circles and passes through their centers. Then let a straight line be drawn through the three centers of these circles. This line is [accordingly] divided into four equal parts, as is manifest in the figure.

[IV.] The fourth conclusion: TO CONSTRUCT A SQUARE FROM FOUR EQUAL STRAIGHT LINES.

This is indeed manifest, but nevertheless it can be demonstrated as follows. Let two straight lines be drawn which meet at the head [of each] so that at the point of contact a right angle is formed. Then let the first foot of a compass be placed in the point of contact of these lines and the

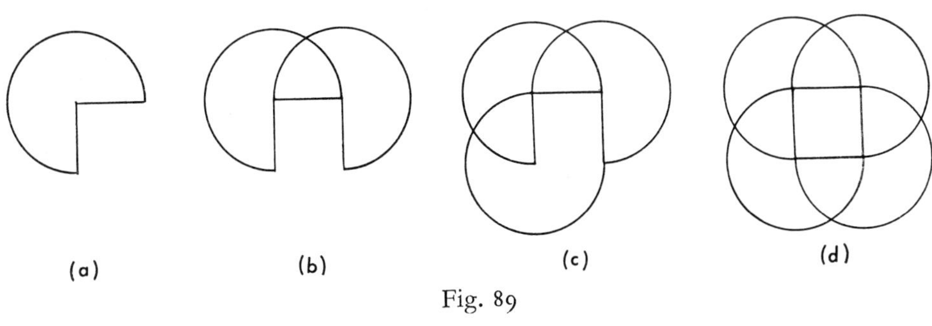

Fig. 89

other foot [of the compass] at the head of one of the aforesaid lines, and let it be revolved up to the head of the other one of the lines [see Fig. 89*a*]. Do not let the circle actually be completed but only done so conceptually. Then let [one] foot of the compass which has not been changed be placed at the head of the other of the aforesaid lines where it meets the circumference. These two above-mentioned lines are evidently two radii of the afore-designated circle. Now let the other foot be placed in the center of the aforesaid circle and revolved so as to form a circle intersecting the

in loco uno usque ad locum ad quem ducta de centro linea recta
65 constituat angulum rectum cum semidyametro circuli primi que terminatur in centro huius circuli secundi, ut hic patet [Fig. 89*b*]. Post hec autem ponatur pes circini non diversificati in capite alterius semidyametri circuli primi versus circumferentiam, reliquus vero pes ponatur in centro eiusdem circuli primi. Et ducatur usque ad locum ubi
70 terminatur linea ducta de centro secundi constituens circulum intersecantem primum, et se per illum in loco uno [Fig. 89*c*]. Et tunc linea recta trahatur de centro tertii huius usque ad caput linee procedentis de centro secundi, ut patet hic in figura [Fig. 89*d*]. Deinde ponatur pes
18v circini non mutati in capite predicte / linee procedentis de centro
75 secundi circuli ad circumferentiam. Alter autem pes ponatur in centro tertii et ducatur usque ad centrum secundi constituens circulum intersecantem ipsos, tertium et secundum, quemlibet in loco uno, et se per illos, ut in figura plenius apparet [Fig. 89*d*]. Quatuor igitur linee recte in predictis circulis contente constituunt quadratum equilaterum, sunt
80 enim equales sibi invicem omnes, nam quelibet due sunt in eodem circulo a centro ad circumferentiam protracte. Et nota quod ideo non

64 loco uno *tr. Ed* | ad locum ad: eo *Mi* | de centro: decenter *Bd*
65 constituit *Ed* | cum: et cum *Er* | circuli *om. Ed* | circuli primi *tr. Ga* | que: qui *Q*
65–66 que terminatur *iter. Eb*
66 centro: circulo *Er* | circuli secundi *tr. Ga* circuli *Eb* | hic patet *tr. Bd* patet in figura *Er* patet in hac figura *Ed* patet in 6ª figura *Ga*
67 autem: *om. EbEd* | diversificati: variati *BdErGa*
67–68 dyametri *Eb*
68 circuli: anguli *Er* | circuli primi *tr. EcBdGaEd* | reliquo *Eb*
69 circuli primi *tr. Bd* | ad *om. Bd* | locum: lineam *Mi*
70 de: a *Ed* | secundi: secundi circuli *Q*
70–72 intersecantem....centro *om. Er*
71 loco uno *tr. EcBdEd* | Et: ex *BdEd*
72 tertii *om. Bd* | tertii huius *tr. EcGaEd*
72, 74 precedentis *Q*
73 de: a *Ec* | secundi *om. Er* | patet...figura *Q* in figura patet 2ª *Bd* in figura patet *EbEc* in figura presenti patet *Mi* patet in figura *Er* patet in figura 7ª *Ga* patet in hac figura *Ed*
75 secundi *om. QGa* | circuli *om. ErMi* | autem *QEd om. Er* vero *Mi* | ponatur *om. Er*
76 ad *om. Bd*
76–77 circulos intersecantes *Ec*
77 ipsos *EbEcBdMiGa om. Q* ipsos scilicet *ErEd* | tertium: primum *Ed* | loco: loco sui (?) *Mi* | se per: semper *Ed*
78 figura...apparet: hac figura plenius declaratur *Ed* 8ª figura manifeste patet *Ga* | plenius *om. Ec* palius *Er* | igitur: ergo *MiEr*
79 in *om. Ec* | predictis: 4 istis *GaEc* | circulis: quatuor circulis *Ed*
80 enim *iter. Eb* | omnes *om. EcBd* eius *Er* | due: due linee *Bd*
81 a...protracte: etc. ut prius *Ed* et cetera *Ga* | a: ab eodem *Mi* | ad: usque ad *Er* | protracte: producte *Mi* pertracte *Er*

aforesaid [circle], and intersected by it, in one place. [It is revolved] up to the point where it meets a line drawn from the center at a right angle with the radius of the first circle—which radius is terminated at the center of this second circle, as is evident here [see Fig. 89*b*]. Thereupon, let a foot of the compass, [still] not altered, be placed at the head of the other radius of the first circle where it meets the circumference [of the first circle]. Now let the other foot be placed in the center of the same first circle and revolved up to the place where the line drawn from the center of the second circle ends. Thus it forms a circle intersecting the first [circle], and intersected by it, in one point [see Fig. 89*c*]. And then let a straight line be drawn from the center of the third circle to the head of the line proceeding from the center of the second circle, as is evident in the figure [Fig. 89*d*]. Then let a foot of the unchanged compass be placed at the head of the aforesaid line proceeding from the center of the second circle to [its] circumference. Then let the other foot be placed in the center of the third circle and revolved up to the center of the second circle. Thus a circle is so formed which intersects the second and third circles, and is intersected by them, each in one place, as is fully evident in the figure [Fig. 89*d*]. Therefore, the four straight lines contained in the aforesaid circles form a square, for they are mutually equal, any two of them being radii of the same circle. And note that hence the aforesaid circles are not

complentur actu dicti circuli quia completi actu tollerent eiusdem sensibilitatem quadrati sub eis constituti.

[V.] Quinta conclusio: OMNIS FIGURA PLANA UNICA LINEA 85 ORBICULARITER DUCTA CONTENTA CUIUS DYAMETER TRANSCENDIT PRECISE QUARTAM EIUSDEM FIGURE IN SEMIPARTIBUS TRIBUS EST EQUALIS QUADRATO CUIUS LATUS EIUSDEM CIRCULI DYAMETER TRANSCENDIT PRECISE IN SEMIPARTIBUS TRIBUS.

90 Huius veritas sic patet. Nam quecunque ab eodem superantur equaliter inter se sunt equalia. Si enim tetracubitum aureum et tetracubitum argenteum a pentacubito ligneo equaliter superantur, quia in cubito uno, tetracubitum aureum et tetracubitum argenteum necessario equantur. Quia igitur quelibet quarta et quodlibet latus huius quadrati a 95 dyametro circuli equaliter superantur, quia in semipartibus tribus,

82 completur *Q* / actu[1]: in actu *Bd et tr. Bd post* circuli / actu [1, 2]: arcus *Q* / dicti *EbErEdGa* isti *Ec* (*sed Ec habet* circuli dicti) predicti *QMi* / predicti circuli *tr. Mi post* ideo / quia: quia scilicet ut *Er* / actu[2] *om. Bd* / tollentur *Eb* tolerent *Mi* / eiusdem *om. Er* evidentem *BdEd* eius *Mi*

83 quadrati: istius quadrati *Er* / sub... constituti *om. Er hic, sed tr. Er* sub eis contenti *inter* li- *et* -nea *in linea* 84 / *post* constituti *add. Ga* sequitur demonstracio

84 Quinta conclusio *mg.Q mg.Bd Ed, om. EbEcErGa* Prima propositio sive maior *Mi* / *ante* Omnis *add. Ed* rem novam mirabilem quadraturam circuli. Velud inscrutabilem apud doctores populi. Olim siscibilem puri cernunt oculi. Vere demonstrabilem nunc in fine seculi (*Cf. var. post lineam 121*) / plana *om. Q*

86 transendit *Mi hic et ubique* / precise *om. Mi* / figure *om. Bd*

86, 89 in *om. Ed*

89 *post* tribus *add. Mi* Commentum / *post* tribus *add. Ed* omnis circulus est figura plana etc. Conclusio ergo omnis circulus est equalis quadrato cuius latus eiusdem circuli diameter transcendit precise semipartibus tribus / *post* tribus *add. Ga* omnis circulus est figura plana unica linea orbiculariter ducta contenta cuius diameter transcendit quartam eiusdem figure in semipartibus tribus. Ergo omnis circulus est equalis quadrato cuius latus eusdem circuli diameter transcendit precise in semipartibus tribus.

90 Huius....Nam: Maior sic patet *Ed* / Huius...sic: Veritas manifeste *Ga* / seperantur *EbEcEr*

91[1, 2] 93 -cubicum *EbQEd*

92 -cubico *EbQEd*

92 ligneo: lineo *Er* / superentur *Mi* separantur *Er* / quia *om. Bd*

92–93 in...uno: minimo cubico igitur *Ed*

93 tetracubitum[2] *om. EcEd*

93–94 equabuntur *Ed*

94 quarta: quarta circuli *GaEd* natura *Q*

95 seperantur *Eb Er*

actually completed, for if actually completed they would sensibly obscure the square formed under them.

[V.] The fifth conclusion: EVERY PLANE FIGURE CONTAINED BY A SINGLE LINE DRAWN CIRCULARLY WHOSE DIAMETER EXCEEDS A FOURTH OF [THE PERIMETER] OF THE SAME FIGURE BY PRECISELY THREE HALF-PARTS IS EQUAL [IN RESPECT TO PERIMETERS] TO A SQUARE WHOSE SIDE THE DIAMETER OF THE CIRCULAR FIGURE EXCEEDS BY PRECISELY THREE HALF-PARTS.

The truth of this is evident as follows. For whatever things are equally exceeded by the same thing are equal to each other. Thus if four cubits of gold and four cubits of silver are equally exceeded by five cubits of wood, namely, by the amount of one cubit, the four cubits of gold and the four cubits of silver are necessarily equal. Therefore, since any quarter [of the circular perimeter] and one side of this square are equally exceeded by the diameter of the circle, namely, by three half-parts, [hence] the quarter

quelibet quarta circuli et quodlibet latus quadrati huius sunt equales, et circulus et huiusmodi quadratum sunt equalia.

[VI. Sexta conclusio:] OMNIS CIRCULUS EST FIGURA PLANA UNICA LINEA ORBICULARITER DUCTA CONTENTA CUIUS 100 DYAMETER TRANSCENDIT PRECISE QUARTAM PARTEM EIUSDEM FIGURE IN SEMIPARTIBUS TRIBUS.

Huius declaratio patet in secunda propositione. Si enim secundum quod plerique mathematici scripserunt iuxta physicam veritatem circulus dividitur in 22 partes equales et remota una, scilicet 22ª parte, 105 tertia remanentis, scilicet 7, est dyameter circuli et quarta circuli continet 5 partes et dimidiam, et dyameter, scilicet 7, transcendit precise quartam circuli, scilicet 5 partes et dimidiam, in semipartibus tribus, id est in tribus dimidiis partibus. Ex premissa igitur 5ª maiore

96 quelibet: igitur quelibet *Ed* / quarta *om. Ga* / quadrati huius *tr. Mi* / huius: necessario *Ga* huiusmodi necessario *Ed* / *ante* sunt *add. Mi* a dyametro circuli

97 et circulus *tr. Er* / et [1]: et sic *BdEd* / et huiusmodi *EcBd tr. Q* huius et *EbErMi* / huiusmodi *tr. GaEd ante* sunt / quadratus *ErEd* / equalia *QEbEcEr* equales vel equalia *Mi* equales. nam quorumcunque omnes partes sibi inter se sunt equales et ipsa inter se sunt equalia *GaEd*

98 Sexta conclusio *mg.Q mg.Bd, om. EbEcErGaEd* propositio minor *Mi* minor propositio quae est sextra conclusio Ed_2

98–101 Omnis...tribus *om. Ed hic, cf. var. linea* 89

98 Omnis: Sed omnis *Er* / plana *om. Q*

100 *post* dyameter *add. Bd* in septem divisa/ transcendit *tr. Q. post* figure *in linea 101* / partem *om. Er*

101 figura *Er*

102 Huius...propositione: minor propositio *(om. Ga)* etiam vera est, ut (et *Ga*) apparet ex hiis que dicta sunt in secunda conclusione *GaEd* / in: ex *Q* / secunda *correxi ex* quarta *in QEbEcErMi et* quinta *in Bd (Cf. GaEd supra)*

104 equales *om. BdGaEd, iter. Eb* / et *om. Ed* / remota una *tr. Q* / una scilicet *om. Ga* / parte *om. Eb*

105 tertia: tercia pars *Er* / scilicet 7: scilicet 7ª *Mi*

106 partes *om. Q* / dimidiam: dimidium unius *Ed* / et² *om. Ec* / et dyameter: nam quarta 22 partium est 5 cum dimidio sive 5 partes et dimidium unius partis diameter igitur *Ed* nam 4ᵗᵃ 22ᵃʳᵘᵐ parcium est 5 partes et dimidium diameter *Ga*

106–107 et²...dimidiam *om. BdEr*

106 7: 7ª *EcMi*

107 partes: partes eius *Ed* / dimidium *GaEd* / semipartibus tribus *tr. Ga*

108 dimidiis *om. Er* / dimidiis partibus *tr. EcBd* / partibus: partibus circuli *Ed*

108–110 Ex...quod: premissis ergo universalibus propositionibus veris recte dispositis in primo modo prime figure sequitur necessario conclusio universalis vera que iam dicta est *Ga* premissis ergo propositionibus universalibus veris recta dispositis in primo modo prime figure sequitur necessario universalis conclusio vera secundum quod *Ed*

108 premissa: predicta *Ec*

108–109 5ª...propositione *QEbMi* regula maiore *Er* quinta propositione maiore *BdEcEr*

of the circular figure and the side of this square are equal. And [so] the circular figure and the square of this kind are equal [as to perimeters].

[VI. The sixth conclusion:] EVERY CIRCLE IS A PLANE FIGURE CONTAINED BY A SINGLE CIRCULARLY DRAWN LINE WHOSE DIAMETER EXCEEDS THE QUARTER PART [OF THE PERIMETER] OF THE SAME FIGURE BY PRECISELY THREE HALF-PARTS.

The clarification of this is evident in the second proposition. For if, following what many mathematicians have written in regard to physical truth, a circle [i.e., a circumference] is divided into 22 equal parts, then, with one part subtracted, i.e., the 22nd part, a third of the remainder, i.e., 7, is the diameter of the circle, and a quarter of the circle [i.e., circumference] contains $5\frac{1}{2}$ parts, and [its] diameter, i.e., 7, exceeds a quarter of the circle, i.e., $5\frac{1}{2}$ parts, by precisely three half-parts, i.e., by $3/2$ parts. Hence, with the fifth proposition as the major premise and the following sixth

propositione et sequente 6ª minore sequitur conclusio universalis in
110 primo modo prime figure, scilicet quod [septima conclusio]:

OMNIS CIRCULUS EST EQUALIS QUADRATO CUIUS LATUS EIUSDEM CIRCULI DYAMETER TRANSCENDIT PRECISE IN TRIBUS SEMIPARTIBUS.

Singulis autem huius rei evidentia fiet hoc modo. Constituatur circulus
115 cuiusvis magnitudinis. Eiusdem dyameter dividatur in 7 partes equales per doctrinam datam in tertia conclusione. Deinde constituatur quadratum equilaterum per artem quarte conclusionis cuius quadrati pars precise contineat 5 partes et dimidiam dyametri supradicte, ut patet in figura [Fig. 90]. Sicque premissis diligenter inspectis patebit

109 et sequente 6ª *BdEbEcMi* consequente 6ª *Q* et sequente *Er* et sequitur sexta minor Ed_2
110 prime: 5ᵉ/ *Bd* / scilicet quod *BdEcMi* scilicet quod ergo *Eb* scilicet illa conclusio *Er* secundum quod ergo omnis et cetera *Q et cf. var. lin.* 108–110 / septima conclusio *mg. Q* (*et mg. Mi?*) *om.* $EbEcBdGaErEdEd_2$
111–113 Omnis...tribus *om. Ga*
111 Omnis: Ergo omnis *Er* / est *om. Bd*
113 tribus semipartibus *tr. Q* semipartibus *Er*
114 *ante* Singulis *add. Mi* R (Ratio ?) / Singulis: scibilis *Ed* sensibilis *Ga* / *ante* fiet *add. GaEd* et facilis intelligencia
114–115 circulus: circulus unus *Mi*
115 cuiusvis...Eiusdem *EcEbBdMi* cuius eiusdem magnitudinis *Q* cuius magnitudinis eiusdem *Er* cuiusvis magnitudinis eiusdemque *GaEd* / dyameter *om. Eb* / dividatur *tr. Bd post* conclusione
116 datam: habitam *Ec* / *post* conclusione *add. Er* supra / Deinde: de hinc *Ed* dictum *Er* de hinc deinde *Bd*
117 conclusionis: propositionis *Mi* / cuius: cuius in quanto *Q*
118 pars: latus *EcBdGaEd* / supradicti *Bd*
119 ut...figura *om. Ed* / figura: figura sequenti *Ga* / Sicque: et sic *Er*
119–20 diligenter...quod: omnibus perspectisque diligenter et intellectis prudenter cognoscietur indubitanter quoniam *Ed* omnibus diligenter perspectis intellectisque prudenter cognoscitur indubitanter quoniam *Ga*

proposition as the minor, a universal conclusion in the first mode of the first figure follows, namely [the seventh conclusion]:

EVERY CIRCLE IS EQUAL TO A SQUARE WHOSE SIDE THE DIAMETER OF THE SAME CIRCLE EXCEEDS BY PRECISELY THREE HALF-PARTS.

The evidence of this statement is produced in respect to particulars as follows. Let a circle of any magnitude be constructed. Let the diameter of this same circle be divided into 7 parts by the doctrine of the third conclusion. Then let a square be constructed by the technique of the fourth conclusion—a square whose side contains precisely $5\frac{1}{2}$ of the parts of

120 quod hic circulus huic quadrato est equalis et talis ac tantus circulus, tali ac tanto quadrato. Explicit quadratura circuli edita a Campano.

120 huic quadrato *tr. Ed post* equalis / est equalis *QEcEd* erit equalis *Eb* erit equalis. Amen. *Mi* equaliter *Er*
120–21 et...quadrato *om. Mi* / et...tanto: Omnis ergo circulus est equalis *Ga*
120 ac: et *Ed*
121 tali...quadrato: est qualis et quantus est quadratus sicut ex premissis est manifestum patet etiam per sensum in hac figura *Ed* / ac: et *QEr* / *post* quadrato *add. Er* Amen / Explicit...Campano *om. Ed* / edita a Campano *om. Ga* secundum Campanum *Bd* / Campano *QEb* magistro campano Rome *Ec* Campano. Do gracias. Amen. Campanus nunquam edidit quia tunc esset oblitus sui comentum *Mi* magistro campano qui commentavit geometriam euclidis *Er* / *Et in fine add. EbEcGaErMi (et post* quadrato *add. Bd*; *cf. Ed. var. linea 84)*:
Rem novam [et, *Bd*] mirabilem... quadraturam circuli
Velud (Velut *Ga*, Velicet *Bd* Velud *Mi*) inscruptibilem (inscrutabilem *BdEcGa* inscriptibilem *Eb*)... apud doctos populi
Olim licet (dictam, *Bd*) scibilem... puri (Plures *Er*, quam modo *Bd*) cernunt (cernant *Ec*) oculi
Vere demonstrabilem... nunc (ut, *Bb*) in fine seculi. Amen (*Er*, *om. EbEcBdGa* Finis est huius operis. Laus deo. Amen. *Mi*)
[This whole poem is omitted by *Q*] / [In *Ga* there are two added marginal glosses: in reference to *puri* we read "immo lustri" and for "demonstrabilem" we read "Nondum demonstratum."] / [In *Ec* there is a poem added to the above poem, expressing, skepticism:
Tu qui te presumpseris hic quadrasse circulum
Verissime *(?)* deciperis tecum gerens oculum
Igorans qui loquens te faceris et populum
Antequam quadraveris finietur seculum.] /
[In *Bd* there is added a reference to the figures on folio 90v and 91r: Vide quadraturum in altera parte folii et circulum divisum in viginti duas partes.]

the aforesaid diameter, as is evident in the figure [Fig. 90]. And thus with the premises diligently observed, it will be evident that this circle is equal to this square [in respect to perimeters], i.e., a circle of some given amount to a square of the same amount [the comparison being of perimeters]. [Here] ends the Quadrature of the Circle composed by Campanus.

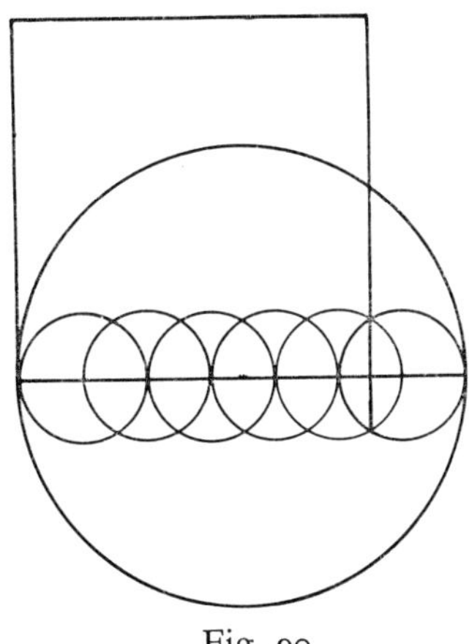

Fig. 90

COMMENTARY

2–3 "Aristotiles...est." Aristotle mentions this in the *Predicamenta*, 7, 7b. We can quote this in the text that accompanied Boethius' commentary on this work: *In categorias Aristotelis* (ed. in Migne, *Patrologia latina*, *64*, cc. 230–31): "Nam si scibile non sit, non est scientia; scientia vero si non sit, nihil prohibet esse scibile, velut circuli quadratura si modo est scibilis, scientia siquidem eius nondum est, ipsa vero scibilis est."* Boethius comments: "Eodem quoque modo quaesitum est si sit propositum circulo aequum fieri quadratum. Quadratum ergo est quod aequalibus lateribus omnes quatuor angulos aequos habet, id est rectos, et Aristotelis quidem temporibus non fuisse inventum videtur. Post vero repertum est, cuius quoniam longa demonstratio est, praetermit-

* The passage in the new edition of Boethius' translation of the *Categoriae* of L. Minio-Paluello (Bruges, Paris, 1961, p. 21) reads: "nam, si scibile non sit, non est scientia, si scientia vero non sit, nihil prohibet esse scibile; ut circuli quadratura si est scibile, scientia quidem eius nondum est, illud vero scibile est." Cf. also pp. 61, 99, for the *editio composita* and the translation of William Moerbeke.

tenda est. Atque hoc est quod ait, velut circuli quadratura: nam sicut manente quadrato, linea per obliquum ducta triangula figura producitur; ita circulo non mutato circumpositis angulis, qui et ipsius circuli lateribus aequaliter diriguntur, quadrati forma consurgit, quod (ut potuimus) conjectura depinximus. Cum enim alicui circulo aequum quadratum constituitur, in quadraturam circuli illius mensura redigitur. Nunc ergo hoc est quod dicit, ut circuli quadratura, id est aequi quadrati ad circulum constitutio id fieri potest, et si res est quae sciri possit, scientia quidem eius nondum inventa est. Nondum enim quisquam sub Aristotele aequum quadratum circulo constituerat. Quod si est aliqua eius scientia quae nondum reperta est, certe prius est quod sciri possit, post vero scientia. Nam cum posset Aristotele vivo sciri circuli quadratura, nulla tamen adhuc eius scientia reperta est, atque ideo prius erat quod sciri posset, quam ipsius rei ulla notitia." Incidentally, this passage of Aristotle's *Predicamenta* was also commented upon by Simplicius in his commentary on the *Predicamenta*. This text became available to medieval Latin Schoolmen with its translation into Latin by William of Moerbeke in 1266. For the sake of completeness, we can cite Simplicius' remarks from the edition of Venice, 1516, 29v, c.2: "ipse autem ostendit hoc in quadratura circuli: nondum enim tunc inventa ipsa dubie dicens: si est scibilis: scientia quidem ipsius nondum est. ipsum autem scibile est. Est autem quadratura circuli: quando dato circulo: equale quadratum constituerimus. hoc autem Aristo. quidem ut videtur nondum novit. Apud pithagoricos autem inventum fuisse ait Iamblicus sicut palam est ex demonstrationibus Sexti Pithagorici qui desuper secundem successionem susceperat artem demonstrationis et posterius ait Archimedes per lineam λυκομήδου[ς] et Nicodemus per lineam proprie tetragonicantem vocatam: et Appollonius per quandam lineam quam ipse vocat sorerem κοχλιοειδοῦς: est autem eadem cum ea que dicitur Νικομήδους et Carpus autem per quandam lineam quam simpliciter ex duplici motu vocat. Alii autem multi varie problema probaverunt sicut narrat Iamblicus: et mirabile est quod hoc latuit plurimum studiosum Porphyrium: qui ait quod est quedam demonstratio secundum quod est figuram tetragonam circulo comparare: sicut et alias figuras: nondum autem comprehensa est: neque inventa est. Dicunt autem (ait) quidam eorum quidem post Aristo. invenire. Forte igitur organica quedam inventio facta est ex theorematis: sed non demonstrativa." This passage is, for the most part, translated and discussed by I. Thomas, *Selections Illustrating the History of Greek Mathematics*, vol. *1* (London, Cambridge, Mass., 1951), pp. 334–35. I have added accents to the Greek words. (Cf. Thomas Heath, *Mathematics in Aristotle* [Oxford, 1949], pp. 18–19.)

Appendix II

The *Quadratura circuli per lunulas*

Another treatment of the quadrature problem, quite unlike any of those presented in Appendix I, went under the title of *Quadratura circuli per lunulas* and, although translated from Greek rather than Arabic, circulated quite widely in the thirteenth and fourteenth centuries in codexes including treatises in the Arabic tradition. In reality, two versions of this short tract can be distinguished. The first (Version I) is simply a faithful translation of a passage of Simplicius' *In Aristotelis Physicorum commentaria*, I.iii (ed. of H. Diels, Berlin, 1882, pp. 56–57). The second of the versions is a paraphrase of the first done into considerably altered Latin, with the letters on the diagrams changed; it was first published by H. Suter in 1884 on the basis of a single fifteenth-century manuscript (*Z*).[1] The first version (together with a new edition of Version II based on earlier manuscripts) I published in 1955.[2] I repeat here my earlier texts (slightly corrected) of both versions but now include English translations as well.

Of the two versions of the *Quadratura*, the second was by far the most popular, for we have at least ten extant manuscripts of it, while we have only one manuscript of the first version. We cannot be absolutely certain of the translator of Version I, but it seems probable that it was Robert Grosseteste since the scribe of its unique manuscript tells us that he found "this demonstration at Oxford in a certain document of the lord Lincoln." My guess is that Grosseteste made this translation sometime during the 1240's when he was engaged in Aristotelian translations.[3] As to who made the second version, we cannot be sure. MS *Ka* ascribes it in the title to

[1] H. Suter, "Der Tractatus 'De quadratura circuli' des Albertus de Saxonia," *Zeitschrift für Mathematik und Physik*, vol. 29 (1884), Hist.-liter. Abt., pp. 85–87.

[2] *Essays in Medieval Life and Thought Presented in Honor of Austin Patterson Evans* (New York, 1955), pp. 99–108.

[3] S. H. Thomson, *The Writings of Robert Grosseteste* (Cambridge, 1940), pp. 42–71.

Grosseteste (*Lincolniensis de quadratura çirculi*), but of course this ascription may have arisen from the fact that the original paraphraser was being honest about his source—namely, Version I as translated by "Lincoln." The other manuscripts are without ascription, but it is evident from the fact that Version II was included in the thirteenth-century sections of MSS *B*, *J*, and *Ha* that it was composed in the thirteenth century.

This short piece is reported by Simplicius from Alexander of Aphrodisias; it contains an attempted (but incorrect) quadrature of the circle by means of the quadrature of a lune, and the first part of the discussion of Simplicius on the quadrature of lunes appears to go back ultimately to Hippocrates of Chios.[4] The most important difference between the two Latin versions is that Version I faithfully gives Simplicius' comment that this demonstration, being based on the assumption that the quadrature of a particular lune holds also for other lunes, is false, while the comment is omitted in the second version, thus leaving us with the impression that the quadrature of the circle can be demonstrated in this way. The later copy of Version II in manuscript *DZa* attempts to justify the proof by an appeal to Campanus and Euclid.[5]

In giving the text of the first version I have indicated the principal variants of the Greek text as established by Diels. I have also included the common Greek words which the translator renders with the hope that someone who studies Grosseteste's style of translation can decide whether this work was translated by him.[6] The last sentence I have included in

[4] T. L. Heath, *A History of Greek Mathematics*, vol. *1* (Oxford, 1921), pp. 183–200, gives an English summary of the Greek text, evaluates the relative value of Alexander's and Eudemus' discussions for getting at the original treatment of Hippocrates of Chios, and criticizes the modern literature on this passage since Bretschneider. Incidentally, the passage represented by Version I here is, so far as I know, the only part of Simplicius' *Commentary on the Physics* translated in the Middle Ages. However, it should be observed that Jordanus de Nemore may have had some contact with still another part of Simplicius' *Commentary*; see M. Clagett, *The Science of Mechanics in the Middle Ages* (Madison, 1959), pp. 258–61. Incidentally, the paraphraser of Archimedes, Francesco Maurolico, adds to his version of the *Measurement of the Circle* a reworking of the *Quadratura per lunulas* which he may have taken from one of the medieval versions, although it is more likely that he took it directly from Simplicius since he calls it *Hippocratis tetragonismus* (see MS Paris, BN lat. 7465, 29r–v).

Finally, we can note that Nicole Oresme in his *Quaestiones super geometriam Euclidis*, dating from about 1350, ed. of H. L. Busard (Leiden, 1961), p. 58, lines 10–27, gives a paraphrase of the first part of the tract (probably from Version II), showing how it is possible to construct a lune equal to a given triangle (dato triangulo equalem lunulam describere).

[5] See variant readings, Version II, lines 14–19.

[6] From the variant readings it will be

brackets since it was obviously added by the scribe. Version II has been prepared principally on the basis of *B*, but variant readings for the other manuscripts have also been given. The late MS *Z* is the only manuscript that varies significantly.[7] The marginal folio references are to *Ca* in Version I and *B* in Version II.

Sigla of Manuscripts

Version I

Ca = Oxford, Corpus Christi College 251, 83v–84r, 13c.
Gr = Diels Greek text, as indicated above.

Version II

B = Oxford, Bodleian Library, Auct. F.5.28, 116r, 13c.
J = Berlin, Deutsche Staatsbibliothek (MS now at Marburg, Westdeutsche Bibliothek), Q. 510, 94v, 13c.
Ha = Oxford, Bodl. Libr., Digby 190, 87v, 13c.
V = London, Brit. Museum, Royal 12 E. 25, 150v, ca. 1300.
Ka = Oxford, Bodl. Libr., Digby 153, 184r, 14c.
Z = Bern, Bürgerbibliothek A. 50, 168r–169r, early 15c. Has *linula* for *lunula*.

[A further version, which is like the free rendering of Version II but which resembles Version I in recognizing that the proof is erroneous, is found in Glasgow University Library, MS BE 8–y.18, 209v–210r. The text of this version will be published in an appendix to Volume II.]

Additional Manuscripts, Not Used for Text

Florence, Bibl. Naz., Conv. Soppr. J.V.18, 33r, 14c.
Dresden, Stadtbibliothek Db, 86, 213r, 14c.
Erfurt, Stadtbibliothek, Amplon. Q. 234, 113r, ca. 1325.
Erfurt, Stadtbibliothek, Amplon. Q. 385, 207r, 14c.

obvious that the translator renders δὲ by both *quoque* and *vero*, ὥστε by *sicque* and *sic itaque*, ἄρα by *ergo* or *igitur*, ἕκαστον by *unisquisque*, and πρὸς ἄλληλα by *ad invicem*.

[7] The text given in *Z* has several more references to Euclid's *Elements* than does the pristine text. It also has a reference by name to Campanus' commentary on the *Elements* and so no doubt was composed at some time posterior to Campanus. For the additions present in *Z*, see the variant readings.

Version I and Version II
Quadrature of the Circle by Lunes

[Versio I]

83v

/Quadratura circuli

Esto circa lineam rectam *AB* semicirculus descriptus *AGB* [Fig. 91*a*]. Et dividatur per medium *AB* linea in puncto *D*. Et a *D* puncto ortogonaliter ducatur super *AB* lineam linea recta *DG*. Et a *G* subtendatur recta *GA*, que videlicet *GA* linea est latus quadrati inscripti in circulo cuius semicirculus est *AGB*. Et super rectam *AG* describatur semicirculus *AEG*. Et quia quadratum *AB* linee est equale quadrato *AG* linee et quadrato alterius tetragoni inscripti in circulo *AGB*, hoc est, quadrato linee *GB*—est enim *AB* linea subtensa angulo recto trianguli rectanguli—sicut vero quadrata descripta a diametris circulorum se habent adinvicem sic et circuli circa ipsos diametros descripti se habent adinvicem; et sic etiam semicirculi, sicut demonstratum est in XII libro Elementorum, duplus est igitur *AGB* semicirculus ad semicirculum *AEG*. Est quoque *AGB* semicirculus duplus ad *AGD* quartam circuli totius particularem. Equalis est igitur *AGD* quarta pars circuli *AEG* semicirculo. Commune itaque tollatur, videlicet quod continetur sub latere quadrati et sub arcu *AG*. Residuum ergo, videlicet *AEG* lunula, equale est trigono *AGD*. Trigonus vero quadratur. Propter hoc iam demonstrata est lunule equalis ipsi trigono quadratura. Postea itaque temptabitur per istud exemplum lunule iam quadrate quadrare totum circulum sic.

2 *post* Esto *hab. Gr* φησί (*the subject of this verb being Alexander of Aphrodisias*)
3 linea *om. Gr* / puncto[1, 2] *om. Gr*
4 lineam linea *om. Gr*
4–5 subtendatur: ἐπεζεύχθω *Gr*
5 videlicet GA linea *om. Gr*
6 rectam *om. Gr*
7 quadratum AB linee: ἀπὸ τῆς AB (*and similarly elsewhere*)
9–10 AB...recto: ὑποτείνουσα ἡ AB *Gr*
11 circa...diametros *om. Gr*
14 quoque: δὲ *Gr*
15 quartam...particularem: τετρατημόριον
16 itaque *om. Gr*
17 quod: τὸ τμῆμα *Gr*
18 vero: δὲ *Gr*
20 Postea itaque: ἑξῆς *Gr*
20–21 per...quadrate: διὰ τοῦ προδεδειγμένου *Gr*

[Version I]

Quadrature of the Circle [by Lunes]

Let semicircle AGB be described on straight line AB [see Fig. 91*a*]. And let line AB be bisected at point D. From point D let line DG be erected perpendicularly on line AB. And let chord GA be drawn from G, which chord is obviously the side of a square inscribed in the circle whose half is AGB. And let semicircle AEG be described on straight line AG. And because AB^2 is equal to AG^2 plus the square of the other [side of the] square inscribed in circle AGB, that is, GB^2—line AB being subtended by a right angle of a right triangle—and because the squares described by the diameters of circles are related to each other as are the circles described

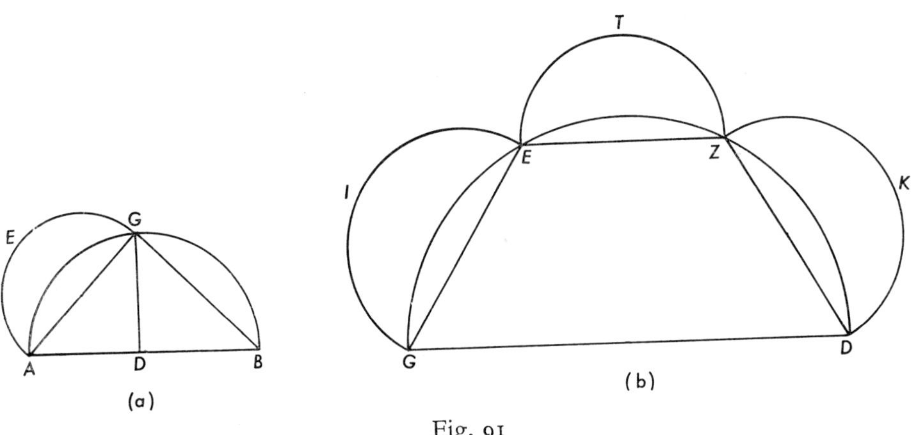

Fig. 91

on these diameters, and also as are the semicircles, just as has been demonstrated in the twelfth book of the *Elements*, therefore semicircle AGB $= 2$ semicircle AEG. But also semicircle $AGB = 2$ quadrant AGD. Therefore, quadrant $AGD =$ semicircle AEG. And so let their common part be subtracted, evidently that part which is contained by the side of the square and by arc AG. Hence, the remainder, evidently the lune AEG, is equal to $\triangle AGD$. Meanwhile, the triangle is squared. Accordingly, the quadrature of the lune equal to this triangle has now been demonstrated. And so afterwards the squaring of the whole circle will be attempted by that example of the lune just squared, as follows:

Sit recta *AB* et circa ipsam semicirculus describatur. Et proponatur ipsi *AB* dupla recta *GD* [Fig. 91*b*]. Et super ipsam rectam *GD* semicirculus describatur. Et inscribantur in ipso semicirculo latera exagoni
25 inscripti in circulo, ista videlicet *GE* et *EZ* et *ZD*. Et super ipsa latera exagoni describantur semicirculi *GIE*, *ETZ*, *ZKD*. Unusquisque ergo semicirculorum super latera exagoni descriptorum equalis est semicirculo descripto super *AB*. Et enim *AB* linea equalis est cuilibet lateri exagoni, quia diameter dupla est cuilibet linee exeunti a centro,
30 latera vero exagoni equalia sunt rectis a centro exeuntibus, et ipsius *AB* linee dupla est linea *GD*. Sic itaque quatuor semicirculi equales sunt adinvicem. Quatuor igitur semicirculi sunt quadruplum ad semicirculum *AB*. Est quoque semicirculus super *GD* quadruplus ad semicirculum super *AB*; quia enim *GD* linea dupla est ad *AB* lineam; quad-
35 ruplum est quod fit a *GD* linea in se ei quod fit ab *AB* linea in se, sicut vero quadrata diametrorum sic et circuli eorum diametrorum se habent adinvicem, et sic etiam semicirculi se habent adinvicem. Sicque quadruplus semicirculus *GD* ad semicirculum *AB*. Equalis est igitur semicirculus *GD* quatuor semicirculis, videlicet semicirculo *AB* et tribus
40 semicirculis descriptis super tria latera exagoni. Communia igitur tollantur semicirculorum super latera exagoni descriptorum et semicirculi descripti super *GD* lineam, que communia sunt portiones contente sub lateribus exagoni et arcubus periferie *GD* quos subtendunt eadem latera exagoni. Residua igitur tres, videlicet lunule *GIE*, *ETZ*, *ZKD*,
45 cum semicirculo *AB*, sunt equalia quadrilatero contento [sub] tribus lateribus exagoni et diametro *GD*. Itaque ab isto quadrilatero rectilineo tollatur pars equalis istis lunulis—demonstratum est enim equale rectilineum lunule invente. Relinquitur itaque residuum de quadrilatero equale semicirculo *AB*. Illud autem relictum de quadrilatero,
50 cum sit figura rectilinea, scimus duplare, et duplatum quadrare; quod

23 super...rectam GD: GD *Gr*
26 GIE: ΓΗΕ *Gr* / Unusquisque: ἕκαστον *Gr*
27 descriptorum *om. Gr*
28 descripto super AB: τῷ AB *Gr*
29 quia: γὰρ *Gr* / cuilibet linee...centro: τῶν ἐκ τοῦ κέντρου *Gr*
30 vero: δὲ *Gr*
31 Sic itaque: ὥστε *Gr*
32 igitur: ἄρα *Gr*
33 quoque: δὲ *Gr*
34 quia: ἐπεὶ *Gr*
37 et sic etiam: καὶ *Gr* / Sicque: ὥστε *Gr*
44 GIE: ΓΗΕ *Gr*
45–46 quadrilatero...GD: ΓΕΖΔ τραπεζίῳ *Gr*
46 diametro *corr. ex* semidiametro
46–47 quadrilatero rectilineo: τραπεζίου *Gr*
47 pars equalis: τὴν ὑπεροχὴν...τουτέστι τὸ ἴσον *Gr*
48 invente *om. Gr* / itaque: δὲ *Gr*
48, 49 de quadrilatero *om. Gr*
50 *post* quadrare *add. Gr* τουτέστιν ἴσον αὐτῷ τετράγωνον λάβωμεν *Gr*
50–51 quod dupla[tum]: τετράγωνον *Gr*

Let there be a straight line AB and a semicircle described on it. And let $GD = 2\ AB$ [see Fig. 91*b*]. And let a semicircle be described on straight line GD. In this semicircle let three sides of a hexagon inscribed in a circle be inscribed: to whit, GE, EZ, and ZD. And let semicircles GIE, ETZ, and ZKD be described upon these sides of the hexagon. Hence each one of these semicircles described on the sides of the hexagon is equal to the semicircle described on AB: for line AB is equal to each side of the hexagon—the diameter being double its radius and the sides of the hexagon being equal to radii, and $GD = 2\ AB$. And so the four semicircles are mutually equal. Therefore, the four semicircles [together] are quadruple semicircle AB. And also the semicircle on $GD = 4$ semicircle on AB: for, since $GD = 2\ AB$, $GD^2 = 4\ AB^2$, the squares of the diameters being related to each other as are the circles on the diameters and also as are the semicircles. And so semicircle $GD =$ 4 semicircle AB. Therefore, semicircle GD is equal to four semicircles, namely, semicircle AB and the three semicircles described on the three sides of the hexagon. Therefore, let us subtract the parts common to the semicircles described on the sides of the hexagon and to the semicircle described on line GD, which common parts are the segments contained by the sides of the hexagon and the arcs of the semicircumference GD which these same sides of the hexagon subtend. Therefore, the three remainders, namely, the lunes GIE, ETZ, and ZKD, together with the semicircle AB, are equal to the quadrilateral contained by the three sides of the hexagon and by diameter GD. And so from this quadrilateral let the parts equal to these lunes be subtracted, for it has been demonstrated that there exists a rectilinear figure equal to the lune found. And so it results that the remainder of the quadrilateral is equal to semicircle AB. But we know how to double that remainder of the quadrilateral since it is a rectilinear figure, and we know how to square the double, which

dupla[tum] equale est totali circulo descripto super rectam *AB*. Et sic circulus descriptus super *AB* quadratur. Hoc autem falsigraphia est ex eo quod non universaliter demonstratum sumitur sicut demonstratum universaliter. Non enim demonstratur omnis lunule quadratio
55 sed tantummodum lunule cui subtenditur latus quadrati. [Hanc demonstrationem inveni Oxonie in quadam cedula domini Lincolniensis.]

[Versio II]

116r /Quadratura circuli per lunulas hoc modo est. Ponatur circulus quadrandus cuius semicirculus sit *ADC* [Fig. 92*a*]. Et in eo protrahantur duo latera quadrati inscriptibilis in eo, que sint *AD*, *DC*. Et super lineam *AD* describatur semicirculus *AID*. Probo ergo quod
5 scis quadrare lunulum *AQDI*. Quadratum enim *AC* est duplum ad quadratum *AD*. Sed *AC* est diameter semicirculi *ADC* et *AD* est diameter semicirculi *AID*. Ergo per secundam duodecimi Euclidis semicirculus *ADC* est duplus ad semicirculum *AID*. Ergo semicirculus *AID* est equalis portioni *ADB*, que est medietas semicirculi
10 *ADC*. Ergo cum semicirculus *AID* et portio *ADB* habeant quoddam commune, portionem que continetur a corda *AD* et arcu *AQD*,

52 descriptus super AB *om. Gr*
52–53 Hoc...est: καὶ ἔστι μὲν εὐφυὴς ἡ ἐπιχείρησις· τὸ δὲ ψευδογράφημα γέγονε *Gr*
55 sed...quadrati: ἀλλ' εἰ ἄρα, ὁ περὶ τὴν τοῦ τετραγώνου πλευρὰν τοῦ εἰς τὸν κύκλον ἐγγραφομένου· οὗτοι δὲ οἱ μηνίσκοι περὶ τὰς τοῦ ἑξαγώνου πλευράς εἰσι τοῦ εἰς τὸν κύκλον ἐγγραφομένου.
1 Quadratura...est: Incipiunt demonstrationes de quadratura circuli. Circulum quadrare est possible *Z* Ecce quadratura circuli per lunulas *Ha* / circuli *KaHa om. BVJ*
2 quadrandus: quadratus *Ha*
2–3 in eo protrahantur: describantur in eo *Z* in eo describantur *Ha*
3 in eo *om. ZKa* / sint *BJ* sunt *VHaKaZ* / DC: et DC *ZV* / DC: AC *Ha*
4 AID: AD *Z*
4–5 Probo...AQDI: lunula ergo AD *(Z*, AQDI *Ha)* sic potest quadrari *HaZ*
5 AQDI: AQD *V* AD *Z* / enim *om. V* / AC: AI *Ka* / est: est per penultimam primi Euclidis *Z* per penultimam primi Euclidis et per propositionem que dicit quod angulus in semicirculo est rectus ergo quadratum *V*
6 diameter: diametrum *Ka* / AD: AD sed *Ha*
7, 8, 9, 10 AID: AD *Z*
7 secundam...Euclidis: 12 Euclidis secundam *Ha* / duodecimi: id est duodecimi *V*
9 equalis...est *om. V*
10 habent *Ha*
10–11 quoddam commune: commune quoddam scilicet *V* commune scilicet *Z*
11 portioni *Ka* / que continetur: continentem *Z*
11 a...AQD: ab arcu et a corda AD *Z* / AD: AB *V*

double is equal to the whole circle described on straight line *AB*. And thus the circle described on line *AB* is squared. This, however, has been falsely reasoned, since that which was not universally demonstrated is taken as universally demonstrated. For the quadrature of every lune is not demonstrated but only that of the lune which is subtended by the side of an [inscribed] square. [I have found this demonstration at Oxford in a certain document of the lord Lincoln.]

[Version II]

The quadrature of a circle by means of lunes is done in this way: Let there be a circle to be squared whose semicircle is *ADC* [see Fig. 92*a*]. And draw in it two sides of a square inscribable in it, the sides being *AD* and

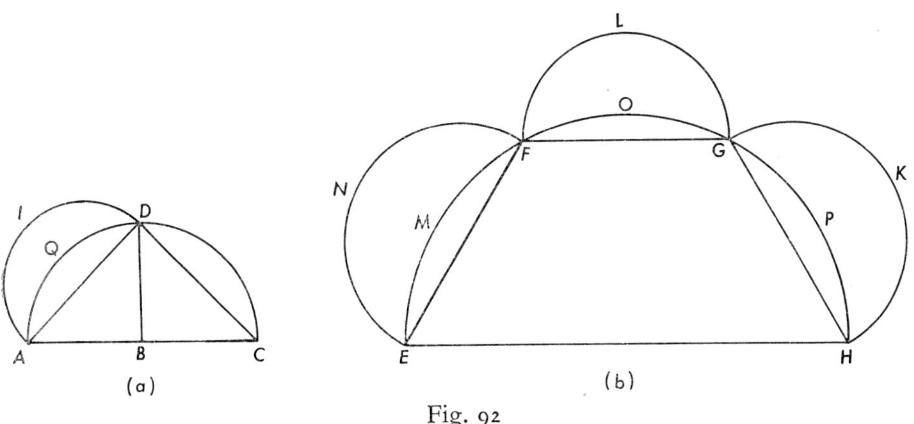

Fig. 92

DC. Let semicircle *AID* be described on line *AD*. I prove, therefore, that you know how to square lune *AQDI*. For $AC^2 = 2\ AD^2$. But *AC* is the diameter of semicircle *ADC* and *AD* is the diameter of semicircle *AID*. Therefore, by XII.2 [of the *Elements*] of Euclid, semicircle *ADC* = 2 semicircle *AID*. Therefore, semicircle *AID* is equal to sector *ADB*, which is one half of semicircle *ADC*. Therefore, since semicircle *AID* and sector *ADB* have a certain area in common, [namely,] the segment contained by chord *AD* and arc *AQD*, when that segment is subtracted

dempta illa portione ab utroque, erit *AIDQ* equalis triangulo *ADB*. Sed triangulum scis quadrare; ergo et lunulam.

Supponas ergo quod sicut contingit quadrare lunulam super latus
15 quadrati descriptam, ita contingat quamlibet lunulam quadrare super cuiuscunque figure inscriptibilis circulo latus descriptam, ut supra latus exagoni. Sequetur demonstrative quod scias quemlibet circulum quadrare.

Probatio: Sit, ut positum est, circulus quadrandus cuius semicir-
20 culus sit *ADC*. Et sumatur linea dupla ad eius diametrum, que sit linea *EH* [Fig. 92*b*] Et super eam fiat semicirculus *EFGH*. Et in ipso distinguantur tria latera exagoni, que sint *HG*, *GF*, *FE*. Et super illa tria latera describantur tres semicirculi : *ENF*, *FLG*, *GKH*. Cum igitur linea *EH* sit dupla ad lineam *AC*, erit quadratum linee *EH*
25 quadruplum ad quadratum linee *AC*. Ergo cum ea sit proportio circulorum adinvicem que quadrati diametri unius ad quadratum diametri alterius, per secundam duodecimi Euclidis, erit circulus cuius diameter

12 illa: ista *Z* / illa portione *tr. V* / erit *om. J* / erit AIDQ *tr. Ha* / AIDQ: lunula AD *Z* ADIQ *V* / equalis: quare equales *V*
13 *post* Sed *add. V* per XXII V[ti] et per ultimam secundi / triangulum...lunulam: trianguli quadratura scitur per ultimam 2[i] Euclidis, quare et lunule *Z* trianguli quadratura scitur quare et lunule *Ha* / et *om. V*
14–19 Supponas...quandrandus: Supponatur ergo quod sit possibile quadrare lunulam super latus quadrati descriptam, et sicut hoc contingit, ita contingit quadrare quamlibet lunulam super cuiuscunque figure circulo inscriptibilis latis descriptam, ut super latus exagoni. Que suppositio confirmari potest per illud principium, quod premittit Campanus primo elementorum Euclidis, et quo utitur in demonstrando 2[am] 12[i] Euclidis, quia sicut se habet lunula super latus quadrati descripta ad lunulam super latus exagoni, ita se habet quadratum quodcunque ad aliquod allud principium: Quanta est quelibet magnitudo ad aliquam 2[am], tantam necesse est esse quamlibet tertiam ad quartam. Et tunc ex 7[a] et 12[a] et 21[a] 5[ti] Euclidis facile concludes hoc quadratum esse equale lunule descripte super latus exagoni, facilius tamen concludes illud ex permutata proportionalitate. Hoc autem presupposito demonstrabitur quemcunque circulum posse quadrari. Sit enim circulus quadrandus ut prius *Z*
14 Supponas: Supponatur *HaZ* / ergo: igitur *V* / lunulam: lunualas *V*
15 descriptam: descripti *V* / contingit *HaKa*
16 cuiuscunque: utrumcunque latus *V*/descriptam *Z* fuerit descripta *BVHaJ* fuerit descriptam *Ka*
19 Sit ut: sicud *Ka* / quadrandus: quad[a] ne' *Ka*
20 sit[1] *om.Ka*
21 fiat: describatur *ZHa*
22 distinguantur: ducantur *Z* / sint *BJ* sunt *VHaKaZ* / HG, GF, FE: EF, FG, GH *Z*
23 ENF: scilicet ENF *Z* ENG *V* EF *Ka* / FLG, GKH: FHG, GHK *V* FG, GH *Ka*
25 linee *om. Ka* / *post* AC *add. Z* per 4[am] 2[i] Euclidis et 18[am] 6[i] eiusdem / Ergo: igitur *Z* / ea sit: sit eadem *Z*
26 que: que est *VZ* / diametri[1] *om. V*
27 duodecimi: X *V* / erit *om. Z*

from each, [then lune] $AIDQ$ will be equal to $\triangle ADB$. But you know how to square the triangle; therefore, [you know how to square] the lune as well.

You suppose, therefore, that just as it is possible to square the lune described on the side of a square, so one may square any lune described on the side of any figure inscribable in a circle, as for example on the side of a hexagon. It will follow demonstratively that you know how to square any circle.

Proof: As has been posited, let there be a circle to be squared whose semicircle is ADC. And let there be taken a line double its diameter, namely, line EH [see Fig. 92*b*]. And let semicircle $EFGH$ be constructed on it. And in that semicircle let three sides of a hexagon be determined, the sides HG, GF, and FE. Then upon these three sides let there be described the three semicircles ENF, FLG, and GKH. Therefore, since line $EH = 2$ line AC, $EH^2 = 4\ AC^2$. Therefore, since the ratio of circles to each other is as that of the squares of their diameters, by XII.2 of Euclid,

est *EH* quadruplus ad circulum cuius diameter est *AC*. Ergo et semicirculus *EFGH* est quadruplus ad semicirculum *ADC*. Sed unusquis-
30 que semicirculorum *ENF*, *FLG*, *GKH* est equalis semicirculo *ADC*, quia omnium illorum diametri sunt equales. Ergo quatuor semicirculi *ADC*, *ENF*, *FLG*, *GKH* sunt equales semicirculo *EFGH*. Ergo demptis tribus portionibus *EMF*, *FOG*, *GPH*, que sunt communes tribus semicirculis *ENF*, *FLG*, *GKH* et semicirculo *EFGH*, relin-
35 quitur quod semicirculus *ADC* cum tribus lunulis, que sunt *ENFM*, *FLGO*, *GKHP*, sit equalis illi quod residuum est de semicirculo *EFGH* post demptionem trium portionum communium. Et est illud residuum figura quadrilatera que continetur quatuor rectis lineis *EF*, *FG*, *GH*, *HE*. Sed illam figuram quadrilateram scis quadrare. Ergo et
40 eius equale scis quadrare. Sed eius equale sunt semicirculus *ADC* et lunule dicte. Ergo ipsis simul iunctis scis quadratum equale designare.

28 est[1] *om. Ka* EH: EH est *Z* / diameter *om. J* / Ergo *om. Z*
29 EFGH: EFG *V* / est quadruplus *tr. Z* / *post* ADC *add.* per 15 [m] 5[i] Euclidis *Z*
30 semiciculorum *HaJ* circulorum *BVKaZ*
31 illorum: istorum semicirculorum *Z* / quatuor *om. Z*
32 GKH *om. V* / EFGH: EFG *V*
32–33 Ergo demptis: demptis igitur *Z*
32–34 Ergo...EFGH *om. Ha*
33 EMF: EMG *V* / FOG: FTG *Z* / *in loco* communes *habet V lacunam*
34 semicirculis: circulis *V*
34–35 relinquet *Ka*
35 cum...lunulis: et tres lunule *Z* / que sunt *om. Z*
35–36 ENFM, FLGO, GKHP: ENF, FLG, GKH *Z*
36 sit equalis: sunt equales *Z* / residuum est *tr. Z*
37 portionum *om. Z* / communium *om. V*
37–38 Et...residuum: Est autem residuum illud *Z*
38 quatuor: a quatuor *Z* / rectis *om. Z*
39 HE: EH *Z* / illam....Ergo: eius quadratura scitur quare *Ha* / illam...quadrare: quadratum illius figure scitur per ultimam 2[i] Euclidis, quare et quadratum ei equale *Z*
40 scis quadrare *om. VHa* / eius equale *BJ* eius equales *V* equales *Z*
41 lunule dicte: tres lunule predicte *Z* / Ergo: igitur *Z* / ipsis: hiis *V*
41–45 scis....lunulas: scimus equale quadratum posse signare. Sed tres lunule, ut prehabitum est, possunt quadrari, quare et totus circulus cuius dyameter est AC; consequentia ultima patet: Si enim tribus lunulis per se est dare tria quadrata equalia, quibus divisis in 6 triangulos datur unum quadratum equale per demonstrationem ultime 2[i] Euclidis; dempto ergo isto quadrato equali tribus lunulis de quadrato totali equali lunulis ipsis et semicirculo ADC residuum, quod est equale semicirculo, remanet figura rectilinea AC et per consequens talia quadrata ut docetur super ultimam secundi Euclidis *Z*
41, 42, 43, 44 scis: scimus *Ha*

the circle whose diameter is *EH* will be quadruple the circle whose diameter is *AC*. Therefore, semicircle *EFGH* = 4 semicircle *ADC*. But any one of the semicircles *ENF*, *FLG*, and *GKH* is equal to semicircle *ADC*, since the diameters of all of them are equal. Therefore, the four semicircles *ADC*, *ENF*, *FLG*, *GKH* are [together] equal to the semicircle *EFGH*. Therefore, when the three segments *EMF*, *FOG*, *GPH*, which are common to the three semicircles *ENF*, *FLG*, *GKH* and to semicircle *EFGH*, have been subtracted, the result is that the semicircle *ADC*, plus the three lunes *ENFM*, *FLGO*, and *GKHP*, is equal to that which remains out of semicircle *EFGH* after the subtraction of the three common segments. And that remainder is the quadrilateral figure contained by the four straight lines *EF*, *FG*, *GH*, and *HE*. But you know how to square its equal, and its equal is the sum of semicircle *ADC* and the said lunes. Therefore, you know how to construct a square equal to that sum.

Sed ipsas tres lunulas, ut suppostitum est et videtur probatum, scis quadrare. Ergo et quartum scis quadrare. Ergo semicirculo *ADC* scis equale quadratum facere; igitur et toti circulo. Hec igitur est
45 quadratura per lunulas.

42 ipsas *om. Ha* / suppositum est et *om. Ha* / et: et ut *Ka*
43 Ergo[1]...quadrare: Et scis subtrahere quadrata tribus lunulis equalia, et quod relictum erit quadratum semicirculi et positi primo. Et illud quadratum scis duplare; ergo et quadratum *V* / Ergo[2]: quare *Ha*
44 igitur *BJ* ergo *VaHaKa*
44–45 Hec...lunulas: et per consequens sequitur intentum quod quadrare circulum est possibile *Ka*
44 igitur est: est ergo *VHa*

But you [also] know how to square these three lunes, as has been supposed and seems to have been proved. Therefore, you know how to square the fourth term. Therefore, you know how to construct as square equal to semicircle *ADC*, and therefore [a square equal] to the whole circle. This, therefore, is the quadrature by means of lunes.

COMMENTARY

Version I

13 "XII...Elementorum." The reference is to Proposition XII.2 of the *Elements*. See the Commentary, Appendix I, Section 1, line 5.

52–55 "Hoc....quadrati." This admonition by Simplicius of the falseness of proceeding from a proof concerning a particular lune to all lunes was, as I have said, omitted by the author of Version II.

55 "sed...quadrati." It will be noticed by consulting the variant reading for this sentence that the Greek text adds the following phrase: "while these lunes [considered here] are upon the sides of a hexagon described in the circle."

Version II

6, 27 "per...Euclidis." See above, the Commentary to Version I, line 13.

14–18 "Supponas....quadrare." The reader should note (variant readings, lines 14–19) that the author of the copy of Version II appearing in MS *Z* attempts to prove the case of quadrature of a lune on the side of a hexagon in the following way: "Therefore, let it be supposed that it is possible to square a lune described on the side of a square and, just as this is so, that it is possible also to square any lune described on the side of any figure inscribable in a circle, as for example on the side of a hexagon. This supposition can be confirmed by that principle which Campanus postulates in the first book of the *Elements* of Euclid and which he uses in demonstrating XII.2 of Euclid. For just as the lune described on the side of a square is related to the lune described on the side of a hexagon, so any square is related to some other square, by that principle, [which is this:] 'Any magnitude is to some second magnitude as any third magnitude is to a fourth' [cf. ed. of the *Elements*, Basel, 1546, p. 3]. And then you easily deduce by V.7, V.12, and

V.21 of Euclid that this square is equal to the lune described on the side of a hexagon; however you more easily deduce that by the use of alternate ratios. With this [quadrature of the lune on the side of a hexagon] presupposed, it will be demonstrated that any circle can be squared." The author of this copy of Version II apparently feels that this so-called proof justifies the phrase "videtur probatum" appearing in line 42 of the pristine form of Version II. It is evident that this added comment constitutes not a "construction" proof but only an "existence" proof. Briefly we can say that the author, following Campanus, holds that for any ratio of lunes (L_1/L_2), where L_1 is the lune on the side of a square and L_2 is the lune on the side of a hexagon, there must exist an equal ratio of squares (Q_1/Q_2), where Q_1 is any given square and Q_2 is some other square, or $L_1/L_2 = Q_1/Q_2$. By the alternation of ratios, $L_1/Q_1 = L_2/Q_2$. But in the first part of the tract it was proved that there is a Q_1 equal to L_1; therefore, there must be a Q_2 equal to L_2. But it should be clear that we are not told how to construct such a square Q_2. The basic similarity of this added comment in Z to the "proofs" given in Section 1 of Appendix I should be evident.

Appendix III

Some Medieval Latin Citations of Archimedes

I have included in this appendix some further references to Archimedean ideas beyond those already included in the various works edited in the course of this volume. All but those in passage number 9 appear in works that were translated in the twelfth or thirteenth century and all appear in tracts which circulated fairly widely with the corpus of Arabo-Latin mathematical works.

1. *Twelfth century*. Anaritius, *In decem libros priores Elementorum commentarii*, the translation of Gerard of Cremona, edited by M. Curtze as a *Supplementum* to J. L. Heiberg and H. Menge, *Euclidis opera omnia* (*Leipzig*, 1899), p. 5, lines 21–23: "Ac si vellet dicere illud, quod Aximethes (*Ed*, aximetes *corr ex*. Eximetes *in Vr*)[1] intellexit, hoc est: 'brevior dimensio, que coniungit (*Vr*, contingit *in Ed*) illud, quod est inter duo puncta.'" ("And if he means that which Archimedes understands [as to the definition of a straight line], it is this: 'the shortest measure that joins the interval between two points.'") While the form of this statement makes it appear as if the actual words of Archimedes are being used, it is evident on consulting the second citation below that the author is merely paraphrasing Archimedes' assumption in the *On the Sphere and the Cylinder* that the straight line is the least of all lines having the same extremities. On definitions of a straight line, consult T. L. Heath, *Euclid, The Elements*, vol. *1* (Annapolis, 1947), pp. 165–69.

2. *Twelfth century*. Anaritius, *op. cit.*, p. 6, lines 1–5: "Et ideo diffinivit eam Asamithes (*Ed*, assamites *in Vr*) dicens: 'Linea recta est brevior lineis,

[1] The variant readings in parentheses are from Vat. Reg. suev. 1268, 145r (abbreviated *Vr*), and were very kindly supplied to me by John Murdoch.

quarum extremitates sunt eedem,' et vult dicere, quod sit (*Ed*, fit *in Vr*) brevior linea, que coniungit, quod (*Ed*, que *in Vr*) est inter duo puncta." ("And therefore Archimedes has defined it, saying: 'A straight line is the least line among those whose extremities are the same,' and he means that it is the least line that joins the interval between two points.") Compare Archimedes, *De sphaera et cylindro*, *Opera omnia*, ed. of J. L. Heiberg, vol. *1* (Leipzig, 1910), p. 8. Actually, Archimedes' statement is not a definition but an assumption (see the reference to H. G. Zeuthen in Chapter Three, Section 1, note 2).

3. *Twelfth century*. Anaritius, *op. cit.*, p. 24, lines 29–31: "Supra hoc Sambelichius: Figure iste vocantur trapezie (*Ed*, trapetie *in Vr*), eo quod sunt (*Ed*, sint *in Vr*) inordinate, quas Asamithes (*Ed*, assamites *in Vr*) similiter nominavit." ("In regard to this, Simplicius says: These figures are called *trapezie* because they are irregular. Archimedes has similarly named them.") Compare Archimedes, *De sphaera et cylindro*, *ed. cit.*, p. 36, line 15, and elsewhere.

4. *Twelfth century*. Anaritius, *op. cit.*, p. 28, lines 16–26: "Et hec radix aut erit impossibilis, sicut illud, quod Asamithes (*Ed*, assamites *in Vr*) premissit et petiit, ut concederetur ei, scilicet, ut esset extra mundum (dixit enim, quod, si illud [*Ed*, istud *in Vr*] concederetur ei, ipse ostenderet, quod moveret terram, ubi dixit: 'Puer concede mihi, quod sit possibile, me elevari et manere extra mundum, et ego faciam te videre, quod ego movebo terram'). Et hoc fuit, cum iactavit se invenisse 'virtutem geometricam.' Et petiit, ut premitteretur istud, et poneretur sic esse, licet sit impossibile." ("Or the principle will [sometimes] be impossible, as in the case where Archimedes has premised and postulated that it be conceded to him that he would be outside of the world—for he has said that if that were conceded to him he would prove that he could move the world. His statement is this: 'Boy, concede to me that I can be lifted up and remain outside of the world, and I shall [then] make you see that I could[2] move the world.' And this was when he had boasted that he had found a 'geometric power.' And [for the sake of proof] he postulated that 'if that were permitted' and 'if it were posited to be so,' even though it is impossible.") This famous story associated with Archimedes is given by the author to illustrate the practice sometimes followed in mathematics of postulating what is known to be impossible. The story appears in Plutarch (*Marcellus*, 14)—"if there were another world and he could go to it, he would move

[2] Following the subjunctive in the Arabic text; this is also the evident sense of the phrase in Latin in spite of the use of the future.

this one"—and in Pappus (*Collectio*, ed. of Hultsch, vol. *3*, p. 1060, lines 1–4)—"Give me a place to stand on and I shall move the earth." Note that the phrase "geometric power" translates the Arabic القوة الهندسية (cf. R. O. Besthorn and J. L. Heiberg, *Codex Leidensis 399.1*, Part I, Fasc. 1 [Copenhagen, 1893], p. 12). Simplicius no doubt originally had the Greek word *dynamis*, by which the five basic machines (including the lever) were known.

5. *Twelfth century*. Anaritius, *op. cit.*, p. 162, lines 23–24: "Asamithes (*Ed*, assamites *in Vr*) vero vocat eas [quantitates homogeneas] quantitates, quarum alie aliis comparantur." ("While Archimedes calls them quantities which are mutually compared.") I do not believe that Archimedes gives a definition of "homogenous" magnitudes, but presumably it is just such magnitudes that are involved in the fifth of the *lambanomena*, or postulates, in the *De sphaera et cylindro*, *ed. cit.*, p. 8.

6. *Twelfth century*. Ametus filius Iosephi, *Epistola de proportione et proportionalitate* (see the text of this passage edited in M. Clagett, *The Science of Mechanics in the Middle Ages* [Madison, 1959], pp. 69–70, on the basis of BN lat. 9335, 66r; cf. Curtze's text, based on Vienna, Nat. bibl. cod. 5277, 309v; a critical text of the whole treatise has been prepared by my student, Miss Dorothy Schrader): "Arsamides quoque ponderum proportionalitatem diffinivit dicens: 'Pondera proportionalia diversa sunt, que uno ponderantur angulo.' Per quod voluit intelligi, ut, cum primum ponderum ponitur in lance trutine et secundum eorum in altera lance, et suspenditur trutina suspensorio suo, erit angulus quem circumdat statera et suspensorium trutine unus ad tertium et quartum cum tertium fuerit positum in loco primi et quartum in loco secundi, et similiter si quintum in loco primi et tertii ponatur et sextum in loco secundi et quarti. Et cum primum etiam et secundum in duabus lancibus ponuntur, et terium et quartum in duabus lancibus alterius trutine, et quintum et sextum in duabus lancibus trutine tertie, anguli qui sunt inter suspensoria trutinarum et stateras earum sunt etiam uni. Per hoc autem quod in verbis eius invenitur, scilicet ex diversis, voluit intelligi quod, cum primum ponderum fuerit equale secundo, statera trutine erit cum suspensorio ipsius coniuncta, neque erit inter ea angulus. Yrinus autem diffinivit proportionalitatem dicens: 'Pondera proportionalia diversa sunt ea que, cum appensa fuerint, erunt linee ordinate unicuique antecedenti earum et consequenti super equales angulos superficiei orizontis.' Quod etiam in nullo separatur ab eo secundum quod Arsamides ponderum proportionalitatem diffinivit." ("Archimedes also defined the proportionality of weights, saying: 'Diverse proportional weights are those which are weighed at the same angle.' By this he meant

that when the first of the weights is placed in a pan of the balance and the second of them in the other pan with the balance suspended by a suspension cord, the angle between the suspension cord and the needle is the same as when, in the case of a third and fourth weight, the third weight is substituted for the first and the fourth for the second, and similarly if the fifth is substituted for the first and third and the sixth for the second and the fourth. And also when the first and second are placed in two pans [of one balance], the third and fourth in the two pans of a second balance, and the fifth and sixth in the two pans of a third balance, the angles made by the suspension cords of the balances and their needles are also the same. When he uses the expression 'diverse' he wants it to be understood that when the first of the weights is equal to the second, the needle of the balance will coincide with its suspension cord and there will be no angle between them. Hero also defined proportionality, saying: 'Diverse proportional weights are those which, when suspended, will respectively orient the balance beam in equal angles with the plane of the horizon.' This does not differ from Archimedes' definition of the proportionality of weights.") I am unable to find either the definition of Archimedes or that of Hero in any of their extant works. Perhaps Archimedes' definition appeared in his lost work *On Balances*. Incidentally, the Arabic author goes on to show the basic identity of the two definitions, as the fuller quotation from the passage in *The Science of Mechanics* reveals.

Another reference to Archimedes by Ametus (BN lat. 9335, 68r, c. 2) is most general and seems to reflect no particular Archimedean passage: "Et Arsamides in proportione que est aliarum in superficiebus et corporibus ad alias ostendit illud quo probatio constat non indigens positione." The purport of this passage with what goes before seems to be that Archimedes demonstrated propositions regarding the ratios of surfaces and volumes. And indeed Archimedes does this in most of his works.

7. *Thirteenth century*. Anonymous, *De ysoperimetris*, translated from the Greek, Bodleian MS, Auct. F.5.28, 106v, ex Proposition VI: "Quoniam vero quod sub ea que ex centro et perimetro circuli duplum circuli demonstratum est Archimenidi in mensuratione circuli; demonstravit enim quoniam omnis circulus equalis est trigono orthogonio, cuius que e centro equalis est uni earum que circa rectum, reliqua vero perimetro circuli." ("That the product of the radius and the circumference of a circle is double the circle has been demonstrated by Archimedes in the *Measurement of the Circle*, for he has demonstrated that every circle is equal to a right triangle,

one of whose sides about the right angle is equal to the radius while the other is equal to the circumference of the circle.")

8. *Ibid.*, ex Proposition VII: "Intelligatur primum solidum contentum sub conicis superficiebus sicut sumebatur et in eis que Archimenidi, cuius generatio erat poligonii inscripti in circulo, cuius latera sub tetrados mensurantur, et delati circa manentem circuli diametrum.... Adiaceat igitur circulus equalis superficiei solidi *AB* et intelligatur ab *AB* conus altitudinem habens eam que ex centro inscripte solido spere. Equalis est ergo solido. Hoc enim demonstratum est Archimenidi.... Et est *GDT* conus equalis spere, velud colligitur ex eis que Archimenidi.... Quoniam vero conus basim habens circulum equalem superficiei spere, altitudinemque equalem ei que e centro spere equalis est spere colligitur ex eis que Archimenidi ita. Quoniam enim demonstravit quod chilindrus basim habens maximum circulum altitudinemque diametrum spere sexquialter est spere, talis vero chilindrus sexcuplus est coni basim quidem habentis eandem, altitudinem vero eandem que e centro, quadrupla erit spera talis coni, est autem et eiusdem quadruplus et conus quidem altitudinem habens eandem, basim vero superficiei spere equalem, sub eadem enim altitudine existentes adinvidem sunt sicut bases. Superficies autem spere quadrupla est maximi circuli. Quare equalis spera dicto cono." ("Let the first solid be understood as one contained by conical surfaces, like the one assumed in the statements of Archimedes [in the *On the Sphere and the Cylinder*, *passim*, Propositions I.23–I.35]. The generation of this solid was by means of the rotation about a fixed diameter of a polygon inscribed in a circle, a polygon the number of whose sides is divisible by four [*On the Sphere and the Cylinder*, Proposition I.23].... Therefore, let there be a circle equal to the surface of solid *AB*, and *AB* is to be understood as a cone having as its altitude the radius of the sphere inscribed in the solid. Therefore, it is equal to the solid. For this has been demonstrated by Archimedes [*On the Sphere and the Cylinder*, Proposition I.26].... And cone *GDT* is equal to the sphere, as is inferred from the statements of Archimedes.... That a cone—having as its base a circle equal to the surface of a sphere and an altitude equal to the radius of the sphere—is itself equal to the sphere is inferred thus from the statements of Archimedes. Since he has demonstrated that a cylinder having as its base the greatest circle [of the sphere] and an altitude equal to the diameter of the sphere is three halves of the sphere [*On the Sphere and the Cylinder*, Proposition I.34, Cor.], while such a cylinder is six times a cone having the same base and an altitude equal to the radius [of the sphere], [hence] the sphere will be quadruple such a cone, and also the

cone having the same altitude but a base equal to the surface of the sphere will be quadruple it—for those things having the same altitude are related as their bases. Moreover, the surface of the sphere is quadruple the greatest circle [of the sphere] [*On the Sphere and the Cylinder*, Proposition I,33]. Therefore, the sphere is equal to the said cone.") As the references in brackets indicate, the author of this tract has used the *On the Sphere and the Cylinder*, giving the results of material that extends approximately from Proposition I.26 through the Corollary of Proposition I.34. Let me emphasize that although this work was translated from the Greek, it appears in many of the codexes that contain mathematical works of the Arabo-Latin tradition; hence the inclusion of these references in this volume.

9. *Fourteenth or fifteenth century*. Anonymous, *De inquisicione capacitatis figurarum*, ed. of M. Curtze, *Abhandlungen zur Geschichte der Mathematik*, *8* Heft (1898), Proposition 18, lines 44–45: "Dato circulo duplum circulum depingere. Sit circulus *abcd*, cuius dyameter *ac*, cui per 7am quarti circumscribatur quadratum *efgh*. Item per 9am quarti Euclidis eidem quadrato *efgh* transcribam circulum *efgh*. Quia igitur quadratum lineae *fh*, quae est dyameter circuli maioris, per penultimam primi est duplum ad quadratum lineae *fg*, quae est aequalis dyametro circuli minoris: ergo per 2am duodecimi circulus, cuius dyameter est *fh*, est duplus ad circulum, cuius dyameter est *ca*." ("To draw a circle double a given circle. Let there be circle *abcd*, whose diameter is *ac*. By IV.7[3] [of the *Elements*] let square *efgh* be circumscribed about it. Also, by IV.9[4] of Euclid, I shall describe about this same square *efgh* circle *efgh*. Hence, because, by the penultimate[5] [proposition] of the first [book of the *Elements*], the square of line *fh*, which is the diameter of the larger circle, is double the square of line *fg*, which is equal to the diameter of the lesser circle, therefore, by XII.2[6] [of the *Elements*], the circle whose diameter is *fh* is double the circle whose diameter is *ca*"). This proposition is equivalent to Lemma VII of the *Liber assumptorum* (or *Lemmata*) attributed to Archimedes, which exists only in an Arabic text (see Archimedes, *Opera omnia*, ed., of J. L. Heiberg, vol. *2* [Leipzig, 1913], pp. 517–18). However, there is no evidence that this work was translated into Latin, and in fact the work in its present form can scarcely be by Archimedes, whom it quotes. Furthermore, it is most doubtful that this partic-

[3] See the Commentary to Chapter Three, Section 3, Florence Version F.IB, line 33.

[4] Edition of the *Elements* (Basel, 1546), p. 92.

[5] See the Commentary to Chapter Three, Section 1, Cambridge Version, line 25.

[6] See the Commentary to Appendix I, Section 1, line 5.

ular lemma is based on an original proposition of Archimede. Still, for the sake of completeness I have included it in this appendix.

10. Two other citations of Archimedes may be mentioned, although neither constitutes a fragment. The first appears to be a simple confusion with a certain Basilides of Tyre mentioned by Hypsicles in his preface to Book XIV of the *Elements*. The customary preface (as translated by Thomas Heath, *Euclid, the Elements*, vol. *1*, p. 5) opens: "Basilides of Trye, O Protarchus, when he came to Alexandria and met my father, spent the greater part of his sojourn with him on account of their common interest. And once, when examining the treatise written by Apollonius about the comparison between the dodecahedron and the icosahedron inscribed in the same sphere, (showing) what ratio they have to one another, they thought that Apollonius had not expounded this matter properly, and accordingly they emended the exposition, as I was able to learn from my father...." In the manuscript containing the medieval Latin translation from the Greek of the *Elements* (Bibl. Nat. lat. 7373, 167v), discovered by John Murdoch, there is added a version of Book XIV, containing a preface which opens as follows: "Acefalus in commento super euclidem de archimede siro scribit: dum esset alexandrie in studio forte ad manus eius pervenisse duos apollonii libros de habitudine figurarum ad invicem in eadem spera constructarum (constructarunt, MS), quos cum sumo affectu pertractaret...." However interesting this statement is for the Euclid text, it obviously has little significance for the knowledge of Archimedes in the Middle Ages. Note the "Greek" spelling "archimede" rather than the customary "Archimenide."

The second reference to Archimedes occurs in the Latin translation of Alhazen's *Liber de speculis comburentibus* and is the attribution to Archimedes of the invention of a burning mirror that would concentrate the rays at a point so that combustion would be stronger: "et ex eis fuerunt quidam, qui assumpserunt specula plurima sperica, quorum radii converterentur ad punctum unum, ut combustio fortior esset, et illi, qui invenerunt specula ista, famosi fuerunt, sicut Archimenides et Anthimus et alii ab istis duobus" (ed. of J. L. Heiberg and E. Wiedemann in *Bibliotheca Mathematica*, 3. Folge, vol. *10* [1909–10], p. 219).

11. Finally, we can briefly refer to the medieval survival of an experiment on refraction that may well go back to Archimedes. This experiment is most fully described in Ptolemy's *Optics*, V, 5, translated into Latin from the Arabic by Eugene the Amīr (A. Lejeune, *L'Optique de Claude Ptolémée*, Louvain, 1956, p. 225): "[5] Quod autem est apparens et manifestum, pos-

sibile est nobis intelligere per se ex nummo qui fit in vase quod vocatur baptistir. Visus enim cum steterit fixus in loco quo radius qui transit per marginem vasis, efficitur sublimior nummo, et manente situ in statu suo, effundetur aqua in vase moderate, quousque radius qui transit per marginem vasis frangatur ad interiora et ceciderit super nummum, accidit inde res que prius non videbantur, videri tunc super lineam rectam protractam a visu ad locum subliorem vero loco, et existimabitur radius non esse refractus ad eas, sed quod ipse natent et eleventur ad radium. Et hac de causa apparebunt tunc secundum rectitudinem visibilis radii, et ex ali a parte apparebunt secundum perpendicularem cadentem super aque superficiem, iuxta principia que prius explicavimus." ("That this is apparent and manifest is possible for us to understand immediately from a coin placed in a vessel called a *baptistir*. For when the line of sight has been fixed positionally by a ray which transits the edge of the vessel so that the line of sight is higher than the coin, then, with the coin remaining in its place, let the vessel be carefully filled with water until the ray which transits the edge of the vessel is bent to the interior and falls on the coin. Thus is happens that things which were not seen before are seen on a straight line projected from the sight to the position higher than the true position [of the objects], and it will be judged that the ray will not be refracted to the objects but rather that they float and are lifted to the ray. And so for this reason they will appear along the straight path of the visible ray, and in another direction they will appear along the perpendicular falling on the surface of the water, following the principles which we have explained before.") Briefer descriptions of this same experiment were also available in Latin to medieval readers in the *Quaestions naturales* (I, vi, 5) of Seneca and in the sixth postulate of the popular Pseudo-Euclidian *Catoptrics* (*De speculis*). For the association of this experiment with Archimedes and reference to it by Greek authors, see A. Lejeune, *op. cit.*, p. 225n, and A. Lejeune, *Recherches sur la catoptrique grecque* (Brussels, Paris, 1957), pp. 55, 143, 153.

[12. In the final stages of printing, there came to light an additional brief citation of Archimedes in Averroes' commentary on Aristotle's *De caelo*, Bk. II, text no. 112 (Junta e.d., Venice, 1562, f. 172r, D): "et ex mensura maximi circuli inveniunt mensuram sphaerae secundum Archimenidem" ("and from the measure of the largest circle they find the measure of the sphere, according to Archimedes").]

Appendix IV

A Medieval Treatment of Hero's Theorem on the Area of a Triangle

The theorem for the area of a triangle as a function of its sides (namely, $A = \sqrt{s(s-a)(s-b)(s-c)}$, where s is the semiperimeter and a, b, c, are the sides) has had a long history since its enunciation by Hero of Alexandria.[1] The most important early study of this history was done by F. Hultsch in 1864.[2] Much of the recent historical investigation of this theorem was summarized succinctly by S. Gandz.[3] In brief, we can note that this theorem was given without proof by the author of the Mishnat ha-Middot (*ca.* 150);[4] by one of the *agrimensores* (*ca.* 400?), also without proof;[5] by the Indian mathematician Brahmagupta (*ca.* 628),[6] who extends it to a quadrilateral but has no proof; by the Banū Mūsā accompanied by a proof (see Chapter

[1] Hero of Alexandria, *Metrica*, I, viii, ed. of H. Schöne in *Heronis Alexandrini opera... omnia*, vol. *3* (Leipzig, 1903), pp. 18–24; *Dioptra*, xxix, *ibid.*, pp. 280–84; *Geometrica*, ed. of J. L. Heiberg, *ibid.*, vol. *4* (Leipzig, 1912), p. 248. Incidentally, al-Bīrūnī (see note 7 below) assigns the theorem to Archimedes. While we have no antique evidence of this, it certainly makes sense that the theorem is earlier than Hero.

[2] F. Hultsch, "Der Heronische Lehrsatz über die Fläche des Dreieckes als Function der drei Seiten," *Zeitschrift für Mathematik und Physik*, vol. *9* (1864), pp. 225–49.

[3] S. Gandz, ed., *Mishnat ha-Middot*, *Quellen und Studien zur Geschichte der Mathematik, Astronomie und Physik*, Abt. A: *Quellen*, vol. *2* (1932), p. 45, note 40.

[4] *Ibid.*, pp. 45–46.

[5] M. Cantor, *Die römischen Agrimensores und ihre Stellung in der Geschichte der Feldmesskunst* (Leipzig, 1875), p. 107. See the text of the theorem by Marcus Junius Nipsus in his so-called *Podismus*, F. Blume, K. Lachmann, and A. Rudorff, *Die Schriften der römischen Feldmesser*, vol. *1* (Berlin, 1848), pp. 300–301.

[6] H. T. Colebrooke, *Algebra with Arithmetic and Mensuration from the Sanscrit of Brahmegupta and Bhascara* (London, 1817), pp. 72, 295–96. Cf. Hultsch, *op. cit.* in note 2 above, p. 239. Other later Indian authors took up the theorem.

Four, Proposition VII); by that superb polymath al-Bīrūnī (*ca.* 1000),[7] who assigns the enunciation to Archimedes and takes his proof from one Abū 'Abdallāh al-Shannī; by al-Karkhī (*ca.* 1020);[8] by Savasorda (twelfth century) without proof;[9] by Leonardo of Pisa, from the Banū Mūsā;[10] by Luca Pacioli in 1494 with a proof from Leonardo;[11] by Widmann without proof in 1489;[12] by Leonardo of Cremona in the same century, again without proof;[13] and in the sixteenth century by Pierre de la Ramée[14] (with a proof similar to that of the Banū Mūsā), and no doubt by others in that century. Incidentally, while the theorem itself is not given by Campanus, its possibility of development is suggested by the thirteenth-century mathematician in his commentary on the *Elements* (see note 22 below).

As I pointed out in Chapter Four, the first contact of Latin scholars with a proof of the theorem came with the translation by Gerard of Cremona of the *Verba filiorum* of the Banū Mūsā, and it was this proof that was reflected in the subsequent treatments by Leonardo of Pisa, Pacioli, and Pierre de la Ramée. A quite different proof of this theorem also circulated during the Middle Ages, a proof associated with the name of Jordanus. It is this proof, somewhat closer to the proof by Hero, that is the object of our discussion in this appendix. It existed in two versions, both of which have here been edited.

The first of these two versions was previously published by Curtze on

[7] H. Suter, "Das Buch der Auffindung der Sehnen im Kreise von Abū 'l-Raiḥān Muḥ. el-Bīrūnī," *Bibliotheca Mathematica*, 3. Folge, vol. *11* (1910–11), pp. 39–40, 70.

[8] Gandz, *loc. cit.* in note 3 above.

[9] "Der 'Liber embadorum' des Savasorda in der Übersetzung des Plato von Tivoli," in M. Curtze, *Urkunden zur Geschichte der Mathematik im Mittelalter und der Renaissance*, *Abhandlungen zur Geschichte der mathematischen Wissenschaften*, *12.* Heft (Leipzig, 1902), p. 72.

[10] *Practica geometrie*, in *Scritti di Leonardo Pisano*, ed. by B. Boncompagni, vol. *2* (Rome, 1862), pp. 40–42.

[11] Hultsch, *op. cit.* in note 2 above, pp. 242–46, gives the citation to Pacioli's *Summa*, and he translates and discusses the proof given by Pacioli.

[12] See the note by G. Eneström, *Bibliotheca Mathematica*, 3. Folge, vol. *5* (1904), p. 311.

[13] "Die 'Practica geometriae' des Leonardo Mainardi aus Cremona," in M. Curtze, *op. cit.* in note 9 above, *13.* Heft, pp. 386–87.

[14] P. de la Ramée. *Scholarum mathematicarum libri unus et triginta* (Frankfort, 1599), p. 313. Ramée's proof is substantially the same as that of the Banū Mūsā. He appears to claim that the proof comes from Jordanus and Tartaglia and that it is lacking in logic. My guess is that he found the proof in a work of Tartaglia, who perhaps got it from Pacioli. I suspect that Tartaglia also saw the different proof that circulated with the *De ratione ponderis* of Jordanus and that perhaps he mentioned such a proof without reproducing it, reproducing rather the proof of the Banū Mūsā as given by Pacioli.

the basis of a single manuscript, namely, *I*;[15] his suggestion as to authorship was based on a table of contents at the beginning of the manuscript, where we read *Theoremata Cratili*.[16] He thought that the theorem might be one of those included under that title, but as to who Cratilus was, Curtze had no idea. This proof, discovered by Curtze, was associated for the first time with the name of Jordanus by Pierre Duhem for the following reasons:[17] (1) The theorem appears in the manuscripts in close proximity to (or even as a part of) works attributed to Jordanus.[18] (2) There is a reference in the Vatican manuscript (*Xa*) to the effect that the theorem "is a part of the phyloteigni and ought to be joined to it" (cf. Version I, variant reading to line 1); but Jordanus refers in his *Elementa de ponderibus* to a work of his entitled *Philotegni*, a work which can now confidently be identified with his *De triangulis*.[19] Still, the accuracy of this marginal reference may well be questioned since none of the four copies of the *De triangulis* that I have examined contains the proof in question, while in manuscript *I* the proof presented here is separated from the *De triangulis* by more than one hundred folio pages. Furthermore, manuscripts *Xa* and *Q* of Version I of the theorem say that "this rule is said to have been written in Arabic" (cf. Version I, lines 7–9), while *Ya*, the unique manuscript of Version II,

[15] M. Curtze, "Über eine Handschrift der Königl. öffentl. Bibliothek zu Dresden, *Zeitschrift für Mathematik und Physik*, vol. *28* (1883), Hist.-lit. Abt., pp. 5–6, 78.

[16] *Ibid.*, pp. 4, 6. The hand giving this table of contents is that of Valentinus Thaus (Thaw) and is dated 1580 (see p. 1). Incidentally, it is Thaus' hand that adds the combined Greek and Latin marginal note opposite the theorem under consideration (see Version I, variant readings, line 1).

[17] P. Duhem, "Un ouvrage perdu cité par Jordanus de Nemore: le Philotechnes," *Bibliotheca Mathematica*, 3. Folge, vol. *5* (1904), pp. 323–25.

[18] In manuscript *Q* (40v) it comes at the end of the *Liber de ratione ponderis* attributed to Jordanus and actually precedes the explicit of that work (40v): "Explicit liber quartus Iordani de ponderibus." In manuscript *Xa* it follows two folios after the *Liber de ratione ponderis*, which occupies folios 50r–55v. In manuscript *I* there are a number of works attributed to Jordanus, including the *De triangulis* (50r–61v), an *Arithmetica* (61v–110v), *Elementa de ponderibus* (186r–187v), *De forma spere in plano* (224r–225v), *De numeris datis* (228r–242v), *De ratione ponderis* (243r–249v). Our theorem is not particularly close to any of these, occupying folios 178r–v. But the association of the theorem with Jordanus, or at least with one of the tracts *De ponderibus*, is suggested by a statement in Leonardo de Cremona's *Practica geometriae* (*ed. cit.* in note 13 above, p. 386), where he notes that he has found the theorem in a book of mechanics (*libro de mechanici*). This could very possibly have been a manuscript of one of the tracts *De ponderibus* attributed to Jordanus, as Curtze asserts.

[19] See E. A. Moody and M. Clagett, *The Medieval Science of Weights*, 2d printing (Madison, 1960), pp. 130, 134–36, 379, 381. Note further that in MS Bruges, Stadsbibliotheek 530, 1v–8v, the *De triangulis* is specifically entitled *Phylotegni Iordani de triangulis liber primus* (1v; cf. 8v).

states more categorically: "this rule concerning the triangle was written in Arabic" (cf. Version II, line 150).

These remarks seem to throw some doubt on Jordanus' authorship of the theorem. But let us examine them more closely, speculating as to their meaning. At least three possibilities suggest themselves, the first of which is incompatible with the idea of Jordanus as the original author of the theorem. The first interpretation of the remarks is that the whole theorem (enunciation and proof) was composed in Arabic and merely translated into Latin. In support of such a theory we recognize that the proof, unlike that of the Banū Mūsā, was fairly close to an Arabic version of Hero's proof that circulated in the Middle Ages,[20] although to be sure we can readily see that the Arabic version was not itself the text from which our proof was translated in spite of the general similarity of the two. Supposing this interpretation to be the correct one, then, if Jordanus had any connection at all with the theorem, it was merely to transmit it or possibly to modify if somewhat. We know that Jordanus did on occasion take certain theorems almost verbatim from other authors, as for example when he drew the theorems and proofs concerning the trisection of an angle and the finding of two mean proportionals from the Banū Mūsā (see Appendixes V and VI).

A second possible interpretation of the remarks on the Arabic writing of this theorem is that they refer only to the enunciation and not to the proof. This seems to have been Duhem's opinion.[21] In support of this theory is the fact that the comment follows immediately after the enunciation in manuscripts *Xa* and *Q* and that it is added in manuscript *Ya* only in connection with the full, formal enunciation as it is given at the end. But one could object to this interpretation (and as a matter of fact to the first one as well) by pointing out that the remark in *Xa* and *Q* is by no means an assertive statement. If the author of that remark really knew that this rule was written in Arabic (and translated therefrom), why did he use the rather tentative verb *dicitur* ("is said"). This objection, of course, does not hold for the statement in *Ya*, which flatly asserts that the rule was written in Arabic. Still, there is considerable evidence that Version II is merely a rewrite of Version I in the *Xa* and *Q* tradition. Hence, the change in tone in the remark as found in *Ya* ought perhaps to have no significance.

A third interpretation would deny the Arabic origin of either enuncia-

[20] This version of Hero's proof was added to the end of al-Ṭūsī's edition of the geometry of the Banū Mūsā (*Majmū' al-Rasā 'il*, vol. *2* [Hyderabad, 1940]). Cf. H. Suter, "Über die Geometrie der Söhne des Mûsâ ben Schâkir," *Bibliotheca Mathematica*, 3. Folge, vol. *3* (1902), pp. 271–72.

[21] Duhem, *op. cit.* in note 17 above, pp. 323–24.

tion or proof. It would hold that the remark is only a vague, general statement that the same rule can be found in writing of Arabic origin (as, for example, in the *Verba filiorum*, known to be of Arabic origin), but that both the enunciation and proof as presented here are independent of Arabic sources. That is, they are either Greek in origin or original with Jordanus or some other Latin geometer. There is no sure way to decide between these varying interpretations and I suspect that no decision can be made until further evidence appears.

The line of argument followed in both of the versions is clearly enough indicated through the accompanying translations to demand little additional comment. As I indicated earlier, it resembles the proof of the theorem given in Hero's *Metrica* and *Dioptra* (see note 1) more than it does that found in the *Verba filiorum* of the Banū Mūsā. It is evident, however, that the medieval proof is less economical than Hero's proof. This is sharply brought out by contrasting Version II with Hero's proof, since Version II has added geometrical steps lacking in Version I. But even in Version I the author takes an excessive number of steps to draw his obvious conclusions after showing the similarity of triangles *PAC* and *DBF* and that of triangles *DFH* and *CPQ*. I do not know when the elaboration represented by Version II was composed, but its unique manuscript dates from the fifteenth century. It seems probable to me that it was composed from some copy of Version I that was closer to the tradition of manuscripts *Xa* and *Q* than *I*, since many of the variant readings of *Xa* and *Q* have been incorporated in Version II. That Version II postdated Campanus' commentary on the *Elements* of Euclid is possible since in Version II the enunciation is changed to a form similar to that found in Campanus' commentary.[22] Incidentally, the only author cited in either version is Euclid, and he is cited only once in Version II (line 94)—a very general citation to Book XI.

My text of Version I has been constructed on the bais of the three manuscripts listed under Version I in the Sigla below. It is clear from an

[22] Thaw points out in a comment on the margin of manuscript *I* opposite the theorem (see variant readings, Version I, line 1) that there is another way of proof in Campanus' treatment of Proposition II.13 of the *Elements*, a remark that puzzled Curtze since he could find no such reference. However, in the edition of Basel, 1546, p. 51, we read after the proof of II.13: "Notandum autem per hanc et precedentem et penultimam primi quod cognitis lateribus omnis trianguli, cognoscitur area ipsius, et auxiliantibus tabulis de chorda et arcu, cognoscitur omnis eius angulus." Compare line 1 of Version II, where the long enunciation of Version I has been abandoned in favor of a short problematic statement of the theorem, in the manner of the brief statement of Campanus.

examination of the variant readings that *Xa* and *Q* again and again agree with each other against *I*, but sometimes circumstances of style and meaning have demanded that I follow *I* and sometimes *Xa* and *Q*. As between *Xa* and *Q*, *Xa* is to be preferred since on occasion *Q* omits necessary material (e.g., see variants to lines 46–48 and 61–63) and often alters correct forms (e.g., see *medietasque* instead of *mediatatisque* in lines 1–2, *superadditus L* instead of *Sed super AC* in line 32, and so on). *Xa* has no figure. The bottom part of the triangle is missing in the figure in *I*, as are the lines *R*, *S*, and *T*; but solids *X* and *Y* are represented as rectangular parallelopipeds. My text of Version II follows *Ya* with only an occasional change, the pristine readings having been included in the variant readings. The marginal folio numbers for Version I refer to manuscript *I*, those for Version II to manuscript *Ya*.

Sigla of Manuscripts

Version I

I = Dresden, Sächs. Landesbibliothek, Db. 86, 178r–v, early 14c. Cf. the text of M. Curtze, cited in footnote 15.
Xa = Rome, Vat. Reg. suev. 1261, 57v–58r, *ca.* 1350–75.
Q = Paris, BN lat. 7378A, 40r–v, 14c.

Version II

Ya = Munich, Bay. Staatsbibliothek, cod. 234, 105v–108v, 15c.
[I have published the Latin text of both of these versions in S. Prete, ed., *Didascaliae* (New York, 1961), pp. 79–95.]

[The Area of a Triangle Measured in Terms of Its Three Sides]

[Area trianguli tribus lateribus mensurata]

[Versio I]

178r / SI TRIA TRIANGULI LATERA COACERVENTUR, MEDIETATISQUE COMPOSITI AD SINGULA LATERA DIFFERENTIE SUMANTUR, PRIMAQUE IN SECUNDAM DUCATUR ET IN PRODUCTUM TERTIA, ITEMQUE QUOD INDE PRO-
5 VENIT IN PREDICTAM MEDIETATEM, ILLIUS ULTIMO PRODUCTI RADIX ERIT AREA TRIANGULI.

[Regula hec in arabico conscripta dicitur, in qua quoniam tertia multiplicatio, videlicet linee in solidum, que in continuis non habetur, rationem in ea sumere oportet numerorum.]

10 Sit itaque datus triangulus *ABC* [Fig. 93] et medietas coacervati ex lateribus ipsius sit linea *KLM*, differentieque ipsius sint ad *AB* linea *T*, et ad *AC S*, et ad *BC R*. Sitque solidum quod continetur *R*, *S*, *T* designatum nota *Z*. Intelligamus itaque in dato triangulo circulum contingentem latera notis *F*, *G*, *E*, cuius centrum *D*, a quo prodeant
15 ad puncta contactus linee *DF*, *DE*, *DG* singulis lateribus perpendiculares, atque inter se equales. Sed et a *D* ad tres angulos linee pro-

1 *mg.* *Xa* Hec est pars Phyloteigni et debet ei subiungi / *mg.* *1 6c m. I* σφάληρον θεώρημα. vide hac de re Campanum Prop. 13, lib. 2, Euclidis aliter idem / tria trianguli *tr.* *XaQ*
1–2 medietasque *Q*
4 quod *XaQ om. I*
4–5 provenit *XaQ* productum *I*
5 illius ultimo *XaQ om. I*
6 producti radix *tr. I*
7–9 [Regula...numerorum] *XaQ om. I*
10 itaque *XaQ om. I*
11 -que *I om. XaQ* / linea[2] *om. Xa*
12 et[1] *om. Q* / S: scilicet (?) *Xa* / et[2] *om. Q* / Sitque *XaQ* sed *I* / continetur *XaQ* continent *I* / R, S, T *XaQ* idest *I*
13 itaque *XaQ* igitur *I* / triangulo *XaQ* trigono *I*
14 latera *I om. XaQ*
15 *ante* ad *add.* *XaQ* D / DG *I* GD *XaQ* / lateribus *XaQ* lineis *I*

[The Area of a Triangle Measured in Terms of Its Three Sides]

[Version I]

IF THE THREE SIDES OF A TRIANGLE ARE ADDED TOGETHER, AND THE EXCESSES OF ONE HALF OF THIS SUM OVER EACH OF THE INDIVIDUAL SIDES ARE TAKEN, AND THE FIRST EXCESS IS MULTIPLIED BY THE SECOND, AND THE THIRD IS MULTIPLIED BY THIS PRODUCT, AND THEN THE RESULT OF THIS IS MULTIPLIED BY THE AFORESAID HALF [PERIMETER], THE ROOT OF THE FINAL PRODUCT WILL BE THE AREA OF THE TRIANGLE.

[This rule is said to have been written in Arabic. Since in the rule the third multiplication, being of a line by a solid, is one which is not had among continua, it is necessary to assume in it the nature of numbers.]

And so let the given triangle be ABC [see Fig. 93] and let one half the

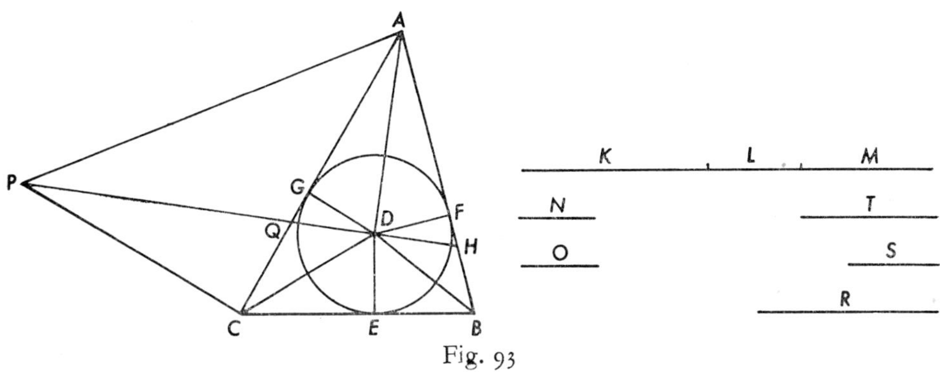

Fig. 93

sum of its sides be line KLM, and let $KLM - AB = T$, and $KLM - AC = S$, and $KLM - BC = R$. Let $R \cdot S \cdot T =$ solid Z. And so let a circle be imagined as in the given triangle, and let it be tangent to the sides at points F, G, and E. Let the center of the circle be D, and from D let lines DF, DE, DG, perpendicular to the separate sides, proceed to the points of tangency, and they will be mutually equal. But also from D let

trahantur, que singulos angulos per equa partientur, quorum omnium medietates quoniam equantur uni recto et quia angulus *DBF* et angulus *BDF* equantur recto, erit angulus *BDF* tanquam angulus *DAF*
20 et angulus *GCD*. Ducatur itaque linea *PQDH* ita ut angulus *HDF* equatur angulo *DAF*, eritque angulus *HDA* rectus, et similiter angulus *QDA*, et ob id etiam angulus *GDQ* equalis angulo *GAD*, et reliquus, scilicet *QDC*, equalis angulo *DBH*, quoniam totus angulus *GDC* est equalis duobus, scilicet *GAD* et *DBE*. Erigatur itaque
25 perpendicularis a *C* donec concurrat cum linea *HDQP*, que sit *CP*, et continuetur *P* cum *A*. Deinde, quoniam *AF* est equalis *AG*, et *BF* equalis *BE*, atque *CE* equalis *CG*, linee enim ab eodem puncto exeuntes usque ad contactus circuli sunt equales, tunc *AF* et *FB* et *CE* sunt tanquam tres relique. Erunt ergo medietas trium laterum
30 componentque lineam *KLM*, sitque *K* equalis *AF* et *L* equalis *FB* et *M* equatur *CE*. Manifestum est etiam quod *KLM* addit super *AB* quantum est *M*, quare *M* equalis est *T*. Sed super *AC* addit *L*; quare *L* equatur *S*. Atque super *BC* addit *K*, itaque *K* est equalis *R*. Sint etiam linee *N* et *O* equales perpendicularibus, quarum una *DF*, soli-
35 dumque sub lineis *KLM* et *N* et *O* contentum sit *Y*. Quia item anguli *ADP* et *ACP* sunt recti et equales, puncta *A*, *D*, *C*, *P* in eodem semicirculo consistent; quare anguli *PAC* et *PDC* sunt equales. Erunt ergo anguli *PAC* et *DBF* equales, itaque trianguli *PAC* et *DBF*
178v sunt similes, / eritque *BF* ad *FD* sicut *AC* ad *CP*. Sed etiam *FD* ad *FH*
40 sicut *PC* ad *CQ*, quoniam trianguli *DFH* et *CPQ* sunt similes. Itaque *BF* ad *FH* tanquam *AC* ad *QC*. Ergo coniunctim *BF* et *AC* ad *FH*

18 medietas *Q* / uni *I* cum *XaQ* / DBF *I* BDF *XaQ*
19 BDF[1] *I* DBS *XaQ* / BDF[2] *I* DBF *XaQ*
20 GCD *I* DCG *XaQ* / itaque *XaQ* ergo *I*
21 equatur *IQ* equetur *Xa*
22 id etiam *I* hoc *XaQ* / GDQ *Xa* QDG *Q* GQD *I*
23 QDC *XaI* QDE *Q*
24 GDC *XaQ* QDC *I* / duobus scilicet *I* angulo *XaQ*
25 HDQP *I* HDPQ *XaQ*
27–28 enim...exeuntes *I* siquidem exeuntes ab eodem puncto *Xa*
28 contactus *I* contactum *XaQ*
30 -que *XaQ* ergo *I*
31 equatur *I* equetur *XaQ*
32 equalis est *I* est equale *XaQ* / Sed super AC *Xa* superadditus L *Q* sed et similiter AC *I* / quare[2] *I* itaque *XaQ*
33 L *I* B *XaQ* / super *XaQ* similter *I* / itaque *I* ergo *XaQ* / equalis *I* equale *XaQ* / R *I* Y *XaQ*
34 O *I* D *XaQ*
35 item *Xa* idem *I* tunc *Q* / anguli *I* trianguli *XaQ*
36 et equales *I om. XaQ*
37 consistent *I* consistunt *XaQ* / quare *I* et *XaQ*
38 DBF[1] *XaQ* FBD *I* / PAC[2] et DBF *XaQ* FBD et ACP *I*
40 PC *XaQ* PO (*?*) *I* / CPQ *I* PQC *Xa* PAC *Q*
41 QC *I* CQ *XaQ* / FH[2] *XaQ* HF *I*

lines be drawn to the three angles, thus bisecting the individual angles. Since the halves of all the angles [together] equal one right angle, and since $\angle DBF + \angle BDF = 1$ right angle, therefore $\angle BDF = \angle DAF + \angle GCD$. And so let line $PQDH$ be drawn so that $\angle HDF = \angle DAF$, and $\angle HDA$ will be a right angle, and similarly $\angle QDA$ will be a right angle. Consequently $\angle GDQ = \angle GAD$, and $\angle QDC = \angle DBH$ since the whole angle $GDC = \angle GAD + \angle DBE$.* And so from C let a perpendicular be erected until it meets line $HDQP$, and this perpendicular is CP. Let P be joined with A. Then, since $AF = AG$, and $BF = BE$, and $CE = CG$—for lines proceeding from the same point [outside of a circle] which are tangent to the circle are equal—then $AF + FB + CE = AG + BE + CG$. Therefore, $(AF + BF + CE)$ equals half the sum of the three sides and [so] comprises line KLM. And let $K = AF$, $L = FB$, and $M = CE$. Also it is evident that $KLM - AB = M$, and so $M = T$. But $KLM - AC = L$, and so $L = S$; and $KLM - BC = K$, and so $K = R$. Further, let lines N and O be equal to the perpendiculars, of which one is DF. And let $KLM \cdot N \cdot O = Y$. Also, since angles ADP and

* Since $\angle GDC + \angle GCD = 1$ right angle, and $\angle GAD + \angle DBE + \angle GCD = 1$ right angle.

et *CQ* sicut *BF* ad *FH*. Sed *FH* est tanquam *QG* propter triangulos similes et quia *FD* equalis est *DG*, [et itaque *FH* et *QC* sunt tanquam *GC*.] Itemque *BF* et *AC* sunt tanquam *KLM*. Est ergo *KLM* ad *GC*
45 sicut *BF* ad *FH*. Sed proportio *BF* ad *FH* aggregatur ex proportione *BF* ad *DF* et ex proportione *DF* ad *FH*. Sed *AF* ad *FD* tanquam *FD* ad *FH*. Proportio ergo *BF* ad *FH*, constat ex proportionibus *BF* ad *DF* et *AF* ad *DF*. Sed proportio *BF* ad *DF* tanquam linee *S* ad *N*, et *AF* ad *DF* sicut differentie *R* ad *O* proportio. Itaque
50 *KLM* ad *T* aggregatur ex proportione *S* ad *N* et *R* ad *O*. Quia igitur solidi *Y* ad solidum *Z* aggregatur proportio ex proportionibus *KLM* ad *T* et *N* ad *S* et *O* ad *R*, et proportiones *N* ad *S* et *O* ad *R* faciunt proportionem *T* ad *KLM*, erit solidum *Y* equale solido *Z*. Producit autem *KLM* in *O* superficiem equalem triangulo dato. Divisus est
55 enim triangulus *ABC* in tres triangulos *ADB* et *BDC* et *ADC*. Sed ex dimidio *AB* in *DF* fit equale triangulo *ADB* et ex dimidio *AC* in *DG* equum est trigono *ADC* et quod ex dimidio *BC* in *DE* est equale triangulo *BDC*. Et quia omnes perpendiculares sunt equales linee *O*, ideo quod fit ex dimidio omnium laterum, et ipsum est *KLM*, in *O* est
60 tanquam area trianguli *ABC*. Cum ergo solidum *Y* habeat tria latera *N*, *O*, *KLM*, et ex *O* in *KLM* fiat area trianguli, tunc *N* ducta in eandem aream perficit solidum *Y*. Quare cum *O* sit equalis *N*, erit area trianguli inter lineam *KLM* et solidum *Y* [medium] proportionale; quare inter eandem et solidum *Z*. Si ergo, ut proponitur, *KLM* ducatur
65 in *Z*, radix producti erit area trianguli.

42 FH[2] *correxi ex* BF *IXa et* BH BF *Q*
43 *post* similes *add. XaQ* FDB, DGQ / DG *I* GD *XaQ*
43–44 [et...GC] *supplevi*
44 Itemque: tuncque *Q* / sunt *I om. XaQ*
46 ex *I om. XaQ*
46–48 tanquam...DF[2] *om. Q*
47 Proportio ergo *tr. Xa* / proportionibus: proportione *Xa*
48 DF[1]: FD *Xa*
51 aggregatur proportio *tr. XaQ*
54 equalem...dato *I* dato triangulo equalem *XaQ* / Divisus *I* quia divisus *XaQ*
55 enim *I om. XaQ* / ADB *I* scilicet ABD *XaQ* / et[1, 2] *I om. XaQ*
56 fit *I* sit *XaQ* / triangulo *I* trigono *XaQ* / ADB: ABD *Q*
57 est[1] *I om. XaQ* / est equale *I* equum est *XaQ*
58 BDC *I* DBC *XaQ* / sunt...O *I* linee O sunt equales *XaQ*
59 fit *I* fit P *XaQ* / ipsum *I* ipse *XaQ* / in O *tr. I ante* et
60 ergo *I* igitur *XaQ* / habeat tria *XaQ* habet *I*
61 N, O *XaQ* O, N *I*
61–63 tunc...trianguli *om. Q*
62 perficit *I* producit *Xa* / equalis *Xa* equale *I*
63 [medium] *supplevi*
64 proponitur *I* proponatur *XaQ*
65 trianguli: trianguli explicit *Xa*

ACP are right angles and equal, points A, D, C, and P will lie on the same semicircle. Therefore, $\angle PAC = \angle PDC$.* Therefore, $\angle PAC = \angle DBF$.** And so $\triangle PAC$ is similar to $\triangle DBF$, and $BF/FD = AC/CP$. But also $FD/FH = PC/CQ$ since $\triangle DFH$ is similar to $\triangle CPQ$.*** And so $BF/FH = AC/CQ$ [by inverting the last proportion and dividing it into the first proportion]. Therefore, $(BF + AC)/(FH + CQ) = BF/FH$. But $FH = QG$ because of the similarity of triangles and because $FD = DG$. [And so $FH + CQ = GC$,] and also $BF + AC = KLM$. Therefore, $KLM/GC = BF/FH$. But $BF/FH = (BF/DF) \cdot (DF/FH)$, and $AF/FD = FD/FH$ [by similar triangles]. Therefore, $BF/FH = (BF/DF) \cdot (AF/DF)$. But $BF/DF = S/N$, and $AF/DF = R/O$. And so [GC being equal to T], $KLM/T = (S/N) \cdot (R/O)$. Therefore, since $Y/Z = (KLM/T) \cdot (N/S) \cdot (O/R)$, and $T/KLM = (N/S) \cdot (O/R)$, therefore, solid Y = solid Z. Further, $KLM \cdot O$ = the given triangle. For $\triangle ABC$ has been divided into three triangles: ADB, BDC, and ADC. But $\triangle ADB = \frac{1}{2}(AB \cdot DF)$, and $\triangle ADC = \frac{1}{2}(AC \cdot DG)$, and $\triangle BDC = \frac{1}{2}(BC \cdot DE)$. And since all the perpendiculars [DF, DG, DE] are [each] equal to line O, hence $\triangle ABC = KLM \cdot O$. Hence, since solid $Y = N \cdot O \cdot KLM$ and $O \cdot KLM$ = area of triangle, then ($N \cdot$ area triangle) = solid Y. Hence, since $O = N$, KLM/area triangle = area triangle/Y. Therefore, KLM/area triangle = area triangle/Z. If, therefore, KLM is multiplied by Z, as is proposed, the square root of the product will be the area of the triangle.

* Since they are angles on a circumference subtended by the same chord (PC).

** Since $\angle PDC$ is identical with $\angle QDC$, and $\angle QDC = \angle DBE = \angle DBF$.

*** Since $\angle HDF = \angle DAF$, $\triangle DFH$ is similar to $\triangle DAH$, and $\triangle DAH$ is similar to $\triangle ADQ$. But $\triangle ADQ$ is similar to $\triangle CPQ$. Therefore, $\triangle DFH$ is similar to $\triangle CPQ$.

[Versio II]

105v / Cognitis tribus lateribus cuiusque trianguli eius aream invenire.
Propositum est cognitis tribus lateribus trianguli invenire quanta
106r sit superficies que ab eis ambitur. Ad huius itaque / exemplum ponam
ut sint latera trianguli *ABG* propostiti nota [Fig. 94], et per hoc inve-
5 niam eius aream. Inscribam autem intra propositum triangulum circulum *CFE* supra centrum *D* et a centro ducam lineas tres ad tria
puncta in quibus circulus contingit latera trianguli, que sint *DC*, *DF*,
DE, et ab eodem centro ducam lineas tres ad tres angulos eius, que
sunt *DA*, *DB*, *DG*. Ex hoc igitur declaratur nobis quod quilibet illo-
10 rum angulorum divisus est in duos angulos equales, propter hoc quod
latera trianguli *ADE* sunt equalia lateribus trianguli *ADC*, et latera
trianguli *BDC* lateribus trianguli *BDF*, et similiter latera trianguli
GDF lateribus trianguli *GDE*. Cum igitur tres anguli trianguli *ABG*
sint equales duobus rectis, tres medietates eorum, que sunt anguli *DAE*,
15 *DBF*, et *DGE*, erunt equales uni recto. Set angulus *DGE* cum angulo
EDG valet unum rectum, quoniam tertius, scilicet *GED*, est rectus.
Ergo duo anguli *DGE* et *EDG* sunt equales tribus angulis qui sunt
DAE, *DBF*, et *DGE*. Ergo remoto angulo *DGE* remanebit angulus
EDG equalis duobus angulis *DAE* et *DBF*. Ergo ipse erit maior
20 angulo *DAE*. Faciam itaque ipsum sibi equalem et protraham lineam
HD que cum *ED* faciat angulum equale angulo *DAE*, quam etiam
producam in continuum usque ad *P*, et ponam notam *Q* ubi interse-
cabit lineam *AB*. Et quoniam medietas trium laterum cuiusque trian-

14 anguli *corr. ex* angulus
16 valet *corr. ex* valent
22 producam *corr. ex* productam

[Version II]

With three sides of any triangle known, to find its area.

It has been proposed—with three sides of a triangle known, to find how great is the area enclosed by them. And so for the exemplification of this, I posit that the sides of the proposed triangle ABG are known [see Fig. 94], and by means of this I shall find its area. Moreover, I shall inscribe within the proposed triangle a circle CFE about center D. And from the center I shall draw three lines to the three points where the circle touches the sides of the triangle; these lines we let be DC, DF, and DE. And from the same center I shall draw three lines to its three angles, the lines being

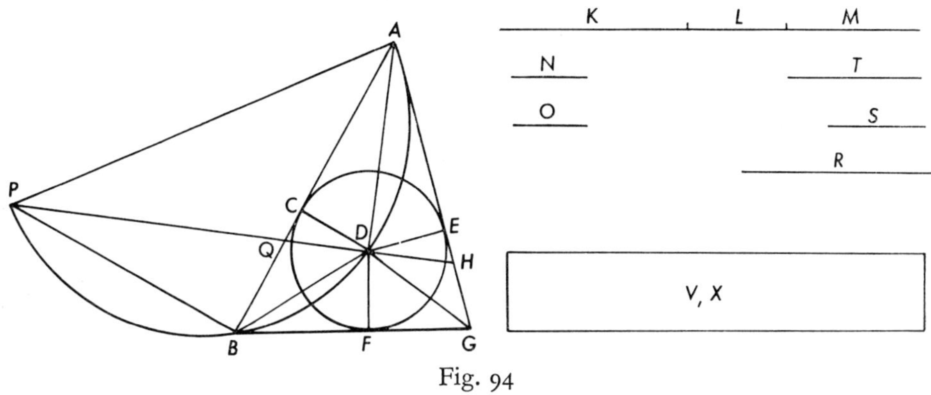

Fig. 94

DA, DB, DG. Therefore, from this it is evident to us that each of those angles is bisected, since the sides of $\triangle ADC$ are equal to the sides of $\triangle ADE$, and the sides of $\triangle BDC$ to the sides of $\triangle BDF$, and similarly the sides of $\triangle GDF$ to the sides of $\triangle GDE$. Since, therefore, the three angles of $\triangle ABG$ are equal to two right angles, then their three halves, which are $\angle DAE$, $\angle DBF$, $\angle DGE$, will be equal to one right angle. But $\angle DGE + \angle EDG = 1$ right angle, since the third angle, i.e., GED, is a right angle. Hence, $\angle DGE + \angle EDG = \angle DAE + \angle DBF + \angle DGE$. Therefore, with $\angle DGE$ subtracted [from each side], $\angle EDG = \angle DAE + \angle DBF$. Therefore, $\angle EDG > \angle DAE$. And so I shall make an angle equal to it [i.e., make an angle equal to DAE], and [to do so] I shall draw line HD, which with ED forms an angle equal to $\angle DAE$. I shall also extend this line [HD] continuously to P. And I shall mark point Q where it will intersect line AB. And since half of the three sides of any triangle is greater than any one of its sides, the whole

guli est maior quolibet latere ipsius, erit tota linea *AG* cum *BF* maior
25 quolibet latere propositi trianguli, nam *AE* est medietas *AE* et *AC*, et *EG* est medietas *EG* et *GF*, et *FB* est medietas *BF* et *BC*. Ergo
106v *AG* et *FB* simul sunt medietates omnium laterum / predicti trianguli. Sit itaque linea *KLM*, quod agregatur ex *AG* et *FB*, et sumatur differentia eius ad quodlibet trium laterum. Et sit eius differentia ad latus
30 quidem *AG T* et ad latus *AB* sit *S* et ad latus *BG* sit *R*. Manifestum est igitur quod *T* erit equalis *CB*, eo quod medietas trium laterum excedit *AG* in *CB*, et *S* erit equalis *GE* quoniam medietas trium laterum excedit *AB* in *GE*, et *R* erit equalis *AE* quoniam medietas trium laterum excedit *BG* in *AE*. Sit itaque solidum quod fit ex *T*, *S*, et *R*
35 solidum *Z*. Sumantur etiam due linee equales perpendiculari *DE*, que sunt *N* et *O*, et fiat ex eis et linea *KLM* solidum *Y*. Dico igitur tunc quod solidum *Y* est equale solido *Z*, quod sic ostendam.

Angulus *HDE* est equalis angulo *DAE*. Set angulus *DAE* cum angulo *EDA* valet unum rectum quoniam tertius est rectus. Ergo
40 angulus *HDE* cum angulo *EDA* valet unum rectum. Ergo totus angulus *ADH* est rectus. Ergo etiam reliquus *ADP* erit rectus, propter hoc quod linea *HDP* est linea una. Ergo ipsi sunt ambo equales. Si igitur ab eis demantur duo anguli *ADC* et *ADE*, qui propter hoc quod reliqui duo sunt equales reliquis duobus remanebit angulus
45 *EDH* equalis angulo *CDQ*. Et iterum quoniam anguli *DAC* et *DBC* et *DGE* sunt equales uni recto, eo quod ipsi sunt medietas trium angulorum trianguli, et angulus *QDC* est equalis angulo *DAC*, erunt tres anguli *QDC*, *DBC*, et *DGE* equales uni recto. Set etiam anguli *CDB* et *CBD* equantur uni recto propter quod alter est rectus. Ergo
50 tres anguli *QDC*, *CBD*, et *DGE* equantur duobus qui sunt *CDB* et *CBD*. Quare demptis duobus qui sunt *QDC* et *CBD* remanebit *QDB* equalis *DGE*. Producam itaque a puncto *B* perpendicularem a linea
107r *AB* / et protraham ipsam quousque concurrat cum *HQP* et ponam *P* in loco concursus. Concurrent autem quoniam angulus *BQD* est maior
55 recto, eo quod ipse est extrinsecus ad *QCD*, qui est rectus. Ergo reliquus *BQP* est minor recto et *PBQ* est rectus. Ergo isti duo simul sunt minores duobus rectis. Igitur in eandem partem protracte concurrent. Deinde a puncto *P* ducam lineam ad punctum *A* super eum. Igitur describam semicirculum supra quem sunt *A*, *D*, *B*, *P*; et quo-

26 BC *corr. ex* FC
34 *corr. ex* scilicet
48 anguli[2] *corr. ex* angulus
51 QDB *corr. ex* QBD
52 Producam *corr. ex* productam

line AG plus BF is greater than any one side of the proposed triangle; for $AE = \frac{1}{2}(AE + AC)$, and $EG = \frac{1}{2}(EG + GF)$, and $FB = \frac{1}{2}(BF + BC)$. Hence, $AG + FB =$ the semiperimeter of the aforesaid triangle. And so let there be a line KLM, which is made up from AG and FB, and let the excess of it over each of the three sides be taken; and let $KLM - AG = T$, $KLM - AB = S$, and $KLM - BG = R$. It is evident, therefore, that T will be equal to CB because the semiperimeter exceeds AG by CB, and that S will be equal to GE since the semiperimeter exceeds AB by GE, and that R will equal AE since the semiperimeter exceeds BG by AE. And so let $T \cdot S \cdot R =$ solid Z. Also, let two lines, [each] equal to perpendicular DE, be taken, namely, lines N and O, and let $N \cdot O \cdot KLM =$ solid Y. I say, therefore, that solid $Y =$ solid Z, which I shall demonstrate as follows.

$\angle HDE = \angle DAE$. But $\angle DAE + \angle EDA =$ 1 right angle, since the third angle is a right angle. Therefore, $\angle HDE + \angle EDA =$ 1 right angle. Therefore, the whole angle ADH is a right angle. Therefore, the adjacent angle ADP will be a right angle since line HDP is a single straight line. Therefore, these angles are equal to each other. If, therefore, the two angles ADC and ADE are subtracted from them [respectively], the angles ADC and ADE being equal because the other two angles [of each triangle] are equal [respectively], thus $\angle EDH = \angle CDQ$. And again, since $\angle DAC + \angle DBC + \angle DGE =$ 1 right angle, because they are [together] half of the three angles of the triangle and $\angle QDC = \angle DAC$, [hence] $\angle QDC + \angle DBC + \angle DGE =$ 1 right angle. But also $\angle CDB + \angle CBD =$ 1 right angle, because the third angle is a right angle. Therefore, $\angle QDC + \angle CBD + \angle DGE = \angle CDB + \angle CBD$. Therefore, with angles QDC and CBD subtracted, $\angle QDB = \angle DGE$. And so I shall draw from point B a perpendicular to line AB, extending it until it meets HQP, and I shall place P in the point of meeting. Moreover, the lines will meet since $\angle BQD$ is greater than a right angle. This is so because $\angle BQD$ is extrinsic to $\angle QCD$, which is a right angle. Therefore the other angle, BQP, is less than a right angle, while $\angle PBQ$ is a right angle. Therefore, these two angles together are less than two right angles. Consequently when the two lines are extended in the same direction, they will meet. Then I shall draw a line from point P to point A. Therefore, I shall describe a semicircle, on which [points] A, D, B, and P lie. And since it [i.e., line AP] is opposite right $\angle ABP$

60 niam ipsa opponitur angulo recto qui est *ABP* et angulo recto qui est *ADP*, erit uterque illorum in circumferentia semicirculi descripta supra eam. Ergo puncta *A*, *D*, *B*, et *P* sunt in circumferentia eiusdem circuli. Propter hoc igitur erit angulus *PDB* equalis angulo *PAB*, eo quod ambo sunt super unam portionem circuli que est *PB*. Qua-
65 propter angulus *PAB* est equalis angulo *DGE*, erunt igitur duo trianguli *APB* et *DGE* equianguli et eorum latera proportionalia. Set et similiter erunt duo trianguli *PQB* et *DQC*, quoniam angulus *DQC* est equalis angulo *PQB* quia sunt contra se positi et reliquus utriusque est rectus; et quia triangulus *QDC* fuit equiangulus et equilaterus cum
70 triangulo *HDE*, erit *HDE* similis *QPB* et eorum latera proportionalia. Erit itaque proportio *AB* ad *BP* sicut proportio *GE* ad *ED*. Set etiam proportio *PB* ad *BQ* est sicut proportio *DE* ad *EH*. Ergo a primo proportio *AB* ad *BQ* est sicut proportio *GE* ad *HE*. Ergo coniunctim proportio *AB* et *GE* simul ad *BQ* et *HE* simul est sicut
75 proportio *GE* ad *HE*. Set *HE* est equalis *QC*, ut ostensum est prius. Ergo proportio *AB* et *GE* simul ad *BQ* et *CQ* simul ita quod *CQ* accipiatur loco *HE* est sicut proportio *GE* ad *HE*. Set *AB* et *GE*
107v simul est equalis *KLM*, eo quod ipsa est medietas trium laterum; / et *BC* est equalis *T*, quoniam *T* fuit differentia medietatis trium laterum
80 ad latus *AG*. Ergo proportio *KLM* ad *T* est sicut proportio *EG* ad *HE*. Proportio autem *EG* ad *HE* est agregata ex duplici proportione, scilicet *GE* ad *ED* et *DE* ad *EH*. Set proportio *DE* ad *EH* est sicut proportio *AE* ad *ED*, propter hoc quod ab angulo *ADH* recto cadit *DE* perpendiculariter super latus quod opponitur recto angulo. Ergo
85 facit duos triangulos similes sibi invicem. Et totalis ergo proportio *GE* ad *EH* est agregata ex proportione *GE* ad *ED* et *AE* ad *ED*. Sed iam fuit proportio *KLM* ad *T* sicut *GE* ad *EH*. Ergo proportio *KLM* ad *T* est agregata ex proportione *GE* ad *ED* et *AE* ad *ED*. Set *GE* est equalis *S* et *ED* est equalis *N*, et iterum *AE* est equalis *R*, et
90 *ED* est equalis *O*. Ergo proportio *KLM* ad *T* est agregata ex proportione *S* ad *N* et *R* ad *O*. Ergo everse proportio *T* ad *KLM* est agregata ex proportione *N* ad *S* et *O* ad *R*. Set rursus proportio solidi *Y* ad solidum *Z* est agregata ex proportione laterum suorum, sicut patet per undecimum Euclidis. Ex proportione itaque *KLM* ad *T* et *N* ad

63 PDB *corr. ex* PDP / PAB *corr. ex* PAP
91 *mg. c.2 107v add. Ya* Esto memor ponere ut sit unum latus 21 et aliud 20 et tertium 19 et tunc sit area trianguli fere 172 partes quadrate quarum longitudo et latidudo erit secundum magnitudinis unius partium positarum.
92 Y *corr. ex* X

and opposite right angle ADP, hence each of these angles will lie on the semicircle described on line AP. Therefore, points A, D, B, and P are in the circumference of the same circle. Accordingly, $\angle PDB = \angle PAB$, since both [are on the same circumference and] are subtended by the same segment of the circle, namely, PB. Since $\angle PAB = \angle DGE$, therefore triangles APB and DGE are equiangular, and their sides are proportional. But this is true in the same way of triangles PQB and DQC, since $\angle DQC = \angle PQB$, these angles being vertical angles, and another angle of each triangle is a right angle. And since $\triangle QDC$ is similar to $\triangle HDE$, then $\triangle HDE$ is similar to $\triangle QPB$ and their sides are proportional. And so $AB/BP = GE/ED$. But also $PB/BQ = DE/EH$. Therefore, [inverting the last proportion and dividing it into the first] $AB/BQ = GE/HE$. Therefore, $(AB + GE)/(BQ + HE) = GE/HE$. But $HE = QC$, as was shown before. Therefore, $(AB + GE)/(BQ + CQ) = GE/HE$. But $AB + GE = KLM$, because KLM is the semiperimeter; and $BC = T$, since T was the excess of the semiperimeter over side AG. Hence, $KLM/T = EG/HE$. Further, $EG/HE = (GE/ED) \cdot (DE/EH)$. But $DE/EH = AE/ED$, since DE descends perpendicularly from right angle ADH to the side which is opposite the right angle. Therefore, it produces two mutually similar triangles. Therefore, $GE/EH = (GE/ED) \cdot (AE/ED)$. But $KLM/T = GE/EH$. Therefore, $KLM/T = (GE/ED) \cdot (AE/ED)$. But $GE = S$, and $ED = N$, and further $AE = R$ and $ED = O$. Hence $KLM/T = (S/N) \cdot (R/O)$. Hence, by inversion, $T/KLM = (N/S) \cdot (O/R)$. But, further, the ratio of solid Y to solid Z is compounded of the ratios of their various sides, as is evident [by Book] XI of Euclid. And so, $Y/Z = (KLM/T) \cdot (N/S) \cdot (O/R)$. But we saw that $(N/S) \cdot (O/R) = T/KLM$. Therefore, $Y/Z = (KLM/T) \cdot (T/KLM)$. Since this ratio is

95 *S* et *O* ad *R* fit proportio solidi *Y* ad solidum *Z*. Set iam fuit proportio *N* ad *S* ducta in proportionem *O* ad *R* sicut proportio *T* ad *KLM*. Ergo proportio solidi *Y* ad solidum *Z* fit ex proportione *KLM* ad *T* et *T* [ad] *KLM*. Cum itaque fiat ex proportione primi ad secundum et secundi ad primum, ipsa erit proportio equalitatis. Ergo solidum *Y* 100 est equale solido *Z*. Quare cum sit solidum *Z* notum quia omnia eius 108r latera sunt / nota, erit propter hoc solidum *Y* notum.

Set ex ductu medietatis laterum in perpendicularem fit area trianguli, quoniam ex ductu *AE* in *ED* fit superficies quedam dupla ad triangulum *AED* propter quod ipse est orthogonius. Ergo ipsa superficies 105 est equalis toti superficiei *ACDE*. Et similter ex ductu *GE* in *DE* fit quedam superficies equalis *GEDF*. Et ex ductu *BF* in *FD*, que est equalis *CD*, fit superficies equalis *BFDC*. Ergo ex ductu medietatis trium laterum in perpendicularem fit area trianguli. Ergo ex ductu *KLM* in *N* fit area trianguli. Set ex ductu *O* in hoc productum fit soli- 110 dum *Y*. Ergo ex ductu perpendicularis in aream trianguli fit solidum *Y*; ergo et solidum *Z*. Ponam itaque ut area trianguli sit *V* et accipiam aliam sibi equalem que sit *X* et accipiam [*N*] in loco *O* propter hoc quod ipse sunt equales. Dico ergo quod ex ductu *N* in *KLM* fit *V*, quod est ipsa area trianguli, et ex ductu *N*, quod est *O*, in *X*, quod 115 etiam est area trianguli, fit solidum *Y*. Hic una et eadem quantitas, scilicet *N*, multiplicat duas, videlicet *KLM* et *X*. Ergo eadem est proportio multiplicatorum et productorum. Ergo que est proportio *KLM* ad *X* eadem *V* ad *Y*. Set que est *KLM* ad *X* eadem est *KLM* ad *V*, eo quod *X* et *V* sunt equalia. Ergo que est proportio *KLM* ad 120 *V* eadem est *V* ad *Y*. Ergo *V* est medio loco proportionalis inter lineam *KLM* et solidum *Y*. Set solidum *Y* est equale solido *Z*. Ergo area trianguli est medio loco proportionalis inter *KLM* et *Z*, quod est, inter medietatem trium suorum laterum et solidum quod fit ex tribus differentiis medietatis trium laterum ad tria latera. Ergo quod 125 fit ex ductu medietatis laterum [in] illud solidum valet illud quod fit ex ductu aree in se. Si igitur coacervemus tria latera trianguli et agregati sumpserimus medietatem et illius ad tria eius latera sumpserimus differentias et multiplicaverimus primam in secundam et tertiam in pro- 108v ductum fiet solidum / illud inter quod et medietatem laterum est area 130 trianguli proportionalis, et si in ipsum multiplicaverimus medietatem illorum trium laterum quod provenerit erit equale ei quod fit ex ductu

108–10 Ergo.... trianguli *mg.* *Ya*

compounded of a ratio of the first to the second and a ratio of the second to the first, it will be a ratio of equality. Therefore, solid Y = solid Z. Therefore, since solid Z is known because all of its sides are known, so accordingly will solid Z be known.

But the product of the semiperimeter and a perpendicular equals the area of the triangle, since $AE \cdot ED = 2 \triangle AED$, the triangle being a right triangle. Hence, $2 \triangle AED = ACDE$. Similarly, $GE \cdot DE = GEDF$, and $BF \cdot FD = BFDC$, FD being equal to CD. Therefore, the product of the semiperimeter and a perpendicular equals the area of the triangle. Hence, $KLM \cdot N$ = area of triangle. But $KLM \cdot N \cdot O$ = solid Y. Hence, the product of the area of the triangle and a perpendicular equals solid Y, and therefore it equals solid Z. And so I posit that the area of the triangle is V, and I assume another area equal to it, namely, X. And I assume N in place of O since they are equal. I say, hence, that $N \cdot KLM = V$, V being the area of the triangle, and that $N \cdot X$ = solid Y, N being equal to O, and X to the area of the triangle. Hence this same quantity N is multiplied by two quantities, namely, KLM and X. Therefore, since the ratios of the multipliers and their products are the same, $KLM/X = V/Y$. But $KLM/X = KLM/V$, X and V being equal. Hence, $KLM/V = V/Y$. Therefore, V is the mean proportional between KLM and solid Y. But solid Y = solid Z. Therefore, the area of the triangle is the mean proportional between KLM and Z, i.e., between the semiperimeter and the product of the excesses of the semiperimeter over [each of the] three sides. Therefore, the multiplication of the semiperimeter by that product is equal to the square of the area. Therefore, if we add the three sides of the triangle together, then take half of that sum, then take the excesses of the semiperimeter over [each of] the three sides, multiplying the first excess by the second, and then their product by the third, a solid results such that the area of the triangle is the mean proportional between this solid and the semiperimeter. And if we multiply the semiperimeter by that solid, that which results will be equal to the square of the area. Therefore, if we

aree in se. Si igitur illius sumpserimus radicem, ipsa area trianguli et hoc est quod voluimus demo[n]strare.

Est etiam alter modus inveniendi per hanc eandem demo[n]stratio-
135 nem in cognitione aree trianguli suppositis notis lateribus. Demo[n]-stratum enim est quod solidum quod fit ex tribus differentiis que sunt medietatis laterum ad tria latera valet solidum quod fit ex medietate trium laterum in perpendicularem vel in semidiametrum circuli in-scripti et ex ductu eiusdem semidiametri in productum. Item demo[n]-
140 stratum est quod illud quod fit ex semidiametro in medietatem trium laterum est equale aree trianguli. Si ergo semidiameter circuli esset nobis nota cum medietas trium laterum sit nota quia omnia latera nota, multiplicaremus semidiametrum in illam medietatem laterum et fieret productum area trianguli. Set semidiameter sic fiet nota. Solidum
145 quod fit ex tribus differentiis in se divide per medietatem laterum et exibit superficies que fit ex semidiametro in se. Quare ergo illius radicem [sume], et ipsa erit semidiameter, quam multiplica in medie-tatem laterum et habebis aream. Et iste modus est omnino idem cum priori.

150 Hec autem regula de triangulo fuit scripta in arabico: Si tria latera trianguli coacerventur, medietatisque compositi ad singula latera dif-ferentie sumantur, primaque in secundam ducatur et in productum tertia, itemque quod inde provenerit in predictam medietatem, illius ultimo producti radix erit area trianguli.

140 medietatem *corr. ex* medietate
147 [sume] *supplevi*
152 *post* productum *habet Ya et delevi* in
154 producti *corr. ex* producta

take the root of that, it will be the area of the triangle, and this is what we wished to demonstrate.

There is also another method, by means of this same demonstration, of finding the area of a triangle when the sides are known. For it has been demonstrated that the solid arising from the mutual multiplication of the three excesses of the semiperimeter over [each of] the three sides is equal to the solid arising from the multiplication of the semiperimeter by a perpendicular, i.e., by the radius of an inscribed circle, and the multiplication of that product by the radius. Also it has been demonstrated that the product of the radius and the semiperimeter is equal to the area of the triangle. If, therefore, the radius of the circle were known to us since the semiperimeter is [also] known by the fact that all of the sides are known, we would multiply the radius by that semiperimeter and the resulting product would be the area of the triangle*. But the radius may be known in the following way. Divide the product arising from the mutual multiplication of the three excesses by the semiperimeter, and the quotient is equal to the square of the radius. Therefore, take the root of this and it will be the radius, which you multiply by the semiperimeter and you will have the area. And this method is completely the same as the preceding one.

This rule for a triangle was written in Arabic: If the three sides of a triangle are added together, and the excesses of one half of this sum over each of the individual sides are taken, and the first excess is multiplied by the second, and the third is multiplied by the product, and the result of this is multiplied by the aforesaid half [perimeter], the root of the final product will be the area of the triangle.

* We can illustrate the procedure of this second method by utilizing the quantities employed in the first part of the proof: We know that $NO \cdot KLM$ = area of triangle. Knowing the sides, we know KLM. But we can also find NO as follows. $Y = Z$, or $R \cdot S \cdot T = N \cdot N \cdot KLM$. Thus $N = \sqrt{(R \cdot S \cdot T)/KLM}$.

Appendix V

A Version of Philo's Solution of the Problem of Two Mean Proportionals

In Propositions XVI and XVII of their *Verba filiorum*, the Banū Mūsā presented two of the antique solutions of the problem of finding two continually proportional means between two given quantities. These were the solutions ascribed to Archytas and Plato by Eutocius (see the Commentary to Propositions XVI and XVII in Chapter Four). Both of these solutions were taken over by Leonardo Pisano in his famous *Practica geometrie*,[1] and the first of them appeared in the *Liber de triangulis* of Jordanus de Nemore.[2] In addition, each of these thirteenth-century mathematicians included another solution of the problem not present in the *Verba filiorum*. This additional solution is the one that Eutocius in his commentary to Archimedes' *On the Sphere and the Cylinder* assigns to Philo of Byzantium.[3] It is essentially the same as the solution given by Hero in his *Mechanica* and his *Belopoeica* (and repeated by Eutocius and Pappus).[4] It also bears close resemblance to the solution attributed to Apollonius by Eutocius.[5] Since the only known source of the solution by Philo is the Eutocius commentary, it is probable that the source of the Latin text we are here considering was ultimately the commentary of Eutocius. Whether the commentary was

[1] *Scritti di Leonardo Pisano*, ed. of B. Boncompagni, vol. *2* (Rome, 1862), pp. 153–55.

[2] Jordanus de Nemore, *Geometria vel de triangulis libri iv*, ed. of M. Curtze, *Mitteilungen des Coppernicus-Verein für Wissenschaft und Kunst zu Thorn*, *6*. Heft (Thorn, 1887), pp. 40–41.

[3] Archimedes, *Opera omnia*, ed. of J. L. Heiberg, vol. *3* (Leipzig, 1915), pp. 60–62. Cf. note 7 below.

[4] *Ibid.*, pp. 58–60 (for the Eutocius reproduction of Hero). For the citations to the *Mechanica* and the *Belopoeica* of Hero and to Pappus, see I. Thomas, *Greek Mathematical Works*, vol. *1* (London, 1951), pp. 266–71, who gives the Greek text (from Pappus) and an English translation.

[5] *Op. cit.* in note 3 above, pp. 64–66. Incidentally in addition to this solution associated with the name of Apollonius there is another proof based on the *Conics* included by Maḥmūd ibn ᶜOmar abū

familiar in Arabic or in Greek to the author (or translator) of this piece is not certainly known. While I am not sure of the existence of an Arabic text of the whole of Eutocius' commentary, the section on the problem of the mean proportionals is extant,[6] and it appears to have been the source of the "Platonic" solution of the problem given by the Banū Mūsā (see the Commentary to Chapter Four, Proposition XVII, above). One possible indication that our Latin piece stems from the Arabic rather than from the Greek is that the letters employed on the diagram and in the text to mark magnitudes differ markedly from the Greek,[7] thus suggesting the Greek text had gone through the Arabic before being turned into Latin.

It will be evident to the reader that the texts of the solution presented by Jordanus and Leonardo are virtually the same, although the Jordanus text in the four copies I have read omitted one crucial phrase (added by

'l-Fath al-Isfahānī in his *Talkhīs* of the *Conics* in 1119 (cf. the Latin translation of Abraham Ecchellensis, Florence, 1661: Lemma VII, pp. 30–31), and by al-Ṭūsī in his version of the *On the Sphere and the Cylinder*, Bk. II, Proposition 1 (Arabic edition of Hyderabad, 1940, pp. 80–82). So far as I know, al-Ṭūsī's version of this proof has not been hitherto noted. He remarks that most of the methods of finding the proportional means involve the moving of instruments or devices (al-ālāt), while this is a theoretical method "based on some of the elements of Apollonius mentioned in the Book of Conics." Actually, Pappus mentions that Apollonius composed a proof based on conic sections (*Collectio*, ed. of Hultsch, vol. III, 21, p. 56), and apparently the Arabic mathematicians had recourse to some version of it, as the existence of the two above-mentioned versions of the proof based on the use of an hyperbola seems to indicate. Note that Paul Ver Eecke has given a French rendering of Abraham Ecchellensis' Latin text of the proof (*Les Coniques d'Apollonius de Perge*, Bruges, 1923, p. 425n). Cf. T. L. Heath, *Apollonius of Perga...*, pp. cxxv-cxxvii.

[6] See the specimen from and description of the Eutocius commentary in Arabic manuscript, Escurial 960, 22v–38v, given by F. J. Carmody *The Astronomical Works of Thabit B. Qurra*, pp. 219, 220. I have not had a chance to examine this manuscript but I would judge from the specimen given by Carmody that it starts off immediately with the section on the solutions of the problem of the proportional means. Cf. MS Instanbul, Fatih Mosque Library, Ar. 3414, 60v–66v, 73r–v, which appears to be a similar extract. See M. Krause "Stambuler Handschriften islamischer Mathematiker," *Quellen und Studien zur Geschichte der Mathematik, Astronomie und Physik*. Abt. B, vol. *3* (1936), p. 441.

[7] The Greek text was translated in 1269 by William Moerbeke, and we can give his translation of Philo's solution for the sake of comparison with the text being here treated (MS Vat. Ottob. lat. 1850, 36v): "Ut Filion bisanteus. Sint date due recte que *AB*, *BG*, quarum oportet duas medias proportionales invenire. Iaceant ita ut rectum angulum contineant qui apud *B*, et copulata *AG* describatur circa ipsam semicirculus qui *ABEG*, et ducantur ad rectos angulos ipsi quidem *BA* que *AD*, ipsi autem *BG* que *GZ*, et secusponatur regula que moveatur apud *B* secans lineas *AD*, *GZ* et sit mota circa *B* quidem donec utique que a *B* ad *D* equalis fiat ei que ab *E* ad *Z*, hoc est, intermedie periferie circuli et ipsius *GZ*. Intelligatur igitur canonium habens positionem qualem habet que *DBEZ*, ipsa *DB* equali existente, ut dic-

me to lines 11–12). It may be that this phrase was in the original version of Jordanus but somewhere along the line was dropped out of the text (since none of my copies of the Jordanus text appears to be earlier than the fourteenth century). At any rate, I would hesitate to say that Jordanus copied the text from Leonardo, omitting the necessary phrase. But if he did take our present piece from Leonardo, he made certain alterations, such as replacing Leonardo's peculiar use of *ex* in the statements of ratios by *ad* (see the various uses of *ad* in lines 13, 14, and 15 in the Jordanus text and compare them with the corresponding usage of *ex* in the Leonardo text). On the whole, I think it likely that both authors copied from some original piece that circulated separately. It is perhaps such a piece that is present in two manuscripts that are in great part devoted to translations made by Gerard of Cremona. The first of these, manuscript *Zm* (see Sigla), includes the proof among several propositions added to the end of the text of the *Verba filiorum* of the Banū Mūsā. It occupies a similar position in manuscript *S*. In neither case is an author mentioned; nor is it given as a part of a treatise but rather as an isolated proposition. It is thus quite possible that the text of the proposition, which agrees very closely with Leonardo's version of the proof, was translated from the Arabic by Gerard of Cremona and that it was the source of both Jordanus' and Leonardo's versions, with Leonardo copying it more faithfully. Of course, it must be granted that both manuscripts are much later than the time of the compo-

tum est, ipsi *EZ*. Dico quod que *AD*, *GZ* sunt medie proportionales ipsarum *AB*, *BG*. Intelligantur enim educte que *DA*, *ZG* et concidentes apud *T*; manifestum

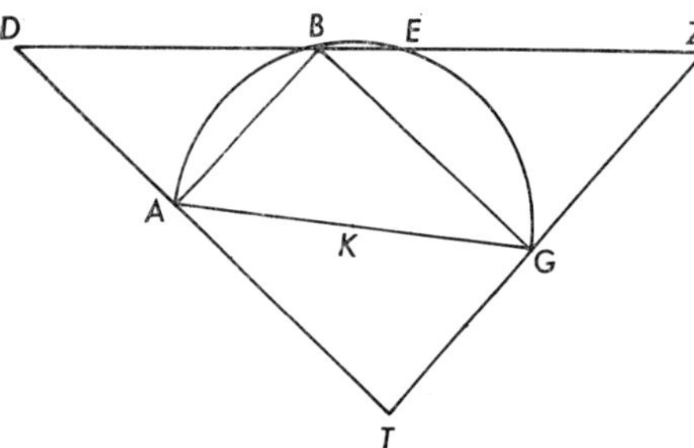

itaque quod equedistantibus existentibus ipsis *BA*, *ZT* angulus qui apud *T* est rectus et circulus *AEG* completus transibit et per *T*. Quoniam igitur equalis est que *DB* ipsi *EZ*, ergo et quod sub *EDB* est equale ei quod sub *BZE*. Sed quod quidem sub *EDB* est equale ei quod sub *TDA*; utrumque enim est equale ei quod contingente a *D*. Quod autem sub *BZE* est equale ei quod sub *TZG*; utrumque enim similiter est equale ei quod a contingente a *Z*. Quare et quod sub *TDA* est equale ei quod sub *TZG*, et propter hoc est ut que *DT* ad *TZ* ita que *GZ* ad *DA*. Sed ut que *TD* ad *TZ* ita que *BG* ad *GZ* et que *DA* ad *AB*; trigoni enim *DTZ* ipsi quidem *DT* equedistanter protracta est que *BG*, ipsi autem *TZ* equedistanter que *BA*. Est ergo ut que *BG* ad *GZ* que *GZ* ad *DA* et que *DA* ad *AB*, quod quidem propositum erat demonstrate." Notice that sometimes Moerbeke preserves the syncopated Greek forms of writing quantities with a common letter (e.g., *EDB* for *ED*, *DB*; *BZE* for *BZ*, *ZE*, etc.).

sition of the *Practica geometrie* of Leonardo and thus they might have used the *Practica*. I think this less likely since, as I have noted, Leonardo, and Jordanus as well, did extract whole propositions from the *Verba filiorum* and both authors might easily have found this proposition in the manuscripts of the *Verba filiorum* which they were using.

My extract from the Jordanus text is based on manuscript *Oa*, manuscript *E*, Curtze's edition *I* (*Ed*), and manuscript *Sl*. In Curtze's edition the extract is found in Proposition 22 of the fourth book. In manuscript *E*, where the propositions are collected in three books, our piece comes from the thirty-fourth proposition of the third book. But in manuscript *Sl*, where all the propositions are numbered successively, the proposition is numbered 67. The propositions are not numbered in *Oa*. The extract from Leonardo is taken from Boncompagni's edition. I have added variant readings from the aforementioned Gerardus manuscripts, *Zm* and *S*.

Since both extracts are virtually the same, I have translated only the first of them. The drawing is identical for both extracts and is as given in *E* and *I*(*Ed*) of the Jordanus text as well as in the Boncompagni edition of the Leonardo text (except for the letter *B* omitted from the Leonardo edition). The drawing is badly made in *Sl* and *O*.

Sigla of Manuscripts

Jordanus Proof

I(*Ed*) = Dresden, Säch. Landesbibliothek Db. 86, 60r, early 14c. I have used Curtze's edition cited in note 2, above, pp. 41–42.
Oa = Paris, BN. lat. 7434, 86r–v, 14c.
Sl = London, Brit. Museum, Sloane 285, 91r, 14c.
E = London, Brit. Museum, Harleian 625, 129v, 14c.
For other manuscripts of the *De triangulis*, see the Sigla of Appendix I, Section 1.

Leonardo Proof

Ed = B. Boncompagni, *Scritti di Leonardo Pisano*, vol. 2 (Rome, 1862), Nac. 10010, p. 154.
Zm = Madrid, Bibl. Nac. 10010, 83v–84r, 14c.
S = Oxford, Bodl. Libr., Digby 168, 122r (old page 121r), 14c.

[Probatio ex libro Jordani de triangulis]

86r / Aliter ad idem. Sint due linee *AB*, *BG* [Fig. 95], et angulus earum sit rectus. Deinde continuabo *AG* et faciam super lineam *AG* trianguli *ABG* circulum *ABHG*, fiatque perpendicularis *AD* a puncto *A* super lineam *AB* et ex puncto *G* fiat perpendicularis *DG* super *BG*. Et
5 protraham utramque, hanc in *Z* et illam in *E*. Deinde faciam transire regulam que moveatur super punctum *B* et non separetur ab eo et abscindat duas lineas *DZ*, *DE* et non cesset moveri donec illa pars regule que est inter *Z* et *B* sit equalis illi parti regule que est inter arcum *BG* et *E*. Sit ergo illud in hoc situ, ut sit regula *EHBZ*, et sit *EH*
10 equalis *BZ*. Ergo quod fit ex *ZH* in *ZB* est equum ei quod fit ex *BE* in *HE*. Verum *EB* in *HE* equalis est *DE* in *GE* [et *HZ* in *ZB* est equalis *DZ* in *ZA*]. Ergo *DZ* est in *ZA* sicud *DE* in *EG*. Ergo proporcio *DE* ad *DZ* est sicud *AZ* ad *GE*. Sed proporcio *DE* ad *DZ* est sicud *BA* ad *AZ*. Ergo proporcio *BA* ad *AZ* est sicud pro-
15 porcio *AZ* ad *GE* et est etiam sicud proporcio *GE* ad *GB*. Stat ergo propositum.

Probatio Jordani

1 et...earum: quarum angulus B *E*
2, 5 Deinde: demum *I(Ed)*
2 AG *E om. SlI(Ed)Oa*
3 ABHG *corr. ex* ABGH *in MSS*
5 hanc *om. Oa* / et *om. Oa*
6 separatur *E*
7 abscindat *OaI(Ed)* abscindet *E* abcindit *Sl*
8 illi *E* ei *SlI(Ed)Oa* / est *om. I(Ed)*
9 ergo: igitur *E hic et ubique* / hoc: H *Ed* / situ *om. Oa et habet lacunam, sed scr.* situ *in marg.*
10 est equum *tr. E*
11 Verum...est[1] *I(Ed)* (*cf. Leon.*) que est equalis ei quod fit *E* et que est *Sl* / equalis: equum *Oa*
11–12 [et...ZA] *supplevi* (*cf. Leon.*)
13 DE[1]: D *Sl* ED *E* / DE[2] *I(Ed)* ED *EOaSl*
14 AZ[1]: Z *SlOa*
15 AZ: AG *SlOa* / est etiam *tr. Oa* / etiam *om. E* / Stat: constat *E*
16 propositum: propositio *Sl*

[The Solution from Jordanus' *Book on Triangles*]

The same thing [can be found] in another way. Let there be two lines AB, BG [Fig. 95], joined at a right angle. Then I shall draw AG and construct circle $ABHG$* on line AG of $\triangle\ ABG$. Let perpendicular AD be constructed on line AB from point A and perpendicular DG on line

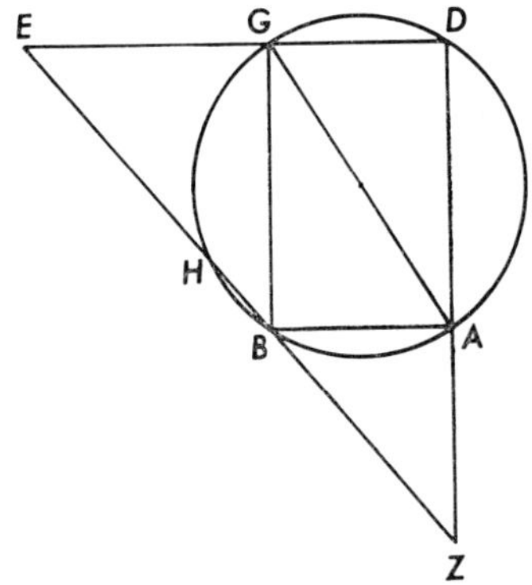

Fig. 95

BG from point G. I shall extend each of them [i.e., DG and AD] to E and Z respectively. Then I shall lay out a rule to be moved about point B without ever being separated from it, to intersect the two lines DZ and DE, and to continue to be moved until that part of the rule between Z and B is equal to that part of the rule between the arc BG and E. Therefore, let the rule be in this position so that it is $EHBZ$, EH being equal to BZ. Therefore, $ZH \cdot ZB = BE \cdot HE$. Now $BE \cdot HE = DE \cdot GE$** [and $HZ \cdot ZB = DZ \cdot ZA$]. Therefore, $DZ \cdot ZA = DE \cdot EG$. Therefore, $DE/DZ = AZ/GE$. But $DE/DZ = BA/AZ$ [by similar triangles]. Hence $BA/AZ = AZ/GE = GE/GB$. Hence that which was proposed stands [AZ and GE being the mean proportionals between BA and GB].

* Strictly speaking H is not in fact introduced until later. Hence in proper mathematical procedure we should have ABG (cf. Leonardo text, line 5). But all manuscripts have $ABHG$ and thus this seems to be Jordanus' error.

** This can be easily proved by using Proposition II.12 of the *Elements*. See the proof of Hero summarized by T. L. Heath in connection with Proposition III.8 of the *Elements* (*Euclid, the Elements*, vol. 2 [Anapolis, 1947], p. 20).

[Probatio ex Leonardi practica geometrie]

Ostendam aliter qualiter inveniam inter duas lineas duas lineas, ita ut continuentur omnes secundum proportionem unam. Ponam ergo duas lineas *AB*, *BG*; et angulus earum sit rectus. Deinde continuabo *AG*; et faciam super lineam *AG* trianguli *ABG* circulum
5 *ABG*. Deinde protraham ex puncto *A* perpendicularem *AD* super lineam *AB*; et ex puncto *G* super *BG* perpendicularem *DG*; et protraham utramque secundum rectitudinem usque ad *Z* et *E*. Deinde faciam transire regulam, que moveatur super punctum *B*, et non separetur ab eo, et sic abscidens duas lineas *DA* et *DE*, et non cesset
10 moveri donec sit illud quod cadit ex eis inter *Z*, *B* equale ei quod cadit inter *E*, *H*. Ergo linea *ZB* est equalis linee *EH*. Ergo multiplicatio *ZH* in *ZB* est equalis multiplicationi *BE* in *HE*: verum *EB* in *HE* est equalis *DE* in *GE*; et *HZ* in *ZB* est equalis *DZ* in *ZA*. Ergo multiplicatio *DZ* in *ZA* est sicut multiplicatio *DE* in *EG*. Ergo proportio
15 *DE* ex *DZ* est sicut proportio *AZ* ex *GE*. Sed proportio *ED* ex *DZ* est sicut proportio *BA* ex *AZ*; et proportio *BA* ex *AZ* est sicut proportio *AZ* ex *GE*, et etiam sicut proportio *GE* ex *BG*. Et illud est quod demonstrare volumus.

Probatio Leonardi
1 Ostendam aliter *Ed* Volo ostendere *ZmS*
4 *AG om. ZmS*
5 ABG *Ed* super quem sit ABGH *ZmS*
9 DA *Ed* DZ *ZmS* / DE *Ed* ED *ZmS*
11 E, H *Ed* arcum BH et E *ZmS* / Ergo *Ed* Sit ergo illud ita ergo *ZmS*
15 ex[1]: ad *Zm*
15, 16, 17 *pro* ex *ubique habet S* ad
17 etiam *Ed* est etiam *ZmS* / BG *Ed* GB *ZmS*

[The Solution from Leonardo's *Handbook of Geometry*

[Virtually the same as the above Jordanus text]

Appendix VI

Jordanus and Campanus on the Trisection of an Angle

1. Jordanus' Solution

In the *Verba filiorum* of the Banū Mūsā presented in Chapter Four, we noticed that Proposition XVIII outlined a mechanical method of solving the *neusis* problem to which the trisection of an angle had been reduced, a method equivalent to tracing a conchoid referred to a circular base.[1] While we have no sure evidence of this particular method having been employed in antiquity, it is well known that Nicomedes and others treated conchoids of a line (rather than of a circle) and used them specifically to solve problems that reduce to *neuseis*.[2] Of particular interest to us is the

[1] The conchoid of a circle (called by Roberval the limaçon of Pascal) has the general equation in rectangular coordinates of $(x^2 + y^2 - 2ax)^2 = b^2 (x^2 + y^2)$ [or in polar coordinates, $r = b + 2a \cos \theta$]. This curve is described as follows: when a rod of length $2b$ moves in such a way that its midpoint describes a circle of diameter $2a$ while the rod is always directed to a fixed point on the circle, the ends of the rod will describe the conchoid of the circle. The particular conchoid traced by the movement of the rule in the Banū Mūsā proposition is such that $a = b$ (see Fig. 96 for such a conchoid). Roberval used this conchoid for the solution of the trisection problem (see "Curves, Special," *Encyclopaedia Britannica*, 14th ed., vol. *6* [Chicago, 1960], p. 889).

[2] For a brief history of *neusis* problems and their connections with the trisection of an angle in antiquity, see T. L. Heath, *A History of Greek Mathematics*, vol. *1* (Oxford, 1921), pp. 235–41. See his more extended treatment in *The Works of Archimedes* (Cambridge, 1897), pp. c–cxxii. In connection with the *neusis* involved in the Banū Mūsā proposition, we should note the statement defining and illustrating *neusis* given by Pappus (Heath, *ibid.*, p. c): "Two lines being given in position, to place between them a straight line given in length and verging towards a given point. If there be given in position (1) a semicircle and a straight line at right angles to the base, or (2) two semicircles with

reduction of trisection to a *neusis* found in the eighth lemma of the *Liber assumptorum* (or *Lemmata*), existing only in Arabic but attributed to Archimedes. While there is little doubt that the extant *Liber assumptorum* was not composed by Archimedes, Heiberg and Heath think it reasonable that this particular lemma is derived ultimately from Archimedes, since the *neusis* to which the trisection is reduced in this lemma is exactly similar to the *neuseis* assumed as possible in Propositions VI and VII of the *On Spiral Lines* of Archimedes.[3] Because of its similarities with the solution found in the *Verba filiorum*, let us examine this lemma in more detail. We can

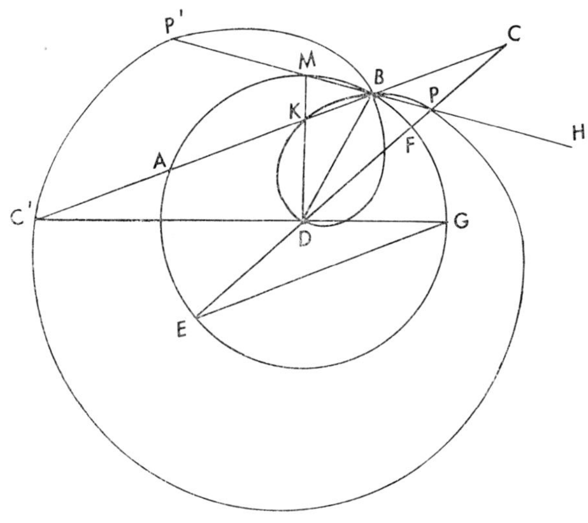

Fig. 96

translate it as follows:[4] "If we let line AB be led everywhere in the circle and extended rectilinearly [see Fig. 96], and if BC is posited as equal to the radius of the circle, and C is connected to the center of the circle D, and the line (CD) is produced to E, arc AE will be triple arc BF. Therefore, let us draw EG parallel to AB and join DB and DG. And because the two angles DEG, DGE are equal, $\angle\ GDC = 2 \angle\ DEG$. And be-

their bases in a straight line, to place between the two lines a straight line given in length and verging towards a corner *(gōnian)* of a semicircle." It is the first case, with the perpendicular a radius and the given line equal to a radius, that is involved in the Banū Mūsā proposition. Note that most of the pertinent passages from Proclus, Pappus, and Eutocius on *neusis* problems have been collected together and discussed by M. Curtze, *Reliquae Copernicanae* (Leipzig, 1875), pp. 7–21.

[3] Heath, *The Works of Archimedes*, pp. ci–cii; J. L. Heiberg, *Archimedis opera omnia*, vol. 2 (Leipzig, 1913), p. 518.

[4] Heiberg, *ibid.*

cause $\angle BDC = \angle BCD$ and $\angle CEG = \angle ACE$, $\angle GDC = 2 \angle CDB$ and $\angle BDG = 3 \angle BDC$, and arc BG = arc AE, and arc AE = 3 arc BF; and this is what we wished."

This proposition shows, then, that if one finds the position of line ABC such that it is drawn through A, meets the circle again in B, and its extension BC equals the radius, this will give the trisection of the angle. It thus demonstrates the equivalence of a *neusis* and the trisection problem—but without solving the *neusis*. This proposition can be related to the Banū Mūsā solution if we draw the line DM perpendicular to DG [Fig. 96]. Then the point K is equivalent to point S in the Banū Mūsā proof and AK is equal to the radius (or TS in the Banū Mūsā proof). Point A can be found by finding C', and C' can be found by the intersection of an outer part of the conchoid with the extension of DG. If then line $C'AB$ were extended so that BC also equals the radius, the *neusis* in the form here presented is solved. That is to say, if we had an undetermined line MH (equivalent to ZH in the Banū Mūsā proof) on which was laid in one direction segment MP equal to the radius and in the other direction segment MP' also equal to the radius, and M continually moved on the circumference of the circle while line $P'MPH$ always passed through point B, then the motion of P would trace the inner loop of the conchoid whose intersection with DM determines point K and the motion of P' would trace the outer part of the conchoid whose intersection with the extension of DG determines point C'.

Although it is evident, as we have seen, that no solution of the *neusis* was given in the *Liber assumptorum*, a procedure similar to that employed by the Banū Mūsā was attributed by a later Arabic author to an Ancient (i.e., a Greek).[5] Whatever its origin, the technique of the Banū Mūsā was not without influence among the Arabic mathematicians.[6] But our particular concern in this section is its use by Jordanus in the Latin West. If the reader examines the proposition here presented from the *Liber de triangulis*, he will notice that actually Jordanus presents three "solutions" of the trisections (all reducing to the same *neusis*): (1) the solution taken substantially from the *Verba filiorum* (see text below, lines 1–26); (2) a slight modification of that solution (lines 27–31) where, instead of the mobile line being ZQH and point Z moving to its final position at T [Fig. 97], an equivalent line LON moves from the contrary direction so that L comes to the same final position at T: in either case the line TSE (with TS equal

[5] E. Woepcke, *L'Algèbre d'Omar Alkhayyâmî* (Paris, 1851), p. 120.

[6] *Ibid.*, p. 117–26. Cf. Curtze, *op. cit.* in note 2 above, page 25.

to the radius) is the objective; and (3) a solution where the required line *TSE* is constructed according to the method given in Proposition V.19 of a work entitled *Perspective* (see lines 32–47). In Alhazen's *Optics*, which had already been translated into Latin, Proposition V.34 gives a solution of a *neusis* of the kind represented by the construction of *TSE*,[7] and I suspect it is to this proposition that Jordanus is referring in spite of the difference in proposition numbers. Incidentally, Alhazen's solution is one based on conic sections and would clearly appeal to a geometer more than the mechanical solution of the Banū Mūsā. Hence we are not surprised at Jordanus' preference for the third solution.

As in Appendix V, my text of the Jordanus proposition is based on four copies: *I*(*Ed*), *E*, *Oa*, and *Sl*, and the same procedures for the establishment of my text used there were followed here. I have had to depend on Curtze for the reading of *I* (since I did not have films of the part of the Dresden manuscript containing the Jordanus work). Sometimes I have questioned the reading of Curtze by adding a question mark in the variant readings where it seemed likely that Curtze had misread the manuscript (e.g., where he reads *demum* when all the other manuscripts have *deinde*). I have made only one addition to the text, namely, the insertion of *non* before *ponam* in line 6, since the whole point of the mechanical procedure used depends on line *ZH* being indeterminate in length so that, regardless of where point *Q* is, *ZH* will still pass through point *E*. The *non* was in the Banū Mūsā text (Chapter Four, Proposition XVIII, line 9), and so we must assume either that all of our manuscripts stem ultimately from a copy which had carelessly omitted the *non* present in the Jordanus text or that Jordanus in the preparation of his text from the Banū Mūsā original had omitted it.

Manuscripts *Sl* and *Oa* have three figures, one for each of the three solutions. Manuscripts *I* has two (at least Curtze reproduces only figures for the first and third solutions), and manuscript *E* economically combines all three figures. Since all three solutions reduce to precisely the same *neusis*, for the sake of economy I have followed *E* in combining all three

[7] Alhazen, *Opticae thesaurus* (Basel, 1572), Book V, Proposition 34, p. 144: "A puncto peripheriae circuli extra datam diametrum dato, ducere lineam rectam, ita sectam data diametro, ut segmentum inter diametrum et punctum peripheriae dato puncto oppositum, aequetur datae rectae, minori circuli diametro."

figures in a single figure (Figure 97).[8] The marginal folio references are to manuscript *I*, as given by Curtze.

Sigla of Manuscripts

I(Ed) = Dresden, Säch. Landesbibliothek, Db. 86, 59r–v, early 14c. *Ed. cit.* in note 2 above of Appendix V, pp. 38–39.
Oa = Paris, BN. lat. 7434, 85v–86r, 14c.
Sl = London, Brit. Museum, Sloane 285, 90r–v, 14c.
E = London, Brit. Museum, Harleian 625, 129r, 14c.

[8] In *Sl*, *Oa*, and *I(Ed)* the first figure is simply my Figure 97 without line *LON* and point *F*. The second figure consists only of semicircle *DZL* with lines *BE*, *BZ*, and *LON* drawn in a manner similar to their disposition on Fig. 97. And the third figure is essentially my Figure 97 without lines *ZH* and *LON*.

Proposition IV.20
from Jordanus' *On Triangles*

[Propositio IV.20
ex libro Jordani de triangulis]

59r /QUEMLIBET ANGULUM RECTILINEUM IN TRIA EQUA DIVIDERE.

Sit angulus *ABG* acutus in tria dividendus [Fig. 97]. Super *B* sumpto centro describatur circulus *DZM* et protrahatur *DB* in *L*,
5 erigaturque *BZ* perpendicularis ad *DL*, protrahamque lineam *ZE* in *H*. Et ⟨non⟩ ponam linee *ZH* finem determinatam, et resecabo de *ZH*, *ZQ* equalem *DB* semidiametro. Ymaginemur igitur quod linea *ZEH* moveatur versus *L*, ita quod *Z* motu suo non recedat a circumferencia et linea *ZH* non cesset transire super *E* sed semper inhereat
10 puncto *E*, et non cesset *Z* moveri quousque *Q* sit super *BZ*; sitque terminus illius motus *T*; erit ergo et quasi pars linee *ZH*, vel ut aliter dicam *ZH*, iacebit super *TE*. Eritque *TS* equalis *ZQ* sive *BD* semidiametro. Dico autem *TL* esse terciam arcus *DE*. Protrahamus a
59v puncto *B BM* equedistantem *TE* linee, et protraham / *BM* in *K*, et
15 coniungantur puncta *T*, *M*. Age: *TS* equalis est *MB* et equedistans; ergo *MT*, *BS* sunt equales et equedistantes; et *BZ* est perpendicularis

1 rectilineum: rectum *Oa*
4 et *E* *om. I(Ed)OaSl*
5 ad DL: super BL *E* / *ante* lineam *add. Sl* Z
6 ⟨non⟩ *supplevi*; *sed cf. Verba filiorum, Prop. XVIII, linea 9* / determinatum *Oa*
7 Yamginemur *EOa* imaginemur *Sl* imaginemus *I(Ed)*
8 L *E* A *I(Ed)OaSl*
9 E: E quiescens *E*
10 et non: neque *E* / moveri: motu (?) *I(Ed)*
11 terminus...motus: Z illius motus in *Oa*
11–12 erit...TE: ita quod tunc iaceat ZH quasi super lineam TE *E*
13 autem: igitur *E* / terciam: terciam partem *E*
14 equidistantem *habet Oa hic et ubique* / BM: MB *E*
15 Age: cum igitur *E* argue (?) *I(Ed)*
16–17 ergo...ergo[1]: igitur *E* ergo *Sl*
16 BS *Oa* BG *I(Ed)*

[Proposition IV.20 from Jordanus' *On Triangles*]

TO DIVIDE ANY RECTILINEAR ANGLE INTO THREE EQUAL PARTS

Let the acute ∠ *ABG* be the one to be trisected [see Fig. 97]. With *B* assumed as the center let circle *DZM** be described. Let *DB* be extended to *L*, and let *BZ* be erected as a perpendicular to *DL*. Then let line *ZE* be extended to *H*. And I do not pose any determined limit to line *ZH*. And I shall cut *ZQ* equal to radius *DB* from *ZH*. Therefore, let us imagine

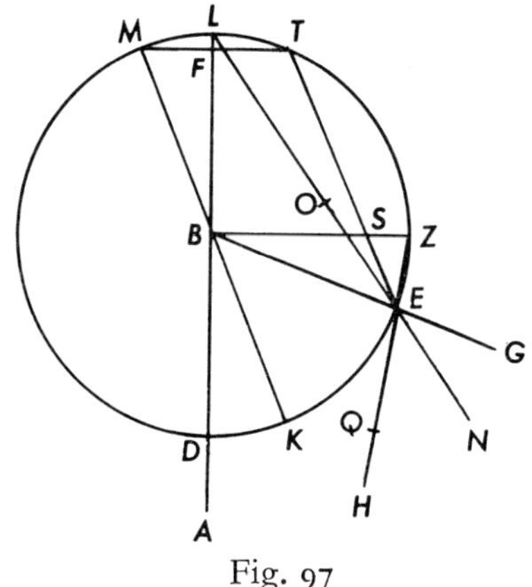

Fig. 97

that line *ZEH* is moved toward *L* in such a way that *Z* during its motion is not separated from the circumference and line *ZH* continues to pass through and adhere to *E*, and *Z* continues to be moved until *Q* falls on *BZ*. And let the terminus of the motion [of *Z*] be *T*. Therefore, a part of line *ZH*—or in other words *ZH*—will coincide with *TE*, and *TS* = *ZQ* = radius *BD*. I say, moreover, that arc *TL* = 1/3 arc *DE*. From point *B* let us draw *BM* parallel to line *TE*, extending *BM* to *K*, and let points *T* and *M* be connected. Proceed: *TS* = *MB* and the two lines are

* Points *Z* and *M* are in fact to be determined later. It is not good geometrical form to use these points to designate the circle. Cf. the better procedure of the Banū Mūsā in Proposition XVIII of their *Verba filiorum*.

ad *DL*, ergo *MT* secat *DL* ad angulos rectos. Ergo *DL* secat *MT* cordam in duo equa. Ergo *ML*, *LT* arcus sunt equales. Item *ML*, *DK* arcus sunt equales, quia *MK*, *DL* sese intersecantes in centro *B*
20 faciunt angulos ad invicem equales. Ergo a duplici pari *KE* arcus ad *DK* est duplus. Ergo angulus *KBE* ad *KBD* est duplus. Diviso ergo *KBE* in duo equa erit angulus propositus *ABG* in tria equalia divisus.

Si vero proponatur angulus maior acuto in tria equalia dividendus, dividatur primo ille in duas medietates, quarum utraque pars erit
25 angulus acutus. Deinde dividatur utraque illarum medietatum in tria equa secundum dictum modum. Constat ergo propositum.

Paululum quoque apercius idem probabitur, hoc solo variato, quod pro *HZ* protrahatur *LEN*; et cum *LBZ* sit rectus, sit *OL* equalis *BL* linee. Ymaginemur ergo *NL* sic moveri versus *Z* ut *LN* semper per-
30 transeat super *E*, moveaturque *NL* quousque *O* sit in *BZ*, et cetera ut prius.

De divisione anguli in tres partes equales mihi nequaquam sufficit dicta demonstracio, eo quod nihil in ea certum reperio. Ut autem mihi me sufficientem faciam, hoc idem sic demonstro. Datus angulus acutus
35 sit *ABG*. Igitur in *B* posito pede circini| describatur circulus et protrahatur *AB* ad *L* in periferia, et a centro super *DL* extrahatur perpendicularis *BZ*, et a puncto *E* per *BZ* semidiametrum ducatur linea per figuram 19 quinti perspective, ut *TS* sit equalis semidiametro *BL*. Ducatur ergo *BM* equedistanter linee *TSE*, et protrahatur ad *K*. Quia

17 *post* rectos *add.* *Oa* per 29 primi / Ergo[2]: propter hoc quod ZB ita facit igitur *E*
18 cordam *om.* *E* / equa: equalia *E* / *post* equa *add.* *Oa* per terciam tercii et 4 primi et 27 tercii / Ergo: quare et *E* / Item: Sed et *E*
19 quia: eo quod *E* / centro: puncto *E*
20 Ergo: igitur *E* / KE: KL *I(Ed)*
21 Ergo: igitur *E* / KBD: angulum DBK *E* / est *om.* *E* / duplus *tr.* *E ante* ad / ergo[2]: igitur *E* /
22 equa: equalia *Oa*
23 proponatur: ponatur *E*
24 ille *om.* *Oa*
25 Deinde: demum *(?)* *I(Ed)*
26 Constat ergo: et constat *E*
27 quoque: quidem *Oa* / idem: hoc *Oa* / probatur *I(Ed)* / solum *E*
28 protrahatur: variatur *Oa* / LEN: LON *E* / OL: omni *Oa*
29 Ymaginemur *OaE* imaginemur *Sl* imaginemus *(?)* *I(Ed)* / ergo: igitur *E* / ut *I(Ed)* et *OaSl* quod *E*
29–30 pertranseat: transeat *E*
30 moveaturque: et moveatur *E* moveatur *Oa* / NL: LN *I(Ed)* / et *E*, *om.* *I(Ed)* *OaSl* / ut: sicud *E*
32–33 De...demonstracio *om.* *Oa*
33 eo quod: ita quod *Sl* / in ea certum *ESl* certum in eo *I(Ed)*
35 Igitur...circini: Posito igitur centro in B *E* / describatur *I(Ed)Sl* circumscribatur *EOa*
36 a *Oa*, *om.* *Sl* e *I(Ed)*
38 TS *I(Ed)Sl* ST *E* IS *Oa*
39–47 Ducatur....propositum: et cetera sicud prius *E*
39 equedistanter *OaSl* equedistans *I(Ed)*

parallel. Therefore, *MT* and *BS* are equal and parallel. And *BZ* is perpendicular to *DL*. Therefore, *MT* will cut *DL* at right angles. Therefore, *DL* will bisect chord *MT*. Therefore, arcs *ML* and *LT* are equal. Also, arcs *ML* and *DK* are equal because *MK* and *DL*, intersecting each other at the center *B*, form mutually equal [vertical] angles. Therefore, by equality twice, arc *KE* = 2 arc *DK*. Therefore, ∠ *KBE* = 2 ∠ *KBD*. Therefore, when ∠ *KBE* has been bisected, the proposed ∠ *ABG* will have been trisected.

Now if an obtuse angle is to be trisected, let it first be bisected so that each of its halves will be an acute angle. Then let each of these halves be trisected by the said method. Therefore, that which was proposed is clear.

The same thing will also be proved a little more clearly with only one change made, namely, that instead of *HZ*, let line *LEN* be drawn. And since *LBZ* is a right angle, let *OL* = line *BL*. Therefore, let us imagine that *NL* is so moved toward *Z* that it always passes through *E* and that it continues to move until *O* falls on *BZ*, and so on as before.

The said demonstration concerning the trisection of an angle does not at all suffice for me, for I can find no certainty in it. To make it suffice for me, I demonstrate the same thing as follows. Let the given acute angle be *ABG*. Hence, with the foot of a compass placed at *B*, let a circle be described, and let *AB* be extended to *L* in the circumference. And from the center let *BZ* be erected as a perpendicular to *DL*. Then by Proposition V.19 of the *Perspective** let a line be drawn from point *E* through radius *BZ* so that *TS* = radius *BL*. Hence let *BM* be drawn parallel to line *TSE*, and [then] extended to *K*. Since *BM* and *TS* are equal and parallel, *BS* and *MT* will be equal and parallel. Therefore, since ∠ *LBZ* is a right angle,

* For a discussion of this work, see the Introduction to the text.

40 ergo *BM*, *TS* sunt equales et equedistantes, erunt *BS* et *MT* equales et equedistantes. Ergo, cum angulus *LBZ* sit rectus, erit *BFT* rectus. Ergo *MT* secatur per lineam *BF* per equalia. Ergo arcus *TM* est duplus ad arcum *ML*. Sed arcus *MT* est equalis arcui *KE* propter equedistantes. Ergo arcus *KE* est duplus ad arcum *ML*, ergo et ad
45 arcum *DK*. Dividatur ergo *KE* per equalia, habetur propositum. Si fuerit angulus maior acuto, dividatur in duos acutos, et utriusque sumatur pars tercia, et habetur propositum.

40 ergo: igitur *I(Ed)*
41 *post* rectus[2] *add. Oa* per 29 primi euclidis
42 *post* BF *add. Oa* per terciam tercii euclidis
43 *post* ML *add. Oa* per 27 tercii euclidis
44 equedistantes *OaSl* equedistanciam *I(Ed)*
45 habetur *Sl* et habetur *Oa* habebitur *I(Ed)*
46 acutos: a centro *Sl*
47 habebitur *I(Ed)*

$\angle BFT$ will be a right angle. Therefore, MT will be bisected by line BF. Therefore, arc TM = 2 arc ML. But arc MT = arc KE, because of the parallels. Therefore, arc KE = 2 arc ML. Therefore, arc KE = 2 arc DK. Therefore, if KE is bisected, that which was proposed is had. If the angle is obtuse, let it be bisected into two acute angles, and let a third part of each be taken, and then that which was proposed is had.

2. The Solution Attributed to Campanus

In the *editio princeps* of Campanus' version of the *Elements* of Euclid (Venice, 1482) there is added to the end of the fourth book (no pagination, but see signature d2 v) a solution of the trisection problem that seems to stem from Jordanus' treatment of the problem, and that probably arises from what we have called the third solution of Jordanus.[1] While I have not been able to find this solution in four manuscript copies of the Campanus commentary available to me (Columbia Univ. Plimpton MS. 156; Vat. reg. suev. 1261; Vat. Urb. lat. 506; Vat. Urb. lat. 507), it *is* present in one manuscript recently examined,[2] and so for the present we can hold to the Campanus attribution. It does, however, have a serious geometrical defect, as Copernicus recognized in a note to his personal copy of the *editio princeps*.[3] For Campanus loosely tells us to construct from point E "a line equal to CB in such a way that it cuts the circumference of the circle in point F" [see Fig. 98]. It is only after this has been done that he adds the instruction to extend the line from E to A. Thus the *neusis* is not properly stated, for the *neusis* must include the requirement that the intercept between CD and the circumference of the circle be on a line verging towards A as well as the requirement that the intercept be equal to the radius. It should be further noted that if Campanus did draw this proof from the third solution of Jordanus, he not only completely altered the lettering but also omitted reference to the proposition in the *Perspective* that would permit the solution of the *neusis*. It may be, of course, that the future discovery

[1] I have suggested its possible origin in Jordanus' third solution, because, like that solution, it assumes the *neusis* as done and thus does not contain the mobile lines (HZ or LN) used in the first two solutions to solve the *neusis*.

[2] As this volume went to press, Mr. John Murdoch informed me that Naples, Bibl. Naz. VIII. C. 21 (15c ?), contains the proposition relative to the trisection of an angle in the margin at the end of Book IV of the Campanus *Euclid* (f. 20v). I have collated this manuscript with my text and it agrees exactly with it except that *equidistantes* is spelled throughout as *equedistantes* and the scribe has written and deleted *ad equalitatem CO* after *duco lineam* in line 6, obviously arising from homoeoteleuton.

[3] M. Curtze, *Reliquae Copernicanae* (Leipzig, 1875), p. 6. Copernicus' note reads: "Datum angulum (intellige, qui non fuerit maior recto) trifarium secare et in linea *cd* etc. Id aptius explanatum fuisset hoc modo: Et ducatur recta linea *aef* scans *cd* in *e* et circumferentiam in *f*, ita ut *ef* aequalis sit ipsi *cb*. De quo qide Nicomedem de conchoidibus." One is not to infer from the last sentence that Copernicus had access to Nicomedes' tract, long since lost, but rather that he had learned from Pappus that Nicomedes had employed a conchoid for the trisection of an angle.

of manuscript copies of this Campanus solution will contain such a reference.

I have drawn the text of this proposition almost completely from the *editio princeps*, only correcting it in two places by references to a later edition (Ed_2, Basel, 1546, p. 586).

[Ex Commentario Campani in libros elementarum]

DATUM ANGULUM IN TRIA EQUA DIVIDERE

Sit angulus datus *C* [Fig. 98]. Volo ipsum dividere in tres equales angulos, quod sic facio. Pono primo *C* centrum circuli describendo circulum qualitercunque contingat, et protraho latera continentia da-
5 tum angulum usque quo secent circumferentiam in punctis *A* et *B*. Tunc a puncto *C*, quod est centrum circuli, duco lineam *CD* perpendiculariter ad lineam *CB*, et in linea *CD* assigno punctum *E*, a quo duco lineam ad equalitatem *CB* usque quo secet circumferentiam circuli in puncto *F*. Et produco *E* usque ad *A*. Deinde protraho lineam
10 *GH* equidistantem *FA*, que scilicet *GH* transeat per centrum; et duco lineam *FG* equidistantem linee *EC*; et protraho lineam *CB* in continuum et directum usque ad *L*, que secat lineam *FG* orthogonaliter in puncto *O* et per equalia. Dico ergo quod arcus *LQ* est equalis arcui *HB* propter hoc quod angulus *LCG* est equalis angulo *HCB* cum sint con-
15 tra se positi. Cum igitur arcus *FG* sit duplus arcui *LG*, erit etiam duplus arcui *HB*. Sed arcus *FG* est equalis arcui *HA* cum sint inter duas lineas equidistantes, que sunt *FA* et *GH*. Ergo arcus *HA* est duplus arcui *HB*; ergo et angulus *ACH* est duplus angulo *HCB*. Dividam ergo angulum *ACH* per equalia per lineam *CM*, et patet
20 propositum.

14 LCG *Ed*$_2$ LGC *Ed* 16 HA *Ed*$_2$ AH *Ed* [cf. *lineam 17*]

[From the Commentary of Campanus on the *Elements*]

TO DIVIDE A GIVEN ANGLE INTO THREE EQUAL PARTS

Let the given angle be C [see Fig. 98]. I wish to divide it into three equal angles, which I do in this way. First I posit C as the center of a circle by describing a circle of any sort, and I extend the sides containing the given angle until they cut the circumference in points A and B. Then from point C, which is the center of the circle, I draw line CD perpendicular to line CB. And in line CD I assign point E, from which I draw a line equal to CB in such a way that it cuts the circumference of the circle in point F. And I extend [the line from] E to A. Then I draw line GH parallel to FA, GH evidently passing through the center. And I draw

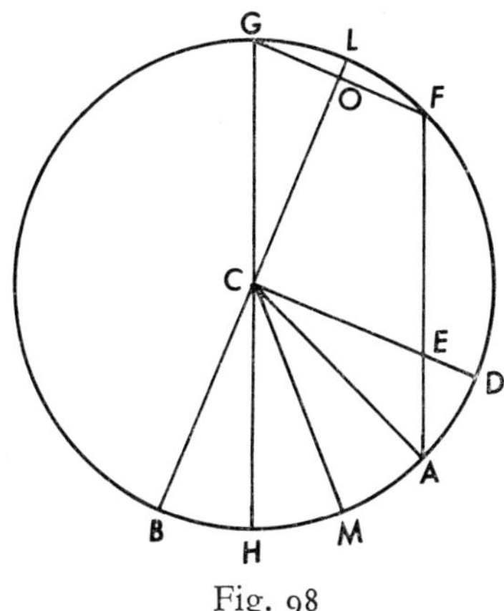

Fig. 98

line FG parallel to line EC, and I extend line CB continuously and in a straight line to L, which extension cuts line FG orthogonally in point O and [hence] bisects it. I say, therefore, that arc LG is equal to arc HB, since $\angle LCG = \angle HCB$, the angles being vertical angles. Since, therefore, arc $FG = 2$ arc LG, also arc $FG = 2$ arc HB. But arc $FG =$ arc HA, since they are between two parallel lines, FA and GH. Therefore, arc $HA = 2$ arc HB. Therefore, $\angle ACH = 2 \angle HCB$. Therefore, I shall bisect $\angle ACH$ by line CM, and that which was proposed is evident.

BIBLIOGRAPHY and INDEXES

Bibliography

This bibliography is primarily of works bearing on ancient and medieval mathematics. It does not include the manuscript catalogues cited in the introductory section or the standard Greek texts of non-mathematical authors like Aristotle and his commentators which are occasionally cited in the footnotes and commentaries.

Abraham bar Ḥiyya. *Liber embadorum*. Latin translation of Plato of Tivoli. Ed. of M. Curtze in *Abhandlungen zur Geschichte der mathematischen Wissenschaften*, 12. Heft (Leipzig, 1902), pp. 1–183.

Albert of Saxony. *Questio de quadratura circuli*. (Text and English translation in Chapter Five, Section 3.)

Albert of Saxony (?or Nicole Oresme?), *Utrum dyameter alicuius quadrati sit commensurabilis costae eiusdem*. Ed. by H. Suter in *Zeitschrift für Mathematik und Physik*, vol. 32 (1887), Hist.-lit. Abt., pp. 43–54.

Algazel. See al-Ghazzālī.

Alhazen. *Opticae thesaurus*. Basel, 1572.

Ametus filius Iosephi (Ahmad ibn Yūsuf). *Epistola de proportione et proportionalitate*. Latin translation of Gerard of Cremona. MS Paris, BN lat. 9335, 64r–75v. See forthcoming edition of Dorothy Schrader. Arabic text in MS Cairo, Bibl. Nat., no. 39 mathematique M.

Anaritius (al-Nairizī). *In decem libros priores Elementorum commentarii*. Ed. of M. Curtze as Supplementum to Euclid, *Opera omnia*, Leipzig, 1899. See also the Arabic text edited by R. O. Besthorn, J. L. Heiberg, G. Junge, J. Raeder, and W. Thomson, *Codex Leidensis 399.1*, Copenhagen, 1893–1932.

Anonymous. *De inquisicione capacitatis figurarum*. Ed. of M. Curtze in *Abhandlungen zur Geschichte der Mathematik*, 8. Heft (1898). Cf. MS Vienna, Nat.-bibli. 5277, 101r–110r.

———. *De ysoperimetris*. MS Oxford, Bodl. Auct. F. 5. 28, 105v–106v.

———. *Mishnat ha-Middot*. Ed. of S. Gandz in *Quellen und Studien zur Geschichte der Mathematik, Astronomie und Physik*, Abt. A: Quellen, vol. 2 (1932), pp. 1–96.

———. *Practica geometrie*. MS Vat. Reg. suev. 1261.

Apollonius of Perga. *Opera quae Graece exstant.* Ed. of J. L. Heiberg, Leipzig, 1891–93.

———. *Apollonius of Perga: The Treatise on Conic Sections.* Tr. of T. L. Heath, Cambridge, 1896. Reprinted 1961.

———. *Les Coniques d'Apollonius de Perge.* Tr. of Paul ver Eecke, Bruges, 1923. Reprinted 1961.

Archimedes. *De quadratura paraboles et alia opera.* In translation of Antonio de Albertis, MS Vienna, Nationalbibliothek, cod. 10701. (See Chapter One, note 30.)

———. *De mensura circuli.* In medieval Latin versions of Plato of Tivoli and Gerard of Cremona. (See Chapter Two for manuscripts and Chapters Three and Five for paraphrases and emendations.)

———. *De sphaera et cylindro.* In a medieval Latin fragment. (See Chapter Six, Section 1.)

———. *Monumenta omnia mathematica quae extant.* In translation of F. Maurolicus, Palermo, 1685.

———. *Opera nonulla in latinum conversa.* Translation with commentaries by F. Commandino, Venice, 1558.

———. *Opera omnia.* In the medieval Latin translation of William of Moerbeke, MS Rome, Vat. Ottob. lat. 1850, 1r–60r.

———. *Opera omnia.* In the modern edition of J. L. Heiberg, 2d ed., 3 vols., Leipzig, 1910–15.

———. *Opera quae quidem exstant omnia.* Basel, 1544. First Greek edition with translation of Jacobus Cremonensis.

———. *The Works of Archimedes.* Translated and paraphrased by T. L. Heath, Cambridge, 1897 (reissued by Dover Publications, New York, 1950).

———. See also Dijksterhuis and Gaurico.

Archimedes (Pseudo-). *De insidentibus in humidum* (or *De ponderibus*). For Latin text and English translation, see Moody and Clagett, *The Medieval Science of Weights*, Madison, 1952.

Aristotle, *Physica.* In the medieval *nova translatio* of Moerbeke, published with Walter Burley's *Commentaria* on the *Physics* (Venice, 1589), and with the exposition of Aquinas on the *Physics* (Rome, 1954).

Bachon Alardus. *In 10 Euclidis.* (See Chapter Five, Section 2, note 1.)

Bacon, Roger. *Communia mathematica.* Ed. of R. Steele, Oxford, 1940.

———. *Compendium studii philosophiae.* In *Opera quaedam hactenus inedita*, ed. of J. S. Brewer, London, 1859.

Banū Mūsā. *Verba filiorum* (*Liber de geometria*). Translation of Gerard of Cremona. (For new edition, see Chapter Four, where comments are made on the older edition of Curtze [Halle, 1885].)

Becker, O. "Eudoxus-Studien II," *Quellen und Studien zur Geschichte der Mathematik, Astronomie und Physik*, B. Studien, vol. 2 (1933), pp. 369–87.

Bernardus Pitoiensis. *Geometrie liber*. MS Brit. Mus. Addit. 17420, 1r–30r.

al-Bīrūnī. See Suter, H.

Björnbo, A. A., and S. Vogl. "Alkindi, Tideus, und Pseudo-Euklid," *Abhandlungen zur Geschichte der mathematischen Wissenschaften*, 26. Heft (1912), pp. 1–176; "Handschriften-Beschreibung," pp. 123–47.

———. "Gerhard von Cremonas Übersetzung von Alkhwarizmis Algebra und von Euklids Elementen," *Bibliotheca Mathematica*, 3. Folge, vol. 6 (1905), pp. 239–48.

———. "Über zwei mathematische Handschriften aus dem vierzehnten Jahrhundert," *Bibliotheca Mathematica*, 3. Folge, vol. 3 (1902), pp. 63–75.

Blume, F., K. Lachmann, and A. Rudorff. *Die Schriften der römischen Feldmesser*, vol. 1, Berlin, 1848.

Boethius. *In Categorias Aristotelis*. Ed. in Migne, *Patrologia latina*, vol. 64.

Boncompagni, B. "Della vita e delle opere di Gherardo cremonese etc.," *Atti dell' Accademia Pontificia de' Nuovi Lincei*, vol. 4 (1851), pp. 387–493.

———. "Delle versioni fatte da Platone Tiburtino etc.," *Atti dell' Accademia Pontificia de' Nuovi Lincei*, vol. 4 (1851), pp. 249–86.

Bradwardine (Pseudo-). *Quadratura circuli*. (Text and translation in Chapter Five, Section 1.)

Bradwardine, Thomas. *De continuo*. MSS Thorn, R 4° 2, 153–92; Erfurt, Stadtbibl. Amplon. Q 385, 17r–48r. A critical edition will soon be published by John Murdoch.

———. *Geometria speculativa*, Paris, 1495. (For manuscripts, see Chapter Two, Section 2, note 11.)

Campanus of Novara (?). *Quadratura circuli*. (For new edition and manuscripts, see Appendix I, Section 3.)

———. *Theorica planetarum*. MS Oxford, Bodl. Auct. F.3.13.

———. See also Euclid.

Cantor, M. *Die römischen Agrimensores und ihre Stellung in der Geschichte der Feldmesskunst*. Leipzig, 1875.

———. *Vorlesungen über Geschichte der Mathematik*. Vol. 2, 2d ed., Leipzig, 1899–1900.

Carmody, F. J. *The Astronomical Works of Thabit B. Qurra*. Berkeley, 1960.

Casiri, M. *Bibliotheca Arabico-Hispana Escurialensis*. Vol. 1, Madrid, 1790.

Chasle, M. *Aperçu historique sur l'origine et le développement des méthodes en géométrie*. 3d ed., Paris, 1889.

Clagett, M. "A Medieval Fragment of the *De Sphaera et Cylindro* of Archimedes," *Isis*, vol. 43 (1952), pp. 36–38.

———. "Archimedes in the Middle Ages: The *De mensura circuli*," *Osiris*, vol. 10 (1952), pp. 587–618.

———. *Giovanni Marliani and Late Medieval Physics*. New York, 1941.

———. "King Alfred and the *Elements* of Euclid," *Isis*, vol. 45 (1954), pp. 269–77.

———. "The *Liber de motu* of Gerard of Brussels and the Origins of Kinematics in the West," *Osiris*, vol. 12 (1956), pp. 73–175.

———. "The Medieval Latin Translations from the Arabic of the *Elements* of Euclid," *Isis*, vol. 44 (1953), pp. 16–42.

———. "The *Quadratura per lunulas*, A Thirteenth Century Fragment of Simplicius' Commentary on the *Physics* of Aristotle," *Essays in Medieval Life and Thought Presented in Honor of Austin Patterson Evans*, New York, 1955, pp. 99–108.

———. *The Science of Mechanics in the Middle Ages*, Madison, 1959; reprinted with minor additions, 1961.

———. "Three Notes," *Isis*, vol. 48 (1957), pp. 82–83.

———. "The Use of the Moerbeke Translations of Archimedes in the Works of Johannes de Muris," *Isis*, vol. 43 (1952), pp. 236–42.

———. See also Moody, E. A.

Colebrooke, H. T. *Algebra with Arithmetic and Mensuration from the Sanscrit of Brahmegupta and Bahascara*. London, 1817.

Commandino, F. See Archimedes.

Crombie, A. C. See Murdoch, J.

Curtze, M. *Reliquae Copernicanae*. Leipzig, 1875.

———. "Über eine Handschrift der Königl. öffentl. Bibliothek zu Dresden," *Zeitschrift für Mathematik und Physik*, vol. 28 (1883), Hist.-lit. Abt., pp. 1–13.

Dijksterhuis, E. J. *Archimedes*. Copenhagen, 1956.

Duhem, P. "Un ouvrage perdu cité par Jordanus de Nemore: le Philotechnes," *Bibliotheca Mathematica*, 3. Folge, vol. 5 (1904), pp. 321–25.

Euclid. *Elementa*. Ed. of J. L. Heiberg with a Latin translation, 4 vols., Leipzig, 1883–1888.

———. *Elementa*. In Latin translation designated as Adelard II. MS British Mus. Addit. 34018. For other MSS see Clagett, "The Medieval Latin Translations...."

———. In Latin translation of Gerard of Cremona. MSS Vat. lat. 7299, Vat. Ross. 579, BN lat. 7216.

———. With Latin commentary of Campanus, Venice, 1482; also ed. of Basel, 1546.

———. *The Elements*. English translation of T. L. Heath, 3 vols., 2d ed., Cambridge, 1926; reprinted Annapolis, 1947.

Eutocius. *Commentarius in dimensionem circuli*. Ed. of J. L. Heiberg in Archimedes, *Opera omnia*, vol. 3, Leipzig, 1915.

Francischus de Ferraria. *Questio de proportionibus*. MS Oxford, Bodl. Canon. Misc. 226, 58r–63r.

Franco of Liège. *De quadratura circuli*. Ed. by Dr. Winterberg, *Abhandlungen zur Geschichte der Mathematik*, 4. Heft (1882), pp. 137–90.

Galilei, Galileo. *Opere*. Ed. Naz. 23 vols., Florence, 1891–1909.

Gaurico, L., ed. *Tetragonismus, id est circuli quadratura, per Campanum, Archimedem Syracusanum atque Boetium...adinventa*. Venice, 1503.

al-Ghazzālī. *Algazel's Metaphysics*. Medieval Latin translation, ed. by J. T. Muckle, Toronto, 1933.

Gerard of Brussels. *Liber de motu*. See M. Clagett.

Gordanus. *Compilacio quorundam canonum in practicis astronomie et geometrie*. MS Vat. Pal. lat. 1389. (See Chapter Three, Section 4.)

Grabmann, M. *Guglielmo Moerbeke, O.P. il traduttore delle opere di Aristotele*. Rome, 1946.

Grant, E. "Nicole Oresme and his *De proportione proportionum*," *Isis*, vol. 51 (1960), pp. 293–314.

Haskins, C. H. *Studies in the History of Medieval Science*. 2d ed., Cambridge, Mass., 1927.

Heath, T. L. *A History of Greek Mathematics*. 2 vols., Oxford, 1921.

———. *Mathematics in Aristotle*. Oxford, 1949.

Heiberg, J. L. "Beiträge zur Geschichte der Mathematik im Mittelalter," *Zeitschrift für Mathematik und Physik*, vol. 35 (1890), Hist.-lit. Abt., pp. 41–48.

———. "Le Rôle d'Archimède dans le développement des sciences exactes," *Scientia*, vol. 20 (1916), pp. 81–89.

———. "Neue Studien zu Archimedes," *Zeitschrift für Mathematik und Physik*, vol. 34 (1890), Suppl., p. 83.

———, and Wiedemann, E. "Ibn al Haitams Schrift über parabolische Hohlspiegel," *Bibliotheca Mathematica*, 3. Folge, vol. 10 (1910–11), pp. 201–37.

Hero of Alexandria. *Opera quae supersunt omnia*. Ed. of W. Schmidt, L. Nix, H. Schöne, and J. L. Heiberg, 5 vols. and 1 suppl., Leipzig, 1899–1914.

Hultsch, F. "Der heronische Lehrsatz über die Fläche des Dreieckes als Function der drei Seiten," *Zeitschrift für Mathematik und Physik*, vol. 9 (1864), pp. 225–49.

———. See also Pappus of Alexandria.

Johannes de Tinemue. *De curvis superficiebus*. (See Chapter Six for MSS and edition.)

Jordanus. *Liber de triangulis*. Ed. of M. Curtze, Thorn, 1887. (For full bibliographical reference and manuscripts, see Sigla in Appendix I, Section 1.)

Kapp, A. G. "Arabische Übersetzer und Kommentatoren Euklids," *Isis*, vol. 22 (1934–35), pp. 150–72; vol. 23 (1935), pp. 54–99; vol. 24 (1935–36), pp. 37–79.

Krause, M. "Stambuler Handschriften islamischer Mathematiker," *Quellen und Studien zur Geschichte der Mathematik, Astronomie und Physik*. Abt. B, vol. 3 (1936), p. 441.

Kristeller, O. *Renaissance Thought and Letters*. Rome, 1956.

Lejeune, A. *L'Optique de Claude Ptolémée*. Louvain, 1956.

———. *Recherches sur la catoptrique grecque*. Brussels, Paris, 1957.

Leonardo Pisano. *Practica geometrie*. Ed. by B. Boncompagni in *Scritti di Leonardo Pisano*, vol. 2, Rome, 1862.

Lull, Ramón. *De quadratura et triangulatura circuli*. Ed. of J. E. Hofmann, "Cusanus-Studien VII. Die Quellen der Cusanischen Mathematik I: Ramon Lulls Kreisquadratur," *Sitzungsberichte der Heidelberger Akademie der Wissenschaften*, Philosophisch-historische Klasse, Jahrgang 1941–42, 4. Abhandlung, pp. 1–37.

Macrobius. *Commentarius in somnium Scipionis*. Ed. of F. Eyssenhardt, Leipzig, 1893.

Mainardi, Leonardo, of Cremona. *Practica geometriae*. Ed. of M. Curtze in *Urkunden zur Geschichte der Mathematik im Mittelalter und der Renaissance*, *Abhandlungen zur Geschichte der mathematischen Wissenschaften*, 13. Heft (1902), pp. 339–433.

Maurolico, Francesco. See Archimedes.

Meyerhof, M. "New Light on Hunain ibn Ishaq and his Period," *Isis*, vol. 8 (1926), pp. 687–724.

Millás Vallicrosa, J. M. *Estudios sobre historia de la ciencia española*. Barcelona, 1949.

———. Another but different work with same title, Barcelona, 1960.

Moody, E. A., and M. Clagett. *The Medieval Science of Weights*. Madison, 1952; reprinted in 1960 with some emendations.

Murdoch, J. "The Medieval Language of Proportions," in A. C. Crombie, ed., *Scientific Change*, New York, 1963, pp. 237–71.

Nasīr al-Dīn al-Ṭūsī. *Majmū' al-Rasā'il*. Vol. 2, Hyderabad, 1940. (For the variant readings of his Arabic texts of the *De mensura circuli* and the *Verba filiorum*, see Chapters Two and Four.)

Oresme, Nicole. *De configurationibus*. MS BN lat. 7371. A critical edition based on fourteen extant manuscripts is being prepared by M. Clagett.

———. *Quaestiones super geometriam Euclidis*. Ed. of H. L. Busard, Leiden, 1961.

Pappus of Alexandria. *Collectio*. Ed. of F. Hultsch, 3 vols., Berlin, 1876–78. See also French translation of Paul Ver Eecke, Bruges, 1923.

Ramée, Pierre de la. *Scholarum mathematicarum libri unus et triginta*. Frankfurt, 1599.

Sarton, G. *Introduction to the History of Science*. 3 vols. in 5, Baltimore, 1927–48.

Schmidt, W. M. "Zur Textgeschichte der 'Ochumena' des Archimedes,' *Bibliotheca Mathematica*, 3. Folge, vol. 3 (1902), pp. 176–79.

Simplicius. *In Aristotelis physicorum libros quattuor priores commentaria*. Ed. of H. Diels, *Commentaria in Aristotelem graeca*, vol. 9, Berlin, 1882, pp. 56–57. (For the medieval Latin translation of this section by Grosseteste, see Appendix II.)

Suter, H. "Das Buch der Auffindung der Sehnen im Kreise etc.," *Bibliotheca Mathematica* 3. Folge, vol. 11 (1910–11), pp. 11–78.

———. "Der Tractatus 'De quadratura circuli' des Albertus de Saxonia," *Zeitschrift für Mathematik und Physik*, vol. 29 (1884), Hist.-lit. Abt., pp. 81–101.

———. "Die Mathematiker und Astronomen der Araber," *Abhandlungen zur Geschichte der mathematischen Wissenschaften*. 10. Heft (1900).

———. "Die Mathematiker-Verzeichniss im Fihrist des Ibn Abî Ja'kûb an-Nadîm," *Abhandlungen zur Geschichte der Mathematik*, 6. Heft (1892), pp. 1–87.

———. "Über die Geometrie der Söhne des Mûsâ ben Schâkir," *Bibliotheca Mathematica*, 3. Folge, vol. 3 (1902), pp. 259–72.

Tannery, P. *Mémoires scientifiques*. Vol. 5, Paris, Toulouse, 1922.

———. "Sur le 'liber augmenti et diminutionis' etc.," *Bibliotheca Mathematica*, 3.

Folge, vol. 2 (1901), 45–47. (See also *Mémoires scientifiques*, vol. 5, pp. 304–307).

Themistius. *Commentum super librum posteriorum*. Latin translation of Gerard of Cremona, edited by J. R. O'Donnell, "Themistius' Paraphrasis of the Posterior Analytics in Gerard of Cremona's Translation,' *Mediaeval Studies*, vol. 20 (1958), pp. 239–315.

Thomas, I. *Selections Illustrating the History of Greek Mathematics*. 2 vols., London, Cambridge, Mass., 1951.

Thomson, S. H. *The Writings of Robert Grosseteste*. Cambridge, 1940.

al-Ṭūsī. See Nasīr.

Ver Eecke, Paul. See Apollonius of Perga and Pappus of Alexandria.

Wiedemann, E. "Beiträge zur Geschichte der Naturwissenschaften III," *Sitzungsberichte der Physikalisch-medizinischen Sozietät in Erlangen*, vol. 37 (1905).

Wüstenfeld, F. "Die Übersetzungen arabischer Werke in das Lateinische," *Abhandlungen der K. Gesellschaft der Wissenschaften zu Göttingen*, vol. 22 (1877).

Woepcke, E. *L'Algèbre d'Omar Alkhayyâmî*. Paris, 1851.

Zeuthen, H. G. "Über einige archemedische Postulate," *Archiv für die Geschichte der Naturwissenschaften und der Technik*, vol. 1 (1909), pp. 320–27.

Zoubov, V. P. "Quelques observations sur l'auteur du traité anonyme 'Utrum dyamter alicuius quadrati sit commensurabilis costae ejusdem'," *Isis*, vol. 50 (1959), pp. 130–34.

A Selective Index of Latin Geometrical Terms

The purpose of this index is to provide a guide to the geometrical language of the Archimedean texts included in this volume. Its aim is not primarily philological; and in fact only a few examples of the use of each term have been indexed rather than all instances. However, enough instances have been cited to illustrate the varying meanings of the terms and particularly any special or peculiar usage. While a great many terms might well have been cited with a *passim*, I have reserved *passim* for terms that constantly recur, e.g., *angulus*, *circulus*, *latus*, *maior*, and the like. A number of words that are not ordinarily geometrical in nature have been included because they play play a particular role in geometrical exposition and argument, e.g., *facere*, *ducere*, *probare*, *producere*, *impossibile*, *deinceps*, and the like. I have also tried to include unusual spellings of common terms. In general, the infinitive form of the verb is used in the index to serve for all forms of the verb found in the texts, but on occasion when a participle form has been used extensively or exclusively that form has been included in the index. Similarly adjectival forms are indexed, for the most part, under the masculine form, although occasionally when another form appears extensively or exclusively in the texts it has been given in the index. There is, I believe, no particular ambiguity in this practice since consultation of the passages indexed will provide a ready answer to any question. I also thought it useful to include proper names cited in the texts. The only exception is the name *Euclides*, for I have given a separate Index of Citations to the *Elements* of Euclid which includes all the textual references to Euclid. Neither cardinal nor ordinal numbers have been included in the index. Roman numbers refer to pages, italic numbers to lines; *c* preceding a line number refers to the Commentary to that line, while *v* refers to the variant reading to that line. Text titles have been abbreviated in the index in accordance with the following table:

ATI = *Area trianguli*, Versio I (Appendix IV)
ATII = *Area trianguli*, Versio II (Appendix IV)
CC = The Corpus Christi Version (Chap. 3, Sec. 5)
CQ = *De circulo quadrando* (Appendix I, Sec. 2)
CS = *De curvis superficiebus* (Chap. 6, Sec. 2)
CSA = *Propositiones addite* (Chap. 6, Sec. 4)
CSC = *Commentum in septimam propositionem* (Chap. 6, Sec. 5)
CSI = *Propositiones interposite* (Chap. 6, Sec. 3)
CV = The Cambridge Version (Chap. 3, Sec. 1)
FIA = The First Florence Version (Chap. 3, Sec. 3)
FIB = The Second Florence Version (Chap. 3, Sec. 3)
FII = The Florence Version of Proposition II (Chap. 3, Sec. 3)
FIII = The Florence Version of Proposition III (Chap. 3, Sec. 3)
GV = The Gordanus Version (Chap. 3, Sec. 4)

MV = The Munich Version (Chap. 3, Sec. 6)
NV = The Naples Version (Chap. 3, Sec. 2)
PA = *Propositiones Archimenidis* (Chap. 6, Sec. 1)
PJ = The Proof of Jordanus on Mean Proportionals (Appendix V)
PL = The Proof of Leonardo on Mean Proportionals (Appendix V)
QA = *Questio Alberti* (Chap. 5, Sec. 3)
QC = *Quadratura circuli Campani* (Appendix I, Sec. 3)
QG = The Gerard Translation of the *De mensura circuli* (Chap. 2, Sec. 2)
QJ = Theorem of Jordanus (Appendix I, Sec. 1)
QLI = *Quadratura per lunulas*, Versio I (Appendix II)
QLII = *Quadratura per lunulas*, Versio II (Appendix II)
QT = The Plato Translation of the *De mensura circuli* (Chap. 2, Sec. 1)
TC = Campanus on Trisection (Appendix VI, Sec. 2)
TJ = Jordanus on Trisection (Appendix VI, Sec. 1)
VA = *Versio abbreviata* (Chap. 5, Sec. 2)
VF = *Verba filiorum* (Chap. 4)
VV = *Versio Vaticana* (Chap. 5, Sec. 1)

abscidere: QT 20, *14*; 22, *31*, *36*; VF 342, *36*; PL 664, *9*
abscindere: NV 82, *10*; 90, *c 32*: PJ 662, *7*
abscisio (=abscissio): QT 22, *17*, *44*
abscitio: QT 22, *43*
abstrahere ("draw"): QT 22, *35*
accipere: QG 46, *64*; 54, *v 148*; VG 156, *119*; MV 198, *14*; VF 278, *1*; 346, *10*; 369n; ATII 654, *111*; *and see* pariter accepti
acutus: VF 352, *46*; QA 424, *292*; 430, *c 45*; CS 458, *86*; TJ 672, *3*; 674, *23*, *25*
addere: FIII 114, *31*; MV 198, *27*, *28*; VA 390, *19*; CS 500, *66*; QC 590, *22*
addere super ("exceed"): QG 42, *18*; 48, *84*; FIII 112, *1–2*; 132, *349*; VA 392, *42*; ATI 644, *31*
adducere: 139, *c 15*
adequare: VG 150, *14*; 156, *119*; MV 198, *19*; 206, *127*, *128*
adhere: VF 342, *38*; 346, *12*
adiacere: QT 20, *4*; 22, *24*; 631
adicere: FIII 128, *265*; QJ 572, *2*
adiectio ("addition"): CV 74, *88*, *89*
ad invicem: VF 308, *84*; *and see* equalis
adiungere: FIII 126, *219*; 130, *306*, *313*
agere: NV 84, *27*; CC 178, *109*
aggregare *or* agregare: VF 278, *6*, *7*; *et passim in VF*; 280, *v 18–124*; VA 394, *66*; ATI 646, *45 et seq.*
aggregatum *or* agregatum: QG 42, *26*; VF 248, *17*; 250, *8*; *et passim in VF*; QA 414, *165*; PA 436, *11*
agregatio: VF 258, *16*, *18*; *et passim in VF*
Albertus de Saxonia: QA 406, *1*
alkaydem (=basem): FIA 100, *24*; 135, *c 24*
alternatim: CSA 532, *3*
alternus: 78, *c 33*
Almagestum: MV 210, *179*; VV 380, *79*; *and see General Index*
altitudo: CV 74, *69*; 79, *c 70*; NV 88, *69*; FIA 100, *13*; *VG* 158, *164*; MV 202, *77*; VF 240, *32*, *33*; 242, *52*, *53*; VV 376, *46*; VA 390, *17*; QA 416, *182*; PA 436, *9*
altus: VF 240, *v 35–39*; 242, *48*; 320, *52*; CS 502, *11*; CSI 526, *2*
ambiens: FIA 102, *34*; VG 148, *4*; CC 172, *22*, *37*; MV 198, *8*; VV 374, *6*; 384, *144*; *and see* linea ambiens
ambire: MV 216, *299*; ATII 648, *3*
ambitus (=perimeter): CV 68, *6*; FIB 108, *25*; FIII 122, *148*; CC 172, *37*; CS 456, *59*
ambligonius: VF 352, *46*
amplecti: QT 26, *104*
ampliatus: QC 592, *26*
amplitudo: VF 240, *v 35–95*; 244, *59*
amplissimus: QA 430, *c 45*; 431, *c 45*
angularia: VG 158, *141*
angulata: VF 252, *4*; *and see* figura laterata
angulosa (=angulata): VF 254, *18*, *21*
angulus: QT 20, *5*; 24, *79*; 33n, *25*; QG 40, *4*; CV 68, *10*; NV 82, *3*; *et passim ubique*

angulus contingentie: 431, *c 45*; CSA 538, *72*; 568n
angulus rectilineus *or* rectilineus angulus: 220, *c 56*; QA 408, *44*; 424, *282*; TJ 672, *1*
angulus semicirculi: QA 408, *44*; 431, *c 45*
angustissimus: QA 430, *c 45*; 431, *c 45*
animi, *see* communis conceptio
antecedens ("antecedent term in a proportion"): QG 48, *v 95*; CSA 534, *22*; ("first part of an implication"): VV 374, *19*; 376, *25*; *et passim in VV*; VA 396, *80*; QA 422, *270*
Antifon: QA 406, *3*; 422, *276*
antiqui ("the Ancients, i.e., Greeks"): VF 238, *9*; 240, *18*, *23*
applicare: CV 68, *14*; 74, *91*; CC 186, *231*; CSI 529, *c 29*
applicatio: VF 308, *79*
apponere: FIII 116, *78*; 118, *96*; *et passim in FIII*
Archemides: VF 278, *v 182*
Archimedes: 36n
Archimenides: 34n, *43*, *64*, *75*, *97*; 36n; QG 48, *v 84*; CV 68, *1*; NV 82, *1*; FIA 100, *1*; FIII 112, *v 1*; 141, *c 147*; 144n; VG 148, *2*, *8*; CC 176, *89*; MV 198, *3*; VF 264, *5*; 266, *21*; 278, *182*; 352, *42*, *65*; QA 416, *v 197*; CS 460, *107*; 466, *34*; 482, *v 1*; 502, *22*; 504, *4*, *13*; CSA 534, *16*; 538, *85*; 542, *22*; CSC 550, *1*
Archindi (=Archimedes): 36n
arcus: QT 20, *13*; QG 40, *12*; CV 68, *5*; NV 84, *20*; VF 256, *v 4–40*; 268, *34*; VV 378, *76*; 430, *c 45*; TJ 674, *18*; TC 680, *13*; *et passim ubique*
area: QT 22, *25*; 24, *59*; 33n, *3*; QG 42, *26*; FIA 106, *71*, *74*; 144n; 146n; CC 188, *243*; MV 198, *2*; CS 462, *35*; ATI 642, *6*; 646, *60*, *61*, *62*, *65*; ATII 648, *1*; *et passim in ATII*
arguere: VG 158, *164*
argumentatio: QA 430, *c 45*
argumentum: 34n, *88*, VG 158, *151*; 160, *169*
Aristoteles *or* Aristotiles: 34n, *87*; MV 198, *25*; 210, *179*; QA 412, *119*; 422, *275*; QC 588, *2*
Arsamithes (=Archimedes): QG 40, *1*
assignare: 79, *c 50*; 146n; 193, *c 278*; QA 422, *266*
assignatus: VG 148, *5*, *6*; 222, *c 307*
assumere: MV 198, *2*; VF 260, *44*; 633
astronomia: CC 170, *2*
attingere: MC 202, *77*
auferre: FIA 100, *20*
augere: QA 406, *14*; 424, *292*
augmentum: QG 44, *47*; 54, *153*; VF 282, *47*, *48*, *50*; VA 394, *67*; CS 466, *9*; 468, *11*; 490, *107*, *111*
axis: VF 294, *34*; CS 460, *3*; 472, *38*; CSA 540, *1*

basis: QT 24, *54*; 33n, *24*; QG 46, *60*; CV 70, *31*; 79, *c 70*; NV 84, *15*; VF 292, *2*; VA 390, *17*; QA 416, *185*; PA 436, *7*; *et passim ubique*
Bradwardinus: VV 386, *187*
brevior: MV 206, *130*; VF 252, *2*, *12*; 256, *9*; *et passim in VF*; QA 420, *228*; 627
brevius: VF 252, *10*
Brisso: QA 406, *3*; 422, *276*, *279*, *280*; 428, *c 3–4*
Bryso: QA 429, *c 3–4*

cadere: QC 48, *v 89*; 60n; 164, *c 149*; 221, *c 226*; VF 246, *78*; 280, *19*; VV 376, *34*; QA 416, *177*; 430, *c 45*; PA 436, *3*; CS 460, *12*; PL 664, *10*
Campanus: MV 202, *60*; 216, *310*; QA 408, *37*, *51*, *60*; 412, *107*, *131*; 424, *297*, *316*; 430, *c 45*; QC 588, *1*, *v 1*, *v 2–8*, *v 5–6*; 606, *121*, *v 121*; QLII 620, *v 14–19*
canonium: 659n
capacior: QA 406, *27*
capacissima: QA 414, *136*
capacitas: QA 410, *82*
capere: VV 378, *78*; VA 390, *20*; QA 430, *c 45*
capud: VF 292, *v 13*
caput ("vertex"): VF 292, *1*, *4*, *8*, *11*; *et passim in VF*; ("top" *or* "head"): VF 320, *64*; CS 460, *15*; QC 596, *55*
casus: VV 376, *26*, *46*; 378, *63*
catechus (=cathetus): QG 44, *v 51*
catetus (=cathetus): CV 72, *39*
cathetus: QT 22, *20*; QG 44, *51*; CC 176, *97*; CS 450, *7*
causa: 33n, *4*; *and see* exempli causa
causa brevitatis: FIA 100, *28*; 104, *61*; FIII 126, *236*; CC 180, *137–38*; CS 458, *78*; 480, *24–25*

causa compendii: NV 84, *26*; 88, *75*
causare: VV 374, *15*, *21*; 376, *56*
centralis: CS 476, *104*
centrum: QT 22, *20*; QG 42, *20*; CV 72, *39*; NV 84, *31*; VF 248, *6*; VV 374, *7*; *et passim ubique*
certificare: 33n, *10*
certificatio: 33n, *11*
certissime: 34n, *66*
certitudo: CSC 550, *9*
chilindrus (=cylindrus): 631
circinus: VG 158, *161*, *162*; QC 592, *26*, *27*; *et passim in QC*; TJ 674, *35*
circonferentia: 97n
circuitus: CS 460, *15*
circulari ("to be made into a circle"): QA 406, *22*, *24*; 424, *302*
circularis: 31; QA 406, *27*; 414, *135*; CSC 554, *67*; QC 592, *23*
circulatio: 146n
circulus: QT 20, *2*; 33n, *16*; QG 40, *1*; CV 68, *9*; VF 246, *1*; VV 374, *1*; *et passim ubique*
circulus magnus: VF 316, *23*; 318, *24*
circulus maior *or* maior circulus: 225n; VF 280, *19*; 286, *90*; 328, *2*; 352, *56*; PA 436, *3*; *and* maximus circulus: CS 478, *3–4*; 482, *50–51*
circumdare: QT 22, *19*; QG 42, *22*
circumdatio: QT 22, *26*
circumducere: CS 468, *3*; 470, *5*; 478, *6*; CSA 538, *69*; QC 594, *32*
circumductio: QT 22, *22*; 26, *107*
circumferencialis: CC 172, *21*, *24*, *29*
circumferencialiter: CC 170, *11*
circumferens, *see* linea
circumferentia: QT 24, *57*; 34n, *41*; QG 42, *23*; CV 68, *13*; VA 394, *77*; *et passim ubique*
circumflectere: CC 170, *11*, *13*
circumflexa, *see* linea
circumiactus: QA 410, *95*, *98*
circumponere: 608
circumrotare: CC 170, 15
circumscribere: NV 86, *50*; FIA 104, *50*; VG 150, *20*; CC 188, *247*; VV 376, *28*; 380, *107*; VA 394, *44*; CS 511, *c 14*; CSA 538, *71*
circumscriptibilis: VV 380, *103*; 382, *132*; VA 396, *81*; CS 482, *6*; 550, *3*
circumscriptus: FIII 112, *6*; 132, *323*; VG 158, *138*; CC 176, *84*; VV 382, *135*; QA 406, *7*
circumvolucio: CSC 552, *17*, *18*, *23*, *et seq.*
circumvolutus: CC 170, *16*; CS 472, *39*, *48*; 478, *10–11*; CSC 550, *13*, *15*
circumvolvere: CS 450, *8*; 454, *44*; 498, *36*, *37*; CSC 554, *76*
circunferentia: 430, *c 45*
claudere: 143n; QA 408, *42*
coacervare: NV 84, *38*; ATI 642, *1*, *10*; ATII 654, *126*; 656, *151*
coalternus: CS 472, *26*
collectus: QT 24, *84*; CS 496, *23*
colligere ("to infer"): QT 22, *25*; FIA 102, *44*; 106, *71*; CC 190, *271*; 631
collocare: 78, *c 40*; 163, *c 31*; QC 596, *53*
colocare: VV 380, *111*
columna *or* columpna ("cylinder"): VF 336, *20*, *28*; 338, *31*; PA 436, *8*; CS 460, *3*; 517, *c 22*; CSC 552, *31*, *et seq.*; *and see* piramis columpna
columpnalis: CS 460, *15*; 476, *94*
commentator: FIII 118, *v 112–67*; CSC 554, *60*; 569n
commentum: FIII 122, *v 166–67*; CSC 554, *70*
communis: QG 52, *125*; FIA 102, *46*; VF 282, *53*; 290, *21*; 298, *4*; CS 468, *17*; QLI 616, *40*
communis animi concepcio: CC 178, *105–106*; MV 198, *13*
communis concepcio *or* communis conceptio: VG 158, *150*; MV 198, *25*; 202, *62*
communis sciencia *or* communis scientia: 136, *c 38–69*; MV 204, *103*; 210, *183*, *186*; 216, *284–85*, *318*
comparare: 136, *c 38–69*; VF 244, *60*, *62*; 246, *85*; *et passim in VF*; 629
comparatio: 31
compar: QA 414, *164*
complere: NV 84, *36*; VF 260, *14*; 304, *29*; QC 600, *82*; 660n
componere: QG 54, *146*; VG 150, *43*; 158, *146*; CC 172, *32*; VF 308, *84*; 316, *3*; 318, *27*; ATI 642, *2*; 644, *30*
comprobare: 139, *c 15*
computare: VF 268, *39*
computatio: VF 266, *22*; 350, *22*, *38*; 354, *72*
computator: VF 264, *v 1-24*; 266, *23*

concedere: 143n; VG 152, *65*; 160, *182*; CC 180, *153*
conceptio: 431, *c 45*; *and see* communis
concludere: FIB 112, *56*; 146n; VG 154, *97*; MV 204, *96*; 206, *148*; VV 374, *1*; QLII 620, *v 14–19*; (=includere): QA 429, *c 42–43*
conclusio: 33n, *12*; MV 198, *18*; 210, *204*; 296, *45*; VV 374, *2*, *3*; *et passim in VV*
concurrere: FIB 110, *47*; VG 158, *144*; *CC* 172, *31*; 184, *184*; *et passim in CC*; VF 310, *11*, *12*; 312, *28*; ATI 644, *25*; ATII 650, *53*, *54*
concursus: VF 310, *16*; ATII 650, *54*
confacere: VF 256, *v 4–40*
confirmare: QLII 620, *v 14–19*
congregare (=agregare): VF 286, *v 83*
conicus: CS 470, *4*; 472, *33*; 474, *61*; 482, *5*; *et passim in CS*; CSC 550, *2*; 631
coniunctim: FIII 114, *39*; 116, *84*; VF 314, *73*; 320, *60*; CSA 536, *51*; ATI 644, *41*
coniunctus: QG 48, *92*; FIII 112, *17*; *et passim in FIII*; VF 252, *5*; 278, *11*; 312, *30*; CSA 534, *21*
coniungere: CS 460, *16*; 627–28; TJ 672, *15*
conscribere: 36n; 78, *c 25*; VA 392, *33*
consequens ("consequent term in a proportion"): QG 48, *v 95*; CSA 534, *22*; ("conclusion of an implication"): 33n, *1*; MV 214, *265*; QA 406, *23*; 422, *250*
consequentia *or* consequencia ("implication"): MV 210, *201*, *206*; VV 374, *19*; 376, *24*; *et passim in VV*; QA 406, *22*; 408, *32*; *et passim in QA*; QLII 622, *v 41–45*
consimilis: FIA 104, *56*; CC 184, *202*
constituere: QT 24, *62*; 78, *c 33*; FIA 100, *13*, *29*; FIII 132, *347*; VG 150, *41*, *43*; CC 174, *79*; MV 198, *5*; VF 324, *102*; VA 390, *22*; QA 410, *95*; CS 470, *10*; CSI 529, *c 29*; QC 592, *28*
contactus: CV 70, *30*; 72, *62*; 90, *c 61*; FIA 102, *45*; MV 214, *270*, *273*; VV 382, *136*; VA 396, *84*; QC 596, *55*; ATI 642, *15*; 644, *28*
conterminare: CC 172, *21*
continere: QT 20, *13*; 24, *80*; 34n, *49*; QG 40, *4*; CV 68, *11*; NV 82, *4*; VF 246, *3*, *5*; VA 392, *42*; QA 408, *34*; PA 436, *2*; QC 594, *42*; QLII 618, *11*; *et passim ubique*
conteneri sub ("to be produced by the multiplication of"): 33n, *23*; CS 460, *2*; 482, *4*
contingencia *or* contingentia: MV 212, *221*; 218, *323*; CSA 538, *72*
contingens: QT 22, *35*; QG 48, *88*; CV 74, *66*; 86, *59*; 90, *c 61*; VF 254, *18*; 280, *24*; VV 382, *136*; VA 394, *45*; QA 430, *c 45*; CSA 536, *66*; QC 594, *34*
contingere: QT 24, *79*; QG 42, *35*; 90, *c 61*; VA 394, *48*; CS 454, *39*; CSA 538, *69*
continuare: VF 242, *v 35–95*; 244, *71*; 282, *55*; 302, *18*, *20*; 334, *3*; 336, *13*; 340, *61*, *5*; 348, *2*; ATI 644, *26*; PJ 662, *2*; PL 664, *2*
continuatio: VF 242, *v 35–95*; 246, *81*, *82*
continue: QA 406, *14*; 424, *292*
continuum: ATI 642, *8*; in continuum: ATII 648, *22*; in continuum et directum CC 188, *250–51*; VV 376, *31–32*; 382, *121*; 384, *140*; TC 680, *11–12*
continuus: FIA 100, *27*; 104, *60*; CC 182, *177–78*, VF 334, *8*; QA 418, *206*
contra se positi ("vertical angles"): ATII 652, *68*; TC 680, *14–15*
contradicere: 569n
contrarium: QG 42, *29*; CC 180, *140*; VF 256, *v 4–40*; 258, *20*; *et passim in VF*; CS 460, *9*
conus: 631
convenire: 78, *c 33*; 429, *c 3–4*; CSA 536, *67*
conversa: 138, *c 4–167*; MV 212, *226*; CSI 526, *38*
convertere: QG 54, *v 146–47*; 633
convexitas: CC 172, *21*, *26*, *30*
convexus: CC 182, *175*; 184, *205*
convincere: FIA 102, *31*
cooperire: QA 424, *296*; 430, *c 45*
copulare: QG 40, *13*; VA 392, *37*
corda: CV 68, *5*; NV 84, *44*; FIA 102, *41*; FIB 108, *24*; FIII 112, *9*; VG 154, *v 101*; 156, *109*; CC 170, *5*, *7*; MV 206, *128*, *130*; VF 272, *95*; VV 378, *76*; QLII 618, *11*; TJ 674, *18*
cordans: VV 378, *78*
corellarium: CS 458, *v 104*
corollarium *or* corolarium: VG 158, *149*; CC 188, *237*; VF 292, *v 24*; CS 458, *104*; 462, *39*; 488, *73*

corporale: VF 240, *v 35–95*; 244, *61*
corporeus: MV 210, *181*; 221, *c* 179–*81*; VF 240, *31*; 252, *21*
corpus: VF 238, *5*; 240, *34*, *v 35–95*; *et passim in VF*; CS 470, *4*; 482, *2*; *et passim in CS*; CSC 556, *79*, *80*
correlarium: FIII 112, *v 6*; 138, *c 4–167*; 141, *c 168-352*; 430, *c 45*; CS 458, *v 104*
correlativa: MV 206, *134*
costa: CC 178, *118*
credulitas: VF 352, *48*
cubicari: QA 406, *18*
cubicus: VF 348, *v 1–20*; QA 406, *20*
cubitum: QC 600, *92*
cubus: VF 334, *6*, *7*, *8*, *10*; 348, *6*, *13*; CS 504, *1*, *8*, *et seq.*
curta piramis ("truncated cone"): CSA 540, *1*; 542, *31*
curva: CV 68, *3*; linea curva: 143n; CC 170, *8*; 172, *20*, *25*, *27*, *31–32*, *36*, *37*; CQ 578, *18*; *et passim ubique*
curva superficies: CC 170, *2*; CS 450, *1*; *et passim in CS*

dabilis: VV 380, *87*, *103*
dactus (=dattus=datus): VV 374, *16*, *22*; *et passim in VV*
dare: FIB 106, *6*; VG 156, *117*; MV 204, *101*; VA 392, *28*; QA 424, *282*; 429, *c 3–4*; QC 590, *18*
datus: 33n, *8*; CV 68, *13*; 163, *c 37*; CC 176, *98*; MV 202, *47*; 221, *c 219*; *et passim ubique*
decisio: QA 408, *55*; 569n
declarare: QG 46, *83*; 50, *110*; FIA 104, *70*; FIII 112, *22*; VG 148, *8*; 164, *c 68–69*; VF 242, *52*, *54*; 248, *21*; *et passim in VF*; ATII 648, *9*
declaratio: QG 46, *68*; VF 242, *49*
decursus, *see* ratio
deducere: VA 390, *11*; QA 424, *297*
deductio: 34n, *88*
deesse: FIII 124, *185*; 132, *347*
deforis: QA 428, *c 3–4*
deincepsque: 60n; CC 178, *112*
demere: 60n; NV 86, *57*; CC 178, *112*; 182, *172*; MV 200, *38*; VA 390, *19*; QLII 622, *33*; ATII 650, *43*
demonstrabilis: VV 374, *2*; QA 410, *97*; 422, *265*
demonstrare: 34n, *45*; CC 174, *50*; VF 260, *8*; CSI 528, *14*; QC 588, *4*, *5*; QLI 618, *53*, *54*
demonstratio *or* demostracio: VG 152, *65*; CC 170, *1*; MV 198, *7*; VF 248, *9*; VV 374, *1*; 429, *c 3–4*; CS 452, *15*; CSI 528, *14*; 569n; CQ 578, *16*; TJ 674, *33*; *et passim ubique*
demonstrative: 144n; VG 148, *1*; VV 374, *1*; 382, *125*; QA 410, *79*, *85*
demonstrator: CS 484, *25*; 496, *6*
demptio: QLII 622, *37*
denominatio: FIII 114, *51*; 126, *222*; *et passim in FIII*; 146n; QA 414, *158*, *159*, *161*
dependens: CS 488, *93*; 492, *153*
depingere: 608; 632
deponere: QG 48, *v 95*
descendere: FIA 100, *23*
describere: QT 20, *12*; 26, *109*; QG 42, *30*; CV 74, *v 83*; 78, *c 25*; FIB 106, *8*; FIII 122, *168*; VF 252, *v 11–16*; VV 378, *61*; 428, *c 3–4*; CS 460, *10*; QLI 616, *22*; ATII 650, *59*; TJ 672, *4*; *et passim ubique*
descriptio: VG 150, *44*; 154, *83*; 160, *168*; VF 318, *27*; CS 484, *26*, *33*
designare: QT 26, *103*; 137, *c 33*; VG 150, *32*; 156, *120*; 165, *c 152*; 222, *c 307*; VF 338, *31*; ATI 642, *13*
determinare: VF 242, *53*
determinatus: VF 340, *13*; 342, *35*; 346, *10*; TJ 672, *6*
detractio *or* detraccio: FIA 100, *27*; 104, *61*; VG 152, *63*; 158, *134*; CC 180, *135*; 184, *178*; *et passim in CC*
detrahere: 60n; CV 72, *58*; VG 150, *15*, *36*; CC 178, *111*
diameter *or* dyameter: CV 68, *22*; 97n; FIA 100, *13*; FIII 112, *7*; 146n; CC 176, *95*; 188, *246*; 225n; VF 260, *v 4–5*; 310, *1*; VA 394, *72*; QC 590, *10*; *et passim ubique*
diametros: CV 74, *87*
diametrus *or* dyametrus: QT 24, *61*, *78*; QG 46, *71*; VF 264, *15*
differentia: 220, *c 25*; CS 466, *3*, *4*, *9*; 488, *96*; CSI 524, *3*; ATI 642, *2*
differentia communis: VF 318, *45–46*; 338, *39*
diffinire: 627, 629
diffinitio *or* diffinicio: 136, *c 38–69*; 138, *c 4–167*; 141, *c 168–352*; MV 200, *45*;

202, *55*; 204, *110*, *118*; 206, *136*, *141*; 210, *197*; 214, *278*; CS 468, *22*
dimensio: 627
diminuere: 140, *c 147*; VF 306, *67*
diminuibilis: VV 380, *96*
diminutio: QG 50, *117*; 140, *c 147*
dimittere *or* demittere: CS 450, *8*; 454, *40*, *41*; 468, *16*
dinumerare: 430, *c 45*
directe: 91, *c 63*; FIB 108, *18*; CC 186, *231*
directum, in: CC 170, *11*, *15*; 188, *251*; VV 376, *31–32*; 382, *121*; 384, *140*; VA 390, *14*
disponere: CV 76, *103*; MV 210, *198*
dispositio: VF 262, *33*; 300, *29*; 342, *25*; QA 428, *c 3–4*; CS 454, *38*; 458, *86*; 476, *97*; 480, *30*; 496, *17*; 500, *66*; 502, *7*, *16*; QJ 572, *2*; QC 594, *32*
dissolutio: 169n; CS 454, *48*
distinguere: QLII 620, *22*
diversificare: QC 594, *44*; 598, *67*
dividere: QT 22, *34*; 24, *83*; 33n, *25*; QG 42, *33*; CV 72, *39*; 78, *c 40*; NV 84, *30*; VF 260, *12*; VV 376, *33*; VA 394, *45*; QA 408, *51*; CSI 532, *5*; CSC 552, *31*; QC 590, *12*, *13*; ATI 646, *54*; TJ 672, *2*; *et passim ubique*
divisibilis: VV 380, *96*; QA 418, *208*
divisibiliter: VV 380, *91*, *109*
divisio: FIII 114, *57*; 122, *v 154–63*; 128, *254*; VF 310, *7*; 312, *35*; VV 376, *34*; CS 454, *32*; 570n; TJ 674, *32*
divisor: FIII 122, *v 154-63*
divisum (=dividendum): FIII 122, *v 154–63*
docere: 34n, *94*; 36n; 144n; MV 200, *44*, *45*; 202, *60*; 210, *196*; 212, *213*; QLII 622, *v 41–45*
doctrina: CC 170, *3*; MV 216, *306*; QC 604, *116*
dubitabile: CC 170, *18*; 172, *27*
ducere (=multiplicare): 33n, *20*; 34n, *34*; 78, *c 25*; FIA 102, *33*; CS 496, *17*; CSA 540, *2*; ATI 642, *3*; (=protrahere): 34n, *60*; CV 70, *30*; NV 84, *31*; FIA 102, *31*; FIB 108, *18*; 143n; VG 150, *38*; MV 212, *233*; 221, *c 219*; VV 374, *12*; VA 390, *9*; QA 414, *147*; 430, *c 45*; CS 470, *14*; QC 590, *9*, *17*; ATI 644, *20*; TC 680, *8*; ("to turn an integer into a fraction"): FIII 114, *50*; 116, *95*; *et passim in FIII*
ductus (=multiplicatio): CV 72, *42*; NV 84, *34*; FIB 108, *24*; 137, *c 26*; CC 174, *57*; MV 206, *125*; CS 454, *48*; *et passim in CS*; ATII 654, *103*
dulk (=Euclid, *Elementa*, I.47): QG 48, *v 94*; 97n
duodecagonum: FIII 112, *v 6*
duplare: FIII 122, *157*; VF 242, *v 35–95*; 244, *70*; 246, *80*; QLII 624, *v 43*
duplatio: 246, *77*
duplicare: VF 246, *v 80*; 262, *22*; CS 500, *71*; QJ 572, *v 18–21*; CQ 580, *26*
dyameter, see diameter
dyagonalis: 34n, *71*

econverso: FIII 122, *152*
educere: CSI 526, *26*; CSA 538, *68*
egredi: QG 46, *67*, *79*; MV 206, *124*; VF 294, *34*; *et passim in VF*; PA 436, *5*, *6*
eicere: 221, *c 197*
Elementa: VV 386, *182–83*; *and see separate Index of the Citations of Euclid*
elementum: CS 496, *7*
Elencorum [*liber*]: QA 406, *4*
elevare: VF 294, *21*; 302, *14*; 318, *33*; et *passim in VF*
elevatio: VF 240, *v 35–95*; 244, *65*
eligere: 97n
elongare: VF 280, *35*, *36*, *39*; 310, *26*
embadum: QT 20, *9*; 24, *52*; VF 246, *4*; 248, *8*, *12*, *15*; *et passim in VF*; 568n
enumerare: CS 474, *63*
equalis: QT 20, *7*; 33n, *4*; QG 40, *3*; CV 68, *3*; NV 82, *2*; VF 252, *v 11–36*; *et passim ubique*; ad invicem equales: 138, *c 46–47*; QA 428, *c 3–4*; sibi invicem equales: 91, *c 63*
equalitas: 34n, *61*; 193, *c 275*; VF 306, *60*; QA 408, *36*; 424, *316*; CS 504, *22*; QJ 572, *17*; ATII 654, *99*; TC 680, 8
equaliter: QA 406, *26*; 408, *51*; QC 590, 18; 600, *95*
equare: QT 20, *6*; 34n, *50*; 36n; QG 40, *5*; CV 68, *11*; FIA 100, *5*; FIB 106, *3*; CC 176, *93*; 225n; VF 324, *114*; VA 392, *25*; CS 450, *4*; 460, *109*; CSA 540, *5*
eque: 79, *c 50*; *and see the compound words that follow where* eque *is often written in a compound word without being changed to* equi

equemultiplices: QA 414, *163*
equiangulus: VG 158, *147*; 165, *c 152*; VV 376, *24*; VA 390, *113*; QA 414, *142*; ATII 652, *66*, *69*
equidistans *or* equedistans: 34n, *67*; CV 70, *31*; NV 84, *15*; VG 152, *49*; MV 202 *59*; 212, *226*; VF 298, *3*; VV 376, *47*; VA 390, *17*; CS 468, *15*; 529, *c 29*; TJ 672, *14*; *et passim ubique*
equidistanter: 35n, *v 67*
equidistantia *or* equidistancia: VG 150, *31*; 152, *55*; MV 212, *226*
equidistare: VF 302, *3*
equilaterus: *33*n, *22*; 60n; CV 70, *27*; VG 158, *147*; 165, *c 152*; VV 374, *4*; VA 390, *13*; QA 414, *142*; QC 604, *117*; ATII 652, *69*
equiparens: FIII 116, *78*; 130, *303*
equipollere: VG 152, *68*
equivalentes: FIII 114, *52*; 128, *251*
equivalere: 114, *53*
equus: QT 20, *9*; CV 72, *40*; 78, *c 25*; 90, *c 32*; 164, *c 149*; 222, *c 307*; CS 474, *69*; *et passim ubique*
erectus: VF 310, *9*, *23*; 336, *23*, *27*; CS 428, *2*
erigere: QT 22, *37*; VF 266, *27*; 294, *36*; 336, *20*; CS 460, *11*; CSI 524, *14*; ATI 644, *24*; TJ 672, *5*
errare: FIII 118, *v 112–67*; QC 588, *5*
error: VF 242, *49*; 278, *v 179–82*
Ersemides (=Archimedes): QT 20, *2*
esse sicut (=esse equalis): QG 46, *70*; CV 74, *85*; FII 106, *75*
esse ut (=esse equalis): QT 24, *60*; *et passim in QT*
estimare: VF 242, *39*, *48*; 300, *17*; 342, *20*
Euclides, *see separate Index of the Citations of Euclid*
eversa proportio: CS 490, *106*
eversa proportionalitas: CSA 540, *103*
everse: ATII 652, *91*
eversim: CSA 534, *33*; 536, *51*
evidentia *or* evidencia: FIB 110, *52*; CC 174, *50*; 184, *185*
ex adverso (=oppositus): 163, *c 31*
exagonum (=hexagonum): QG 48, *v 89*; 50, *120*; FIII 112, *12*; 122, *169*; VF 266, *30*; QLI 616, *24*, *26*, *et seq.*
exagonus: CC 172, *40–41*; QA 416, *192*
excedere: QT 22, *18*; FIII 116, *62*; VG 156, *118*; MV 210, *211*; 219, *c 25*; VV 380, *91*, *110*; ATII 650, *34*
excellere: MV 198, *24*, *25*, *26*; 210, *209*; 219, *c 25*
excessio: CV 68, *v 17*
excessus: CV 68, *17*; 72, *56*; FIA 100, *11*; 102, *38*; FIII 132, *351*; VG 150, *13*; CC 178, *107*; MV 198, *27*; VV 380, *91*; QA 418, *207*
excrescere: FIII 114, *51*; 145n; 146n
exempli causa: QG 46, *71*; FIA 106, *76*; VF 292, *13*
exemplum: 36n; VF 270, *59*, *70*, *82*; 350, *34*; CS 466, *6*; 496, *13*; ATII 648, *3*; pro exemplo: MV 214, *270–71*
exire: QLI 616, *29*, *30*
existere: QT 20, *7*; FIA 102, *45*
exitus: VF 300, *30*
extendere: CC 170, *12*, *13*; VF 240, *35*; 242, *37*; 304, *30*; 340, *12*; QA 406, *26*; 408, *60*; 424, *305*
extensio: VF 240, *v 35–95*; 242, *38*, *46*, *47*
exterior: CV 70, *25*; CC 174, *48*; CSC 554, *72*
exterior lunula: NV 88, *72–73*
extra: FIA 104, *54*; MV 212, *243*
extra scribere (=circumscribere): VG 150, *21*, *25*
extra designare (=circumscribere): VG 150, *32*
extrahere: VF 334, *6*; 350, *41*; 352, *64*
extremitas: QG 46, *67*; VF 260, *40*, *42*; 296, *57*; CSC 554, *63*; QJ 572, *16*; 628
extremum: QA 406, *11*; CSC 554, *75*; QC 592, *30*
extrinsecus: FIB 110, *50*; FIII 132, *325*; ATII 650, *55*

facere: QT 22, *17*; QG 40, *10*; CV 72, *40*; NV 84, *25*; FIB 108, *20*; FIII 118, *120*; VF 252, *13*, *17*; 258, *14*; *et passim in VF*; VA 392, *36*; QLII 624, *44*; PJ 662, *2*
falsigraphia: CC 186, *v 218*; QLI 618, *51*
falsigraphus: 169n; CC 186, *218*; CS 453, *33*, *46*; 456, *72*, *76*; 458, *80*, *84*, *94*; 464, *13*; 480, *39*; 500, *65*, *81*; 568n
falsitas: QA 406, *23*; 422, *249*
falsus: NV 84, *30*; FIA 102, *31*; FIII 122, *v 166-67*; VG 154, *81*; 160, *187*; MV 216, *302*; QA 406, *12*; 418, *220*; CS 464, *20*

fere ("approximately"): QT 24, *71*; QG 46, *81*; FIA 106, *83*; FIII 116, *61*, *62*; ATII 652, *v 91*; per fere ("by approximation"): CS 504, *5*
fieri: QG 42, *24*; CV 68, *22*; NV *84*, 21; FIII 116, *82*; VF 252, *3*; CS 460, *16*; et *passim ubique*
figere: VF 336, *25*, *29*
figura: QT 22, *19*; 26, *103*, *107*; 33n, *3*; CV 68, *6*; NV 84, *36*; FIII 120, *140*; 220, *c 45*; VF 238, *5*; PA 436, *13*; CS 500, *63*; QC 588, *8*; 600, *84*; 633; *et passim ubique*
figura altera parte longior(=rectangulum): 33n, *17*, *18–19*
figura laterata [et angulata] ("a regular polygon"): VF 246, *1*, *4*; 252, *4*, *8–9*; *et passim in VF*
figura poligonia equalium laterum et angulorum: CS 454, *38–39*
figura poligonia equilatera [et equiangula]: VV 374, *4*; VA 390, *1*
figura rectilinea poligonia ("a regular polygon"): QG 42, *18*
figuraliter: 546, *c 13*
finis: VF 266, *21*; 340, *13*; 342, *35*; 346, *10*; 348, *13*; 350, *22*; TJ 672, *6*
fixus: QC 592, *v 27*; 594, *46*
forma: CV 68, *18*; VF 254, *36*; 260, *45*; et *passim in VF*
formatum: MV 214, *258*
fortitudine, in ("squared"): QT 26, *86*; *and see* potentia
fractio ("fraction"): VF 350, *25*
Franco scolasticus: QC 588, *v 1*
frequenter: QT 22, *17*, *44*
fundata: CS 458, *83*; 490, *102*

gemini [anguli]: CS 470, *23*; 476, *104*
generatio: 631
geometer: 169n
geometria: FIII 112, *15*; CC 178, *110*; 180, *125*, *130*; *et passim in CC*; VF 240, *29*; 334, *5*; 336, *12*; 348, *4*; 352, *54*; 354, *72*; VV 386, *187*; 569n
geometricus: VF 352, *60*; 428, *c 3–4*; 568n, 628
gnomicus: 34n, *38*
gratia brevitatis: VG 160, *181*
gratia compendii: VG 152, *64*
gratia consorcii: FIII 128, *266*
gratia exempli: MV 206, *124*; 214, *268*; VA 396, *83*

habere, se: FIII 126, *237*; QA 416, *166*; CSI 526, *33*; QLI 614, *11*; *et passim ubique*
habitudo: 633
hortogonius (=orthogonius): VF 246, *v 92*
hypothesis: VF 256, *v 4–40*; CS 452, *15*

iacere: 659n; TJ 672, *12*
id quod fit ex ductu (*or* ex *or* ex multiplicatione)—in—("the product of—and—"): NV 84, *34*; 88, *80*; VG 154, *72–73*; CC 174, *56–57*; VF 322, *75*; CS 468, *32*; *et passim in CS*
idem (=equalis): VA 390, *19*
illatio: VG 152, *68*; 154, *72*
illud quod fit ex ductu (*or* ex multiplicatione)—in—("the product of—and—"); CV 72, *41–42*, *43*; FIB 108, *23*; 137, *c 26*; VG 154, *96*; 191, *c 66*; VF 304, *46–47*
imaginari: VF 342, *28*, *32*, *33*; 430, *c 45*
immota: CS 450, *8*
impar: CS 476, *93*; 494, *175*; CSC 550, *6–7*, *10–11*
imperceptibilis: FIA 106, *84*
implicare ("to imply a contradiction"): VG 156, *109*; VV 382, *127*, *128*, *131*; 384, *152*, *153*, *173*
implicatio ("contradiction"): VV 384, *173*
impossibile: QT 22, *29*; QG 42, *29*; VG 154, *103*; 156, *110*; CC 180, *154*; MV 200, *31*, *et seq.*; VF 258, *21*; QA 406, *26*; *et passim ubique*; per impossibile: NV 82, *6*; 86, *47*; VF 256, *v 4–40*
improbare: NV 88, *79*; CS 466, *32*
improbatio: CS 458, *85*
improportionalis: CC 186, *v 218*
includens: CV 68, *6*; FIA 100, *18*; VV 380, *106*; CS 458, *97*; 480, *40*; CSA 538, *95*
includere: VF 252, *v 11–36*; VV 380, *106*; CSA 536, *60*; 552, *49*
inclusus: CV 68, *7*; VV 380, *106*; CS 456, *70*; 458, *96*; 460, *18*; 480, *40*; CSA 536, *61*, *67*; 538, *95*; CSC 550, *14*
incompossibile: CC 186, *218*
inconcussum: CS 466, *33*; 500, *66*
inconveniens: VG 156, *108*; 162, *190*

incurvare: VV 378, *75*
indistincte: CSC 550, *2, 4*
indivisus: FIII 122, *162*; 191, *c 66*
indubitabile: CC 170, *6*
inducere: QC 588, *7*
inequalis: 33n, *28*; 60n; NV 82, *9*; 90, *c 32*; CC 178, *110*; 222, *c 307*; 369n; VV 378, *78*; *et passim ubique*
inequalitas: QA 424, *317–18*
inferioris: VF 308, *85*; VV 378, *77*
inferre: VF 328, *v 4–36*
infinitum, in: VV 380, *96*
infra: MV 200, *30, 43*
ingenium ("device"): VF 344, *2*; 346, *33*
inherere: TJ 672, *9*
iniungere: MV 202, *66*
inordinatus: 628
inproportionalitas: 138, *c 4–167*; 141, *c 168–352*
inscribere: CV 68, *21*; NV 82, *12*; FIA 100, *11*; VG 150, *18*; CS 454, *38*; CSA 534, *12*; *et passim ubique*
inscripcio: MV 202, *55*
inscriptibilis: VV 380, *88, 92, 98*; VA 396, *92*; CS 482, *5–6*; CSC 550, *3*; QLII 618, *3*
inscriptus: CV 68, *24*; FIA 100, *19*; VV 378, *77*; 382, *138*; QA 406, *8*
inseparabilis: VF 342, *22, 30*; 348, *36*
integra: FIII 114, *52*; 116, *60*; *et passim in FIII*; CS 504, *6*
inter se ("mutually"): VG 154, *91*
interiacere: NV 86, *64*
interior: 220, *c 45*; CS 460, *17*; CSA 534, *14*
intersecare: QC 590, *13*; 592, *28*; *et passim in QC*; ATII 648, *22*; TJ 674, *19*
intra: 78, *c 40*; MV 204, *101*; 220, *c 45*
intrinsecus: QA 430, *c 45*; CSI 529, *c 29*
invenire: 33n, *4*; 36n; CV 74, *82*; FIII 112, *23*; 122, *v 154–63*; CC 176, *88*; 188, *241*; VF 264, *8*; QA 406, *25*; CS 456, *55*; 474, *59*; CQ 578, *22*; ATII 648, *1*; PL 664, *1*
inventio: 33n, *19–20*; QA 408, *58*; 568n
invicem ("mutually"): MV 202, *55*
ipothesis (=ypothesis): CSA 536, *56*
ire: VF 342, *31*
iungere: QA 410, *72*; 412, *111*; *and see* simul iuncti
lateralis: MV 206, *125, 126, 135*
laterata: VF 246, *1, 4*; 252, *4, 9*; 256, *v 4–40*; *et passim in VF*; *and see* figura laterata
latitudinaliter: VF 240, v *35–95*; 242, *37*
latitudo: VF 240, *31, 32, 33*; 242, *39, 42, 46, 47, 48, 51, 53*; VV 380, *91, 97, 110*; ATII 652, *v 91*
latus: QT 20, *4*; 33n, *28*; QG 40, *3*; CV 68, *10*; NV 82, *3*; VF 246, *3*; VV 374, *5*; *et passim ubique*
latus cubi: VF 334, *6*; 348, *3, 6, 10*
linea: QT 20, *15*; 33n, *24*; QG 42, *21*; VF 242, *40*; *et passim ubique*
linea ambiens (=perimeter): FIA 102, *33*
linea ambiens octogonium: CC 180, *145*
linea ambiens poligonium: FIA 104, *64*; FIII 132, *349*
linea ambiens quadratum : CC 182, *163–64*
linea circumdans (=circumferentia): VF 322, *79*
linea circumdans poligonium (=perimeter): QG 42, *22, 26–27*
linea circumducta (=circumferentia): VF 306, *58*, *et seq.*
linea circumferens (=circumferentia): QT 20, *6*; 24, *70*
linea circumflexa (=circumferentia): VF 278, *178*
linea continens circulum (=circumferentia): 34n, *51*; 36n; QG 40, *5–6*; FIII 132, *351*; 134, *352–53*; 140, *c 147*; CC 176, *93–94*; VF 278, *v 179–82*; 292, *12*; *et passim in VF*; PA 436, 5
linea continens poligonium (=perimeter): QG 44, *51*
linea curva, *see* curva
linea recta: 90, *c 62*; FIII 120, *139*; CC 182, *174*; 627; *et passim ubique*
lineare (" to draw" *or* "to construct"): VF 266, *25*; 342, *42*; 346, *9*
linula (*an error for* lunula): CV 70, *v 29*, *v 34*
locabilis: VV 380, *91, 94, 97*
locare: VV 380, *96, 98, 100*
locus: VF 300, *30*; 336, *25*; 346, *16, 18*; QC 592, *28*; 598, *64*, *et seq.*
longior: 33n, *17, 19*; VF 256, *9*; 258, *22*; *et passim in VF*; VV 378, *71*; QA 422, *250*
longius: 221, *c 218*; *et passim ubique*
longitudo: VF 240, *30, 32, 34*; 242, *36, 37, 40, 42, 44, 47, 51, 53*; ATII 652, *v 91*; in

longitudine ("in the first power" *as opposed to* "squared"): QT 26, *89*, *98*; QG 48, *95*; FIII 118, *122*
lunaris, figura: 569n
lunula ("lune"): QLI 614, *18–20*; 616, *44*, *47*, *et seq.*; 618, *54*, *55*; QLII 618, *1*; 622, *35*, *41*, *et seq.*; (*misused as* "segment of a circle" *or as* "a figure contained by curved and straight lines"): CV 70, *29*, *34*; 74, *72*; NV 84, *17*, *18*, *19*; 86, *57*; 88, *70*, *72*, *et seq.*; FIA 100, *20*; FIB 108, *2*; CC 180, *134*, *135*; *et passim in CC*

magnitudo: VF 238, *5*; 240, *33*; 242, *53*; 244, *56*; *et passim in VF*; VV 380, *93*; QA 428, *c 3–4*; QC 604, *115*; QLII 620, *v 14–19*; ATII 652, *v 91*
magnus, maior, maius: *passim ubique*
maior ("major premise"): QA 414, *161*; QC 602, *108*, *v 108–109*
maiuratura: 145n–46n
manere: VV 378, *75*; CS 458, *85*; *et passim ubique*
margo: VF 346, *13*; 348, *36*
mathematica: QC 590, *19*
mathematicus *(n)*; QC 590, *v 19*; 602, *103*
maximus: 33n, *31*; FIII 126, *238*; 128, *268*; 221, *c 179–81*; VV 408, *29*
medietas: QT 20, *16*; 33n, *24*; QG 40, *12*; 60n; CV 68, *19*; *et passim ubique*
meditas diametri (=radius): 34n, *41–42*; QG 40, *4*; CV 68, *11*; FIA 100, *4*; 225n; VF 246, *2*; *et passim ubique*
medium *(n)* ("middle"): 33n *27*; NV 84, *35*; VG 154, *75*; 163, *c 31*; CC 176, *83*; VF 296, *66*; VV 378, *61*; VA 392, *37*; CSA 536 *66*; ("half"): QG 40, *13*; FIII 112, *14*; 124, *192*; MV 204, *104*; VF 266, *31*; 310, *5*
medium proportionale: 568n; QJ 572, *12*; 574, *19*, *21*; 659n; *and see* proportionalis
medius ("middle"): NV 84, *31*; CC 174, *55*, *62*; VV 374, *7*; ("mean"): 33n, *19*; VF 288, *114*; 334, *10*; 430, *c 45*
meguar ("axis"): VF 336, *29*, *30*
mensura: 33n, *6*; 34n, *44*; QG 40, *1*; VF 238, *4*; 264, *v 18*; 348, *13*; 350, *16*; CS 466, *34–35*; 502, *23*; QC 590, *15*, *16*; 608
mensurare: VF 240, *v 35–95*; 244, *70*, *72*; 246, *78*, *83*; *et passim in VF*; 631
mensuratio: 36n; 225n; VF 240, *v 35–95*; 244, *60*, *73*; 246, *78*; 286, *85*; 288, *98*; 630
Mileus (=Menelaus): VF 336, *12*; 340, *1*; 352, *68*
minimus: FIII 126, *239*; 128, *268*
minor: QT 22, *17*; 40, *10*; 60n; NV 82, *7*; *et passim ubique*
minor ("minor premise"): 34n, *96*; QA 416, *166*; QC 602, *102*; 604, *109*
minuere: 36n; VF 306, *65*
minutum ("fraction"): FIII 114, *51*; *et passim in FIII*; ("sexagesimal fraction of the first order"): VF 264, *19*; 350, *16*, *31*
modus: QT 26, *90*; 60n; NV 88, *75*; FIA 100, *24*; VG 152, *62*; MV 210, *189*; VF 250, *13*; 264, *5*; 348, *11*, *12*; 352, *45*; ATII 656, *148*; TJ 674, *26*
monstrare: QT 22, *29*
motus: VF 342, *22*, *23*, *30*; 346, *13*; 348, *36*, *38*; TJ 672, *8*
movere: VF 342, *20*, *24*, *28*; 346, *12*, *15*; 348, *34*; QA 424, *295*; 430, *c 45*; 659n; PJ 662, *6*; PL 664, *8*; TJ 672, *8*
Macrobius: CV 74, *87*
multiangula: QT 22, *22*; 26, *103*; *and see* rectilinea
multiplicare: 36n; QG 54, *v 148*; FIII 114, *24*; 225n; ATII 654, *116*, *117*
multiplicatio: QT 22, *23*; 24, *53*; QG 42, *24*; 79, *c 50*; FIA 106, *72*; FIII 116, *82*; VF 246, *2*; *et passim in VF*; ATI 642, *8*; PL 664, *11*; *et seq.*; *et passim ubique*
multiplices: 79, *c 50*, *c 107*; CC 190, *280*; 193, *c 278*
multo a forciori: MV 212, *240*; *all of the phrases with* multo *are a part of the exhaustion procedure and are used passim*
multo fortius: FIII 122, *155*; VG 160, *171–72*; MV 210, *184*
multo magis: CV 74, *71–72*; NV 88, *70*; FIB 110, *48*; VIII 114, *48*
multo maior: QT 22, *40*; VF 256, *v 4–40*
multo minor: VF 256, *v–40*
mutekefia ("reciprocally proportional"): 193, *c 239*

necessaria: VF 238, *6*; 428, *3–4*
necessario: VG 148, *7*; 156, *114*; VV 382, *129*; QC 600, *93*

necessarium: CC 170, *3*; 172, *43*; 182, *160*
necesse: 60n; FIA 100, *27*; 139, *c 15*; *et passim ubique*
necessitas: VG 152, *63*; 156, *116*; VF *238*, *4*
notum, per se: CC 170, *4*, *6*, *9*; 172, *23*
numerare: 430, *c 45*; CS 456, *54*
numerus: 33n, *10*; QG 52, *134*; 97n; FIII 112, *23*; *et passim in FIII;* VG 152, *69*; 225n; VF 334, *8*, *10*; 428, *c 3–4*; CS 476, *93*; 494, *175*; CSA 550, *4*; 557, *c 41*; ATI 642, *9*
numerus remanens ("the remainder"): FIII 122, *v 154–63*
numerus qui relinquitur ("the remainder"): FIII 122, *v 154–63*

oblique: CS 472, *28*; 476, *97*; per obliquum: 608
obtusus: CS 458, *87*
occupare: VG 158, *161*, *163*
occurrere ("to occur"): CV 68, *20*; NV 84, *25*; FIA 104, *61*; VG 152, *63*; CC 178, *115*; ("to meet"): VF 312, *33*; 336, *19*
occursus: CS 472, *26*; 484, *17*
octogonium: 60n; CV 70, *27*; FIA 100, *19*; CC 180, *142*; MV 206, *122*; *et passim ubique*
octogonius: CC 174, *67*; 180, *139*; MV 210, *185*, *188*; QA 418, *224*
octogonum: NV 84, *28*, *30*, *36*; *et passim in NV*
octogonus: FIB 108, *19*, *25*; *et passim in FIB*; VG 150, *41*, *43*; 152, *66*; *et passim in VG*
operacio: MV 204, *90*; VF 348, *v 1–20*
operari: MV 212, *246*
operatio: VF 336, *11*
opponere: ATII 652, *60*, *84*
oppositum: VV 382, *127*; 384, *161*, *164*
oppositus: 33n, *26*; 78, *c 25*; VG 150, *42*; 164, *c 149*; 192, *c 191*; MV 198, *9*; 212, *217*; 221, *c 218*; CS 470, *14*; 529, *c 29*; CSC 556, *77*; 669n
orbiculariter: QC 590, *9*, *17*; 600, *85*; 602, *99*
ordinare: QC 590, *22*; 629
ordo: CS 462, *21*
orthogonaliter *or* ortogonaliter: 34n, *60*; 78, *c 40*; 164, *c 149*; MC 206, *123*; VF 260, *14*; 266, *28*; 294, *33*; QA 408, *50*; 430, *c 45*; CS 460, *14*; OLI 614, *4*; TC 680, *12*
orthogonius *or* ortogonius: QT 20, *3*; 36n; QG 40, *2*; CV 68, *9*; FIA 100, *2*; FIB 106, *2*; 143n; CC 176, *90*, *96*; 148, *203*; MV 198, *3*; VF 242, *v 35–95*; 246, *86*, *90*, *92*; VV 374, *5*; VA 390, *2*; CS 450, *2*; 468, *13*
orthogonus: QA 414, *143*, *151*; 416, *201*
ostendere: QT 22, *37*; VG 148, *9*; MV 200 *34*; VF 238, *16*; CS 460, *107*; PL 664, *1*

par: CS 550, *4*
pari, a ("by equality"): CS 502, *23*; 504, *17*; CQ 580, *30*; TJ 674, *20*
paralellogramum: 78, *c 33*; VG, 152 *53*; QJ 572, *18*
paralellogrammum: MV 202, *76*; *et passim in MV*
parallellogramum: CV 70, *v 31*
parallelogrammum: 34n, *69*
parallelogramum: CV 70, *30*; NV 84, *14*; FIA 100, *22*; FIB 108, *15*; CC 180, *131*; 220, *c 110*; CQ 578, *14*; *et passim ubique (perhaps some of these are* parallelogrammum; *the abbreviations are ambiguous)*
parallelogramum rectangulum: 220, *c 110*; CS 476, *108*
paralogismus: 222, *c 320*
pariter accepti: NV 84, *39*, *43*; FIA 100, *20*; FIII 114, *58*; QA 414, *165*; CS 456, *65*, *67*; 514, *c 78*
pariter sumpti: CS 456, *69–70*; 458, *88*
pars: QT 24, *63*; QG 50, *106*; CV 74, *88*; NV 84, *27*; VF 310, *11*; 529, *c 29*; *et passim ubique*
partialis: NV 86, *63*, *64*; FIII 120, *143*; CC 172, *39*
partialis [triangulus] ("a triangle which is a part of another triangle"): CV 72, *43*; NV 84, *33*; VG 152, *50*
particulatim: 191, *c 66*
partire: ATI 644, *17*
parvitas: VV 380, *93*, *94*
parvus: CV 70, *36*; 72, *42*; CC 184, *199*; VV 380, *95*; *et passim ubique*
penetrare: VF 300, *12*; 310, *26*
pentacubitum: QC 600, *92*
pentagona: VA 390, *8*
pentagonum: 165, *c 152*

pentagonus: VV 374, *12, 13, 16*; *et passim in VV*; VA 390, *11, 12*; *et passim in VA*
per se, *see* notum
perficere: 34n, *62*; CS 450, *9*; ATI 646, *62*
periferia: NV 82, *4*; 84, *44, 46*; 86, *47*; 144n; 369n; QLI 616, *43*; 659n; TJ 674, *36*
perimeter: 630
peripheria: 669n
permittere: QG 48, *v 95*
permutata proportionalitas: QLII 620, *v 14–19*
permutatim: QG 52, *v 128*; CV 76, *107*; FIII 112, *16*; CS 464, *16, 29*; QJ 572, *6*
perpendicularis: QT 24, *53*; QG 42, *20–21*; NV 84, *31–32*; FIA 100, *23*; CC 174, *54*; 193, *c 237*; VF 250, *9*; 282, *46, 52*; VV 378, *60*; VA 390, *9*; CS 450, *7*; 634; ATI 642, *15*; PJ 662, *3*; TJ 672, *5*; *et passim ubique*
perpendiculariter: QT 22, *36*; 34n, *33*; VF 292, *v 3*; VV 384, *158*; VA 394, *76*; QA 414, *147*; CS 470, *17*; ATII 652, *84*; TC 680, *6*
perscriptiones: 34n, *37*
perspectiva: TJ 674, *38*
pervenire: CV 74, *75*; CC 170, *17*; VF 264, *13*; 290, *2*; QA 406, *12*; 424, *289*; 430, *c 45*
pes: 33n, *8*
pes circini: VG 158, *161*; QC 592, *27, 31*; TJ 674, *35*
petere: CC 170, *1, 3, 6, 9*; 172, *22*; 628
petitio *or* peticio: 34n, *44*; 36n; CV 68, *24*; 72, *49*; CC 174, *45*; 180, *146*; 182, *163*; 184, *214*; QA 408, *43*
petitum: CC 170, *5, 8*; 172, *20, 34*
phalsigraphus (=falsigraphus): CS 458, *v 80*
Philosophus (=Aristoteles): QA 406, *3*
philosophus: QA 412, *108*; CS 452, *22*
physica *or* phisica: QC 590, *19, v 19*; 602, *103*
Physica [*Aristotelis*]: MV 198, *26*; 220, *c 25*; QA 406, *4*; 569n
piramis ("pyramid"): PA 436, *13*; ("cone"): CSC 550, *12, 15*; 552, *44, et seq.*; *and see* pyramis
piramis columpna *or* piramis columna ("cone"): VF 292, *2, 10*; 294, *41, v 41*; *et passim in VF*; PA 436, *16, 18*
piramis rotunda ("cone"): CS 450, *9*; *and see* rotunda piramis
planicies: VF 242, *v 54*
plana superficies: CC 170, *10, 14–16*; 186, *231*; 188, *248*
planum *or* planus: CC 170, *12*; 531n; QC 588, *8*; 600, *84*; 602, *98*
poligonia *or* polygonia: QG 42, *18*; FIII 120, *140*; 140, *c 147*; VV 374, *4, 11*; VA 390, *1*; *et passim in VV et VA* (*VV and VA use* poligonia *for* figura poligonia); QA 418, *220*; 428, *c 3–4*; *et passim in QA*
poligonium: 34n, *34*; QG 42, *22*; CV 70, *36*; FIA 102, *30*; CC 172, *36*; MV 200, *30*; CS 462, *9*; 468, *1*; *et passim in CS*; CSC 554, *74*
polus: VF 316, *10*; 318, *24, 30*
ponere: QT 22, *20*; 33n, *12*; QG 48, *v 94*; 60n; VG 158, *161*; CC 178, *110*; MV 200, *29*; 210, *177*; VF 252, *12*; 268, *38*; 334, *1*; QA 422, *274*; CSI 524, *15*; QC 598, *67*; *et passim ubique*
portio *or* porcio: QT 22, *34*; QG 42, *15*; FIA 104, *58*; VG 150, *35*; CC 178, *121*; MV 202, *67*; VF 302, *1, 15, 16*; *et passim in VF*; VA 390, *21*; QA 424, *282*; PA 436, *4*; 531n; CSA 532, *4*; QLI 616, *42*; *et passim ubique*
portiuncula *or* porciuncula: VG 150, *47*; 152, *62, 65, 70*; 160, *172*; CC 180, *134, 136*
positio *or* posicio: QT 20, *11*; FIB 106, *11*; MV 210, *204*; 214, *264*; VF 252, *v 11–36*; 340, *2*; 348, *1*; 659n
possibile: QG 42, *30*; MV 198, *23*; 206, *150*; VF 258, *10, 14*
postremus: QG 52, *135*; 193, *c 275*
postulare: 163, *c 37*; CS 452, *17*
potentia, in ("squared"): QG 48, *94, v 94*; *see* fortitudine
practica: 225n
precise: QC 600, *86, 89*; 602, *100, 107*; 604, *112, 118*
preconcessum: CS 480, *22*
Predicamenta [*Aristotelis*]: QA 412, *119*
premittere: QG 52, *132*; VG 148, *2*; MV 204, *97*; VF 288, *109*; 320, *61*; VV 374, *2*; CS 496, *23*; QLII 620, *v 14–19*; 628
preostendere: CS 488, *89*
presupponere: QLII 620, *v 14–19*
presumptio: VV 378, *68*

principale: FIII 116, *65*
principia communia: QA 429, *c 3–4*
principium ("principle"): CC 170, *9*; 172, *23*, *34*, *35*; 568n; QLII 620, *v 14–19*; a principio: CC 190, *274*; ("beginning"): MV 200, *45*; 206, *137*
probabiliter: VF 350, *38*
probare: QT 22, *44*; QG 54, *v 148*; CV 70, *27*; NV 82, *6*; FIB 106, *9*; FIII 114, *48*; 144n; VG 150, *45*; *et passim ubique*
probatio: 36n; 143n; VF 252, *v 11–36*; CSC 554, *60*, *70*
problema: 531n
procedere: QT 24, *53*; FIII 122, *v 154–63*; 132, *330*; VG 156, *130*; CC 180, *156*; MV 206, *133*; VA 390, *12*; QC 598, *72*
procedere in infinitum: FIB 108, *22*
procreare: FIII 118, *106*; 120, *126*; 132, *316*
prodire: ATI 642, *14*
producens ("a multiplier"): CS 460, *113*; 462, *31*
producere: ("to draw" *or* "to extend"): QT 20, *15*; QG 42, *20*; MV 212, *232*; VF 250, *10*; 310, *10*; 336, *19*; VV 376, *55*; VA 390, *15*; ATII 650, *52*; TC 680, *9*; ("to produce" or "to result in"): FIA 102, *34*; FIII 116, *70*; 118, *97*; 130, *302*; 191, *c 66*; CSA 540, *2*; 557, *c 41*; 608; ATI 646, *53*
productum ("product" *in multiplication*): 145n; VG 154, *89*, *95*; CS 460, *113*; 462, *31*; ATI 642, *4*; 646, *65*; ATII 654, *109*; 656, *154*
propinquior: VF 350, *27*, *28*
propinquitas ("approximation" *in finding roots*): *VF* 264, *12*, *17*, *20*; 278, *179*; 348, 7, *11*, *12*; 350, *15*
proponere: CV 74, *83*; CC 176, *86*; VF 252, *v 15*; CS 458, *102*; 490, *112*; 496, *8*; QJ 572, *13*
propositum: FIB 112, *57*; FIII 116, *65*; 118, *100*; 126, *227*; VG 148, *10*; *et passim ubique*
propositus: NV 82, *8*; 137, *c 33*; 146n; VG 162, *200*; 164, *c 82*; VV 374, *3*; CS 458, *79*; *et passim ubique*
proportio: QT 24, *59*; QG 46, *69*; 79, *c 50*; NV 88, *67*; VF 260, *1*; CSA 532, *3*; ATI 646, *45*; PL 664, *2*; *et passim ubique*
proportio equalitatis: ATII 654, *99*
proportionum nexus: CS 452, *21*
proportionabilis: VF 288, *111*
proportionalis: FIII 124, *204*; 126, *240*; 128, *270*; 139, *c 15*; VF 300, *27*; CS 462, *2*; 518, *c 104*; 557, *c 41*; 629; ATII 652, *66*; medio loco proportionalis: CC 188, *238*; QJ 574, *20–21*; CQ 578, *11*, *12*; 580, *23*; ATII 654, *120*, *122*; medium proportionale: 568n; QJ 572, *12*; 574, *19*, *21*; ATI 646, *63*; 659n
proportionalitas: 193, *c 275*; 629; *and see* conversa, eversa, *and* permutata
proportionaliter: VF 352, *62*
proportionare: QG 54, *v 148*; VF 240, *26*
propositio *or* proposicio: 34n, *43*; FIA 106, *81*, *85*; FIII 112, *15*; VG 150, *15*; MV 198, *3*, *14*, *17*; 212, *219*; 216, *306*; VF 280, *v 18–124*; 428, *c 3–4*
protendere: QC 594, *45*
protrahere: QT 22, *20*; QG 48, *v 90*; 91, *c 63*; FIA 100, *18*; CC 172, *25*; 180, *126*; VF 248, *10*; 250, *9*; *et passim in VF*; VV 374, *8*; VA 390, *14*; QC 598, *81*; 634; ATII 650, *53*; TC 680, *4*
provenire: QT 24, *57*; QG 54, *v 148*; NV 88, *75*; FIII 114, *32*; VG 154, *77*; MV 206, *135*; 225n; VF 242, *38*; 286, *v 83*; 320, *v 74*; CS 462, *35*; ATI 642, *4–5*; *et passim ubique*
proximus: FIB 110, *42*; FIII 114, *30*; CC 188, *260*
Ptholomeus: VV 380, *97*
Ptolemeus: MV 210, *179*
punctum: FIA 100, *17*; CC 174, *72*; VF 248, *6*; 256, *4*; 430, *c 45*; *et passim ubique*
punctus: QT 24, *78*; 430, *c 45*
pyramis: 447; CS 450, *v 9–10*; 452, *v 11–12*, *v 25*; CSI 524, *1*, *et seq.*; *and see* piramis

quadrangulum: FIA 106, *72*; CC 186, *229*; CS 460, *6*, *9*
quadrangulum rectangulum: CC 186, *226*; CS 478, *2*
quadrangulus rectangulus: CS 460, *16*
quadrare: 33n, *3*; 34n, *84*, *et seq.*; CV 72, *53*; FIB 106, *1*; VG 148, *1*, *6*, *7*; 162, *201*; VV 382, *127*; 386, *180*; VA 394, *79*; QA 406, *2*, *4*; *et passim in QA*; 428, *c 3–4*; 432, *c 49–53*; 569n; QJ 574, *23*; CQ 578, *1*; 580, *24*, *25*; QLI 616, *50*; QLII 618, *2*, *5*; 622, *39*, *40*

quadraticus: QA 406, *16*
quadratio: QLI 618, *54*
quadratum ("square"): FIA 100, *11*; FIB 106, *8*; FIII 112, *22*; *et passim in FIII*; VF 268, *40*; 284, *66*; ("multiplication" *or* "product"): VF 284, *65*; ("quadrilateral"): 60n; VF 312, *38*, *39*
quadratura ("squaring"): 31; 33n, *2*, *5*, *13*; 34n, *36*; CV 68, *1*; NV 82, *1*; FIA 100, *1*; FIII 112, *v 1*; 144n; CC 170, *1*; 176, *87*; MV 198, *1*; VF 242, *v 35–39*; VV 374, *1*; 382, *117*; QA 406, *1*; 408, *46*, *48*, *52*; *et passim in QA*; 429, *c 3–4*; CS 460, *107*; 482, *47*; 504, *13*; 568n; QC 588, *1*, *2*, *et seq.*; 608; ("squareness"): VF 246, *81*
quadratus ("square"): QT 20, *2*, *12*; QG 40, *11*; 42, *32*; CV 68, *24*; NV 84, *19*; VF 242, *v 35–39*; *et passim ubique*; ("squared" *or* "square-like"): 33n, *3*, *5*; VF 246, *89*
quadrilaterum: QLI 616, *45*, *46*, *48*, *49*; figura quadrilatera: QLII 622, *38*, *39*
quantitas: QT 22, *32*; 26, *110*; 36n; QG 42, *17*; 60n; NV 82, *8*; 86, *49*; FIA 100, *11*; 225n; VF 240, *33*, *34*; VV 374, *17*; VA 392, *41*; CS 452, *21*; 529, *c 29*; *et passim ubique*
quantitas rectilinea: CQ 578, *3*
quantitas curvilinea: CQ 578, *3*
quantitas in quam cum multiplicatur diameter erit illud quod agregatur ipsa linea circumdans ("π"): VF 322, *83–84*; *86–87*; 88-89, *et passim in VF*
quantitas in quam cum multiplicatur diameter provenit linea circumdans circuli (*or* ——— provenit circumferentia)["π"]: VF 324, *98–99*, *113–14*
quantum, in: FIII 116, *61*; 118, *99*; 132, *347*
quod fit ex ductu——in——("the multiplication of ——and——"): NV 84, *45*; VG 152, *55*; CC, 174, *72*

raciocinare: VF 264, *9*; 266, *23*; 278, *178*, *180*; 348, *5*, *6*, *7*, *8*; CS 478, *113*
radius: 634
radix: FIII 114, *29*; 146n; VG 150, *27*; ATI 642, *6*; ATII 656, *147*, *154*; *and see* vera radix
radix surda: VF 266, *23*; 348, *5*, *8–9*
ratio: QG 54, *v 157*; NV 84, *17*; 97n; FIA 100, *21*; 146n; VG 150, *31*; ATI 642, *9*; *et passim ubique*
rationis causa: CS 452, *15*
rationis decursus *or* decursus rationis: VG 154, *106*; 156, *129*
recedere: TJ 672, *8*
rectangulum: CC 188, *238*; 191, *c 66*; 222, *c 307*; 557, *c 41*
rectangulus: 78, *c 25*; CC 186, *226*; 193, *c 237*; 220, *c 110*
rectificata: VV 378, *72*; 382, *138*; 384, *161*
rectilinea: 31; 34n, *90*; QG 42, *18*; VV 376, *35*; PA 436, *13*; QLI 616, *50*
rectilinea figura multiangula ("regular polygon"): QT 22, *18–19*
rectilinee superficies equidistantium laterum ("parallelograms"): 79, *c 70*
rectilineus: CC 184, *203*; 220, *c 56*
rectitudo: VF 240, *35*, *v 35–95*; 242, *38*; 282, *56*; 294, *18*; *et passim in VF*; 634
recta (=recta linea): VB 150, *38*; *et passim ubique*
recta linea: CV 68, *3*; 90, *c 61*; *et passim ubique*; *and see* linea recta
rectum, in: QA 408, *59*, *63*; QC 590, *22*
rectus: QT 20, *4*; QG 40, *3*; CV 68, *10*; NV 82, *3*; VF 266, *27*; CS 452, *12*; *et passim ubique*
reducere: 33n, *15*, *21*; FIII 132, *344*; QA 408, *55*
regula ("rule" *or* "law"): CS 488, *95*; 494, *154*; 500, *75*; QC 602, *v 108–109*; ATI 642, *7*; ATII 656, *150*; ("ruler"): 659n; PJ 662, *6*, *8*; PL 664, *8*
relativum: QG 52, *135*
relatus: 40, *v 13*
relictus: CV 70, *37*; FIA 104, *62*; FIB 108, *21*; MV 214 *254*
relinquere: 60n; CV 70, *36*; 72, *59*; 74, *81*; FIB 106, *11*; FIII 124, *173*; VG 152, *63*; CC 178, *112*; MV 200, *40*; CS 456, *72–73*; QLII 622, *34*
reliquus: QT 20, *15*; QG 42, *23*; 60n; CV 72, *43*; NV 82, *4*; VF 306, *68*; VV 384, *145*; QA 418, *202*; ATII 650, *44*; *et passim ubique*
remanere: QT 22, *18*; QG 42, *17*; 54, *v 148*; FIII 114, *25*, *57*; VG 150, *47*; CC 178, *121*; VF 282, *60*; 304 *43*; VA 392, *41*; QC 590, *20–21*; 602, *105*; ATII 648, *18*

reperire: FIA 102, *47*; QA 430, *c 45*; CS 474, *64*; TJ 674, *33*
repugnare: VV 382, *113*, *127*
resecare: FIB 106, *10*; 110, *39*; MV 210, *191*; VA 390, *15*; QA 418, *207*, *220*; 420, *237*, *238*, *243*; CS 468, *15*; CSI 524, *19*; QJ 572, *17*; TJ 672, *6*
residuum: CV 68, *19*; FIII 124, *184*; CC 178, *114*; MV 200, *39*; QLI 616, *44*; QLII 622, *36*
residuus: QG 42, *14*; NV 84, *19*; FIA 102, *30*; FIB 110, *39*; VG 150, *35*; MV 216, *317*; VA 392, *39*
resolutio: 34n, *35*
resolvere: MV 206, *143*; VF 246, *v 5–21*; QA 416, *176*
respectus: MV 208, *159*; QA 406, *16*
respicientia: 90, *c 36*; FIII 120, *143*; 132, *329*
respicere: CV 74, *65*; NV 86, *65*; FIA 102, *32*; FIII 114, *42*; 164, *c 149*; QA 418, *204*
restrictus: QC 592, *26*
resumere: MV 210, *207*
revolutio: VF 336, *27*; 338, *32*, *35*
revolvere: VF 280, *19*; 290, *15*; 336, *15*, *22*; 348, *35*
rotunda columna ("cylinder"): CSI 526, *1*
rotunda piramis *or* rotunda pyramis ("cone"): CS 450, *1*, *v 9–10*; *et passim in CS*; 518, *c 104*; CSI 524, *1*

scientia: VF 238, *4*, *12*, *16*; 240, *17*, *18*; *et passim in VF;* VV 386, *182*; *and see* communis sciencia
scribere ("describe"): QA 429, *c 3–4*
secare: QG 40, *12*; 78, *c 40*; NV 84, *20*; FIA 100, *16*; FIB 108, *20*; MV 212, *216*; 221, *c 197*; 222, *c 307*; VF 298, *9*; 300, *18*; VA 392, *36*; CS 454, *43*; TC 680, *5*, *et seq.*
sectio: QT 20, *16*; QG 42, *35*; 46, *63*; 60n; FIB 108, *18*; 193, *c 237*; 222, *c 307*; VF 298, *3*, *6*, *10*; *et passim in VF*; VA 394, *47*
sector: 60n; VF 352, *51*; PA 436, *18*; CSA 540, *10*, *11*, *14*
secusponere: 659n
segmentum: 669n
semicirculus: FIA 100, *14*; VG 158, *161*, *163*; 220, *c 56*; QA 408, *34*; 430, *c 45*; CS 474, *82*; 476, *105*; QLI 614, *2*
semicircumferentia: 34n, *83*; CC 186, *235*; 430, *c 45*
semidyameter *or* semidiameter: 34n, *50*; 36n; CV 68, *13*; NV 82, *5*; FIA 100, *8*; 102, *32*; FIB 106, *3*; 143n; VG 148, *5*; 154, *96*; CC 176, *97*; MV 198, *4*; VF 256, *v 4–40*; VV 382, *120*; VA 394, *75*; 396, *85*; CS 460, *106*; AITI 656, *138*; *et passim ubique*
semipars: QC 600, *87*, *89*, *95*; 602, *101*
semipoligonium: CS 472, *32*; CSC 554, *72*
semper, et sic ("sucessively" *or* "continuously"; *equivalent to* deincepsque): VG 150, *16*; 160, *180*; et hoc semper: MV 200, *39*
separare ("to separate out" *or* "extract" *in the exhaustion procedure*): QT 20, *16*; QG 40, *11*; 42, *32*; FIA 102, *40*; VA 392, *38*; 394, *53*; ("to be separated from"): PJ 662, *6*; PL 664, *9*; ("to divide"): FIA 102, *42*; ("to intersect"): VF 336, *28*
separatio: VA 394, *53*
sequi: QG 42, *16*; 50, *112*; FIA 102, *36*; VG 148, *7*; *et passim ubique*
serratile: PA 436, *14*, *15*
sibi invicem equales: 91, *c 63*; 193, *c 277*; MV 216, *315*
signare: VG 158, *141*; CC 184, *181*; MV 202, *70*; VF 300, *15*; 304, *37*; 338, *33*
signatus: 138, *c 46–47*; QJ 572, *15*
significare: VF 264, *10*; 348, *12*
similis: QT 26, *90*; 33n, *14*; QG 40, *13*; CC 174, *79*; VF 254, *32*; 288, *120*; VA 392, *38*; CS 468, *23*; CQ 580, *29*; *et passim ubique*
similitudo: QG 48, *97*; FIII 126, *235*; CSA 442, *29*
simul ("at one time" *or* "added together"): ATII, 652, *74*, *76*, *et seq.*
simul collecti: QT 24, *84*
simul iuncti: QA 408, *59*; QLII 622, *41*
simul sumpti: VG 150, *14*; 156, *123*; *et passim in VG*; MV 202, *67*; VV 374, *9–10*; 376, *33*; VA 390, *7*
singularitas: VF 244, *76*
singulus: VG 150, *41*, *42*; MV 214, *267*, *270*; QA 414, *163*; ATI 642, *2*
situs: QA 408, *55*; 430, *c 45*; CS 456, *74*, *76*; 502, *11*
soliditas: 225n

solidum: 221, *c 179–81*; CS 482, *5, 7, 8*; 490, *120*; CSA 538, *69*; ATI 642, *8, 12*; ATII 650, *34, 35*; *et passim in ATI, ATII*
solidum contentum sub conicis superficiebus: 631
spacium: MV 216, *289*
spera (=sphaera): MV 210, *181*; 221, *c 179–81*; 225n; VF 248, *20*; 252, *20, 21*; *et passim in VF*; QA 406, *17*; PA 436, *2, 3, et seq.*; CS 478, *118*; 502, *1*; *et passim in CS*; CSA 532, *1, 5, 7*; *et passim in CSA*; CSC 550, *3, 11*; 554, *63*; 631
spericus (sphaericus): CC 170, *2*; QA 406, *19*; CSA 540, *99, 102, 104*; 633
stare ("to stand"): 78, *c 40*; 90, *62*; VF 260, *14*; VV 376, *36*; CS 450, *8*; 468, *3*; ("to be valid logically"): FIA 102, *36*
submultiplices: 79, *c 50*; CC 190, *280*; 193, *c 278*
subscribere: QT 24, *73*
subtendere: CV 68, *14*; NV 86, *61*; FIII 112, *v 6*; 122, *170*; VF 266, *29*; 268, *35*; QLI 616, *43*; 618, *55*
subtraccio: CC 180, *123*; 184, *202*
subtrahere: CV 68, *19*; NV 82, *11*; FIA 104, *60*; FIII 114, *25*; VG 150, *35*; CC 178, *120*; QLII 624, *v 43*
successive: VA 394, *65*
Sudor Archimenidis: QG 48, *v 84*
sumere: 79, *107*; NV 84, *24*; FIB 110, *31*; FIII 112, *22*; VG 150, *14*; 156, *123*; 164, *c 149*; MV 204, *103*; VF 256, *v 4–40*; 260, *39*; 430, *c 45*; CS 484, *25*; 502, *7–8*; 504, *5*; CSA 532, *3*; QLI 618, *53*; ATI 642, *3, 9*; TJ 676, *47*; *and see* simul sumpti
summa: 145n–46n; MV 198, *3*
summitas: CSI 529, *c 29*
sumptio: VV 378, *65*
superaddere: QT 24, *71, 75*
superare: QT 22, *46*; 97n; QC 600, *92, 95*
supereminens: CS 482, *1*
superesse: QT 22, *47*; NV 82, *10*; 86, *58*; FIII 114, *29*; MV 214, *265*
superficialis: MV 210, *180*; 221, *c 179–81*; VF 238, *5*; 240, *v 35–95*; 244, *61*
superficies: 33n, *8*; CV 70, *36*; 225n; VF 240, *30*; 262, *19*; VA 392, *2–8*; PA 436, *2, 4, et seq.*; *et passim ubique*
superficies equidistantium laterum ("parallelogram" *or* "rectangle"): 79, *c 70*; FIA 100, *12*; 193, *c 239*; 222, *c 307*; QA 416, *184*
superfluitas: VA 278, *1*; 304, *43–44*; *et passim in VF, Prop. 7*
superfluum: VF 304, *41*; 306, *53, 68*
superior: FIII 124, *182, 185*; VF 302, *4, 5, 8*; *et passim in VF*; CSA 540, *5*; 542, *33*
superius: NV 88, *84*; VF 302, *2, 13*
supervenire: CSI 529, *c 29*
supponere: 34n, *42*; VA 392, *27*; QA 412, *115*; QLII 624, *42*
suppositio: VV 382, *134*; 384, *140*; VA 396, *82*; QLII 620, *v 14–19*
suppositus: FIII 124, *185*; VV 384, *143*
supraponere: FIA 104, *57*
surda, *see* radix surda
suscepta: QA 428, *c 3–4*

tangere: 33n, *1*; VF 252, *v 11–36*; 254, *32*; 280, *21*; 330, *17, 30*; CSA 534, *13*; CSC 554, *63*
tempus: QA 428, *c 3–4*
teorema (=theorema): CS 452, *v 16*; 458, *v 103*
terminare: VF 240, *33, 35*; 244, *56*; 252, *1, 2, 6*; *et passim in VF*; CS 468, *3*; QC 592, *31*; 598, *65*
terminus: CV 72, *41, 61*; VG 150, *38*; 164, *c 149*; CC 172, *20*; 430, *c 45*; CS 468, *2*; TJ 672, *11*
tetracubitum: QC 600, *91, 93*
tetragonismum *or* thetragonismum (=quadrangulum): 35n, *v 68*; QA 406, *v 1–47*; tetragonismus ("squaring"): 32; 429, *c 3–4*; QC 588, *v 1, v 2*
tetragonizare ("to make a quadrangle"): 35n, *v 82*
tetragonum: 33n, *23, 29*; CV 76, *v 101*; CS 460, *2*; CSI 526, *9*; 528, *11*; QLI 614, *8*
Theodosius: VF 292, *v 24*
theorema: FIII 122, *167*; VF 292, *v 24*; CS 452, *16*; 454, *49*; 458, *103*; 460, *5*; 462, *40*
theoreuma: 146n; CC 176, *82*; 178, *109*; 180, *137*; 188, *237*
totalis: CV 74, *73*; NV 86, *62*; FIB 108, *25*; FIII 112, *11*; VG 150, *43*; CSA 532, *8*; ATII 652, *85*
totum: QT 22, *42*; 34n, *46*; VG 160, *167*; CC 178, *119*; MV 202, *63*; *et passim ubique*

totus: QT 22, *25*; QG 44, *42*; CV, 72, *44*; FIA 106, *78*; VG 152, *54*; *et passim ubique*
trahere. FIII 112, *v 6*; VG 158, *143*; CSI 526, *27*; QC 594, *34*
transcendere: QC 600, *86*, *88*; 602, *100*, *106*; 604, *112*
transire: QG 42, *34*; CC 172, *26*; VF 298, *6*; 304, *30*; 318, *40*, *51*; 342, *23*; VA 394, *50*; QA 406, *11*; 424, *288*; 430, *c 45*; CSA 532, *2*; QC 592, *29*; 634; 660n; PJ 662, *5*; PL 664, *8*; TJ 672, *9*; TC 680, *10*
transitus: QA 406, *10*, *16*; 424, *288*
transpositio: QA 408, *55*; 412, *102*
trapezie, figure: 628
triangulus: QT 20, *3*; 33n, *22*; QG 40, *2*; CV 68, *15*; VF 246, *v 5–21*; ATI 642, *1*; *et passim ubique*
triangulus orthogonius: QG 40, *2*; VV 374, *5*; *et passim ubique*
triangulus rectanglus: 78, *c 25*; *et passim ubique*
trigonus: CV 68, *9*; NV 82, *2*; 164, *c 149*; QA 422, *272*; CS 468, *13*, *14*; QLI 614, *18*; 660n
triplare: QC 590, *21*
triplicata: CV 74, *87*, *89*

unica: QC 600, *84*; 602, *99*
uniformiter: QA 406, *14*
unitas: CV 76, *93*; VF 334, *7*
universalis: QC 602, *v 108–110*; 604, *109*
universaliter: MV 216, *295–96*; VV 376, *51*, *52*; 380, *97*; QA 424, *281*, *286*; QLI 618, *53*, *54*

vacuitas: VF 242, *v 35–95*; 244, *71*, *v 71*; 246, *82*
vacuum: MV 198, *26*; 220, *c 24*
valere: FIII 112, *v 6*; 114, *42*; *et passim in FIII*; VG 150, *26*; VV 380, *100*; 430, *c 45*; ATII 650, *39*; 654, *125*
variare: QC 594, *45*
vera radix: FIII 114, *31*; *et passim in FIII*
verificare: QG 46, *81*; VF 246, *93*; 428, *c 3–4*; CSC 554, *65*
veritas ("truth" *as applied to roots*): VF 264, *10*, *14*, *18*; 278, *180*; 348, *11*; 350, *16*, *17*
verus: FIII 126, *240*; MV 212, *238*; 216, *297*; QA 424, *282*, *284*, *286*; 429, *c 3–4*
vicis ("times" in multiplication): CC 188, *257*, *258*, *264*, *265*; 190, *271*
virtus: 628

ymaginare *or* ymaginari: VF 346, *11*, *14*; TJ 672, *7*; 674, *29*
ypothenusa ("hypotenuse"): CC 176, *99*; ("slant height"): CS, 450, *4*; *et passim in CS*; CSI 524, *2*; CSA 540, *2*
ypothenusalis: CS 456, *57*, *65*, *67*, *69*; *et passim in CS*
ypothesis: FIII 124, *203*; VG 152, *67*, *71*; CC 182, *166*; 188, *255*; 190, *268*; MV 204, *112*; 206, *148*; 208, *164*; 210, *207*; 216, *301*, *302*; CS 452, *15*; 464, *13*
ysochelis: 34n, *v 22*
ysoperimetrus: 221, *c 179–81*; QA 408, *28*, *29*
ysozelis: 33n, *22*

An Index of the Citations of Euclid's *Elements*

Roman numbers refer to pages, italicized numbers to lines. When the italicized number is preceded by *v*, the reference is to the variant reading to the line; when by a *c*, to the Commentary to the line. In the case of Latin texts accompanied by English translations, reference is made only to the citation in the Latin text.

Book I

Def. 2: 242, *45*
Def. 7: 242, *45–46*
Def. 15 (circle): 136, *c 38*
Def. 17 (diameter): 138, *c 4*
Ax. 1: 198: *13*; 376, *25*
Ax. 2 and 3: 390, *19–20*; 398, *c 19–20*
Pseudo-Post.-Ax.: 408, *42–43*; 429, *c 42–43*
Addit. Ax. 1: 136, *c 38*
Addit. Ax 2: 620, *v 14–19*; 625, *c 14–18*
Prop. 3: 84, *32*; 90, *c 32*
Prop. 4: 48, *v 90*; 84, *36*; 90, *c 36*; 136, *c 38*; 150, *44*; 152, *50*; 154, *84*; 158, *149*; 184, *194*; 387, *c 25*; 674, *v 18*
Prop. 5: 86, *63*; 90, *c 63*; 136, *c 38*; 158, *150*; 216, *316*
Prop. 6: 48, *v 90*; 86, *65*; 91, *c 65*; 136, *c 38*; 184, *193*, *195*; 202, *50*, *56*; 218, *319*
Prop. 7: 54, *v 157*
Prop, 8: 84, *32*; 90, *c 32*; 136, *c 38*; 150, *44*; 152, *49*; 154, *83*; 158, *149*
Prop. 9: 138, *c 4*; 141, *c 168*
Prop. 10: 154, *82*; 186, *233*
Prop. 13: 86, *62*; 90, *c 62*
Prop. 14: 212, *218*
Prop. 15, cor.: 138, *c 4*
Prop. 16: 138, *c 4*
Prop. 18: 135, *c 7*; 136, *c 38*; 192, *c 191*; 212, *222–23*
Prop. 19: 184, *191*; 192, *c 191*
Prop. 26: 136, *c 38*; 158, *149*; 164, *c 149*
Prop. 28: 526, *29*; 529, *c 29*
Prop. 29: 674, *v 17*; 676, *v 41*
Prop. 32: 48, *v 90*; 138, *c 4*; 141, *c 168*
Prop. 33: 526, *29*; 529, *c 29*
Prop. 34: 150, *31*; 163, *c 31*
Prop. 37: *526*, *29*; 529, *c 29*
Prop. 38: 212, *226*, *236*; 221, *c 226*; 376, *48*
Prop. 41: 70, *33*; 72, *44*; 78, *c 33*; 108, *14*, *21*, *29*; 135, *c 7*; 136, *c 38*; 137, *c 74*; 152, *52*, *56*; 154, *86*; 174, *66*, *73*, *75*; 180, *130*, *151*; 182, *167*; 186, *227*; 202, *82*; 206, *120*, *139*; 208, *164*, *167*; 212, *236*; 214, *275*, *277*; 454, *49*
Prop. 42: 216, *307*; 221–22, *c 307*
Prop. 44: 216, *307*; 222, *c 307*
Prop. 46 (=Greek 47): 48, *v 94*; 70, *25*; 78, *c 25*; 110, *48*; 138, *c 4*; 141, *c 168*; 178, *118*; 202, *51*; 618, *v 5*

Book II

Def. 1: 204, *111*, *118*; 206, *137*, *141*; 214, *278*; 220, *c 110*
Prop. 1: 135, *c 7*; 136, *c 38*; 174, *66*, *75*; 176, *81*; 191, *c 66*; 208, *167*; 214, *282*; 484, *27*; 486, *50*
Prop. 2: 108, *26*; 137, *c 26*
Prop. 4: 620, *v 25*
Prop. 5: 216, *307*; 222, *c 307*
Prop. 10: 454, *v 40*
Prop. 13: 639n; 642, *v 1*
Prop. 14: 74, *82*; 79, *c 82*; 112, *56*; 148, *6*; 198, *11*; 216, *306*; 386, *182*; 403; 422, *271*; 574, *23*; 620, *v 13*; 622, *v 41–45*

Book III

Def. 2: 210, *197*; 221, *c 197*
Prop. 3: 72, *40*; 78, *c 40*; 84, *32*; 174, *56*, *71*; 180, *143*; 674, *v 18*; 676, *v 42*
Prop. 15 and cor. (=Greek 16, Por.): 136, *c 38*; 164, *c 149*; 403; 408, *45*; 424, *297*; 430–31; 568n
Prop. 16 (=Greek 17): 143n; 212, *219*; 216, *308*; 221, *c 219*
Prop. 17 (=Greek 18): 86, *61*; 90, *c 61*; 136, *c 38*; 138, *c 4*; 158, *142*, *149*; 212, *222*; 214, *272*; 216, *314*; 218, *323*
Prop. 25 (=Greek 26): 136, *c 38*
Prop. 26 (=Greek 27): 158, *149*; 164, *c 149*
Prop. 27 (=Greek 28): 674, *v 18*; 676, *v 43*
Prop. 28 (=Greek 29): 150, *44*; 163, *c 44*
Prop. 29 (=Greek 30): 150, *37*; 163, *c 37*; 180, *125*
Prop. 30 (=Greek 31): 138, *c 4*; 141, *c 168*; 202, *56*; 220, *c 56*; 618, *v 5*
Campanus Comment to Prop. 35: 138, *c 46*; 216, *311*; 222, *c 308*
Prop. 36 (=Greek 37): 216, *308*; 222, *c 308*

Book IV

Def. 1: 200, *45*; 220, *c 45*
Prop. 5: 138, *c 4*
Prop. 6: 106, *9*; 135, *c 7*; 137, *c 9*; 150, *18*; 200, 44
Prop. 7: 110, *33*; 136, *c 38*; 137, *c 33*; 150, *20*; 156, *120*; 632
Prop. 9: 632
Prop. 12: 158, *152*; 165, *c 152*
Prop. 15, cor.: 138, *c 4*; 141, *c 168*

Book V

Campanus Comment to Def. 1: 138, *c 4*; 141, *c 168*
Campanus Comment to Def. 3: 138, *c 4*; 141, *c 168*
Prop. 1: 414, *162*; 432, *c 161–66*
Prop. 3: 76, *107*; 79, *c 107*
Prop. 7: 620, *v 14–19*; 625, *c 14*
Prop. 8: 48, *v 94*; 138, *c 4*; 141, *c 168*
Prop. 11: 76, *106*; 79, *c 106*; 190, *277*; 193, *c 277*
Prop. 12 (=Greek 13): 620, *v 14–19*; 625, *c 14*
Prop. 13 (=Greek 12): 138, *c 4*
Prop. 15: 72, *50*; 79, *c 50*; 138, *c 4*; 190, *278*; 193, *c 278*; 552, *42–43*; 622, *v 29*
Prop. 16: 138, *c 4*; 141, *c 168*
Prop. 18: 138, *c 4*
Prop. 20 (not in Greek): 190, *275*; 193, *c 275*
Prop. 21 (not in Greek): 620, *v 14–19*; 625, *c 14*
Prop. 24: 306, *57*, *64*; 474, *78–79*; 504, *21*; 514, *c 78*; 520, *c 21*; 526, *36*; 529, *c 35*

Book VI

Prop. 1: 74, *70*; 79, *c 70*; 88, *68*; 104, *52–53*; 110, *45*; 136, *c 38*, *c 71*; 137, *c 74*; 158, *165*; 184, *196*; 188, *261–62*; 416, *178*, *182–83*; 432, *c 178*; 536, *48*
Prop. 3: 48, *v 91*; 52, *v 128*; 112, *15*; 138, *c 4*; 139, *c 15*; 141, *c 168*
Prop. 4: 141, *c 168*
Prop. 8, cor.: 188, *237*; 192, *c 237*; 344, *v 55–57*
Prop. 13 (=Greek 14): 188, *239*; 193, *c 239*
Prop. 15 (=Greek 16): 54, *v 157*; 552, *41*; 557, *c 41*
Prop. 18 (=Greek 20): 620, *v 25*
Prop. 33: 138, *c 4*; 141, *c 168*

Book VII

Prop. 19 (=Campanus 20): 141, *c 168*; 552, *41*; 557, *c 41*

Book X

Prop. 1: 5; 60n; 68, *20*; 78, *c 19–21*; 82, *11*; 100, *28*; 110, *39–40*; 135, *c 7*; 136, *c 38*; 150, *15*; 158, *134*; 178, *109*; 180, *137*; 182, *176*; 184, *208*; 200, *41*; 202, *66*; 212, *214*; 357, *c 17–19*; 368–69; 389; 402; 561

Book XII

Prop. 1: 464, *14*; 500, *72*; 511, *c 14*
Prop. 2: 5; 60n; 202, *61*, 220, *c 60–61*; 254, *v 18*; 262, *25–28*; 263v; 357, *c 1–39*; 572, *5*, *11*; 578, *6*; 618, *7*; 620, *27*, *v 14–19*; 625, *c 14*; 632
Prop. 9 (=Greek 10): 441n; 444; 484, *22*, *37*; 486, *44–45*; 488, *83*; 492, *142*, *145*; 502, *14*; 517, *c 22*; 540, *5*; 552, *51*
Prop. 11: 444; 490, *104*, *110*; 494, *168*; 504, *25*; 518, *c 104*
Prop. 14 (=Greek 17): 534, *13*; 546, *c 13*
Greek Prop. 16 (=Adelard 13): 357, *c 17*; 369

An Index of Latin Manuscripts Cited

Basel, Bibl. Univ.
F.II, 33: xxiii, 228n–229n, 235, 448, 568n, 570
Berlin, Deutsche Staatsbibl. (MMS now at Marburg, Westdeutsche Bibl.)
Q.150: xxiv, 37, 447, 612
Bern, Bürgerbibl.
A.50: xxviii, 404, 612
Bruges, Stadsbibl.
530: 96n, 570, 637n
Cambridge
Cambridge Univ. Libr.
Ee. III. 61 (1017): xxiii, 568n, 586
Mm.III.11 (2327): xxii, 449
1572: 587
Gonville and Caius College,
504/271: xxv, 66
Dresden, Sächs. Landesbibl.
Db.86: xxiii, 37, 448, 569, 612, 637n, 640, 661, 670
Dublin, Trinity College Libr.
D.2.9: xxviii, 18
Erfurt, Stadtbibl.
Amplon. F.178: xxii, 586
Amplon. Q.234: 612
Amplon. Q.361: xxii, 586
Amplon. Q.385: xxii, 586, 612
Florence, Biblioteca Nazionale
Conv. Soppr.
J.I.32: 570
J.IV.29: 33n
J.V.18: xxv, 65–66, 448, 530, 612
J.V.30: xxi, 6, 91–93, 96–98, 135–41 *passim*, 448, 520
J.IX.26: xxi, 389
J.X.40: 570
Bibl. Medicea-Laurenziana
San Marco 184: 235
Bibl. Riccard.
106: 12n
107: 14n
Glasgow, Univ. Libr.
Be 8-y.18: 38, 97n, 569, 587, 612
London, British Museum
Addit. 17368: xxvii, 37
Addit. 17420: 36n
Addit. 34018: 60n, 78, 90, 137, 163, 191, 511, 514, 518, 529, 557
Harleian 625: xxi, 449, 547–48, 569, 661, 670
Royal 12E.25: xxvii, 65–66, 612
Sloane 285: xxvii, 569, 661, 670
Madrid, Bibl. Nac.
9119: 11, 13
10010: xxix, 234-35, 433, 435, 661
Marburg, see Berlin
Milan, Bibl. Ambros.
H.144 Inf.: xxv, 586
Munich, Bay. Staatsbibl.
56: xxviii, 195–96
234: xxviii, 640
4377: 587
Naples, Bibl. Naz.
VIII.C.21: 678
VIII.C.22: xix, 37, 80–81, 449
New York, Columbia Univ. Libr.
Plimpton 156: 678
Nuremberg, Stadtbibl.
Cent. V. 15: 12n

Oxford, Bodl. Libr.
Arch. Seld. B.13: xxiv, 37
Auct. F.3.13: 97n, 432
Auct. F.5.28: xx, 37, 438, 446–47, 612, 630
Canon. Misc. 226: 445n
Digby 147: xx, 586
Digby 153: xxiv, 612
Digby 168: xxvi, 236, 434–35, 661
Digby 174: xx, 37, 447, 577
Digby 190: xxiii, 612
Corpus Christi College
234: xx, 165, 169
251: xxi, 569, 612
Paris, Bibl. Mazarine
3637 (1256): xxv, 235
Bibl. Nat.
Fonds latin
7216: 60n
7220: 12n
7221: 12n
7224: xxv, 17–18
7225A: xxvi, 236
7371: 445n
7373: 60n, 633
7377B: 36n, 225n
7378A: xxvi, 37, 570, 586, 637n, 640
7380: 369n
7434: xxvi, 569, 661, 670
7465: 611n
9335: xxvi, 4, 17n, 30, 37, 227n–28n, 433, 629–30
11246: xxiv, 17–18
11247: xxi, 448, 521
16089: 587
16649: 37–38

Nouv. acquis. lat. 1538: 12
Thorn (Torún), Gymn.-bibl.
R 4° 2: xxvii, 236, 387, 444n
Vatican Libr.
Vat. lat.
3102: xix, 33n, 372
4275: 38, 80n, 97n
7299: 60n
Ottob. lat.
1850: 8, 11–12, 369n, 403n, 438–39, 531n, 659n
1870: xxix, 92–93, 98, 135–41 *passim*
Pal. lat. 1389: xxviii, 142–43, 146
Reg. seuv.
1253: 11
1261: xxviii, 36n, 637n, 640, 678
1268: 60n, 627n
Ross. 579: 60n, 228n
Urb. lat.
261: 12n
506: 678
507: 678
Venice, Bibl. Marc.
f.a. lat. 327: 12n
Vienna, Nationalbibl.
5257: xxii, 143, 146, 404
5277: 236, 629
5303: xxii, 37, 448, 577
10701: 14n

General Index

Place names in connection with manuscripts have not been indexed here (but see the separate Index of Latin Manuscripts Cited). I have also reserved for a separate index the geometrical terms employed in the Latin texts themselves. The proper names employed in the Latin texts are generally included in the Selective Index of Latin Geometrical Terms and cross-references (to *see* Selective Index) are made here to that index. I have not indexed the Bibliography since it is arranged alphabetically and may be easily consulted. Nor have I indexed here the contents of the manuscripts in the General List of Manuscripts at the front of the volume, since the material is repeated in the separate Sigla before each text, and the manuscript numbers are indexed in the separate Index of Latin Manuscripts Cited. But the manuscript catalogues are indexed. The running page number form (e.g., 125-26) does not necessarily mean that there is continuous discussion of the indexed item; it may merely indicate that the item appears on each page of the series.

Abbot, T. K., xxviii
Abbreviated Version of the *De mensura circuli*, 388–98
Abraham bar Ḥiyya, 17, *and see* Savasorda
Abū ᶜAbdallāh al-Shannī, 636
Abū Jaᶜfar Muḥammad, 226–27, 229n, 238–39
Abū 'l-Kāsim Aḥmad, 226–27, 229n, 238–39
Acefalus, 633
Adelard of Bath, 60n, 64, 135, 388n, 441n, 569, *and see the copious citations of the* Index of the Citations of Euclid *where the Adelard II version of the* Elements *is the one usually cited*
agrimensores, 15, 359, 635
Alardus Bachon (=Alardus Bathoniensis), 388n, 569
Albert of Saxony: his "betweenness" postulate, 357, 369, 402, 562; his *Question on Quadrature*, 7, 36, 60, 64, 167–69, 368, 371, 398–432, 560–62, 568–69n; mentioned, 6, 10, 561, 582
Albertis, Antonius de, 13–14
Alexander of Aphrodisias, 611, 614n
Alexander, Pseudo-, 427
Algazel, 167
algorism, 97
Alhazen, *see* Ibn al-Haitham
alkayda (alkaydis), 135, *and see* Selective Index
Alkindi, *see* al-Kindī
Almagest, 195, 210, 221, 371, 379n–81, 387
Alverny, M.-T. d, x, xxiv
Amelli, A. M., xxv
Ametus filius Iosephi, 9n, 17n, 235, 629–30
Anaritius (=al-Nairīzī), 16n, 63n, 627–29
Anthimus (=Anthemios), 633
Antiphon, 60n, 399, 407, 423, 426–27, 432
Apollonius of Perga, 608, 633, 658–59n
Aquinas, Thomas, 219, 429, 443
Arabic mathematicians: their calendaric tables, 142: their knowledge of Archimedes, 3–4, 38, *and see* Chap. 4; their terminology in Latin, 434; their use of Euclid, 60n, 139; *and see* Banu Mūsā
Archimedes: his *Balances*, 630; his *Bovine Problem*, 3, 7; his *Conoids and Spheroids*, 4, 8, 12, 13n, 14n,; his *Equilibium of Planes*, 3–4, 11, 14n, 362; his *Floating Bodies*, 3,

11, 13; his *Measurement of Circle*, 3–7, 9, 10n, 11–14n, 223–24, 264–65, 278–79, 357–59, 364, 439, 441n, 444, 505, 511, 520, 558–62, 570, 583, 630, *and particularly see* Chaps. 2, 3, and 5, *as well as index entries for* Gerard of Cremona, Plato of Tivoli; his *Method*, 3, 4, 10n; his *Quadrature of Parabola*, 3–4, 8, 11, 13, 14n; his *Sandreckoner*, 4, 7, 13n–14n; his *Sphere and Cylinder*, 3, 6–7, 12, 14n, 31, 63–64, 165–68, 223, 363–66, 558–60, 627–29, 631–32, 658, *and see* Chaps. 4, 6, *passim*; his *Spiral Lines*, 3–4, 8, 11–12, 13n–14n, 369n, 667; his *Stomachion*, 3; knowledge of among the Arabs, 3–4, 38, among the Latin Schoolmen, Chaps. 1 and 7; lemma of, 402n–403n; Principle of, 8, 10, 563; statics of, 9–11, 562, 628–29; techniques used by Galileo, 1, by Gerard of Brussels, 10, 560; *and see* exhaustion, method of: *and consult* Selective Index

Archimedes, Pseudo-: the *De curvis superficiebus* and, 6–7, Chap. 6, *and see* Johannes de Tinemue; the *De insidentibus in humidum* and, 8, 10, 563; the *Liber assumptorum* (or *Lemmata*) and, 4, 224, 367, 559, 632, 667–68; tracts of, in Arabic, 4, 38

Archimenides (=Archimedes), 17n, 31, 225n, *and see* Selective Index

Archytas, 224, 365, 559, 658

arcs: postulated as greater than their chords, in Cambridge version, 63, 68, 166; in Corpus Christi version, 63, 166, 170; in Gordanus version, 63, 156, 166; in Munich version 63, 195; in Naples version, 84, 90

area, *see* cone, area and volume of; polygon; sphere; triangle, area of; *etc.*

Aristophanes, 400, 431

Aristotle: his *Categories*, 15, 401, 413, 432, 586, 589, 607–608: his *De caelo*, 195, 210, 634; his *Physics*, 16n, 62, 192, 194, 198, 219–20, 399, 407, 426; his *Prior Analytics*, 35; his *Posterior Analytics*, 426, 428; his *Sophistical Refutations*, 169, 407, 399, 426–27; on density, 10; *and see* Selective Index

Arsamides (=Archimedes), 16–17n, 629–30

Arsamithes (=Archimedes), 16, 31

Arsamites (=Archimedes), 17n

Arsimenides (=Archimedes), 17n

Asamithes (=Archimedes), 627–29

Assamites (=Archimedes), 627–29

astronomy, 142–43

Averroes, 62, 569n, 634

Aximethes (or Aximetes) (=Archimedes), 627

axioms, 139, 195, *and see* postulates

Bacon, Roger, 7–8, 225, 444, 561

Baghdad, 225–26

balances 9, 629–30

Bale, J., 581n

Banū Mūsā: biography of, 225–27; their *Verba filiorum*, 7, 10, 17n, 29, 33, 59–60, 95–96, 97n, 168, 368–71, 433, 507, 510, 514–15, 519, 558–59, 635–36, 638–39, 658–61, 666–68, 673n, mentioned 6, 562; *and see particularly* Chap. 4, *passim*

bapistir, 634

Barozzi, Francesco, xxviii, 18

Basilides of Tyre, 633

Bateson, M., 581n

bent lines, 167, 191

Bernardus Pistoiensis, 36

Besthorn, R. O., 629

al-Bīrūnī, 7n, 365, 635n, 636

Björnbo, A. A., xxi, xxiii, xxv–xxvi, xxviii, 30n–31n, 227n, 443

Blasius of Parma, 10

Blume, F., 635n

Boethius: as translator of Archimedes, 2n; his *Commentary on the Categories*, 15, 587, 607–608

Boncompagni, B., 17n, 30n, 224n, 357–58, 636n 658n, 661

Boulliau, I., 227n

Boyer, C., x

Bradwardine, Thomas: his *De continuo*, 220 387–88; his *De proportionibus*, 444–45n; his *Geometria*, 33, 62, 94, 163, 370, 387, 586–87; on quadrature, 33-36; mentioned, 6, 225, 561

Bradwardine, Pseudo-, 7, 36, 168, 357, 370–89, 402

Brahmagupta, 635

Bryson, 399, 403, 407, 423, 426–29, 432

Bubnov, N., xxiv

Burley, Walter, 220

Byzantine, 3, 32, 226, 441-43

calendaric tables, 142–43
Cambridge Version of the *De mensura circuli*: cited in comparison, 59, 61–62, 80–81, 89–91, 94, 137–38, 144–45, 163–67, 191–92, 195, 220–21, 371, 388, 402, 432, 507, 508, 546; text and translation of, 63–79; mentioned, 530
Campanus of Novara: and the fourth proportional term postulate, 625–26; his *Commentary on the Elements*, 59, 135–36, 138–39, 141, 192, 194, 196, 220–22, 225, 400, 403, 409, 425, 429–31, 529, 546, 581, 606n, 612n, 639, 678–81; his *Theory of the Planets*, 409, 431–32; on Hero's formula for area of triangle, 636, 639; on quadrature, 15n, 33, 36, 401, 409, 413, 581–609; on trisection of an angle, 678–8; mentioned, 443, 561, 583; *and see* Selective Index
Cantor, M., 582n, 635n
Carmody, F. J., 659n
Carpus, 608
Cassiodorus, 2n
Cereo de Burgo Sancti Sepulchri, Francesco, 12
Chasles, M., 582n
Chinemue, 15n, 440, 578v, *and see* Johannes de Tinemue
circle, *see* quadrature of a circle
Clagett, M.; "Archimedes in the Middle Ages," 31n, 441n; *Giovanni Marliani*, 140; "King Alfred," 169n, 444n; "Liber de motu," 8n–9, 80n, 440n–41n; "Medieval Latin Translations," 135, 219, 388n, 429; *Medieval Science of Weights*, 8n–9n, 445n, 637n; *Science of Mechanics*, 1n–2n, 8n–10n, 164, 166n, 398n, 629–30; "Three Notes," 12n; "Use of Moerbeke," 11n, 13n, 369n
Colebrooke, H. T., 635n
Commandino, Federigo, 13
compass, *see* circinus *in* Selective Index
Copernicus, 678
componendo, 529
conchoid, 366, 666–68, 678n
cone, area and volume of, 224, 451–61, 467–69, 507–508, 511–12, 520, 524–27, 531, 540–43, 546–47, 559
continuity and continuous magnitudes, 357, 369, 380–81, 387, 389, 402, 418–19, 427, 562
Corpus Christi Version of the *De mensura circuli*: cited in comparison, 60, 62, 98, 144–45, 195, 370–71, 401–402, 442n, 517, 557, 562; text and translation of, 165–93
Coxe, H. O., xx–xxi
Craster, H. H. E., xx, xxv
Cratilus, 637
Cremonensis, *see* Gerard of Cremona, Jacobus Cremonensis, *and* Leonardo Mainardi of Cremona
Creton, Mich., 14n
Crombie, A. C., 61n, 369, 431
cube roots, *see* roots
Curtze, M., x, xxiii–xxiv, xxvi–xxvii, 17n, 195–96, 225n, 229–30, 233, 235, 387. 443, 445n, 567–69, 627, 629, 632, 636n, 640, 658n, 661, 667n–70, 678n
curved lines, postulated as equal to straight lines, *see* rectification
cylinder, area, of, 461–63, 509–10, 520–21, 526–29, 559

De canonio, 9, 563
De curvis superficiebus, *see* Johannes de Tinemue
De inquisicione capacitatis figurarum, 225, 445, 632
De pyramidibus rotundis, 440, *and see* Johannes de Tinemue
De proportionibus, anonymous, 92n
De replentibus locuum, 225n
De ysoperimetris, 32, 92, 221, 225n, 365, 409, 518–19, 630–32, *and see* ysoperimetrus *in* Selective Index
definitions, 136, 139, 141, 627–28, *and see* diffinito *in* Selective Index
Delisle, L., xxiv, xxvi
denomination of fractions or ratios, 140
Diacetto, Francesco da, 13–14
Diels, H., 426, 610
Dijksterhuis, E. J., x, 3n
dispositio, 444
Dozy, R., 249n
Duhem, P., 637
dulk or *dulkarnon* (=Pythagorean theorem), 48v, 97n, *and see* Selective Index
Durnheimer, Wigandus, 143, 146, 404, 561
dyscolus, 442

Ecchellensis, Abraham, 659n
ekthesis, 444

elementum (=a principle), 469, 518
Eneström, E., 582n, 636n
Ersemides (=Archimedes), 16, 17n, 18, 20
Essex, *see* Gervasius de Essexta
Euclid: Adelard I version of his *Elements*, 135; Adelard II version of his *Elements*, 60n, 64, 511, 514, 517, *and see* Index of the Citations of *where citations are usually to the Adelard II version*; a tract *On the Balance* attributed to, 4; Anaritius commentary on his *Elements*, 16n, 63n, 627–29; translation from the Greek of his *Elements*, 60n, 633; Gerard of Cremona translation of his *Elements*, 60n, 228n; Prop. X.1 of his *Elements*, 60–62, 81, 144, 168, 195, 368–69, 561, *and see* Index of the Citations of Euclid; use of his *Elements*, 5, 59–61, 64, 94, 96, 144, 168, 194–95, 242–43, 355, 371, 400, 402–403, 444, 561, 612n, 639, *and see* Index of the Citations of Euclid; mentioned, 2, 226, 427, 445n
Euclid, Pseudo-: *Catoptrics* of, 634
Eudemus, 365, 427
Eudoxus, 427, 429
Eugene the Amīr, 633
Eutocius: his *On the Equilibrium of Planes*, 11; his *On the Measurement of the Circle*, 3, 7, 63, 90, 95, 139–42, 224, 358–59, 560; his *On the Sphere and Cylinder*, 3, 365–66, 658–60; mentioned, 667n
exemplum, 444
exhaustion, method of; 5–6, 60–62, 81, 94–95, 195, 223, 531, 559–61; "remaindering" and, 137
Eximetes (=Archimedes), 627
Exter, 443
Eyssenhardt, F., 79

falsigraphus, 169, 192, 442, 500; *and see* Selective Index
Ficino, Marsiglio, 13
Fihrist, 226, 365
Florence Versions of the *De mensura circuli*, 59, 62, 91–142, 144, 163–64, 358–59
Flügel, G., 226
fractions, use of in computing *pi*, 95–96, 140
Franciscan Order, 582, 587
Franciscus de Ferraria, 445, 561
Franco of Liège, 15n, 582, 586
Fredricus, 568n–69n

Galileo, 1–2
Gandz, S., 635–36n
Gaurico, L., 13, 582n, 587
geometry: Euclid's Book of (=*Liber geometrie*, or *Elementa*), 96, 168, 177–91 *passim*, *and see* Index of the Citations of Euclid; Greek, 7, 223, *and passim*
Gerard of Brussels, 9–10, 32, 80, 440–41n, 444–46n, 560
Gerard of Cremona: his translation of *De mensura circuli*, 4–7, 13, 17n, 28–29, 30–58, 59, 64, 80, 89, 93, 95, 136–37, 139, 165–66, 194, 223, 402, 433–34. 509, 558, 576; his translation of *Verba filiorum*, 7, 17n, 30n, 95, 224, 227, 228n, 229–33, 245v, 433, 636, *and* Chap. 4 *passim*; his translation of the *Elements*, 60n–61n, 228n; other translations of, 9, 16, 17n, 31, 433, 627, 660–61; translating techniques of, 30, 232–33, 434
Gerbert, 224
Gervasius de Essexta *or* de Assasia, 440, 443, 445, 449, 506v
Gordanus: his *Compilacio*, 142–43; his version of the *De mensura circuli*, 60, 62–63, 142–65, 371, 561
Grabmann, M., 443n
Grant, E., 140
Greek: translations of Archimedes from, 7–8, 11–14; translation of *De curvis superficiebus* from, 442–43; translation of other works from, 9, 32, 226
Grosseteste, Robert, 16n, 568n, 610–11
Guilhelmus Anglus, 581n

al-Ḥassan, 22n, 226–27, 238–39
Haskins, C. H., 17n, 31n
Heath, T. L., x, 167n, 366, 426–28, 431, 530, 609, 627, 659n, 663n, 666n–67n
Heiberg, J. L., x, 3n, 11–12, 28–29, 31, 37, 56–57, 90, 139, 358, 365–66, 434n, 441, 443, 627–29, 632–33, 635n, 658n, 667n
Hermann, archbishop, 582
Hero of Alexandria: his *Belopoeica*, 658; his citation of corollary to Prop. I of *De mensura circuli*, 5, 32, 57; his definition of proportional weights, 629–30; his *Dioptra*, 635n, 639; his formula for area of

triangle, 7, 224, 274–89, 359–62, 559–60, 635–57; his *Geometria*, 635n; his *Mechanics*, 4, 166, 441–42, 658; his *Metrica*, 635n, 639; *Stereometry* attributed to, 441, 443; mentioned, 4
Hippocrates of Chios, 611
horn angle, 403, 423–25, 430–31
Hugo, 446n
Hugh of St. Victor, 224
Hultsch, F., 357, 359, 629, 635–36n, 659n
Hunt, R., x, xxiv
hydrostatics, 2, 164
Hypsicles, 633

Iamblicus, 608
Iannelius, C., xix
Ibn al-Haitham (=Alhazen), 560, 633, 669
improbatio (=refutation), 444
including and included figures, 63–64, 68, 172–74, 402
inertia, 1n
al-Iṣfahānī, 658n–59n
Islam, 223
Isoperimeters, *see De ysoperimetris*

Jacobus Cremonensis, 12
James, M. R., xxv
Johannes de Muris, 6, 10, 12, 162, 369n
Johannes de Tinemue: his *De curvis superficiebus*, 6, 63, 65, 169n, 191, 225n, 363–65, 558–60, *and see* Chap. 6, Sects. 2–5, *passim*; quadrature tract attributed to, 15n, 576–77
Johannes Gervasius, 443, *and see* Gervasius de Essexta
Jordanus de Nemore: Hero's formula for area of triangle and, 636–38; his *De triangulis*, 96n–97n, 224–25, 366, 387, 567, 572–75, 637, 658–63, 670–77; his *Philotegni*, 637; on quadrature, 15n, 33, 567–75, 577; on trisection of an angle, 668–77, 678; Philo's solution of mean proportionals problem and, 366, 560, 658, 660–63; statics of, 9–11; mentioned 142, 561, 637n
Juliacensis, Johannes Caesarius, 582n, 587
Junius Nipsus, Marcus, 635n

kataskue, 444
al-Kharkhī, 636
al-Kindī, 226, 228n, 433
kinematics, 9–10, 32
Krause, M., 659n
Kristeller, O., 14

lambanomena, 629
Lachmann, K., 635n
Lejeune, A., xxiii, 633–34
Lemmata, *see Liber assumptorum*
Leonardo da Vinci, 13
Leonardo Fibonacci Pisano: Philo's solution of mean proportionals problem and, 660–61, 664–65; his *Practica geometrie*, 33, 224, 357–58, 363–66, 519, 636, 658–61, 664–65; mentioned, 7, 560–61
Leonardo Mainardi of Cremona, 636–37n
lever, law of, 9, 628–29
Liber assumptorum, 4, 224, 367, 559, 632, 667–68
Liber de curvis superociebus, Chap. 4, Sec. 2, *and see* Johannes de Tinnemue
Liber embadorum, 17n
Liber trium fratrum (=Verba filiorum), *see* Banū Mūsā
Libri, G., xxiv–xxv, 443
limaçon, 666
logic and medival geometry, 62, 144–45, 169, 193–95, 222, 388, 398–99, 561, 582–83
Lull, Ramón, 15n–16n
lunes, 62, 65, 81, 92n, 94, 145, 165, 169, 569n, 610–26 *passim*, *and see* lunula *in* Selective Index

al-Ma'mūn, 226
Macray, G. D., xx, xxiii–xxiv
Macrobius, 59, 64, 74, 79
Madan, F., xx, xxv
Mahumed, xxi
Mainardi, *see* Leonardo Mainardi of Cremona
Maurolico, Francesco, 13
means: single proportional, 15n, 62, 169, 186–88 two continuously proportional, 7, 224, 335–45, 365-66, 559–60, 658–65
mechanical methods: in solution of problem of mean proportionals, 334-35, 366, 658–65 in solution of problem of trisection of an angle, 344–48, 366, 666–81
Menelaus, 224, 228n, 336–37, 340–41, 352–53, 365–66
Menge, H., 627

Meyerhof, M., 226n
Michael Ephesius, 427
Migne, L'Abbé, 607
Mileus, *see* Menelaus
Millás Vallicrosa, J. M., xxix, 16n–17n
Minio-Paluello, L., 607n
Mishnat ha-Middot, 635
Moerbeke, William of, 4, 6–8, 11–13, 29, 63, 165, 167, 219, 369n, 433, 443, 531n, 558, 563, 607n, 608
Molinier, A., xxv
Moody, E., 8n–9n, 445n, 637n
Muckle, J. T., 167
Munich, 143n
Munich Version of *De mensura circuli:* cited in comparison, 60, 98, 387, 399, 429, 561; text and translation of, 193–222
Murdoch, J., x, 37, 60n–61n, 139, 369, 402n, 429, 431, 627n, 633, 678n
Mūsā ibn Shākir, 7, 226, *and see* Banū Mūsā
al-Mu^ctaṣim, 226

al-Nadīm, 365, *and see Fihrist*
Naples Version of *De mensura circuli:* cited in comparison, 59, 62, 144, 164–65, 220–21, 387, 402, 429; text and translation of, 80–91
Nasīr al-Dīn al-Ṭūsī, *see* al-Ṭūsī
Neugebauer, O., 442n
neusis, 366–67, 559–60, 666–69, 678
New Testament, 442
Nicholaus, Magister, xxiv
Nicomedes, 608, 666, 678n
Nix, L., 166n, 441n–42n
numerals: Indo-Arabic, 18, 36, 65, 97; Roman, 18, 65, 98
numbers, their use to simplify argument, 144, 164

O'Donnell, J. R., 428
Oresme, Nicole, 12, 140, 399, 445, 561
Oxford, 610

Pacioli, Luca, 636
Pallas, 442
Pappus, 357, 363, 510, 629, 658–59n, 666n–67n, 678n
Paris: Archimedes at, 11–12; statics at, 10; mentioned, 398–99
Pascal, 666n
pentagon, 374–77, 390–93
per impossibile, 444, *and see* impossibile *in* Selective Index
Perspective, 669, 675, 678
Philo of Byzantium, 658–65
Philoponus, John, 427–28
philosophy and geometry, 561–62
pi: determination of by Banū Mūsā, 97n, 223–24, 264–79, 358–59; in Florence version, 95–96, 112–34, 138, 560; in Tivoli's translation, 17, 25–27, 29; in Gerard's translation, 48–55, 402; other versions of Prop. III of *De mensura circuli* and, 80n, 96n–97n, 140–41; rhetorical expression of, 224, 322–25, 364; value of in Gordanus' *Compilacio*, 143–44n; in Campanus' tract, 584; mentioned, 6, 441, 520, 559
Pines, S., x
Plato, 224, 559, 658
Plato of Tivoli: as possibile translator of *De mensura circuli*, 4, 16–29, 40v, 358; as translator of *Liber embadorum*, 17, 636n
Plutarch, 628
polygon: area of regular, 168, 174–77, 246–49, 356, 370–71, 374–79, 390–94, 401, 414–17; area and volume of solid formed by rotation of, 469–79, 483–95, 512–17, 547–57, 631
polyhedron, volume of regular, 248–49, 356
Poole, R. L., 581n
Porphyry, 608
portiuncula, 62, 145, 164, 169, *and see* Selective Index
postulates: betweenness, 369 for various versions of *De mensura circuli*, 63–64, 68, 90, 137, 143n, 166–68, 170–72, 368–69, 402, 416–19, 562; other, 628–29; use of physical, 166, 562
power, geometric (=mechanical advantage), 628–29
Practica geometrie, anonymous, 36
Prete, S., 640
prism, 434, 437–38
Proclus, 428, 667n
proportionals, problem of mean: *see* means
pseudographos, 159, *and see falsigraphus*
Ptolemy: *Almagest* of, 195, 210–11, 221, 371, 379n–81, 387; *Optics*, 633–34
pyramid, volume of, 249, 434, 356

Pythagorean theorem, 78, 97n, *and see dulk*
Pythagoreans, 608

al-Qifṭī, 4n
quadrature of a circle: in Archimedean sense, 7, 224, 256–79, 324–25, 357–59, 382–87, 394–97, 398–432, 630–31, *and see* Chaps. 2, 3, 4, (Prop. I–VI), 5, *and* Appendixes I and II, *passim*; in the sense of quartering a circle, 400, 409; *and see* quadratura *in* Selective Index

Ramée, Pierre de la, 636
Ratdolt, E., 220–21
ratio (=*apodeixis*), 444
rectification (=the equation of curved and straight lines), 63–64, 69, 143n, 166, 168, 170, 371, 378–79, 411, 413
reduction to absurdity, 80, 90, 95, 144–45, 444, 455, 561, *and see per impossibile*
refraction, 633–34
Regiomontanus, 12
Renaissance, 3, 12–14, 224
Roberval, 666n
Roman balance, 9
Rome, 606n
roots: cube, 349–51, 367 square, 95, 139–42
Rudorff, A., 635n
Ruska, J., 226n

Sambelichius (=Simplicius), 628
Sarton, G., 434n
Savasorda, 17, 359, 636
Schöne, H., 5n, 357
scholastic proofs, 59–60, 194
Schum, W., xxii
Scott, E. J. L., xxvii
Seneca, 634
separando, 529
serratile, 434, 437
Sextus Pithagoricus, 608
Shrader, D., 629
Simplicius, 16n, 62, 426–27, 608, 610–26, 628
Simson, R., 530
specific gravity, 8, 10
sphere: area and volume of, 223-25, 317–35, 437–39, 479–83, 497–507, 514–15, 518–20; segment of, 531, 532–41, 544–46, 560
statics, 8–11, 562, 629–30
Steele, R., 225n
steelyard, 9
Stereometry, *see* Hero of Alexandria
Sturm, A., 582n
Suter, H., xxviii, 3n, 7n, 17n, 226n–28n, 344v, 365, 582n, 587, 610, 638n

Tanner, T., 443n
Tannery, P., xxvi, 15n, 227n, 568
Tartaglia, Niccolò, 13, 636n
Thābit ibn Qurra: his *Liber karastonis*, 9, 563; his translation of *Measurement of Circle*, 17n, 29; other tracts and translations of, 38, 226
Thaus, Valentinus, 637n
Themistius, 426–28
Theodosius, 228n, 292v, 363
Thomas, I., 365–66, 608, 658n
Thorndike, L. , xx–xxi, xxvi–xxvii
Thomson, S. H., 610n
Tinemue (*or* Tinnenie), 440, *and see* Johannes de Tinemue
tiphis (=Tiphys), 442, 507
Torre, A. della, 14
trapezium, 628
triangle, area of, 35, 43, 356, *and passim*, *and see* Hero of Alexandria
trisection of an angle, 7, 224, 344–49, 366–67, 560, 666–81
Troianus, Curtius, 13
al-Ṭūsī: his version of the *Measurement of the Circle*, 17, 29, 38, *and see* Chap. 2, Sec. 2 (Arabic variants) *passim*; his version of the *Verba filiorum*, 231–33, 355–56, 361–63, 638n, *and see* Chap. 4 (Arabic variants) *passim*; on mean proportionals, 659n

Vacca, G., 583n
Van Ryzin, Sister St. Martin, 228n
Venatorius, Thomas Gechauff, 12n
Verba filiorum, *see* Banū Mūsā
Vienna, 143n, 398
Viterbo, 443
Vogl, S., xxiii, xxvi, xxviii, 30n
volume, *see* polygon; polyhedron, volume of regular; pyramid, volume of; sphere; *etc.*

Widmann, Johannes, 636

Wiedemann, E., 3n, 4n, 366, 633
Willis, J., 79
Winterberg, Dr., 15n
Winton, 443
Witelo, 443
Woepcke, E., 668n
Wüstenfeld, F., 30, 223n

Yaḥya ibn Abī Manṣūr, 226
York, 443
Yrinus (=Hero), 629

Zeuthen, H. G., 63n, 628
Zoubov, V. P., 399n

ADDENDA AND CORRIGENDA

Page xxix: line 11. *For* Millas *read* Millás.

Page 5: line 26. *For* page 605 *read* page 60.

Page 12: line 14. After *Cylinder* insert *Floating Bodies.* Note 21, *add* Oresme made numerous citations to Moerbeke's translation of the genuine *Floating Bodies*, presumably from MS Vat. Ottob. lat. 1850, in his *Questiones in libros de caelo* (MS Erfurt, Amplon. Q. 299, 47v-48v). These citations will be discussed in Volume Two.

Page 287: line 6. *For* trianle equals to *read* triangle equals.

Page 369: line 10 of note. *For* property *read* property of.

Page 442: line 23. *For* sources. *read* sources, although there is an English canonist, Johannes de Tynemuth, or Tinemue, who died about 1221 and who could have had something to do with the translation or rewriting of the text (see A. B. Emden, *A Biographical Register of the University of Oxford to A.D. 1500*, vol. 3, Oxford, 1959, p. 1923). The possibility of his authorship of the *De curvis superficiebus* will be discussed in Volume Two.

Page 443: line 9. *For* Winton *read* Winchester; *for* 1261 *read* 1262. Line 11, *for* 1261 *read* 1262. Note 15, *before* Th. Tanner *add* A. B. Emden, *A Biographical Register of the University of Oxford to A.D. 1500*, vol. 2, Oxford, 1958, p. 757.

Page 467: Note to Figure 66. Note should appear under Figure 67, page 471.

Page 531: line 32. For *DEF* read *DLF*.

Page 541: line 1. *For* two, surfaces *read* two surfaces.

Page 611: line 17. For *DZa* read Z.

Page 648: numbered line 4. *For* propostiti *read* propositi.

Page 692. *For* adhere *read* adherere.

Page 713. *For* Alverny, M.-T. d *read* Alverny, M.-T. d'.

Page 717: col. 1, line 1. *After* 274-89 *add* 353v. *For the entry* al-Kharkhī *read* al-Karkhī. *Add new entry* al-Khāzin, 353v. Col. 2. *For the entry Liber de curvis superociebus*, Chap. 4, *read Liber de curvis superficiebus*, Chap. 6.